CASE STUDIES IN STRATEGIC BOMBARDMENT

R. Cargill Hall
Editor

AIR FORCE History and Museums PROGRAM
1998

To those in World War II who fought for their countries high above the Earth, where aerial armadas moved amidst the din and frost for the first and, perhaps, last time.

Library of Congress Cataloging-in-Publication Data

Case studies in strategic bombardment / edited by R. Cargill Hall.
p. cm. — (Special studies)
Includes bibliographical references and index.
1. Bombardment — Case studies. I. Hall, R. Cargill. II. Series: Special studies (Air Force History and Museums Program (U.S.))
UG700.C378 1998 98–23457
358.4′2 — dc21 CIP

For sale by the U.S. Government Printing Office
Superintendent of Documents, Mail Stop: SSOP, Washington, DC 20402-9328
ISBN 0-16-049781-7

Foreword

This volume, the third in a series of historical case studies of important air power missions, addresses the most controversial (and arguably most significant) air power mission of all — strategic bombardment. The ability of aircraft and missiles to destroy or disrupt an enemy's war-making potential and to break or weaken his will to resist, independent of the actions of ground and naval forces, has served as the central theme of air power theory and as the rallying point of air advocates, who made it the *raison d'être* for independent air forces. Written by well-known military historians, each chapter stands alone as a case study of an important stage in strategic air operations; combined, the chapters provide a comprehensive and insightful analysis of the theory and practice of strategic bombardment from its inception in World War I through the Gulf War of 1991.

From "Boom" Trenchard and "Billy" Mitchell to John Warden and Charles Horner, the vision of air power prophets and airmen is tested against the reality of bureaucratic inertia, aircraft capability, technological advances, and bombing accuracy. Seldom in the twentieth century has technology fully met the demands of air power theory. Yet in each era a practitioner of the art appears, such as Harris, Spaatz, LeMay, or Horner, who in some measure modifies prevailing doctrine and stretches the paradigm of his time and circumstances to achieve extraordinary results. Technology, of course, is the prime determinant of doctrine and operations. This exceptional volume surveys the entire history of strategic bombardment and its technology, from the Zeppelin and Gotha of the Great War to the F–117 and the penetrating precision guided bomb of the Gulf War. The reader will find technological advances — such as radar bombing and range-extending air-to-air refueling — that answer one problem only to produce new requirements and expectations that demand more advanced technology. Guided munitions, while offering remarkable precision, have underscored the problems of strategic intelligence collection and dissemination, and of locating and attacking both fixed and mobile targets.

This volume also examines the changes in the public's perception of strategic bombardment. The exaggerated fears of a "knockout blow" and near total destruction bruited about before World War II gradually were replaced by a general acceptance of area bombing and contemporary satisfaction at the casualties inflicted on the foe during the war itself. That public tolerance evaporated soon after the appearance of the atom bomb. The ever-present threat of a civilization-ending global nuclear war continued for another forty years and made the B–52 bomber and intercontinental ballistic missile both symbols of ultimate destruction and potent tools of nuclear stability. Indeed, public acceptance of combat casualties for both friend and foe has steadily declined in the latter half of the twentieth century. If the World War II bombing of Berlin and Tokyo resonated favorably with public opinion at the time, changing attitudes by the 1990s would never have condoned such an approach to the Gulf conflict.

Finally, later chapters in this volume consider some of the most significant missions and accomplishments of the United States Air Force during the Cold War. They examine the actions and events associated with strategic air power that helped repel or deter communist aggression and protect the vital interests of Western democracies throughout the world. The U.S. Air Force's commitment to strategic air power has been consistent and seamless since the days of the B–17 to the era of the B–2. Through the development and elaboration of strategic air warfare capabilities and thought, it has created a modern world in which global air power will be the strategic instrument of choice for power projection and presence in the twenty-first century.

RICHARD P. HALLION
The Air Force Historian

Preface

No subject provokes greater emotion or more strongly held views than does strategic bombing. One can find numerous histories of specific strategic bombing campaigns and of strategic bombardment during a particular war or period. Some of them, as is increasingly the fashion, have measured strategic bombardment against the ethos of another era and found it driven by dubious technical or ideological imperatives. Others have superimposed strategic bombardment on military hypotheses and portrayed it evolving or unfolding to become the decisive instrument of warfare, making all other forms of combat obsolescent. Balanced, reasonably dispassionate histories of this subject are difficult to find.

Case Studies in Strategic Bombardment is not only a balanced work, but a comprehensive one. It covers the subject from its conceptual beginnings near the turn of the twentieth century through an assessment of its most recent application in the 1991 Gulf War.* In scope and content, there is no other anthology like it. Although commissioned by Maj. Gen. Perry McCoy Smith, then U.S. Air Force Director of Plans, the contract authors approached their respective topics independently and without constraints, except in the space allotted by their sponsor. Collectively, their history of strategic bombardment is reflective and interpretive, yet it has the ring of truth. Their analyses and assessments address hard military and moral issues, and their contribution to this arena of military history is significant.

The contract authors had completed a review of their respective contributions in 1990 when the advent of the Gulf War prompted another chapter, prepared by Richard Davis of this office. This supplement and the intervening

---

* Other historical case studies in this series are Benjamin Franklin Cooling, ed., *Case Studies in the Development of Close Air Support* (Washington, D.C.: Office of Air Force History, 1990), and Benjamin Franklin Cooling, ed., *Case Studies in the Achievement of Air Superiority* (Washington, D.C.: Center for Air Force History, 1994).

demands of World War II and service anniversaries, regrettably, delayed publication of their work. If some of the bibliographic essays have become dated, the fault lies here, not with our authors who ably discharged their commissions.

Besides the authors, we are indebted to a number of people who organized this work and brought it to press. Herman S. Wolk, B. Franklin Cooling, and the late Col. John F. Shiner (USAF, Ret.) structured this volume, and it was undertaken at the direction of Richard H. Kohn, then the Air Force Historian. Barbara Wittig, the editor, rendered the manuscript uniform mechanically and stylistically, executed the layout, and otherwise attended to all of the myriad details in publication. Graphics support was provided by Veronica Williams and SSgt. Dee Blake of 11th Communications Squadron, Headquarters Air Force Graphics. Evelyn Buhl of the same organization executed the cover design of Robert E. Bell. Special thanks are owed to the 1st Combat Camera Squadron, Office of the Secretary of Defense Public Affairs, and to Lee Kennett, who, in response to a request at the last minute, contributed a provocative concluding analysis of strategic bombing from its inception through the demise of the Soviet Union and the inactivation of the Strategic Air Command.

R. CARGILL HALL
Air Force History Support Office

Contents

Photographs

The photographs of Giovanni Caproni and Giulio Douhet appear courtesy of the Italian Air Force History Office. Those of Lt. Gen. Walther Wever and General Adolf Galland are courtesy of the Hergestellt um Bundesarchiv. Air Chief Marshals Sir Cyril Newall, Sir Arthur Harris, Sholto Douglas with Maj. Gen. Ira Eaker, and Sir Charles Portal with General H. H. Arnold, and the Fairey Battle aircraft all appear courtesy of the British Imperial War Museum. President Kennedy and Ambassador Lodge appear courtesy of The John F. Kennedy Library. The following images are from the Air Force Historical Foundation: Portal and Eaker; *Memphis Belle*; and Vandenberg, Goering, and Spaatz. The two-page schematic drawing of the B–29 appears courtesy of The Boeing Company.

Maps

Charts

Tables

Introduction

David MacIsaac

Not least among the problems facing the student of military aviation is the absence of an accepted, or agreed upon, vocabulary among those who have chosen to write on the subject. As Professor I. B. Holley, Jr., observed, this problem begins with the practitioners themselves. "One is struck by the absence of standard or stable terminology. In each new resurgence of interest . . ., those involved seem to have coined virtually a whole new vocabulary."[1] Subsequent commentators, historians, and analysts struggle with the difficulties thus generated, sometimes despairing of any hope of progress in defining the "vocabulary from which on the basis of observed usage the grammar of air power may eventually be compiled."[2]

With regard to strategic bombing specifically, practitioners have, on the whole, a better record of semantic consistency than commentators, some of whom have felt no compunction to distinguish strategic bombing from any other form or forms of aerial bombardment. Yet practitioners themselves are not absolved in this respect; they too have been happy on occasion to label as strategic bombing any kind of aerial bombardment that they think worked in the circumstances of the moment. In short, strategic bombing has been the term applied over the years to at least three distinct visions of long-range aerial bombardment: one aimed at the devastation and destruction of industrial and population centers on a vast scale; another aimed at assisting surface forces; and a third form involved the systematic attack on selected targets considered critical to the enemy's war-making potential. Looking back to the late summer and early fall of 1917, one is reminded that each of these three types was clearly distinguishable from the very beginning.

On July 11, 1917, the British Cabinet met to discuss the German aerial attacks of May and June on London. A committee of two — Prime Minister Lloyd George and General Jan Smuts — was appointed to make recommendations.

> This was a crucial moment, perhaps the crucial moment, in the history of strategic bombing. Apprehension had laid hold of men's minds and many were certainly driven by the strong motive of fear. The climate was sympathetic to radical proposals. General Smuts, to whom Lloyd George left the work, was a visionary and far-sighted man.[3]

The resulting Smuts Memorandum of August 17th, the paper that led directly to the establishment of the Royal Air Force (RAF), discussed air warfare in the following terms:

> As far as can at present be foreseen there is absolutely no limit to the scale of its future independent war use. And the day may not be far off when aerial operations with their devastation of enemy lands and destruction of populous centres on a vast scale may become the principal operations of war, to which the older forms of military and naval operations may become secondary and subordinate.[4]

At the time these lines were written, neither General Douglas Haig, the British commander on the Western Front, nor his loyal air commander, Air Marshal Hugh Trenchard, found them relevant to the existing military situation in France (although Trenchard, to be sure, would come eventually to embrace them wholeheartedly). And yet it is to this vision expounded by General Smuts that we can trace the roots of what others would later call area attack, morale bombing, terror bombing, fire bombing, indiscriminate bombing, and the like.

A second vision, less cataclysmic and tied more closely to the war on the surface, was offered by Winston Churchill when he was the British Minister of Munitions in October 1917.

> It is improbable that any terrorization of the civil population which could be achieved by air attack would compel the Government of a great nation to surrender . . . . Nothing that we have learned of the capacity of the German population to endure suffering justifies us in assuming that they could be cowed into submission by such methods, or, indeed, that they would not be rendered more desperately resolved by them. Therefore our air offensive should consistently be directed at striking at the bases and communications upon whose structure the fighting power of his armies and his fleets of the sea and the air depends. Any injury which comes to the civil population from this process of attack must be regarded as incidental and inevitable.[5]

This view of Churchill's — which he felt obliged to abandon early in World War II owing to the operational limitations of the RAF — speaks to what we might today call long-range interdiction and to what, in fact, occupied the "strategic" air forces frequently during World War II, as in the bombing immediately before and after the assault on Normandy.

A third vision is that of precision attack against carefully selected targets chosen from among potential "bottlenecks" in enemy war production. Although normally attributed almost exclusively to the work of the Air Corps Tactical School in the 1930s, this idea can also be traced to late 1917, in the plan drawn up by Lt. Col. Edgar S. Gorrell, head of Strategical Aviation, Zone of Advance,

American Expeditionary Force. Although the plan was approved, the means to effect it could not be found before the war ended. Still, since it remains the "earliest, clearest, and least known statement of the American conception of the employment of air power,"[6] it deserves mention.

"When we come to analyze the targets," Colonel Gorrell wrote, "we find that there are a few certain indispensable targets without which Germany cannot carry on the war." A few paragraphs earlier Colonel Gorrell had offered some specific examples:

> The manufacture of [German] shells is dependent upon the output of a few specific, well-known factories turning out the chemicals for them, so we can readily see that if the chemical factories can be blown up, the shell output will cease . . . . The same is true of airplane output; the large Mercedes engine plants and the Bosch magneto factories are [both located in Stuttgart], and if bombing airplanes . . . can inflict damage on one or both of these plants, the output of German airplanes will cease in proportion to the damage done.[7]

Colonel Gorrell's detailed plan even included a section on difficulties to be encountered that reads today like a listing of the actual problems encountered in World War II! Among other things, his plan embraced navigation training and adequate aerial maps; effects of weather on operations; force structure (not only aircraft but crews, bombs, logistics, and so on); basing considerations (size, location, and proximity to the enemy, among others); intelligence requirements; communications; and the significance of concentration of effort. In this final respect, Gorrell wrote what could be considered the first draft of the British and American 1943 Casablanca Conference Directive:

> There are two large systems of aerial bomb-dropping. One is what is known as daylight bombing; the other is night-time bombardment. There are many arguments both in favor of and against each kind of bomb-dropping, but it goes without saying that our efforts should be directed against the German objectives *both by day and by night*, giving the Germans no rest from our aerial activities and no time to repair the damage inflicted. [Emphasis added.][8]

Despite such early theorizing, World War I ended before strategic bombing received any real test. As one result, during "The Long Armistice" of 1919–1939 virtually no progress was made toward a consensus of what the term itself implied. The absence of significant data on the effects of such bombing as did occur left everyone free to define strategic bombing as one might. The only matter on which widespread agreement could be found was that *tactical* aviation had come to play an increasingly significant combat role in each succeeding year of the war.

More important than the relative effects of military aviation on the outcome of World War I was the impact of that war on the future of air power. The interminable trench stalemate on the Western Front led inevitably to a determined search for the means to restore mobility — and hence, it was hoped, to restore decisiveness — to warfare. Here, advancing technology made air power appear particularly promising. In this respect, the war's appalling length and death toll led others to see in the emerging potential of air power — in particular strategic bombing — a means to attain a quicker — and hence, some argued, a more humane — route to victory.

Foremost among the airmen who fostered such ideas were Brig. Gen. Giulio Douhet in Italy, Sir Hugh Trenchard in Great Britain, and Brig. Gen. William "Billy" Mitchell in the United States. Douhet was the most strident of the three, seeing a long-range air force composed entirely of bombers (called battleplanes) as both necessary and sufficient to secure command of the air, that is, air superiority from which, he argued, victory in war must inevitably follow. Trenchard's view was broader and included the concept of air control, in which military aviation was applied to the functions of policing the Empire. Mitchell's approach was the most flexible of the three, emphasizing that all future warfare would be dominated from the air, and hence a separate air force, consisting of all aircraft types and led by airmen, should be organized to capitalize on what he saw as the virtually unlimited potential of emerging aircraft capabilities. The vision of such prophets invariably outpaced the actual technical capabilities of their air arms, despite continuing advances in speed, range, altitude, and munitions-carrying capacity.

Most of the matters thus far addressed are treated in rich detail in Richard Overy's opening chapter to this volume, especially the extent to which developments in theory and force structure varied from nation to nation *and* in almost every case devolved from compromise among conflicting viewpoints within each nation. As Overy observes, "the imaginative and intellectual foundations of strategic bombardment were laid long before such a technique became technically and organizationally feasible on any large scale."

The debate over bombardment doctrine proceeded at several levels. One, among political leaders, addressed overall national strategy within the affected nations; another occurred within the various air arms, among proponents and detractors; and a third involved a part of the wider public debate. At each level, the debate reflected geographic considerations of the nations involved and the organizational setting of each nation's air elements. The particular strength of Overy's work is the detail with which he illustrates this thesis with reference to Great Britain, the United States, France, the Soviet Union, and Germany. An important element of his detail centers as well on the technical and practical problems associated with creating a strategic bombardment capability. Overy reminds us that the development of such capabilities to the level we witnessed during World War II was in no sense inevitable, especially given the disappointing results of aerial bombardment in Spain, Ethiopia, and China on the eve of that war.

William Jacobs's chapter on RAF Bomber Command's offensive against Germany stresses the operational considerations that drove the course of decision and policy. It has been thirty years since Noble Frankland decried the extent to which critics of strategic bombing have focused their attention on the level of policy, at the expense of the determining operational factors,[9] a difficulty still faced by some historians today if we are to judge by Ronald Schaffer's *Wings of Judgment* or Michael Sherry's *Rise of American Air Power*. So far as Bomber Command is concerned, the relevant operational and tactical considerations have never been more succinctly set forth than they are here.

Jacobs ranges confidently from early, naive assumptions governing the bombers' accuracy and defensive needs to such matters as aircrew rotational policies and the important leadership roles of station, squadron, and individual aircraft commanders. His tables and illustrations are models of their kind. When he concludes that "rarely in military history has there been a greater disparity between strategic intentions and the means to carry them out," the reader is reminded of one of the perennial questions too little addressed by today's (and tomorrow's?) enthusiasts for one or another all-encompassing answer to emerging problems.

Most American readers will find themselves on more familiar ground in Chapter 3, where Stephen McFarland and Wesley Newton tell the story of the U.S. Strategic Air Forces (USSTAF) between 1942 and 1945. Following brief treatments of the background to and early operations of the U.S. effort, they move on in more detail to the fall crisis of 1943, the winning of air superiority early in 1944, the OVERLORD diversion of mid-1944, and, ultimately, the final campaign of September 1944 through April 1945 when the combined efforts of USSTAF and Bomber Command, finally loosed in full force, brought German war production to a virtual halt. An unusual aspect of their treatment is the less than flattering picture presented of General Henry H. Arnold, Army Air Forces Commanding General, who, they show, was not entirely fair in his dealings with subordinate commanders, in particular, with Lt. Gen. Ira C. Eaker.

Like Jacobs in Chapter 2, their concluding analysis reminds us that the major weight of the Combined Bomber Offensive was not loosed on Germany until after the invasion at Normandy, a point often lost on earlier writers (J.F.C. Fuller, for one) who often complained that important elements of German production peaked during the summer and fall of 1944 "despite the heavy bombing."[10] More than 70 percent of the bombs dropped on Germany fell after July 1, 1944, a fact too little appreciated over the years; both of these chapters are welcome accounts in this respect.

In Chapter 4, Alvin Coox presents an exceptionally well-balanced treatment of the bombing of Japan, one that marks a distinct advance over most earlier accounts in two important respects: first, he takes particular care from the opening pages to treat the bombardment campaign in the context of the Pacific War as a whole; second, he relates the Japanese side of the air campaign with remarkable thoroughness and insight. Nor does he in any way slight the less well-known aspects of the effort (Operation MATTERHORN, the unsuccessful attempt to bomb Japan from China with B–29s, for one example), or the ticklish questions regarding command arrangements, or the important role played from afar by the seemingly ubiquitous General Arnold.

As the first four chapters make abundantly clear, the role of individual personalities in the story of strategic bombing far outweighs any other single factor — including, I would argue, even technology. The roles played by men like Smuts, Trenchard, Churchill, Roosevelt, Mitchell, Douhet, Arnold, Spaatz, Eaker, LeMay, Hansell, Marshall, Harris, Portal, Zuckerman, Lovett, and Tedder (to cite only the most obvious) were at various points decisive in the evolution of both policy and tactics affecting the bombing campaigns. To argue that overarching trends were at work — that, in effect, the game would have been played out pretty much the same with other players — is to my mind a conceit into which historians and other commentators too often have fallen.[11]

However that may be, as the essays by Jacobs, McFarland and Newton, and Coox clearly reveal, the actual conduct of the war did little to resolve the dilemma of finding a precise definition of strategic bombing. Each of the major participants employed long-range bombardment aircraft in multiple roles, driven to do so as circumstances and personalities determined. Looking back from today's vantage, one may suggest that strategic bombing came to mean different things to different people, as well as different things to the same people at different times.

If there is any one source to which today's historian might hope to refer for a precise definition of strategic bombing, it would have to be the United States Strategic Bombing Survey. Its 323 constituent reports provide the most thorough data base existing on the subject.[12] Alas, one will look there in vain. The survey's chairman, Franklin D'Olier, refused all pleas to seek a precise definition. For the survey's purposes, strategic bombing would be all bombing conducted by air forces designated as strategic, regardless of the targets attacked.

D'Olier was encouraged in this respect by the senior air commanders who wanted assurance that no part of their effort be ignored in the final reckoning. Also affecting his decision was a growing awareness that a measurement of the effectiveness — as opposed to the effects — of the bombing that occurred in Europe would not prove possible in the end.[13]

Almost twenty years later Kent Roberts Greenfield, in his essay "Air Power and Strategy," addressed this question of definition as well as if not better than anyone before or since. "What did [the airmen] mean by 'strategic bombing'?" he asked.

> The term requires definition because it is inexact. It carries a charge of aspiration, if not of boastfulness. It implies that the kind of air offensive to which it refers is the only kind of offensive that is truly strategic. What the term, as used in World War II, actually meant was massive and systematic bombing of the enemy's war economy and of the enemy population's will to resist. . . . The term "strategic" bombing was used to designate this unprecedented kind of warfare because no other acceptable term for it was available.[14]

In the years since World War II, strategic bombing in its classic sense has not recurred. Certainly *one* of the reasons it has not is that the United States built up and maintained strategic forces of such magnitude as to deter any nation with a similar though lesser capability. That story is the subject of Chapter 5, Steven Rearden's "U.S. Strategic Bombardment Doctrine since 1945." Rearden divides the postwar era into four periods: 1945–1953, which saw the emergence of what was then called the "air-atomic age"; 1953–1961, the era of Strategic Air Command dominance, countervalue targeting, and massive retaliation; 1961–1969, the McNamara era and a shift in emphasis toward assured destruction and, increasingly, counterforce targeting; and the period since the early 1970s that featured strategic arms limitation talks, the waning of America's previous superiority in the weapons of nuclear warfare, and, ultimately, the demise of the Soviet Union. The latter event, concluding a forty-year Cold War, also eliminated the *raison d'être* of the Strategic Air Command and of the strategic bomber configured exclusively as a nuclear weapons carrier.

Thomas Hone's chapter, "Strategic Bombing Constrained: Korea and Vietnam," includes the significant narrative details, of course, but goes well beyond that in what is in fact a penetrating essay on the U.S. Air Force's continuing struggle to come to grips with the uses of strategic air power in limited wars, especially those when the specter of escalation to the nuclear level is a possibility. The extent to which presidents and their advisers, both military and civilian, can become entrapped in old (not to say wishful) thinking and language when approaching new problems is his persistent, if muted, theme, along with the idea that limited strategic bombing in an air-pressure campaign might be in fact a new (or at least special or unique) form of warfare deserving special attention.

Richard Davis's chapter on "Strategic Bombardment in the Gulf War" presents a detailed account and analysis of the special concepts, equipment, and innovative approaches that marked that conflict of early 1991. It provides the context for the Coalition air campaign, stressing the technological and doctrinal developments over almost two decades following the Indochina experience, the onset of the Kuwait crisis, and the initial U.S. deployment and war plans. The actual conduct of the air campaign is described and analyzed in considerable detail, informed by Davis's special access to many sources unavailable to academic historians and by numerous interviews with the principals involved.

The many successes of the Gulf War strategic air campaign, some unprecedented, are spelled out, with emphasis on the new techniques made possible by the marriage of stealth aircraft with precision-guided munitions, aerial refueling, and space-based navigation and communication aids. These were widely noticed at the time, to be sure, but for the reader who takes the long view of strategic air campaigns, several problems appear to have dogged such efforts from the very beginning. In this respect, historians will notice in particular the difficulties prompted by force diversions occasioned in the shifting of objectives and by the seemingly perennial tensions that bedevil communication and cooperation between the intelligence and operations communities, sometimes reflecting little more than differences in viewpoint between and among determined personalities.

The seven chapters in this volume, taken together, constitute a truly exceptional introduction to the history and theory of strategic bombardment from initial conception through its classical application during World War II, to its apparent transformation into a term now used somewhat loosely to describe any form of aerial bombardment undertaken to achieve a strategic purpose. Surely, we can all hope that Noble Frankland was correct when he wrote that "the introduction of nuclear weapons and guided missiles has . . . rendered a repetition of a long drawn-out campaign of attritional strategic bombing in a major war inconceivable."[15] Even if this is correct, it does not likely mean that we have seen the end of aerial bombardment — via aircraft, ballistic missiles, or cruise missiles — directed to attain specific strategic objectives. In fact, as Hone argues, today's and tomorrow's airmen must come fully to grips with a wholly new set of conceptual problems, among which advancing technology is only the most obvious.

Consider just one example: that technology today gives weapons of all kinds the ability to strike targets accurately from great distances, thereby blurring former distinctions between land, sea, and air power.[16] It could even be (again as Hone suggests) that the older concepts of air power have become outmoded and have outlived their usefulness. The truly strategic questions now, as Sir Michael Howard has observed, are What should be attacked to fulfill the purpose of the war — and from what platform — in the air, in space, at sea, or on land? and How can this be accomplished with the greatest effectiveness, efficiency, and prospects for success?[17] If that is in fact the emergent issue, it could well be that airmen, along with those who write about them, will be forced to abandon their long-cherished hope for the establishment of an outlook, along with an attendant vocabulary, that permits some measure of purity in the concept of air power, or of war in the third dimension.

NOTES

1. Benjamin Franklin Cooling, ed., *Case Studies in the Development of Close Air Support* (Washington, D.C.: Office of Air Force History, 1990), pp 544–545.

2. Noble Frankland, *The Bombing Offensive against Germany: Outlines and Perspectives* (London: Faber & Faber, 1965), p 16. For the most thorough analysis of this problem yet to become available, see Air Vice-Marshal Tony Mason, *Air Power: A Centennial Appraisal* (London: Brassey's, 1994), esp. chaps 1, 8.

3. Frankland, *Bombing Offensive against Germany*, p 33.

4. For the full text of this memorandum, see H. A. Jones, *The War in the Air: Appendices Volume* (Oxford: Clarendon Press, 1937), pp 8–14. For the background, see Mason, *Air Power*, pp 20–27.

5. Quoted in Frankland, *Bombing Offensive against Germany*, p 17.

6. The assessment of Maj. Gen. Laurence S. Kuter in 1943 is quoted in Maurer Maurer, *The U.S. Air Service in World War I*, vol 2, *Early Concepts of Military Aviation,* 4 vols (Washington, D.C.: Albert F. Simpson Historical Research Center & Office of Air Force History, 1978), vol 2, p 141.

7. *Ibid.*, pp 142–143.

8. *Ibid.*, pp 146–149.

9. Frankland, *Bombing Offensive against Germany*; Ronald Schaffer, *Wings of Judgment: American Bombing in World War II* (New York & Oxford: Oxford Univeristy Press, 1985); Michael S. Sherry, *The Rise of American Air Power* (New Haven, Conn.: Yale University Press, 1987).

10. J. F. C. Fuller, *The Second World War, 1939–1945: A Strategical and Tactical History* (New York: Duell, Sloan, & Pearce, 1949).

11. For a recent example of what I term an overarching trend, see the emphasis placed on "technological fanaticism" as a driving force in Sherry, *Rise of American Air Power*, esp. chap 9.

12. For an introduction to the survey, along with reproductions of its thirty-one most important reports, see David MacIsaac, ed., *The United States Strategic Bombing Survey: Selected Reports in Ten Volumes* (New York & London: Garland Publishing, 1976).

13. See David MacIsaac, *Strategic Bombing in World War II: The Story of the U.S. Strategic Bombing Survey* (New York & London: Garland Publishing, 1976), esp. pp 161–164.

14. Kent Roberts Greenfield, *American Strategy in World War II: A Reconsideration* (Baltimore: Johns Hopkins Press, 1963), p 88.

15. Frankland, *Bombing Offensive against Germany*, p 14.

16. For evidence that air forces worldwide are beginning to address these implications, see either of the following two examples from the Indian Air Force: Wg Comdr P. Khokhar, "Is the IAF a Tactical Air Force?" *USI Journal* [New Delhi], vol 119, no 496 (Apr–Jun 1989), pp 158–162, esp. p 161; or Jasjit Singh, "The Strategic Deterrent Option," *Strategic Analysis*, vol 12, no 6, pp 585–604, esp. p 601. For further comments by this writer, see his contributions in R. A. Mason, ed., *War in the Third Dimension: Essays in Contemporary Air Power* (London: Brassey's, 1986), pp 11–31, and in Samuel Sarkesian and John Williams, eds., *The United States Army in a New Security Era* (Boulder & London: Lynne Reinner Publishers, 1990), pp 236–245.

17. Lecture, Michael Howard, spring 1987, The Pentagon, Washington, D.C.

1

Strategic Bombardment before 1939 Doctrine, Planning, and Operations

Richard J. Overy

Strategic bombardment lies at the very center of the history of air power. "The bomb," according to the Royal Air Force *War Manual* of 1936, "is the chief weapon of an air force and the principal means by which it may attain its aim in war."[1] Early views of air power as a form of warfare with distinct objectives and operational features, and independent in a real sense of the other fighting services, were based largely on the argument that bombardment could be an autonomous *strategic* activity. The strategic use of air power differed from tactical bombardment in two important ways. First, it formed part of the grand strategic aim by directly attacking the enemy's will to resist, bypassing the surface campaign and independent of its immediate objectives. Second, strategic bombardment focused on complex target systems chosen not because of any direct or necessary relationship with the enemy's forces in the field, but because their destruction would undermine the enemy nation's willingness and capability to wage war at all. Tactical bombardment prepared, supported, and expedited ground initiatives, even with long-range attacks, but it did not constitute a separate air strategy.

In the interwar years, the idea of strategic bombardment gained wide currency in military circles and among the public at large. The strategic use of air forces gave the contemporary and widespread concept of *total war* a direct foundation. It was argued that only strategic bombardment brought war directly to civilian populations, attacking their morale and economic base of their war effort, thus breaking national resistance. When, in 1936, U.S. Army Maj.

Harold George lectured on air power to young American airmen at the Air Corps Tactical School (ACTS) at Maxwell Field, Alabama, he asserted that the very nature of war itself was changing as a result of the bomber and its power to attack not just armed forces, but whole nations as well: "It appears that nations are susceptible to defeat by the interruption of their economic web. It is possible that the moral collapse brought about by the breaking of this closely-knit web will be sufficient, but closely connected therewith is the industrial fabric which is absolutely essential for modern war."[2] The "national objective" in war, announced a senior RAF officer that same year, is "the demoralization of enemy people."[3] These views reflected the growing awareness of a fundamental change in the nature of warfare in the early twentieth century, from a war between opposing armies and navies to a war of attrition involving the material resources and morale of whole nations. This change afforded strategic bombardment a special place in the evolution and popular perception of air power.

In practice, strategic bombardment was limited before 1939; it was just one of many roles assigned to national air forces. Resources devoted to strategic bombardment air forces were, in the case of most major powers, negligible. For much of the interwar period, the technology for major strategic air operations was woefully deficient. The acceptance of strategic bombardment doctrine and the size and nature of bomber forces varied widely from country to country, according to circumstance. In Britain and the United States, strategic bombardment aviation was accepted by their air forces as a necessary component of air power, but in the Soviet Union, France, Germany, Italy, and Japan, bombardment, whether long- or short-range, was closely related to the immediate aims and objectives of the other services. Why such a disparity should emerge is one of the central questions addressed here, though it would be wise to recall that no air force was capable of mounting an effective strategic bombing campaign before 1942, including that of the United States and Great Britain. In 1941, Lt. Gen. Henry H. "Hap" Arnold, commander of the Army Air Corps,* rated America's offensive capability in the air "at zero strength."[4]

All bombardment operations before the outbreak of war in 1939, even those of World War I, were small in scale and largely tactical in design. Strategic bombardment was an unredeemed promise, "an article of faith."[5] It had but a limited and relatively ineffective life at the end of the First World War. After 1918, however, general ideas about the strategic use of aircraft developed together with specific bombardment doctrine, and the foundations were laid for organized bombing forces. Major technological advances made possible the evolution and ultimately large-scale production of heavy bombers, and hurried efforts in the late 1930s helped translate this evolution into effective strategic bombardment forces with serious operational potential.

---

* Renamed the Army Air Forces after July 1941.

The Genesis of Strategic Bombardment

The powerful and ambitious conviction that a future war might be decided from the air had its roots in the dramatic transformation of aviation in the ten years preceding 1918. When, in November 1918, Germany signed the Armistice ending World War I, British and American forces were preparing a major air offensive against German industry and communications with the use heavy bombers capable of flying 1,300 miles with a maximum bombload of 7,500 pounds.[6] Just ten years earlier, powered flight was in its infancy, "a form of sport which consists in providing a spectacle to draw a crowd," according to the British Admiralty.[7] Military skepticism about the future of aviation in war was considerable, justifiably based on the unreliability and poor lifting power of early aircraft. Serious military and naval interest in aviation only began in the two years before the outbreak of war, and that interest was divided between airships and airplanes. The main role expected of these aircraft was to gather intelligence, for they promised a more economical and effective way to obtain reconnaissance information over both land and sea.[8] Neither air combat nor aerial bombing was regarded as a significant element of use in aviation. In 1910 the Admiralty held the view that "direct attacks by dropping explosives, though possible, is not likely to have so much effect . . . ."[9]

Nonetheless, the coming of effective powered flight immediately opened the prospect of bombing from the air. In 1910 Glenn Curtiss demonstrated this in America with an attack using simulated bombs on a dummy battleship. A year later the U.S. Army used live bombs in tests in San Francisco. In 1911 Italian aviators dropped small bombs on Turkish forces in the Libyan war. Even the British navy carried out bombing experiments "with some considerable measure of success" before 1914.[10] It was an obvious development, and none of the major powers ignored its potential before 1914. A lively campaign in France to expand French air power in 1912 highlighted not only the ability of aircraft to attack troop formations with bombs, but their ability to attack rear areas, convoys, and even arsenals and munitions stores as well.[11] The German airship was regarded as a major threat to the Western Powers if its aviators ever chose to drop bombs rather than take photographs. The Russian designer Igor Sikorsky produced the first four-engine bomber in 1913, forerunner of a generation of Russian heavy aircraft. Though thirty years would elapse before major air campaigns along these lines could actually be launched, early discussions of bombing operations and their probable effect helped frame the evolution of aerial bombardment. The imaginative and intellectual foundations of strategic bombardment were laid long before such a technique became technically and organizationally feasible on any large scale.[12]

World War I almost immediately ushered in an important period of experimentation with all forms of aerial warfare. This included occasional and sporadic attacks by bomb-dropping aircraft, largely in pursuit of objectives closely connected with the land battle. Such attacks, neither coordinated nor sustained, were carried out by small numbers of aircraft suffering high losses.

The first aerial operations that could be regarded as strategic in any real sense of the term were conducted, not by airplanes, but by Zeppelin airships. At the outbreak of war, the German deputy chief of naval staff, Paul Behncke, argued that airships, each carrying a 2,200 pound bombload over a range of 400 miles, should be used to attack the British mainland. The strategic result he aimed for was not material destruction that, with the force at his disposal, would be slight, but was panic "which may possibly render it doubtful that the war can be continued."[13] The commander of the German Naval Airship Division, Peter Strasser, was an even greater enthusiast. At the height of the airship campaign in 1916, he urged the naval high command to make available as many airships as possible "in the interest of a prompt and victorious ending of the war." Through the proper conduct of bombing operations, Britain could "be deprived of the means of existence through increasingly extensive destruction of cities, factory complexes, dockyards, harbor works with war and merchant ships lying therein, railroads, etc. . . . ."[14] Though the German high command approved of the attacks, the number of airships produced could not match Strasser's strategic conception; airship losses proved unacceptably high, and the amount of damage done (196 tons of bombs, 557 deaths, £1.5 million in material destruction) was of little strategic significance.[15]

The vulnerability of the airship fleet was demonstrated by the immediate and largely successful response of Allied forces to its threat. The only successful independent strategic air operations mounted by the Allies before 1918 were directed at the airship pens and at airship construction and repair facilities. These attacks, undertaken mainly by the Royal Naval Air Service (RNAS) based in France, anticipated the counterforce strategy of later years. At the time, these air attacks were regarded as an extension of the naval war, designed with the support of the First Lord of the Admiralty, Winston Churchill, as a defense against the Zeppelin threat to the British fleet.[16]

Attempts to extend the frontline battle account for the isolated attacks made by the French Army Air Force and the Royal Flying Corps (RFC) on positions in Germany deep behind the German lines, or Italian attacks on the Trieste and Pola naval bases and British attacks on Turkish military targets in the Middle East. The British directive "Bomb Dropping Attacks," issued February 1915, specifically identified troop concentrations and rail links with the German front as vital targets.[17]

French Army aviators, with extensive public support, planned a more wide-ranging campaign against particular industrial targets within range of French aircraft, including the chemical works at Mannheim, the Krupp works at Essen, the Mauser factory at Oberndorf, and an explosives factory at Rottweil. British bombing campaigns, meanwhile, proved ineffective; the RFC and RNAS lacked adequate aircraft, a satisfactory bombsight, trained bomber personnel, and a core of tactical doctrine. By mid-1915, the British Army concluded that "aerial attack has not proved to be a serious operation of war."[18] French forces, too, lacked sufficient numbers of aircraft designed for bombing to do much serious damage, and French Army leaders, disenchanted with bombing attacks far from the front line, gave priority to direct support on the Western Front.[19]

Much the same thing happened with German plans for bombing operations against England from airplanes or with German air attacks on Paris. As with the airship campaign, German plans for long-range operations had a clearly strategic intent, but they shared the same fate as the early French and British attempts when bombing distant targets. German operations broke down because of a lack of effective equipment, the demands of aerial combat directly over the battlefield (where Germany enjoyed considerable success), and the relatively junior and marginal organization of the infant air forces — a problem not exclusively German. At the senior command level of all major belligerents, bombing was seen as a minor diversion. In November 1916 Sir Douglas Haig, British commander on the Western Front, told his War Office, "Long-range bombing as a means of defeating the enemy is entirely secondary . . . its results are comparatively unimportant."[20] French Army Marshals Henri Philippe Pétain and Ferdinand Foch later echoed this view.

The problem of developing a bombing strategy was not simply one of technology, it was also one of politics. Without the active support of senior military leaders or influential politicians, promoting strategic bombardment was difficult. This political problem largely stemmed from a growing fear among leaders in both Germany and France that an extension of bombing in civilian urban areas might produce a domestic political crisis, a view that explains the French General Staff's limited support for strategic bombing for the remainder of the war.[21] The initial bombing successes of the RNAS in 1914 and 1915 depended on the enthusiastic support of First Lord of the Admiralty Winston Churchill for long-range bombardment operations, and German bombing attacks needed not just the support of the German Navy Commander in Chief (C-in-C) Admiral Alfred von Tirpitz and the German Army Quartermaster General, General Erich Ludendorff, but that of the Kaiser himself.

Yet the failure of the army campaigns of 1916 and the strategic stalemate on the Western Front promoted greater interest in long-range bombardment in 1917. For the first time, discussions about bombardment strategy attracted the attention of politicians and military leaders at the highest levels. This shift was partly due to changes in technology — better aircraft, bombs, and bombsights — and in bombardment training — almost nonexistent at the beginning of the war, now a familiar part of the training program. The major combatants gradually accepted that targets distant from the battlefield were legitimate objectives of war.

Discussions on the British side about the nature of long-range bombing, though in the short term putting an end to the RNAS's operations, helped

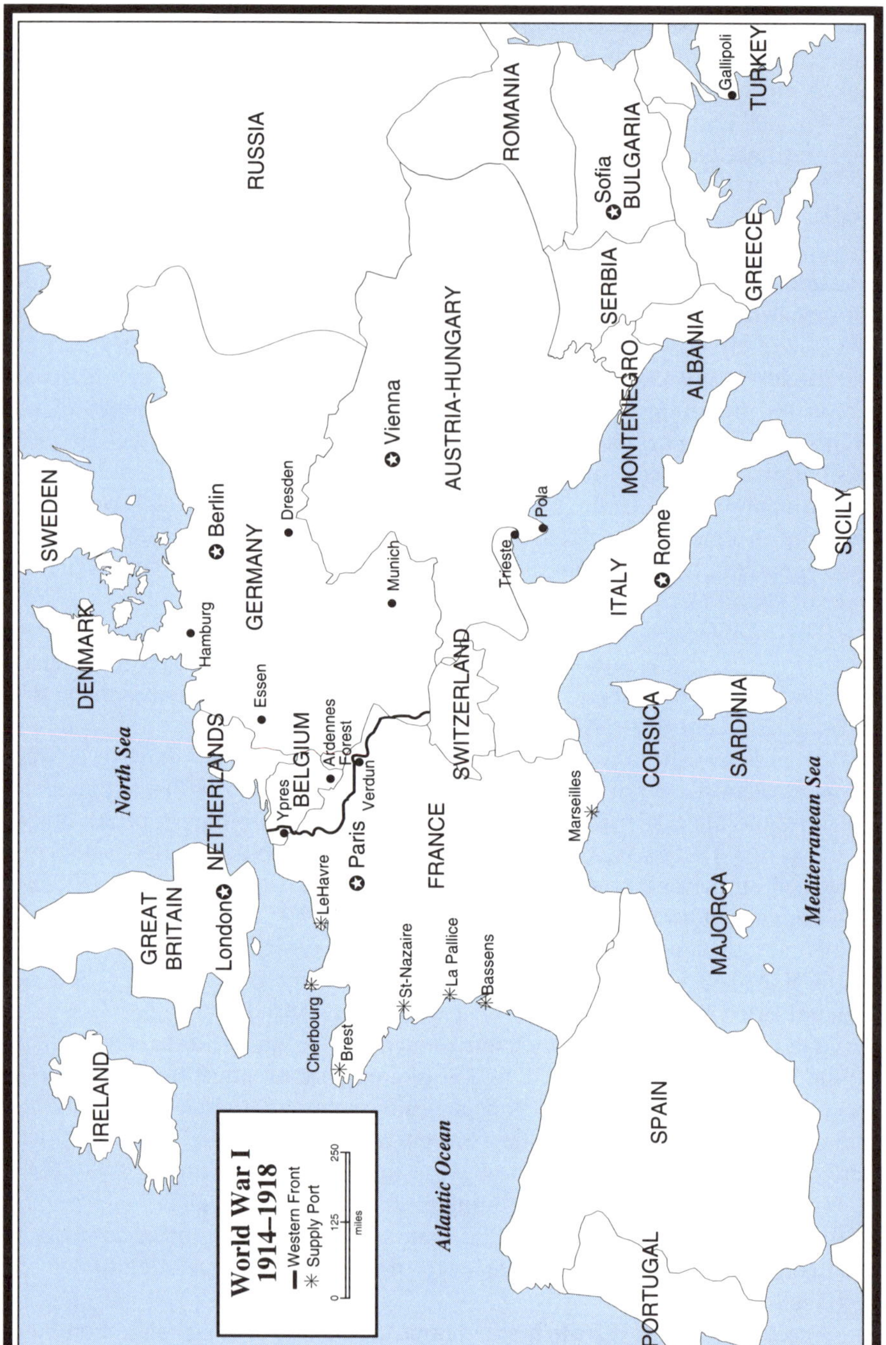
World War I
1914–1918
Western Front
Supply Port
0
125
250
miles
RUSSIA
SWEDEN
DENMARK
GERMANY
Berlin
Hamburg
Essen
Dresden
Munich
NETHERLANDS
BELGIUM
Ardennes Forest
Ypres
Verdun
Paris
FRANCE
LeHavre
Cherbourg
Brest
St-Nazaire
La Pallice
Bassens
Marseilles
SWITZERLAND
AUSTRIA-HUNGARY
Vienna
Trieste
Pola
ROMANIA
BULGARIA
Sofia
SERBIA
MONTENEGRO
ALBANIA
GREECE
TURKEY
Gallipoli
ITALY
Rome
SICILY
CORSICA
SARDINIA
MAJORCA
SPAIN
PORTUGAL
GREAT BRITAIN
London
IRELAND
North Sea
Atlantic Ocean
Mediterranean Sea

define what the purpose of such operations might be. The naval squadrons were "to attack the enemy's fleets, dockyards, arsenals, factories, air sheds . . . (i.e., long-distance bombing)."[22] For the RFC, the statement of duties included preparation "to undertake long range offensive operations against Military or National objectives," though this came at the bottom of the list behind "fighting in the air," "reconnaissance," and "bombing in connection with military operations."[23] In November 1916, Lord George Curzon, head of a newly appointed British Air Board, informed the War Cabinet that "we do not recede from the view that a long-range offensive is in itself a most desirable thing and should be systematically pursued when the force is available for the purpose."[24] French supporters of strategic bombardment went even further. Late in 1916 Col. Maurice Barres, a French liaison officer, visited London to persuade the British to join a campaign in which "the end of the war would be brought about by the effective bombing of open towns."[25]

Perceptions in Germany moved in the same direction in 1917. The new Gotha IV bomber could reach British targets with a larger payload (formerly 660 but now 1,000 pounds of bombs, according to range). In the summer of 1917, with the United States committed to the Allied cause, Germany was forced to find more drastic ways to break the stalemate. With the active support of General Ludendorff, a renewed strategic air campaign was launched at England, partly in hopes that it would create a morale crisis and weaken England's willingness to prolong the conflict. A special England Squadron was established and charged with providing "a basis for peace" by destroying "the morale of the British people."[26] With military and military-industrial targets in Britain as objectives, Gotha bombers undertook 27 raids and dropped 110 tons of bombs largely on residential and dockland areas.[27] Though the raids did little material damage, they created panic in the populations under threat, but it was nothing like the collapse of morale intended. In 1918 the Gothas were accompanied by the larger R-Gigant aircraft, each capable of carrying 3,900 pounds of bombs, but by then German industry was unable to provide bomber aircraft in the necessary quantities because of the demand for tactical aircraft. Moreover, the lessons of the Gotha campaign demonstrated that bombing was not a quick war-winning strategy, and since no adequate defense existed against bombing, it was a strategy that could rebound if and when the enemy sought to combat air offensive with air offensive.[28] Above all, it was a strategy that withdrew aircraft from what frontline generals saw as their most important function: protecting the armies in the field.

This was a view fully shared across the trenches. Although in 1916 and 1917 French air forces continued occasional attacks on major industrial cities in Germany within range, and even as late as 1918 some air officers argued for strategic attacks "to paralyze all the time German economical life and their war industries," during 1917 the French Army high command tightened its control over its air forces and concentrated its efforts on winning air superiority over

The new German Gotha IV bombers developed in 1917 made twenty-seven raids on British residential and dock areas during the following year.

the Western Front.[29] Colonel Barres, the major champion of strategic bombardment, resigned, and Marshal Pétain, worried by the prospect of further unrest at home in 1917 should the Germans take up city bombing seriously, cut back on long-range air raids, even for retaliation. Priority in the French forces went to combined operations and tactical support bombardment on or just behind the front line. By 1918 the French Air Force began to lay the foundation of a sophisticated theory of tactical aviation; as a result, tactical aviation, which had proved so much more successful in practice, dominated French strategic thinking until 1939.[30]

This change in doctrinal emphasis was important for it colored German and French perceptions of air power in the 1920s and 1930s. But the lesson that the British drew from the Gotha campaign in World War I was the reverse: the only way to combat bombing was to launch an independent bombing offensive in return. This response by the British to the Gotha attacks on London and in southeast England is the real root of strategic bombardment. Two elements in this story are particularly significant.

First, support for a bombing strategy appeared at the highest level. Prime Minister Lloyd George used the Gotha raids as an opportunity to examine the whole question of British air power, and he had the support of Parliament and public opinion in doing so.[31] This brought into the open submerged, ongoing arguments about the functions of the RFC and RNAS and the most effective use of air forces. The South African statesman, Jan Smuts, visiting London in 1917 for the Imperial Conference, was invited to produce a report on the organization of the air forces. His conclusions closely reflected the arguments of many in the ministerial and military establishment who favored strategic operations, and they led directly to the formation of a separate Air Ministry and, in 1918, a unified and independent service, the Royal Air Force. This development was of the greatest importance for those who supported strategic bombardment because it created an administrative and organizational structure within which Smuts's vision of air power "as an independent means of war operation" could be promoted.[32]

The second element was that the British government decided against a strategy based simply on reprisal. Indeed, the direct reprisal raid hastily ordered by Lloyd George after the second attack on London was later canceled. During the summer of 1917, inspired by Smuts's arguments that "continuous and intense pressure against the chief industrial centres of the enemy . . . may form an important factor in bringing about peace,"[33] British leaders took the first practical steps to develop a clear strategy for long-range air bombardment and to make it operationally effective. This change in strategic outlook was closely identified with the new Air Minister, industrialist Sir William Weir, and the new Chief of Air Staff, Sir Frederick Sykes, both firm advocates of strategic bombardment and both enjoying the support of Lloyd George and the War Cabinet. In August 1917, Weir authorized Lord Tiverton, a senior air staff officer, to prepare an aerial bombardment plan assuming a bombing force of 2,000 aircraft and a principal aim of achieving "the systematic destruction of the German munitions industry." The target systems chosen — iron and steel, chemicals, aero engines, and magneto works — were attacked from October 1917 onward by the RFC VIII Brigade under Lt. Col. Cyril Newall, though he operated with only a fraction of the aircraft on which the plan had been based.[34]

If the air force now had a clearer strategic guideline, the gap between plan and execution remained as wide as ever. British strategic forces were too small and too hindered by poor weather with its attendant, limited visibility, and the air crews lacked sufficient training in bombing or in navigation. The average flying experience of the bomber pilots was only 17½ hours. British bombers, including the new DH–9, were simply inadequate to the difficult task of flying long distances over hostile territory to destroy precision targets. Furthermore, the 250- and 500-pound bombs they carried were later found to have had little effect either on factories or on communications. The result was similar to the later experience of Bomber Command in 1940: a move away from specific, precision target bombardment to an indiscriminate bombing of cities. The Air Policy Committee advocated operations "to attack the important German towns systematically . . . until the target is thoroughly destroyed," with the object of disrupting industry and undermining civilian morale.[35] Given the available aircraft, this was something of a recognition of operational reality, though it produced a campaign that was neither systematic nor effectively strategic.

On the basis of the work accomplished during 1917 and the early months of 1918, the Air Ministry pressed ahead with the establishment of a bombardment force with a properly defined function, better equipment, and clearer objectives. On May 13, 1918, the cabinet agreed with an Air Council proposal to "constitute an independent force of the Royal Air Force for the purpose of carrying out bombing raids on Germany on a large scale."[36] On June 5, the

Independent Air Force (IAF) was established. Based in France, its commander was General Hugh Trenchard, formerly commander of the RFC and a previous critic of the policy of independent bombardment.[37] Two weeks later, Chief of the Air Staff General Frederick Sykes produced a paper for the War Cabinet outlining the nature of the independent air strategy.[38] It so closely matched the course pursued by the RAF for the next twenty years that it is worth considering in some detail.

The Air Staff started from the assumption that long-range bombardment and offensive aviation were the most important functions of air power. Aircraft performed a task of "strategic interception," attacking the "root industries" of the enemy and "the moral [*sic*] of his nation" thereby undermining war willingness at home and draining away supplies from the enemy's naval and land forces. If properly organized, air power was considered "the most probable determining factor for peace" and "affords the best and most rapid return for the expenditure of national resources of man-power, material and money."[39] The "Strategic Striking Arm" was to be employed in two complementary ways: first, in bombing specific target systems such as munitions, submarine construction, chemicals, iron and steel, machine engineering, aero engines, and magnetos; and second, in bombing "densely populated industrial centres" in

Sir Hugh Trenchard, commander of the Royal Flying Corps in the First World War and known as the Father of the Royal Air Force, initially criticized independent bombardment.

order "to destroy the morale of the operatives."[40] The geographical center of attack was the Ruhr, where much of Germany's heavy industry was concentrated. Sykes's memorandum plainly expressed the strategic principles on which independent bombardment should be based, and it established the foundation for the postwar commitment of the RAF to the independent exercise of air power.[41]

At this stage of the war, American soldiers and airmen became introduced to Allied air strategy. It presented an opportunity to those in the American forces who also argued for strategic operations to contribute directly to the planning and operation of the first major strategic air campaign. This was an important event, since most American airmen were of junior Army rank and would have found great difficulty in promoting the wider use of air power on their own. General John J. Pershing, the American Expeditionary Force C-in-C, like Haig, the British commander on the Western Front, saw long-range operations as "of secondary importance" and insisted that battlefront aviation was the primary object of air forces.[42] This remained the dominant view of air power among senior U.S. officers throughout the war. Cooperation with the British allowed the Americans to introduce independent air operations by the back door.

At the time, much of American bombardment aviation, like that of Britain and France, was confined to large-scale tactical bombing of the battlefield and rear areas of the German lines. Its organizer was the energetic Brig. Gen. William "Billy" Mitchell, most prominent among the new corps of Army aviators. Mitchell's contribution to the development of air power in the war was arguably as significant for tactical aviation as it was for strategic aviation. His main ambition was to establish concentration and independence of air forces and to use the flexibility of the new weapon to the full. Like Trenchard, whom he admired, Mitchell was an advocate of bombardment to support the battle on the ground, using air forces where he felt they were most successful — in massed offensives of fighters and bombers against troop concentrations, supplies, and rear communications, distant from the front.[43]

If any have claim as the originator of American strategic bombardment, two names stand out: Maj. Raynal Bolling and Lt. Col. Edgar Gorrell. Secretary of War Newton Baker appointed Bolling to head the June 1917 aeronautical mission to Europe charged with drawing up recommendations on the size and purpose of America's air forces. Gorrell, accompanying Bolling, was named Chief of the Air Services Technical Section in Paris and, in November 1917, Chief of the Strategical Section. Both men were closely involved in discussions on the merits of strategic bombardment among British, French, and Italian military and industrial leaders. Bolling recommended in an August 1917 report that the United States should produce large numbers of bombardment aircraft, particularly for night bombing when success depended on the sheer numbers of aircraft. "Could night bombing be conducted," he wrote, "on a sufficiently great

Brig. Gen. William "Billy" Mitchell advocated bombardment against troop concentrations, supplies, and rear communications distant from front lines.

scale and kept up continuously for a sufficient time, there seems good reason to believe that it might determine the whole outcome of military operations."[44] It was Gorrell, however, influenced strongly by Italian aircraft designer Giovanni Caproni and British naval officer Spenser Gray who spelled out in November 1917 how this might be achieved. Gorrell argued for sustained, concentrated attacks by day and night against major industrial targets in Germany "with a view to causing the cessation of supplies to the front."[45] The effect would be to undermine civilian morale as well. The impact on morale and material would destroy the flow of resources to the front armies and end the war.

To achieve these objectives, the Bolling mission called for equipping 260 squadrons of aircraft by 1919, including 60 for bombardment. In the summer of 1918 Gorrell, now Aviation Officer at the American General Headquarters (GHQ), increased the number of squadrons planned to 358; 110 were bombardment squadrons scheduled for July 1919.[46] In Washington, D.C., Congress appropriated $1.6 billion for an aircraft procurement program, but as Maj. Gen. Mason M. Patrick, Commander of the Air Service, American Expeditionary Force, later remarked, it was based "on frantic boast and foolish word."[47] Little of this program materialized. In France General Pershing diverted almost all American aircraft to tactical aviation; Gorrell was removed

Maj. Raynal Bolling (shown here as a lieutenant colonel at the *left*), headed the aeronautical mission to Europe in 1917 which recommended the production of a large number of bombers especially for night bombardment. Lt. Col. Edgar Gorrell (*right*, as a colonel, being decorated by General Sir D. Henderson) argued for sustained, concentrated bomber attacks against major industrial targets in Germany.

in February 1918 as head of the Strategical Section (significantly rechristened simply GHQ Reserve a few months later, in June); and Bolling was killed in an ambush on the ground. The production of large numbers of technically advanced bombers and trained pilots, the major ingredient of the whole enterprise, had only begun when the war ended.

Much the same could be said of the British IAF. The supply of new bombers and aero engines increased more slowly than had been hoped, and aerial priority continued to be given to demands from the front. The French high command remained hostile to the whole idea of an independent air force. Trenchard later complained that they "put every difficulty in my way," though he himself was something of an obstacle to the strategic employment of aircraft, giving way to demands from his own army to attack tactical objectives.[48] A September 1918 survey conducted by Lord Tiverton showed that in August, of all IAF sorties, those against enemy aerodromes accounted for 49 percent of the activity, but those directed against iron and steel works amounted to only 7 percent and those against chemical works could claim only 8 percent. In June

1918, the bulk of the aerial raids (55 percent) was directed against rail communications behind the German front lines. The first major attacks on German cities did not occur until September, almost a year after the strategic campaign was first authorized.[49] In October, after much argument, an interallied force was established to combine British, French, American, and Italian bombers in a single strategic force, but it did not see action before the Armistice brought the war to an end.[50]

Postwar investigation of the impact of the bombing campaign showed that the raids were small in scale and that the material damage was slight. The greatest problems for the Germans were caused by air raid alarms disrupting their activity and the civilians' fear of bombing in the major target areas. Throughout its existence the IAF dropped only 540 tons of bombs (much of them on tactical targets), while British bombardment aircraft as a whole dropped 6,402 tons of bombs between 1916 and 1918 on the Western Front alone.[51] At its peak the IAF comprised only 120 aircraft. Trenchard later reflected candidly about the force: "I told them that this high-sounding name was all moonshine.... What I commanded was a few squadrons. I was not anybody very much."[52]

The experiences of 1918 highlighted the considerable amount of time and industrial effort required to mount a bombing campaign. At war's end, Allied air forces had completed plans for a large-scale strategic campaign during the spring and summer of 1919. In June 1918, the American War Department authorized the 358-squadron program to provide a total of 110 squadrons of day and night bombardment aircraft in 1919. Domestic aircraft production between July 1918 and June 1919 was to provide 15,000 DH–4 light day-bombers, 1,350 Handley-Page heavy bombers, and 1,115 Caproni bombers.[53] When these figures proved unrealistic, the General Staff reduced the number of bombardment squadrons planned for mid-1919 from 110 to 42. By the Armistice in November 1918, only 60 American bombers had reached the front.[54]

The IAF had also planned for rapid expansion, from ten squadrons in the autumn of 1918 to forty-eight by May 1919, when the IAF was to be used for a systematic attack on German chemical, iron, and steel works in the Ruhr. In June 1918 Britain had 360 bombers in service for tactical and strategic bombardment; by November the number was 500.[55] Moreover, American and British airmen were preparing the operational and intelligence details necessary to conduct the 1919 strategic campaign, including producing target maps and folders and collecting detailed information on the German industrial economy.[56] At the end of 1918, both forces were poised to conduct the first, large-scale strategic air campaign against the economy and morale of an enemy power, independent of, though related to, the operations of Allied ground forces.

The end of the war rudely disrupted these preparations and left a strong sense of unfulfilled expectation among the airmen and officials involved. This was an important legacy, for it allowed supporters of strategic operations to claim they never had a real opportunity to prove the efficacy of strategic bombardment. The whole strategy, left suspended in midpreparation, also allowed an attitude of might-have-been to survive among the air officers, and it created, certainly in the RAF, a reluctance to approach air power with any other strategic vision. General Sykes later argued that had 500 bombers been available in the summer of 1918, "there can be no reasonable doubt that the Germans must have collapsed...owing to the disorganization of their munitions factories and industries."[57] It is tempting to see the development of strategic bombardment doctrine in Britain and the United States in the interwar years as a renewed attempt to resolve the unfought campaign of 1919.

For all of their drawbacks and evident ineffectiveness, the German attacks on London in 1917 and the Allied attacks on Germany in 1918 were crucial reference points in future discussions of air strategy. The obvious evidence that German bombing had done relatively little material damage and had caused only temporary and local panic did nothing to blunt RAF claims that the effect of the limited raids of the IAF on Germany "can hardly be over-estimated," or dilute the view of General Percy R. C. Groves, RAF Director of Flying

The Italian aircraft designer and manufacturer, Giovanni Caproni, is shown here with his Caproni bomber that was planned for production in late 1918.

Operations, that "never before in the annals of warfare has so small an application of force produced such immensely disproportionate results."[58]

This perception remained even after a commission sent to Germany to study the effects of the bombing placed the Allied achievement in a much more sober perspective.[59] Nor did the experience of war provide clear lessons about the relative merits of attacking military and economic targets with those of attacking home morale, though the RAF accepted uncritically Trenchard's view that morale effects were twenty times greater than material ones.[60] Early attempts at bombardment first made clear to RAF and American Air Service leaders that strategic campaigns could only be mounted by large and operationally independent air forces and that, despite much political hostility and public misgivings, the civilian population and domestic economy could be regarded as legitimate targets in any future major war. This marked a fundamental change of attitude toward warfare and brought the issue of strategic bombardment into the center of subsequent discussions on total war. This became the single most important issue in the arguments, after 1918, among airmen, soldiers, and sailors about basic questions of strategic doctrine.

The Development of Bombardment Doctrine

In evaluating the evolution of strategic bombardment doctrine in the interwar years, it is essential to distinguish between the popular debate on bombing and air power and the development of doctrine within the armed forces. A significant gap always existed between the often exaggerated, unproven assertions of military writers who fueled popular fears about air power and the actual strategic intentions of air planners. That is not to say that this debate, particularly when it touched on the principles of warfare, did not influence the military and civilian authorities, for abundant evidence exists that it did so. But conclusions from public debate on doctrine do not necessarily reflect official air doctrine, and they should not be confused with it. Official doctrine was rooted much more firmly in the operations of World War I, and the pattern established then in terms of air power theory is what largely colored the subsequent development of air doctrine among the major belligerent states. *No* major doctrinal breakthroughs occurred after 1918.

What aerial bombardment could do covered a relatively wide spectrum from direct battlefield attack to major strategic operations against enemy cities, but it was an easily defined spectrum. After the war, no air force remained ignorant of this form of warfare, but each adopted a different part of the spectrum according to its own view of air power or the prevailing view held by its civilian and military authorities. Nor did ideas about air power produce any general principles on which all were agreed. Bombardment doctrine was never fixed over this period; it was often poorly defined and open to the most ambiguous interpretation. It was "a picture," wrote a German air historian in 1926, "of general uncertainty and the greatest conflicts."[61] This remained true even for Britain, the one country that in 1939 had an air strategy in which strategic bombardment was a major component. If doctrine remained volatile before 1939, however, it was also genuinely international. Ideas on strategic bombardment enjoyed wide currency, and its defenders drew regularly on a fund of foreign opinion to strengthen their own claims.[62]

Strategic bombardment doctrine thus developed at a number of levels. At the highest level it had to be integrated into the general doctrine of the armed forces and into the grand strategy pursued by national authorities. Below this, it developed within the individual air forces. Finally, it was influenced by and related to the wider public debate on the nature of air power and total war. In general, strategic bombardment enjoyed greater prominence in the second of these levels than in the first; indeed, a wide disparity often existed between the role of air power in the general plans of the armed forces and an air force's own perception of that role. At the level of public debate, strategic bombardment had an influence out of all proportion to the role actually allotted to it in future warfare, and out of all proportion to the actual experience of bombardment during the Great War.

At the highest level — integrating strategic bombardment doctrine into the overall mission of the armed forces — few developments during the interwar period devolved from the experience of 1918. In almost all cases the prevailing general aim remained the immediate destruction of the enemy armed forces on land, at sea, and in the air. For this the air forces remained a vital but auxiliary service. The U.S. Army Field Service Regulations in 1923 concluded that "the mission of the infantry is the general mission of the entire force" and that pursuit aviation was "the most vital element of the air service," views that were echoed elsewhere.[63] In France, the army insisted that the general strategy was one of defense in depth on a "continuous front" and that the major role of the air forces was in direct ground support and in providing an umbrella (*coverture*) over the fighting front.[64] Even when, in the 1930s, the armies in America and France accepted that an air force might be capable of conducting independent operations, these operations were only acceptable as an extension of the general aim of defeating the enemy ground forces and of protecting the home army from air attack.[65]

Such views of air power highlighted the important elements conditioning the development of strategic bombardment: the attitude, often hostile, of the traditional services toward air forces, and the enormous influence exerted by the other services at the level of national policy. In most cases during the interwar period, the armies had powerful and entrenched positions in the national military establishments, which gave them much greater influence than the air forces in deciding questions of national importance. In the United States, the War Department was really an Army ministry, and the Air Corps, a subordinate arm of the Army.

In France, the *Conseil Superieur de la Guerre* was dominated by army spokesmen; in the Soviet Union, the air forces were entirely subordinate to the army and the army view of war. In Japan, the army and navy each had their own subordinate air services. Even in countries where an air force had a separate existence — in Britain and Italy — the highest military councils were still dominated by the two senior services. Under these circumstances it was inherently unlikely that the general mission of the armed forces would be influenced to any great degree by doctrines of independent air strategy.

Two other conditions also deeply affected the reception of strategic bombardment doctrine. The first was the political reaction against bombing. After 1918, repeated efforts were made to outlaw the bombardment of civilians and open towns, a proposal that would have reduced strategic bombardment to impotence. Though no general agreement was ever reached, governments were left in no doubt about the general public revulsion against long-range aerial bombardment. In the 1920s and 1930s, the RAF had to produce arguments for the legality of its proposed air offensives before the Committee of Imperial Defense (CID) would accept the strategy.[66] In both Britain and the United States, formal directives from the government forbade bombardment of civilians when hostilities opened.[67] Nor was the attitude confined only to politicians. In May 1939, General Maxime Weygand, French forces C-in-C in the Middle East, told a British audience: "There is something in these bombardments of defenseless people behind the front that smacks of cowardice, which is repugnant to the soldier."[68] Political sensitivity compelled air forces, right up to the outbreak of war, to work with a hidden agenda of strategic bombardment plans that were not able to be fully integrated into the general military mission.

The second condition was geopolitical. Given the limited range of bombers in the 1920s, incorporating a major strategic capability into the general military plan made little sense. This was certainly the argument used in America against strategic bombing (though it changed in the 1930s as aircraft range rapidly increased). With reference to Japan and the Soviet Union, questions of geography precluded the development of strategic air doctrine and long-range forces until at least the mid-1930s. In France and Germany, where range was less of a problem because of the proximate common frontier between the industrialized Ruhr and the French industrial regions of the northeast, the crucial issue remained the need for large land armies to physically repel invasion. In neither country was air power ever seriously considered as a substitute for the military shield.

The one country that, from a geopolitical view, had the most to gain by promoting an independent air strategy was Britain, whose special position as an island naval power close to Europe was compromised by aviation. Lord John Fisher, First Sea Lord during World War I, warned his countrymen that "[the] approaching aircraft development knocks out the present fleet, makes invasion

The first attempt at air-to-air refueling to extend the limited range of bombers in the 1920s came with the flight of the *Question Mark*, accomplished by the four men flanking Maj. Gen J. E. Fechet, Chief of the Air Corps, and shown here (*left* to *right*): Lt. Elwood Quesada, Capt. Ira C. Eaker, Maj. Carl Spaatz, and Sgt. Roy Hooe.

practicable, [and] cancels our being an island . . . ."[69] Air power was greeted as a new form of deterrent, a means to compensate for the loss of full naval protection. Prime Minister Neville Chamberlain in 1936 thought that aviation was "of first-rate, if not decisive importance" in British military strength.[70] This capability would also have the merit of reducing Britain's need to maintain a large conscript army, a requirement the continental powers could not avoid. In the 1920s strategic bombardment was presented as one way to defeat the French, should war ever occur between these powers, and in the 1930s it was embraced as a major deterrent or instrument for a counteroffensive against Hitler's Germany.[71] It formed the core of a new strategic conception of defense based on rapid offensive air action against an enemy's vital centers and morale, a view that was promoted with equal vigor by the U.S. Army Air Corps after 1939, when the political brakes on air power were finally removed by President Franklin D. Roosevelt.[72]

Yet even in the British case, strategic bombardment doctrine remained only a limited component of overall strategy, one that was never clearly defined at the highest level until the very onset of World War II, when its position as a doctrinal point became much less significant than airmen expected. Nevertheless, after World War I British strategy did not reject strategic bombardment as an important component of future warfare. This can be explained not only by

Agreement on the basic strategy for the course of the World War II was hammered out in the Grand Alliance between Roosevelt and Churchill aboard the HMS *Prince of Wales* in mid-August 1941. Shown above are General George C. Marshall (in conversation with FDR) and Admirals Ernest J. King and Harold R. Stark.

Britain's insular geopolitical position, but by the relative weakness of her army in the military establishment — Britain had, for only the first time, produced a major conscript army between 1915 and 1918 — and also by the support for long-range aviation in navy circles during and after this war. Moreover, a strategic bombardment strategy had powerful support among sections of the British administrative and political elite, for it promised cheaper wars, an end to attrition warfare in the trenches, and independence from continental allies.

Some of these conditions contributed to the survival of the RAF as an autonomous branch of the armed forces after 1919, which made it easier to pursue directly the idea of independent air strategies.[73] Despite the support it received, little attempt was made to define exactly what such a strategy entailed, or indeed to define grand strategy at all in the face of sharp cuts in defense expenditures and defense forces. Trenchard's view that "It is on the bomber offensive that we must rely for defence" was promoted, nonetheless, with great force and certainty.[74] At War Office staff exercises in 1923, Trenchard argued that the other services were essentially redundant: "It is probable that any war on the European continent in which we might be involved in the future would resolve itself, virtually, into a contest of morale between the respective civilian populations."[75] Not until 1928 were these arguments finally scrutinized by the Chiefs of Staff when they initiated a major debate, "The War Object of an Air Force."

The British Navy was particularly critical of the loosely worded and poorly evaluated doctrine of the aerial offensive suggested by Trenchard and the RAF Plans Department in the 1920s. "The time is now ripe," wrote the First Sea Lord, "to lay down explicitly the doctrine of the Air Staff as to the object to be pursued by an Air Force in war."[76] Some naval leaders held that the RAF doctrine of "direct air attack on the centres of production, transportation and communication" had no basis in existing operational evidence, was contrary to international law, and departed from the traditional principles of war, the central one being the concentration of effort against enemy armed forces. In the opinion of the navy and the army, an air force should properly attack enemy air forces.[77] Trenchard rejected this idea, and it was not included in the compromise formula for the air doctrine finally agreed to in 1928 by the Chiefs of Staff:

> The aim of the Air Force in concert with the Navy and Army is to break down the enemy's resistance. The Air Force will contribute to this aim by attacks on objectives calculated to achieve this end in addition to direct co-operation with the Navy and Army.[78]

This remained the major definition of the function of air power until the outbreak of war in 1939.

The definition established in 1928 left open what were the exact nature of the objectives for attack and how the enemy's resistance was to be broken. Trenchard insisted that it was not through attacking enemy air power but was through "air attacks which are to be delivered against the enemy's vital centres."[79] He now conceded that air power was just one of many ways to defeat an enemy: "I do not for a moment contend that any of the three services could act alone."[80] Strategic bombardment "will be one of the many means of exercising pressure on the enemy, in conjunction with seapower and blockade, and the defeat of his armies . . . ."[81] These views of the place of strategic bombardment in general strategy were not fully explored again until 1937 when the Cabinet and the CID began to plan seriously for the prospect of a war with Germany. What resulted was the first full evaluation of air doctrine since 1918, and its product was more relevant to that earlier conflict than to the one the RAF would fight in World War II. To the dismay of air force leaders, the CID relegated bombing forces mainly to a subsidiary role: to defend Britain against attack or support British surface operations.[82]

One reason for this view was the vastly different international situation the

British government faced in the middle to late 1930s. The aerial threat posed by Germany underscored a fundamental ambiguity in British strategic thinking. Belief that the bomber will always get through created a powerful fear that Germany would be able to inflict great destruction on Britain because of her superior air strength, and it would be foolish to invite such a prospect by deliberately planning to attack Germany from the air. This fear proved a strong deterrent against British action. "We thought of air warfare," Prime Minister Harold Macmillan later wrote, "rather as people think of nuclear warfare today." Until parity was reached and mutual deterrence appeared possible, the government gave offensive air power a more modest role and diverted financial resources to aerial defense.[83]

In the 1937 guidelines prepared by the CID, even offensive aviation was given a defensive purpose — to attack the enemy air forces to prevent their mounting a terror campaign against Britain. Second, the RAF was detailed to cooperate closely with the British Navy in preventing invasion; third, it was to support any army expeditionary force in Europe through attacks on enemy troop concentrations and supply columns. These functions were largely tactical rather than strategic in the sense understood by the RAF.[84] Only *after* these functions had been fulfilled would bomber forces be permitted to launch a limited offensive against the Ruhr, and that, as it turned out, occurred only after the government had issued a clear directive permitting attacks on civilian areas.[85] Though operational independence was retained, little was left of the central doctrinal conception of strategic bombardment fostered in the 1920s. It was not fully revived until Winston Churchill promoted it in late 1940.[86]

In many respects the American experience was the reverse of the British. Beginning in the early 1920s with a doctrine that subordinated bombardment entirely to the requirements of the ground battle, the U.S. Army Air Corps by 1939 had reached a position where independent strategic bombing operations were formally, if rather unclearly, recognized as part of the general strategic plan. In the 1920s, general strategy remained based on the assumption, firmly expressed by the Menoher Board report of 1919, that "an air force acting independently cannot win a war against a civilized nation, nor, by itself, accomplish a decision against forces on the ground."[87]

Air strategy was a priori subsidiary to the general aim of defeating an enemy's armed forces. That view was formulated in the 1923 U.S. Army Field Service Regulations, and despite widespread efforts by the Air Service to promote ideas of independent aviation, it was repeated more forcefully in the 1926 War Department's *Fundamental Principles for the Employment of the Air Service*. According to these regulations, "the organization and training of all air units is based on the fundamental doctrine that their mission is to aid the ground forces to gain decisive success."[88] This did not preclude all independent operations, though an attempt to include these words in the 1923 Army regulations precipitated such a storm of criticism that it was withdrawn.[89] So-called distant operations were justified only to the extent they directly promoted the aims of the ground forces; that is, they served as a long-range extension to tactical bombardment. This War Department view of the role of bombardment remained in effect throughout the 1930s.

The problem facing American airmen involved the difficulty of demonstrating that offensive, strategic aviation made sense in a situation when the nearest potential enemy was 3,000 miles away and could not be reached, even remotely, with existing bombardment aircraft. Furthermore, the American political climate of isolationism and widespread pacifism made it difficult to propose overtly offensive doctrine or gainsay the U.S. Army argument that the employment of light bombers and pursuit aircraft in cooperation with the Army and Navy was the most sensible course to pursue.

Some idea of just how bitterly the U.S. Army resented aviation claims to a larger slice of the strategic pie can be seen in the arguments, during the first Roosevelt administration, about the proper place of air power that marked a fresh attempt to assert the primacy of ground operations. In 1934 the War Department Special Committee on the Army Air Corps concluded that "the Army, with its own Air Forces, remains the ultimate decisive factor in war."[90] Maj. Gen. Hugh Drum, Army Deputy Chief of Staff, told the board of inquiry that there should be "no air operations not contributing to the success of the ground campaign" and that independent air operations "would be largely wasted and might be entirely ineffective."[91] Brig. Gen. Charles Kilbourne, Chief of the Air Plans Division, could mordantly dismiss Mitchell's defense of strategic aviation as "sensational fiction."[92] But the most hostile and persistent critic of strategic aviation was Brig. Gen. Stanley Embick, Assistant Chief of Staff, whose attack on air power in 1935 is worth recording in full:

> During the World War the role of aviation was substantially that of an auxiliary to the ground and sea forces. Though presented with innumerable opportunities, neither on land nor sea was it able to accomplish by independent operations anything of a decisive nature. It will be objected that the aircraft of today is a vastly improved and much more formidable instrument of war. Only to an extent is this true. They are larger, faster, and have a greater range, but they are not nor can they be made free from some fundamental inherent limitations. They cannot occupy nor control permanently either land or sea areas, they are impotent and helpless save in flight and must depend largely upon land and sea forces in their protection. They are fragile, vulnerable to the smallest missile, inoperable in bad weather, and exceedingly costly.[93]

While these views dominated the War Department and continued to do so during the mid-1930s under Chief of Staff Maj. Gen. Malin Craig, strategic bombardment played virtually no part in American strategy. Nor was the Navy Department helpful. Navy leaders resented the Army air arm's claims to autonomy in the 1920s and remained anxious to keep naval control over long-range aviation for coastal defense and attacks on the enemy fleet. A compromise reached on this issue in 1931 gave some responsibility for coastal defense to the Air Corps, but it left a residue of mistrust and hostility between the two services that hampered efforts to apply ideas of strategic bombardment even to defense of America's coastline.[94]

Between 1935 and 1939, American general war plans and field regulations reflected this more conservative view of air power. Strategic bombardment was not a component of those plans although the 1935 establishment of a GHQ Air Force as a separate U.S. Army Air Forces component did permit the possibility of independent operations in support of "the general plan of the Commander-in-Chief . . . and must be designed to support the general strategic purpose which he desires to attain."[95] Air arm leaders hoped to exploit this loophole fully, though the primary mission remained defensive support for the Army and Navy and attacks on enemy air power and the enemy's immediate rear areas. The prospect of independent, or strategic, operations was held in reserve by the Air Corps itself, but the concept could not be fully promoted until the political climate became favorable.

This occurred decisively in 1939 when President Roosevelt was able to commit America to higher levels of rearmament and General George C. Marshall became Assistant Chief of Staff. In January 1939, Roosevelt announced that "the Baker Board report of a few years ago is completely out of date."*[96] He authorized major air programs that included, for the first time, a large complement of heavy bombers. Roosevelt, like Churchill, had a sympathetic, if impressionistic, view of aviation which he brought with him to the presidency. (His symbolic flight to Chicago in 1932 to accept the Democratic Party's nomination flowed naturally from his previous interest in nascent naval aviation as Assistant Secretary of the Navy in World War I.) By the time of the Munich Conference in 1938, Roosevelt thought air power might actually end a war on its own. As Secretary of the Interior Harold Ickes confided to his diary: "the President would make war principally one of the air. He believes that with England, France, and Russia all pounding away at Germany from the air, Germany would find it difficult to protect itself . . . . It is his opinion that the morale of the German people would crack under aerial attacks."[97]

Helped by the changing mood of Congress in the late 1930s, Roosevelt and Marshall willingly supported a larger role for American air forces. By September 1939, the Secretary of War had approved new regulations on the employment of air power, and in October Marshall approved them and sent them on to the President. The new regulations placed aerial bombardment at the center of the Air Force mission and strongly implied that strategic bombardment was part of this center: "The strategic function of the military services is to defeat the enemy by the destruction of his means of waging war or by overcoming his will to resist."[98] These objectives in war, central to the strategic conception of Air Corps leaders, provided a framework for the evolution of the

---

* The Baker Board had been set up by the War Department in 1934 to report on and guide development of the Air Corps.

strategic offensives of World War II.

Outside Britain and the United States, strategic bombardment doctrine evolved quite differently, a consequence to large extent of the different lessons drawn from the air experience of World War I. In France, where strategic bombardment had enjoyed early support from prominent advocates, the army continued to regard air forces as entirely subordinate to the requirements of the ground forces. Not until 1933 did the French Air Force win any kind of organizational autonomy; even then it remained tightly within the army's strategic conception of combined operations. The army view of air power turned on the considerable success enjoyed by French tactical air forces in winning air superiority over the Western Front in 1918. The lesson of this conflict dominated its strategic choices for the next twenty years. French air forces would be used for tactical bombardment to disrupt the enemy's mobilization and advance in the crucial early stages of the war and would then provide an aerial cover for the army along the whole of the front in support of its major offensives.

These French views of air power, laid down in a general *Instruction* of 1921, remained in most important respects the major statement on air doctrine until 1939. It was modified, only briefly, between 1936 and 1938 when French Popular Front Air Minister Pierre Cot attempted to create operationally independent air forces in addition to the air forces that were assigned to individual armies. Even in this case, however, the 1936 *Instruction* did not commit the air forces to strategic bombardment. Rather it committed them to concentrated, massed attacks at critical points on the front which were supplemented by selective attacks on rear areas and communications to effect a breakthrough for the land forces. This more limited view of offensive air power was still difficult to promote in the face of army and navy hostility, and both of these services retained direct control over 118 of 134 squadrons. After 1938 the General Staff broke up Cot's independent force and divided it along the continuous front once again.[99]

This attitude toward air power was not, as is sometimes argued, mere conservatism on the part of French generals who dominated the French military establishment between the wars. In the first place, their views on air power were based on a firm conviction that this represented the most militarily effective use to which aircraft could be put. During the war, strategic bombardment achieved little of significance, whereas tactical bombardment clearly had. Moreover, by the 1930s, prospects for a more effective defense were growing. The rise of Nazi Germany again shifted the emphasis in French military planning to preventing a successful German breakthrough (*attaque brusquee*) by the concentration of air attacks (*une offensive aerienne*) on the enemy's mobilization. The French regarded this kind of air power as strategic because it would weaken the enemy's army and contribute in a major way to a victory on land for the combined services.

A second consideration was not doctrinal; it was political. Strategic bombardment was bound up with the historic tension in France between the army and civilian authorities. During World War I public and parliamentary pressure for bombing Germany contributed to soured relations between army leaders and politicians because the army regarded such attacks as a wasteful diversion from the main effort on the Western Front. This conflict of opinion in France, as in Britain, led to demands for an air ministry and an independent air force. Army leaders interpreted this as another attempt by Republican ministers to dictate military policy, and they rejected strategic bombardment, both then and later, as a strategy for amateurs, one poorly rooted in the classic principles of warfare.

In the 1930s these issues resurfaced. The Popular Front government's efforts to restructure French air power and its favorable view of independent bombardment were seen as interference in army affairs. Worse still from the French Army's point of view, Air Minister Cot was strongly influenced by the example of aviation in Communist Russia. He had visited the Soviet Union himself and had offered the Russians examples of French aviation technology. Strategic bombardment was seen as undermining army influence and challenging its conception of strategy with a doctrine its leaders regarded as left-wing, even communist. Cot's deliberate promotion of younger officers with Republican and strategic bombardment sympathies and the forced resignation of army generals hostile to his views gave this doctrinal conflict an overtly political character. When the Popular Front fell, the army, from doctrinal conviction and political self-interest, insisted on reversing Cot's achievement. By 1939, French aviation was governed once again by traditional instructions on combined operations and the aerial umbrella.[100]

Much the same thing happened to bombardment aviation in the Soviet Union, which Cot tried to imitate. During the mid-1930s the Red Army began for the first time to develop a separate heavy bombardment force designed, like its putative French counterpart, to achieve concentration of aerial firepower at critical points in the ground battle and directed toward the enemy's reserves and battle supplies. The ideas of Soviet Army Chief of Staff Marshal Mikhail Tukhachevsky, who favored mobile, mechanized warfare and concentration of air forces, were echoed by air advocates A. N. Lapchinskii and V. V. Khripin, who argued for independent bombardment aviation to conduct heavy attacks of limited range against hostile forces and force a path for advancing Soviet armies. Discussions in the mid-1930s also postulated strategic air attacks on Germany and Japan, but these ideas never became a part of Soviet aviation doctrine, a consequence, in part, of the politics of the Stalinist purges. In 1937, Tukhachevsky and half of the Soviet Air Force officer corps were purged by Stalin, who ordered a shift in priorities to attack aviation and small tactical bombardment planes as a consequence of the operational experience gained in Spain. In 1940, heavy bomber forces were broken up and sent to support individual military districts and army groups, leaving just a small headquarters air force to be used for emergencies on the front line.[101]

The situation in Germany, like that in France, was dominated by the lessons of World War I. The German Army never subscribed to the view that an independent bombardment campaign was strategically desirable. During the 1920s Germany was effectively disarmed in the air, and when secret rearmament began in the late Weimar period, it was the German Army that laid the foundations for the revival of German tactical air power. As a result, the army view of aviation strongly influenced the new German Air Force that Hitler created after 1933 and formally constituted in 1935. As in France, the German Army favored the doctrine of combined operations, with air forces detailed to perform three vital tasks: attack enemy air forces, defend German territory from air attack, and provide an aerial spearhead and protect ground armies. The primary task of the armed forces, according to the *Conduct of Air Warfare*, first published in 1936, was "to defeat the enemy armed forces," and it was the task of the Air Force "to serve this aim."[102]

This tactical requirement was underlined as the German high command gradually came to develop the strategy of blitzkrieg in the late 1930s, which massed air power with armor to pierce the enemy's forces in one decisive blow. German military leaders, drawing on their operational experience to support this view and relating air power to the basic principles of warfare in which German soldiers were steeped, were convinced that this was the most effective use of aircraft. Despite some arguments in favor of long-range heavy bombers to the contrary, in their judgement this method of using aircraft made operational sense, particularly since it favored concentrating the force on attainable objectives. The campaigns in Poland and France convincingly demonstrated that this approach could be successfully applied. Moreover, until the late 1930s it was not clear to German military planners who their potential enemies were, since Adolf Hitler's strategic guidelines were loosely worded and changed during the course of the period. Consequently, tactical aviation fitted more closely with the demands of warfare against the major land enemies that the German generals expected to fight.

Not until May 1939 did Hitler suggest that his army leaders evaluate the prospects of strategic air warfare of a different kind: "Is air attack against a factory more important than against a battleship?"[103] Though both Hitler and his *Luftwaffe* chief, Hermann Goering, had been happy to use air power for diplomatic ends by threatening bombardment from the air and were, at times, attracted to the idea of using a knockout blow, these views of air power were not incorporated in any substantial way into the general doctrine of the armed forces. Terror attacks against civilian morale were, according to prevailing instructions for the German Air Force at that time, "fundamentally rejected."[104] In Germany, as in France, geopolitical considerations, the weight of the army in the development of a central doctrine of combined operations, and a widespread aversion to war against civilians left little room for a serious doctrinal commitment to strategic bombardment.

Air Force Views

At the second level in the evolution of strategic doctrine — the national air forces themselves — ideas about the nature and employment of bombardment forces appeared much more prominently. To be sure, a significant gap existed between what the air forces thought their role should be and what was actually assigned to them by the general campaign plan. The U.S. Army Air Corps in some cases best exemplifies this because its refinement of strategic bombardment doctrine continued in defiance of the official view on aircraft employment. This doctrine also developed well in advance of operational capability and bombardment technology, and a gap existed between doctrine and operational reality. Nor did all airmen subscribe to the view that strategic bombardment was a major element in the exercise of air power. For much of the period, prevailing technology made tactical bombardment a much more realistic course, and combat experience, whether in Spain or China, underlined this.

Where the various Air Force or corps manuals or instructions included long-range aerial attacks, they generally specified a tactical objective: a long-range extension of battlefield bombing aimed at weakening the enemy armed forces at the front. In France, Germany, and the Soviet Union, many airmen assumed this to be the most effective use of bombardment aircraft and, on this basis, willingly accepted their integration into the general plan of the armed forces. Perhaps the most important internal division within air forces developed in the 1930s when improvements came in active and passive defenses against air attack. Champions of strategic bombardment depended on the correctness of their assertion that against the bomber there was no real defense. This had been entirely true during World War I, but it was demonstrably less so in the 1930s. Improvements in effective defense aviation (as the fighter again overtook the bomber in the 1930s), the development of radar and better communications, and the growth of static antiaircraft forces all created a situation in which one arm of aviation threatened to undermine the strategies promoted by the other.

Nonetheless, strategic bombardment was promoted widely after 1918, and with particular force after the American and British experiences. First among the reasons for this was the powerful sense that bombardment had not been used effectively during the war — the "supreme blunder" as General Groves, RAF Director of Flying Operations in 1918 put it — and just at the point when it was to prove itself, the Armistice intervened. "We realize," said Major George of the ACTS, "that air power has not proven itself under the actual test of war."[105] General Mitchell campaigned for American air power on this potential: "Just as the war was ending, the vital centers of the enemy were brought within range of the airplane's action. Heavy bombing of the Rhine cities was one of the main reasons for the Germans giving up the contest. They knew . . . if they did not give up the struggle, their cities would be laid in ruins."[106] Mitchell reflected the widespread view that the airplane was the one modern weapon of war not yet tested to the full, that any future war would realize its awful potential in large-scale, strategic bombardment attacks.

The second reason for promoting this concept was the link established in the popular mind between strategic bombardment and total war. The most important influence here was the Italian general, Giulio Douhet, who argued that future wars would be won by massing a huge air force, "an offensive power so great that it defies human imagination,"[107] to attack persistently the economic and population centers of the enemy until "the people themselves, driven by the instinct of self-preservation, would rise up and demand an end to the war."[108] Douhet believed that the central air weapon was the heavily armed strategic bomber, against which there was no effective defense, and by using this weapon to attack the enemy air force and the major industries supplying the enemy armed forces, conditions could rapidly be created for a complete collapse of the enemy's will to resist.[109] It was a vivid, overdrawn account based on flimsy operational evidence, but it fitted well with the new view of war promoted by military thinkers everywhere that stressed the importance of the economic and moral dimensions of warfare and the need to mobilize all the

General Giulio Douhet, an architect of early military aviation, believed that the central air weapon was the heavily armed strategic bomber, against which no effective defenses could be made.

resources of a nation. By implication, the total war theory made any part of the national structure a legitimate objective, and "Douhetism" provided the strategic guidelines for attacking that structure. The doctrine of strategic bombardment, for all the moral scruples on which it trampled, was perceived as the central doctrine of modern industrialized warfare.

The third reason for promoting strategic bombardment was the link between claims for air force independence and air strategy. Strategic bombardment was the only major strategic course that could manifestly justify operational autonomy and concentration of force and, most important of all, exploit the offensive character of the airplane to the full. Here, two figures whose influence on air power thinking was as great, if not greater than Douhet's stand out: General Mitchell and Marshal Trenchard.[110] By no means the only airmen who argued the case for independent, strategic air operations, these men had more success than their colleagues in promoting the cause with force and in public. Trenchard urged the "supremacy of the offensive" in the use of air power. Like Douhet, he placed his faith in the strategic bomber to attack any enemy repeatedly and indiscriminately at the outset of hostilities in the hope the enemy would give in first.[111] Mitchell's view was more sophisticated — he emphasized selection and destruction of the vital centers on which the enemy's war-making capacity depended, but he and Trenchard stressed that only an organizationally independent air force, free to pursue its own strategy, could produce the "proper use of air power."[112] Neither man argued that armies and navies had no role to play, but that the condition for achieving a decision on the ground could only be created by an air force, and indeed might be achieved without ground forces having to fire a shot.[113]

Strategic bombardment theory was, by its nature, closely bound up with these efforts to create an independent, unitary air force by giving air forces a distinct function, the chance to develop highly specialized and technically advanced weapons, and the opportunity (real or imagined) to win a war. If these arguments were sometimes carried to extremes, it must be remembered that they were conducted in an atmosphere of interservice rivalry, personal hostility, and profound skepticism.

It is no coincidence, therefore, that the only major, independent air force to be established during World War I, the RAF, was also the only major force in which strategic bombardment took firm doctrinal root from the start. The RAF *War Manual*, issued in 1935, summarized the commitment to strategic air power. The main aim of an air force was the counteroffensive carried out by strategic bombardment against an enemy nation. Air fighting and air defense were subsidiary duties — means, not ends. Subject to the obvious restrictions of geography and technology, strategic bombardment should be directed at the "nerve centres, main arteries, heart and brain" of the enemy nation; "If they are exposed to air attack, the continual interruption and disorganization of the activities of those vital centres by sustained air bombardment will usually be the most effective contribution which can be made by air power towards breaking down the enemy's resistance."[114] The most important effect of bombardment would occur on enemy morale, or war willingness: "the morale effect... is always severe and is usually cumulative, a proportionately greater effect being obtained by continuous bombing especially of the enemy's vital centres...."[115] The *War Manual* also detailed the tactical conduct of strategic operations. It emphasized the need for sound preliminary intelligence and for repeated attacks, concentrated on one target or target system at a time, carried out by waves of bombers until the objective was completely destroyed.[116] These two arguments — attacks on morale and vital centers and the tactical necessity of massed continuous attacks — were both well established in the RAF by 1923, became orthodoxy in the late 1920s, and continued to dominate RAF thinking during the period of rearmament and war planning.[117]

Much of this view of air power was shared by the American air arm in the 1920s, despite an official policy that saw air power in quite a different light. An RAF officer, involved in the CID survey of air strategy in 1928, quoted approvingly from the 1926 American Manual of Combined Air Tactics: "The objective is selected with a view to undermining the enemy's morale.... Such employment of air forces is a method of imposing will by terrorizing the whole population of a belligerent country."[118] Addressing the Army War College in 1923, Mitchell, then Assistant Chief of the Air Service, underlined the need for a "vigorous offense" and discussed in detail the nature of bombardment aviation aimed at distant "transportation centers [and] industrial centers" with attacks by day and night with the use of high-explosive, gas, and incendiary bombs.[119] The Chief of the Air Service, General Patrick, argued regularly for offensive aviation based on the British idea that "decisive blows from the air on rear areas" might end future conflicts.[120] At least some of this enthusiasm for strategic bombardment could be attributed to Mitchell's publicity or the influence of Douhet, but it is clear that men like Patrick and Benjamin Foulois were profoundly influenced by what they saw of German bombing in London in 1917 and 1918 and by their close association with the combined planning for the great unfought campaign of 1919.[121]

The U.S. Army Air Service in the 1920s was all too conscious of having to justify its existence and its strategic claims in terms that both reflected the uniqueness and novelty of air power and reduced its operational dependence on the Army and Navy. While not officially recognized, the air arm, by the 1930s, privately preached the view that "bombardment is the basic arm."[122] Maj. Gen. Frank Andrews, who became Commander of the GHQ Air Force in 1935, typified this approach, paying lip service in public to a more limited view of bombardment while arguing in private for "independent air operations" against factories, refineries, power plants and utilities, and centers of population, a strategy very similar to that eventually pursued during World War II.[123] The GHQ Air Force was used as a Trojan horse by bombardment aviators hoping

Maj. Gen. Benjamin D. Foulois, Chief of the Air Corps, stands in front of the air route map.

to turn it from an instrument of the army into a force for offensive, long-range bombardment. General Andrews's view was that the "backbone of this air power is Bombardment Aviation, for that is the type which can exert direct pressure on an enemy through the destructive effect of its bombing . . . ."[124]

Unlike the RAF, the American air arm devoted much more effort to evaluating the nature of independent air strategy and relating means to ends. Much of this work was undertaken in the 1930s at the ACTS at Maxwell Field, Alabama. It was here that the doctrine of strategic bombardment was developed, refined, and taught with little interference from the War Department.[125] Also unlike the RAF, the Army Air Corps had a case to prove in the face of strong resistance to strategic bombardment doctrine, not only from the Army itself but also from pursuit aviation which had a vested interest in rejecting the notion that a "bombardment attack once launched, cannot be stopped."[126] The ACTS bombardment staff started from the basic principles that air power was essentially offensive and could be directed at the enemy's will to resist independently of engaging his armed forces. The nature of modern industrial systems made them particularly vulnerable to disruption from the air, but this disruption could be effected, not by attacking morale, but through "destruction of carefully selected elements of the industrial system."[127]

This growing emphasis on selective or precision bombing of key targets, rather than on indirect attacks of ill-defined targets such as morale, reflected not only the influence of changes in bombardment technology but also the sensitivity of plans that involved civilian bombardment. Moreover, during the 1930s, ACTS instructors and other Army Air Corps thinkers believed pursuit airplanes were incapable of stopping a well-armed, massed bomber force. The new B–17 possessed excellent speed, which existing Air Corps fighters did not. Pursuit advocates, unaware of radar developments, remained a weak group in terms of their influence on Air Corps thinking. Not until after 1939, when the Air Corps had an opportunity to observe the war in Europe, did its leaders begin to take aerial defenses more seriously and view the enemy air force as an "intermediate target" whose defeat through bombing and air fighting was an important preliminary in the conduct of major strategic campaigns. In this regard they differed significantly from the RAF, whose leaders maintained, even in the face of radar and fast fighter development, that defeat of the enemy air force was a wasteful diversion from the central objective.[128]

By the early 1940s American aviators worked out a doctrine for strategic bombardment that greatly influenced the eventual campaigns waged against Germany and Italy. Despite, or perhaps because of, the official attitude toward strategic bombardment, the Army Air Corps produced a more thorough, detailed, and well-defined doctrine than any other air force. By contrast, the RAF, with much more doctrinal room for maneuver, depended far too long on loosely worded assertions about the nature of air power, assertions that had to be hastily reevaluated on the very eve of war, with disillusioning consequences.

In France, tension between the air force and the army certainly existed, but it was based more on organizational issues and less on doctrinal grounds. French airmen in general subscribed to the views of the army — that the air force was most usefully employed in protecting ground forces, attacking rear areas and communications, and counterforce operations designed to win air superiority over the battlefront. This could only be achieved by a direct conflict between air forces, not through an indirect campaign of bombing the sources of production. This represented the fundamental difference between Continental and Anglo-Saxon views of air power. French air theory did not preclude long-range attacks, but Frenchmen regarded such attacks as extensions of the main battle, a development of the theory of tactical bombardment developed in World War I. The main statement of French air theory, Camile Rougeron's *Aviation de Bombardement* published in 1937, argued that the ideal bombing aircraft should be derivative of fighter designs and used mainly against military targets, particularly the enemy air force.[129] Even airmen who urged attacks on "*des points sensibles du territoire ennemi*" [sensitive points in enemy territory] did so only to the extent that this contributed to weakening enemy mobilization, delaying his attack, or preparing the ground for major army offensives.[130]

Within the French military establishment it was generally believed — even by those like General Paul Armengaud who wanted a wider role for bombard-

Strategic bombardment was "kept alive" in the United States with a few B–17s produced in the mid-1930s.

ment — that Douhet's picture of air warfare was exaggerated and that a country with effective antiaircraft and air defenses could withstand strategic bombardment. It was the winning of air superiority that mattered, and arguments within the French Army turned on questions of how best to organize air units rather than questions of doctrine. French air leaders wanted large, operationally independent air forces that could be moved from one part of the front to another in massed offensives executed with a combination of fighter and bomber aircraft, whereas army leaders wanted to divide and assign air units among their own individual units. Though the army officially got its way in 1938, until 1940 French airmen continued to think in terms of air forces and massed offensives.[131]

In Germany, too, the role assigned bombardment in the plans of the *Luftwaffe* reflected the army view of combined operations. Many among the first generation of German air force leaders came directly from an army background and thought instinctively in terms of fighting enemy armed forces rather than in terms of attacking national structures. Bombardment was primarily intended to contribute to the general task of subduing an enemy's armed forces through defeat of his air force, intervention in ground and naval actions, and disruption of the supply of men and weapons to the front line. Even Lt. Gen. Walther Wever, usually regarded as the only *Luftwaffe* chief of staff to favor strategic air warfare, argued that "in the war of the future, the destruction of the armed forces will be of primary importance."[132] In the 1936 issue of the *Luftwaffe* Service Manual 16, the general statement of air doctrine and operational conduct specifically excluded the use of terror raids on cities, but it did under special conditions permit retaliatory raids provided their retaliatory nature was clear. The manual included no general plan for a strategic air offensive nor did it give particular prominence to the bombardment airplane, but it did not rule out the possibility of attacks to disrupt "the enemy's materiel supply" in the event of a stalemate on the ground.[133]

Some German airmen wanted a more serious commitment to strategic bombardment strategy, among them Robert Knauss who taught the *Luftwaffe* Air Academy cadets the virtues of the strategic offensive and the probability that an enemy's morale would collapse.[134] But stronger pressures prevailed. First, the Chief of the German Air Staff after 1939, Hans Jeschonnek (previously he had served as Chief of Operations), strongly favored tactical dive-bombing aviation after his early experience as commander of *Luftwaffe* training units. This preference for ground-attack aviation, reinforced by the campaign in Poland, was shared by other leading *Luftwaffe* generals from an army or fighter-pilot background. No prominent bombardment officers were to be found in the higher echelons of the air force.[135] Second, the *Luftwaffe* placed great emphasis on defense against bombing attacks, in particular on static antiaircraft forces which, after much interservice wrangling, remained under *Luftwaffe* control. The new generation of fast German fighters, the outstanding 88-mm antiaircraft gun, and improvements in communications all argued in favor of an effective defense. German air officers drew the conclusion the RAF refused to draw: that it made more operational sense to fight the enemy air force and protect the ground army than it did to waste men and machines on long-range bombardment.

Only the *Luftwaffe* C-in-C Hermann Goering might have changed this situation had he prevailed with his vague ideas about the knockout blow and

Lt. Gen. Walther Wever, the only *Luftwaffe* chief of staff to favor strategic air warfare.

attacks on enemy morale. But he failed to impress these views on Hitler and the German Army, and he refused to cooperate with the German Navy to plan air attacks on British trade and shipping, which might have produced genuinely strategic results. Nor did Goering direct his own staff to seriously consider strategic warfare until 1939, when four-engine heavy bombers were ordered in quantity for the first time.[136]

Some Italian and Russian air force officers also saw the merit of strategic bombardment, but in neither service was the conception thought through fully. Furthermore, the lessons of Spain and China inclined them to accept what the armies had argued all along: fighting an enemy's armed forces was the proper role of an air force. In Japan, the question never seems to have arisen; there, the air forces remained entirely subordinate to the army and navy, which used aviation to support surface operations.[137]

Proponents of strategic bombardment attempted as best they could to use or allay popular concerns about this form of warfare. At the public level, essayists expressed genuine fears about the possible effects of aerial bombing on civilian life. Throughout the interwar years a stream of books and articles appeared urging the need for air disarmament, effective aerial defenses, or large, deterrent bombing fleets. The basic assumption underlying most of this tension was expressed by French General Maurice Duval, Commander of the French Air Service in World War I: "The decision will be reached in the air." Future warfare was widely expected to produce aerial "frightfulness" in the form of mass destruction of urban areas by bombs and gas, which would constitute the knockout blow that haunted the popular imagination before 1939. Some of this writing had considerable influence on professional airmen as well, particularly Douhet's book on war or the British military writer Capt. Basil Liddell Hart's *Paris, or the Future of War*. But much was couched in alarmist and apocalyptic language, for example, *War on Great Cities* and *Air Power and Civilization*.[138]

This side of the debate was, from an air force point of view, a mixed blessing, for it encouraged a backlash against exaggerated claims for air power and provided critics of strategic bombardment with ample ammunition for their cause. Army General Charles E. Kilbourne, Chief of the War Plans Division, caustically dismissed Mitchell as a man "infected with the virus of that fatal disease, love of publicity" who resorted, according to Admiral William Moffett, Chief of the U.S. Navy's Bureau of Aeronautics, "to the revolutionary methods of the Communists."[139] Public perceptions also had the effect of restraining governments from pursuing a strategic bombardment capability because of fears of retaliation by the enemy or because such a course was judged morally repugnant. As a consequence, efforts devoted to aerial and static defenses were increased at the expense of bombardment. The disparate views held by airmen, the other military services, and the public all help to explain the relatively modest part strategic bombardment played in overall strategy before 1939.

The Establishment of Strategic Bombardment Forces

A great number of practical problems affected the evolution of strategic bombardment between the wars. In particular, those related to technology and the size and nature of the aeronautical industry ruled out a number of nations from the start. Creation of a strategic bombardment force required not only an organizational structure and training program, but also a satisfactory bombardment technology and the commitment of a great amount of productive and scientific resources to create an effective force. These elements did not necessarily develop in tandem with doctrine (as evidenced by the disparity between the number and performance of bombing aircraft available in the 1920s and the doctrine they were designed to support), but all were elements essential for turning theory into practice. As war in Europe approached, the relationship between theory and practice drew much closer and became, in the end, entirely one of interdependence.

The problems of formulating doctrine and of obtaining its acceptance at the highest levels were reflected throughout the interwar years in the very limited resources specifically devoted to a strategic bombardment force. Indeed, in terms of the strategic offensives waged during the war, no organization was satisfactorily equipped before 1939, and only one, the RAF, had a bombardment force primarily organized for major strategic operations. This situation resulted in part from the nature of bombardment aircraft of the period, which were in the main small, twin-engine aircraft capable of performing both tactical and strategic roles. Some, such as the multipurpose French aircraft, the Bleriot 127, were even designed to perform armed reconnaissance as well.[140] This tendency to group bombardment aviation, whether for performing tactical or strategic tasks, into a single bombardment organization obscured the degree to which strategic bombardment was a component of any individual air force. In the 1920s such distinctions were in any case academic, for bombardment aviation was small in scale everywhere, either divided up in direct support of major ground units or grouped, like the RAF, into a general air defense organization.

Not until the onset of large-scale rearmament and increased funding in the 1930s did Britain establish the first independent bombardment aviation force. In July 1936, Bomber Command and Fighter Command were set up as separate organizations, each with its own C-in-C and separate network of air bases. The bombers were placed under group commanders who kept in close communication with the central commander, permitting the organization of large-scale operations from scattered bases in the south and east of England. What the force lacked, even by 1939, was an adequate number of aircraft. The frontline bombardment force in August of that year comprised some twenty-one squadrons of twin-engine bombers.[141]

In the United States, an organization for independent operations emerged more by accident than by design. In the 1920s and early 1930s, American bombardment aviation was insignificant. Only fifty bombardment aircraft were produced in 1930, and the total bomber strength of the Air Corps in 1932 was ninety-six aircraft.[142] During arguments in 1934 and 1935 about the nature of air organization, the major concession the Army made was to establish a formal GHQ Air Force consisting of reconnaissance, pursuit, and bombardment groups directly under the Army high command. The establishment consisted of twelve active bombardment squadrons with a primarily defensive role.[143] This allowed greater concentration and a more flexible use of bombardment aviation to support ground forces in repelling attacks. Army Air Corps leaders saw this establishment as a potential base for the development of long-range independent bombardment aviation. But not until 1939, when the GHQ Air Force became the organizational umbrella under which the Air Corps could build up the offensive aviation force that Roosevelt's directives authorized, did this potential base become a remote possibility.

Outside Britain, decentralized organization prevailed, with bombardment units attached to particular geographical zones or military units. In Germany, *Luftwaffe* forces were divided into four major air fleets, or *Luftflotten*, composed of all major aircraft types. These fleets operated from seven air districts, or *Luftkreise*, each having from two to five bombardment groups attached.[144] No central strategic reserve existed, but bombardment aircraft could be organized for massed attacks, as during the Battle of France and the Battle of Britain. Yet the overall structure, like the French *Armee de l'air*, was designed to facilitate the general battle plan. In France, bombers were attached to individual army corps and divisions throughout the country. Even after the French Air Force won a measure of organizational independence in 1933, over 86 percent of all aerial forces remained tied to individual army units. Only during 1936–1937 was an attempt made to establish an independent bombardment organization whose forces could strike en masse with concentrated firepower at particular parts of the front. But this organization, disbanded in 1938, never comprised more than a few squadrons. In 1938 the French nominally had an establishment of 66 bombardment squadrons, although in April 1938 the actual bombardment strength consisted of 68 heavy and medium bombers and 125 light attack bombers.[145]

Much the same pattern developed in the Soviet Union. The Red Army air forces were first to organize independent bombardment units, though not a complete command, in the force reorganization of 1935. These units were grouped together in a special strategic reserve, *Aviatsya osobovo naznachenya* (AON), composed largely of the heavy Tupolev four-engine bombers. These forces were designed for support of major Army campaigns, but in 1936 Stalin ordered a halt to the development of heavy bombers because of their poor safety record, and in the following year he threw his support behind medium-bomber and ground-attack aviation on the basis of the Soviet experience in Spain. In 1940, AON was broken up and its bombardment forces were distributed among major army units for direct support of the army in combined operations.[146] For the most part, European bombardment units were organized to work closely in tactical support of the other services. At most, they could be massed to provide large-scale shock attacks at decisive points in the land campaign.

The question of training went hand in hand with the development of bombardment aviation. Two areas, bombing accuracy and navigation, related directly to strategic bombardment, though only British flying personnel trained with a strategic offensive specifically in mind. Once again, however, something of a gap existed between strategy and performance. Operational and tactical questions took second place to organizing and equipping the force prior to 1939. Until 1937, it was even assumed that a pilot could fly *and* navigate his own aircraft. A second pilot, or copilot, was introduced on bomber aircraft in

1938, but on bombers with cabin space for only one person, the observer/gunner was given a short course in navigation so he could assist the pilot. Not until 1939 was it finally realized that a fully trained navigator was essential for any aircrew engaged on long-range bombardment operations. Training in navigation was then altered to produce specially trained navigator officers, and the pilot was relieved of this broader responsibility. Nevertheless, training in night and blind flying was still rudimentary at the outbreak of war. The hours of night-flying training amounted to only 10 percent of the hours spent on day flying in 1938. With existing technology and training, Bomber Command could only attack on clear days and moonlit nights with any hope of even reaching the target! Thus it was inevitable that Bomber Command could not launch a large-scale attack with any accuracy against Germany's vital centers when war began in 1939.[147]

These same problems affected bombing accuracy. The British used bombsights in 1939 that differed little from those available in 1918. Accuracy in high-level bombing, thought to be the only tactic viable for strategic attacks, simply could not be obtained with available methods.[148] In March 1938, a Bombing Policy Subcommittee examined bombardment tactics, and, at its first meeting, members admitted that these tactics needed to be completely reevaluated, "as the effort required for the destruction of certain targets by existing methods under war conditions would appear to be of considerable magnitude."[149] Prior assumptions held that the problem was not technical — that there would be no "very marked improvement" in bomb aiming and navigational aids — and it could be solved simply through better training.[150] Bombing exercises, however, showed that even at altitudes of 2,500 feet against undefended targets, only 15 percent of the bombs hit the target.[151] Exercises later in 1938 showed that accuracy of high-level bombing was only three hits per hundred.[152]

During the course of 1938, RAF leaders decided to establish a Bomber Development Unit as a "center for the formulation of tactical and strategical doctrine" with facilities for "practical trials to assist in the determination of bombing errors."[153] This unit, however, did not become activated until two days before the outbreak of war. Equipped with small numbers of light and medium bombers, it lost even these at the end of September 1939 when they were transferred to operations. "For the time being," concluded Bomber Command C-in-C, "we shall have to rely upon getting our experience on active service . . . ."[154] The poor level of preparation in bomb aiming and navigation led by default to the situation in which Bomber Command was compelled in 1941 to abandon precision bombing and adopt area bombing.

The U.S. Army Air Corps' approach to training and tactics was much different. Accurate, tactical bombardment training went back to the early 1920s and contributed in the long run to the American preference for a strategy of precision bombardment attacks. This view was reinforced by a belief that the

American public would not accept a bombing strategy based on massed attacks against civilian targets. Furthermore, the bomb-aiming equipment of the Air Corps was superior to that of the RAF, and with the development of the Norden stabilized bombsight, the American equipment had a level of technical achievement that permitted degrees of accuracy well beyond anything available to the RAF at the outbreak of war.[155] The U.S. air forces also taught a higher standard of navigation, and the training was more specialized. From the early 1920s, the American forces pioneered long-range display and training flights that encouraged high standards for navigational equipment and training.[156]

The operational evaluation of bombardment accuracy and bombardment tactics was also well advanced, and at least part of the 1920's argument for a separate and larger Army air arm was based on the demonstration of bombing accuracy and destructive power.[157] Bombing probability charts, drawn up in 1933, recorded bombing accuracy for the exercises held from 1924 to 1932. The probability of hitting a precision target the size of a small factory was 64 percent from 3,000 feet and 19 percent from 10,000 feet. Moreover, bombing accuracy was improving yearly.[158] By 1939, the GHQ Air Force commander was confident his force could hit targets through clouds and during bad weather.[159]

A restrictive factor in the United States was the small number of trainees in bombardment aviation. In 1941, Army Air Forces Commander General Arnold viewed this as the major constraint on the expansion of strategic bombardment forces.[160] Yet the element that really counted in the evolution of bombardment aviation was not men, it was machines. Neither a satisfactory strategy nor an effective organization could compensate for two related and vital considerations: the quality of bombardment aircraft available and the quantity that could be procured. The relationship between technology and strategy was complicated and depended on a number of important variables — the range needed to reach potential enemies, the state and size of the aircraft industry, the quantity of state funds available, and the ability or willingness of the air staff to specify clearly the type and number of aircraft required. Of all these variables, the state of existing aviation technology and productive capacity was arguably the most significant, for heavy bombardment aircraft were immense consumers of economic and scientific resources.

The development of strategic bombardment forces depended on planning for the required industrial capacity and providing generous funding well in advance — up to four years — of actual operational capability. Nations that first made this commitment enjoyed a lead that made it difficult, if not impossible, for others to catch up in the short term, even if the competitors could afford the cost. Indeed, it would not be an exaggeration say that the recognition in Britain and the United States of the importance of a purposefully built heavy bomber and the eventual success in securing its production in quantity are the most important differences in explaining the use of air power during World War II. In contrast, the failure of heavy bombardment aircraft to materialize in Germany in the numbers and quality required limited the *Luftwaffe*'s strategic options. The situations in these three cases demonstrate not only the importance of technology and production capacity, but the wide range of factors that influence the choice of weapon systems.

During the interwar years, however, air forces were compelled to work mainly with small twin- or three-engine medium and light bombers. These were, by the 1930s, very different in performance from the bombers of 1918 in terms of range, altitude, and speed; however in terms of bomb-carrying capacity, developments transpired much more slowly. Most of the medium bombers operated by the major belligerents in 1939 could carry little more than the amount carried by the largest bombers of 1918 and considerably less than that of the so-called Barling Bomber developed in the United States in the early 1920s, which could carry up to 8,000 pounds of bombs.[161] The exception was the Soviet four-engine ANT–6 that saw service after 1930, but its fixed undercarriage, flimsy construction, and slow speed made it an easy target, and its bombload was only 5,000 pounds. Production after 1937 was halted in favor of more modern medium bombers such as the Ilyushin Il–4, which could more satisfactorily fulfill the tactical bombardment requirement.[162]

In Britain, the light and medium bombers were, for better or worse, the aircraft allocated for strategic operations throughout the interwar period. Their inadequacies were clear, for most could barely reach northern Europe from British bases with more than a modest bombload. Yet British airmen argued from their experience in World War I that strategic operations required, at a minimum, large numbers of bombers capable of carrying as large a bombload as possible. This question of the quantity of bombing capacity, expressed in terms of the potential daily bomb tonnage the force could deliver, became a major element in the bombing debate in the 1930s, for it was *the* measure of how limited Britain's bombardment capability actually was. This realization, coupled with exaggerated reports of Germany's bombardment potential, left RAF planners with no doubt that they had to find more bomb-carrying capacity. By the mid-1930s, aviation technology — spurred on by commercial developments, improved construction materials, and better aero engines — had reached a stage where very large, all-metal monoplane bombers of high performance had become a technical possibility.

Early in 1936, the British Air Staff drew up plans for a major long-range, four-engine bomber force. This force was needed, reasoning held, not only to face the danger of German air rearmament by achieving parity of bomb lift and avoiding the risk of obsolescence, but to use as a potential weapon against the Soviet Union, whose aviation was a threat to India and the Empire.[163] The Air Staff agreed to issue specification B12/36 for a four-engine aircraft with a maximum range of 3,000 miles, the ability to carry 8,000 pounds of bombs, a service ceiling of 28,000 feet, and a speed of 230 mph.[164] Though some favored procuring large numbers of medium bombers instead, Group Captain Arthur Harris's argument that the RAF needed "to get the maximum offensive potential (which in bombers meant weight-carrying capacity and range) out of any given number of first line aircraft" carried the day.[165] The specification, issued to industry later in 1936, resulted in the Stirling heavy bomber.[166]

This was by no means the end of the story. During 1936 and 1937, rapid developments in fighter aircraft and antiaircraft defenses would make the heavy-bomber project vulnerable to strong objections. Moreover, the situation in the British aircraft industry, which had only just begun to expand in response to increases in armament expenditures in 1936, made it unlikely that large numbers of heavy bombers could be produced until 1941 or 1942 at the earliest. Powerful financial and industrial interests favored abandoning heavy bombers and concentrating production instead on new medium bombers such as the Hampden and Wellington already in the pipeline.[167] Undeterred, the Air Staff initiated a search in 1938 for what it called the Ideal Bomber. The search would take account of technical advances in bomber design and also provide much improved armament in anticipation of strong fighter resistance. The bomber was specified as the B9/38 with "a high cruising speed, powerful gun defense, long range, and a substantial bomb load." Its range was set at 2,000 miles with a bombload of at least 12,000 pounds and a speed of 300 mph. Bombload and range could be compromised for shorter or longer journeys, but the craft had to be capable of rapidly reaching distant parts of the Empire as a reinforcement.[168]

To meet RAF specifications for bombload capacity and range in 1936, British industry produced the Stirling heavy bomber, shown here in these front, side, and rear views.

Once again, objections to this proposal turned on the view that, for the same expenditure, a greater number of medium bombers could be procured, and more quickly. The Air Staff in December 1938, however, produced a detailed defense of the plan, which showed that for a 4,000-ton daily bomb lift, the force needed 3,584 Wellington medium bombers but only 896 Stirling heavy bombers, at a respective cost of £79 million and £47 million. The most telling argument of all was that the heavy aircraft needed only 27 percent of the flight and maintenance personnel for an equivalent medium bomber capability, and only half the quantity of aluminum.[169] These arguments, communicated directly to the Cabinet, tipped the scales.

By this time, serious delays in British bomber development and production arose owing to shortages of labor and industrial capacity and the difficulty of finding a satisfactory engine. Without question, Bomber Command would have liked to have had some of these aircraft available in 1939 or 1940, when they were sorely needed, but heavy bombardment aircraft would not be available *in quantity* until 1942. Regardless, their cost-effectiveness and greater striking power made them the central weapon for the RAF in the 1940s. As it turned out, the "Ideal Bomber" failed to materialize with the outbreak of war, but two projects for heavy twin-engine bombers issued as specification P13/36 were converted to four-engine types in 1939 and 1940 to meet the Ideal requirements. These became the Halifax and the Lancaster heavy bombers.[170]

Only the size of the commitment remained to be determined. The industrial effort was enormous, but serious industrial preparation only began in 1936 under Lord James Weir, a champion of strategic aviation in World War I. Recognizing that industrial capacity was an important constraint on the eventual size of the bomber force, Weir embarked on a program of industrial mobilization with great urgency: "The country may be at peace," he wrote, but "the whole spirit and atmosphere in the supply departments must be one of war pressure on every individual. . . ."[171] A system of shadow factories was set up to produce aero engines and components in wartime; programs of labor retraining began; and aircraft firms, in close cooperation with the Air Ministry, greatly expanded their own productive capacity.

The purpose of this British effort was to create in peacetime the industrial infrastructure necessary to produce much greater numbers of modern aircraft, including heavy bombers, if war came. A commitment to forge a strategic striking force was built into these plans from the start. The program planned in 1939 was to produce a force of 2,250 heavy bombers by the autumn of 1941, with nine weeks' reserves and a total bomb-lift capacity of 3,795 tons. These figures were designed to match what Germany was expected to achieve with her heavy bomber force by the same date.[172] Although the British figures were never fully achieved, enough was done to secure the technical means to embark on the bombing offensive in earnest in 1942, six years after the initial commitments to the heavy bomber. By that date, however, Germany still had

During the mid-1930s, financial and industrial interest favored production of new medium bombers, such as the Hampden (*top*) and Wellington (*bottom*).

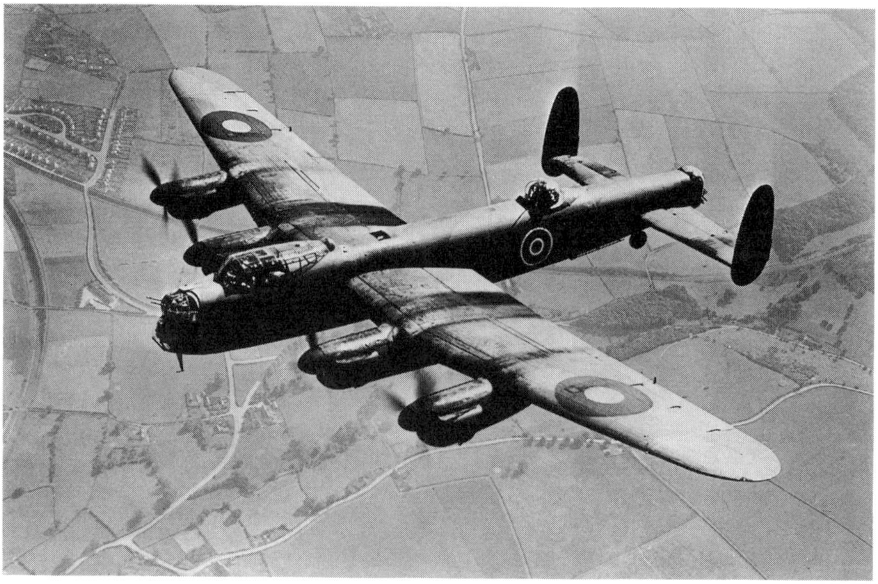

British Halifaxes (*top*) and Lancasters (*below*) were developed as heavy bombers in 1939 and 1940 to meet "Ideal Bomber" requirements at the outbreak of World War II.

no heavy bomber force.

U.S. Army Air Corps efforts to produce a heavy bombardment force met with much greater resistance. This was partly a consequence of the insistence in the 1930s on a defensive strategy and the absence of any obvious targets for offensive planes of great range and payload, and partly a by-product of the general hostility of the War Department to Air Corps planning that did not conform with the Army view of tactical aviation. Certainly, the standard of performance of the new generation of monoplane medium bombers in the 1930s fulfilled the Army's requirements and undermined the Air Corps' case. Yet in defending the need for a heavy bomber, the Air Corps was also defending its potential to develop an independent strategic force. Both sides well understood that this was the principle at stake.

The roots of the American heavy bomber lay in the late 1920s and early 1930s when American bombardment aviation reached something of an impasse. The great improvements in aircraft performance in commercial airlines were not reflected sufficiently in military aircraft, but the low priority given to bombardment aviation and the low level of military spending made it difficult to embark on a major redevelopment of the weapon. As a result, obsolescent types were kept in service longer than Air Corps leaders thought desirable.[173] In 1931 the Chief of the Air Corps authorized a thorough revision of bombardment aviation, which led in 1933 to recommendations that the corps proceed to develop four-engine heavy bombardment aircraft.

The Army Air Corps laid down two specifications for these aircraft. One was for a bomber capable of at least a bombload of 2,000 pounds carried over a distance of 1,000 miles at speeds of 200 to 250 mph. The other called for a more ambitious project, a bomber with a 5,000-mile range, capable of hitting targets in Hawaii or Alaska from mainland bases. The Boeing Airplane Company won both design competitions, producing in 1935 a prototype for the first (popularly known as the XB–17) and embarking that same year on the development of the second (the XB–15). Since both projects were in the experimental stages, the General Staff approved research on both. The Air Corps was delighted with Boeing's product, for the heavy bomber was regarded as more cost-effective than the medium bomber in terms of resources and personnel, and it was an aircraft that could ultimately defend Hawaii, Alaska, and Panama. General Oscar Westover, who became Chief of the Air Corps in 1936, considered the Boeing B–17 "the most outstanding airplane development of modern times."[174]

During 1936, the War Department attitude toward heavy bombers changed sharply, coinciding with the transfer of General Douglas MacArthur (who had been sympathetic to the project) and his replacement as Chief of Staff by General Craig. Both Craig and his deputy, General Embick, were hostile to efforts at procuring strategic bombers, since their priority was to obtain for the Army as many ground support bombers as possible from a limited congressional appropriation. In April 1936, the Secretary for War approved twenty-six B–17s for the Fiscal Year 1937 program instead of the sixty the Air Corps had requested. In August, Westover was told to reschedule these for 1938, and in December the order was canceled entirely in favor of medium twin-engine bombers.[175] By early 1937, the Boeing company faced financial difficulties because of unpredictable shifts in the procurement program, and the Air Corps budget ran a deficit, even without the heavy bombers, because of the Baker Board's recommendation for additional spending for aircraft.[176]

Both of these problems weakened the support for heavy bombers and fueled the hostility of the Army's leadership to the whole concept of long-range bombardment. The War Department took the view that heavy long-range aircraft were an unnecessary waste of resources when they could only be used overseas to perform tasks for which the Navy was better equipped. Moreover, what the Army wanted was more small bombers, such as those that had proved their utility in the Spanish Civil War and promised "greater efficiency, lessened complexity, and decreased cost."[177] Given the likely conflicts facing the United States and the prevailing military strategy, Craig believed that planes as heavy as the B–17 were "not justified." "Aside from their undemonstrated military utility," Craig told the Army-Navy Joint Board in 1938, "the relatively high cost of the large, long-range planes must be considered in relation to the effect of that cost on the other requirements of the Army Air Corps."[178] Development of the heavy aircraft was permitted to continue, though starved of adequate research and development resources. But during the course of 1938, even the modest plans for heavy bombers in 1939–1941 were scrapped in favor of a large number of smaller attack bombers.[179]

The Army assault on the heavy bomber project at least had the welcome effect of forcing senior airmen to think hard about why they wanted a weapon of this kind. "The basic problem," Arnold wrote to the Adjutant General in 1937, "is 'What is the role and employment of the GHQ Air Force?' The answer has never yet been defined." Arnold's own definition included the view that the force was rapidly becoming obsolescent and that it should be provided, in principle, with "bombardment aviation at least the equal in numbers, range and speed performance, and striking power, to that of any other nation."[180] General Andrews, commander of the GHQ Air Force and a powerful advocate of the heavy bomber, defined the need in bolder strategic terms. On his assumption that "bombardment aviation is the basic element of air power," he argued that America should concentrate development and procurement on the very best bombardment aircraft available, aircraft capable of attacking enemies in the Pacific, reinforcing American possessions, and ranging far out over the Atlantic. Offense was the best form of defense.[181]

Arguments such as these merely confirmed what the General Staff already suspected: that the heavy bomber was a back door to strategic bombardment and Air Corps independence. Despite Air Corps protests, Secretary of War

Harry H. Woodring endorsed the Army view and cut from the plan the small number of B–17s already scheduled for Fiscal Years 1939 and 1940. The Army defended the decision by demonstrating that, with the money saved from a few B–17s, 154 smaller bombardment aircraft could be procured. In the terms in which air strategy was officially cast in 1938, this choice made considerable sense and it satisfied the demand of Congress for rapid results in expanding the number of frontline aircraft.[182] With only a few B–17s available, strategic bombardment was kept alive in the Air Corps by the slenderest of margins.

President Roosevelt's decision to expand American rearmament in response to the Munich crisis of September 1938 transformed the situation. He appointed an Air Board in March 1939 to investigate procurement, and the board came out strongly in favor of the heavy bomber. Procurement of B–17s in quantity was authorized, and development of a larger bomber, previously carried on with a single prototype aircraft up to 1939, was accelerated. In January 1940 permission was given to issue specifications for a new superbomber with a range of 4,000 miles, capable of reaching Europe from the United States. This became the forerunner of the B–29 and the B–36.[183] A substantial complement of heavy bombers was included in the production programs; the initial objective was 498 by 1941, with 1,520 by the end of 1942.[184]

The sudden expansion of aircraft production placed great demands on the aviation industry. Though the new appropriations covered plans for heavy bombers, the additional factory capacity first had to be created. The President was asked to authorize a major program of industrial expansion at a cost of over $2 billion.[185] This included a considerable amount of direct government investment in the aircraft industry to provide additional floor space, similar to the shadow factory scheme operated in Britain. And, as in the British case, the American plan involved industrial leaders and engineers directly in organizing the effort.[186]

Preparations for just such an industrial mobilization had in fact begun long before 1939. The Air Corps recognized, in the 1920s, that the real cause of America's disappointing aviation record in the 1917–1918 period was the almost complete lack of industrial preparation. General Arnold later wrote that "after World War I the lessons of the failure of aircraft production in 1917 remained uppermost in my mind." During the 1920s the Air Corps drew up detailed plans to convert industry to this purpose. Projections included lists of firms where key components and materials could be procured in wartime. By the early 1930s, mobilization planning was based on the ability to produce 24,000 aircraft in the first year of war, including 3,250 bombers.[187] From 1936 onward, industrial flesh began to cover the planning bones, and by 1939, industry already had a firm foundation for large-scale mobilization. Without this preparation of a manufacturing base capable of mass-producing heavy four-engine aircraft, America's strategic offensive in 1942 and 1943 would scarcely have been possible.

German efforts to create a long-range strategic bombing force broke down over just such questions of production and development. In 1935 the *Luftwaffe* began to search for a heavy bomber with long range, able, like its British counterpart, to attack targets in the Soviet Union. Technical deficiencies led to the cancellation of the initial designs, the Dornier Do 19 and Junkers Ju 89, both of which had flown by 1936. These design efforts were replaced with a Bomber A project for a very long-range bomber aircraft, which went to the Heinkel firm.[188] The subsequent design, the He 177, was built into *Luftwaffe* production planning from January 1939 onward, to be produced in quantity in 1941 and 1942.[189] Yet this effort suffered difficulties from the start.

The head of *Luftwaffe* aircraft procurement, Col. Ernst Udet, was a fighter ace turned stunt man and film star. He understood very little about aircraft production and even less about large aircraft. Udet, supported by Chief of Operations Jeschonnek and other key members of the air staff, now insisted that all German bombers, irrespective of size or purpose, have a dive-bombing capability. This was partly to emphasize their tactical potential, because it was assumed that dive-bombing would achieve much greater bombing accuracy and therefore reduce the number of bombers needed to obtain a given level of destruction.[190] This effectively forced Heinkel designers to work within very limiting technical constraints. Two engines were coupled together on each wing to make diving more practical, but no way could be found to make these engines technically secure, and the aircraft throughout its life had a very poor safety and performance record. Nor, with dive-bombing capability, was it possible to attain the kind of bombloads or range that the original *Luftwaffe* specification had called for.[191]

The general hostility to heavy, level-flying, long-range bombers gave the German project a low development priority. Moreover, arguments about the relative merits of heavy versus medium bombers went in favor of the smaller aircraft in Germany. "We do not want these expensive, heavy machines," Udet told Heinkel, "which eat up more in material than a medium, twin-engine dive bomber costs."[192] This attitude was encouraged by Germany's shortages of materials and skilled labor in the late 1930s, but it owed something to the development of a most successful medium bomber, the Junkers Ju 88, whose promised range, lifting power, and short development time made it a much more attractive proposition for the immediate future. Goering diverted all available resources in the aircraft industry to building up an intermediate force of high-performance medium bombers capable, it was believed, of strategic as well as tactical tasks.[193]

By 1939, a significant portion of Germany's aircraft industrial capacity was already mortgaged to medium bomber output and could not be switched to heavy bombers without serious disruption. The irony was that Germany, alone among the major powers, had developed radio navigational aids capable of locating targets in poor weather and at night with a greater degree of accuracy

Col. Ernst Udet, a World War I ace turned stunt man and film star, insisted that all German bombers have a dive-bombing capability.

The He 177 (side and bottom views) that went into *Luftwaffe* production in 1939 featured this dive-bombing requirement and was beset with development difficulties.

The Ju 88, the *Luftwaffe*'s most successful medium bomber.

than was available to either the RAF or the U.S. Army Air Corps; further, the *Luftwaffe* was also well aware of the tactical advantages of mixing high-explosive and incendiary bombs to achieve the maximum destructive effect in urban areas.[194] Had arguments over the type of bomber in the mid-1930s gone in favor of the heavy four-engine aircraft for immediate development, Germany not only would have possessed a strategic bombing capability sooner than Britain or the United States, but it would have had one with a high degree of operational effectiveness.

Bombardment in Practice

The German emphasis on tactical bombardment and smaller aircraft was also a consequence of its participation in the Spanish Civil War. Because the lessons of that conflict confirmed assumptions already held by many German airmen, the impact of that experience should not be exaggerated. It was a factor that decisively tipped the balance at an important time in favor of ground support aviation. This was true as well for other powers. All the air forces closely examined the limited experience of air fighting between the World Wars — in Ethiopia, China, and Spain — to see what clues they supplied to the practical use of air power in the future.

German airmen drew a number of lessons in Spain. The first and most important was that "strategic" attacks on morale and economic targets were much harder to execute and less militarily effective than had been previously assumed. Such attacks did not defeat enemy forces. Morale was, if anything, strengthened through bombing, increasing the enemy's determination to resist. Attacks on distant targets were not decisive, whereas dive-bombing attacks on enemy formations and enemy air installations could be. Neutralizing enemy air power permitted the ground forces to operate more freely.[195] As one German writer put it: "The decision in war devolves on the ground and on the forces that fight on the ground, not in the air or from the air. . . ."[196] The results of the air war in Spain also demonstrated how difficult it was to mount a persistent bombing campaign in the face of well-organized resistance that produced a steady loss rate of aircraft and crews. The need for highly specialized flying personnel coupled with relatively high levels of attrition made bombing a costly operation.[197] It is a curious historical twist that the air force responsible for the notorious bombing attack on the Basque town of Guernica in fact drew conclusions from the conflict unfavorable to independent, long-range bombardment.

The French and Russian experience in Spain produced similar conclusions. An article in *La Science et la Vie* in 1937 enumerated three major lessons from the conflict: numerous defending fighters could prevent heavy bombers from reaching their targets; heavy bombers of the kind envisaged by Douhet "have not justified their existence"; and light bombers performed well in support of ground forces.[198] During the Spanish conflict in 1937 and 1938, the fact that the balance in performance and destructive power was moving in favor of the fighter became generally clear. This contributed to the view that tactical bombardment was more viable, since bombers could be protected more easily by fighter cover and could be used as a counterforce instrument. The experience in Ethiopia and China was broadly similar. When General Embick attacked Army Air Corps calls for autonomy in 1935, he recalled that in Ethiopia the "Italian progress from day to day is measured solely by the slow advance of the men in the mud . . . the role of military aviation must by its inherent nature be essentially of an auxiliary character."[199] In China, the Japanese army and navy began by attacking military targets but they moved on to attack urban residential districts from the air. Though such attacks were widely condemned by international opinion, observers recalled that these attacks were coordinated with movements of ground forces and that the small aerial defense forces of the Chinese experienced greater success against bombardment aviation than had been anticipated. American fighter pilots who fought on the Chinese side demonstrated that Douhet's view of the bomber always getting through was no longer tenable.[200]

The same was true for Soviet forces that fought with the Chinese in the two major battles against Japan at Chang-ku Feng and Nomonhan in 1938 and 1939. Evidence from these conflicts showed the need for good ground attack aircraft supported by large numbers of fast, light support bombers to protect ground units and destroy enemy air power. The combat lessons learned by Soviet airmen between 1936 and 1939 only further underscored their commitment to tactical aviation, massed in short penetration attacks in support of major ground operations, and to counterattacks by enemy bombers and fighters. In the Soviet Union, the emphasis placed on the large, level-flying bomber gave way to an emphasis on tactical, attack aviation. This brought the Soviet Air Force very close to the position that the French and German Air Forces had reached in 1918 when combat experience underlined the importance of battlefront support and the winning of air superiority.[201]

If the RAF had been involved in this kind of active combat in the three years before 1939 it, too, might have drawn some of the same conclusions. It was not, however, involved in major combat until 1940, although it kept close watch on what other air forces were doing. The only active air fighting that the RAF faced during the interwar years, far from dampening enthusiasm for strategic bombardment, had the effect of giving it a spurious validity. During the 1920s the RAF practiced what it called Imperial Air Control, a cheap means of maintaining internal security in certain areas in the Empire and mandated territories.[202] This approach consisted of small-scale bomb attacks on largely undefended native villages with the purpose of subduing rebel tribesmen cheaply and quickly. The Air Staff saw these attacks as a microcosm of the larger strategic campaign, since detailed intelligence on targets was required and attacks on "vital centers" were involved. Sir John Slessor, Director of Plans in the Air Ministry in the late 1930s, later wrote:

> Whether the offender concerned was an Indian Frontier tribesman, a nomad Arab of the northern deserts, a Moreilli slaver on the border of Kenya, or a web-footed savage of the swamps in the Southern Sudan, there are almost always some essentials without which he cannot maintain his livelihood.[203]

These "essentials" were the vital centers whose destruction ended the natives' will to resist through demoralization as much as physical damage.

Whatever French General Weygand might think of it, Imperial Air Control helped sustain the RAF view that morale was the vital objective, not only against "semi-civilized peoples," as the RAF *War Manual* put it, but against major industrial powers as well.[204] In the 1920s, an explicit assumption in air planning was that the French would yield sooner than the British under bombing. The same naive racialism was directed at the Germans in the 1930s. One senior RAF officer, faced with the awkward evidence of mass panic among Londoners during the World War I bombing raids, blamed it on "aliens from the East and Northeast end of London."[205] Imperial arrogance remained undaunted by the observation of other wars in the 1930s — indeed it was argued that the morale impact of bombing was evident in Ethiopia and China — so that British air strategy, though it involved some concessions to defense after the invention of radar and the development of the Spitfire, was still committed to strategic bombardment when other nations were turning to tactical aviation.

Planning for War

Only two air forces, the RAF and the U.S. Army Air Corps, made systematic operational plans for a strategic air offensive before the onset of hostilities. Other forces either fully committed to tactical aviation or, in the German case, postponed serious preparation until a heavy bomber force became available. German armed forces as a whole had not begun operational planning for a war in the west even by the summer of 1939, so convinced was Hitler that the Polish conflict could be localized. The only preparation undertaken by the *Luftwaffe* was a preliminary study by General Hans Felmy on the prospects of an air war against Britain, and he concluded that little success could be expected until 1942 when heavy bombers would reach the squadrons and the training program would be complete.[206] Though the German Navy wanted to plan a coordinated air/sea blockade of Britain, Goering was too jealous of the naval service to divert air resources to assist it. Air intelligence did produce target folders for major military installations in Britain and France, but no evidence exists that German planners thought in terms of establishing targeting systems designed independently of the other services to achieve strategic aims, which may well explain the haphazard pattern of objectives chosen during the Battle of Britain and the Blitz.[207]

The important thing about British and American preparations is that they were conducted with particular enemies in mind and a particular kind of strategy. As early as 1937 the Cabinet accepted that Germany was Britain's key enemy and that war with her was highly probable. The U.S. Air Force — whose own contingency plans in 1936 had identified Germany, Italy, Japan, and the Soviet Union as the most likely enemies — was directed to concentrate its efforts on the assumption of a war with Germany. For the RAF, the resultant Air Ministry directive issued at the end of 1937 opened the door to active preparation for strategic operations against Germany's economy and home morale, although the terms of the document made it clear that combined operations and attack on the enemy air force should receive priority.[208]

The Air Ministry directive, issued at the end of 1937, was based on the CID's more limited view of air power. According to this approach, the RAF had to give effective help to the defense of the British Isles in cooperation with the Royal Navy. Members of the Air Staff accepted these limitations with poor grace. They wanted operations that would directly attack Germany's domestic economy and home population since this would weaken the whole of the German air effort and destroy German air power at its source. The argument about priorities continued until the outbreak of war itself, even after Bomber Command realized that it simply lacked the resources to mount the more ambitious strategic campaign it favored.[209]

The December 1937 Air Ministry directive ordered Bomber Command to begin detailed planning for a bombing campaign. Earlier in the year, the Air

Staff drew up a list of important objectives and targets that it organized into four main groups in order of priority. Group I, *Western Air Plan W5*, included mainly defensive and tactical tasks. Only the fifth and final objective in this group was strategic in the RAF sense of the term: "Plans for Attacking Enemy's Manufacturing Resources in the Ruhr, Rhineland, and Saar." Group II included attacks on the enemy aircraft industry, but general attacks on German industry, government offices, and administrative centers were included only at the very end of the list, in Group IV, *Plans W11* and *W13*.[210] The Air Ministry directive did not include attacks on civilian morale or communications distant from the front. Top priority instead went to counterforce activity — attacking German striking power with the British striking force. Bomber Command C-in-C Air Marshal Sir Edgar Ludlow-Hewitt protested strongly against this choice since in his view and that of most of his staff, attacks on the enemy striking force were wasteful and ineffective. He favored concentrating on *Plan W5*, the attack on the Ruhr: "I am much impressed with the high importance of this plan as a practicable operation." He calculated that Bomber Command could cripple the electric power system of the Ruhr "in about a fortnight" with the use of 1,500 sorties and an estimated loss of 88 aircraft. For an attack on German air bases, his force lacked bombers of sufficient range and adequate intelligence on base locations.[211]

The argument remained unresolved in the months leading up to the Munich crisis in September 1938. When it appeared likely that the Czech issue would lead to war with Germany, Bomber Command wanted to be free to ignore the counterforce instruction and attack the enemy's economy instead: "When ready, go for the RUHR as intensively as possible — primary targets, power generating stations and coking plants. Bound to cause heavy loss of civilian life."[212] This was precisely the kind of attack the British and French governments wished to avoid for fear of massive retaliation from Germany, whose air striking power was thought to be much greater. The threat of a knockout blow was far too serious to risk. In April 1938, the French and British staffs formally agreed to avoid "the intentional bombing of civilians." This agreement was endorsed on September 19 when Bomber Command was directed to attack only clearly identifiable military targets and conserve its forces for defense of the United Kingdom and British shipping.[213] These instructions remained in force until the outbreak of war a year later.

Not only did this decision make sound strategic sense, but Bomber Command knew all too well that it stemmed from its own weaknesses. When the Command confronted the problem of actually demonstrating how it would conduct a strategic campaign in 1937, a whole host of deficiencies was exposed. Most obvious was the failure until then to evaluate the nature of the vital centers and award them some sort of priority as targets, or to assess how they might be attacked. Using material provided by the Industrial Intelligence Unit's Air Targets Sub-Committee that had been established in 1936, the Air

Ministry prepared a series of detailed evaluations of major German target systems in the first half of 1938. Six major targets were selected: electric power, fuels, the chemical industry, the metallurgical industry, engineering and armament facilities, and transportation. Each evaluation was accompanied by detailed appendices that listed individual targets, provided maps of each system, and analyzed each system's disposition and vulnerability to attack.[214] Of these, the Air Staff selected electric power and fuel oil, the basic energy supply of the German economy, as the key targets, and the Ruhr as the area in which these selective attacks should be concentrated, with the object to cause the maximum possible destruction of the German war industry.

Selection of the Ruhr was justified, not only because the largest part of German heavy industry was concentrated there but because air leaders realized Bomber Command could not reach more distant targets with the aircraft then available.[215] Indeed, when operational preparations began in the summer of 1938, it was soon evident that not even the Ruhr targets could be reached with any effect, nor could they be hit with any acceptable degree of accuracy. In August 1938 Air Marshal Ludlow-Hewitt wrote to the Air Ministry explaining that attacks on German targets from British bases could barely reach the fringes of northwest Germany for at least the foreseeable future, and that his command would face unacceptably high losses even on these attacks. Britain should rely instead on the North Sea and strong air defenses to mitigate the force of a German bombing campaign. For the British at that point to launch a similar campaign "might end in major disaster."[216]

If the RAF had identified vulnerable targets by 1939, its operational capacity for systematically attacking them was almost nonexistent and would remain so until larger aircraft became available. In September 1939, the frontline force still had only 488 medium and light bombers, mainly Blenheims and Whitleys, with short range and poor lifting power.[217] Moreover, too few air bases allowed the operation of large aircraft. Discussions about the production of bombs in sizes larger than five pounds only began in the summer of 1938, but by April 1939 the force still had nothing larger.[218] The central tactical questions about whether or not to provide escort fighters for bomber formations and what the ratio of defensive to offensive armament on bombers should be were only addressed during the spring and summer of 1939, and then only inconclusively.[219] In a whole range of more minor tactical and operational questions, Bomber Command was quite unprepared; it even lacked adequate maps for navigating flights in northwest Europe. Air Marshal Slessor, head of Air Ministry Plans in 1939, acknowledged these failures in his memoirs:

> It must be admitted that our imagination was not sufficiently flexible and our experience too limited to comprehend quickly enough the very far-reaching technical requirements of a modern striking force, capable of operating — of finding and hitting its targets — at long range in bad weather . . . . We attached insufficient importance to things which afterwards became a commonplace, like bombing and navigational aids, signals equipment, D/F [direction finding] homing beacons and blind landing systems.[220]

These cumulative deficiencies fundamentally reduced Bomber Command to ineffectiveness during 1938 and 1939 and left a large gap in British grand strategy when war finally broke out in September 1939. Bomber Command had no choice but to accept the gap between intention and reality — and the diversion of resources to the buildup of British air defense and a new continental expeditionary force. In discussions with the French General Staff during 1939, besides agreeing that bombing of civilian targets should be avoided, the two sides concurred that aerial bombardment should be directed against German troop formations and the areas immediately behind German lines. The French also urged counterforce attacks and raids on communications in the main area of German mobilization, very much along the lines laid down in French air doctrine. Bomber Command agreed that, in the event of a German attack on France, "the object of all the available bombers would be to contribute to the success of the battle on land."[221] Not only did Bomber Command have to admit its own weaknesses, but the RAF was forced, very

In September 1939, Britain's frontline bomber force consisted of 488 Blenheims (shown *above*) and Whitleys (*below*).

much against its will, to subscribe to French views of air power. Nor could Bomber Command even carry out the more limited tasks that the French requested.

Operational research showed that Blenheim and Battle medium bombers used at full strength in the first week of a German attack would only be able at most to keep three German railway lines closed![222] Anglo-French plans did not rule out an eventual attack on the Ruhr, although French generals remained skeptical of its value, but for the early stages of the war any general strategic bombardment plan was ruled out. Though the RAF planners still talked openly of the Ruhr plan, both before and after the outbreak of war, Bomber Command was directed in September 1939 to confine its activities to reconnaissance, propaganda flights, and attacks on North Sea targets. "Unrestricted air war," the Air Ministry declared, "is not in the interests of Great Britain."[223] This was the French view too. French Air Force Commander General Joseph Vuillemin told the French high command that weaknesses in bombardment aviation called for "*une tactique d'emploi tres prudente*" [a most prudent tactic of employment] confined to occasional attacks against close and poorly defended enemy targets or in support of the land battle, where it would be really effective. French bomber forces were even smaller than the British ones and amounted to only 440 mainly obsolescent bombers in metropolitan France.[224]

This failure on the part of the one air force formally committed to strategic air warfare can be explained in various ways — a shortage of funds during the mid-1930s, the demands of reorganization and force expansion during 1936–1938, or the fact that the major potential enemy was not identified with certainty until 1937. Nevertheless, it is difficult not to see some merit in the April 1938 complaint of the Permanent Secretary to the Treasury Sir Warren Fisher that "for some years we have had from the Air Ministry soothing-syrup and incompetence in equal measure."[225] The failure to evaluate strategy in operational terms showed both the strengths and weaknesses of the British military establishment. Much of the development of bombing doctrine and the force itself was left in the hands of civilian officials or ministerial officers, whose real talents were exploited in defining and promoting strategic thinking and in examining questions of organization and procurement but who often lacked sufficient practical or technical knowledge to turn thinking into operational reality.

Modern bombardment aviation employed a sophisticated weapon system that required special skills to understand and operate, skills that people at the top of the ministerial tree patently lacked. Operational questions were regarded as the province of regular airmen who were victims of poorly defined guidelines and inadequate resources, and of a certain discrimination on the part of the politicians and civil servants against the practical side of military effort. Thus there developed something of a gulf between "passive" and "active" parts of the force that only began to be bridged in the last months of peace. This situation contrasted sharply with what held in Germany. By 1939, Germany's bombardment forces had the high standards of operational effectiveness characteristic of all her forces, but they lacked clear strategic guidelines and satisfactory procurement policies. Britain had the strategy and procurement policies in place; she lacked the operational competence and tactical imagination necessary to make the effort work.[226]

Although it could be argued that the United States benefited from the mistakes made by the British and that the Army Air Corps learned much from closely observing the course of air combat after 1939, the American military establishment was able to strike a more effective balance between strategy and operations. Because of its late start in preparing its force, the Air Corps was no better placed than the RAF was to carry out strategic bombardment when war came in 1941, but its background preparations were more comprehensive and more thorough, and showed a more professional understanding of key tactical and operational issues. The shift in strategy in 1939 that favored building up large-scale offensive bombardment aviation and developing heavy bomber production served the Army Air Corps well. Between 1939 and 1941 the American service concentrated its efforts on the training program, on tactical appreciations of bombardment, and on securing sufficient detailed target information from the British to permit the drawing up of a full operational plan for a strategic bombardment campaign.[227]

The question of intelligence information was vital. In 1940, General Arnold established an air intelligence section with responsibilities for providing information and analysis of foreign air forces and bases and for detailing the economic systems of potential enemy nations, "culminating in analysis and description of the vulnerable systems and, finally, target selection and target folders."[228] Much material was secured from the British, who were better informed on the German aircraft industry and German communications. Other information came from industries and banks with investment and trading links in Germany and from individuals with detailed knowledge of German conditions. The final target systems selected for major evaluation were electric power, steel, gasoline products, the aircraft industry, and transportation. When the British and Americans agreed in 1941 that part of any joint strategy would be a "sustained air offensive against German military power" and Roosevelt authorized the so-called Victory Program for the defeat of Germany, the Air Corps already possessed all the material needed to produce a strategic plan.[229]

The Army Air Forces Air Staff produced an air plan as a separate annex to the general war plan, and the subsequent outcome in August 1941 — i.e., AWPD–1 — owed its authorship to officers who had championed strategic bombardment at the ACTS in the 1930s: Lieutenant Colonels Harold George and Kenneth Walker and Majors Haywood Hansell and Laurence Kuter. The air plan bore all the characteristics of the school's teaching. The target systems selected were electric power, transportation, and German fuel oils, all vital to sustaining every aspect of the German war effort, but all requiring precision bombardment for effective destruction.[230] Moreover, and in marked contrast to the RAF, the plan emphasized the importance of the enemy air force as an intermediate target, whose destruction was a necessary condition for the successful attack on the other target systems. Morale was rejected as a proper objective on its own; the assumption held that the interruption of the economic system would have a secondary impact on German morale since Germany's economy served both military and civilian needs. Altogether, some 154 targets were selected for repeated daytime attacks, with the use of large numbers of bombers.[231]

AWPD–1 called for an aggregate force of 251 groups (including B–36 bombers that would not be ready until 1945). This force would be composed of 61,800 combat aircraft including 37,051 trainers and 11,800 heavy bombers. These were numbers much greater than those included in the air plans produced earlier, in 1940 and 1941.[232] They reflected the evidence available from Britain that strategic operations required "greater numbers of long range airplanes."[233] The strategy depended on daytime precision bombing of the targets selected, always the American view, even in World War I. Again, it was vindicated by British experience. When Major Hansell met Group Captain Slessor in December 1940, the latter told him: "We feel convinced that bombardment aviation is effective *only* when it is so employed. Area bombardment does not produce effective results. You must do precision bombing."[234] This was something of an irony, for in 1941 the RAF gave up daytime precision bombing because of poor accuracy and strong German defenses and returned to the strategy of area bombing and attacks on morale by night — little different from the stage that strategic bombardment had attained in 1918.

The differences between British and American planning — the emphasis in the Air Corps on attacks against enemy air power, the rejection of morale as a target,[235] and the selection of a limited range of vital but precise targets — conditioned to a considerable extent the sort of campaign the two forces would fight during the war. These differences also resulted from the different circumstances facing the two forces between 1939 and 1941. After the defeat of France, the RAF was forced to produce an air strategy in 1940 as thc only effective way of striking Germany, and do so with little combat experience and a small force of short-range, twin-engine bombers. Nighttime attacks were imperative with forces of this kind, and accurate bombing at night was technically impossible with the equipment at hand. Moreover, establishing a large bombardment organization under conditions of war was quite a different proposition from preparing one in peacetime.

American airmen had the advantages not only of better equipment (a result of the closer attention paid to technical and scientific questions in the 1930s) but also of having learned from the earlier European war. These lessons furnished another profound irony. Observing German air attacks in 1939 and

1940, Army Air Corps observers arrived at the erroneous conclusion that the Germans had practiced massed attacks with long-range bombers for strategic purposes and that strategic bombardment would be a key component of the German war effort to come.[236] Grossly exaggerated, though not deliberately so, intelligence reports stressed the enormous air power potential of Germany and its commitment to large, heavy bomber production. The Air Corps maintained that Germany had produced more than 9,000 long-range bombers between July 1939 and December 1940 and could produce 12,000 more in 1941 (the true figures were 4,200 medium bombers in 1939 and 1940 and only 2,800 in 1941).[237]

The American strategic bombardment campaign was designed not only to complement the British effort, but to match the large aircraft production expected from Germany and the resultant German aerial campaign. It is easy to see how, with German forces victorious across the whole of Continental Europe, this picture of German air strength was derived but without this fear of German power, American production and preparation might well have been slower, or directed into different channels. When war came to America in December 1941, this was the one major element of strategy on which the two Allies agreed completely.

Conclusion

The great German strategic air offensive never materialized and, after the failure of the Blitz, was never seriously pursued until Hitler asked industrial leaders in 1943 about his heavy bombers.[238] The RAF and the U.S. Army Air Forces — the one overcoming its lack of operational preparation; the other, its relatively late start — did, however, realize their strategic ambitions by 1943. The campaign of 1919, postponed for a generation, was renewed with weapons that dwarfed all those of previous aerial bombings. Campaigns on this scale could only be mounted by major industrial powers. This explains at least some of the reluctance of countries like Italy and Japan to develop strategic bombardment forces before 1939. Even for the more industrially developed powers such as Germany or the United States, economic constraints could be used as an argument in favor of smaller bombers. To be effective, bombardment strategy required an enormous economic commitment as well as a massive military effort. The heavy bomber was so large and technically complex that plans had to be prepared well in advance for its production in quantity. This occurred in Britain and the United States and gave these powers technical and industrial leads that made it difficult for other nations to catch up in wartime. The choices that mattered had been made well before war broke out.

What governed these choices in a great number of cases was the political condition at the time. Bombardment aviation in the Soviet Union was restricted to tactical uses after 1937 largely because of the hostility displayed by the Stalinists toward strategic bombardment theory and the heavy bomber. In France, strategic aviation was identified in the eyes of the Army leadership with the left-wing Popular Front and its presumed collaboration with Bolshevism, and rejected as part of a conservative backlash in 1938. In the United States, isolationist and pacifist opinion strongly influenced the development of the armed forces in the 1930s and the attitude of the President toward air power. A strong case can be made for arguing that a receptive political climate, in its broadest sense, was essential for the development of strategic bombardment as a central feature of general strategy. Widespread support for bombardment strategy among officials and within military circles in Britain transcended any constraints that moral or practical objections might have exercised. Support from the President in 1939 and 1940 rescued American air strategy from the War Department's view of air power. If either Roosevelt or Churchill had strongly opposed strategic bombardment, it is unlikely that the combined offensive would have taken place.

Reasons for these differing political and strategic preferences also lie in the specific experiences of the Great War. American airmen were impressed by what they saw of the bombing in London in 1917, as were British politicians clustered on the balconies of the Savoy and Cecil Hotels to watch the German aircraft attacks.[239] Many junior officers and officials involved in planning the great retaliatory aerial campaign scheduled for 1919, Hap Arnold among them, later rose to more senior rank in the air organizations they served during the 1930s. By disseminating ideas about bombing and recruiting to those of similar views, each force adopted a profile dominated by the strategic use of air power. This did not mean that the choice of a strategic bombardment offensive was a foregone conclusion, for it still depended on a number of crucial decisions including, above all, the choice of a heavy, four-engine bomber. Without this, the campaigns would have been impossible. Yet the eventual success of the effort to develop a heavy bomber in the United States and the failure of that effort in Germany says much about the contrast in air power development of the major powers.

British leaders saw strategic bombardment as a way of overcoming the decline of sea power and the growing vulnerability of the Empire. It was a grand, new kind of deterrent, one to replace the dreadnought and the gunboat. It was "a striking and offensive air weapon," according to Lord Weir, "so powerful as to compel the most wholesome respect from friend and foe."[240] Air power promised to end dependence on European alliances and avoid another war of trench-based attrition. Bombers could bring war to any enemy (even the Soviet Union), blockade enemy powers, and undermine the enemy's economic capacity to carry on fighting. The British tradition of war-making depended to a greater than normal degree on these forms of economic and moral pressures and limited military commitment, to which World War I had been an unhappy exception.

The French and German experiences were just the opposite, and their military traditions correspondingly different. Neither power could make war without an army, and neither power was at home with indirect strategies of blockade, aerial or otherwise. Both countries' armed forces were dominated by army officers hostile to the idea of fighting civilians rather than other soldiers, and both forces shared a common fear born in World War I that diversion of aircraft to independent bombing would allow the opening the enemy army wanted to begin a ground offensive. The outcome of the Great War was clearly a victory for ground armies. Long-range bombing achieved very little for either the French or the Germans, but for the British and Americans it represented a campaign that might have broken the stalemate and won the war in 1919.

For American airmen, too, the experience of the war was most important. The birth of American air power was closely linked to the Europeans' arguments about the use of aircraft. American airmen were attracted to the idea of bombardment as a way to end the deadlock on the Western Front and avoid the high casualties of the trenches. They also had more confidence than their European colleagues in the war-winning potential of the new weapon. They came into the conflict at a time when bombardment on this scale was at last technically possibile. Moreover, strategic aerial bombardment highlighted the strength of America's war capability, which required a strong industrial base and advanced technology. This continued to be true after 1918. The development in aviation technology and air transport in the United States and the emphasis it placed on scientific research, technology, and mass production made bombardment not only a practical proposition, but one that matched America's distinctive modern stance and materialist outlook.

In the absence of official permission to study firsthand the industrial centers of any foreign powers, Army Air Corps officers in the 1930s studied their own, drawing conclusions about the vulnerability of their country's economic structure by looking at American industry.[241] The strategic air offensive reflected their deeper interest in seeing warfare as an extension of economic power and vitality and in terms of industry and technology, not simply seeing it in terms of combat. Perhaps it is no coincidence that strategic bombardment was embraced wholeheartedly before the Second World War by the two states in which industrial society was most fully developed and in which the planning and preparation for war relied much more on civilian initiative and a close collaboration of the politicians and industrialists.

NOTES

1. Lecture, Air Vice-Marshal A. S. Barratt, "Air Policy and Strategy," Mar 23, 1936, p 1, AIR 9/39, Public Record Office, Kew (PRO).

2. Lecture, Maj Harold L. George, "An Inquiry into the Subject of War," 1936, p 17, Andrews Papers, Official Correspondence, box 11, Library of Congress, Washington, D.C. (LC).

3. Barratt lecture, p 5.

4. Jean H. DuBuque and Robert F. Gleckner, *The Development of the Heavy Bomber, 1918–1944* (USAF Hist Study 6, Maxwell AFB Ala., 1951), p 53.

5. J. Slessor, *The Central Blue: Recollections and Reflections* (London, 1956), p 166.

6. C. H. Barnes, *Handley Page Aircraft since 1907* (London, 1976), p 147. The aircraft was the Handley Page 15 V/1500, of which three were in service by the Armistice.

7. Ltr, Admiralty to Foreign Office, Jul 21, 1910, in S. W. Roskill, ed., *Documents Relating to the Naval Air Service*, 2 vols (London, 1969), vol 1, p 21 (hereafter NAS Docs).

8. J. Cuneo, *Winged Mars: The German Air Weapon, 1870–1914* (Harrisburg, Pa., 1942), pp 128–133; J. A. Hennessy, *The United States Army Air Arm* (Washington, D.C., 1985), esp chaps 4–6.

9. See note 7 above; see also R. F. Futrell, *Ideas, Concepts, Doctrine: A History of Basic Thinking in the United States Air Force* (Maxwell AFB, Ala., 1971), p 8. Futrell records the view of the Army's Chief Signal Officer in 1907: "For the purpose of dropping explosives on an enemy, a high-speed aeroplane is hardly suitable."

10. B. Collier, *A History of Air Power* (London, 1974), pp 40–41; "First Annual Report of the Air Committee on the Progress of the Royal Flying Corps," Jun 7, 1913, NAS Docs, vol 1, pp 97, 101; "Board Minutes and Regulations for the Royal Naval Air Service," Apr 6, 1914, NAS Docs, vol 1, p 136.

11. *Notre avenir est dans l'air* (Paris, Feb 1, 1912), copy in Imperial War Museum, London, pp 11–12, 20, 24.

12. R. Kilmarx, *A History of Soviet Air Power* (London, 1962), p 13; D. H. Robinson, *The Zeppelin in Combat: A History of the German Naval Airship Division* (London, 1971), p 17 ff; C. Graham-White and H. Harper, *The Aeroplane in War* (London, 1912), pp 32–41, 202–211; R. P. Hearne, *Aerial Warfare* (London, 1909), chap 12, "Can England Be Raided?"

13. Robinson, *Zeppelin*, p 50.

14. Ltr, Peter Strasser to Imperial Command, High Scas Fleet, Aug 10, 1916, in Robinson, *Zeppelin*, p xv. Admiral von Tirpitz suggested that "a really effective concentrated bombardment of London by all available means . . . would have been thoroughly justified as one way of shortening this inhuman war." Von Tirpitz, *My Memoirs*, 2 vols (London, 1920), vol 2, p 306.

15. Robinson, *Zeppelin*, pp 345–351, 404. Zeppelins carried out a total of 366 sorties; 177 reached England. For details of the Zeppelin raids, see J. Morris, *The German Air Raids on Great Britain, 1914–1918* (London, 1925), pp 265–272.

16. "Report from Commander Spenser Gray to Director of Air Dept., Admiralty, on the Raids on Cologne and Dusseldorf," Oct 17, 1914, NAS Docs, vol 1, pp 179–181; Memo, Winston Churchill, "Aerial Defence," Oct 22, 1914, NAS Docs, vol 1, pp 181–183; Paper, Joint War Air Committee, AIR 4, Mar 3, 1916, NAS Docs, vol 1, pp 309–311. For operational details, see W. Raleigh, *The War in the Air: Being the Story . . .* (London, 1922), vol 1, pp 389–390, 395–401.

17. C. C. Turner, *The Struggle in the Air, 1914–1918* (London, 1919), pp 161–169; N. Jones, *The Origins of Strategic Bombing* (London, 1973), pp 54–61.

18. Jones, *Origins*, pp 64, 75–77; see also P. Bernard, "*A propos de la* STRATÉGIE AÉRIENNE *pendant la Première Guerre Mondiale: Mythes et realités*," *Revue d'Histoire Moderne et Contempor-*

aine 16 (1969), pp 357–361 (hereafter Bernard, "*STRATÉGIE AÉRIENNE*"). British raids are discussed in H. A. Jones, *The War in the Air: Being the Story*... (London, 1928), vol 2, pp 426–451.

19. Details in R. Martel, *L'aviation française de bombardement: Des origines au Nov 11, 1918* (Paris, 1939), pp 49–63.

20. Gen Sir Douglas Haig to War Office, Nov 1, 1916, NAS Docs, vol 1, pp 405–407. See also pp 336–337 for the hostile view of bombing circulated by Lord Derby to members of the Joint Air War Committee on Mar 31, 1916: "Opinion has been misled by the air raids against towns, munitions factories, aerodromes &c., which are really secondary operations . . . ."

21. Jones, *Origins*, pp 148–149; Bernard, "*STRATÉGIE AÉRIENNE*," p 362.

22. Memo, Lord Curzon for the War Committee, "Long-range Bombing Operations by Aircraft," Nov 9, 1916, NAS Docs, vol 1, p 420.

23. *Ibid.*; Jones, *Origins*, p 94.

24. Curzon Memo, p 422.

25. Jones, *Origins*, p 91.

26. R. H. Fredette, *The Sky on Fire: The First Battle of Britain, 1917–1918, and the Birth of the Royal Air Force* (London, 1966), pp 14–40 (hereafter Fredette, *First Battle of Britain*).

27. *Ibid.*, p 196. On the Gotha bomber see P. Grey and O. Thetford, *German Aircraft of the First World War* (London, 1970), p 32.

28. Fredette, *First Battle of Britain*, pp 160–172.

29. Memo, French Aviation Mission, "New French Program of Aircraft Production," Aug 1918, pp 4–5, Arnold Papers, box 222, LC. On the organizational and political conflicts in France, see the account in J. M. Spaight, *Beginnings of Organised Air Power* (London, 1927), pp 223–235.

30. Bernard, "*STRATÉGIE AÉRIENNE*," pp 361–366.

31. D. Lloyd George, *War Memoirs*, 6 vols (London, 1934), vol 4, pp 1857–1863.

32. J. Sweetman, "The Smuts Report of 1917: Merely Political Window Dressing?" *Journal of Strategic Studies*, vol 4 (1981), pp 152–156, 164–166; M. Cooper, "A House Divided: Policy, Rivalry and Administration in Britain's Military Air Command, 1914–1918," *Journal of Strategic Studies*, vol 3 (1980), pp 178–201.

33. Sweetman, "Smuts Report," p 165; Lloyd George, *War Memoirs*, vol 4, p 1860.

34. Memo, Lt Comdr Lord Tiverton to Capt Vyvyan, Air Board, "Scheme for Bombing German Industrial Centres," Sep 3, 1917, AIR 1/462 15/312/121, PRO; Memo, for Supreme War Council, "Bombing Operations," Jan 1918, pp 1–4, AIR 1/463 15/312/737, PRO. On Weir, see W. J. Reader, *Architect of Air Power: The Life of the First Viscount Weir of Eastwood, 1877–1959* (London, 1968), pp 73–75.

35. Supreme War Council memo, p 3; Memo, Sir Henry Norman to Sec of State for Air, "Long-range Bombing," Mar 25, 1918, p 9, WEIR 1/2, Weir Collection, Churchill College, Cambridge. Norman wrote, "It is evident that these raids do not even approach an offensive sufficient to have an appreciable effect on the duration of the war."

36. Jones, *Origins*, p 161; "Memorandum for the War Cabinet on Certain Lines of Main Policy," May 14, 1918, and "Memorandum on Independent Air Force Command for Long-range Bombing of Germany," May 23, 1918, both WEIR 1/2, Churchill College.

37. Air Ministry, Comnd 100, *Synopsis of British Air Effort During the War* (London, April 1919), pp 5–6 (hereafter Comnd 100, *Synopsis*).

38. Memo, Chief of Air Staff, "Review of Air Situation and Strategy for the Information of the Imperial War Cabinet," Jun 27, 1918, AIR 9/8, PRO.

39. *Ibid.*, pp 2–3.

40. *Ibid.*, pp 5–6. This judgment was based on earlier intelligence work in assessing suitable targets. The "main objectives" selected by intelligence were chemical factories, iron and steel industry, aero engine and magneto works, submarine bases and yards, gun shops, and engine repair shops. War Ministry D.F.O., "Strategic Bombing Objectives in Order of Importance," [autumn 1917], AIR 1/460 15/312/97, PRO.

41. Jones, *Origins*, pp 186–187; F. Sykes, *From Many Angles: An Auto-Biography* (London, 1942), pp 224–230 (hereafter Sykes, *Many Angles*).

42. T. H. Greer, *The Development of Air Doctrine in the Army Air Arm* (Maxwell AFB, Ala., 1955), p 3 (hereafter Greer, *Air Doctrine*). Tiverton reported his conversations with the American military leaders: "They will not be ready to do any serious bombing next summer and they do not intend to undertake it." Ltr, Tiverton to Vyvyan, Sep 15, 1917, AIR 1/462 15/312/121, PRO.

43. Memo, Maj William Mitchell, Signal Corps, for Chief of Staff, USAF, "Air Policy," Jun 10,

1917, in M. Maurer, ed., *The U.S. Air Service in World War I*, 4 vols (Washington, D.C., 1978), vol 2, pp 107–108 (hereafter Maurer, *Air Service*); memo, Mitchell for Chief of Staff, "Aeronautical Organization," Jun 13, 1917. It is evident from these two early papers by Mitchell that his main concern was to organize large-scale attacks with bombers against "enemy aircraft and enemy material." See also memo, Wm Mitchell for Director of Air Service, "Air Service Program Report," Apr 16, 1919, in Maurer, *Air Service*, vol 2, p 111. Mitchell argued that the primary objective of an air force was to defeat the air service of the enemy.

44. Rpt, Maj R. C. Bolling, Aeronautical Commission, to Chief Signal Officer, Aug 15, 1917, in Maurer, *Air Service*, vol 2, p 132.

45. Lt Col E. Gorrell, "Plan for Strategical Bombardment," Nov 28, 1917, in Maurer, *Air Service*, vol 2, pp 141–145. See also E. S. Gorrell, *The Measure of America's World War Aeronautical Effort* (Norwich University Pub 6, Northfield, Vt., 1940), pp 1–4 (hereafter Gorrell, *Aeronautical Effort*); T. H. Greer, "Air Arm Doctrinal Roots, 1917–1918," *Military Affairs*, vol 20 (1956), pp 213–215.

46. Memo, Operations Section, HQAEF, for Chief of Staff, Sep 6, 1917, in Maurer, *Air Service*, vol 2, pp 135–136.

47. General Patrick, "Our Wartime Expenditure for Aeronautics," prepared for *Current History*, Apr 1922, Patrick Papers, RG 18/229, National Archives, Washington, D.C. (NA). Patrick demonstrated that the net sum actually expended by the Treasury was $569 million, not the full $1.6 billion appropriated by Congress. Gorrell put the sum at $868 million. Gorrell, *Aeronautical Effort*, p 7.

48. Intvw, with Lord Trenchard, on the Independent Air Force, Apr 11, 1934, pp 2–3, AIR 8/179, PRO; Bernard, "*STRATÉGIE AÉRIENNE*," pp 366–367.

49. Ltr, Gen Groves for CAS, "I.F.R.A.F. Policy," Sep 11, 1918, AIR 1/460 15/312/97, PRO. Groves objected that "the policy pursued at present amounts to the diversion of maximum effort against targets of subsidiary importance." Even the city attacks were small in scale and faced heavy odds. For example, the attack on Mainz on July 31 involved only nine aircraft of the IAF. Four were shot down on the way at Saarbrucken, the original objective was abandoned, and three more aircraft were lost on the return journey. Turner, *Struggle in the Air*, pp 183–185.

50. Notes, CAS, Aug 7, 1918, in Sykes, *Many Angles*, app 6, pp 555–558. These notes address the Independent Royal Air Force and the proposed interallied strategic bombing force.

51. See note 37 above and A. Morris, *The First of the Many: The Story of the Independent Force RAF* (London, 1968), app A, "Missions." The force lost 458 aircraft in total and flew 650 raids, 172 on the German homeland. United States Bombing Survey: Narrative Summary, n.d. [1919], in Maurer, *Air Service*, vol 4, pp 500–502. Gorrell drafted much of the American report and was critical of the British failure to prepare a plan for using successive raids to destroy the industries most vital to the maintenance of Germany's fighting forces.

52. Trenchard intvw, Apr 11, 1934, p 6.

53. Bureau of Aircraft Production, Planning Dept, "Service Plane and Engine Production Program," Jul 12, 1918, program D, RG 18/11, NA; ltr, Gorrell to Patrick, Jul 29, 1918, in Maurer, *Air Service*, vol 2, p 220.

54. Division of Military Aeronautics, "Accomplishments and Program Requirements under Maj Gen Wm. L. Kenly, Apr–Nov 1918," Jul 1, 1919, p 8, RG 18/11, NA; AEF History Charts, "Comparison of Airplane Strengths of Allied and Enemy Air Services," Nov 11, 1918, RG 18/109, box 1, NA; Memo, Gen Patrick for Chief of Staff, G–3, Aug 16, 1918, in Maurer, *Air Service*, vol 2, p 227.

55. Gorrell, *Aeronautical Effort*, pp 40–41; Comnd 100, *Synopsis*, p 6.

56. "The Operation of the Independent Air Force," pp 6–12, file, Bombing of Land Objectives, 1919–1937, AIR 9/6, PRO; "Objectives—Germany," n.d., and ltrs, Lord Tiverton to Rear Adm Mark Kerr, Dec 1917, encl "Military Industries of Eastern Germany," and War Trade Intelligence Department to Lord Tiverton, Jul 13, 1918, all in AIR 1/460 15/312/97, PRO. On the Air Service, see Greer, *Air Doctrine*, p 10. The Office of Air Intelligence was established in March 1918.

57. Sykes, *Many Angles*, p 932.

58. For German attitudes to bombing, see G. P. Neumann, *The German Air Force in the Great War* (London, 1921), pp 157–159; on France see Bernard, "*STRATEGIE ARRIENNE*," pp 371–375. The effect of raids is quoted in Comnd 100, *Synopsis*, p 5. Groves is quoted in P. R. C. Groves, *Behind the Smoke Screen* (London, 1934), p 119.

59. United States Bombing Survey, [1919], in Maurer, *Air Service*, vol 4, pp 495–503; rpt, British Bombing Commission, AIR 1/2104 207/36 PRO.

60. "Operations of the Independent Air Force," p 12, AIR 9/6 PRO; United States Bombing Survey, [1919], in Maurer, *Air Service*, vol 4, p 498. The survey claimed: "It is certain that air raids

had a tremendous moral effect on the morale of the entire people."

61. H. Ritter, *Der Luftkrieg* (Berlin, 1926), p 215.

62. This is evident in the primary material. The views of Groves, Douhet, Liddell Hart, Fuller, Sykes, Mitchell, and others were widely circulated, read, and requoted.

63. H. S. Hansell, *The Air Plan That Defeated Hitler* (Atlanta, Ga., 1972), p 7.

64. R. J. Young, "The Strategic Dream: French Air Doctrine in the Inter-War Period, 1919–1939," *Journal of Contemporary History*, vol 9 (1974), pp 59–61; C. Messenger, *The Art of Blitzkrieg* (London, 1976), pp 91–93.

65. For a general discussion of armies and air power, see R. J. Overy, *The Air War 1939–1945* (London, 1980), pp 8–12.

66. Chiefs of Staff, "The War Object of an Air Force," May 22, 1928, AIR 9/8, PRO.

67. "Note on the Employment of the AASF," Sep 7, 1939, AIR 14/170, PRO; R. Schaffer, *Wings of Judgement: American Bombing in World War II* (New York, 1985), p 26. The "Note" also states that "the objectives for attack have been confined to purely military targets as defined in the instructions governing naval and air bombardment . . . ."

68. M. Weygand, "How France Is Defended," *International Affairs*, vol 18 (1939), pp 470–471.

69. R. Burlinghame, *General Billy Mitchell: Champion of Air Defense* (New York, 1952), p 114. See also Sir C. Burney, *The World, the Air, and the Future* (London, 1929), pp 54–140, where one can find a general discussion of the changing nature of Britain's international position with the advent of the airplane.

70. N. Gibbs, *Grand Strategy*, vol 1, *Rearmament Policy* (London, 1976), p 534.

71. Note, Air Ministry, "A Locarno War," May 2, 1929, file item 28, Preparations for a Possible War with France in 1929, AIR 9/8, PRO. On the development of British air strategy, see M. Smith, "A Matter of Faith: British Strategic Air Doctrine in the Thirties," *Journal of Contemporary History*, vol 15 (1980) (hereafter Smith, "British Strategy"); B. Powers, *Strategy without Slide Rule* (London, 1976), pp 185–205.

72. Futrell, *Ideas*, pp 49–51.

73. J. Sweetman, "Crucial Months for Survival: The Royal Air Force, 1918–1919," *Journal of Contemporary History*, vol 19 (1984), pp 530–540; Viscount Templewood, *Empire of the Air: The Advent of the Air Age, 1922–1929* (London, 1957) pp 54–84; M. Dean, *The Royal Air Force and Two World Wars* (London, 1979), pp 33–37. Powers provides a full account of the spread of "airmindedness" among Britain's political and administrative elites. Powers, *Strategy*, pp 115–157.

74. Fredette, *Battle of Britain*, p 243. On the "supremacy of the offensive" in the 1920s, see Slessor, *Central Blue*, p 41.

75. Address, CAS, War Office Staff Exercise, Apr 9, 1923, p 3, AIR 9/8, PRO. Sir Henry Wilson, Chief of the Imperial General Staff in 1921, had a different view of the air force: "Aircraft are at present and for many years to come must act as an auxiliary arm to the naval and military services." "The Unified Air Force: Part II, 1918–1937," p 8, SWIN I 2/6, Swinton Papers, Churchill College, Cambridge.

76. Chiefs of Staff (COS) Paper 156, Note by the First Sea Lord, May 21, 1928, p 1, AIR 9/8, PRO. Page 3 of the note states: "No evidence has so far been produced that such bombing in the face of counter attack will have such a result . . . ."

77. Chief of the Air Staff (CAS), "Note upon the Memorandum of the Chief of the Naval Staff, Paper No COS 156," n.d. [May 1928], AIR 9/8, PRO.

78. "Notes for Address by CAS to the Imperial Defence College on the War Aims of an Air Force," Oct 9, 1928, p 1, AIR 9/8, PRO.

79. *Ibid.*, p 5. Trenchard thought aircraft should attack "vital centres at which the enemy is at his weakest."

80. "The War Object of an Air Force: Note upon the Memorandum of the CIGS, Paper No COS 156," n.d. [May 1928], p 10, AIR 9/8, PRO.

81. A. Boyle, *Trenchard* (London, 1962), p 577.

82. Charles Webster and Noble Frankland, *The Strategic Air Offensive against Germany, 1939–1945*, 4 vols (London, 1961), vol 1, pp 76–77, 89–90.

83. U. Bialer, "Elite Opinion and Defence Policy: Air Power Advocacy and British Rearmament in the 1930s," *British Journal of International Studies*, vol 6 (1980), p 37.

84. Ltr, R. Pierce, Air Ministry, to C-in-C, Bomber Command, Dec 13, 1937, app A, pp 1–2, AIR 14/225, PRO.

85. Ltr, Deputy Director Plans to CAS, Sep 9, 1938, pp 1–3, AIR 8/251, PRO.

86. "The Air Programme," n.d. [late 1940], file, Bombing Policy, AIR 8/258, PRO. According to Churchill's directive, "Our aim is to win the war by building up a crushing measure of air superiority . . . the Army has no primary offensive role."

87. Futrell, *Ideas*, p 17.

88. *Ibid.*, p 28.

89. Lecture, Asst Chief Air Service to the Air War College, "Air Force Tactics," Nov 1923, pp 2, 4–6, 13–14, and "Fort Leavenworth Lecture as Delivered by Gen Patrick, March 27, 1924," pp 1–2, both in Patrick Papers, RG 18/229, NA. On the second occasion, Patrick reiterated his view: "The mission of the Air Service is to assist the ground forces to gain strategical and tactical successes by destroying enemy aviation, by attacking enemy ground forces and other enemy objectives on land and sea, and, in conjunction with other agencies, protecting our own ground forces from hostile aerial observation and attack."

90. Final Rpt, War Dept Special Committee on Army Air Corps, Jul 18, 1934, p 12, RG 165/888, NA.

91. Memo, Maj Gen Hugh Drum for WPD, "Information on Aviation and Department of National Defense," May 1, 1934, p 3, RG 165/888, NA. "Battle," General Drum continued, "is the decisive element in warfare. Only by defeat of the enemy's armies can his morale be broken and vital centers occupied, thus forcing him to consent to yield."

92. War Dept Statement, Brig Gen C. E. Kilbourne, for the Federal Aviation Commission, "Truth about the General Staff and the Army Air Corps," Aug 3, 1934, p 5, RG 165/888.86, NA.

93. Memo, Gen S. Embick, Asst COS, WPD, "Aviation versus Coast Fortifications," Dec 6, 1935, RG 165/888.96, NA.

94. J. F. Shiner, *Foulois and the US Army Air Corps, 1931–1935* (Washington, D.C., 1983), pp 52–55, 220–226. On Jan 7, 1931, the so-called Macarthur-Pratt Agreement was reached, giving responsibility for coastal defense to the Air Corps in the first instance.

95. Futrell, *Ideas*, p 44.

96. DuBuque and Gleckner, *Development of the Heavy Bomber*, p 35. For the role of the President, see H. H. Arnold, *Global Mission* (New York, 1949), p 177. The vital meeting at which Roosevelt decided to promote air power was held shortly before Munich, on September 26, 1938.

97. H. L. Ickes, *The Secret Diary of Harold L. Ickes*, 3 vols (London, 1955), vol 3, p 469.

98. Memo, Gen George C. Marshall for Gen E. M. Watson, Sec to the President, Oct 10, 1939, p 1, RG 165/888.103, NA. "Air Power," wrote Marshall, "is based upon the offensive fire power of the bombardment aeroplane."

99. P. Cot, *Triumph of Treason* (Chicago, 1944), pp 274–275. According to Cot (pp 302–303), the major influence on French air doctrine in the 1930s was Marshal Pétain, during his period as Inspector General of Aerial Defense, 1931–1934. On the *Instructions*, see Young, "Strategic Dream," pp 58–59, 66–67, 70; P. Le Goyet, "*Evolution de la doctrine d'emploi de l'aviation française êntre 1919 et 1939,*" *Revue d'histoire de la Deuxième Guerre Mondiale* 19 (1969), pp 22–23; and Messenger, *Blitzkrieg*, pp 91–93.

100. Cot, *Triumph*, pp 303–304; R. J. Young, *In Command of France: French Foreign Policy and Military Planning, 1933–1940* (Cambridge, Mass., 1978), pp 174–176; Le Goyet, "*Evolution,*" pp 9–14, 22–3, 29–30.

101. J. T. Greenwood and Von Hardesty, "Soviet Air Force in World War II," in R. J. Murphy, ed., *The Soviet Air Forces* (London, 1984), pp 35–39; J. Erickson, *The Soviet High Command* (London, 1962), pp 382–383, 500–501; K. R. Whiting, "Soviet Aviation and Air Power under Stalin, 1928–1941," in R. Higham and J. W. Kipp, eds., *Soviet Aviation and Air Power: A Historical View* (London, 1978), pp 50–57.

102. *Luftwaffendienstvorschrift*, "*Luftkriegführung*," Mar 1940 ed, in K. H. Völker, *Dokumente und Dokumentarfotos zur Geschichte der Deutschen Luftwaffe* (Stuttgart, 1968), pp 469–471; P. Deichmann, *German Air Force Operations in Support of the Army* (USAF Hist Study 163, New York, 1968), pp 9–15; H. Boog, "Higher Command and Leadership in the German *Luftwaffe*, 1935–1945," in A. F. Hurley and R. C. Ehrhart, eds., *Air Power and Warfare*, Proceedings of the 8th Military History Symposium, USAF Academy, Oct 18–20, 1978 (Washington, D.C., 1979), pp 134–135.

103. "Conference with Hitler, May 23, 1939," in H. Michaelis and E. Schraepler, eds., *Ursachen und Folgen von deutschen Zusammenbruch 1918 und 1945*, 26 vols (Berlin, 1968), vol 13, p 296.

104. Völker, *Dokumente*, p 482. On the general background of German air strategy, see K. Maier, "Total War and German Air Doctrine before the Second World War," in W. Deist, ed, *The German Military in the Age of Total War* (Leamington Spa, 1985), pp 213–215; W. Murray, "British and

German Air Doctrine between the Wars," *Air University Review*, vol 31 (1980), pp 49–55 (hereafter Murray, "Air Doctrine"); and R. J. Overy, "Hitler and Air Strategy," *Journal of Contemporary History*, vol 15 (1980), pp 408–412.

105. Groves, *Behind the Smoke Screen*, p 123; George lecture, p 3.

106. "Give America Airplanes or We Shall Perish as a Nation," [*ca.* 1934], p 4, Mitchell Papers, box 27, LC.

107. G. Douhet, *The Command of the Air* (London, 1943), p 24. On the influence of Douhet, see C. G. Segre, "Douhet in Italy: Prophet Without Honor?" *Aerospace Historian*, vol 26 (1979), pp 72–79.

108. Quoted in J. W. Shiner, "Reflections on Douhet," *Air University Review*, vol 37 (1986), p 95.

109. Douhet, *Command of the Air*, pp 14–22. Compare these views with Sir F. Sykes, *Aviation in Peace and War* (London, 1922), pp 96–100.

110. Futrell, *Ideas*, pp 15–22. Slessor claims that Douhet had not influenced RAF thinking to any extent. Slessor, *Central Blue*, p 41. On French criticisms of Douhet, see Le Goyet, "*Evolution*," pp 9–11, 13–14; on the general reception of Douhet abroad, see B. Brodie, *Strategy in the Missile Age* (Princeton, N.J., 1959), pp 71–95.

111. Slessor, *Central Blue*, pp 50–51; Smith, "British Strategy," pp 58–63.

112. "Aviation in the Next War" and "Give America Airplanes," both n.d., Mitchell Papers, box 27, LC; W. Mitchell, *Winged Defense* (New York, 1925), pp 4–6, 214–216.

113. Boyle, *Trenchard*, p 577; Smith, "British Strategy," pp 57–60. Even Douhet argued only that air power "should lead armies and navies to reorganize on a new basis" rather than to replace them altogether. Douhet, *Command of the Air*, p 161.

114. Royal Air Force, *War Manual: Part I, Operations* (Air Ministry Pub 1300, May 1935), p 57, copy in RG 18/223 box 1, NA.

115. *Ibid.*, p 59.

116. *Ibid.*, pp 62–64. The RAF *War Manual* also states, "Air bombardment should be as sustained as possible" and "Attacks should be concentrated against one kind of target."

117. M. Smith, "The RAF and Counter-Force Strategy before World War II," *Journal of the Royal United Services Institution* (*JRUSI*), vol 121 (1976), pp 69–72 (hereafter Smith, "Counter-Force Strategy"); Webster and Frankland, *Strategic Air Offensive*, vol 1, pp 86–89; A. Verrier, *The Bomber Offensive* (London, 1968), pp 31–35.

118. Group Capt Foster, "Notes on a Memorandum by CIGS," May 2, 1928, p 2, AIR 9/8, PRO.

119. Asst Chief Air Service lecture, pp 1, 14–15.

120. "Fort Leavenworth Lecture," pp 7–8. In this lecture, General Patrick argued that an air force could be used for "distant operations against enemy communications and industrial centers."

121. Memo, Brig Gen B. Foulois for the Chief of Staff, "Report on Air Service Problems," Dec 23, 1917, in Maurer, *Air Service*, vol 2, pp 161–163; see also Greer, "Air Arm Doctrinal Roots," pp 205, 212–213.

122. Memo, Maj Chaney, "The Employment of Aviation at the Beginning of the World War, at the Close of the War, and at the Present Time," [*ca.* 1932], p 7, RG 18/231, NA; Gen Andrews, "The Airplane in National Defense," n.d., pp 2–4, Andrews Papers, RG 18/231, NA.

123. Lecture, Gen Frank Andrews to the Newport Rotary Club, "The GHQ Air Force," Mar 26, 1935, p 4, and Memo, Gen Andrews, on the GHQ Air Force, n.d., both Andrews Papers, box 9, LC. In the second of these papers, Andrews added to the list of possible targets "attack of factories, refineries, etc.; Attack of power plants and other utilities; Attack upon centers of population."

124. HQAF, G–2 Section, "The GHQ Air Force," 1929, p 2, RG 18/258, box 4, NA; see also Futrell, *Ideas*, pp 39–41.

125. Hansell, *Air Plan*, pp 37–48; H. S. Hansell, *The Strategic Air War against Germany and Japan: A Memoir*, (Washington, D.C., 1986) pp 10–19; Greer, *Air Doctrine*, pp 115–116; W. F. Craven and J. L. Cate, eds., *The Army Air Forces in World War II*, 7 vols (Chicago, Ill., 1948), vol 1, pp 51–52.

126. Hansell, *Air Plan*, p 15.

127. *Ibid.*, pp 37, 40.

128. *Ibid.*, p 40; Smith, "Counter-Force Strategy"; ACTS, "Air Force: Air Operations, Counter Air Force," May 5, 1927, Air Force History Support Office, Washington, D.C. (AFHSO).

129. "Rougeron's '*Aviation de Bombardement*,' Part II," *Royal Air Force Quarterly*, vol 10 (1939), pp 39–44; Bernard, "*STRATÉGIE AÉRIENNE*," pp 372–375.

130. General de Goys, preface to *L'aviation française*, by Martel, p x. Raids on German cities

could only be taken in retaliation against similar German raids. Young, "Strategic Dream," pp 68–69.

131. General Armengaud, *"L'armée de l'air et la defense nationale: I aux premières heures de guerre," Revue des deux mondes*, vol 103 (1933), pp 284–299; *"II dans la bataille genérale," Revue des deux mondes*, vol 103 (1933), pp 545–552; General Voisin, *La doctrine de l'aviation française de combat au cours de la guerre, 1915–1918* (Paris, 1932), pp 153–154.

132. C-in-C, *Luftwaffe*, to Supreme Commander, *"Weisungen fur die Führung der Operationen in der ersten Zeit eines Krieges: Beitrag zur Wehrmachtstudie,"* Nov 18, 1935, in Völker, *Dokumente*, pp 445–446. The Wever quotation is in E. M. Emme, "Technical Change and Western Military Thought, 1914–1945," *Military Affairs*, vol 24 (1960), p 15.

133. *Luftwaffendienstvorschrift, "Luftkriegführung,"* in Völker, *Dokumente*, pp 479–482.

134. Maier, "Total War," pp 212–213; Ritter, *Luftkrieg*, pp 143–144. Ritter states, "Bombardment has above all a great morale effect."

135. R. Suchenwirth, *Historical Turning Points in the German Air Force War Effort* (USAF Hist Study 189, New York, 1968), pp 36–44 (hereafter Suchenwirth, *Turning Points*). According to Suchenwirth's sources, bombardment officers feared they might be thought old-fashioned if they supported level-bombing against dive-bombing. See also Deichmann, *German Air Force Operations*, pp 10–12, 39–56, for the development of tactical bombardment, and Boog, "Higher Command and Leadership in the German *Luftwaffe*," in Hurley and Ehrhart, eds., *Air Power and Warfare*, pp 140–148.

136. R. J. Overy, "From '*Uralbomber*' to '*Amerikabomber*': The *Luftwaffe* and Strategic Bombing," *Journal of Strategic Studies*, vol 1 (1979), pp 154–159 (hereafter Overy, "*Luftwaffe*").

137. On Italy, see Segre, "Douhet in Italy," pp 74–79; F. P. Dunworth, "Italian Aviation," *JRUSI*, vol 80 (1935), pp 393–404; and M. Knox, *Mussolini Unleashed: Politics and Strategy in Fascist Italy's Late War, 1939–1941* (Cambridge, Mass., 1982), pp 23–25. On Japan, see United States Strategic Bombing Survey, *Japanese Air Power*, Rpt no. 62 (Washington, D.C., 1946), copy in Imperial War Museum, London, pp 1–2; and R. J. Francillon, *Japanese Aircraft of the Pacific War* (London, 1970), pp 29–33.

138. Duval's view is quoted in C. E. Wheeler, "Mitchell, Moffet and Air Power," *Air Historian*, vol 8 (1961), p 83. See also B. Liddell Hart, *Paris, or the Future of War* (London, 1925); M. J. B. Davy, *Air Power and Civilization* (London, 1941); F. Morison, *War on Great Cities* (London, 1937).

139. Kilbourne statement, p 2; Burlinghame, *General Billy Mitchell*, p 153.

140. Young, "Strategic Dream," p 61; see also W. R. Kirkland, "The French Air Force in 1940: Was It Defeated by the *Luftwaffe* or Politics?" *Air University Review*, vol 36 (1985), p 102.

141. Gibbs, *Grand Strategy*, vol 1, pp 598–599. The squadrons were composed of 320 Blenheims and Whitleys, and 168 Wellingtons.

142. Memo, J. P. Tracy for the Chief of Staff, "Wartime Allocations of Airplanes to the Army," Apr 4, 1932, p 9, RG 18/223, box 4, NA.

143. M. Maurer, *Aviation in the U.S. Army, 1919–1939* (Washington, D.C., 1987), pp 325–329, 467–470; A. Goldberg, ed., *A History of the United States Air Force, 1907–1957* (Princeton, N.J., 1957), pp 40–41.

144. H. Schliephake, *The Birth of the Luftwaffe* (London, 1971), app 6, "Activation Survey of Operational Units of the *Luftwaffe*," pp 70–77.

145. Cot, *Triumph*, pp 282–286, 302, 330–335; J. Truelle, *"La production aéronautique militaire française jusqu'en Juin 1940," Revue d'histoire de la Deuxième Guerre Mondiale*, vol 19 (1969), p 97; Kirkland, "French Air Force," p 105. By August 1939 this number had risen to 429 frontline bombers of all kinds.

146. Whiting, "Soviet Aviation and Air Power," in Higham and Kipp, eds., *Soviet Aviation and Air Power*, pp 52–63; A. Boyd, *The Soviet Air Force since 1917* (London, 1977), pp 56–58; Greenwood and Hardesty, "Soviet Air Force," in Murphy, ed., *Soviet Air Forces*, pp 38–39.

147. Webster and Frankland, *Strategic Air Offensive*, vol 1, pp 107–117.

148. Jones, *Origins*, pp 165–166. The sight designed by Wimperis in 1917 was still in use in a modified form twenty years later.

149. Minutes, Bombing Policy Sub-Committee of the Bombing Committee, First Meeting, Mar 22, 1938, p 1, AIR 9/92, PRO.

150. *Ibid.*, p 6.

151. RAF Staff College, 16th Comnd, "Bombing: The Assessment of the Probability of Hits," Feb 23, 1938, app, p 3, AIR 9/92, PRO.

152. "Note on A.T.S. Results," [*ca.* Sep 1938], AIR 9/92, PRO.

153. Minutes, Bombing Committee, Air Ministry, Ninteenth Meeting, Oct 4, 1938, pp 2, 4, AIR 14/189, PRO. This concept dated from earlier in the year. Ludlow-Hewitt wrote the Air Minister in June to strongly support the idea of a bomber establishment: "I am convinced that the many tactical and technical problems of bomber operations which now confront us would be more expeditiously, more searchingly and more accurately solved if some such establishment were formed at an early date." Ltr, C-in-C, Bomber Command, to Air Min, Jun 30, 1938, p 4, AIR 14/189, PRO.

154. Ltr, Air Ministry to C-in-C, Bomber Command, Jul 31, 1939, AIR 14/189, PRO; Ltr, C-in-C, Bomber Command, to ACAS, Air Ministry, Sep 23, 1939, AIR 14/189, PRO.

155. Craven and Cate, eds., *Army Air Forces in World War II*, vol 1, pp 598–599. The British were well aware of this deficiency and attempted to procure the sight from the American Air Corps in 1939, without success.

156. See, for example, House of Representatives (Lampert Committee) to Chief of the Air Service, Aug 2, 1924, encl, "Topic No 15, Operations and Activities," RG 18/91, NA. General details can be found in Maurer, *Aviation*, pp 165–188.

157. Maurer, *Aviation*, pp 113–127.

158. Rpt, War Plans Div, G–3, "No 16: Bombing Probability," tables 1–6, RG 165/888, NA.

159. Memo, Chief of the Air Corps, "Bombing within and above Overcast on Large Area Targets," Mar 27, 1939, RG 18/241A, box 3, NA.

160. Memo, Maj Gen B. Emmons to R. Lovett, War Dept, "Training 4-engine Pilots," RG 18/241A, box 3, NA. Like the RAF, the Air Corps did not have a specialized Bombardment School in 1939, nor a published bombardment manual. Emmons to Arnold, "Report on Bombing Training, 1937–1938," and ltr, Emmons to Arnold, Sep 7, 1939, both RG 18/241A, box 3, NA.

161. On the Barling Bomber, see Lampert Committee, "Engineering Division: A Consideration of Its Status with Respect to the Air Service," Nov 20, 1924, pp 71–72, RG 18/91, NA; F. G. Swanborough and P. M. Bowers, *United States Military Aircraft since 1909* (London, 1971), p 562.

162. Whiting, "Soviet Aviation and Air Power," in Higham and Kipp, eds., *Soviet Aviation and Air Power*, p 50; Boyd, *Soviet Air Force*, pp 55–56, 62–68; R. Wagner, ed., *The Soviet Air Force in World War II: The Official History* (New York, 1973), pp 18–19.

163. Note, D.D. Plans to CAS, Sep 24, 1936, pp 1–2, AIR 9/8, PRO. The note states, "Russian progress as forecast for the next say ten years, must entirely alter our conception of the problem of the defence of India . . . on strategic considerations alone, India will soon be at the mercy of the Russian Air Force." See also memo, Air Staff, "The Potential Dangers to the Security of the British Empire and Our Consequent Defence Requirements," Jan 15, 1936, AIR 9/8, PRO.

164. Draft Air Staff Requirement for a Four-engine Bomber Land-plane, Specification B12/36, pp 1–2, AIR 9/77, PRO.

165. Minutes, Operational Requirement Committee Meeting, May 27, 1936, p 3, AIR 9/77, PRO.

166. M. M. Postan, D. Hay, and J. D. Scott, *Design and Development of Weapons* (London, 1964), pp 124–125.

167. Smith, "British Strategy," pp 264–267; J. Terraine, *The Right of the Line: The Royal Air Force and the European War, 1939–1945* (London, 1985), pp 50–51. Sir Thomas Inskip, Minister for the Coordination of Defence, wanted more light and medium bombers to accompany the expensive heavy bombers.

168. Draft Air Staff Requirement for a Bomber Land-plane, Specification B19/38, AIR 9/77, PRO; Slessor, *Central Blue*, pp 174–177.

169. Note, Air Staff, "On the Size of Bomber Aircraft, No 170," Dec 22, 1938, pp 1–7, Air 8/244, PRO. Aluminum was estimated for the Wellingtons at 26,000 tons, and for the Stirlings, at 13,860 tons. Total flying and maintenance personnel for the two forces were calculated to be 56,791 and 15,354, respectively. See also ltr, Air Staff for the Prime Minister, Dec 1938, and memo, Air Staff, "Considerations Affecting the Design of the Ideal Bomber Aircraft," Mar 1938, both AIR 8/244, PRO.

170. Postan, Hay, and Scott, *Design and Development*, pp 93–94, 124–125; Slessor, *Central Blue*, pp 176–177.

171. Memos, Lord Weir, "Air Policy," [1935], and "Acceleration of Defence Measures," Jun 10, 1936, and ltr, Lord Swinton (Air Minister) to Lord Weir, "War Potential," Nov 20, 1936, all in WEIR 19/2–3, Churchill College. See also memo, Sec of State for Air, "Expansion of the Royal Air Force," Mar 12, 1938, pp 1–7, Cabinet Paper 65 (38), SWIN I 2/7, Churchill College.

172. Memo, Sec of State for Air, "Relative Air Strengths and Proposals for the Improvement of the Country's Position," Oct 25, 1938, p 3, AIR 8/250, PRO. The new bomber program included 1,500 Manchester, 1,500 Stirling, and 500 Halifax heavy bombers. See also "Striking Power of the Metro-

politan Bomber Force on Completion of Scheme L Accelerated," Apr 15, 1939, pp 1–4, Cabinet Paper 218 (38), AIR 8/250, PRO.

173. Proceedings, Procurement Planning Board, 1927–1930, esp those of Mar 13, 1929, Mar 10, 1930, and Sep 19, 1930, box 1, RG 18/222, NA. In 1930 the Air Corps procured only fifty bombardment planes in total. See intvw, Brig Gen George W. Goddard with Maj Gen Hugh J. Knerr, Nov 24, 1971, K239.0512–1012, p 1; DuBuque and Gleckner, *Development of the Heavy Bomber*, pp 8–10.

174. R. W. Krauskopf, "The Army and the Strategic Bomber, 1930–1939," part 1, *Military Affairs*, vol 22 (1958/59), p 94. See also memo, Carl Spaatz for Executive, "Comments on Doctrine of the Army Air Corps," Jan 5, 1935, p 1, and draft ltr, Andrews to Col Watson, aide to the President, n.d., p 2, both in Andrews Papers, box 11, LC. Andrews thought the B–17 the "finest military airplane in the world." See also Shiner, *Foulois*, pp 230–233, and DuBuque and Gleckner, *Development of the Heavy Bomber*, pp 15–17.

175. Ltr, Gen Westover to General Craig, COS, Dec 4, 1936, RG 94/452.1, NA.

176. Ltrs, Arnold to Adjutant-General, Sep 11, 1936, pp 1–2, and Air Corps Materiel Division to CAC, Apr 14, 1937, pp 1–2, both RG 94/452.1, NA. According to the second letter, "some uncertainty exists in connection with the affairs of the Boeing Company" because of high overhead costs per unit produced and the reduced orders for B–17s, which eliminated all of Boeing's anticipated profits from the project.

177. Memo, Gen Embick for COS, "Changes in Fiscal Year 1938 Airplane Requirements Program," May 16, 1938, RG 94/452.1, NA.

178. Ltrs, Gen Craig to Sec of War, "Army Bombardment and Reconnaissance Aviation — Limitations of Development of," Jun 20, 1938, pp 1–2, and Craig to Joint Board, "Army Bombardment Aviation," Jun 2, 1938, p 1, both RG 94/452.1, NA.

179. Memo, for COS, "Review of Present Approved Airplane Program," Jul 19, 1938, p 2, and ltr, Louis Johnson, Asst Sec of War, to CAC, Jun 8, 1938, both RG 94/452.1, NA.

180. Arnold, Acting CAC, to Adjutant General, Jun 9, 1937, pp 1–3, RG 94/452.1, NA.

181. Ltrs, Andrews to Adjutant General, Jun 1, 1937, pp 1–3, and to Sec of War, "Air Corps Procurement Program," Jan 24, 1938, p 4, both Andrews Papers, boxes 9 and 11, respectively, LC. "The trend of development," wrote Andrews, "of the modern bombardment airplane, already a powerful engine of war, is toward even larger machines, with ever greater range, greater bomb capacity, and greater powers of self-defense."

182. Memo, for COS, "Review of Present Approved Airplane Program," Jul 19, 1938, RG 94/452.1, NA; R. W. Krauskopf, "The Army and the Strategic Bomber, 1930–1939," part 2, *Military Affairs*, vol 22 (1958/59), pp 212–213; DuBuque and Gleckner, *Development of the Heavy Bomber*, pp 28–33.

183. Memo, for COS, "Airplane Replacement and Development Program," Sep 21, 1939, RG 94/452.1, NA; memo, Gen George Strong for COS, May 10, 1940, RG 94/580, NA; Krauskopf, "Army and the Strategic Bomber," part 2, pp 211–214; Greer, *Air Doctrine*, pp 118–119.

184. Memos, for COS, "Army's Second Aviation Objective," Feb 28, 1941, p 2, and "Army's Second Aviation Objective, Table D, Comparison of First and Second Objective," Feb 27, 1941, both RG 94/580, NA.

185. Air Defense Board, "Summary of the Important Questions to Be Studied in Connection with the Expansion of Aircraft Production," Nov 23, 1938, RG 94/580, 12–14–39, sec 2, NA. The costs of expanding aircraft output were set out as follows:

Utilization of existing plant	$400 million
Expansion to 15,000 planes a year	$555 million
Creation of government plants	$155 million
Operation of government plants	$1,280 million
Total	$2,390 million

See also I. B. Holley, *Buying Aircraft: Materiel Procurement for the Army Air Forces* (Washington, D.C., 1964), pp 169–176.

186. Ltr, Johnson for the President, "Aircraft Procurement Planning, 10,000 Program," Nov 23, 1938, Arnold Papers, box 222, LC; see also F. Walton, *Miracle of World War II* (New York, 1956), pp 362–365. The number of people employed in the aircraft industry increased from 28,000 in 1939 to 200,000 in 1940.

187. Memos, for COS, "Wartime Allocation of Airplanes to the Army," Apr 4, 1932, p 1, and Arnold to CAC, "Cumulative Production of Airplanes of Mobilization Planning," Mar 24, 1931, both

RG 18/223, box 4, NA; Arnold, *Global Mission*, p 62.

188. Suchenwirth, *Turning Points*, pp 40–41; W. Green, *Warplanes of the Third Reich* (London, 1970), pp 128, 223–227; Overy, "*Luftwaffe,*" pp 154–156; E. Heinkel, *He 1000* (London, 1954), pp 215–218; A. Kesselring, *Memoirs* (London, 1955), chap 5.

189. *Flugzeugsbeschaffungs-Programm* No 10, Jan 1, 1939, p 3, RL3/158, Bundesarchiv-Militärarchiv, Freiburg, Federal Republic of Germany (BA-MA); *Vortragsunterlagen fur den Vortrag vor dem Generalfeldmarschall Goering*, Dec 13, 1938, p 65, Milch Documents, 7409–10, BA-MA.

190. "Statement Issued on March 17th 1954 by Field Marshal Kesselring on the Subject of *Luftwaffe* Policy and the Question of a German Four-engine Bomber," in C. Bekker, The *Luftwaffe* War Diaries (London, 1966), pp 478–479; Suchenwirth, *Turning Points*, pp 40–41; R. Suchenwirth, *Command and Leadership in the German Air Force* (USAF Hist Study 160, New York, 1969), pp 33–35; A. Nielsen, *The German Air Force General Staff* (USAF Hist Study 173, New York, 1968), pp 156–157.

191. Rechlin Research Station to Goering, Aug 13, 1942, pp 1–3, He 177 *Handakten*, and memo, Ernst Heinkel to German Air Ministry, on the He 177 program, Sep 12, 1942, pp 1–10, both RL3 16, BA-MA; ltr, *Luftwaffe Erprobungsstelle* to State Sec Erhard Milch, May 27, 1943, Milch Documents, 50.3, pp 745–746, BA-MA; Heinkel, *He 1000*, pp 222–230.

192. Heinkel, *He 1000*, p 215.

193. Overy, "*Luftwaffe,*" pp 156–158; rpt, H. Koppenberg (Junkers Flugzeug-und Motorenwerke) to Goering, Aug 11, 1939, pp 1–12, RL3 246, BA-MA; "*Aktennotiz betr. Dringlichkeit der Rüstungs-Programm 2.9.1939,*" in G. Thomas, *Geschichte der deutschen Wehr-und Rüstungswirtschaft 1918–1943/5*, ed. W. Birkenfeld (Boppard am Rhein, 1966), pp 403–404.

194. R. V. Jones, *Most Secret War* (London, 1978), pp 92–105; Overy, "*Luftwaffe,*" pp 161–162.

195. H. Klotz, *Militärische Lehren des Burgerkrieges in Spanien* (self-published, 1937), pp 50 ff; Murray, "Air Doctrine," pp 50–55; D. Poulain, "The Role of Aircraft in the Spanish Civil War," *JRUSI*, vol 83 (1938), pp 581–586.

196.. Klotz, *Militärische Lehren*, p 52.

197. F. A. Fischer von Poturzyn, *Luftmacht* (Heidelberg, 1938), pp 57–65, 88. On the Italian lessons, see "Tendencies and Aims of Italian Aviation," *Journal of the Royal Aeronautical Society* (*JRAeS*), vol 43 (1939), pp 716–717.

198. "What We Have Learnt So Far from the Spanish Civil War," translated from *La Science et la Vie*, Nov 1937, reprinted in *JRAeS*, vol 42 (1938), p 94; see also "Air Force on Trial — Lessons of Abyssinia, Spain and China," from *L'Air*, Mar 1938, in *JRAeS*, vol 42 (1938), p 462; "Experience with Heavy Bombers in Spain," from *Les Ailes*, 1937, in *JRAeS*, vol 41 (1937), p 732.

199. Embick memo, Dec 6, 1935, p 3.

200. "The Sino-Japanese War," *JRUSI*, vol 82 (1937), pp 864–865: HQ Air Comnd, Southeast Asia, RAF, *Notes on the Japanese Army Air Force*, Nov 1943, pp 34–40; Shuhsi Hsu, *The War Conduct of the Japanese* (Shanghai: Kelly & Walsh, 1938), pp 2–20.

201. Whiting, "Soviet Aviation and Air Power," in Higham and Kipp, eds., *Soviet Aviation and Air Power*, pp 52–58; Kilmarx, *History of Soviet Air Power*, pp 145–148. On the border incidents, see A. D. Coox, *The Anatomy of a Small War: The Soviet-Japanese Struggle for Changkufeng-Khasan, 1938* (Westport, Conn., 1977); A. D. Coox, *Nomonhan: Japan against Russia, 1939*, 2 vols (Stanford, Calif., 1985).

202. Smith, "British Strategy," pp 28–31. For examples of this "cost-effective" view of bombardment, see H. G. Wilmott, "Air Control in Ovamboland," *JRUSI*, vol 83 (1938), pp 23–29, and N. H. Bottomley, "The Work of the Royal Air Force on the Northwest Frontier," *JRUSI*, vol 84 (1939), pp 769–78.

203. Slessor, *Central Blue*, pp 65–66.

204. AF *War Manual*, part 1, "Characteristics of Operations against a Semi-Civilized Enemy," p xiv, copy in RG 18/223, NA.

205. Barratt lecture, app L, "Some Effects of Air Bombardment," p 6.

206. "*Schlussbesprechung des Planspiels des Luftflottenkommanados 2 zur Frage der Luftkriegführung gegen England und über See,*" May 13, 1939, in Völker, *Dokumente*, pp 460–466.

207. F. H. Hinsley, *Hitler's Strategy* (Cambridge, 1951), pp 16–18; M. Salewski, *Die deutsche Seekriegsleitung 1939–1945*, 2 vols (Frankfurt am Main, 1970), vol 1, pp 213–216, 251–257; Overy, "*Luftwaffe,*" pp 163–164.

208. Directive, Air Ministry to C-in-C, Bomber Command, "Air Ministry *Directif,*" Dec 13, 1937, pp 1–2, AIR 14/225, PRO. On the choice of an enemy, see memo, Air Staff, "The Potential Dangers

to the Security of the British Empire," Jan 15, 1936, AIR 9/8, PRO; for attacks on the enemy air force, see Plan W1, "Appreciation of the Employment of the British Air Striking Force against the German Air Striking Force and Its Ancillary Industries and Reserves," Apr 1938, AIR 14/381, PRO.

209. Draft ltr, Ludlow-Hewitt to Air Ministry, [*ca.* Aug 1938], and ltr, Ludlow-Hewitt to Sec of State for Air, Aug 3, 1938, pp 1–7, both AIR 14/225, PRO. See esp *ibid.*, p 7: "We cannot place much reliance on *Plan W1* as a means of halting the German air offensive." See also "Aide-Memoire by Sir Thomas Inskip, Minister for Co-ordination of Defence, for the Secretary of State for Air, 9th December 1937," in Webster and Frankland, *Strategic Air Offensive*, vol 4, app 5, pp 96–98: "The role of our Air Force is not an early knock-out blow . . . but to prevent the Germans from knocking us out."

210. Air Ministry directive, app A, pp 1–2. The list was revised and an additional target added in September 1939. Webster and Frankland, *Strategic Air Offensive*, vol 4, app 6, "Western Air Plans, 1st September 1939," pp 99–102.

211. Ltr, Ludlow-Hewitt to Sec of State for Air, Aug 30, 1938, AIR 14/225, PRO.

212. "Short Note on Action of Air Force at the Outset," [*ca.* Sep 1938], AIR 8/251, PRO.

213. Ltrs, D.D. Plans to CAS, Sep 9, 1938, CAS to Ludlow-Hewitt, Sep 19, 1938, and Air Council to Ludlow-Hewitt, Sep 15, 1938, and memo, Air Ministry, "Points for CAS to Discuss with General Vuillemin," n.d., all AIR 8/251, PRO. On the 1939 agreement, see Anglo-French Staff Conversations, "Preparation of Joint Plans for Action of Franco-British Air Force," Apr 19, 1939, pp 2–3, AIR 9/105, PRO.

214. Air Ministry, Air (Targets) Intelligence, country file, Germany: 1, Electricity Supplies, Apr 28, 1938; 2, Fuels, May 5, 1938; 3, Chemical Industry, May 20, 1938; 4, Metallurgical Industries, May 20, 1938; 5, Engineering and Armaments, Jun 14, 1938; and 6, Transportation, Jun 14, 1938, all AIR 9/89, PRO.

215. Air Targets Intelligence, Germany file, Oct 13, 1938, pp 4–8, and apps A, C, AIR 9/102, PRO; memo, Air Min Plans (Ops) for CAS, "Appreciation on the Attack of German War Industry (W.A.5 1st Revise)," pp 6–7, and rev app A, "Economic and War Importance of the Ruhr," pp 1–2, Jan 1939, AIR 9/102, PRO.

216. Ltr, Ludlow-Hewitt to Under Sec of State, Air Ministry, Aug 30, 1938, AIR 14/225, PRO; Webster and Frankland, *Strategic Air Offensive*, vol 1, p 100.

217. Gibbs, *Grand Strategy*, vol 1, p 598.

218. Ltr, Ludlow-Hewitt to Air Ministry, "Provision of a One Thousand Pounds Bomb," Jul 27, 1938, and rpt, Air Ministry (Plans), "Bomb Stocks at Apr 26, 1939," May 17, 1939, both AIR 9/92, PRO. It was reported that Britain in April 1939 had no large-caliber bombs in stock at all. The largest items in stock (39 percent) were old-fashioned, 250-pound bombs.

219. Conference, Air Ministry, "Bomber Command War-Readiness," Apr 6, 1939, pp 3–4, AIR 8/270, PRO.

220. Slessor, *Central Blue*, pp 203–205. Regarding maps, see page 4 of the minutes of the April 6, 1939, conference cited in note 219 above.

221. Anglo-French Staff Conversations, "The Employment of British Bombers in the Event of German Invasion of the Low Countries," Apr 21, 1939, p 2, and "Preparation of Joint Plans of Action for Franco-British Air Force," Apr 19, 1939, both AIR 9/105, PRO; Anglo-French Staff Conversations, "Inter-Service Discussions 1/2 May 1939," pp 1–3, AIR 9/117, PRO; Meeting at HQ Advanced Air Striking Force, Sep 13, 1939, AIR 14/170, PRO.

222. Anglo-French Staff Conversations, "The Attack of German Railway Communications East of the Ruhr (Plan 2)," Apr 26, 1939, AIR 9/117, PRO.

223. Memo, Air Ministry, "The Employment of the Air Striking Force on the Outbreak of War," [*ca.* Aug 1939], p 10, AIR 9/131, PRO; ltr, Air Ministry to HQ AASF, Sep 7, 1939, AIR 14/170, PRO. According to the letter, "Objectives should not involve unduly deep penetration . . . . Night operations are not to be undertaken at present."

224. Rpt, First French Air Army, 3eme Bureau, for HQ AASF, "Participation of British Bombers in Eventual Operations between the Rhine and Moselle," Oct 14, 1939, AIR 14/170, PRO; rpt, Gen Vuillemin, French CAS, to French Air Minister, Aug 20, 1939, in *Revue d'histoire de la Deuxième Guerre Mondiale*, vol 19 (1969), pp 73–74.

225. Ltr, Sir Warren Fisher to Neville Chamberlain, Apr 2, 1938, FO 800–309 5537, PRO.

226. C. Messenger, *"Bomber" Harris and the Strategic Bombing Offensive 1939–1945* (London, 1984), pp 22–24; Webster and Frankland, *Strategic Air Offensive*, vol 1, pp 107–126; Boog, "Higher Command and Leadership in the German *Luftwaffe*," in Hurley and Ehrhart, eds., *Air Power and*

Warfare, pp 140–146; R. E. Vintras, "The German Air Force," *JRUSI*, vol 92 (1947), pp 91–95.

227. Memo, for COS, "Second Meeting of the Joint Air Advisory Committee," Jun 5, 1940, and memo, Arnold, "Army and Navy Aviation Objectives and Procurement Plans," Jun 5, 1940, both RG 165/888.103, NA; Futrell, *Ideas*, pp 51–54. For the influence of European operations, see Craven and Cate, eds., *Army Air Forces in World War II*, vol 1, pp 104–116.

228. Hansell, *Air Plan*, p 50.

229. *Ibid.*, pp 53–63.

230. *Ibid.*, pp 298–307; Futrell, *Ideas*, pp 59–61.

231. Hansell, *Air Plan*, pp 81–83.

232. *Ibid.*, pp 92–93.

233. Memo, for Deputy COS for Air, "Army's Second Aviation Objective, Strength of Units," May 17, 1941, p 1, RG 94/580, NA.

234. Memo, Hansell to Col Candee, War Dept, "Conversation on Bombardment Aviation between Captain H. S. Hansell and Air Commodore Schlesser [*sic*]," Dec 4, 1940, 167.6–51(C), AFHSO.

235. Memo, for Deputy COS, WPD, "German Air Operations," May 23, 1940, RG 94/580, NA.

236. *Ibid.*

237. Memo, General Staff, G–2, for Chief of Intelligence, "Estimates of German Air Strength," Jan 21, 1941, p 1, and encl, "Germany, Domestic Production, Capacity and Sources of Aviation Equipment," pp 1–9, Arnold Papers, box 246, LC.

238. Heinkel, *He 1000*, p 233. Hitler told Heinkel that he wanted forty to fifty heavy bombers attacking London by day and by night: "Such continuous attacks," he enthused, "would bring life there to a standstill." When the technical difficulties of developing the He 177 with a dive-bombing capability were explained to him, he reportedly exclaimed, "But that's madness! I've heard nothing of this until today. Is it possible that there could be so many idiots?"

239. Fredette, *First Battle of Britain*, p 125.

240. Reader, *Architect of Air Power*, p 231.

241. Hansell, *Strategic Air War*, pp 12–13. Studying the U.S. economy drew the ACTS to electric power, transportation, fuel, food, and steel as major target systems, in that order of priority: "Loss of any of these systems would be a crippling blow. Loss of several or all of them would bring national paralysis."

Bibliographic Essay

Until the last decade, a serious historical record of air power before 1939 was limited, and substantial gaps remain. Studies of strategic bombardment have been scarcer still. In the more general histories of air power, the period preceding 1939 tends to be overshadowed by the experiences of combat that occurred between 1939 and 1945 and those in Korea and Vietnam. The most useful introductions are L. Kennett, *The History of Strategic Bombing* (New York, 1982); R. Higham, *Air Power: A Concise History* (London, 1972); B. Collier, *A History of Air Power* (London, 1974); and A. Stephens, ed., *The War in the Air, 1914–1994* (Canberra, 1994). Although official histories concentrate on the war period, their opening volumes contain good detailed material. C. Webster and N. Frankland in their four-volume *The Strategic Air Offensive Against Germany* (London, 1961) discuss strategic bombing before 1939 in volume 1 and append a number of key documents for the period in volume 4. On the American side are the seven volumes edited by W. F. Craven and J. L. Cate, *The Army Air Forces World War II* (University of Chicago Press, 1948), of which volume 1 on preparation and volume 6 on economic and technical questions are the most important. The German air force is discussed in the semiofficial history edited by W. Deist *et al.*, *Das Deutsche Reich und der Zweite Weltkrieg: Band I, Ursachen und Voraussetzungen der deuteschen Kriegspolitik* (Stuttgart, 1979), section 3. Detail on RAF Bomber Command is to be found in D. Richards, *The Royal Air Force 1939–1945*, volume 1, *The Fight at Odds* (London, 1953). For the First World War, there is W. Raleigh's and H. A. Jones' *The War in the Air* published in 6 volumes (Oxford, 1922–1937).

More important are the primary documents themselves. For this essay, the British and

American records constitute an indispensable source. For the RAF, the following collections were consulted at the Public Record Office, Kew, London: AIR 1 and 2 for World War I; AIR 8, Papers of the Chief of Air Staff; AIR 9, Director of Plans; AIR 14, Bomber Command; and AIR 19, Secretary of State for Air. Useful supplementary material can be found in the Churchill College Archive, Cambridge, in the records of Lord Weir, Britain's first Air Minister, and those of Lord Swinton, Air Minister during the key period of British air rearmament in the 1930s. From the National Archives in Washington, D.C., the following collections were used: Record Group (RG) 18, Army Air Forces, particularly Group II, General Records 1914–1919; Group IV, Office of Chief of the Air Service; Group V, Office of the Chief of the Air Corps; Group VI, Office of the Commanding General GHQAF; and Group VII, HQ Army Air Forces; RG 94, Office of the Adjutant-General, central decimal files 1926–1939; RG 107, Office of the Secretary of War: Assistant Secretary of War (Air) 1926–1933; and RG 165, War Plans Division, general correspondence, 1920–1942. At the Library of Congress, use was made of the papers of Generals H. H. Arnold, William Mitchell, and Frank Andrews. The last of these proved a particularly valuable source. Primary German sources include the papers of State Secretary Erhard Milch, German Air Ministry, originally housed at the Imperial War Museum, London, now located at the Bundesarchiv-Militärarchiv (BA-MA), Freiburg, Germany; the records of the German Air Ministry, Groups RL2 and RL3, including the so-called He–177 *Handakten*, are also at the BA-MA. In addition, a number of important published collections of documents include S. W. Roskill, ed., *Documents Relating to the Naval Air Service*, 2 vols (Navy Records Society, London, 1969); M. Maurer, ed., *U.S. Air Service in World War I*, 4 vols (Office of Air Force History, Washington, D.C., 1978); and K. H. Volker, ed., *Dokumente und Dokumentarfotos zur Geschichte der Deutschen Luftwaffe* (Stuttgart, 1968).

A number of books touch on the early development of military aviation in relation to strategic bombardment. Very suggestive is the book by A. Gollin, *No Longer an Island: Britain and the Wright Brothers* (London, 1984). On the American experience, one can consult the thorough study by J. A. Hennessy, *The United States Army Air Arm* (Washington, D.C., 1985); on Germany, available sources include J. Cuneo, *Winged Mars: The German Air Weapon, 1870–1914* (Harrisburg, Pa., 1942), and J. H. Morrow, *German Airpower 1909–1914* (Knoxville, Tenn., 1976); and on Britain, we have R. Worcester's interesting book, *Roots of British Air Policy* (London, 1966). For World War I there are two outstanding studies: N. Jones, *The Origins of Strategic Bombing* (London, 1973), and J. Morrow, *The Great War in the Air: Military Aviation from 1909 to 1921* (Washington, 1993). The best account of French bombardment remains R. Martel, *L'aviation française de bombardement: Des origines au Novembre 11, 1918* (Paris, 1939). Details of the German air offensives can be found in E. von Hoeppner, *Deutschlands Krieg in der Luft* (Leipzig, 1921), and in G. P. Neumann, *The German Air Force in the First World War* (London, 1921). On the bombing of England from airships, the definitive work is D. Robinson, *The Zeppelin in Combat: A History of the German Naval Airship Division* (London, 1971); on bombing from airships and aircraft, it is J. Morris, *The German Air Raids on Great Britain, 1914–1918* (London, 1925). Good detail on airship operations is available by one who participated, T. von Buttlar Brandenfels, in his *Zeppelins over England* (London, 1931).

The British response to German bombing is well documented in R. H. Fredette, *The Sky on Fire: The First Battle of Britain, 1917–1918, and the Birth of the Royal Air Force* (London, 1966). This can be supplemented with articles by J. Sweetman, "The Smuts Report of 1917," *Journal of Strategic Studies* 4 (1981), and by M. Cooper, "A House Divided: Policy, Rivalry, and Administration in Britain's Military Air Command, 1914–1918," *Journal of Strategic Studies* 3 (1980). The British bombing counteroffensive is described in A. Morris, *The First of the Many: The Story of the Independent Force RAF*

(London, 1968). The memoirs of P.R.C. Groves, *Behind the Smoke Screen* (London, 1934), and F. Sykes, *From Many Angles: An Autobiography* (London, 1942), are particularly useful, as is W. J. Reader, *Architect of Airpower: The Life of the First Viscount Weir of Eastwood, 1877–1959* (London, 1968). On the different experience in French aviation, there are the valuable study by P. Bernard, "*A propos de la stratégie aérienne pendant la Première Guerre Mondiale: Mythes et realités*," *Revue d'Histoire Moderne et Contemporaine* 16 (1969), and the work C. Christienne, P. Lissarague, *et al.*, *Histoire de l'aviation militaire française* (Paris, 1980). The American dimension is well covered in T. Greer, *The Development of Air Doctrine in the Army Air Arm* (Air University, Maxwell AFB, Ala., 1955), which should be read with Edgar Gorrell's own account of the war period, *The Measure of America's World War Aeronautical Effort* (Norwich University Pub 6, Northfield, Vt., 1940) and the article by L. Kuter, "Edgar Gorrell's Concept of Air War," *Air Force* (Apr 1978).

Little of a general nature on strategic bombardment is available for the interwar period. Two useful books that address the wider strategic implications of air power are B. Brodie, *Strategy in the Missile Age* (Princeton University Press, 1959), and G. Quester, *Deterrence before Hiroshima* (New York, 1966). Much of value on the different approaches to the use of air power, and in particular, of bombardment aviation, is to be found in C. Messenger, *The Art of Blitzkrieg* (London, 1976), and in A. F. Hurley and R. C. Ehrhart, eds., *Air Power and Warfare*, Proceedings of the 8th Military History Symposium, USAF Academy, Oct 18–20, 1978 (Washington, D.C. 1979). An interesting comparative study can be found in W. Murray, "British and German Air Doctrine between the Wars," *Air University Review* 31 (1980). On the influence of General Douhet, there is C. G. Segre, "Douhet in Italy: Prophet without Honor?" *Aerospace Historian* 26 (1979). Giulio Douhet's own book, *The Command of the Air* (London, 1943), is essential. Other important contemporary discussions of strategic air power are F. Sykes, *Aviation in Peace and War* (London, 1922); C. Burney, *The World, the Air and the Future* (London, 1929); E. Schuddekopf, *Britische Gedanken über den Einsatz des Luftheeres* (Berlin, 1939); and B. Liddell Hart, *Paris, or the Future of War* (London, 1925).

Much more has been written on the history of individual air forces, particularly on the RAF and the Army Air Corps. Some of it is directly concerned with strategic bombardment, and some of it, indirectly so. The best general surveys of British air strategy are M. Smith, *British Air Strategy between the Wars* (Oxford, 1984), and B. Powers, *Strategy without Slide Rule* (London, 1976). The official history by N. Gibbs, *Grand Strategy*, volume 1, *Rearmament Policy* (London, HMSO, 1976), also contains useful detail. On strategic bombardment as such, there is very little. The introduction to A. Verrier, *The Bomber Offensive* (London, 1968), supplements the official history. H. M. Hyde's *British Air Policy between the Wars, 1918–1939* (London, 1976) has little to add; more can be found in J. Terraine, *The Right of the Line: The Royal Air Force and the European War, 1939–1945* (London, 1985), but this is much more concerned with the war period. The memoirs of J. Slessor, *The Central Blue: Recollections and Reflections* (London, 1956), detail the views of an airman close to the center of decision-making in the 1930s. A. Boyle's *Trenchard* (London, 1962) is more than a biography; it should be read in conjunction with M. Dean, *The Royal Air Force and Two World Wars* (London, 1979), and C. Messenger, *"Bomber" Harris and the Strategic Bombing Offensive, 1939–1945* (London, 1984). For the public debate on bombing, the pioneering work is U. Bialer, *The Shadow of the Bomber* (London, 1980).

The debate on bombardment in the U.S. Army Air Corps is comprehensively discussed in R. Futrell, *Ideas, Concepts, Doctrine: A History of Basic Thinking in the United States Air Force* (Air University, Maxwell AFB, Ala., 1971), and in Greer, as cited above. Bombardment also features prominently in the major general histories prepared by A. Goldberg, ed., *A History of the United States Air Force, 1907–1957* (Princeton

University Press, 1957), and M. Maurer, *Aviation in the U.S. Army, 1919–1939* (Office of Air Force History, Washington, D.C., 1987). On strategic bombardment specifically, there are the excellent general histories: R. Schaffer's *Wings of Judgement: American Bombing in World War II* (New York, 1985); M. E. Sherry's *The Rise of American Air Power: The Creation of Armageddon* (New Haven, 1987); C. C. Crane's *Bombs, Cities, and Civilians: American Airpower Strategy in world War II* (Lawrence Kans., 1993); H. S. Hansell's *The Air Plan That Defeated Hitler* (Atlanta, 1972); R. W. Krauskopf's "The Army and the Strategic Bomber, 1930–1939," parts 1 and 2, *Military Affairs* 22 (1958/59); and J. H. DuBuque and R. F. Glackner's thoroughly documented *Development of the Heavy Bomber, 1918–1944* (USAF Hist Study 6, Air University, Maxwell AFB, Ala., 1951). On the key personalities involved, see J. F. Shiner's excellent study, *Foulois and the U.S. Army Air Corps, 1931–1935* (Office of Air Force History, Washington, D.C., 1983); R. Burlinghame's *General Billy Mitchell: Champion of Air Defense* (New York, 1952); and H. H. Arnold's *Global Mission* (New York, 1949), a lively and critical account written by a major figure in the bombardment debate. Mitchell's *Winged Defense* (New York, 1925) is the Army Air Service equivalent of Douhet's *Command of the Air.*

The very different approach to air strategy taken in France, Germany, and the Soviet Union is covered in a large number of general studies. On the German side, the most useful are H. Boog, *Die deutsche Luftwaffenführung, 1935–1945* (Stuttgart, 1982); R. Suchenwirth, *Historical Turning Points in the German Air Force War Effort* (USAF Hist Study 189, New York, 1968); A. Nielsen, *The German Air Force General Staff* (USAF Hist Study 173, New York, 1968); W. Murray, *Luftwaffe* (London, 1985); and R. J. Overy, *Goering: The "Iron Man"* (London, 1984). A number of articles address strategic bombing in the German Air Force. The fullest account is in R. J. Overy, "From '*Uralbomber*' to '*Amerikabomber*': The *Luftwaffe* and Strategic Bombing," *Journal of Strategic Studies* 1 (1979), but more information on this subject can be found in K. Maier, "Total War and German Air Doctrine before the Second World War," in W. Deist, ed. *The German Military in the Age of Total War* (Leamington Spa, 1985), and W. Murray, "The *Luftwaffe* before the Second World War: A Mission, a Strategy?" *Journal of Strategic Studies* 4 (1981). Regarding France, a number of important articles deal with, among other things, strategic bombardment. They are P. Le Goyet,"*Evolution de la doctrine d'emploi de l'aviation française êntre 1919 et 1939*," and J. Truelle, "*La production aéronautique militaire française jusqu'en juin 1940*," both in *Revue d'histoire de la Deuxième Guerre Mondiale* 19 (1969); R. J. Young, "The Strategic Dream: French Air Doctrine in the Inter-war Period, 1919–1939," *Journal of Contemporary History* 9 (1974); and F. R. Kirkland, "The French Air Force in 1940: Was It Defeated by the *Luftwaffe* or by Politics?" *Air University Review* 36 (1985). On the political conflicts between civil and military authorities within the armed forces, good detail is found in P. Cot, *Triumph of Treason* (Chicago, 1944), and an interesting analysis is in A. Horne, *The French Army and Politics, 1870–1970* (London, 1984). On the Soviet Union, the most informative general studies are R. Kilmarx, *A History of Soviet Air Power* (London, 1962); A. Boyd, *The Soviet Air Force since 1917* (London, 1977); and Von Hardesty, *Red Phoenix: The Rise of Soviet Air Power, 1941–1945* (London, 1982). R. Wagner, ed., gives detail on organization in *The Soviet Air Force in World War II: The Official History* (London, 1974). Two important articles on shifts in Soviet strategy in the 1930s are K. R. Whiting, "Soviet Aviation and Air Power under Stalin," in R. Higham and J. W. Kipp, eds., *Soviet Aviation and Air Power: A Historical View* (London, 1978), and J. T. Greenwood and Von Hardesty, "Soviet Air Forces in World War II," in P. J. Murphy, ed., *The Soviet Air Forces* (London, 1984).

No bibliography on strategic bombardment would be complete without acknowledging the importance of technical and economic factors in its evolution. On bombardment aircraft, few general studies exist. The book by P. Lewis, *The British Bomber since 1914*, 2d ed. (London, 1980), is one of them. Good technical information can be found in F. G.

Swanborough and P. M. Bowers, *United States Military Aircraft since 1909* (London, 1971); W. Green, *Warplanes of the Third Reich* (London, 1970); J. W. Thompson, *Italian Civil and Military Aircraft* (London, 1964); J. J. Sadkovich, "The Development of the Italian Air Force prior to World War II," *Military Affairs*, vol 51, no. 3 (1987), pp 128–136; and R. Francillon, *Japanese Aircraft of the Pacific War* (London, 1970). A number of distinguished monographs characterize the aircraft industry and air rearmament. Among them are E. Homze, *Arming the Luftwaffe* (Nebraska University Press, 1976); DuBuque and Glackner, as cited above; I. B. Holley, *Buying Aircraft: Materiel Procurement for the Army Air Forces* (Washington, D.C. 1964); and R. Frankenstein, *Le prix du rearmement francaise, 1935–1939* (Paris, 1982). Books that treat economic or development questions including bombardment aviation are available for Britain and the United States. They include H. Penrose, *British Aviation: The Ominous Skies, 1935–1939* (London, 1980); M. Postan, D. Hay, and J. D. Scott, *Design and Development of Weapons* (London, HMSO, 1964); J. B. Rae, *Climb to Greatness* (Cambridge, Mass., 1968); and G. Simonson, *History of the American Aircraft Industry* (MIT Press, 1968). On the development of communications and radio navigation, see R. V. Jones, *Most Secret War* (London, 1978); on the factory program, G. T. White, "Financing Industrial Expansion for War," *Journal of Economic History* 9 (1949).

Finally, there is a wealth of contemporary journal and popular book material on the strategic bombing debate and the changes in the nature of warfare that this implied. Among the most useful journals recording this exchange are the *Journal of the Royal United Services Institution*, the *RAF Quarterly*, the *Journal of the Royal Aeronautical Society*, *Revue des deux mondes*, *Aviation*, *Military Affairs*, and *Flight*. On the popular bombardment debate, the books by Bialer and Powers cited above contain good bibliographies. See also P. Fritzsche, "Machine Dreams: Airmindedness and the Reinvention of Germany," *American Historical Review* 98, no. 3 (1993), pp 685–709.

2

The British Strategic Air Offensive Against Germany in World War II

W. A. Jacobs

On September 1, 1939, the armed forces of Nazi Germany invaded Poland. Despite Reich Chancellor Adolf Hitler's belief that the conflict would be limited, on September 3 the British and French governments, having pledged to assist the Poles, declared war on Germany. Eight and one-half months later the Royal Air Force (RAF) attacked industrial targets in Germany's Ruhr valley. The British bombing offensive would continue for five years, kill or maim hundreds of thousands, devastate entire cities, and lay waste huge amounts of industry.[1] It would cost the lives of almost 48,000 aircrew killed on operations. As the the British official history records, "it was probably the most continuous and gruelling operation of war ever carried out."[2] It was also the most controversial British military campaign of World War II. More than five decades after the last bomb fell, scholars and the informed public remain divided over its utility and morality.[3]

RAF Bomber Command was the main weapon of the British offensive. First formed in 1936, it was one of three major operational commands of the RAF based in the United Kingdom.[4] The first wartime commander-in-chief (C-in-C) was Air Marshal Sir Edgar Ludlow-Hewitt, who served from 1937 through April 1940. His successor, Air Marshal Sir Charles Portal, left to become Chief of the Air Staff (CAS) in October 1940. Air Marshal Sir Richard Peirse held the post until early 1942 when he was replaced by Air Marshal Sir Arthur Harris, the officer with whom the British strategic air offensive was, and still is, most closely identified. The Bomber Command C-in-C answered to the

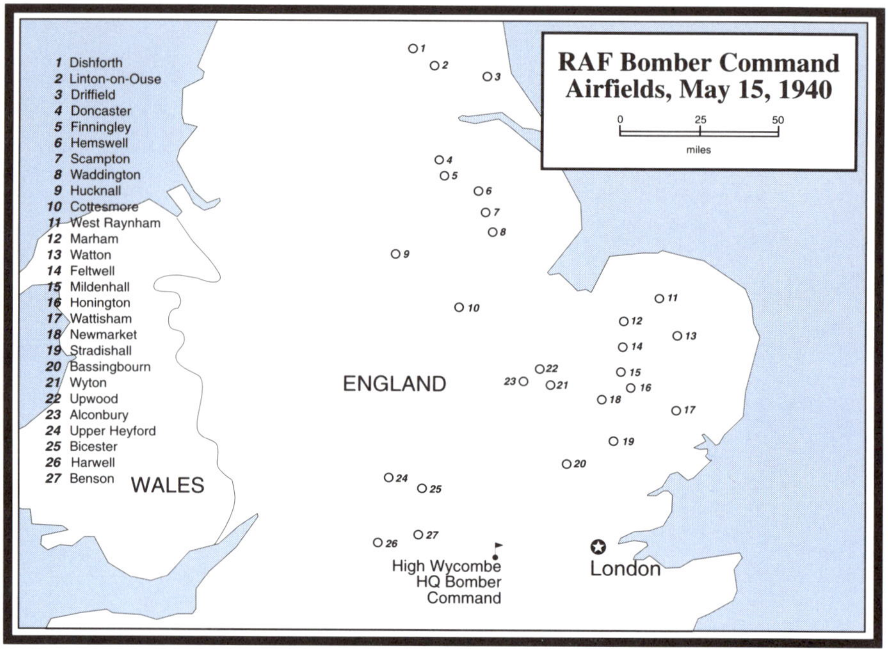

CAS, the highest ranking officer in the RAF who served as administrative head of the service and de facto C-in-C of the home commands. During World War II this post was held by only two individuals: Air Chief Marshal Sir Cyril Newall from 1937 to October 1940, and Sir Charles Portal from October 1940 to the end of the war.

For most of the first three years of the war, Bomber Command was a small force consisting of largely twin-engine aircraft with limited range, small bombloads, and weak defensive armament. At the outset, the Air Staff intended that the command be used to attack German industrial targets in daylight, but its limited destructive power, vulnerability to German air defenses, and aircrews' inability to find and destroy targets at night obliged the British gradually to abandon their original intention.

Beginning in October 1940, the Air Ministry issued a series of directives ordering Bomber Command to attack cities as secondary targets. In February 1942, the Ministry directed Bomber Command to destroy large urban areas with the aim of undermining the morale of the German industrial work force. A few days later, Sir Arthur Harris became C-in-C of Bomber Command, and in May he staged the first thousand-bomber attack (on Cologne). For the next two years Bomber Command, while rapidly changing into a force composed largely of four-engine heavy bombers, focused its efforts almost entirely on what came to be called the night area offensive against German cities. During this period

The highest ranking officer in the RAF was the Chief of Air Staff, a post held by only two officers during the war: Air Chief Marshal Sir Cyril Newall (1937–1940) (*top*) with troops on an airfield in France, and Air Chief Marshal Sir Charles Portal (1940–1945) (*center*), here appearing with Maj. Gen. Ira Eaker, top U.S. airman in England. Sir Arthur "Bomber" Harris (*lower left*) became C-in-C of Bomber Command, staging massive strategic bombardment attacks on German cities. Heading the Fighter Command after the Battle of Britain was Sholto Douglas (*bottom right*), responsible for the defense of home islands.

Allied Command Structure, September 1939–April 1944

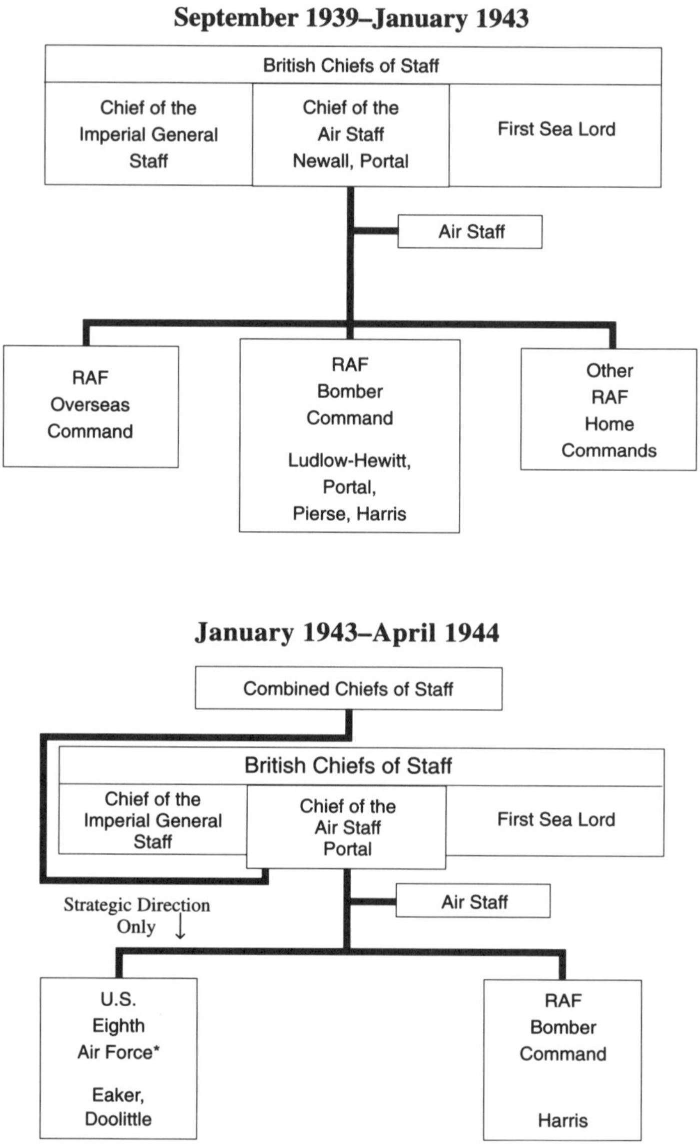

*After January 1944, United States Strategic Air Forces, Europe replaced Eighth Air Force as the highest American air force headquarters in Europe.

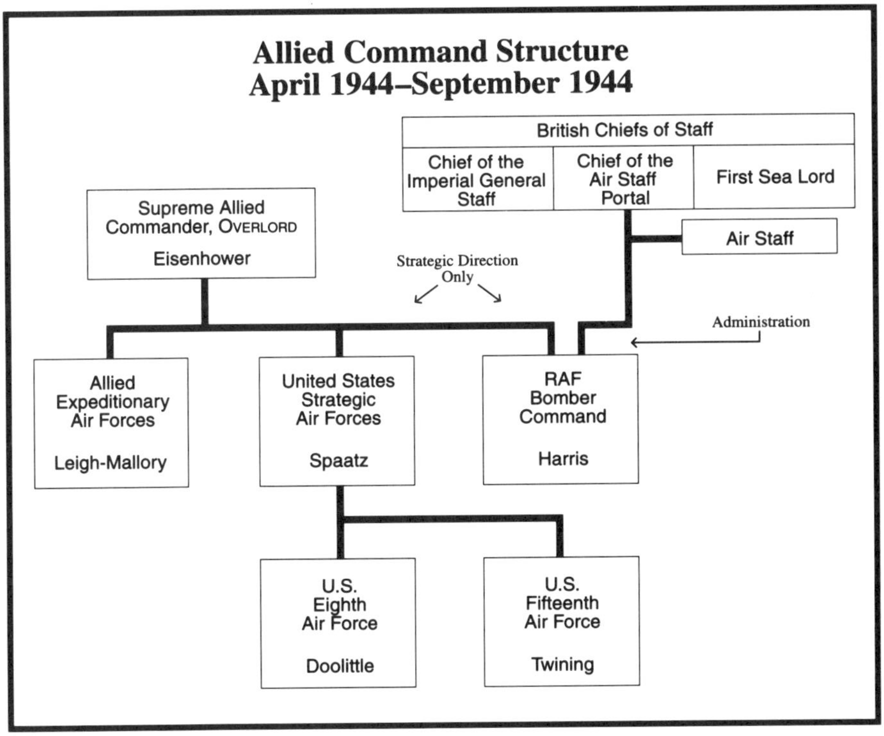

Sir Arthur Harris directed three great aerial campaigns: the Battle of the Ruhr in the late winter and spring of 1943; the Battle of Hamburg in July 1943; and the grueling, five-month-long Battle of Berlin in the autumn and winter of 1943–1944.

The third major phase of Bomber Command's wartime history began in April 1944 when the force was placed under the strategic direction of the Supreme Allied Commander, General Dwight D. Eisenhower, to prepare for and support Operation OVERLORD, the Allied invasion of France. The emphasis on area attack against German cities gave way to attacks on rail centers, launch sites for V–1 buzz bombs, coastal batteries, and enemy forces on the battlefield. The requirement to support the Allied invasion accelerated development of techniques to use in precision attacks at night, and in June 1944 the command resumed attacks on German synthetic oil production, attacks it had been forced to abandon early in the war. The daylight air superiority over Europe created by the American strategic air offensive and the powerful Allied tactical air forces subsequently allowed Bomber Command to resume operations by day. Bomber Command returned to the control of the CAS in September 1944 and, for the remainder of the war, largely devoted its efforts to bombing industrial cities, oil production, and transportation. This final phase in the air offensive was marked by fierce disputes over strategic policy and refinements in Bomber Command's operational techniques and its considerable destructive powers.

Aircraft and Aircrews

During the first three years of the war, Bomber Command did not possess adequate numbers of the aircraft needed to carry out a serious strategic bombing offensive. When the war began in September 1939, as Table 2–1 shows, it possessed only 515 aircraft ready for bombing operations, of which no less than 264 were light bombers (Battles and Blenheims); the rest were mediums (Hampdens, Whitleys, and Wellingtons). The force possessed no heavy bombers. A year and a half later, in January 1941, the total number of bombers ready for operations had risen only to 608. That figure increased to 878 by midsummer 1941, but owing to attrition it fell back to 670 in the following 12 months. From July 1942, however, the trend moved steadily upward, peaking at 1,977 aircraft on V–E Day in May 1945. By midsummer 1942, new four-engine heavy bombers made up about two-fifths of the force; by January 1943 that proportion had increased to two-thirds. The medium bombers disappeared entirely from the command in 1943, some going to the Tactical Air Force and some, overseas. By July 1944, Bomber Command consisted almost entirely of Lancasters, Halifaxes, and light Mosquito bombers.[5]

By July 1944, the Bomber Command consisted almost entirely of Lancasters (*top left*), Halifaxes (*top right*), and Mosquitos (*bottom*).

TABLE 2–1
Force Size and Composition, 1939–1945
(Aircraft Available for Operations)

Date	Aircraft Type Light[a]	Medium[b]	Heavy[c]
Sep 1939	264	251	0
Jan 1940	210	228	0
Jul 1940	307	360	0
Jan 1941	176	432	0
Jul 1941	160	665	31
Jan 1942	63	603	136
Jul 1942	86	293	278
Jan 1943	96	192	551
Jul 1943	51	124	978
Jan 1944	72	15	1,139
Jul 1944	138	0	1,463
Jan 1945	206	0	1,617
V–E Day	269	0	1,708

Source: BBSU, "Strategic Air War against Germany," table 5, Aircraft Available for Operations.

[a]Light = Battles, Blenheims, Bostons, Venturas, Mosquitos.

[b]Medium = Mitchells, Hampdens, Whitleys, Wellingtons.

[c]Heavy = Manchesters, Stirlings, Halifaxes, Lancasters.

Changes in force size and composition had a considerable effect on the command's destructive power. As Table 2–2 demonstrates, in 1939 the average bombload for each aircraft was a puny 204 pounds. By 1944 it had risen to 8,250 pounds. In sharp contrast, the average bombload for an American Eighth Air Force bomber in 1945 was 4,750 pounds.[6] As Table 2–3 shows, more than 70 percent of Bomber Command's tonnage was dropped in the last sixteen months of the war,[7] which underscores the importance of the large force of four-engine heavy bombers.

In 1933 when Hitler came to power in Germany, the standard RAF bombers were wire and fabric biplanes like the single-engine Hawker Hart and the twin-engine Handley-Page Heyford. These venerable aircraft still equipped frontline squadrons as late as 1937. Between 1932 and 1935 the Air Ministry issued new specifications that resulted in the production of monoplane aircraft, including the single-engine Vickers Wellesley and Fairey Battle and the twin-engine Bristol Blenheim.[8] The new twin-engine mediums (or heavies, as they were known at that time) were the Handley-Page Hampden, the Armstrong-Whitworth Whitley, and the Vickers Wellington. These aircraft equipped Bom-

TABLE 2–2
Average Bomblift per Aircraft Dispatched, 1939–1945

	Bomblift (lbs)	
Year	RAF Bomber Command	U.S. Eighth Air Force
1939	204	0
1940	1,457	0
1941	2,324	0
1942	3,405	2,600
1943	6,903	3,220
1944	8,250	3,980
1945	7,385	4,750

Source: BBSU, "Strategic Air War against Germany," pp 42–43.

ber Command squadrons during the rapid rearmament of the late 1930s. Both the underpowered Wellesley and the Battle were obsolete and recognized as such long before the outbreak of war. The Wellesley was used only in the Middle East, and the Battle squadrons were decimated in France in 1940. The Wellington, best of this lot, constituted a large part of Bomber Command until well into 1942.[9] The damage that could be inflicted on Germany by these aircraft, however, was very limited. Over the course of the war the Blenheim's average bombload was only about 600 pounds; the Hampden's, 1,556 pounds; and the Whitley's and Wellington's, 2,441 and 2,502 pounds, respectively.[10]

TABLE 2–3
Annual Tonnages Dropped by Bomber Command

Year	Short Tons[a]	Percent
1940	14,631	1.40
1941	35,509	3.39
1942	51,028	4.87
1943	176,352	16.84
1944	571,057	54.52
1945	198,835	18.98
Total	1,047,412	100.00

Source: BBSU, "Strategic Air War against Germany," table 10, Annual Tonnages Claimed Dropped by RAF Bomber Command. For slightly different figures expressed in long tons, see Webster and Frankland, *Strategic Air Offensive against Germany*, vol 4, app 44.

[a]A short ton is 2,000 pounds.

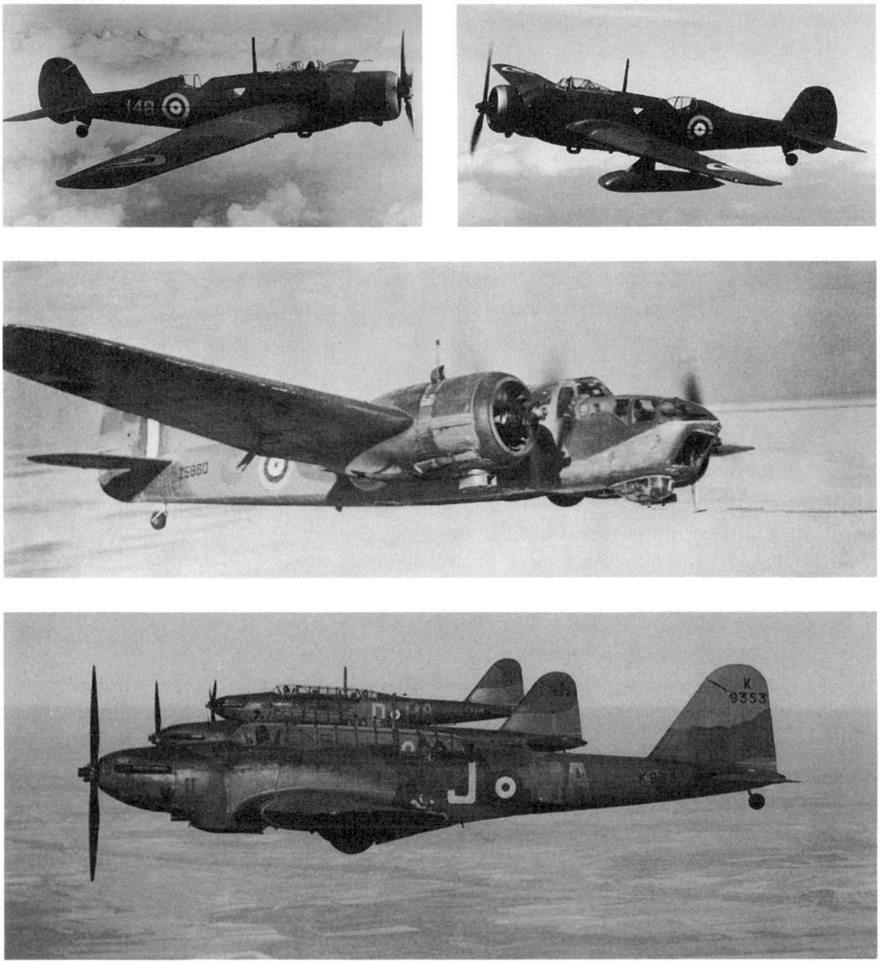

Between 1932 and 1935 the Air Ministry issued new specifications that resulted in the production of monoplane aircraft, including the single-engine Vickers Wellesley (*top left* and *top right*), the twin-engine Bristol Blenheim (*center*), and the Fairey Battle (*bottom*).

These bombers possessed inadequate defenses, mounting few guns and having too many blind spots that constrained the guns' radius of fire. The main armament consisted of .303-cal. Browning machine guns. Though reliable and capable of a high rate of fire, these guns were increasingly outmatched in range and striking power by contemporary fighter aircraft armament: the .50-cal. machine gun and 20-mm cannon. In the absence of a fighter escort, bombers with this defensive armament had little hope of surviving in daylight operations over Europe. RAF leaders, moreover, like their counterparts in the other major air forces of the late 1930s, did not think it possible to develop long-range escort fighters with combat performance good enough to hold their own against short-range fighters.[11]

The Air Ministry issued specifications for the true heavies in 1936. Pressure to arm rapidly and ensure against the failure of any one design led to the purchase of three different models.[12] Because they were the largest and most complex aircraft of the period, their manufacturers experienced a host of development problems, and the heavies did not even begin trickling into squadrons until 1940 and 1941 (Table 2–4). In order to save money on facilities, the four-engine Short Stirling featured a wingspan designed to fit into existing hangars. This economy had grave consequences for performance. The first Stirlings had an operational ceiling of just over 10,000 feet, totally inadequate for flying over flak-defended areas of Germany.[13] Later improvements raised its ceiling to about 17,000 feet, which was still too low. The attrition of Stirlings caused them to be restricted to easier targets in 1943, and they were withdrawn from operations altogether in 1944.

TABLE 2–4
British Aircraft Production
Medium and Heavy Bombers

	Bombers	
Year	Medium	Heavy
1939	758	0
1940	1,926	41
1941	2,777	498
1942	3,463	1,976
1943	2,737	4,615
1944	2,396	5,507
1945	648	1,669

Source: BBSU, "Strategic Air War against Germany," table 2, British Aircraft Production.

The Handley-Page Halifax began as a twin-engine heavy bomber designed to take the new and untried Rolls-Royce Vulture engine. While in prototype, however, it appeared that supplies of the temperamental Vulture engine would be limited, and the Air Ministry ordered the Halifax redesigned to accomodate four Merlin engines. The 20,000- to 21,000-foot service ceiling of the Halifax was much better than the Stirling's, but it was still inadequate in some weather conditions (the need to climb above cloud cover that threatened icing of the wings always constrained operations), and it still left the aircraft vulnerable to

German heavy flak. The Halifax also suffered from high loss rates that owed something to its instability during maneuvers when fully loaded.[14] For a time, in the 1943 to 1944 period, some versions of the Halifax were also confined to easier targets.

The twin-engine Avro Manchester, like the Halifax, was designed to use the Vulture engine, but service trials showed the bomber to be badly underpowered. Worse still, the Vulture exhibited a distressing tendency toward bearing failures and engine fires.[15] In 1940, when assured of a better supply of Merlins, Avro and the Air Ministry agreed to redesign the aircraft to accept four of these proven engines. The result was the most successful of the heavies — the Lancaster.[16] Despite an inadequate service ceiling (20,000 feet), it had good speed and no major handling vices, was easily serviced, and carried a large bombload. It first entered squadron service early in 1942 and by late 1943 formed the backbone of Bomber Command. In 1944 and 1945 it was the RAF weapon of choice against difficult German targets. (Table 2–5)

The Lancaster, Halifax, and Stirling all shared one shortcoming with the aircraft they replaced: they mounted .303-cal. machine guns. The Air Staff recognized the need for heavier armament as early as 1938, but plans were to install it in the next generation of aircraft, not in the 1936 heavies.[17] A proven design combined with the need to rearm rapidly made the Air Ministry reluctant to change defensive armament. Moreover, an increase in gun size involved numerous design considerations including additional electrical power for the turrets, with an increase in weight and consequent adverse effects on the aircraft's range and stability. Bomber Command tried repeatedly to procure .50-cal. gun turrets during World War II, but it succeeded only in fitting a few specially manufactured by the firm of Rose Brothers to its bombers toward war's end. Sir Arthur Harris would later describe the armament branches of the Air Ministry as "incompetent."[18]

Reality was a bit more complicated. British aircraft companies had small design staffs, compared with those of American firms, that were unable to handle a large volume of major changes effectively.[19] The temporary restriction of aircraft production to a limited number of current models in 1940 also inhibited redesign efforts. Bomber Command itself was not without fault, issuing mutually incompatible requirements at different times. In response to Harris's charges, an Air Staff officer wrote:

> Bomber Command found themselves unable to frame a consistent turret policy . . . the direct result was a severe repercussion both material and psychological throughout the M.A.P. [Ministry of Aircraft Production] and the turret industry. Clearly Bomber Command, in spite of many warnings, failed to appreciate the inevitable time lag that there must be between the placing of a requirement and the meeting of it. Thus they allowed themselves to fall into the fatal error . . . of altering their minds before effect could be given to their previous requirements.[20]

TABLE 2–5
Examples of Aircraft in Bomber Command, 1939–1945

		Cruising Speed				
		Most Economical		Performance at That Speed		
Aircraft	Service Ceiling (ft)	MPH	Altitude	Range (miles)	Bombload (lbs)	Fuel (imp gals[a])
Blenheim IV	22,000	180	15,000			
w/max bombload				1,460	1,000	466
Whitley V	17,600	165	15,000			
w/max bombload				630	8,000	370
w/perm tank(s) full				1,370	5,500	705
w/aux tank(s) full				1,930	3,500	969
Hampden	20,000	155	15,000			
w/max bombload				1,200	4,000	438
w/perm tank(s) full				1,885	2,000	654
Wellington III	19,500	180	15,000			
w/max bombload				1,200	4,500	638
w/perm tank(s) full				1,440	3,500	750
w/aux tank(s) full				2,040	1,500	1,030
Halifax III	20,000	225	20,000			
w/max bombload				980	3,000	1,077
w/perm tank(s) full				2,005	6,250	1,998
w/aux tank(s) full				2,785	500	2,688
Stirling III	17,000	200	15,000			
w/max bombload				590	14,000	980
w/perm tank(s) full				2,010	3,575	2,254
w/aux tank(s) full				2,440	None	2,694
Lancaster I and III	20,000	216[b]	20,000			
w/max bombload				1,660	14,000	1,625
w/perm tank(s) full				2,250	10,000	2,150
w/aux tank(s) full				3,150	None	2,950
Mosquito IV	33,000	265	15,000			
w/perm tank(s) full				1,620	2,000[c]	536
w/aux tank(s) full				2,040		657

Table 2–5 — *Continued*

Mosquito XVI	36,000	245	15,000			
w/max bombload				1,370	5,000	500
w/perm tank(s) full				1,485	3,000	536
w/aux tank(s) full				1,795	2,000	657

Source: Webster and Frankland, *Strategic Air Offensive against Germany*, vol 4, app 43.

[a]An imperial gallon is equivalent to 4.545 liters; a U.S. gallon is 3.785 liters.
[b]At mean weight.
[c]Some of this model were modified for a 4,000-lb bomb.

The Air Ministry claimed, furthermore, that the specially manufactured gun turret that Harris demanded could not be mass-produced.[21] Wherever the blame may lie, throughout the war Bomber Command aircraft flew into the teeth of the enemy's defenses with utterly inadequate defensive weaponry.

One bright spot in this materiel procurement story was the Mosquito light bomber, and it possessed no defensive armament at all. In the late 1930s, Ludlow-Hewitt of Bomber Command asked the Air Ministry for a "speed bomber" to perform "harassing" operations and reconnaissance. But the Ministry was not persuaded and refused to issue a specification for such a machine. Sir Geoffrey de Havilland produced a design in 1938 for a fast, twin-engine, plywood-sheathed, unarmed day bomber that relied entirely on its ceiling and speed to evade defenders. De Havilland, who consulted informally with officials in the Air Ministry and with RAF officers, succeeded in winning the support of Sir Wilfred Freeman, the Ministry's chief of research and development. Thus was born Freeman's Folly, the Wooden Wonder. The Mosquito's combination of speed (cruising, 245 mph; maximum, 397 mph at 26,000 ft), service ceiling (36,000 ft), and payload (4,000 pounds to Berlin) was arguably superior to any other Allied light bomber produced during the war. Only toward war's end would German turbojet fighters give it any trouble. Indeed, the Mosquito proved to be so versatile in combat that virtually every Allied air force wanted it.[22]

Various circumstances help explain the slow expansion and reequipment of Bomber Command with heavy bombers. Decisions to concentrate on island air defense, made stagewise between 1936 and 1938, slowed the rate at which Bomber Command received its "interim force" (the twin-engine Hampdens, Whitleys, and Wellingtons). This had little impact on the procurement of heavy bombers because the Air Ministry issued the first orders for them in 1937, and each suffered from a variety of engine and airframe development problems. The heavy bomber prototypes did not fly until 1939, and the first flights of production aircraft did not occur until 1940.[23] In May of that year, however, Lord Beaverbrook, Minister of Aircraft Production, ordered all production concentrated on just five military aircraft to close the gap between German and British air strength: two fighters, the Hurricane and Spitfire; and three medium bombers, the Blenheim, Whitley, and Wellington. Resources that could not be effectively employed on these high-priority programs could be used on aircraft expected to be available by 1941. Work continued with less emphasis, and amid various difficulties, on the heavies.[24]

German bombing of Great Britain and the subsequent dispersal of portions of the British aircraft industry disrupted production in 1940 and 1941. Shortages of key raw materials and manufactured components also affected the expansion of these programs. Insufficient numbers of skilled laborers, a situation that has been described as a "labour famine," became perhaps the most critical problem of all.[25] Taken together, the many difficulties that beset heavy bomber production caused the government to expand output of the older twin-engine Wellington and keep it in Bomber Command's front line for almost a year longer than were originally planned.[26]

The aircraft production record reflected these difficulties. Medium bomber output rose from 758 aircraft in 1939 to a peak of 3,463 in 1942; thereafter it declined to 2,396 in 1944. No heavy bombers were produced in 1939 or in the first quarter of 1940. Only 41 left the factories in all of 1940. Heavy bomber production rose to 498 in 1941 and climbed rapidly and steadily to a peak of 5,507 in 1944 (see Table 2–4). Bomber Command, meanwhile, had to acquire and train the personnel to fly and maintain these complex aircraft, a requirement that the available population of British males could not meet. The Dominions and the Empire provided just over 39 percent of all the aircrew employed by the RAF in the war. Of the 487 squadrons of all types in the RAF in mid-1944, Canada, South Africa, India, Australia, and New Zealand provided no fewer than 100. In addition, some 31 squadrons of Allied forces (Polish, French, Dutch, Belgian, Czech, and Norwegian) were based in the United Kingdom, and a further 20 (French, Greek, Dutch, Polish, and Yugoslav) were based overseas.[27] Within Bomber Command itself, about a third of the personnel by 1944 came from outside the British Isles. Bomber Command, if not the RAF, was in many respects an international force.[28]

These external supplies of men still proved insufficient to meet RAF requirements. Large numbers of British women, therefore, were recruited for the Women's Auxiliary Air Force, and in late 1944 they constituted almost 15 percent of the entire RAF establishment and no less than 20 percent of the personnel based in the United Kingdom.[29] (Table 2–6) At the beginning of the war, women were restricted to only five service trades (Air Force Specialty Codes, in American parlance) because conventional wisdom held them devoid of "inherent mechanical instinct." Wartime needs and demonstrated abilities ended this foolishness; by V–E Day women served in some eighty trades including intelligence, armament, and flight control.[30]

Of 487 squadrons in the RAF in mid-1944, Allied forces provided more than 50 that were based in England and its overseas possessions. Sir John Slessor, C-in-C in the Mediterranean and Middle East confers with the Wing Commander and one of the Flight Commanders of a Yugoslav Spitfire Squadron in Italy.

TABLE 2–6
Air Force Personnel Strength and Composition, 1939–1945

Date	RAF Officers and Other Ranks	Percent of Whole	WAAF Officers and Other Ranks	Percent of Whole	Overall Total
Sep 1, 1939	173,958	99.01	1,734	0.99	175,692
Oct 1, 1940	420,109	96.03	17,364	3.97	437,473
Oct 1, 1941	772,607	92.32	64,309	7.68	836,916
Oct 1, 1942	900,548	86.42	141,467	13.58	1,042,015
Oct 2, 1943	988,396	84.57	180,339	15.43	1,168,735
Oct 1, 1944	1,000,177	85.38	171,244	14.62	1,171,421
Sep 1, 1945	940,867	86.92	141,529	13.08	1,082,396

Source: Terraine, *Right of the Line*, app E.

Securing trained aircrews depended on the careful allocation of experienced personnel among frontline units to serve in combat and among training organizations as instructors. If frontline units expanded too rapidly, too few experienced personnel would return to the training organization as instructors and less competent aircrews would graduate into the operational squadrons. Flying training of Bomber Command aircrews for service in all frontline squadrons in 1939 was relatively poor. Inadequate training facilities, political demands for a rapidly expanded force, and an Air Staff failure to anticipate correctly bomber aircrew requirements all conspired to produce a poorly trained command and a badly equipped and staffed training organization.[31]

With the declaration of war in September 1939, corrective steps were taken. Expecting a protracted struggle, the Air Ministry withdrew seventeen squadrons from the front line for reserve and training duties.[32] The British government also turned to the Dominions and ultimately to the United States for assistance in aircrew training. By the end of 1940, Great Britain had secured agreements with Dominion governments to provide elementary and advanced flying training schools, with the bulk of the advanced schools to be located in Canada. Meanwhile, the U.S. government agreed to train replacements for the Eagle Squadrons (American volunteers serving in the RAF). By the summer of 1941, the U.S. Army Air Forces was operating six flying training schools that also served the RAF. Over the course of the war, overseas training proved absolutely vital to the maintenance and expansion of Bomber Command.[33]

Other problems affected the training program in the early and middle years of the war. Too few operational aircraft were available for training aircrew members in the final stages of preparation before they were to join the frontline squadrons. The machines that were available were often worn out and plagued by poor serviceability rates.[34] The increasing complexity of aircraft equipment, especially electronic aids, further complicated the process, inevitably throwing a large training burden on the frontline squadrons themselves, despite every effort to relieve them of that duty.

One final difficulty, the persistence of the prewar idea that aircrew members should be generalists, also contributed to dilute the quality of training. During the first years of the war, some gunners doubled as wireless operators, pilots were expected to function as navigators, and air observers were to assist in navigation and in aiming and releasing bombs. But specialists were required for the increasingly difficult jobs of navigating at night, using the growing number of complicated electronic devices, aiming bombs, and piloting the aircraft. Not until 1942 did the separation of navigating from piloting, bomb aiming from air observing, and wireless operator from gunner duties become firmly established. The air observer disappeared; in his place emerged a specialist navigator and a bomb-aimer (bombardier in American usage). In 1941, a bottleneck in advanced pilot training led the RAF to eliminate the co-pilot from all heavy bomber crews. For the better part of four years in the strategic air war, RAF pilots flew their aircraft to the farthest reaches of Germany with no assistance from a second pilot.[35]

The combat "tour policy" was closely linked to the eventual success of the training program. It was necessary to relieve men from operational duty before they became inefficient as a result of combat fatigue and became a danger to themselves and their colleagues. It was also crucial to ensure that an adequate number of experienced crewmen made their way back into the training organization.[36] Early in the war, the general policy was to relieve men from combat after 200 hours of combat flying, the time that comprised one tour of duty.[37] By the middle of the war, aircrews were expected to do thirty combat missions in a first tour. If they survived, they were posted to an Operational Training Unit or to other training duties for a six-month period. Afterward, they would return for another combat tour of twenty missions. Bomber Command granted considerable latitude in establishing tour policy to its group commanders. In the Pathfinder Force (later No. 8 Group), the specialist organization created in 1942 for target finding and marking, a first tour was set at no less than forty-five missions.[38] Combat attrition figures bespoke the risks. In 1943, only 33 percent of new crews could expect to survive their first tour, and only 16 percent would survive training duty (itself, a risky proposition) *and* a second combat tour.[39]

Strategy and Operations, to February 1942

How did the British government and RAF leaders expect to use these aircraft and their aircrews in an aerial war against Germany?[40] Early rearmament policy was based on the fear that the refashioned German Air Force, if employed with great intensity and ruthlessness in the early stages of a war, might inflict a knockout blow against Britain. For many airmen, the only possible counter to such a threat was to deter the enemy by developing a similar capability. If deterrence failed, a bomber force could be employed as a counteroffensive, striking the enemy in his homeland. A powerful strategic bomber force also would make unnecessary a large army on the model of the Great War. There would be neither losses like those at the Somme and Paschendaele, nor any need to deal with difficult allies.[41]

These ideas, no matter how popular with some RAF officers, politicians, and sections of the public in the 1930s, did not remain the official policy of the British government. The new aerial policy, established in 1938, gave first priority in air armament to creating an air defense system that could protect the island nation while reducing to second priority the use of an offensive bomber force for deterrence and counterattack. This reflected the Cabinet's conviction that the country could not afford to develop both forces at a maximum pace simultaneously and its belief that air defense technologies held greater immediate promise. The bomber force, when it became available, was to be used in concert with the Royal Navy and the French Army in a complementary strategy.[42] Later, in the spring of 1939, the British government reluctantly added a large British Army to the formula. Should war come, it was expected to pass through three stages analogous to what had occurred in World War I: an initial defensive period when the Germans possessed the initiative, a middle phase in which both sides built up their resources while continuing the fight, and a final period in which the combined effort of British and French arms would defeat the enemy.[43]

In the last year before the outbreak of the war, RAF thinking about strategic bombing focused on three of the sixteen Western Air Plans developed to implement the official strategy: *W.A. 1*, *W.A. 4*, and *W.A. 5*. The first plan called for attacks on the German air force to aid Allied air defenses and blunt the much-feared knockout blow from the *Luftwaffe* in the earliest and most defensive stage of the war. By 1939, however, Air Staff and Bomber Command leaders concluded that the front line of the German Air Force would be dispersed on too many airdromes to be attacked effectively and that attacks on aircraft production would take too long to have any effect on the scale of German air operations early in the war.[44]

The Air Staff was also skeptical about *W.A. 4*, the air plan aimed at destroying key elements in the German transportation system. Staff members assumed that road, rail, and water transportation, while vital to the German industrial system, presented too many targets to be attacked effectively with reasonable economy of effort. They were no more certain of succeeding if they attacked the German Army's lines of communications, partly because the targets were difficult and partly because they feared such an effort would lead to the British Army having authority over Bomber Command.[45]

W.A. 5 was decidedly the prewar favorite of the Air Staff and of Bomber Command. In its original form, it envisioned daylight attacks on the industrial districts of the Ruhr, the Rhineland, and the Saar. As more consideration was given to the campaign, *W.A. 5* soon developed into three major plans — one to attack electricity-generating and coking plants, another to attack transportation in the Ruhr, and a third to attack synthetic oil production.[46]

Several key assumptions lay behind these plans. The first judged the Ruhr to be the "industrial nerve"[47] of Germany for which no other area could substitute. It was thought to account for 65 to 75 percent of German production of bituminous coal, coke, coal tar and its by-products, pig iron, and raw steel. The second assumption held that four 500-lb bombs would be sufficient to disable any factory or plant measuring 100 yards by 100 yards. The average bombing accuracy for high-level day attacks was expected to be 300 yards circular error probable, and for those at low level, a remarkable 75 yards. Eighty percent of the sorties dispatched were expected to reach and bomb the objective if penetration into the enemy's defensive zones did not exceed 80 miles; the figure dropped to 40 percent if the penetration zone exceeded 200 miles. These highly optimistic operational assumptions led the Air Staff to estimate that a successful campaign against key Ruhr industries could be completed within four to six weeks. The only constraints involved the size, composition, and training of the bomber force likely to be available to carry out the ambitious scheme.[48]

Optimism about *W.A. 5* derived from two suppositions made by the intelligence community and widely held throughout the British defense establishment. Modern industrial economies and their associated social orders were believed to be fragile structures that could be made to collapse from one or two massive blows. The evidence or beliefs that supported these assumptions drew on the panicky reaction of the London population to the German Gotha bombardments in World War I, the sudden collapse of Germany in 1918, the dangerous economic crises of the 1920s and 1930s, vague conservative notions about the inferiority of modern social orders, and the contempt for civilian society found in some military circles. Overlooked was the possibility that basic ingredients of industrialized society — high productivity, administrative flexibility and sophistication, a trained and flexible labor force, a large stock of capital resources, good communications, and a highly developed transport network — provided the means with which a strategic bombing campaign could be resisted and a quick victory denied.[49] The intelligence community also believed incorrectly that the German economy possessed few reserves of capacity that could be called into play when war broke out.[50] Consequently, it assumed that the damage caused by strategic bombing would be felt very quickly at the fighting front.

When war came in 1939, the bomber force was incapable of any major contribution to the British war effort, although Sir Edgar Ludlow-Hewitt, C-in-C of Bomber Command in the late 1930s, was at first confident of his ability to execute effective strategic bombing.[51] Nevertheless, as the limitations of his equipment, aircrew training, and reserves became apparent, he became more cautious, even pessimistic. Throughout 1938 and 1939 Ludlow-Hewitt argued that the equipment and training available to his force severely limited what could be accomplished. The more he examined the operational problems involved in a strategic bombing campaign, the more he became convinced that German air defenses would inflict prohibitively high bomber losses in attacks made in daylight. In 1938, for example, he reported that the loss rate in a sustained attack on German targets would eliminate his Blenheims in 3½ weeks and his Hampdens, Whitleys, and Wellingtons in 7½ weeks.[52]

The C-in-C reluctantly abandoned the long-cherished idea among airmen that bomber forces, unlike all other weapons of war, did not have to defeat an intervening enemy armed force before delivering blows to the enemy homeland. Though he continued to share the official view that long-range escort fighters could not contend with the faster, more maneuverable short-range interceptors they would encounter over enemy territory, mounting pessimism over Bomber Command's prospects prompted him to urge the Air Ministry to develop escort fighters.[53] The Air Staff's conviction that a competitive, long-range fighter was a technical contradiction in terms, the nation's commitment to short-range fighter aircraft for air defense, and the belief that the overcommitted British aircraft industry was incapable of handling more design projects combined to thwart Ludlow-Hewitt's suggestion.[54] In the late 1930s, Air Staff leaders also believed that any threat to bombers posed by defending fighters would be, at best, temporary.[55] British airmen were not uniquely shortsighted in this respect. American airmen were also certain of the bomber's invulnerability and, if anything, slower to be convinced of the need for long-range escorts.

Members of the Air Staff and officials in the British government overestimated what the German Air Force was capable of doing at the outset of a war, and leaders of the French government shared their fears.[56] Neither government wanted to commit any act that might trigger reprisal attacks on London and Paris; both were mindful of the need to keep neutral opinion, chiefly in the United States, on their side.[57] In late 1938, Ludlow-Hewitt declared that in light of existing political restrictions, the operational limitations of his force, and a need to conserve strength for the future, he could not foresee an "effective bombardment role" for his command if war was to come soon. If Bomber Command was forced to operate at night, he asserted, about all he could do would be to drop leaflets.[58] These perceptions and fears produced an interim bombing policy: Bomber Command's strength should be conserved for the future and employed only against strictly defined military targets.

Between September 1939 and the German western offensive of the following spring, Bomber Command proceeded accordingly; it carried out limited reconnaissance, dropped propaganda leaflets over Germany at night, and made a few small daylight attacks on German warships in harbor. The command did not attempt any large-scale daylight raids on high-priority targets in the Ruhr of the sort called for in *W.A. 5*. The losses suffered in a few small-scale daylight operations against peripheral targets, moreover, confirmed prewar fears and reinforced the policy of conservation of force.[59] The Air Staff and Sir Edgar Ludlow-Hewitt agreed that major daylight attacks would imperil the long-term future of the command.[60] At Bomber Command, increased attention was paid to training for nighttime operations.[61]

When the Germans invaded France and the Low Countries in May 1940, the War Cabinet authorized aerial operations against the Ruhr. Beginning in May, Bomber Command, hoping to do something useful while conserving the force, attempted to carry out a selective, precision bombing strategy at night, something that had long been thought by many professional airmen to be a contradiction in terms. Increasingly, with the exception of missions flown by the light bombers of No. 2 Group against coastal targets and the occasional foray against small targets in occupied Europe, Bomber Command became a nighttime force.[62] Indeed, Bomber Command did not attempt any large-scale daylight operations again until June 1944, some four years later. Unfortunately, night navigation and bombing proved no more accurate in the early spring of

1940 than it had before the war. As the official British historians observed, science had to come to the aid of the bombers or else the bombing policy embodied in the western air plans had to change.[63]

At first, the Air Ministry stubbornly clung to its prewar philosophy of selective air attack against vital economic targets, emphasizing attacks on synthetic oil production in particular. When the Western Front collapsed and while the British Army extracted itself from the continent as best it could, the Air Ministry ordered Bomber Command to attack only targets most important for an invasion of Britain: airfields, aircraft production, and German railways. When the Battle of Britain began in the summer of 1940, Prime Minister Winston Churchill and the new C-in-C of Bomber Command, Sir Charles Portal, pressed the Air Ministry to abandon selective bombing and attack the German civilian population directly and persistently. The Air Staff, however, continued to favor selective bombing.[64]

The Allied defeat on the continent and the fall of France left strategic bombing the only offensive action that could be taken against Germany. Certainly for the British Army, weakened, in disarray, and hurriedly preparing to repel a German invasion, the prospect of any return to the continent could only be relegated to some distant future. As the Prime Minister put it: "Bombers alone provide the means of victory."[65] In light of what Bomber Command actually could do at that time, this was a considerable rhetorical exaggeration.

Many optimistic assessments were made in the summer of 1940 about Bomber Command's ability to navigate and aim bombs at night. Evidence to the contrary did not appear until late in the year when the immediate crisis of the Battle of Britain passed and the Air Staff again decided to attack Germany's oil production. Photoreconnaissance of two oil plants at Gelsenkirchen showed them to be almost entirely undamaged, despite having been attacked by almost 300 aircraft at one time or another.[66] The Air Staff and Bomber Command nonetheless persisted in nighttime attacks on oil plants for the next two months, with no better results.

At this point other requirements of the war intervened. The increasing success of the U-boat offensive against British sea communications led the Prime Minister to order that Bomber Command's efforts be directed to the antisubmarine campaign for at least four months. Had Bomber Command persisted in its attempts to attack oil targets at night with the existing equipment and operational techniques, the official historians observed, it probably would have "done a great deal more damage to its prestige than to its targets."[67] Bomber Command's operational weaknesses had doomed the prewar strategy of selective, precision bombardment and frustrated Churchill's hopes that bombers would prove the decisive weapons of war.

The belief in selective bombing survived as long as it did because accurate assessments of the nighttime bombing raids could not be made. In the absence of good human intelligence and satisfactory photoreconnaissance from inside Germany in the early days of the war, the Air Staff failed to appreciate the futility of Bomber Command's efforts. Human intelligence from agents and neutral reporters essentially disappeared by mid-1940,[68] and what little remained trickled through reinforced optimistic assessments.[69]

Air Staff and Bomber Command leaders early realized the importance that photoreconnaissance would play in a strategic bombing effort. Ludlow-Hewitt had been especially forceful in his recommendation that specialized aircraft such as the Mosquito be procured for this purpose; however, the Air Ministry had made little progress in acquiring such machines before the war.[70] Having lived so long with austerity, the RAF could not rid itself of a prejudice against expensive specialization. Furthermore, the intelligence assessment of photoreconnaissance remained an exclusive responsibility of the British Army until 1938. In fact, until February 1937, the only posts in the Air Ministry connected with this subject involved the development of cameras. Most airmen believed that any requirements for photoreconnaissance could be met by the existing Blenheim squadrons of No. 2 Group, which were expected simply to add aerial photography to their other duties.[71]

A host of difficulties made the Blenheim photoreconnaissance effort ineffective. These slow-flying aircraft were vulnerable to fighters and flak when operating in daylight, their camera equipment often malfunctioned, and crews were inadequately trained for the precise demands of the mission. From September 1939 through January 1940, No. 2 Group dispatched some forty-eight Blenheims on photo sorties, eight of which were shot down. Five brought back no photographs at all, while many others took photos of little or no value because the cameras froze at the altitudes at which they were designed to operate. Some photographs of good quality were taken, but the success rate was insufficient to warrant continued confidence.[72]

In January 1940, the Air Ministry organized special reconnaissance units using high-performance aircraft.[73] A modest beginning had been made shortly after the war broke out, when the Secret Intelligence Service transferred its three specialized aircraft to the RAF. Supplemented by a few precious Spitfires, this unit, initially known as the Heston Flight, began activities that would make RAF photoreconnaissance one of the most valuable intelligence sources in the war.[74] The revelation of Bomber Command's bombing failures, in particular the lack of damage to oil targets, was one of the first contributions that improved reconnaissance made to the war effort.

Some of the most telling negative evidence came from analysis of a few photographs taken during actual bombing operations. In January 1941, Bomber Command had only 22 cameras in service when the official number was supposed to be 168.[75] By summer, however, enough photographs had been accumulated to make possible a significant scientific assessment of the effectiveness of nighttime precision bombardment. In July, Lord Cherwell

(Professor Frederick A. Lindemann), the Prime Minister's personal scientific adviser, ordered a statistical analysis of accuracy using recent bombing photographs. The result, known as the Butt Report, was devastating. Of all the aircrews claiming to have attacked their targets, the report revealed that only one-third approached any closer than five miles! For short-range targets, such as those in France, the proportion of aircrews that approached within a five-mile radius rose to two-thirds. Over the Ruhr, it fell to a mere one-tenth. During a full moon, the most favorable condition for navigating at night, only two-fifths of all aircrews reporting successful attacks actually came within the five-mile radius. With no moon, the figure fell to one-fifteenth.[76]

By the summer of 1941 this evidence of operational ineffectiveness undermined many of the major premises on which the British strategic bombing offensive had been planned and conducted. Prime Minister Churchill, who one year before termed Bomber Command "the means of victory," now told the CAS and former C-in-C of Bomber Command, Sir Charles Portal: "It is very disputable whether bombing by itself will be a decisive factor in the present war."[77]

The German Defenses, 1940–1942

By the end of 1941, it was increasingly apparent that the darkness to which Bomber Command had retreated in 1940 to conserve its forces no longer provided adequate security against the attentions of the German nighttime defenders. To be sure, the *Luftwaffe* did not at first possess an organized system of air defense that compared with RAF Fighter Command. The whole prewar orientation of the *Luftwaffe* was directed toward offensive counterair action, interdiction, and close air support. An enemy air force could best be defeated over its own territory, and air superiority seized in air combat. Despite this emphasis, many of the key ingredients for an improvised air defense were on hand or under development in 1939.[78] The Germans possessed relatively large numbers of good flak guns of all calibers supported by searchlights, sound detectors, and visual ranging apparatus. They were also deploying an early warning radar (Freya) on the frontiers and coastlines as well as an observer network for reporting the progress of enemy air attacks. They had a first-class day fighter in the Bf 109, but they had no night fighter of any kind. And they were shortly to deploy a new fire-control radar, the Würzburg.

German fighter airfields were disposed along the coast or frontier and linked to a rudimentary early warning system or to the ground observer system, or to both. There was neither centralized control nor any close control of aircraft from ground stations, as was the case with RAF Fighter Command. When Bomber Command began night attacks on the Ruhr in May 1940, the Germans were caught without the organized means to defend themselves

Among superior air defense measures the Germans possessed in 1939 were the Bf 109, a first-class fighter, and early warning radar, the Freya, strategically deployed along the coastlines and frontiers.

effectively. At this stage of the war, they were poorly prepared to defend against even the limited and poorly executed attacks of the Bomber Command.

A number of prewar experiments indicated the main lines of potential development for a night defense system. These trials used a warning service based on sound detectors and a searchlight belt. Night fighters orbited the beacons at altitude outside the illuminated zone until a bomber was caught in the lights. Any focusing of searchlights at altitude signaled the night fighter to leave his beacon and enter the illuminated zone to attack the bomber. Once the fighter-defenders were in the lighted area, flak firing was to cease. There was no ground control of the fighters. These experiments ended in August 1939 when the units involved shifted to day fighting. At that stage, night defense was exclusively in the hands of the searchlight-aided flak guns.[79]

Early in 1940, small, standing night fighter patrols were improvised. Experiments in Denmark seemed to demonstrate that these would be ineffective without radar-based ground control integrated with searchlights and flak.[80] In the middle of 1940, responding to Bomber Command's early attacks, the *Luftwaffe* high command assigned Col. Josef Kammhuber the task of developing a systematic approach to night aerial defense. Kammhuber had no connection with the prewar experiments, but his organizational ability and knowledge of the problems of blind flying were ample qualifications for the task.[81] Over the better part of the next three years, he developed a highly sophisticated

The *Luftwaffe* had not procured a specialized night fighter, so the Germans modified the Messerschmitt Bf 110 twin-engine fighter, which had proved a disappointment in day fighting. Instead of a revolving sweeping antenna, the night fighter was equipped with four fixed arm antennas attached to its nose.

system of controlled night fighting which the British came to call the Kammhuber Line.

Kammhuber began by greatly expanding the illuminated zone, creating searchlight coverage that extended in a line from Denmark down through northern France. Early warning was provided by Freya radar stations backed up by the observations of the sound detector crews. Close control of the night fighters and the flak batteries was provided by Würzburg, a shorter range radar. The *Luftwaffe* had not procured a specialized night fighter, so Kammhuber was compelled to make do with modifications of the Messerschmitt Bf 110 twin-engine fighter, which had proved a disappointment in day fighting, and the Ju 88 twin-engine fast bomber. These two aircraft bore the brunt of night fighting for most of the war. Other aircraft such as the Do 217 and the He 219, the latter a specially designed night fighter, were also used, but not in the same numbers.[82]

Between late 1940 and the middle of 1943 the Germans modified the Kammhuber system in two major ways. First, they introduced airborne interception (AI) radar. Kammhuber asked for the development of this aircraft-mounted equipment at an early date; the first version to be used was the Lichtenstein B in 1941. Several leading night fighter aces opposed this radar

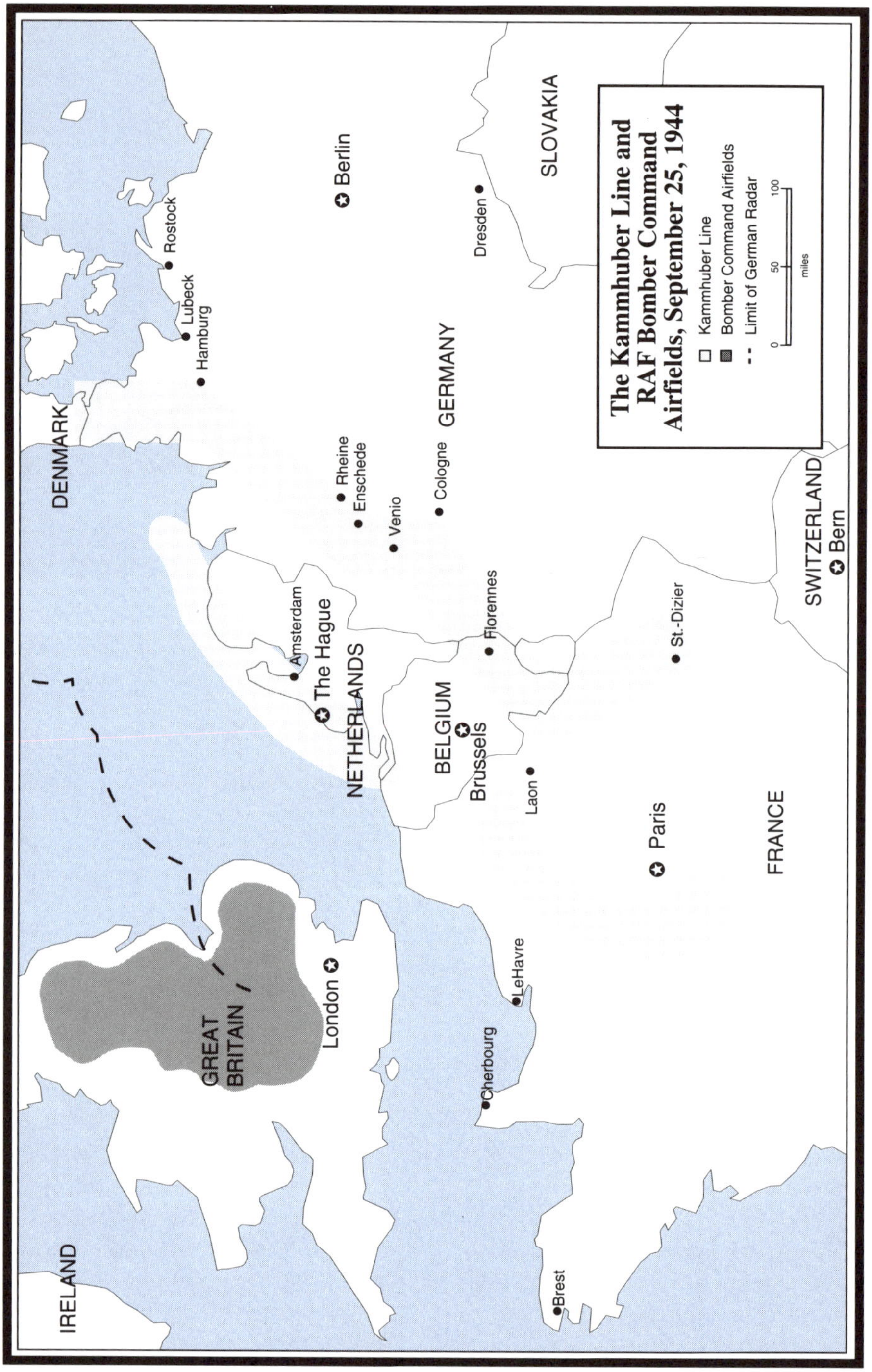
The Kammhuber Line and RAF Bomber Command Airfields, September 25, 1944
Kammhuber Line
Bomber Command Airfields
Limit of German Radar
0
50
100
miles
DENMARK
Rostock
Lubeck
Hamburg
Berlin
SLOVAKIA
Dresden
GERMANY
Rheine
Enschede
Venlo
Cologne
Amsterdam
The Hague
NETHERLANDS
BELGIUM
Brussels
Florennes
Laon
St.-Dizier
SWITZERLAND
Bern
Paris
FRANCE
GREAT BRITAIN
London
LeHavre
Cherbourg
Brest
IRELAND

Prime Minister Winston Churchill.

because its large aerials created drag and decreased the performance of their aircraft; they preferred to acquire the target visually once ground control directed them to its vicinity. The second change involved removing the searchlights and the flak guns from Kammhuber's linear system and grouping them to provide a point defense for individual cities. This was largely a response to RAF raids on Lübeck, Rostock, and Cologne in 1942 that proved sufficiently damaging to create political pressure to improve city defenses. From that point, all controlled aerial fighting along the Kammhuber line occurred in the dark.[83]

However arranged, Kammhuber's aerial defense system had serious weaknesses. To use an industrial term, it had a low "processing rate." The line was composed of a series of contiguous "boxes," the boundaries of which were defined by the range of the Würzburg radars. The awkwardness of the plotting system used within each box and, before 1942, the absence of an aircraft-mounted IFF (Identification, Friend or Foe — an electronic device giving an appropriate response when interrogated by a friendly radar transmitter) meant that only one German night fighter at a time could be controlled from the ground. One Würzburg tracked one bomber, and a second tracked one fighter. The two plots were not represented on a single radarscope; they came to two individual operators, each of whom projected a different colored circle on a plotting table. The controller radioed directions to the fighter on the basis of data portrayed on the plotting table. Until IFF became available, enemy blips could not be separated from friendly blips. When operators lost the fighter, as often happened, the controller would have to order it to return to the beacon in that particular box. In addition, altitude measurements from the two Würzburgs might vary as much as 500 meters, and when compounded by transmission and reading errors, an interception failure often resulted. Airborne intercept radar overcame some of this problem, but the Lichtenstein B had a fairly narrow search angle, and when a bomber employed a radical evasive maneuver, contact could be lost.[84]

Despite these weaknesses, Kammhuber's organization developed a level of sophistication that, combined with the weather, in the autumn and early winter of 1941 succeeded in temporarily halting Bomber Command's nighttime bombing offensive. Bomber loss rates rose sharply and in early November 1941 Prime Minister Churchill ordered operations against German territory restricted, thus preserving the command and allowing it to gather strength for operations in 1942.[85] By the end of 1941, the attempt to combine the prewar strategic emphasis on selective economic targeting with the policy of force conservation by operating at night appeared to have failed on both counts.

The Strategy of the Area Offensive, February 1942 to March 1944

While mass bombing attacks on civilian morale were much discussed after World War I within and beyond the RAF, none of the sixteen Western Air Plans drafted before World War II singled out morale as a target. Moreover, none of the first three Air Ministry directives issued to Bomber Command in early 1940 even mentioned the civilian population or morale in any way. The fourth directive, issued in June, ordered the command to prepare to use incendiary weapons against German crops and forests to achieve some "moral effect" on the enemy. The ninth directive, sent on September 21, 1940, ordered Bomber Command to attack Berlin from time to time to "cause the greatest possible disturbance and dislocation both to the industrial activities and to the civil population in the area."[86] Sir Charles Portal, then C-in-C of Bomber Command, originally proposed this strategy of area "harassing attacks" to compensate for the command aircrews' inability to navigate and bomb accurately at night. As he put it: "The accidents of weather, coupled with the variation in the skill and judgment of individual captains, inevitably produce dispersal which, however, largely increases the moral effect of our operations by the alarm and disturbance created over the wider area."[87]

The desire to retaliate for *Luftwaffe* bombing of British cities also encouraged deliberate aerial attacks on the German civilian population. The Cabinet authorized a major bombing raid on Berlin after a heavy attack on London in late September 1940, and a month later, declare: "While we should adhere to the rule that our objectives should be military targets, at the same time, the [German] civilian population must be made to feel the weight of the war."[88] Churchill urged Portal to attack as many cities in Germany as possible, and when the oil plan was under discussion, he lamented that the targets were not located in heavily populated areas.[89]

The first Air Ministry directive to place any real emphasis on civilian morale appeared in late October 1940, in the document that ordered Bomber Command to focus its primary efforts on the oil plan. When conditions were not favorable for hitting oil targets, the Air Ministry desired "regular concentrated attacks . . . on objectives in large towns and centres of industry, with the primary aim of causing very heavy material destruction which will demonstrate to the enemy the power and severity of air bombardment and the hardship and dislocation which will result from it."[90] The pairing of traditional *selective* economic targets with a secondary emphasis on civilian morale continued to appear in Air Ministry directives until February 1942, when Bomber Command was told to direct its efforts primarily at the morale of industrial workers.

There were dissenters. The Air Staff told Sir Charles Portal back in the

View of London, with smoke from damage caused by a German raid.

summer of 1940 that the "morale effect, although an extremely important subsidiary result of air bombardment, cannot in itself be decisive."[91] Perhaps the strongest criticism of morale attacks appeared in an Air Staff note of January 1941:

> A nation of eighty millions, the conquerors of Europe from the North Cape to Bordeaux, are not fit subjects for moral attacks with the very limited force at our disposal . . . . Any attempt to weaken the fighting spirit of the [German] military services by the indiscriminate attack on their towns and families with our very limited force is doomed to failure.[92]

The criticism nevertheless seemed to reserve final judgement, implying much for the future. If the bomber force were enlarged and if tactical concentration improved, it might prove capable of doing the kind of material damage that would have some effect on German civilian morale.

In February 1942 the Air Ministry directed Bomber Command to focus its operations exclusively "on the morale of the enemy civil population and in particular, of the industrial workers."[93] Over the next three months, the command demonstrated a newfound ability to concentrate sufficient high-explosive bombs and incendiaries to burn out large urban areas in major attacks. This combination of objective and technique, now known as the area offensive, would dominate British bombing operations until March 1944.

In March 1942, one month after the Air Ministry directed Bomber Command to begin its nighttime area offensive, Lord Cherwell, Scientific Advisor to the Prime Minister, advised Mr. Churchill that a sustained, direct attack on German civilian morale had the potential for decisive results. He believed that Bomber Command now could do enough harm to "break the spirit of the [German] people."[94] The Secretary of State for Air, CAS, and Prime Minister agreed. A few days later, Air Marshal Harris, the new C-in-C of Bomber Command, set out to prove that bombardment attacks of the sort Cherwell had in mind could be carried out. On the night of March 28/29, 1942, he sent 234 aircraft against the city of Lübeck on the Baltic Sea. Photographic reconnaissance afterward revealed that 40 to 50 percent of the built-up area had been destroyed, largely by fire. More than 200 acres of buildings appeared to have been knocked down or burned out. To be sure, Lübeck was not an important center of war production, but it was not well-defended, it could be found at night and hit in force, and when hit, it would burn.[95]

Harris's efforts to demonstrate the new destructive powers of his command culminated on the night of May 30/31, 1942, in Operation MILLENNIUM, the 1,000-bomber attack on the Rhine River port of Cologne. Harris had fewer than 400 heavy and medium bombers in his front line at this time. To make so large an effort, he committed to action the whole of his reserves and his entire training organization. He risked the future of his command and his career in this attack, but he was rewarded. Photoreconnaissance revealed that some 600 acres of the ancient Roman city had been completely destroyed. Only 41 aircraft and their crews were lost, not quite 4 percent of the total force dispatched. The British strategic bombing offensive could claim its most important propaganda victory thus far in the war.[96]

From these early experiments, Bomber Command in 1942 and 1943 developed the area offensive more fully. Strictly speaking, the term *area* characterized a tactic in which aiming points were placed right on the center of the built-up sector of the city attacked, not on a specific factory complex usually found along the city's periphery, as was the case earlier when attempts were made to employ selective, precision bombing. Bomber Command organized the area raids to achieve concentration over the target to start large fires, to spread the damage over the largest area possible, and generally to ensure that bombs hit something other than open fields.[97]

Even though the formal promulgation of the area offensive occurred in February 1942 and the first large attacks conforming to the area model were made in April and May of that year, not until early 1943 did Sir Arthur Harris believe his force large enough and well enough equipped for a sustained offensive. From March of 1943 through March of 1944 Bomber Command carried out three major campaigns that constituted the heart of the area offensive: the Battle of the Ruhr, the Battle of Hamburg, and the Battle of Berlin.[98]

The Battle of the Ruhr opened with a major attack against Essen on the night of March 5/6, 1943. This began a period that ended in late July 1943 during which Bomber Command delivered forty-three major area attacks on German cities. No fewer than twenty-one of these attacks were made against six key industrial cities in the Ruhr. Duisburg was hit five times (2,084 sorties) as was Essen (2,074 sorties), while Cologne again suffered four assaults (1,761 sorties). The other seven Ruhr missions were directed against Düsseldorf, Dortmund, and Bochum. The destruction, death, and injury that occurred in the Ruhr and in the other cities was serious and widespread and foreshadowed what was to come.[99]

The second major campaign of the area offensive in 1943 opened with four closely spaced major attacks on the North Sea port of Hamburg, beginning on the night of July 24/25. An armada of more than 700 bombers was dispatched on each of these missions. The second attack on the night of July 27/28 resulted in an almost unimaginable catastrophe. Widespread fires started by the cascade of incendiary bombs merged into a firestorm in the center of the city. Temperatures approaching 1,000 degrees Centigrade created ground winds of 150 mph that sucked everything from people to trees into the inferno. Thousands of men, women, and children perished — incinerated or suffocated in their shelters — and many thousands more were badly burned. Overall, after four attacks within the space of a few days, some 50,000 people died and nearly 1 million were made homeless refugees. A huge area of the city at the mouth of the Elbe River was completely devastated. The oft-quoted Albert Speer, Hitler's Minister of Armaments, wrote in his memoir: "Hamburg . . . put the fear of God in me."[100]

Germans took pictures of Hamburg following one of the air assaults, after which Europe's largest port lay useless as smoke billows from supply dumps at the port and docks and ships go down in water.

Between July 24 and November 18, 1943, when Sir Arthur Harris opened the Battle of Berlin, Bomber Command conducted thirty-three major area attacks. During this period over 600 bombers, almost all four-engine heavies, were dispatched to attack a single target on fourteen occasions. The first all-Lancaster attack of more than 300 aircraft was directed against the German capital on the night of September 3/4.[101] The area offensive against that city, however, did not really begin until November when the long winter nights of northwest Europe offered Bomber Command adequate cover of darkness.

Between November 18/19, 1943, and March 24/25, 1944, Harris sent sixteen major bombing missions, involving more than 9,000 sorties against Berlin.[102] Another nineteen large attacks, more than 11,000 sorties, were made against a dozen other German cities during the same period. The damage to Berlin and the death and injury inflicted on its inhabitants was appalling, but the aerial campaign did not "cost Germany the war" as Sir Arthur Harris had predicted.[103] He had based his prediction on the participation of the American Eighth Air Force in the Berlin bombing campaign, but that organization was wholly committed to a direct attack on German aircraft production and, in any case, could not have staged unescorted daylight attacks against Berlin in November and December 1943 without itself incurring excessive losses. The Americans did attack Berlin in March 1944 when the Eighth Air Force was a much larger force and possessed adequate long-range escort fighters. As it was, however, according to the calculations of the British official historians, Bomber Command on its own actually delivered a weight of attack comparable to what Harris had originally expected from an Allied effort.[104]

Berlin was a much larger target than Hamburg, but attacks on the capital required a flight deep into German airspace. That gave more opportunities to the night fighter defenses. Moreover, winter weather produced an almost constant cloud cover (daylight photoreconnaissance of the city could be made only twice between mid-November 1943 and mid-February 1944). Also, because of the range, the target could not be marked using Oboe, the most accurate of the electronic bombing aids available to Bomber Command. Toward the end of the Battle of Berlin, German aerial defenses became so formidable that Sir Arthur Harris himself began to doubt whether the area offensive could long be carried on with his existing equipment and operational technique.[105]

Tactics and the German Defenses during the Area Offensive, February 1942 through March 1944

Even though area attacks on German cities in campaigns such as the Battle of Hamburg were easier to execute at night than precision ones in daylight, they could not have achieved success without improvements in the nighttime techniques that Bomber Command used to find and hit targets. Indeed, the February 1942 "morale directive" was predicated on the use of the new electronic aids in navigation and bomb aiming.[106] Nor could Bomber Command have ventured into area offensives without a new combination of tactics and devices to reduce casualties inflicted by the continually improving German defenses.

Navigating an aircraft during daylight in peacetime with the available technical aids was an exacting but practicable exercise.[107] One plotted a course from base to destination, taking into account the power settings and estimated wind drift. Once aloft, visual fixes could be used to estimate the actual wind drift, and corrections to the course could then be made. If weather intervened to prevent the taking of visual fixes, then one could rely on dead reckoning or, under the best conditions, home on radio beacons and take direction-finding loop bearings. In wartime, however, once the aircraft left friendly territory, no usable radio beacons were available and, beyond a certain range, loop bearings could not be taken from the aircraft with any accuracy. Everything then relied on visual fixes or dead reckoning.[108]

At night, such fixes could be obtained in the right conditions — plenty of moonlight with no cloud cover. The standard alternative taught to all navigators in the early stages of the war was the star fix, obtained with a sextant through a Perspex bubble in the roof of the aircraft. Unfortunately the Perspex was sometimes covered with frost. When the navigator attempted a fix through an open hatch, he had to wipe the eyepiece between each shot, and at high altitude he ran the risk of freezing his hands.[109]

Good star fixes also required that the pilot fly a straight course (within three degrees of the course plotted) at a constant air speed (the variation could not exceed five knots), something the RAF testing establishment in 1941 assumed could not be done more than half the time by the average pilot.[110] With the equipment then in existence, it was assumed that a navigator could fix his position with less than a ten-mile error 50 percent of the time and less than a 24-mile error 90 percent of the time.[111]

Operational planners usually routed aircraft over easily recognizable points spaced not more than one hundred miles apart if possible. A final identification point was set twenty miles from the target. The run-up from there to the aiming point was done by map-reading, a process complicated in the final approach by bursting flak and searchlight dazzle. In the early days, according to Sir Arthur Harris, the search for a visual fix from which to make the run to the target "almost always took half an hour or so."[112]

On the return trip, once the enemy coast was crossed, loop bearings became more reliable, and direction finding radio fixes could be obtained from friendly stations. After crossing the English coast, an aircraft could home on a light or radio beacon at the airfield. It was necessary to keep a good dead-reckoning plot to guard against wireless failure, which happened all too often. Aircrews also had to keep the IFF turned on when approaching the English coast or else risk being fired upon by friendly flak or being attacked by friendly fighters. If the IFF was down, every attempt had to be made to stay above 2,000 feet (to avoid friendly light flak) and secure an identification on the normal, medium frequencies.[113]

Gee, first among the new radio navigation aids developed for Bomber Command, was initially used early in 1942. The system consisted of a master ground station which transmitted a radio signal simultaneously to the aircraft and two slave stations that relayed it to the aircraft. Using triangulation, equipment in the receiving aircraft measured the time interval between the arrival of the signals, thereby giving the navigator a position fix. Gee's accuracy decreased with distance from the British ground stations, so it rarely found use beyond the Ruhr. Although it could not aid or improve actual bomb aiming, it much improved navigation over the North Sea. Perhaps most important, Gee helped Bomber Command form hundreds of aircraft ascending from different bases into a single bomber stream, a massed formation used against the German defenses from the summer of 1942 through the spring of 1944. Gee also served as a very useful homing aid on the bomber's return.[114]

Oboe, which entered service in December 1942, was a blind-bombing aid which used two ground stations. Signals transmitted from each were received by the aircraft and retransmitted to the originating stations. The range and position of the aircraft in relation to the stations could thus be measured. One station monitored the track of the aircraft and directed it along an arc of constant range, an arc that passed directly over the target. The second followed the progress of the aircraft along the arc, and when the bomber's position matched that of the appropriate target, the station transmitted a signal to the bomber to release its bombs. Because radio pulses travel tangentially to the surface of the earth, Oboe's range was limited by the altitude at which the bomber could fly. Aircraft could be directed to targets in the Ruhr only if the bomber flew at about 28,000 feet. Only one type of aircraft in Bomber Command's inventory — the Mosquito light bomber — could do that. Oboe could not be used against eastern German targets until the autumn of 1944, when mobile transmitters moved to the continent.[115] The accuracy of Oboe varied during the war, but some idea of its utility can be seen in the fact that variations in bomb aiming now could be measured in hundreds of yards rather than in miles.

Oboe had one other significant disadvantage. A pair of stations could control only one aircraft at a time. Because an Oboe run usually required about ten minutes, the number of stations available permitted only six aircraft per hour to be controlled. Even with more stations, Oboe could not be used to control the operations of the entire force. Its most useful employment was as an aid to target-marking in the Pathfinder Force and in target-marking squadrons that sprang up in the other groups in the final year of the war.[116]

Gee-H was a combined navigation and blind-bombing aid. The H component operated rather like Oboe in reverse. A ranging pulse was sent from the aircraft to two ground stations and then retransmitted back to the aircraft. The accuracy of the fix thus obtained was similar to what Oboe could supply, and it far surpassed Gee's. Because H was cumbersome to operate, it was paired with Gee for navigational purposes over the first part of the outward flight. H, which could be operated by about one hundred aircraft simultaneously, was then used in the run-up to the target. Gee-H was first used in the early winter of 1943, but it did not come into its own until the autumn of 1944. No. 3 Group came to specialize in this equipment and used it to great effect in Bomber Command's attacks on oil plants in the Ruhr region.[117]

H2S was an early, crude navigation radar that first entered service in January 1943. Carried onboard the aircraft, it was not limited by distance from base. Its usefulness varied according to the skill of the operator and the signature of the terrain. As an aid to bomb aiming it gave the best performance against targets like Hamburg, where the difference in radar returns from land, water, and built-up areas made it easier to identify aiming points. It did not work nearly as well against Berlin because of the large dispersion of the built-up area and the tendency of radar returns to swamp the screen. Its greatest utility was in getting ground fixes and in keeping the bomber stream together on the way to the target.[118]

When Gee began operational trials in late 1941, the Air Staff proposed the formation of a target-finding force. The idea was to put some of the best aircrews together to share ideas, improve navigation, and operate as a separate force. They would precede the main force, find the target, and use incendiaries to start fires that could be observed by the bombers that followed. When Sir Arthur Harris came to Bomber Command, he resisted this idea. Sharing a general RAF dislike of a *Corps d'Elite*, he knew that his group commanders would not take kindly to giving up their best crews, and he came up with his own proposal for target-finding. In June 1942, after much argument, the Air Ministry ordered Harris to form the Pathfinder Force (subsequently raised to the status of No. 8 Group). Harris chose navigation expert Air Commodore D.C.T. Bennett as its commander.[119]

Ambitious, and ruthless with his subordinates, Bennett was a highly skilled pilot and navigator. His new force did not immediately increase its bombing accuracy in 1942; however, the concentration of the bombs dropped improved significantly. Bennett's work was made more difficult because, as Harris had guessed, group commanders were less than enthusiastic or generous in their assignment of crews to the Pathfinders. Moreover, Bomber Command at that time did not possess the right kind of flares and markers for use by the lead bombers. Worse, the Germans succeeded in jamming the first version of Gee by midsummer 1942.[120] By spring 1943, however, Bennett had made real progress. In good visual conditions, the Pathfinders used ground marking, a technique in which markers ignited near the earth's surface. When the target was obscured in weather, they employed sky marking. These markers were dropped with parachutes on H2S or Oboe indications and ignited in the air. In either case, marking was done from high altitude. The main force, arriving over the target as a stream of bombers, then aimed their bombs on the sky- or ground-marked targets.[121]

Bomber Command's tactics were strongly affected by the weather and the stage of the moon.[122] The command required adequate visibility for takeoff and landing and the absence of icing clouds on route to the target (especially important given the 20,000-foot ceiling of even the best of British heavy bombers). Early in the war, adequate conditions had to exist at the target for visual bomb aiming. Although this situation was always desirable, it was no longer indispensable after sky-marking techniques with H2S or Oboe became available in 1943. Early in the war, clear, moonlit nights were thought to provide the most desirable conditions for navigation and bomb aiming. As the German night fighter defenses improved, however, Bomber Command realized that these conditions favored defenders more than attackers. Again, new radio and radar aids came to the rescue, allowing Bomber Command to shift operations to nights with less moonlight. The command's operations were also determined by the hours of darkness. Long winter nights permitted attacks to be made on the more distant targets in eastern Germany, such as Berlin, although this advantage was partially offset by the more hostile winter weather. Better weather but the shorter nights of summer limited operations to western Germany.[123]

Once over target, the effectiveness of Bomber Command's attack depended on the nature and quality of its ordnance. Throughout the period of the area offensive, Bomber Command's tactics sought to start and spread large fires.[124] High-explosive bombs, dropped in the first waves, were intended to break windows and doors, shatter homes into kindling, and damage the fire-fighting services. The incendiary bombs that followed then became the primary instrument of destruction. In the early days of the war, Bomber Command used as standard ordnance 250- and 500-pound general-purpose bombs. With a low charge-to-weight ratio (only 27 percent), the explosive filling performed poorly, and many were duds. Unfortunately, very large stocks had been accumulated, and they continued to be used long after they were known to be defective, chiefly because too few of the more efficient weapons were available.[125]

The medium-capacity bombs, which first appeared in service toward the end of 1941, improved the charge-to-weight ratio to about 40 percent. Filled with an improved explosive, they were commonly 1,000 and 4,000 pounders. The most famous bombs of this type were immense, special devices created by Barnes Wallis of Vickers. Wallis designed the special 10,000-lb rotating mine that 617 Squadron used to attack major German dams in May 1943. He also conceived the notion of earth-penetrating bombs that exploded underground and destroyed structures by shock waves and ground motion. The crowning results of his work were Tallboy at 12,000 pounds and Grand Slam at 22,000 pounds, each carried in specially modified Lancasters.

Tallboys came into use in the summer of 1944 and were employed against reinforced-concrete submarine pens and, in one spectacular case, the Saumur railway tunnel, a bottleneck on a major rail line leading to the Normandy battle area. A single Tallboy penetrated the roof of the tunnel and collapsed it.[126] The most famous use of the Tallboys was in the final, successful attack in November 1944 against the battleship Tirpitz anchored in a Norwegian fiord.[127] Grand Slams, first used near the end of the war, destroyed the Bielefeld viaduct, a major component of the German railway system.[128]

The most effective pure-blast bombs were in the high-capacity series, manufactured in 2,000-, 4,000-, 8,000-, and 12,000-lb versions that came into service early in 1942. The 4,000-lb weapon was commonly identified in Bomber Command parlance as a "cookie."

Success in fire-starting, however, depended on reliable incendiaries. The version most widely used was the 4-lb magnesium bomb. Though very effective, ordnance designers were slow in solving the problems of aiming and dropping it. Large numbers of the bombs were packaged in canisters which remained onboard the aircraft after the incendiaries dropped. Unfortunately, bomb hang-ups were not uncommon, and most aircrews, sooner or later, found themselves peppered with incendiaries dropped from bombers at higher altitudes.[129] There is no question that some aircraft and crews were lost in this fashion.

During early Bomber Command operations, whether by day or by night, the force divided into small groups to attack several targets. Each crew attempted to navigate its own way to its assigned target. As German aerial defenses became more effective, Bomber Command placed increasing emphasis on narrowing the front along which the force entered German air space, thus overloading the defenses at that particular point. Tactical massing and concentration of the force to saturate defenses also allowed bombing to be more concentrated with respect to time. By the careful management of takeoff times, staggered altitudes, and the use of Gee to navigate to a point on the continental coast, a bomber stream could be formed for night attacks. This was definitely not a formation in the American sense. In the course of a combat mission, most pilots never saw another aircraft except for brief moments. The stream, flying at a designated altitude, was stacked in a formation about 5,000 feet high when Stirlings and Wellingtons were included, or about 2,000 to 3,000 feet high when only Lancasters and Halifaxes were flying. The stream normally was between 25 and 30 miles wide and 100 to 150 miles long when 300 to 400 aircraft comprised the force. The first of the great raids employing these concentration tactics bombed Cologne on the evening of May 30/31, 1942, when some 900 bombers passed over the target in ninety minutes.[130]

To avoid bunching over the target and to ease handling at its bases, Bomber Command divided bomber streams into waves of aircraft of similar performance: Lancasters in one, Halifaxes in another, and Stirlings and Wellingtons in a third. When window (strips of aluminum foil cut to half the wavelength of the German radar selected for jamming) was introduced as a defensive radar countermeasure in July 1943, it reduced the effectiveness of German fire control and interception methods, but it also complicated operations in the British tactical system. For aluminum foil dropped by the leading aircraft to provide the greatest benefit to those following, higher performance aircraft operating at higher altitudes had to be first in the stream. This arrangement, unfortunately, necessitated that the slower and lower-flying Stirlings and Wellingtons following would fall farther and farther behind on the homeward leg, becoming more vulnerable to attack after the bombing run. To reduce this danger, slower aircraft were moved into the leading wave of the attack, but this denied the high-flying Lancasters and Halifaxes the benefit of the window foil. The problem eased somewhat when Bomber Command removed the Stirlings and older Halifaxes from the main force.[131]

The Air Ministry also pursued offensive electronic countermeasures. After its experience in forming an electronic intelligence organization to devise countermeasures against the *Luftwaffe*'s radio navigation aids used during the Blitz in 1940 and 1941,[132] the ministry now proposed to develop methods of airborne, offensive electronic warfare. Bomber Command, however, preferred the maintenance of radio silence and opposed such measures until late 1942. The command's position changed when its loss rate increased and, as a result of a determined effort by the Air Ministry's Directorate of Scientific Intelligence, more details became known about German aerial defenses.[133]

In the subsequent prosecution of the area offensive in late 1942 and 1943, the RAF directed offensive electronic warfare, or radio countermeasures as it became known, at four main targets in German aerial defenses: the early warning system based on the Freya radar, the ground control radars (Würzburg), ground-to-air HF and VHF radio (in both morse and voice), and the AI radars carried in night fighters. Fighter Command began to jam Freya radars in August 1942 with a device known as Moonshine. This had the effect of magnifying the size of the attacking force with the idea of making German defenses vulnerable to feints and decoys by small forces. Mandrel, an electronic jammer employed beginning in December 1942 against German early-warning radars, was used in a variety of forms until the end of the war. In June 1944 the British introduced an improved version to support the Normandy invasion.[134] German modifications to Freya enabled skilled operators to "see" through gaps in the jamming frequencies, and Mandrel radiations served the Germans as a raid warning. But Mandrel jamming reduced the ability of the German early warning system to distinguish between large and small forces, or between the main force and diversions or feint attacks, and it delayed the process of raid assessment.[135]

Unofficial, misguided action by Bomber Command aircrews actually assisted the defenders. Some crews apparently believed that switching on their IFF transponder over the illuminated target zone somehow interfered with the operation of German radar-directed searchlights. The practice became widespread by early 1942, and in the summer of that year Bomber Command actually installed a special switch that allowed the IFF to transmit constantly over enemy territory without being triggered by an electronic interrogation from the ground. The Air Ministry Scientific Intelligence unit vigorously protested. No scientific or technical foundation supported the notion. Moreover, it warned, the IFF could serve the enemy as a locating beacon! Bomber Command ordered a halt to the use of IFF in this way in early 1944, but numerous aircrews, convinced otherwise, continued to do it, and thereby provided the excellent *Luftwaffe* Wireless Intercept Service with important tracking information.[136] Indeed, so valuable was this source to the German defenders and so promiscuous its use by British aircrews that the *Luftwaffe*'s Head of Signals, General Wolfgang Martini, wondered aloud if some undiscovered, surreptitious purpose did not actually lie behind the practice.[137]

Window was by far the most effective jamming weapon against Würzburg radars. The principle was known both in Great Britain and in Germany before the war. Chaff experiments took place in both countries early on; then caution intervened. Hermann Goering, *Luftwaffe* C-in-C, forbade continued development of such jamming devices for fear they would fall into British hands.[138] The British Air Staff also did not allow the use of chaff for almost a year while Fighter Command and the technical research establishment worked on ways to combat its use against British aerial defenses.[139] When introduced in July 1943 in the attacks on Hamburg, it struck at the very heart of the controlled night-fighting system, denying German controllers the ability to identify single aircraft against which to direct fighters. The Germans feverishly attempted to modify their radar to resist window, but they were largely unsuccessful except under special conditions.[140] In March 1944, Bomber Command began to supplement window with active jamming (Carpet II).[141] In the summer of 1944 the command combined Mandrel with window in the Special Window Force to simulate attacks. These provided diversions and kept German defenses in a constant state of activity, even in the absence of a major raid.[142]

The attack on German AI radar started inauspiciously, with ground-based jamming known as Ground Grocer in April 1943. Ground Grocer was abandoned when it became clear to the British that the Lichtenstein B AI radar was as vulnerable to window as was Würzburg.[143] In 1943, the Germans introduced a lower frequency (73–91 MHz) AI radar with a higher performance — the Lichtenstein SN2 — widely used by late spring 1944. When working properly, it was invulnerable to the window then in use. British scientific intelligence reported the appearance of the new AI radar in February 1944, but full details were not known until June 1944.[144]

Bomber Command's tactical concentration of forces, made possible by the electronic navigation aids, and its increased use of electronic countermeasures caused a severe crisis for Kammhuber's system of linear defense. If the attackers could be formed into a compact stream and passed through the defensive line on a fairly narrow front, then only a few defense boxes could come into action, and total British losses would be kept low.[145] In the face of concentration tactics, the Kammhuber system could not increase its number of kills at the same rate that Bomber Command was increasing the number of bombers passing through the boxes. The attack on Cologne at the end of May 1942 illustrated the problem. More than 1,000 bombers took off, with about 900 actually attacking the target. The stream penetrated German air space on a front about 30 km wide and passed through no more than 8 defense boxes. German ground controllers were able to bring only 25 night fighters into action against this enormous British force.[146]

The Germans adopted two alternatives to supplement controlled night-fighting in 1943. The first, proposed by Maj. Hajo Herrmann, came to be known as Wild Boar. Herrmann advocated forming a unit of single-engine

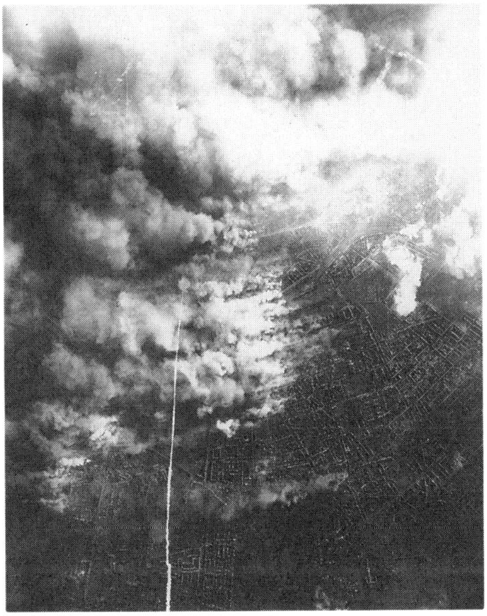

The aerial views above show how repeated bombing assaults on Cologne left this fourth largest city in Germany in devastation.

fighters flown by ex-bomber pilots and other individuals with night flying experience. Once the target was known, Wild Boar fighters would take off and climb to an altitude over the target higher than that flown by the bomber stream. They would then use the illumination from the searchlights and the fires in the burning city to spot individual bombers beneath them. Only visual means would be used to acquire the target and deliver the attack. Against Kammhuber's wishes, Goering gave Herrmann authorization to form such a unit sometime before the disaster at Hamburg.[147] For the remainder of 1943, Wild Boar units were constantly in action.

Kammhuber predicted that Wild Boar would be cheap in terms of organization and equipment but expensive in blood, and he was right. Wild Boar increased the British loss rate, but with worsening weather in the autumn and winter of 1943, the German accident rate rose alarmingly, especially since Wild Boar fighters were not equipped for instrument flying.[148] Several times Wild Boar aircraft attacked friendly conventional night fighters, and Kammhuber complained bitterly that Wild Boar pilots attacked any aircraft with twin stabilizers.[149]

Col. Viktor von Lossberg proposed a second alternative, ultimately known as Tame Boar. Though window hindered the tracking of individual bombers, it did not prevent the plotting of the altitude and the heading of the bomber stream as a whole. The central German controller could put out this information as a running commentary for the night fighters aloft.[150] The fighters could attempt to find the bomber stream by dead reckoning, or they could rely on an alternative to radar for ground control — the Y system. A calibration signal included in the voice transmission from the ground to the fighter and then retransmitted to the ground station could give bearing and distance to the controller. With a second station, the position of the fighter could be fixed.[151] This could only be used with a few aircraft at a time, but it did help infiltrate more night fighters into the bomber stream. Once in the stream, a fighter receiving Y fixes could transmit a homing signal to other fighters on another frequency.[152] Lossberg also expected that the new AI radar, the Lichtenstein SN2, would be less vulnerable to window and would greatly improve target acquisition.[153]

The effectiveness of Tame Boar improved with the addition of a variety of homing devices in night fighters. Naxos, mostly mounted in Ju 88s, homed on the emissions from the H2S radar sets carried in Bomber Command aircraft. A ground-based device known as Korfu could track H2S well enough to give bearings on the bomber stream and aid the running commentary. Flensburg was a device that homed on the British tail-warning radar, Monica, which Bomber Command used for part of 1943 and 1944.[154]

Late in 1942, Bomber Command began to attack the German system of ground-to-air communications with the active jamming reflector, tinsel. Specially equipped aircraft carried a German-speaking operator who, when he detected an active night fighter frequency, broadcast engine noise on that frequency. The *Luftwaffe* countered by increasing the frequency spread and moving increasingly from HF to VHF voice. In response, from midsummer 1943 the British began to employ a series of devices that interjected false information from German-speaking controllers on VHF voice in appropriate frequencies to confuse night fighters and their controllers.[155] The German response to these attacks was to increase again the spread of frequencies used.

British electronic warfare in this area greatly increased the difficulties for all German ground-to-air communications. But, according to one German source, it was never possible entirely to jam all the frequencies simultaneously and continuously, if for no other reason than the British airborne jammers themselves had to avoid being homed on. At one point, the Germans broadcast their running commentary on the bomber stream in voice and morse simultaneously from different command echelons, radar stations, and operational airfields.[156]

On balance, the Germans recovered well from the crisis created by the introduction of window in 1943. The improvised Tame Boar system using Y-direction, the SN2 radar, and passive detection devices brought Bomber Command near the margin of prohibitive losses in the winter of 1943–1944.[157] In a paper prepared shortly before war's end, the British Director of Bomber Operations estimated that an average of 5 percent losses per mission sustained over a three-month period of intensive operations would likely make operational effectiveness "unacceptably low." If the rate climbed to 7 percent over a three-month period, he believed a strategic bombing force would become "relatively ineffective."[158] Sustained losses at any rate above 5 percent would reduce over time the average experience level of the crews and accelerate both combat and accident losses. The flow of experienced men to the training organization would also be gravely impaired.

War records for the entire period between September 1939 and May 1945 indicate that Bomber Command dispatched 297,663 night sorties from which 7,449 aircraft did not return, a loss rate of 2.5 percent.[159] This was well below the 5 percent considered the boundary between acceptable and unacceptable losses. For short periods, however, the defenses imposed prohibitive loss rates on various elements of Bomber Command. For example, squadrons using Halifaxes II and V suffered losses of 6.4 percent in December 1943, 10.1 percent in January 1944, and 5.1 percent in February 1944. Harris took these units off German targets in March 1944. Similarly, in a four-month period beginning in August 1943, Stirlings of No. 3 Group suffered 7.8, 6.3, 3.6, and 5.6 percent losses.[160]

The Tame Boar system on occasion could inflict high casualties on single missions. Three of these, for example, occurred during February and March 1944. Against Leipzig on February 19/20, the missing rate rose to 9.5 percent.[161] Against Berlin on March 24/25, it reached 9.1 percent.[162] Against

Nuremberg on the night of March 30/31, Bomber Command dispatched 795 aircraft and suffered a horrendous loss rate of 13.3 percent.[163]

For Bomber Command, the whole winter of 1943–1944 was a grim business. Beginning in November, the overall loss rates per month were 3.1, 4.1, 5.0, 4.6, and 3.1 percent, each within the limits of acceptability.[164] However, in thirty-five major actions against German targets during this period (sixteen of which were against Berlin) the average loss rate was 5.2 percent.[165] The figures would have been somewhat higher if one included those aircraft damaged beyond economical repair, although in such cases the crews were not entirely lost. When one considers that 2,034 Mosquito sorties dispatched against major German targets in this period are included in the total sortie figures, and they suffered a minimal loss rate averaging only 0.4 percent, one can understand why the British official historians counted the five-month Battle of Berlin a "defeat."[166] Bomber Command C-in-C Sir Arthur Harris found the situation worrisome enough in early April 1944 to observe that if German aerial defenses continued to improve, a point would be reached "at which night bombing attacks by existing methods and types of heavy bomber would involve percentage casualty rates which could not in the long run be sustained."[167]

Strategic Debate during the Area Offensive, 1943–1944

One of the most important policy questions debated among senior British officials in 1943 and early 1944 involved the role Bomber Command should play in the larger Allied strategy for the defeat of Germany.[168] On the face of it, a common policy to rule operations of Bomber Command and the American Eighth Air Force, first deployed to the United Kingdom in 1942, had much to recommend it. At the Casablanca Conference in January 1943, British and American leaders looked forward to what they regarded as the supremely important operation in the European war, the invasion of Northwest Europe, and they saw the strategic bombing effort of Bomber Command and the Eighth Air Force as essential in preparing for that operation. To give effect to that view, the Combined Chiefs of Staff made Sir Charles Portal their agent for the conduct of what was called, after Casablanca, the Combined Bomber Offensive. Portal's authority over the Eighth Air Force in 1943 and early 1944 was vague, but it stood somewhere between identifying the overall aims and primary target systems, which were properly the responsibility of the Combined Chiefs, and the daily choice of targets, which was appropriately a prerogative of the operational commanders. At meetings of the Combined Chiefs, Portal acted as spokesman for both the American and the British operational commanders.[169]

Portal was responsible for executing three directives issued by the Combined Chiefs of Staff for the conduct of the Combined Bomber Offensive — the Casablanca Directive of January 1943 (intended to replace the February

General Arnold and Air Chief Marshal Portal at Casablanca.

1942 document under which Bomber Command had been operating), the POINTBLANK Plan of May 1943, and the February 1944 revision to POINTBLANK. Each of these documents identified a general aim for the offensive and a specific list of target systems ranked according to their strategic importance. The Casablanca Directive declared that the general or "primary" object was "the progressive destruction and dislocation of the German military, industrial and economic system, and the undermining of the morale of the German people to the point where their capacity for armed resistance is fatally weakened."[170] Two aspects of this early statement are significant: It clearly recognized the strategic air offensive as a campaign of attrition, and its wording was sufficiently vague to accommodate different conceptions of how the primary aim should be achieved.[171]

All three documents identified German aircraft production, particularly that part supporting the German fighter force, as a high-priority target. In fact, the POINTBLANK plan approved by the Combined Chiefs in May 1943 termed enemy fighter strength an "intermediate objective second to none."[172] The objective was intermediate in the sense that the German fighter force stood in the way of developing the American daylight strategic bombing offensive. It was a primary objective in the sense that a successful invasion of the continent could not be achieved unless the German Air Force was thoroughly beaten. The POINTBLANK Directive specifically assigned the U.S. Eighth Air Force, which operated in daylight using precision rather than area bombing tactics, the task of defeating the *Luftwaffe*. Bomber Command was to be governed by the following rule:

> While the forces of the British Bomber Command will be employed in accordance with their main aim in the general disorganization of German industry their action will be designed as far as practicable to be complementary to the operations of the Eighth Air Force.[173]

Superficially, it would appear that the American Eighth Air Force and Bomber Command might operate against a common target system on an around-the-clock basis, that is, one in daylight and the other at night. In fact, different tactical considerations involved in operating in daylight and at night made it difficult if not impossible to make coordinated attacks against the same targets within a short period of time. Moreover, Sir Arthur Harris continually argued that the tactical requirements of nighttime operations not only forced the use of area tactics but also prohibited using those tactics to support a selective strategy that struck only those cities in which specific important industries were located according to a strict system of priorities. He made this point often, but perhaps most cogently in a 1944 letter:

> There is . . . an aspect of bombing which it is always difficult to impress on or to keep impressed upon those outside the immediate Command, and that is the decisive effect of weather and tactical factors on what can be done at any given moment. Taking into account the low ceiling of our bombers and the high ceilings of many cloud formations, particularly those associated with high icing indices, it is frequently impossible to go

Major differences between the Allies were ironed out at the Casablanca Conference. General Marshall makes a point to his British counterpart Sir Alan Brooke, Chief of the Imperial General Staff. General Arnold is at Marshall's left. Lord Louis Mountbatten (fist to chin), Sir Dudley Pound, First Sea Lord and Chairman of the British Chiefs of Staff sit with Sir Alan Brooke at his right, while on his left are Sir Charles Portal, Chief of the Air Staff and Field Marshal Sir John Dill.

> where one wants to go and it is as frequently necessary to do anything of some value even though it is not always something near the head of the priority list.[174]

Harris remained skeptical in principle about any selective bombing strategy that resembled the one with which Bomber Command began the war. He simply did not believe that the Ministry of Economic Warfare and the Air Staff could identify a target system that if destroyed would bring the economic collapse of Germany. He denounced all attempts to promote area attacks against selected industrial targets as the work of "panacea mongers and parochial experts." To Harris, the only effective way to get at the German war economy was to conduct area attacks as ruthlessly as possible to reduce the morale of the industrial work force and damage the working economies of the industrial cities that could be reached under the operational conditions of the moment. From 1943 to the end of the war, he and other proponents of the area offensive represented it less as an attack on morale than as an assault on the housing, utilities, communications, and other services that supported the war production effort.[175] Harris also maintained that the general area offensive by itself could defeat Germany. In late 1943 he made a most extravagant claim, declaring that under certain conditions the Lancaster element in his command alone had the power to inflict upon Germany "a state of devastation in which surrender is inevitable."[176]

The latitude of action permitted Bomber Command in the Casablanca and POINTBLANK directives, the ordinary tactical discretion possessed by an operational commander, and Harris's personal belief in the general areaoffensive against the German war economy led to serious disputes between his command and the Air Staff. These arguments arose largely because the Eighth Air Force's selective campaign against German aircraft production at first failed to proceed as expected. The Eighth Air Force had not built up according to schedule; it suffered from maintenance and supply problems; bad weather interfered with American daylight precision tactics; and above all, the casualties inflicted by daylight defenses were exceptionally high on missions flown without fighter escort deep into Germany.[177]

In October 1943, the Air Staff Director of Bomber Operations, Air Commodore Sidney Bufton, and his superior, Air Marshal Sir Norman Bottomley (the Deputy CAS), feared that failure of the Eighth Air Force daylight bombardment campaign might imperil the Allied invasion of France scheduled for the coming year. Bomber Command, they advised, should conduct area attacks against the high-priority, selective economic targets in the POINTBLANK Directive. Bufton and Bottomley believed that the general area offensive was a temporary expedient to be indulged in so long as Bomber Command could do nothing else. When the size of the force increased and its navigation aids improved, it should be returned to the original objectives of the British strategic air offensive: selective attacks on German industry, or at least, selective area attacks on cities associated with vital areas of the German war economy, particularly aircraft production. Bufton declared:

> In our and the Ministry of Economic Warfare's view, [general] area attack only is an unjustifiable and dangerous gamble at the present time. We should direct all possible effort in the immediate future to the reduction of the German fighter force. This will open the way for the full exploitation of area bombing later. It will also open the way for "OVERLORD."
>
> Failure to seize the present opportunity might at the worst prejudice the whole success of "OVERLORD" and in any case result in heavy casualties. It may also result in the failure to demonstrate the power of the strategic air offensive with consequent and dangerous repercussions upon postwar policy.

Bufton concluded that Bomber Command should be ordered to give "full and complete priority . . . to the night attack of cities associated with German fighter production."[178]

The contest between the Air Staff and Sir Arthur Harris in this period became in the end the Schweinfurt question. Bufton and Bottomley believed the Schweinfurt ball bearing plants to be targets of the highest priority. Ball bearings were essential to the entire German war effort, nowhere more so than in aircraft production. And manufacture of this vital commodity was believed to be largely concentrated in Schweinfurt. The Americans attacked it twice, in August and October 1943, both times sustaining very heavy losses. The Ministry of Economic Warfare and the Directorate of Bomber Operations concluded that it must be attacked again, and soon. Given the losses sustained by the Americans, it was imperative that Bomber Command take advantage of the longer winter nights and deliver an area attack on Schweinfurt. Bufton's predecessor also had urged an attack on the ball bearing plants early in 1942, and Bufton himself tried to have Bomber Command follow up the Eighth Air Force's attack in August.[179]

Sir Arthur Harris resisted vigorously, almost to the point of insubordination. When the proposal first landed on his desk, he pointed out the immense tactical difficulties of finding and hitting a small city such as Schweinfurt at night and at great distance. He also disputed the intelligence assessments of Schweinfurt's importance, as he would in almost every other similar case. In a characteristic outburst, he denounced the Air Staff for "stressing of targets which necessarily remove bombing pressure from the German nation as a whole [the area offensive], to concentrate on objectives such as [Schweinfurt]." Any belief in the importance of Schweinfurt, moreover, was the product of disinformation, "in many cases a deliberately engineered A.R.P. [Air Raid Precautions] maneuvre initiated by enemy sources."[180] It took an endless amount of persuasion, a strong letter from the Deputy CAS, one directive, and a specific order to move Harris to attack Schweinfurt, an event that did not occur until February 1944.[181]

That attack became part of Bomber Command's contribution to what the Americans called Big Week, a concentrated seven-day assault on German aircraft production. Harris dispatched 734 aircraft to Schweinfurt on the night of February 24/25; 695 reported they struck the target. Photographs taken

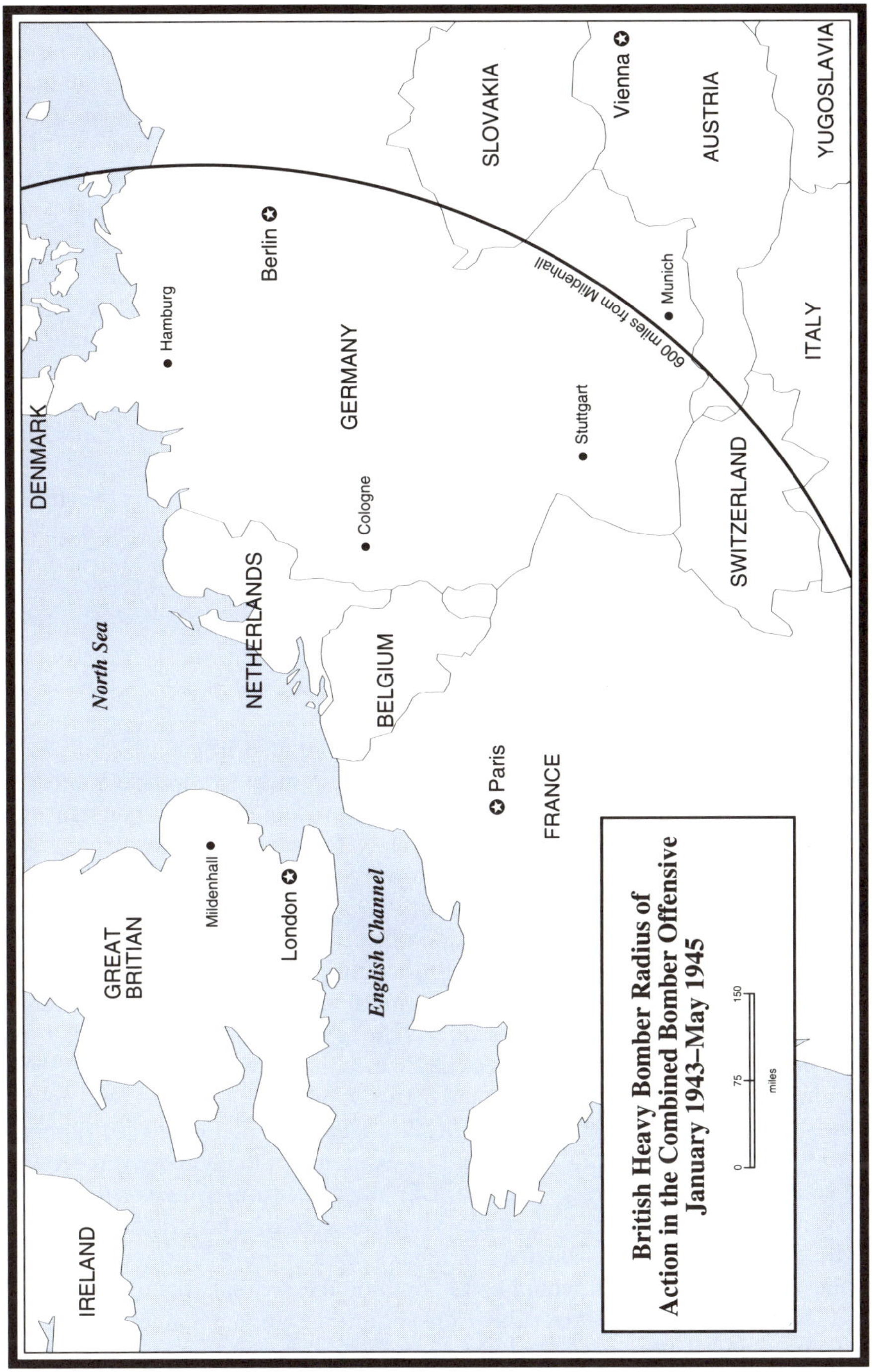

British Heavy Bomber Radius of Action in the Combined Bomber Offensive January 1943–May 1945

during the attack, however, showed that only 298 dropped their bombs within three miles of the aiming point. Of those, only 22 were plotted inside the target area, that is, the boundaries of the built-up districts of the city. Photoreconnaissance revealed that the city and its ball bearing plants were badly damaged, but most of the damage was attributed to the American daylight attack that had preceded Bomber Command's effort. Bomber Command's attack miscarried because of bombing "creepback," the tendency of marker aircraft arriving later to drop markers short, and the equally pronounced tendency of main force aircraft not to aim at the center of the markers, but at the first ones seen.[182] This phenomenon made little difference in strikes on large urban areas, but against a relatively small target like Schweinfurt it meant the bulk of the bombs hit open ground.

This result was not inevitable when Bomber Command aircraft attacked a relatively small city. The following night, on February 25/26, 1944, Harris sent almost 600 aircraft to Augsburg, an important aircraft assembly center. The markers were dropped accurately, they remained concentrated during the attack, and the main force managed to burn out some 60 percent of the unfortunate city.[183] The potential demonstrated in this attack was soon employed in another strategy.[184]

Bomber Command and Operation OVERLORD

The coming of Operation OVERLORD in 1944 ended Bomber Command's emphasis on the general area offensive. The requirement to contribute directly to preparations for OVERLORD — the invasion of France scheduled for June — interacted with the heavy losses caused by the new German Tame Boar night fighters and pushed the command toward a new strategy and to develop the tactics needed for its accomplishment.

The CAS advised Sir Arthur Harris in December 1943 that he would soon have to provide direct support to the forthcoming invasion. Harris responded with a familiar litany of the tactical limitations of his force. The highly specialized aircraft and the operational technique of the force, he insisted, were suitable only for area attacks against German cities. Precise targets could not be reliably attacked even with radio aids. The heavy bomber force could not be brought into action quickly against "fleeting targets," and, above all, Bomber Command could not operate in daylight. A diversion of the bomber force to the uneconomical task of direct support for OVERLORD, furthermore, would give the German nation a six-month respite from the kind of attack most damaging to it. The only effective aid the command could give to assaulting Allied armies, Harris concluded, would be to continue the general area offensive.[185]

The battle was joined over the Transportation Plan, a document produced in January and February 1944 by the organization charged with preparing air plans for the support of Operation OVERLORD, the staff of the Allied Expeditionary Air Force (AEAF).[186] AEAF proposed to cripple the French and Belgian railway systems, thus helping Allied armies to deal with the German counterattack that was certain to follow the invasion. This required an extensive assault on railway centers and marshaling yards beyond the scale of bomblift available to the medium bomber forces under AEAF's direct control. Sir Arthur Harris rejoined that his bomber forces were too inaccurate for attacks against such targets. The Air Staff, however, did not accept his tactical estimates and sent Bomber Command a directive ordering experimental attacks on several targets in France, including a list of marshaling yards.[187] The first of these nighttime raids, against Trappes on March 6/7, was an outstanding success in terms of accuracy and concentration of attack. It heralded a period when, under the stimulus of OVERLORD, the command developed the ability to make precision attacks at night.[188]

In mid-April 1944, Bomber Command and the Eighth Air Force came organizationally under the strategic direction of General Eisenhower. In the space of six weeks, Bomber Command carried out nighttime attacks on radar sites, gun batteries, and launching sites for V–1 weapons as well as executed a massive program of raids on railway marshaling yards. Operationally, the force changed from one that concentrated the whole of its strength against a single target on a given night to one capable of sending out smaller concentrations of bombers against a variety of targets.

Perhaps the most surprising development of all was the resumption of daylight operations. These began on a modest scale at the time of the Normandy invasion in June and, for the most part, were confined to occupied western Europe under cover of escorts provided by RAF Fighter Command. During August, Bomber Command dispatched 10,271 sorties in daylight operations and 10,013 at night. Daylight raids declined in the winter months to a low of 1,304 sorties in January 1945, and day attacks were never used against deep-penetration targets. With these reservations in mind, however, in the last eleven months of the war in Europe, Bomber Command was the only one among the Allied air forces that could operate the bulk of its units by day *and* by night.[189]

After the landings in France, Bomber Command also carried out a variety of attacks in support of the ground forces, including some spectacular close support efforts. Though the utility of these attacks was questioned at the time, there can be no doubt that they represented an advance in technique not thought possible in 1942 when the command had to struggle to concentrate enough force so that it could bomb with sufficient accuracy to produce damage in a large urban area.

Bomber Command's ability to deliver relatively precise nighttime attacks under the optimum conditions in 1944 was the result of improved target marking and the use of an airborne controller (the Master Bomber), who directed the marking and controlled the attack by radio.[190] Alongside the continuing development of marking techniques in the Pathfinder Force, other groups began to work on their own systems. No. 5 Group, for example, possessed 617 Squadron, a unit specializing in what are now called long-range attack missions. This squadron was originally formed to conduct the famous raid on German dams in May 1943 (Operation CHASTISE) and thereafter was used in several highly specialized roles. No. 617 and No. 627 Squadrons pioneered the use of low-level target marking that became a specialty of No. 5 Group.[191] By November 1944, No. 3 Group operated by day and night as Bomber Command's Gee-H force.

On the night of April 22/23, 1944, an attack on Brunswick illustrated the complexity and sophistication of Bomber Command's target-finding and marking techniques. The attacking force, divided into two groups, navigated to the target over different courses. Four Mosquitos and six Lancasters from No. 617 Squadron with the backing of crews from two Pathfinder squadrons marked the target, which was illuminated by hooded flares (to avoid dazzling the marking crews) dropped on H2S indication. When the lead Mosquito (the Master Bomber) identified the aiming point, he warned the other Mosquitos, dove to 3,000 feet, and released two red spot markers. Circling to judge the accuracy of his work, he ordered the remaining Mosquitos to back up his marking with markers of their own and then directed the six Lancasters from No. 617 squadron to add their markers. Forty-four red spot markers now burned in a compact mass over Brunswick, providing an aiming point which the crews of the main force could loose their bombs. This attack heavily damaged the city, but it fell short of its full potential. Some later markers went astray and inevitably attracted too many bombs, errors that could not be corrected because the Master Bomber's radio communication had broken down.[192] Difficulties such as these continued to affect otherwise well-planned and well-executed operations.

Bomber Command's destructive power, harnessed to the support of OVERLORD and using these bombing techniques, promised a great deal — providing the German nighttime aerial defenses that had inflicted such serious losses during the winter of 1943–1944 could be overcome. Fortunately, the Allies mortally wounded the *Luftwaffe* in the spring and summer of 1944. American strategic bombers and their fighter escorts decimated German daylight defenders and secured air superiority over Germany while Allied tactical air forces overwhelmed the *Luftwaffe* in France and Belgium. This allowed Bomber Command to resume daylight operations with improvised tactics and escorts from Fighter Command (renamed the Air Defence of Great Britain). On June 14, 1944, Harris sent 234 Lancasters and Mosquitos in broad daylight against the docks at Le Havre on the left flank of the Allied beachhead in Normandy. During the next two months, in other daylight operations over occupied western Europe, the command enjoyed a loss rate of less than 1 percent of its sorties. In August, Harris extended daylight operations to the Ruhr, supported by the long-range Mustangs in Fighter Command. Near-complete daylight air superiority, won by the American strategic air forces and the Allied tactical air forces, was underscored when only ten Bomber Command aircraft failed to return from five daylight missions totaling 803 sorties against German targets in that month.[193]

The Germans lost crucial parts of their early warning chain and their forward airfields in the west when Allied armies moved out of the Normandy beachhead and drove into France and Belgium.[194] This seriously weakened both day and nighttime aerial defenses, but it did not leave them entirely without the means to detect an incoming attack. According to General Beppo Schmid, Kammhuber's successor, the German Wireless Intercept Service provided warning from a variety of intercepted transmissions, including Allied warnings to British shipping, reports from weather reconnaissance flights, and H2S emissions. To reduce the incidence of casualties from friendly fire, Americans often passed warnings of impending aircraft flights over inner artillery zones to their antiaircraft units. German intercepts of these messages, in Schmid's word, proved "invaluable."[195]

An intensification of British electronic warfare under the auspices of No. 100 Group in Bomber Command delivered a third blow to German defenses in the summer of 1944. The Air Ministry had formed this group in November 1943 to consolidate all offensive electronic countermeasures in one organization and to develop night fighter escort and intruder operations.[196] A shortage of equipment and personnel initially hampered the group's efforts, but in the summer of 1944, upon release from OVERLORD commitments, it achieved much success against the German radars and fighter control organization. The effort degraded the performance of the German AI radar so much that aces like Heinz-Wolfgang Schnaufer stopped using it altogether, relying exclusively on night vision once inside the bomber stream.[197] Other German fighter pilots continued to use devices that homed on H2S and Monica emissions. After midsummer 1944, the British abandoned Monica and restricted the use of H2S. Despite these measures, a recent account suggests that as long as Bomber Command did not maintain absolute radio silence and shut down H2S entirely, the Germans always managed to cobble together some improvisation to track raids as a whole and employ some variant of Tame Boar.[198]

The British also attempted to reduce the effectiveness of German night fighters by using intruders to attack them and by providing fighter escorts for Bomber Command operations. Fighter Command began intruder operations early in the war, but these flights employed inadequate navigation aids and aircraft with relatively poor performance (Bostons). When Bomber Command formed 100 Group in late 1943, Fighter Command intruder squadrons transferred to it. After receiving Mosquitos equipped with Gee and AI radar, intruder activity increased greatly and featured patrols around German night fighter airfields and assembly beacons.[199]

Night escort activity began in 1943, also under Fighter Command. These escorts were Beaufighter aircraft equipped with AI radar and Serrate, a homing device that monitored emissions of the Lichtenstein B. After some early successes, the effort weakened because the Beaufighter IV did not have adequate speed or range. The squadrons changed to Mosquitos when the escort mission moved to Bomber Command's 100 Group in late 1943, but the effort hardly improved. The first installment of aircraft had already seen a lot of action and their engines were worn out. In early 1944 these units were re-engined and began to receive a centimetric AI radar. Later in the year they also used equipment (Perfectos) that could interrogate the German IFF.[200]

British official historians were skeptical about the worth of intruder and escort patrols.[201] Certainly the number of German night fighter aircraft downed was small when compared with the numbers claimed by American day escort fighters. The effect on German operational technique and on the psychology of night fighter crews, however, was out of all proportion to the material damage caused. The Germans had to exercise greater care when orbiting assembly beacons and when landing. A few were attacked while circling airfields, and that led to more panic landings, shaken crews, and smashed undercarriages. In the summer of 1944 a real "Mosquito phobia" took hold. The *Luftwaffe* command forbade assembly over airfields or near the main radio beacons. *Luftwaffe* pilots were ordered to fly directly to the secondary beacons and were forbidden to fire recognition flares, a rule that increased the danger of friendly fire from flak batteries defending the airfields where they attempted to land.

In the allied offensive against German oil production, these three attacking B–17 Flying Fortresses leave clouds of black smoke at Hamburg's oil refineries.

Many pilots were forced to land on darkened runways.[202] Night fighter pilots numbered among the *Luftwaffe*'s most precious assets, and any losses, strain, and inconvenience caused by 100 Group's modest intruding and escort efforts supported Bomber Command's operations.

A fuel crisis created by the Allied air offensive against synthetic oil production dealt the final blow to German night defenses. Attacks, begun by the American Eighth and Fifteenth Air Forces in May 1944, greatly reduced the output of aviation gasoline. Using reserves accumulated during the previous winter and spring, the *Luftwaffe* managed to stave off calamity until August of 1944. Defensive flight operations then had to be curtailed, and the training effort, already grossly inadequate, declined even further. After August, the German night defenses never again could exact serious losses from Bomber Command.[203]

Bomber Command at the Height of Its Powers, September 1944 To May 1945

In September 1944, the Allied Combined Chiefs of Staff returned Bomber Command to the control of the British CAS. In the next eight months the command reached the peak of its strength and destructive ability. The average monthly number of aircraft fit for operations in the squadrons rose from 1,327 in September 1944 to 1,843 in April 1945. In September 1943 and April 1944 the numbers stood at 733 and 944, respectively. Bomber Command now operated at slightly less than twice its strength of the previous winter.[204] Bomb tonnages grew by a still larger margin. From September 1944 to the end of the war, Bomber Command dropped on average more than 2½ times as much high-explosive and incendiaries per month on German targets than it had in the same period a year earlier.[205] This number reflected both the greater size of the force and the increased number of Lancasters within it.

The relative weight of attack against different target systems also changed considerably. In the two years of the area offensive that began in 1942, approximately 84 percent of RAF bomb tonnage was aimed at targets described as "industrial towns."[206] As Table 2–7 shows, in the summer of 1944, after the invasion of Normandy, these targets received only 14 percent of the tonnage. The largest share of the bombing effort, some 33 percent, was dropped on V-weapon sites in France and Belgium. Oil reappeared as a major target system for the first time since 1941, receiving about 11 percent of the command's bombs.

TABLE 2–7
Distribution of Bomb Tonnage Dropped by Bomber Command

	Percentage of Tons Dropped On:				
Period	Oil	V-weapons	Transport	Towns[a]	Other[b]
Sep 3, 1939–Jul 8, 1941	7.96	0	13.54	22.28	56.22
Jul 9, 1941–Feb 13, 1942	0.27	0	23.37	36.95	39.40
Feb 14, 1942–Mar 6, 1944	0.03	1.39	1.65	84.26	12.67
Mar 7, 1944–Jun 6, 1944	0	1.59	38.31	35.01	25.08
Jun 7, 1944–Sep 15, 1944	11.70	33.35	15.67	14.09	25.18
Sep 16, 1944–Dec 31, 1944	11.24	0	12.96	48.19	27.63
Jan 1, 1945–May 4, 1945	26.21	0	13.75	36.78	23.25

Source: BBSU, "Strategic Air War against Germany," table 7, Tonnages Claimed Dropped by RAF Bomber Command in Accordance with Main Directifs. According to the BBSU, the distribution statistics were not very reliable before 1944.

[a]Data are for area attacks.
[b]The term *other* includes aircraft targets, docks and port areas, military installations, specific industrial targets, and "miscellaneous" targets.

The distribution of effort changed again when Bomber Command was removed from the control of the Supreme Allied Commander and when Sir Arthur Harris resumed, to a limited degree, his area campaign. From mid-September 1944 to the end of the year, industrial towns received slightly more than 48 percent of Bomber Command's tonnage, and oil, the target of choice among Allied leaders (except for Harris), received about 11 percent. The V-weapon sites disappeared altogether from target lists when they were overrun by the Allied ground forces. Rail transportation, a system of targets largely ignored between 1941 and the spring of 1944, accounted for about 13 percent of the tonnage.[207] Even though Bomber Command was no longer under the direct control of General Eisenhower, he could ask for its assistance in the land battle at any time. Harris's force dropped slightly more than 21 percent of its tonnage on military installations, and a good part of that was in direct support of the Allied armies; its contribution during the first critical week of the Battle of the Bulge proved particularly helpful.

The command was a much more flexible weapon as a result of the tactical developments wrought in the spring and summer of 1944. It no longer relied exclusively on a stereotypical mass attack using the whole of the main force on one target. No. 8 Group, the Pathfinder Force, possessed between 150 and 200 fast Mosquitos, the bulk of which were employed as the Light Night Striking Force, used largely for harassing raids. Two squadrons (No. 9 and No. 617) served as special attack forces, specializing in the use of 12,000-lb Tallboy earth-penetrating bombs. Toward war's end they employed the larger version of that device, the 22,000-lb Grand Slam. They attacked dams, submarine pens, tunnels, canals, and locks and sank the battleship Tirpitz. No. 100 Group, the command's electronic warfare organization, overcame a great many of its early difficulties and provided night intruder and fighter escorts, jammed the German early warning radar and radio communications, and staged an impressive array of electronic spoofs and diversions. No. 5 Group continued to work as an independent force with an average of 200 to 250 aircraft. Over the summer and autumn of 1944, Harris ordered No. 4, No. 3, and No. 1 Groups to operate in a similar fashion. No. 3 Group specialized in the use of Gee-H and was regularly assigned to the short-range oil targets in the Ruhr. The whole of the main force, it should also be remembered, could now attack both at night and during the day.

Bomber Command's greater power and flexibility did not end the arguments over how it should be used. For a large part of the autumn and early winter of 1944, a dispute raged between the CAS and the C-in-C of Bomber Command. Sir Charles Portal wanted the command to throw the bulk of its available effort against German synthetic oil production. Sir Arthur Harris, again, stoutly resisted.[208] During heated arguments over the Transportation Plan earlier that year, the United States Strategic Air Forces (USSTAF) presented an Oil Plan to General Eisenhower as an alternative which would allow for the strategic bombers to support OVERLORD. He set it aside on grounds that it could be of no immediate help to the invasion. Shortly after his decision, he allowed USSTAF to make "experimental" attacks on oil.[209] In late May 1944, Portal endorsed oil as a target for both American and British strategic air forces — after the OVERLORD invasion forces were firmly established. In June the Air Staff asked Harris for a tactical assessment of proposed attacks on ten synthetic plants in the Ruhr. He observed, in reply, that the attacks would involve the wastage of many bombs, but he was confident they could be made. Harris added that he had already agreed, at the request of Deputy Supreme Allied Commander Sir Arthur Tedder, to devote some effort to the attack on oil.[210] The first of these raids came on the night of June 12/13, 1944.

Sir Charles Portal and many other senior officials of the British government knew that Great Britain, in prosecuting the war effort worldwide, had reached the limit of its material and human resources. The country simply could not manage another eighteen months of total war at the same level of effort. At some time, in the last half of 1944, the British contribution would have to

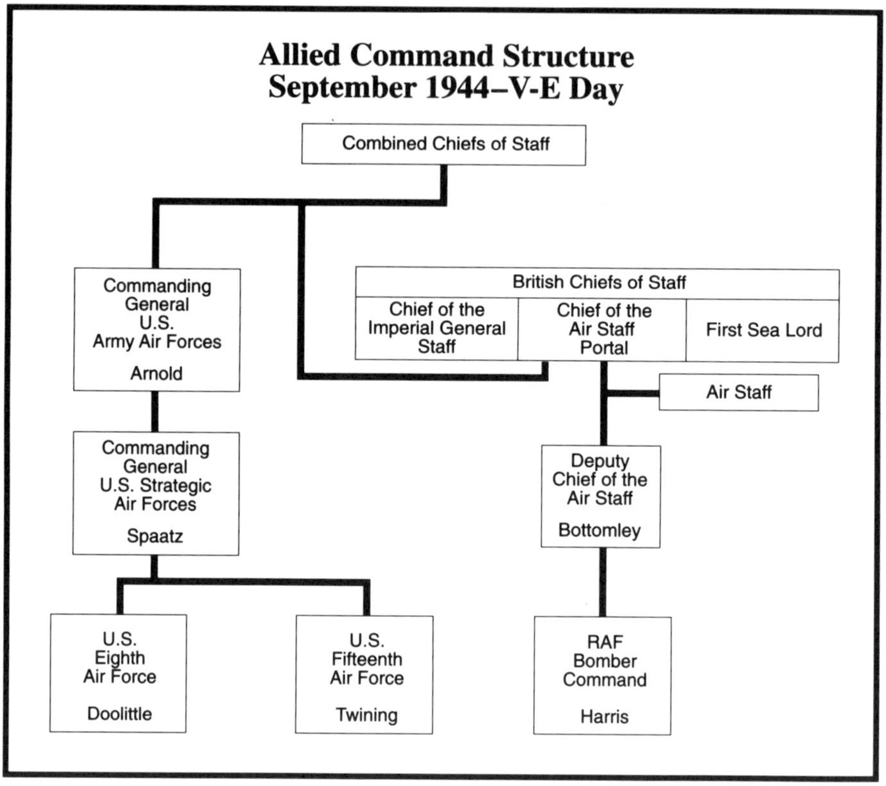

decline, either by withdrawing manpower from the armed forces to feed industry or by contracting the industrial base to keep the battlefield supplied with men. These facts bore heavily on British leaders, never more so than when compelled to resolve difficult problems with their ally, the United States. The British were now the junior power, and many American officials, civilian and military, were quick to take advantage of this fact. Sir Charles Portal, therefore, had good reason to demand that the Allied bombing policy contribute to a quick end to the European war.[211]

The oil plan met his requirements. By late summer, intelligence assessments showed that German oil plants had been hit hard chiefly, but not exclusively, by American strategic air forces, and that petroleum production had fallen precipitously.[212] Oil shortages, intelligence sources reported, were affecting the German armed forces profoundly and adversely. The evidence also showed the Germans were not giving up but instead were devoting vast resources to repairing the damaged plants.[213] Far from being knocked out in a single decisive blow, it appeared that the oil plants, like almost every other economic target, required persistent and protracted bombing to assure positive

Air Chief Marshal Sir Arthur Tedder (*right*) with Maj. Gen. Carl A. Spaatz, commander of the U.S. Strategic Air Forces in Europe.

results, an effort that would become increasingly difficult with the approach of winter weather.[214]

Sir Arthur Tedder shared Portal's sense of urgency. He sought a common denominator by which to combine the efforts of all the Allied air forces. "The various types of operations," he said, "should fit into one comprehensive pattern . . . at present they are more like a patchwork quilt." Though Tedder was not opposed to the oil plan, he thought it should be integrated in a larger program of attacks on western German road, rail, and water communications that he identified as the one common factor in the whole German war effort.[215]

The first Air Ministry directive sent to Bomber Command in September 1944 after the force had been returned to the direct control of the CAS, contained a general mission statement similar to the one in the Casablanca document:

> . . . the overall mission of the Strategic Air Forces remains the progressive destruction and dislocation of the German military, industrial and economic systems and the direct support of land and naval forces.[216]

The directive specifically identified the German petroleum industry as the target of first priority. It placed on an equal footing as second in priority, German rail and water communications, tank and ordnance production and depots, and motor transport production plants and depots. However, it also advised:

> When weather or tactical conditions are unsuitable for operations against specific primary objectives, attacks should be delivered on important industrial areas, using blind bombing technique as necessary.[217]

This last clause, in combination with the general mission statement carried over from previous Combined Bomber Offensive directives, afforded considerable discretion to Sir Arthur Harris.

In the month afterward, according to the official history, two-thirds of the tonnage dropped by Bomber Command fell on industrial cities, and only 6 percent on oil targets. This moved Sir Charles Portal and Sir Norman Bottomley to impress again upon Harris the importance of the oil campaign.[218] Harris responded as he had so often before, explaining that weather and German defenses prevented concentrating on any narrowly defined target system, including oil. If Bomber Command ceased bombing anything other than oil, it would waste a vast amount of effort by attacking in bad conditions or in achieving redundant destruction. If Bomber Command persisted in attacking a specific target system, it immensely simplified the work of the enemy defenders, and the command's casualty rate would rise.[219]

Portal no longer worried about overall efficiency of effort. If halting the production of oil would effect an earlier end to the war, then redundant destruction and otherwise wasted effort mattered not at all. Harris remained unconvinced:

> In the past eighteen months Bomber Command has virtually destroyed forty-five out of the leading sixty German cities. In spite of invasion diversions we have so far managed to keep up and even to exceed our average of two and a half such cities devastated a month. In addition others have been "started on" to an extent where they are already damaged beyond anything experienced in this country. There are not many industrial centers of population now left intact. Are we now to abandon this vast task, which the Germans themselves have long admitted to be their worst headache, just as it nears completion?[220]

This statement, concluding as it did with a rhetorical question, strongly implied that Harris was pursuing — and intended to pursue — a policy no longer favored by his superiors. An exchange ensued between Portal and Harris in which they expressed their respective views directly and sharply. In one note, Harris carped, the misguided oil plan was one of the "panaceas enthusiastically put forward by the amateurish, ignorant, irresponsible and mendacious M.E.W. [Ministry of Economic Warfare]." Sir Charles Portal shot back, saying that it was "an unworthy and inexcusable travesty of our conduct of the war to suggest that our policy is determined on that kind of basis."[221] These exchanges culminated in Harris's inviting Portal to consider removing him from Bomber Command.[222] This Portal declined to do, and Harris remained at his post until the end of the war.

The focus of Bomber Command's operations did change at the beginning of 1945. Oil received more than 26 percent of the bomb tonnage, much more than double the effort of the last four months of 1944. Transportation targets received roughly the same amount of attention as before (12 to 13 percent), while the tonnage devoted to attacking what were called military installations declined from just over 21 percent to about 13 percent. Area attacks on industrial towns still accounted for more than one-third of the command's effort. It is of some interest that the main force did not bomb the German capital at all in 1945, but the city suffered the constant attention of the Light Night Striking Force. Harris sent 4,000 Mosquito sorties to the Reich capital, not quite 1,500 of them carrying 4,000-lb bombs.[223]

No account of Bomber Command's operations, however brief, can ignore the most successful, albeit most controversial of its great area attacks in 1945: the assault on Dresden on the night of February 13/14, 1945.[224] The destruction of Dresden was hardly the result of novel ideas. Strategic air power enthusiasts in the interwar years had sketched apocalyptic pictures of massed air raids on cities, and the Air Staff and Bomber Command, having begun the war as apostles of precision, selective bombing of economic targets, had, as we have seen, come to accept the morale of the civil population as a legitimate and potentially profitable target. Between February 1942 and the spring of 1944 Bomber Command had aimed the bulk of its bombers at the center of German cities and the people living there.

In midsummer 1944, a renewed interest in morale attacks emerged. In July, the British Chiefs of Staff sent a memorandum to the Prime Minister in which they urged consideration of a plan to deliver an "all-out attack by every means at our disposal on German civilian morale."[225] Sir Charles Portal submitted a plan called THUNDERCLAP to meet this requirement in early August. He proposed not to defeat Germany by attacking morale, but rather to use such an attack to push German authorities to surrender once defeat had been ensured by other means. The operation would consist of a 4-day, 3-night attack upon Berlin in which American and British strategic air forces would together drop some 20,000 tons of bombs. In the event that Berlin could not be attacked, consideration might be given to an alternative target. According to Portal:

> Immense devastation could be produced if the entire attack was concentrated on a single big town other than Berlin and the effect would be especially great if the town was one hitherto relatively undamaged.[226]

The authors of the British official history observed that the memorandum in which this passage appears "may be regarded, if only indirectly, as the title deed" for the destruction of Dresden.[227]

The Air Staff argued that THUNDERCLAP would be effective, despite the scale of destruction expected, only if it were carried out in precisely the sort of circumstances laid out in the memorandum. Because neither the Chiefs of Staff nor General Eisenhower thought these conditions existed in the late summer of 1944, the plan was laid aside, only to be revived in January 1945 in a different context. Soviet forces had just begun the major offensive that would carry them to the Oder River, and the British Joint Intelligence Committee urged the high command to assess ways in which the strategic air forces might accelerate the progress of the Red Army.[228] Three days before, Air Commodore Bufton urged Sir Norman Bottomley, Portal's deputy, to consider reviving THUNDERCLAP.[229]

Portal, Bottomley, and Harris were discussing how a modified THUNDERCLAP might be used when Prime Minister Churchill intervened, demanding dramatic action. On January 25, 1945, he asked the Secretary of State for Air what plans he had for attacking the Germans in the east. After consulting with the Air Staff, the Secretary replied on January 26, asserting that the best prospects lay in continuing the attack on oil as long as the weather was favorable. The Air Staff, he added, had "under consideration" a plan to exploit bad weather by attacking cities in eastern Germany, the "administrative centres controlling the military and civilian movements" and the "main communications centres through which the bulk of the traffic moves."[230] The Prime Minister replied in what the official history termed a somewhat preemptory minute:

> I did not ask you last night about plans for harrying the German retreat from Breslau. On the contrary, I asked whether Berlin, and no doubt other large cities in East Germany, should not now be considered especially attractive targets. I am glad this is "under examination." Pray report to me tomorrow what is going to be done.[231]

This retort had an immediate effect. Next day, Bottomley ordered Harris to carry out attacks against eastern German cities where "a severe blitz will not only cause confusion in the evacuation from the East but will also hamper the movement of troops from the West."[232]

A few days later, on their way to the Yalta Conference, the Combined Chiefs of Staff met at Malta in the Mediterranean. They "tacitly approved" a proposal that directed both Allied strategic bombing forces to support the Soviet advance and assigned to several eastern German cities, including Dresden, a priority for attack second only to oil targets.[233] On the night of February 13/14, 1945, nearly 800 RAF heavy bombers attacked Dresden with a combination of high-explosive and incendiaries while the city teemed with refugees, troops on the move, and some Allied prisoners of war. The result was appalling. A firestorm started and, although figures cannot be established with any precision, perhaps as many as 35,000 people perished.[234] The next day more than 300 American bombers again attacked Dresden; they returned to the stricken city twice more in the next few days.

In late March 1945, six weeks after Bomber Command attacked Dresden, a serious controversy broke out between the Prime Minister and the CAS. Churchill, doubtless reacting to unfavorable public discussion of what had happened to Dresden, proposed that Bomber Command area attacks "simply for the sake of increasing the terror . . . should be reviewed." "The destruction of Dresden," he observed, "remains a serious query against the conduct of Allied bombing." This remarkable statement from the man who had so considerable a role in the events leading up to the attack did not go unchallenged. Sir

Charles Portal succeeded in getting the Prime Minister to withdraw his statement and substitute for it another, one not quite so transparent in its attempt to remove the politician from the morally repugnant military business of slaughtering civilians in large numbers.[235]

While this exchange occurred, Germany tottered on the brink of collapse. In the east, the Red Army had closed to the Oder River. In the west, British, Canadian, and American forces were well over the Rhine, and on April 11, elements of the U.S. Ninth Army reached the Elbe River at Magdeburg. Five days later the Red Army began to pound German defenses in front of Berlin. On the 21st, Soviet artillery rounds began to rain down on the German capital where, in a subterranean bunker, Adolf Hitler furtively moved nonexistent Panzer divisions about on a large map. Two days before, the British Chiefs had concluded that "the main mission of the Strategic Air Forces is now to give direct assistance to the land campaign."[236] The strategic air offensive was over. Within three weeks Hitler was dead, and the German armed forces surrendered unconditionally to the Allied powers.

Effects of the Bombing Campaign

The effects of the Allied strategic bombing offensive on the German war effort and the contribution of RAF Bomber Command to them have been a matter of controversy ever since the end of World War II.[237] In assessing the timing and the weight of effects, one must account for the considerable differences in bomb tonnages dropped on German targets in various periods of the war. Although the British strategic offensive began as early as 1940, no less than 70 percent of all the bomb weight dropped by Bomber Command fell after February 1944. Almost 40 percent was dropped in the last 7½ months of the war.[238] Measuring production loss in any period of the war with any kind of precision is most difficult. Does one compare the actual production record to the planned program or to estimates of production possibilities? What does one assume about levels of mobilization and productivity of resources? Separating the effects of one form of attack from another also poses nearly insurmountable difficulties. This is particularly true after early 1944, when very substantial American bomb tonnages began to be dropped. All estimates of the effects of area attack after mid-1944, for example, are highly problematical for this reason.[239] A closely related problem involves disentangling direct and indirect causes of a demonstrated effect. This is especially important because, whether one examines purely military systems or industrial production in an economic web, many causes take time to work their way through to have substantial and measurable effects.

There is also the fundamental problem of establishing the cause and effect relationship between the damage to economic performance and the resulting diminution of fighting power of the armies, navies, and air fleets at the front. In one sense this is no different, for example, than attempting to estimate the contribution of the Eastern Front fighting of 1943 to the success of the Western Allies in 1944. In another sense, however, it is an especially troublesome task because of all the intermediation that occurs in any productive system between the factory and the fighting front.

One must recognize that German munitions output increased substantially while British and American bombers were pounding away at various elements of the German economy. According to figures prepared by the Speer Ministry, total armaments production slightly more than doubled between the base period of January/February 1942 and the beginning of Bomber Command's main offensive against the Ruhr cities in March 1943. (Table 2–8) This early expansion peaked in May 1943, beyond which little advance was made for several months. From the spring of 1944, the armaments index climbed steeply to reach its wartime high in July 1944, when it stood at slightly more than three times the base level. By October 1944 it had fallen back to a slightly lower plateau. In January 1945 the overall production of arms began a steep decline, reflecting a very rapid deterioration in the economy as a whole. By March 1945, production had fallen by 50 percent from the July 1944 peak.[240]

TABLE 2–8
Indices of Actual and Potential Total German Armament Output (January–February 1942 = 100)

Year/Quarter	Actual	Potential
1942/1st	110	—
1942/2d	138	—
1942/3d	154	—
1942/4th	168	—
1943/1st	202	—
1943/2d	224	225
1943/3d	230	235
1943/4th	233	256
1944/1st	248	281
1944/2d	287	307
1944/3d	308	345
1944/4th	270	381
1945/1st	182	406

Source: Webster and Frankland, *Strategic Air Offensive against Germany*, vol 4, app 49, pt 2. Based on BBSU, "Potential and Actual Output of German Armaments in Relation to the Combined Bomber Offensive," table 1.

How were the Germans able to increase production in the face of an increasingly destructive British and, by late 1943, American bombing effort? Two contending interpretations of the German war economy are relevant to this question. The traditional one, established by the United States Strategic Bombing Survey (USSBS), holds that the German government ran only a limited war economy, a *Blitzkriegwirtschaft*, during the first 2½ years of the war. Aiming at short, decisive wars against isolated opponents, it did not adequately mobilize the labor force (particularly women), reduce civilian consumption, or fully exploit its manufacturing capacity until Albert Speer took over a large part of armaments production in early 1942.[241]

The alternative view, formulated principally by Professor Richard J. Overy, holds that Hitler believed a war among the great powers would be a long one, wished to avoid a replay of the catastrophe of 1918, and aimed at a total war economy from the very beginning. The labor supply, including women, was mobilized on that basis early on, and civilian consumption was kept well in hand. The problem was not with state policy but with its execution by the German ministries, the armed forces, and industry; the timing of the outbreak of general war; and the lack of urgency bred by unexpected successes in 1940. Administrative chaos and low productivity initially kept the output of a fully mobilized economy well below its production possibilities. Hence, the Speer regime did not implement a total war economy for the first time; it made the old one work more effectively.[242]

In either case, at the moment Bomber Command's main offensive began to fall on the economy, Germany was expanding and exploiting economic reserves either by mobilizing them for the war effort for the first time or by making more efficient use of them. This meant that direct production losses caused by bombing could be made good by repair and reconstruction supported by basic industrial resources not yet mobilized (or if mobilized, not yet productively used) for war production. In these circumstances, the bombing attacks of 1943 and 1944 could not spread damage heavily enough over the whole industrial economy in a sufficiently short period to force levels of munitions production down. Bomber Command was still too small, the defenses too effective, and the weather too uncooperative for any other result.

Strategic bombing imposed losses of three kinds on the German economy: direct production loss, indirect production loss, and defense opportunity costs. Direct production loss was production that simply did not occur as a result of the material destruction and dislocation caused by bombing. Indirect losses were opportunity costs measured by the sum of those resources devoted to repair and reconstruction made necessary by bombing which, therefore, became unavailable to expand war production. A third form of loss, the strategic opportunity cost, involved building the fighter and flak defenses to defend the towns under attack. The manpower, equipment, supplies, and productive effort invested in this endeavor thus were denied to conventional military requirements at the battlefront.

The USSBS estimated that the combination of direct and indirect production losses from area attack (the computation did not include the opportunity costs of diversions to air defense) amounted to 2.5 percent in 1942, 9.0 percent in 1943, 17.0 percent in 1944, and 6.5 percent in 1945.[243] These were, of course, only educated guesses, and the quality of the estimates was lower for the period after mid-1944 because of the difficulty of separating the effects of area bombing from other bombing activities, and from the injuries the advancing Allied and Soviet armies inflicted.

Official historians concluded that the area offensive of 1943 and early 1944 "sapped some of the reserve within the German war economy" and made it "more vulnerable" to the attacks that followed.[244] This seems a little cautious; the strategic effect of Allied bombing in 1943 and early 1944 in which Bomber Command played the larger role prevented German war production from rising as rapidly as Hitler, Speer, and the rest of the Nazi leadership had hoped.[245] In 1943 and early 1944, Bomber Command hindered German mobilization and the exploitation of resources for war production that Germany might have used to increase its chance of imposing a stalemate. Though not spectacular, it was arguably more than marginally significant.

Strategic bombing did have an impact on morale, but it was never decisive in the way that Lord Cherwell, Sir Charles Portal, Sir Arthur Harris, and others had hoped. Area bombing attacks unquestionably hurt. People do not endure the death or maiming of their children, parents, brothers and sisters, and friends by blast and fire without suffering enormously. They do not lose their houses and belongings without serious depression. Many tens of thousands of people alive today have suffered mentally and physically from these events for more than four decades. To have been militarily decisive, however, area bombing would have had to affect morale sufficiently to produce a breakdown in the political order such as occurred in Russia in 1917 or in Germany and Austria-Hungary in 1918. Nothing like this happened or came close to happening. Failing a political collapse, bombing would have had to depress morale enough to produce a decline in productivity so severe as to affect the fighting efficiency of the armed forces. That did not occur.

What did happen? Morale was badly hurt for short periods in those areas struck by bombing attack.[246] There was evidence of resentment and hatred against the authorities, but those sentiments were not translated politically or economically into significant action. There is some evidence to suggest that industrial productivity suffered from worker morale problems, but never to the extent necessary for decisive effects. Catastrophe is never easy to bear, but when it happens to a great many other people at the same time, some of the depression is eased. German authorities, especially Nazi Party officials, went to great lengths not only to show themselves but to help survivors after the big attacks. Also, the Gestapo and the SS went about, soon afterward, demonstrating the hard face of Nazi authority to the bombed-out population.[247]

In the last ten months of the war, the effect of the strategic bombing offensive as a whole and the contribution of Bomber Command to it changed greatly. American and British air forces achieved the highest totals in numbers of aircraft and bomblift that they attained during the entire war. Strategic air forces attempted to concentrate their efforts against oil and transportation — two target systems directly related to the ability of the German armed forces to fight and the German economy to produce and distribute war materiel.

Oil was the centerpiece of British strategic planning at the beginning of the war. Attacks on oil targets were abandoned only when it became clear that their number was too great and their size too small for Bomber Command to find and hit. The American strategic air forces restored oil to a first-priority target in May 1944 (initially on an experimental basis), and Bomber Command joined in the attack at Sir Arthur Tedder's request in June 1944. The campaign against oil continued well into April 1945. All sources agree that oil production attacks were among the most important contributions made by the air forces to the defeat of Germany. Put simply, the production and import of finished oil products fell from 900,000 metric tons in January 1944 to slightly more than 300,000 in December 1944.[248] Industry was not seriously affected, because it drew the bulk of its power from coal.[249] It was the armed forces that felt the most damaging effect. Aviation fuel shortages first became troublesome for the *Luftwaffe* in 1942 — with the chief consequence, a curtailment of training. From midsummer 1944, operational flying had to be restricted more and more, and fuel for training became almost nonexistent. This, more than almost anything else, led to the relative impotence of the *Luftwaffe* in the last ten months of the war. The mobility of the German ground forces was also seriously hampered by fuel shortages, especially after late summer 1944, and that check dramatically eroded the Army's ability to wage successful counteroffensives.

Bomber Command played a more important role in this campaign than is usually realized or might be assumed from the policy argument between Portal and Harris. According to Table 2–9, the Eighth Air Force delivered some 233 attacks against oil targets in 1944 and 1945; the Fifteenth Air Force delivered 221; and Bomber Command, 200. The Eighth Air Force, led by true believers in the oil offensive, delivered only 16.5 percent more attacks from bases in the United Kingdom than did Bomber Command, even though the latter was led by an officer who denigrated oil as just another panacea target.[250]

The bomb tonnage figures also invite some interesting comparisons. During the course of the oil campaign, the Eighth Air Force aimed 71,042 tons of bombs at oil targets and the Fifteenth, 48,378, for an American total of 119,420 tons. Bomber Command aimed 93,641 tons of bombs at these targets, a sum smaller than the American total but 32 percent higher than the American force operating from the same base area. Overall, Bomber Command aimed more bomb tonnage at oil targets than did the Eighth Air Force in seven of the twelve months beginning in May 1944.[251]

TABLE 2–9
The Allied Attack on Oil

	No. of attacks			Short tons dropped		
Month	8th AF	15th AF	Bomber Command	8th AF	15th AF	Bomber Command
May 1944	11	10	0	2,883	1,540	0
Jun 1944	20	32	10	3,689	5,653	4,562
Jul 1944	9	36	20	5,379	9,313	3,829
Aug 1944	33	23	20	7,116	3,997	1,856
Sep 1944	23	8	14	7,495	1,829	4,488
Oct 1944	18	10	10	4,462	2,515	4,088
Nov 1944	32	19	22	15,884	4,168	16,029
Dec 1944	7	33	15	2,937	6,226	5,722
Jan 1945	17	5	23	3,537	2,023	10,114
Feb 1945	20	20	24	6,161	4,362	15,749
Mar 1945	36	24	33	9,550	6,628	21,211
Apr 1945	7	1	9	1,949	124	5,993

Source: BBSU, "Strategic Air War against Germany," table 49, 1944–1945 Attacks on German Oil Installations.

The relative efficiency — or inefficiency — of the strategic bombardment campaign against oil should also be taken into account. The USSBS examined what it described as "meticulously documented" German records of fifty-seven attacks delivered against three oil–chemical plants: Leuna, Ludwigshafen-Oppau, and Zeitz. When these attacks were examined as a whole without any attempt to distinguish between Eighth Air Force and Bomber Command, just over 87 percent of all bombs dropped landed outside the targeted factories' fences. A very small 1.3 percent hit utilities inside the plant precincts; only 2.2 percent struck and damaged buildings and equipment. All others fell in open spaces within the boundaries of the plant or else did not explode.[252] Oil plants never proved an easy target to find, attack, and hit from high altitudes.

When the USSBS compared the attack records of the two air forces, it found that in good weather and using visual aiming, the Eighth Air Force put 26.8 percent of its bombs inside oil plant boundaries. When the weather was mixed and the force had to use both visual and instrument aiming (with H2X, the American version of H2S), the percentage of bombs that fell within plant boundaries declined to 12.4. When the target was heavily covered with cloud and all bomb aiming was done on instruments, the percentage nosedived to an abysmal 5.4 percent. And measured in tonnage, just over 80 percent of Eighth Air Force's bombardment effort was delivered under partial instrument or full instrument conditions. By comparison, all of Bomber Command's attacks on these targets were made using what the Survey called the Night Pathfinder Technique. These attacks averaged 15.8 percent hits inside the factory fences. On average, in good weather, the Eighth Air Force proved itself much more efficient than Bomber Command as a weapon, but in bad weather — the prevailing condition from October 1944 through March 1945 — Bomber Command was at least the equal of the Eighth, and possibly better.

The average weight of each bomb dropped by Bomber Command against oil targets was some 70 percent higher than that of the corresponding bomb dropped by American strategic forces. Both forces used far too many of the smaller 100-, 250-, and 500-lb iron bombs. The difference was that Bomber Command's Lancasters and Halifaxes carried 4,000-lb cookies in their loads, which the USSBS believed were far more effective than an equivalent weight of the smaller devices.[253]

Some of Bomber Command's edge in winter weather was offset by a large percentage of unexploded bombs. In the attacks on the oil plants at Meerbeck-Homburg, no fewer than 30.6 percent of the bombs dropped by the command did not explode. Overall, at least 18.9 percent of British bombs aimed at oil plants failed to detonate. The corresponding Eighth Air Force figure was 12.2.[254] A large number of these ordnance failures resulted when tail fins separated and the bombs landed flat. The USSBS was absolutely correct when it observed:

> One in every six bombers sent on these attacks might just as effectively have loaded its bomb bays with scrap iron. Missions were wasted and lives of airmen lost flying junk to Germany.[255]

In the last seven months of the war, the Allied strategic bomber forces devoted a considerable effort to attacking the German transportation system, a campaign that had dramatic consequences. In a recent work, Professor Alfred Mierzejewski argues that the spatial distribution of Germany's resource and production centers made the German economy especially vulnerable to a determined attack on its rail communications.[256] For example, the daily average of freight car assignments in Germany fell from 138,000 in June 1944 to 83,000 in December 1944.[257] Waterborne movements of coke and coal from the Ruhr to the rest of Germany declined from a daily average of 76,000 tons in July 1944 to a mere 14,200 in January 1945.[258] Stocks of coal, the main source of power for German industry, rose from a low of 186,000 tons kept at the mineheads in July 1944 to an incredible 2,767,000 tons in February 1945.[259] Raw materials could no longer be moved effectively.

The effect on the German economy was powerful. In the words of the USSBS: "The consequences of the break-down in the transportation system were probably greater than any other single factor in the final collapse of the

German economy."[260] Between May 1944 and January 1945 the total output of finished goods, including armaments, fell by 22 percent. Somewhere between 50 and 60 percent of this decline could be attributed to attacks on inland transport.[261] The effect on the operations of the German armed forces are less clear. The best evidence suggests that essential movements of troops and supplies eventually were subjected to intolerable delays, especially as administrative chaos swept the railway system in the last three months of the war. Some of the great damage to the transportation system after July 1944 was the result of low-altitude attacks by medium bombers and tactical fighter-bombers. It can also be said with reasonable assurance that the combined strategic and tactical aerial attacks on transportation so crippled the German economy that Germany could not have carried on the war past the early summer of 1945.

At the outset of the transportation campaign in March 1944, which at that stage was directed against French railway targets in preparation for OVERLORD, Bomber Command dropped almost 2.4 times more bomb tonnage on that target system than the Eighth Air Force did. During the fighting in France, Bomber Command continued to lead in tonnage by a smaller, but still considerable, margin. For the remainder of the war, however, the Eighth Air Force far outstripped Bomber Command in this area, even though the command's special forces, with the use of large bombs such as Grand Slam and Tallboy, did the bulk of the damage against difficult targets, such as canals, viaducts, and shiplifts in the rail and inland waterway system.[262] In the last four months, for example, the Eighth Air Force dropped 117,158 tons against Bomber Com-

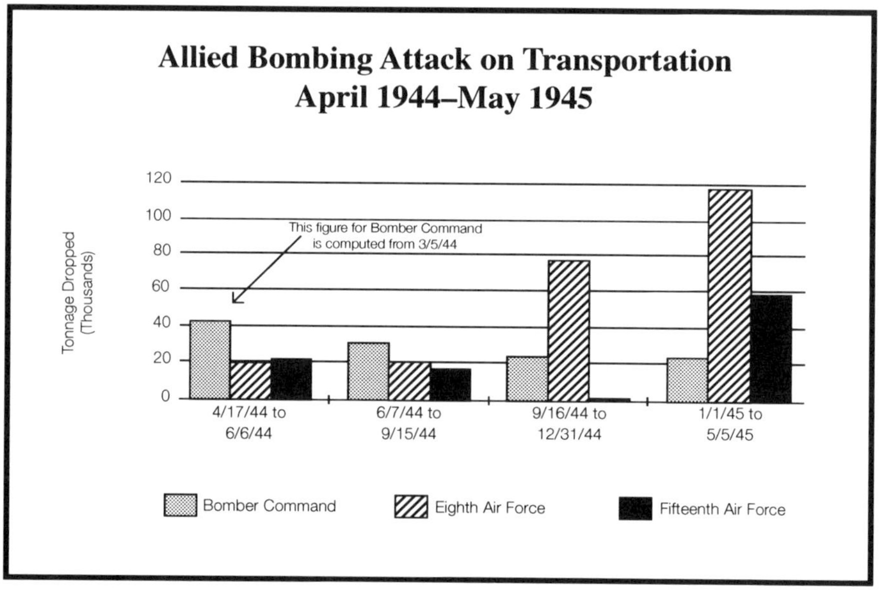

mand's 27,432 tons on transportation targets.[263] Some part of this large difference involved preferred alternative targets in bad weather: the Eighth Air Force chose marshaling yards while Bomber Command aircraft almost always went for an industrial city.[264]

Policy and Leadership

Britain's bomber force in 1939 was not prepared for war. Quite apart from the limitations of their aircraft, bomber aircrews were sadly wanting in navigation and bomb-aiming skills, operational arts fundamental to any successful air campaign. As the official historians put it, "Bomber Command was not trained or equipped either to penetrate into enemy territory by day or to find its target areas, let alone its targets, by night."[265] Many officials were responsible for this state of affairs: politicians, civil servants, and officers of the RAF.[266] Bomber Command lacked adequate facilities for training, but a good training program does not depend exclusively on having enough resources; it depends on making the right choices about how to use them. This must be judged one of the most serious shortcomings of the prewar RAF leadership.

British official historians argued that one of the most important mistakes of the prewar period was assuming that a strategic air campaign could be successfully executed without directly confronting and defeating the enemy air force. Failure to provide adequate defensive armament for bombers, or escort fighters to accompany them, forced Bomber Command into the cover of the night where its acknowledged weaknesses in navigation and bomb aiming, and its want of electronic aids, condemned it to operational impotence for the first 2½ years of the war.

When Bomber Command became capable of doing serious damage in 1942 and 1943, the night defenders became more dangerous, hindering tactical concentration on a particular target. By the winter of 1943–1944 they had become proficient killers of Bomber Command aircrews. By the late summer of 1944, Bomber Command had been relieved of that menace, but not through direct combat. The night fighters were overcome by electronic countermeasures, fuel starvation, declining standards of flying training, and the loss of their early warning and forward airfield positions in occupied Europe.[267] The command's successful daylight operations, begun in 1944, were made possible by the campaign waged against the *Luftwaffe* by the Eighth and Fifteenth Air Forces and the Allied tactical air forces. Flying daylight or nighttime operations, Bomber Command remained a force restricted to a ceiling of 20,000 feet and defenses consisting of .303-cal. machine guns.

RAF leaders were not alone in sharing early optimistic assumptions that daylight bombing could be conducted over Germany without escort fighters. The Americans persisted in a belief that the bomber will always get through —

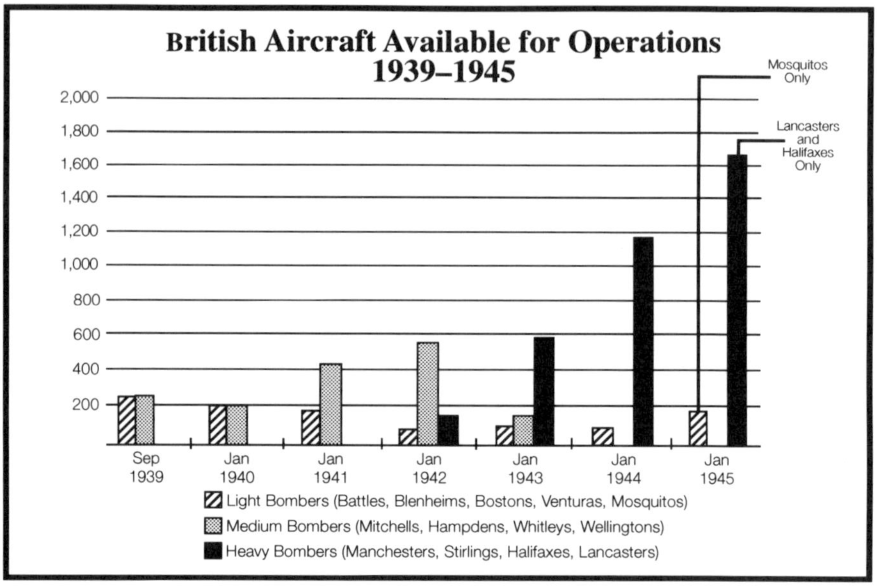

long after the experience should have told them that getting through could often be too costly. They accepted the principal remedy (the P–51 Mustang) both late and over continued objections to its use.[268] The enemy's fighters were formidable, but they were not the principal cause of the inefficiency of strategic bombing at night. Weather determined what could be bombed and how often the attempt could be made. This was true both in daylight and at night. Once over the target, weather played a powerful role in determining how many bombs fell on (or more realistically, within three miles of) the aiming point. Bursting flak and searchlight dazzle added greatly to the problem of getting bombs on target at night.[269] In the age of dumb iron bombs, strategic bombing was bound to be a blunt weapon even if the enemy's fighters offered little resistance.

The official historians also charge, rightly, that British and Allied leadership failed to concentrate strategic bombing on a single objective.[270] The dispute between Sir Arthur Harris and the Air Staff in 1943, and again in the autumn of 1944, was not between area and precision bombing, it was between general area bombing and selective area bombing: between attacking whatever industrial towns could be hit at the moment and attacking those targets thought to be of particular importance to the war effort, whatever the cost to operational efficiency. In the first phase, aircraft production was preferred; in the latter, oil and communication networks were the targets of choice. Sir Arthur Harris was probably right to be skeptical about selective area attack in 1943. He was wrong about it in 1944, when Bomber Command was a much larger force

In the campaign waged against the *Luftwaffe*, these B–24 Liberators of the Fifteenth Air Force leave the Munich airdrome covered with smoke after bombing it on March 24, 1945.

capable of achieving good tactical concentration on its targets.

Much has been written about Sir Arthur Harris. He has had more attackers than defenders, partly because of the moral questions raised over the area offensive. He did not invent area attack on German cities, an established policy when he came to Bomber Command, but he was undoubtedly one of its most enthusiastic advocates. When Harris assumed command, the force had just passed through a winter in which it had been withheld from German targets because it experienced unacceptably high loss rates and produced virtually no significant damage. Some British officials may have still hoped that strategic bombing would be a war-winner; few really believed it would be.

Harris as Bomber Command C-in-C brought considerable practical experience as a flyer, staff officer, and operational commander (at No. 5 Group). Within a year, his professional competence, single-mindedness, determination, willingness to take risks, and energy helped transform the situation. Anyone who has read his memoranda or heard him talk (even in the later years of his very long life) has been impressed by the sheer power of his personality. As a leader, Harris seldom left his headquarters and was almost never seen by any of his men in the operational squadrons. Yet he somehow succeeded in establishing his authority throughout Bomber Command and he left no doubt that his one policy in this total war was to bomb Germany relentlessly and continuously. No one else was quite like him in the RAF.[271]

Harris had a natural gift for sharp, candid, and forceful expression of his views.[272] His candor, however, could easily become gratuitous abuse, with significant consequences. One unfortunate side effect may have been the absence of real debate within Bomber Command headquarters.[273] Relations between the Air Staff and the command were poor, with Harris's own attitudes contributing greatly to that situation. He was, according to the official history, "openly contemptuous" of certain Air Staff officers. His rows with Air Commodore Sidney Bufton, the Director of Bomber Operations, were frequent and legendary. Bufton combined operational experience over Germany, a philosophical belief in the "selective" approach to area bombing, and an unwillingness to be intimidated by the C-in-C. Harris denounced the influence of "junior staff officers" and appears to have allowed or encouraged similar attitudes in his headquarters.[274] Under Harris, the not-invented-here syndrome appears to have characterized much of Bomber Command's attitude toward Air Staff proposals. The bad relations doubtless diverted energy away from joint problem-solving. Harris's greatest contribution to Bomber Command probably occurred in his first eighteen months, when he transformed the command from a relatively ineffective force to a weapon of mass destruction.[275] In this work, he benefited enormously from the long-programmed expansion of the force and the arrival of the four-engine heavies in 1942 and 1943.

Sir Charles Portal, the CAS and Harris's superior, has had something of a mixed press as well. He was a gifted head of service and skilled bureaucratic warrior admired by the Prime Minister, among many others. He probably contributed more than any other senior officer, in Britain or the United States, to the survival of the bomber offensive in Allied strategy in 1941 and 1942. In that difficult time, he kept his attention focused on the potential of strategic bombing, not on its dismal achievements to date. He was an early supporter of the area offensive against German cities and the attack on the morale of the civilian population. Harris was an enthusiast; Portal was an architect.[276]

Should Portal have relieved Sir Arthur Harris? American air commanders were fired for lesser cause, and Portal certainly had sufficient grounds on at least three occasions in 1944. He encouraged Harris to attack Berlin, but he also warned him not to "plan for a sustained and costly series of assaults."[277] When casualties mounted and the attacks continued, Portal might have intervened, but he did not. Harris's belligerent response to Portal's instruction to support Operation OVERLORD was a more important matter. OVERLORD was the critical Allied operation of the war, and the costs of its failure to the fortunes of the Western Allies did not bear thinking about. Portal apparently considered relieving Harris at that time, but thought better of it.[278] Portal recognized that

Harris's views "represented a rigidity of mind fully equal to that of which certain RAF circles accuse the older services," but thought it would "all come right in the wash."[279]

The extraordinary exchanges between Portal and Harris in November and December 1944 unquestionably would have automatically produced a relief of the C-in-C had they occurred in the U.S. Army Air Forces. Indeed, in an early draft of their work, British official historians concluded that Portal "virtually abdicated" his position as CAS when he failed to sack Harris, not only for his persistent opposition to the oil plan, but for the manner in which he expressed it.[280]

Against this view, some argue that Harris was guilty only of taking liberties in official debate. Whatever his views, after some delay, he loyally carried out official policy.[281] Bomber Command gave good support to OVERLORD both in preliminary campaigns against rail communications and in direct support of Allied armies once they had landed. Harris's cooperation with Eisenhower and Tedder was, against expectations, remarkably good. And, it could be said, Bomber Command's attack on oil, though weak in September and October 1944, was persistent and heavy from November onward (see the discussion of statistics above). In short, Portal, whatever the cost to his own authority, may well have been right to keep Harris on.

A considerable share of the credit for the improved performance of Bomber Command after the spring of 1943 must go to group commanders like Air Vice Marshal (AVM) Bennett (No. 8 Group), AVM C. M. McEwen (No. 6 Group from early 1944), AVM, the Hon. Ralph Cochrane (No. 5 Group), AVM Roderick Carr (No. 4 Group), AVM R. Harrison (No. 3 Group), and AVM E.A.B. Rice (No. 1 Group). No list of group commanders would be complete without AVM E. B. Addison, the signals and electronic intelligence expert who headed 100 Group from December 1944. These individuals took the lead in devising the tactical revolution that transformed Bomber Command's operations in 1944, making it a much more flexible and destructive instrument of war.

For most aircrews, the most important commanders were not a Portal or a Harris, or even their group commander. These powerful figures were rarely seen. Most aircrew memoirs of the period either fail to mention men in that position or treat them only in the most perfunctory manner. It is important to remember, so far as night bombing was concerned, that "each crew from take-off to landing was largely a self-determining unit."[282] Each aircrew of seven men did their work out of the sight and hearing of their superiors. And the decisions of the individual aircraft commanders often meant the difference between life and death. In the end, everything depended on the men's collective skill, courage, and persistence.[283]

Capable station, squadron, and, in particular, individual aircraft commanders were critical to the success of Bomber Command's aerial campaigns. Numerous outstanding individuals fortunately staffed these echelons, among them officers such as Wing Commander Leonard Cheshire, Wing Commander Guy Gibson, and Squadron Leader Micky Martin. When the war ended, Cheshire held five major decorations, including the Victoria Cross, the highest honor awarded for armed service to the Crown. Pioneering low-level marking and working as a Master Bomber, he completed two combat tours by 1943 and continued to fly combat missions to the end of the war. Given Bomber Command's aircrew loss rates, the official history noted that one of the most remarkable things about Cheshire was that he survived the war.[284]

The memoir literature focuses on qualities important at the squadron level. Leadership by example stands out. Aircrews tended to respect squadron commanders who flew tough operations with them, and they paid little attention to those who did not. Good luck was also judged immensely important. Aircrews tended to be superstitious in a mild sort of way, and it was thought to be good fortune to fly with a *lucky* pilot or be directed by a *lucky* commander. Bad luck could destroy a man's authority. The officer who succeeded Cheshire at No. 76 Squadron returned early from two missions, and then flew a reciprocal course, lost his way, and had to return early again. It was bad enough to follow a star, but to do so with the curse of bad luck was too much.[285] Technical competence — in particular, an inclination to take great pains with critical detail — appears as another important quality. A commander who flew on dangerous operations and then organized sessions on emergency three-engine flying, methods of improving night vision, and escape and evasion tactics if shot down encouraged his subordinates to take off willingly and press on to the target, even in the face of morale problems caused by sustained aircraft losses.

Conclusion

The British strategic bombing offensive began badly. In 1940 Bomber Command was too small and its aircrews were inadequately trained. Its aircraft were too poorly armed to operate in daylight without escort, and no long-range escort force existed. At night, the smallest target that Bomber Command could find and hit was a large city. Rarely in military history has there been a greater disparity between strategic intentions and the means to carry them out.

In 1943, Bomber Command began to expand rapidly and, with electronic aids, could find large targets at night and achieve sufficient tactical concentration to blast and burn out large urban areas, killing and injuring large numbers of civilians. These successes did not make German capitulation inevitable, as Sir Arthur Harris predicted and many others hoped. Bomber Command shook but did not break civilian morale. The force was still too small, the defenses too strong, the weather too difficult, and the coercive powers of the German authorities too great. But Bomber Command's area offensive did destroy a large part of the cushion in the German economy that might otherwise have been mobilized for war purposes.

During the last year of the war, Bomber Command became a vastly more powerful and flexible instrument than anyone would have dared imagine in the dark days of 1941 or 1942. Its ability to concentrate attacks tactically, its large bombloads and large bombs, its electronic aids to navigation and bombing, and its powerful electronic warfare capability made it a highly flexible instrument of destruction, even in winter weather conditions. In the last ten months of the war, Bomber Command was the only Allied air force capable of operating large units both at night and in daylight.

The Command played a much more important role in the attack on German oil production than is popularly recognized. It is possible (Sir Charles Portal certainly believed so at the time) that Bomber Command could have done even more in this campaign. The aerial attack on oil seriously affected the operations of the German armed forces, and the attack on transportation in the last eight months of the war precipitated, perhaps more than any other factor, a German economic decline that reached proportions of an economic collapse at war's end.

When assessing the importance of strategic bombing, it is well to remember that Allied war aims were extremely ambitious. Not content to restore the *status quo ante* or even to negotiate another Versailles Treaty, the Allies wanted to defeat the German armed forces absolutely, force an unconditional surrender, and extinguish the Nazi political order. They aimed at nothing less than the destruction of the German state and at replacing its sovereignty with their own. War aims that radical required the combined, massive, and unrelenting efforts of the Soviet armed forces employed in the east along a front that once stretched from Leningrad to the Black Sea, the Anglo-American invasion of the European continent in the west, and the widespread destruction wrought by strategic bombing. Without any one of these three elements, and short of using nuclear weapons, victory on the scale achieved and in the time and manner desired was a doubtful proposition.

Acknowledgment

I thank Air Commodore Henry Probert and his staff at the Air Historical Branch of the Ministry of Defence for their kind assistance in the research for this chapter. I am also grateful for the assistance rendered by the late Professor Max Schoenfeld. I remain responsible for any errors of fact or interpretation. Quotations from Crown Copyright documents appear by permission of the Controller of Her Majesty's Stationery Office (HMSO).

NOTES

1. The German Federal Republic's Statistical Office estimated that some 593,000 German civilians died in the Allied (British and American) bombing offensive during World War II. Max Hastings, *Bomber Command* (London: Michael Joseph, 1978), p 352.

2. Sir Charles Webster and Noble Frankland, *The Strategic Air Offensive against Germany*, 4 vols (London: HMSO, 1961), vol 1, p 144.

3. See, for example, John Ellis, *Brute Force: Allied Strategy and Tactics in the Second World War* (New York: Viking, 1990), chap 4; Malcolm Smith, "The Allied Air Offensive," *Journal of Strategic Studies*, vol 13 (Mar 90), pp 67–83; Roger Beaumont, "The Bomber Offensive as a Second Front," *Journal of Contemporary History*, vol 22 (1987), pp 3–14; Dudley Saward, *Victory Denied: The Rise of Air Power and the Defeat of Germany, 1920–1945* (London: Buchan & Enright, 1985), pp 359–363; Richard J. Overy, *The Air War, 1939–1945* (New York: Stein & Day, 1980), pp 152–162; and Hastings, *Bomber Command*, pp 346–352. See also John Terraine, *The Right of the Line: The Royal Air Force in the European War, 1939–1945* (London: Hodder & Stoughton, 1985); Alan J. Levine, *The Strategic Bombing of Germany, 1940–1945* (Westport, Conn.: Praeger, 1992); and Stephen A. Garrett, *Ethics and Airpower in World War II: The British Bombing of German Cities* (New York: St. Martin's, 1993).

4. For the reorganization of the Metropolitan RAF in 1936–1937, see Webster and Frankland, *Strategic Air Offensive against Germany*, vol 1, pp 81–85.

5. British Bombing Survey Unit (BBSU), "The Strategic Air War against Germany, 1939–1945," table 5, Aircraft Available for Operations, U.S. Air Force Historical Research Agency, Maxwell AFB, Ala. (USAFHRA), microfilm reel A5346.

6. *Ibid.*, pp 42–43.

7. *Ibid.*, table 10.

8. In the late 1930s, the Blenheim was designated a "medium." Subsequent descriptions of aircraft are taken from Owen Thetford, *Aircraft of the Royal Air Force Since 1918*, 5th ed (London: Putnam's, 1971).

9. For the development of the prewar bomber force, see Webster and Frankland, *Strategic Air Offensive against Germany*, vol 1, pp 65–81, and Malcolm Smith, "Planning and Building the British Bomber Force, 1934–1939," *Business History Review*, vol 54, no 1 (spring 1980), pp 34–62. See also Neville Jones, *The Beginnings of Strategic Air Power: A History of the British Bomber Force, 1923–1939* (London: Frank Cass, 1987).

10. BBSU, "Strategic Air War against Germany," p 42.

11. Webster and Frankland, *Strategic Air Offensive against Germany*, vol 1, pp 116–117.

12. M. M. Postan *et al.*, *The Design and Development of Weapons* (London: HMSO, 1964), pp 15–24, 142–147.

13. Air Historical Branch (AHB), "The RAF in the Bombing Offensive against Germany: Area Bombing and the Makeshift Force, June 1941 to February 1942" (hereafter AHB, vol 3), USAFHRA, microfilm reel 23355, p 4.

14. The CAS cited the stability problem when replying to a minute from the Prime Minister asking about high Halifax losses. CAS to PM, Aug 12, 1942, AIR 8/330, Public Record Office, Kew (PRO). The Director of Bomber Operations reported to the CAS in December 1942 that there was now "little doubt" that "critical loading" of the aircraft had been responsible for higher wastage of the Halifax. DB Ops to CAS, Dec 8, 1942, AIR 8/409, PRO.

15. AHB, vol 3, p 5.

16. Postan *et al.*, *Design and Development*, pp 93–95.

17. Air Staff Note, "Gun Armament for Bombers," n.d., AIR 9/77, PRO.

18. Harris's case is set out in his "Despatch on War Operations, February 23, 1942, to May 8, 1945," Oct 1945, app C, AIR 14/1252, PRO. I have used the copy held at the Air Historical Branch, Ministry of Defence.

19. M. M. Postan, *British War Production* (London: HMSO, 1952), pp 335–336.

20. "Memorandum by Group Captain E.S.D. Drury in regard to certain of the charges made in the despatch by Air Chief Marshal Sir Arthur Harris on the Operations of Bomber Command from Feb 23, 1942, to May 8, 1945," Aug 1946, AIR 20/6285, PRO. I am indebted for this reference to Professor Vincent Orange of Christchurch University, New Zealand.

21. *Ibid.*

22. Postan *et al.*, *Design and Development*, pp 81–82. Ludlow-Hewitt's requirement was for a fast, defended bomber. See AIR 14/251, PRO. The figures cited are for a Mosquito 16 and are taken from Webster and Frankland, *Strategic Air Offensive against Germany*, vol 4, app 43, pp 449, 453.

23. Thetford, *Aircraft of the RAF*, pp 59, 296–297, 448–449.

24. Postan *et al.*, *Design and Development*, pp 6–8.

25. M. M. Postan, *British War Economy* (London: HMSO 1952; repr 1975), pp 123–126, 145–174. The labor famine is discussed at pp 217–226.

26. *Ibid.*, p 125.

27. Terraine, *Right of the Line*, p 301.

28. Martin Middlebrook, *The Nuremberg Raid* (1973; New York: Morrow, 1974), pp 46–47.

29. Terraine, *Right of the Line*, app E. The figures in this appendix are drawn from AHB, "Manning Plans and Policy," AHB Library, app 3.

30. AHB, "Manning," pp 203–205.

31. On the state of training at the outset of the war, see Webster and Frankland, *Strategic Air Offensive against Germany*, vol 1, pp 107–120. For an overview of wartime training, see Webster and Frankland, *Strategic Air Offensive against Germany*, vol 4, annex 3.

32. AHB, "The RAF in the Bombing Offensive against Germany: Restricted Bombing, September 1939 to May 1941" (hereafter AHB, vol 2), USAFHRA, microfilm reel 23354, pp 5–6.

33. AHB, "Flying Training," AHB Library, pp 75–78, 141–142, app 2.

34. *Ibid.*, p 169.

35. *Ibid.*, pp 169, 191.

36. *Ibid.*, pp 230–231.

37. *Ibid.*, pp 138–139.

38. The evolution of the operational tour policy is described in Terraine, *Right of the Line*, pp 521–527.

39. Martin Middlebrook, *The Battle of Hamburg* (New York: Scribners, 1981), p 39. These calculations were made for Middlebrook by the Phoenix Assurance Co., Ltd., and seem to be a reasonable assessment of the risks. Important Bomber Command papers on the subject of tour policy may be found in AIR 14/1016–18, PRO.

40. For an introduction to early ideas about the use of the bomber, see Lee Kennett, *A History of Strategic Bombing* (New York: Scribners, 1982), chap 3.

41. Malcolm Smith, *British Air Strategy between the Wars* (Oxford: Oxford University Press, 1984), chap 2. Other analyses of British interwar strategy are H. Montgomery Hyde, *British Air Policy between the Wars* (London: Heinemann, 1975); Barry Powers, *Strategy without Slide-Rule: British Air Strategy, 1914–1939* (London: Croom-Helm, 1976), and Uri Bialer, *The Shadow of the Bomber* (London: Royal Historical Society, 1980). See also Webster and Frankland, *Strategic Air Offensive against Germany*, vol 1, pp 34–106.

42. Smith, *British Air Strategy*, chaps 6–7; Webster and Frankland, *Strategic Air Offensive against Germany*, vol 1, pp 86–91.

43. Webster and Frankland, *Strategic Air Offensive against Germany*, vol 1, pp 86–94. This paragraph considerably simplifies a complex policy debate that took place in the British government in 1936 and 1937. See N. H. Gibbs, *Grand Strategy* (London: HMSO, 1976), vol 1, chaps 14 and 15.

44. Smith, *British Air Strategy*, pp 285–291.

45. Webster and Frankland, *Strategic Air Offensive against Germany*, vol 1, pp 96–97, 104.

46. Several numbering systems were employed in the Western Air Plans between 1937 and 1939. I am using the scheme in DD Plans (Op), "Summary of Plans for Action by the Air Striking Force against Enemy War Industry," Sep 17, 1939, AIR 20/22, PRO. The most recent scholarly work on the subject identifies the oil plan as WA 6. Smith, *British Air Strategy*, pp 296–301. The DD Plans

document labels that design "WA 5 (c)."

47. The phrase "industrial nerve" is used in Plans (Ops.) Plan WA 5, app A, in AIR 20/266, PRO.

48. *Ibid.*, app G; Webster and Frankland, *Strategic Air Offensive against Germany*, vol 1, pp 97–98.

49. Smith, *British Air Strategy*, pp 302–305; Smith, "Allied Air Offensive," pp 69–70.

50. F. H. Hinsley *et al.*, *British Intelligence in the Second World War*, 5 vols (London: HMSO, 1979–1981), vol 1, chap 7, app 3.

51. Jones, *Beginnings of Strategic Air Power*, pp 151–152. A good brief portrait of Ludlow-Hewitt's tenure at Bomber Command is Malcolm Smith, "Sir Edgar Ludlow-Hewitt and the Expansion of Bomber Command, 1939–1940," *Journal of the Royal United Services Institution*, vol 126 (Mar 1981), pp 52–57.

52. Webster and Frankland, *Strategic Air Offensive against Germany*, vol 1, p 95.

53. *Ibid.*, pp 95–96.

54. *Ibid.*, pp 116–117.

55. Malcolm Smith, "The RAF and Counter-Force Strategy before World War II," *Journal of the Royal United Services Institution*, vol 112 (June 1976), p 71.

56. See Bialer, *The Shadow of the Bomber*. A good treatment of the prewar intelligence estimates is Wesley Wark, *The Ultimate Enemy: British Intelligence and Nazi Germany, 1933–1939* (Ithaca: Cornell University Press, 1985), pp 35–79.

57. Gibbs, *Grand Strategy*, vol 1, pp 589–595; J.R.M. Butler, *Grand Strategy* (London: HMSO, 1957; repr w/amendments, 1971), vol 2, pp 165–170.

58. Ludlow-Hewitt to Air Ministry, Sep 25, 1938, AIR 20/22, PRO.

59. The early months of the war are treated in Webster and Frankland, *Strategic Air Offensive against Germany*, vol 1, pp 133–143. See also Hastings, *Bomber Command*, pp 15–35.

60. That was the reasoning employed by Sir Edgar Ludlow-Hewitt in a letter to the Air Ministry in January 1940, sent accompanying an appreciation of the Ruhr Plan. Webster and Frankland, *Strategic Air Offensive against Germany*, vol 1, p 139.

61. DD Plans (Op) "Notes for Conference with AOC in C Bomber Command," Feb 13, 1940, AIR 9/96, PRO; Webster and Frankland, *Strategic Air Offensive against Germany*, vol 1, pp 139–140.

62. On the early stages of Bomber Command's development as a night force, see Webster and Frankland, *Strategic Air Offensive against Germany*, vol 1, pp 190–212.

63. *Ibid.*, p 141.

64. See the directives of April 13, 1940, May 30, 1940, June 4, 1940, June 20, 1940, July 4, 1940, July 13, 1940, July 24, 1940, July 30, 1940, September 21, 1940, and September 30, 1940, printed in Webster and Frankland, *Strategic Air Offensive against Germany*, vol 4, app 8, pts 1–10. For an account of the exchanges between Churchill, Portal, and the Air Staff, see Webster and Frankland, *Strategic Air Offensive against Germany*, vol 1, pp 149–154.

65. PM to Chiefs of Staff (COS), Sep 3, 1940, in W. S. Churchill, *Their Finest Hour* (New York: Houghton Miflin; Bantam PB ed, 1962), p 391.

66. Webster and Frankland, *Strategic Air Offensive against Germany*, vol 1, p 163–164.

67. *Ibid.*, pp 165–166.

68. Hinsley *et al.*, *British Intelligence in the Second World War*, vol 1, pp 226–227.

69. Webster and Frankland, *Strategic Air Offensive against Germany*, vol 1, pp 145, 157, 219–220.

70. *Ibid.*, pp 121–123.

71. AHB, "Photographic Reconnaissance," AHB Library, pp 55–57.

72. *Ibid.*, pp 67–68.

73. *Ibid.*, pp 85–86.

74. *Ibid.*, pp 73–78; Hinsley *et al.*, *British Intelligence in the Second World War*, vol 1, app 2.

75. Webster and Frankland, *Strategic Air Offensive against Germany*, vol 1, pp 221–222.

76. The text of this study appears in Webster and Frankland, *Strategic Air Offensive against Germany*, vol 4, as app 13. For a discussion of the Butt Report and its impact, see Webster and Frankland, *Strategic Air Offensive against Germany*, vol 1, pp 178–180.

77. PM to CAS, Sep 27, 1941, quoted in Webster and Frankland, *Strategic Air Offensive against Germany*, vol 1, pp 182–183.

78. For German air strategy and the development of the *Luftwaffe* in the prewar period, see AHB, *The Rise and Fall of the German Air Force* (London: HMSO, 1948; New York: St. Martin's, 1983), pp 1–52, 185; Williamson Murray, *Strategy for Defeat: The Luftwaffe, 1933–1945* (Maxwell AFB: Air University Press, 1983), pp 1–24; Wilhelm Deist, *The Wehrmacht and Nazi Rearmament* (London:

Macmillan, 1981), pp 54–69; Wilhelm Deist, "*Die Aufrüstung der Wehrmacht,*" in *Das Deutsche Reich und der Zweite Weltkrieg*, vol 1 (Stuttgart: *Deutsche Verlags-Anstalt*, 1979); Klaus Maier, "*Totaler Krieg und operativer Luftkrieg,*" in *Das Deutsche Reich und der Zweite Weltkrieg*, vol 2 (1979); and Horst Boog, "*Der angloamerikanische strategische Luftkrieg über Europa und die deutsche Luftverteidigung,*" in *Das Deutsche Reich und der Zweite Weltkrieg*, vol 6 (1990), pp 436–448.

79. Gebhard Aders, *History of the German Night Fighter Force 1917–1945* (London, 1979), pp 7–12. (This work was originally published in Stuttgart in 1978 as *Geschichte der deutschen Nachtjagd 1917–1945*.)

80. *Ibid.*, pp 14–16.

81. AHB, *Rise and Fall*, p 186.

82. The development of the Kammhuber system is described in Aders, *History of the German Night Fighter Force*, chaps 3–5, and AHB, *Rise and Fall*, pp 186–191. See also Murray, *Strategy for Defeat*, pp 132–33; Alfred Price, *Instruments of Darkness* (New York: Scribners, 1977), pp 55–70; and Boog, "*Der angloamerikanische strategische . . .,*" pp 482–492.

83. Aders, *History of the German Night Fighter Force*, pp 40, 57.

84. *Ibid.*, pp 40, 53–57.

85. Webster and Frankland, *Strategic Air Offensive against Germany*, vol 1, p 186.

86. DCAS to C-in-C, Bomber Command, Sep 21, 1940, in Webster and Frankland, *Strategic Air Offensive against Germany*, vol 4, app 8, pt 9.

87. Portal to Air Ministry, Jul 16, 1940, AIR 20/3359, PRO; Webster and Frankland, *Strategic Air Offensive against Germany*, vol 1, pp 144–154.

88. Butler, *Grand Strategy*, vol 2, p 411.

89. Webster and Frankland, *Strategic Air Offensive against Germany*, vol 1, pp 152, 161.

90. Douglas (DCAS) to Peirse (C-in-C, Bomber Command), Oct 30, 1940, in Webster and Frankland, *Strategic Air Offensive against Germany*, vol 4, app 8, pt 11.

91. Minutes, Plans (Op), Jul 7, 1940, AIR 20/23, PRO. See also Webster and Frankland, *Strategic Air Offensive against Germany*, vol 1, pp 150–151, citing a memorandum sent to Portal dated Jul 21, 1941.

92. Plans 2, "Note on Air Bombardment Policy," Jan 1, 1941, AIR 20/24, PRO.

93. DCAS to C-in-C, Bomber Command, Feb 14, 1942, in Webster and Frankland, *Strategic Air Offensive against Germany*, vol 4, app 8, pt 22.

94. Webster and Frankland, *Strategic Air Offensive against Germany*, vol 1, pp 331–332. The text of the minute is printed in full on these pages.

95. *Ibid.*, pp 391–393.

96. *Ibid.*, pp 402–410; Terraine, *Right of the Line*, pp 481–88; see also Ralph Barker, *The Thousand Plan* (London: Chatto & Windus, 1965).

97. For an extended discussion of the development of area bombing tactics in 1942, see Webster and Frankland, *Strategic Air Offensive against Germany*, vol 1, pp 382–417.

98. Sir Arthur Harris describes this period of the air campaign in his *Bomber Offensive* (New York: Macmillan, 1947), pp 144–194.

99. Webster and Frankland, *Strategic Air Offensive against Germany*, vol 2, pp 108–137.

100. Albert Speer, *Inside the Third Reich* (New York: Avon, 1971), p 370. A good account of these events is Middlebrook, *Battle of Hamburg*. See also Webster and Frankland, *Strategic Air Offensive against Germany*, vol 2, pp 147–157; Hans Rumpf, *The Bombing of Germany*, trans. Edward Fitzgerald (London: White Lion, 1961), pp 77–91.

101. Webster and Frankland, *Strategic Air Offensive against Germany*, vol 2, pp 147–148.

102. *Ibid.*, pp 190–211. A good recent account of this part of the area offensive is Martin Middlebrook, *The Berlin Raids: RAF Bomber Command, Winter 1943–1944* (London: Viking, 1988). Middlebrook dates the beginning of the "Battle of Berlin" from the night of Aug 23/24 1943. Hastings provides a highly critical view (*Bomber Command*, pp 257–269). An account more sympathetic to Harris is Dudley Saward, *Bomber Harris: The Story of Marshal of the Royal Air Force Sir Arthur Harris* (Garden City, N.Y.: Doubleday, 1985), pp 214–225.

103. Harris made the prediction in a letter to the Prime Minister on Nov 3, 1943. Webster and Frankland, *Strategic Air Offensive against Germany*, vol 2, p 190.

104. *Ibid.*, p 191.

105. Harris to Air Ministry, April 7, 1944, quoted in *ibid.*, p 193.

106. Webster and Frankland, *Strategic Air Offensive against Germany*, vol 1, pp 320–324.

107. An excellent introduction to the development of aerial navigation in the early years of the Second World War is Monte D. Wright, *Most Probable Position: A History of Aerial Navigation to 1941* (Lawrence: University of Kansas Press, 1972).

108. This description of navigational technique is based on a report submitted by Bomber Command to the Directorate of Bomber Operations on February 2, 1941, a copy of which can be found in AIR 20/283, PRO.

109. "Bomber Command Preliminary Report on Loss of Aircraft during Operations on the Night of Nov 7/8, 1941," Nov 11, 1941, AIR 8/408, PRO.

110. Aeroplane and Armament Establishment, "Accuracy of Target Location by Astro-Navigation," Oct 31, 1941, AIR 20/283, PRO.

111. *Ibid.*

112. Harris, *Bomber Offensive*, p 80.

113. Bomber Command to B Ops, Feb 2, 1941, AIR 20/283, PRO.

114. For a good description of Gee see Webster and Frankland, *Strategic Air Offensive against Germany*, vol 4, annex 1. By August 1942, about 80 percent of the force had been equipped with the device. See also Price, *Instruments of Darkness*, pp 98–101.

115. Webster and Frankland, *Strategic Air Offensive against Germany*, vol 4, annex 1, pp 7–11; Gordon Musgrove, *Pathfinder Force* (London: Macdonald's & Jane's, 1976), app 1.

116. Webster and Frankland, *Strategic Air Offensive against Germany*, vol 4, annex 1.

117. *Ibid.*, pp 15–17.

118. *Ibid.*, annex 1, pp 13–15; Price, *Instruments of Darkness*, pp 121–123; Musgrove, *Pathfinder Force*, app 2.

119. For a good treatment of the argument over the formation of the Pathfinder Force, see Webster and Frankland, *Strategic Air Offensive against Germany*, vol 1, pp 418–436. For the best book on the subject, see Musgrove, *Pathfinder Force*. Harris lays out his views in *Bomber Offensive*, pp 128–130.

120. Webster and Frankland, *Strategic Air Offensive against Germany*, vol 4, annex 1, p 35.

121. Webster and Frankland, *Strategic Air Offensive against Germany*, vol 2, pp 90–107. For an extended discussion of specific operations, see also *ibid.*, pp 108–167. For a good description of the development of the many kinds of marking techniques, see Musgrove, *Pathfinder Force*, app 6.

122. This description of tactics is based on Harris, "Despatch on War Operations," app D. For each major period of the war, Webster and Frankland provide detailed analyses of the tactical problems.

123. Harris, "Despatch on War Operations," app D.

124. Webster and Frankland, *Strategic Air Offensive against Germany*, vol 1, pp 252–253; AHB, "The RAF in the Bombing Offensive against Germany: A Period of Expansion and Experiment, March 1942 to January 1943" (hereafter AHB, vol 4), USAFHRA, microfilm reel 23355, pp 45–46. The policy of incendiarism was influenced, in part, by the *Luftwaffe*'s successes with fire bombs during the Blitz.

125. Webster and Frankland, *Strategic Air Offensive against Germany*, vol 4, annex 4.

126. Webster and Frankland, *Strategic Air Offensive against Germany*, vol 3, p 181.

127. *Ibid.*, pp 191–196.

128. *Ibid.*, pp 203–204.

129. Webster and Frankland, *Strategic Air Offensive against Germany*, vol 4, annex 4, p 34.

130. This description of "concentration" tactics is based on Harris, "Despatch on War Operations," app D.

131. *Ibid.*

132. A good account of the struggle against the German radio navigation is provided in R. V. Jones, *Most Secret War* (London, 1978), pp 92–214, and *passim*. See also Price, *Instruments of Darkness*, pp 19–54. Jones is also the best single source on the development of scientific intelligence in general.

133. For the development of the intelligence picture of the German night defenses, see Hinsley *et al.*, *British Intelligence in the Second World War*, vol 2, pp 244–257; and Jones, *Most Secret War*, pp 264–279.

134. For a brief survey of radio countermeasures, see Martin Streetly, *Confound and Destroy* (London: Jane's PB ed, 1985), pp 1–31. More extended treatment is provided in Price, *Instruments of Darkness*, pp 111–211.

135. Streetly, *Confound and Destroy*, pp 135–138.

136. Jones, *Most Secret War*, pp 210–211, 388–391. See also Hinsley *et al.*, *British Intelligence in the Second World War*, vol 3, pt 1, app 21; Streetly, *Confound and Destroy*, pp 18–19.

137. Streetly, *Confound and Destroy*, p 139.

138. Aders, *History of the German Night Fighter Force*, pp 80–81.

139. Webster and Frankland, *Strategic Air Offensive against Germany*, vol 2, pp 141–145.

140. Aders, *History of the German Night Fighter Force*, p 122.

141. Streetly, *Confound and Destroy*, p 45.

142. *Ibid.*, p 60.

143. *Ibid.*, pp 21–22.

144. Price, *Instruments of Darkness*, p 214; Aders, *History of the German Night Fighter Force*, pp 166–67; Jones, *Most Secret War*, pp 391–393; Hinsley *et al.*, *British Intelligence in the Second World War*, vol 3, pt 1, app 20.

145. Harris, "Despatch on War Opertations," app D.

146. Aders, *History of the German Night Fighter Force*, p 56.

147. *Ibid.*, pp 61–62.

148. *Ibid.*, p 147. For problems of the Wild Boar system, see *ibid.*, pp 97–100.

149. ADI (K) 416/1945, in AIR 40/1394, PRO. This is a report of the postwar interrogations of Kammhuber and his successor, Schmid.

150. Aders, *History of the German Night Fighter Force*, pp 85–86.

151. *Ibid.*, pp 76–77.

152. ADI (K) 416/1945.

153. Aders, *History of the German Night Fighter Force*, p 85.

154. *Ibid.*, pp 125–127; Hinsley *et al.*, *British Intelligence in the Second World War*, vol 3, pt 1, app 21.

155. Aders, *History of the German Night Fighter Force*, pp 146–147.

156. ADI (K) 416/1945.

157. For an account of how (slowly) British intelligence gained a good working picture of the new German defenses, see Hinsley *et al.*, *British Intelligence in the Second World War*, vol 3, pt 1, pp 308–317.

158. DB Ops, "What Is the Highest Percentage of Losses That the Royal Air Force Could Stand over a Period of Three Months of Intensive Operations?" Mar 16, 1945, in Webster and Frankland, *Strategic Air Offensive against Germany*, vol 4, app 42.

159. Webster and Frankland, *Strategic Air Offensive against Germany*, vol 4, app 40. These figures exclude Fighter Command intruder sorties and Bomber Command minelaying, reconnaissance, and secret operations. They do include Mosquito bomber sorties, which had low loss rates.

160. *Ibid.*, app 42.

161. Webster and Frankland, *Strategic Air Offensive against Germany*, vol 2, p 205.

162. *Ibid.*, p 207.

163. *Ibid.*, pp 192–193.

164. Computed from Webster and Frankland, *Strategic Air Offensive against Germany*, vol 4, app 40.

165. Webster and Frankland, *Strategic Air Offensive against Germany*, vol 2, p 194.

166. *Ibid.*, p 193.

167. Harris to Air Ministry, Apr 7, 1944, quoted in Webster and Frankland, *Strategic Air Offensive against Germany*, vol 2, p 193.

168. A stimulating analysis of the British strategic debate about the bombing offensive in the larger Allied war effort is found in Anthony Verrier, *The Bomber Offensive* (London: Batsford, 1968; London: Pan Books, 1974), pp 151–193.

169. W. F. Craven and J. L. Cate, *The Army Air Forces in World War II*, 7 vols (Chicago: University of Chicago Press, 1948–1958; Washington, D.C.: Office of Air Force History, 1983), vol 2, pp 305–307, 373–375.

170. The language is in "Combined Chiefs of Staff Directive for the Bomber Offensive from the United Kingdom," Jan 21, 1943, in Webster and Frankland, *Strategic Air Offensive against Germany*, vol 4, app 8, pt 28. The POINTBLANK Directive, dated June 10, 1943, repeats this text. It is printed as part 32 of the same appendix. This language also occurs in the revision to POINTBLANK. See Air Ministry to Bomber Command, Feb 17, 1944, in *ibid.*, pt 36.

171. Webster and Frankland, *Strategic Air Offensive against Germany*, vol 2, pp 10–21.

172. Webster and Frankland, *Strategic Air Offensive against Germany*, vol 4, app 23.

173. Bottomley to Harris, Jun 10, 1943, in Webster and Frankland, *Strategic Air Offensive against Germany*, vol 4, app 8, pt 32. For a discussion of the exchanges on this matter, see Webster and

Frankland, *Strategic Air Offensive against Germany*, vol 2, pp 27–30.

174. Harris to CAS, Nov 1, 1944, AIR 37/1013, PRO.

175. Webster and Frankland, *Strategic Air Offensive against Germany*, vol 2, pp 64–67, 234–235; Harris, *Bomber Offensive*, p 144.

176. Harris to Air Ministry, Dec 7, 1943, AIR 8/425, PRO. This famous memorandum is printed in full and discussed in Webster and Frankland, *Strategic Air Offensive against Germany*, vol 2, pp 53–61.

177. Craven and Cate, eds., *Army Air Forces in World War II*, vol 2, pp 665–732.

178. DB Ops (Bufton), "A Note on the Conduct of the Combined Bomber Offensive Operations of the Eighth U.S.A.A.F. — October 1/10th, 1943," AIR 20/4764, PRO.

179. For a good account of these debates, see Webster and Frankland, *Strategic Air Offensive against Germany*, vol 2, pp 33–70.

180. Harris to Bottomley, Dec 20, 1943, quoted in Webster and Frankland, *Strategic Air Offensive against Germany*, vol 2, pp 66–67.

181. The letter was dated December 17, 1943; the directive, January 14, 1944; and the order, January 27, 1944. The letter is discussed in Webster and Frankland, *Strategic Air Offensive against Germany*, vol 2, p 65; the order, in *ibid.*, p 70. The directive is printed in Webster and Frankland, *Strategic Air Offensive against Germany*, vol 4, app 8, pt 34.

182. BC Operational Research Section, Night Raid Rpt no. 536, AIR 14/3411, PRO.

183. BC Operational Research Section, Night Raid Rpt no. 537, AIR 14/3411, PRO; Webster and Frankland, *Strategic Air Offensive against Germany*, vol 4, app 49. Table 14, however, shows 29 percent of the area burnt out. This figure is taken from BBSU documents.

184. In fairness, it should also be pointed out that Harris regarded Augsburg as a much easier target than Schweinfurt to find and hit. See the exchange of letters between Harris and Lord Selborne on this subject in Webster and Frankland, *Strategic Air Offensive against Germany*, vol 4, app 15.

185. Harris, "The Employment of the Bomber Force in Connection with the Invasion of the Continent from the United Kingdom," Jan 13, 1944, AIR 2/7080, PRO. Also in this file is a four-column paper containing Harris's assertions, comments by the staff of the Allied Expeditionary Air Force (the organization charged with planning air support for the assault), and comments by the Air Staff. The text of this remarkable paper is printed in Solly Zuckerman, *From Apes to Warlords, 1904–1946* (London: Hamish Hamilton, 1978), app A. For a detailed survey of the issues, see Webster and Frankland, *Strategic Air Offensive against Germany*, vol 3, pp 9–41.

186. A good account of the organizational politics of the fight over the Transportation Plan is found in Alfred Carl Mierzejewski, *The Collapse of the German War Economy, 1944–1945: Allied Air Power and the German National Railway* (Chapel Hill: University of North Carolina Press, 1988), chap 4. For an alternative perspective, see W. W. Rostow, *Pre-Invasion Bombing Strategy: General Eisenhower's Decision of Mar 25, 1944* (Austin: University of Texas Press, 1981).

187. Coryton (ACAS, Operations) to Harris, Mar 4, 1944, AIR 14/780, PRO, and in Webster and Frankland, *Strategic Air Offensive against Germany*, vol 4, app 8, pt 37. For a detailed analysis of Bomber Command's early attacks on pre-OVERLORD targets, see the Operational Research Section reports in AIR 24/280, PRO. Other relevant papers may be found in AIR 14/2687, PRO.

188. For Harris's point of view on these developments, see *Bomber Offensive*, pp 190–205; for a description of Bomber Command's tactical evolution, see Webster and Frankland, *Strategic Air Offensive against Germany*, vol 3, pp 123–162.

189. On the daylight operations, see Webster and Frankland, *Strategic Air Offensive against Germany*, vol 3, pp 163–174.

190. For a discussion of the master bomber's duties, see Musgrove, *Pathfinder Force*, pp 272–274. Documents on the use of the master bomber may be found in AIR 14/791, 2063, and 2325, PRO.

191. The development of No. 617 Squadron's techniques in 1943 and early 1944 is treated in Webster and Frankland, *Strategic Air Offensive against Germany*, vol 2, pp 168–189.

192. This account is taken from the minute, B Ops 2 (a) to DB Ops, Apr 24, 1944, based on telephone conversation with the SASO, No. 5 Group, AIR 20/3208, PRO. For a brief discussion of this attack, see also Webster and Frankland, *Strategic Air Offensive against Germany*, vol 3, p 157. For another description of advanced marking technique during the same period, see DB Ops to DCAS, Apr 19, 1944, AIR 20/283, PRO.

193. Webster and Frankland, *Strategic Air Offensive against Germany*, vol 3, pp 163–172.

194. *Ibid.*, pp 177–178.

195. ADI (K) 416/1945.

196. On the formation of No. 100 Group, see Streetly, *Confound and Destroy*, chap 2.
197. ADI (K) 337/1945, in AIR 40/1397, PRO.
198. Streetly, *Confound and Destroy*, p 148–149.
199. *Ibid.*, chap 2.
200. *Ibid.* For information on Perfectos, see also Price, *Instruments of Darkness*, p 220.
201. Webster and Frankland, *Strategic Air Offensive against Germany*, vol 3, pp 145–150.
202. Aders, *History of the German Night Fighter Force*, p 15.
203. AHB, *Rise and Fall*, pp 347–356.
204. These figures are taken from AHB, "The RAF in the Bombing Offensive against Germany: The Full Offensive, February 1943 to February 1944" (hereafter AHB, vol 5), USAFHRA, microfilm reel 23355, app 11, and AHB, "The RAF in the Bombing against Germay: The Final Phase, March 1944 to May 1945" (vol 6), USAFHRA, microfilm reel 23355, annex A.
205. Computed from Webster and Frankland, *Strategic Air Offensive against Germany*, vol 4, app 44.
206. BBSU, "Strategic Air War against Germany," table 7, Tonnages Claimed Dropped by RAF Bomber Command in Accordance with Main Directifs.
207. One statistical note is important here. From November 1944 Bomber Command often used a marshaling yard rather than the traditional center of the built-up area as its aiming point in area attacks. To the extent that this has not been accounted for in the British statistics, the figures for area attacks should be lowered and those for transportation, raised. Mierzejewski, *Collapse of the German War Economy*, p 103; Bomber Command, Operations Record Book, p 318, AIR 24/206, PRO.
208. Accounts of this debate vary. See Webster and Frankland, *Strategic Air Offensive against Germany*, vol 3, pp 42–94, esp. 65 ff; Denis Richards, *Portal of Hungerford* (London: Heinemann, 1978), pp 317–325; Saward, *Bomber Harris*, pp 260–277; and Hastings, *Bomber Command*, pp 327–336.
209. Craven and Cate, *Army Air Forces in World War II*, vol 3, pp 72–79; Webster and Frankland, *Strategic Air Offensive against Germany*, vol 3, pp 29–35.
210. Webster and Frankland, *Strategic Air Offensive against Germany*, vol 3, p 47.
211. For Portal's desire to focus everything on achieving an early end to the war, see Portal to Harris, Dec 6, 1944, in AIR 20/3246, PRO. The labor problem is discussed in Postan, *British War Production*, pp 217–227.
212. Intelligence on this point was unambiguous. See Hinsley *et al.*, *British Intelligence in the Second World War*, vol 3, pt 2, p 497.
213. *Ibid.*, pp 500–514.
214. Webster and Frankland, *Strategic Air Offensive against Germany*, vol 3, pp 48–52.
215. DSAC, "Notes on Air Policy to Be Adopted with a View to Rapid Defeat of Germany," Oct 25, 1944, in AIR 37/1436, PRO, apps for 10/44. This document is also printed in Webster and Frankland, *Strategic Air Offensive against Germany*, vol 4, app 25. For a discussion of the quarrels over bombing policy from the summer through early winter of 1944, see also Mierzejewski, *Collapse of the German War Economy*, chaps 5–7, and *passim*.
216. Bottomley to Harris, Sep 25, 1944, in Webster and Frankland, *Strategic Air Offensive against Germany*, vol 4, app 8, pt 41.
217. *Ibid.*
218. The long and extremely detailed correspondence between Harris and Portal in this period is well covered in Webster and Frankland, *Strategic Air Offensive against Germany*, vol 3, pp 65–94.
219. For Harris's tactical arguments, see his letters to Portal of Nov 1, 1944, and Nov 6, 1944, both in AIR 8/1745, and of Nov 24, 1944, and Dec 6, 1944, both in AIR 20/3246, all in PRO.
220. Harris to Portal, Nov 1, 1944, AIR 8/1745, PRO.
221. Harris to Portal, Dec 28, 1944, and Portal to Harris, Jan 3, 1944, quoted in Webster and Frankland, *Strategic Air Offensive against Germany*, vol 3, p 88.
222. Harris to Portal, Jan 18, 1945, quoted in Webster and Frankland, *Strategic Air Offensive against Germany*, vol 3, p 93.
223. Webster and Frankland, *Strategic Air Offensive against Germany*, vol 3, p 199.
224. For the official historians' account, see Webster and Frankland, *Strategic Air Offensive against Germany*, vol 3, pp 99–119; see also Saward, *Bomber Harris*, pp 281–297; Hastings, *Bomber Command*, pp 339–345; and Terraine, *Right of the Line*, pp 677–678. Books on the subject include David Irving, *The Destruction of Dresden* (London: Kimber, 1963) and Alexander McKee, *Dresden 1945: The Devil's Tinderbox* (London: Souvenir Press, 1982). For THUNDERCLAP and Dresden from

an American perspective, see Ronald Schaeffer, *Wings of Judgment* (New York: Oxford University Press, 1985), pp 80–103. See also Richard G. Davis, "Operation THUNDERCLAP: The U.S. Army Air Forces and the Bombing of Berlin," *Journal of Strategic Studies*, vol 14, no 1 (Mar 1991), pp 90–111.

225. COS to PM, Jul 5, 1944, quoted in Webster and Frankland, *Strategic Air Offensive against Germany*, vol 3, p 52.

226. Portal to COS, Aug 5, 1944, quoted in Webster and Frankland, *Strategic Air Offensive against Germany*, vol 3, p 55.

227. *Ibid.*

228. JIC Rpt, Jan 25, 1945, cited in Webster and Frankland, *Strategic Air Offensive against Germany*, vol 3, p 98. On the American side, Lt Gen Carl Spaatz, CG, U.S. Strategic Bomber Forces in Europe, ordered Eighth Air Force Commander Lt Gen James Doolittle to begin planning for an American THUNDERCLAP on January 16, 1945. Davis, "Operation THUNDERCLAP."

229. *Ibid.*

230. For the full text of Sir Archibald Sinclair's memorandum, see Webster and Frankland, *Strategic Air Offensive against Germany*, vol 3, p 102.

231. Churchill to Sinclair, Jan 26, 1945, quoted in Webster and Frankland, *Strategic Air Offensive against Germany*, vol 3, p 103.

232. *Ibid.* The cities on the list had been suggested by Sir Arthur Harris in a conversation with Sir Norman Bottomley on January 25. Webster and Frankland, *Strategic Air Offensive against Germany*, vol 3, p 100.

233. The phrase "tacitly approved" appears in Craven and Cate, eds., *Army Air Forces in World War II*, vol 3, p 724. British official historians say the modified THUNDERCLAP plan that was sent to the Malta conferees by Bottomley and his American counterpart General Carl Spaatz was not discussed by the Combined Chiefs. Webster and Frankland, *Strategic Air Offensive against Germany*, vol 3, p 105.

234. Melden E. Smith, Jr., "The Bombing of Dresden Reconsidered: A Study in Wartime Decision-Making," (Ph.D. diss., Boston University, 1971). Some part of the death toll should, of course, be assigned to the American raids that followed the initial Bomber Command blow.

235. This episode is well described in Webster and Frankland, *Strategic Air Offensive against Germany*, vol 3, pp 112–117.

236. *Ibid.*, pp 118–119.

237. For a sampling of interpretations, see the sources cited in note 2 above. The appreciations and results sections in the official history are fundamental: Webster and Frankland, *Strategic Air Offensive against Germany*, vol 1, pp 457–92, vol 2, pp 224–300, and vol 3, pp 207–311. See also Mierzejewski, *Collapse of the German War Economy*, chap 9.

238. The percentages are computed from BBSU, "Strategic Air War against Germany," table 7, Tonnages Claimed Dropped by RAF Bomber Command in Accordance with Main Directifs. The graph is prepared from the monthly tonnage figures in Webster and Frankland, *Strategic Air Offensive against Germany*, vol 4, app 44.

239. United States Strategic Bombing Survey (USSBS), *Area Studies Division Report* (Washington, D.C.: GPO, 1947), p 18.

240. USSBS, *The Effects of Strategic Bombing on the German War Economy* (Washington, D.C.: GPO, 1945), pp 6–11, 139–192, app table 100; BBSU, "Potential and Actual Output of German Armaments in Relation to the Combined Bomber Offensive," USAFHRA, microfilm reel A5346, table 1.

241. The principal sources are USSBS, *Effects of Strategic Bombing on the Germany War Economy*; Nicholas Kaldor, "The German War Economy," *Review of Economic Statistics*, vol 12 (1946); Burton H. Klein, *Germany's Economic Preparations for War* (Cambridge, Mass.: Harvard University Press, 1959); and Alan Milward, *The German War Economy* (London: Athlone, 1965).

242. One may follow the development of Professor Richard J. Overy's views in his "Hitler's War and the German Economy: A Reinterpretation," *Economic History Review*, 2d ser, vol 35 (1982); *Goering: The Iron Man* (London: Routledge & Kegan Paul, 1984); "Mobilization for Total War in Germany, 1939–1941," *English Historical Review*, vol 103, no 408 (Jul 1988); and *"Blitzkriegswirtschaft? Finanzpolitik, Lebensstandard und Arbeitseinsatz in Deutschland, 1939–1943," Vierteljahrschefte für Zeitgeschichte*, vol 36, no 3 (1988).

243. USSBS, *Area Studies Division Report*, p 18.

244. Webster and Frankland, *Strategic Air Offensive against Germany*, vol 3, p 288; see also BBSU, "Report of the Effects of Strategic Air Attacks on German Towns," USAFHRA, microfilm reel A5346.

245. This point is well argued in Overy, *Air War*, pp 158–160.

246. These conclusions are based on USSBS, *The Effects of Strategic Bombing on German Morale* (Washington, D.C.: GPO, [1947]). For an example of recent German work, see Olaf Groehler, "The Strategic Air War and Its Impact on the German Civilian Population," in Horst Boog, ed., *The Conduct of the Air War in the Second World War: An International Comparison* (New York: Berg, 1992).

247. See the description in Middlebrook, *Battle of Hamburg*, pp 334–337.

248. Webster and Frankland, *Strategic Air Offensive against Germany*, vol 4, app 49, table 38 (based on USSBS data).

249. BBSU, "Strategic Air War against Germany," p 151.

250. *Ibid.*, table 49.

251. *Ibid.*

252. USSBS, *Oil Division Final Report* (Washington, D.C.: GPO, 1947), p 121; see also Ronald C. Cooke and Roy Conyers Nesbit, *Target: Hitler's Oil* (London: William Kimber, 1985) pp 202–205.

253. USSBS, *Oil Division Final Report*, pp 125–128.

254. *Ibid.*, table 46.

255. *Ibid.*, p 129.

256. For Professor Mierzejewski's analysis of the economic geography of Germany, the German railways, and the Speer system as they existed before serious Allied air attack in 1944, see his *Collapse of the German War Economy*, pp 1–60.

257. Webster and Frankland, *Strategic Air Offensive against Germany*, vol 4, app 49, table 14 (based on BBSU data).

258. *Ibid.*, table 44 (based on BBSU data).

259. *Ibid.*, table 43 (based on USSBS data). Mierzejewski's analysis of decision-making and operations is contained in his *Collapse of the German War Economy*, chaps 4–7.

260. USSBS, *The Effects of Strategic Bombing on German Transportation* (Washington, D.C.: GPO, 1947), p 4.

261. *Ibid.*; see also Mierzejewski, *Collapse of the German War Economy*, chaps 8–9.

262. Photographs of successful strategic bombing attacks against these very difficult targets with specialist forces are printed in Webster and Frankland, *Strategic Air Offensive against Germany*, vol 3, between pp 238 and 239.

263. BBSU "Strategic Air War against Germany," table 7, Tonnages Claimed Dropped by RAF Bomber Command in Accordance with Main Directifs, and table 8, Tonnages Claimed Dropped by VIIIth USAAF in Accordance with Main Directifs.

264. The difference in the numbers recorded by Bomber Command and Eighth Air Force would lessen if the RAF area attacks that used marshaling yards as aiming points were counted as transportation raids.

265. Webster and Frankland, *Strategic Air Offensive against Germany*, vol 1, p 125. For a particularly withering paragraph, see Webster and Frankland, *Strategic Air Offensive against Germany*, vol 3, p 297.

266. See Terraine, *Right of the Line*, pp 81–92.

267. For the official historians' summary judgment on these points, see Webster and Frankland, *Strategic Air Offensive against Germany*, vol 3, pp 291–295. For a more direct statement of the case, see Noble Frankland, *The Bombing Offensive against Germany* (London: Faber, 1965), pp 105–107.

268. Frankland says "the solution was found largely by chance." *Bombing Offensive against Germany*, p 107. Range-extending drop tanks for fighter aircraft, proposed for development in the United States in 1939, were stoutly resisted in the Army Air Corps. I. B. Holley, "Of Saber Charges, Escort Fighters, and Spacecraft," *Air University Review*, vol 34, no 6 (Sep–Oct 1983), pp 8–10.

269. The interplay between these factors is discussed in Webster and Frankland, *Strategic Air Offensive against Germany*, vol 3, pp 296–302. The official historians do not, as I do, draw the conclusion that fighter defenses, while important, were less significant than other operational factors in blunting the edge of strategic bombing.

270. *Ibid.*, pp 302–306, 309–310.

271. *Ibid.*, pp 78–79. Max Hastings's book on Bomber Command is one of the critical treatments of Harris. Much more sympathetic is Saward, as cited in note 102 above. Charles Messenger has contributed a useful balanced picture of Harris (declining to call it a biography) in *"Bomber" Harris and the Strategic Bombing Offensive, 1939–1945* (New York: St. Martin's, 1984).

272. Max Hastings believes that Harris's failure to cloak himself in what Americans might regard as the stereotypical English "formal dress of circumspection and manners" exposed him to postwar

criticism not suffered by others equally responsible for strategic bombing. Hastings, *Bomber Command*, p 245.

273. *Ibid.*, p 247.
274. *Ibid.*, p 254.
275. The point is Hastings's. *Ibid.*, p 245.
276. For a sympathetic treatment of the CAS, see Richards, *Portal of Hungerford.*
277. AHB, vol 5, p 122.
278. Hastings, *Bomber Command*, p 266.
279. Portal to Bottomley, Jan 28, 1944, AIR 20/5493, PRO.
280. Richards, *Portal of Hungerford*, pp 330–331.
281. *Ibid.*, p 330. See also Saward, *Bomber Harris*, pp 275–277 [this is the authorized biography].
282. Webster and Frankland, *Strategic Air Offensive against Germany*, vol 3, p 297.
283. Terraine, *Right of the Line*, pp 462–468. On the quality of aircrews, see pp 520–537.
284. Webster and Frankland, *Strategic Air Offensive against Germany*, vol 3, p 156.
285. Hastings, *Bomber Command*, pp 218–219.

Bibliographic Essay

Primary Sources

The bulk of the sources for the study of the British strategic bombing campaign are now deposited in the Public Record Office at Kew. Among the many relevant classes of records are the following:

- AIR 2: Registered Papers.
- AIR 8: Chief of the Air Staff.
- AIR 9: Directorate of Plans.
- AIR 14: Bomber Command.
- AIR 16: Fighter Command.
- AIR 20: Unregistered Papers.
- AIR 24: Operations Record Books (Commands).
- AIR 37: Second Tactical Air Force, Allied Expeditionary Air Force, Deputy Supreme Allied Commander and SHAEF (Air).
- AIR 40: Intelligence.

Some files of interest remain classified, for example, AIR 40/2470 which contains prisoner of war interrogation reports concerning German signals intelligence.

Official Histories and Other Government-sponsored Sources

The starting point for any discussion of the historical literature on the subject of British strategic bombing must be the four-volume official history by Sir Charles Webster and Noble Frankland published in 1961 as *The Strategic Air Offensive against Germany* (London: HMSO). Whatever its shortcomings (and they are few rather than many), it is, in this writer's view, the finest of all official histories of the Second World War, British or American. Sir Charles Webster brought to the project a critical independent spirit and a considerable reputation as a diplomatic historian. Frankland, a fine historian, had been a navigator in the command during the war. The balance between the values and perspective of academic scholarship on the one hand and the practical experience of operations on the other produced a first-class account of the British strategic air war. Among the many virtues of the work is one thread that runs throughout: insistence that the results achieved in the bombing campaign were determined by operational considerations.

Any air force officer will profit greatly from reading the opening chapter in volume 1, "The Nature of a Strategic Air Offensive." Volume 4 consists of the best published collection of documents on the strategic air war ever produced to date, including most of the directives sent to Bomber Command during the war, selections from some of the most important correspondence, and a large number of useful tables. It also contains short sections on technical subjects such as radio aids to navigation and bomb aiming.

The Air Historical Branch of the Air Ministry (now the Air Historical Branch of the Ministry of Defence) prepared a series of studies on the strategic air war. Most valuable are the specialized monographs on such subjects as training, maintenance, ordnance, signals, and reconnaissance. Most of these works exist only in typescript and may be consulted at the Air Historical Branch or in the Public Record Office. The Air Historical Branch also produced a six-volume narrative on the evolution of policy and operations. Comprising the "The RAF in the Bombing Offensive against Germany" are the following six volumes, listed in order:

"Pre-War Evolution of Bomber Command, 1917 to 1939."
"Restricted Bombing, September 1939 to May 1941."
"Area Bombing and the Makeshift Force, June 1941 to February 1942."
"A Period of Expansion and Experiment, March 1942 to January 1943."
"The Full Offensive, February 1943 to February 1944."
"The Final Phase, March 1944 to May 1945."

Through the courtesy of the Air Historical Branch, these narratives are available from the U.S. Air Force Historical Research Agency on microfilm reels 23354 and 23355.

Among other important government sources are the reports of the USSBS and the British Bombing Survey Unit (BBSU). The main reports of the USSBS were published shortly after the end of the war and are available in a reprint edition. An evaluation of the USSBS appears in David MacIsaac, *Strategic Bombing in World War Two: The Story of the United States Strategic Bombing Survey* (New York: Garland Press, 1976). For purposes of this study, the five most important USSBS reports are

The Oil Division Final Report,
The Effects of Strategic Bombing on German Transportation,
Area Studies Division Report,
The Effects of Strategic Bombing on the German War Economy, and
The Effects of Strategic Bombing on German Morale.

The reports of the BBSU were never published, and, in fact, remained classified until the early 1960s. The overall report, "The Strategic Air War against Germany, 1939–1945," written chiefly by Professor (later Lord) Solly Zuckerman, is a valuable source largely because Zuckerman, though a civilian, brought a sound understanding of air operations and of strategic bombing to the task of analysis. These reports are also available on microfilm reel A5346 from the U.S. Air Force Historical Research Agency at Maxwell Air Force Base, Alabama:

"The Effects of Strategic Bombing on the Production of U-Boats."
"Report on the Effects of Strategic Air Attacks on German Towns."
"Potential and Actual Output of German Armaments in Relation to the Combined Bomber Offensive."
"The Effects of Air Attack on Inland Communications."
"The Strategic Air War against Germany, 1939–1945."

Sir Arthur Harris's official report, "Despatch on War Operations, February 23, 1942, to May 8, 1945," completed in October 1945, is available in the Public Record Office at AIR 14/1252. Another copy is in the Air Historical Branch. The most valuable parts of his despatch are the appendices which provide a great deal of information on a wide range of technical matters.

The British official history of secret intelligence by F. H. Hinsley *et al.*, is a work of major importance. Five volumes have appeared to date under the title of *British Intelligence in the Second World War* (London: 1978, 1981, 1984, 1988, and 1990). Volume 2 and volume 3, parts 1 and 2, contain much information on the strategic air war.

Nongovernment Sources

Six general histories are recommended. Though Anthony Verrier's *The Bomber Offensive* (London: Batsford, 1968) appeared before the main body of wartime documents was released to the public, his thoughtful and often provocative interpretations continue to deserve careful attention. Equally valuable, though much briefer, is Noble Frankland, *The Bomber Offensive against Germany: Outlines and Perspectives* (London: Faber, 1965). A stimulating, interpretive survey is Lee Kennett's *A History of Strategic Bombing* (New York: Scribners, 1982). Max Hastings's *Bomber Command* (London: Michael Joseph, 1978) is a lively critical work that makes good use of interviews. His portraits of squadrons at various stages of the war provide a valuable perspective not found in other sources. John Terraine's *The Right of the Line: The Royal Air Force in the European War, 1939–1945* (London: Hodder and Stoughton, 1985) sets the strategic offensive in the context of the whole RAF effort. The best general work is Richard J. Overy, *The Air War, 1939–1945* (New York: Stein and Day, 1980). This book is distinguished by its breadth of outlook, its comparative analysis, and the sharpness of its judgment. Recent interpretations are represented in John Ellis, *Brute Force: Allied Strategy and Tactics in the Second World War* (New York: Viking, 1990); Malcolm Smith, "The Allied Air Offensive," *Journal of Strategic Studies* 13 (Mar 90), pp 67–83; and Alan J. Levine, *The Strategic Bombing of Germany* (Westport, Conn.: Praeger, 1992).

Considerable literature exists on special subjects. Malcolm Smith's *British Air Strategy Between the Wars* (Oxford University Press, 1984) is a valuable work based on extensive archival research; it provides a detailed picture of doctrinal development before the outbreak of the war. Smith's book is complemented by Neville Jones, *The Beginnings of Strategic Air Power: A History of the British Bomber Force, 1923–1939* (London: Frank Cass, 1987). Earlier studies of prewar air policy include H. Montgomery Hyde, *British Air Policy Between the Wars* (London: Heinemann, 1975); Barry Powers, *Strategy Without Slide Rule: British Air Strategy, 1914–1939* (London: Croom-Helm, 1976); and Uri Bialer, *The Shadow of the Bomber* (London: Royal Historical Society, 1980).

On prewar intelligence, one should consult, in addition to the official history, Wesley Wark, *The Ultimate Enemy: British Intelligence and Nazi Germany, 1933–1939* (Ithaca: Cornell University Press, 1985). Gordon Musgrove's *Pathfinder Force* (London: MacDonald and Jane's, 1976) contains much useful information about the development of target finding, marking, and bomb aiming in the last three years of the war. An excellent introduction to the subject of electronic warfare, complete with a wealth of illustrations and technical detail, is Martin Streetly's *Confound and Destroy* (London: Jane's, 1978; reprint edition, 1985). Also valuable in this area is *Instruments of Darkness* (London: Kimber, 1967) by Alfred Price. Monte D. Wright's *Most Probable Position: A History of Aerial Navigation to 1941* (Lawrence: Kansas University Press, 1972) is a good introduction to an important aspect of the problem of night bombing. A handy reference to the statistical details of individual Bomber Command operations is Martin Middlebrook and Chris Everitt, *The Bomber Command War Diaries* (London: Viking, 1985).

A number of useful memoirs have been published. Sir Arthur Harris completed *Bomber Offensive* (London: Collins, 1947) shortly after the end of the war. As usual with Harris, it contains a clear and forceful expression of his views. Sir Arthur Tedder's *With Prejudice* (London: Cassell, 1965) is important for the period when Bomber Command operated under Supreme Headquarters Allied Expeditionary Force control. Further insights on that period can be gained from Lord Solly Zuckerman, *From Apes to Warlords* (London: Hamish Hamilton, 1978). D.C.T. Bennett's *Pathfinder* (London: Sphere paperback edition, 1972) provides the perspective of the controversial leader of the Pathfinder Force. *Most Secret War: British Scientific Intelligence, 1939–1945* (London, Hamish Hamilton, 1978), by R. V. Jones, is a valuable memoir by the Director of Scientific Intelligence at the Air Ministry during the war. Another good intelligence memoir, though less well-known, is W. E. Jones, *Bomber Intelligence* (Leicester: Midland Counties, 1983). Jones was a squadron intelligence officer. A number of good aircrew memoirs exist, of which two are this writer's favorite: *The Pilot Walked Home* (London: Collins, 1946) by Denis Hornsey, and *No Moon Tonight* (Angus and Robertson, 1956; reprinted in paperback, London: Goodall Publications, 1984) by Don Charlwood. The latter is a navigator's memoir. Insight into intruder and escort operations can be gained from J.R.D. Braham, *Night Fighter* (New York: Bantam, 1984, originally published 1962). Charles Carrington, historian, biographer of Kipling, and infantry veteran of the First World War, has left a picture of life at Bomber Command headquarters from a unique perspective in his *Soldier at Bomber Command* (London: Leo Cooper, 1987).

Three biographies are worthy of note. Denis Richards in *Portal of Hungerford* (London: Heinemann, 1978) and Dudley Saward in *Bomber Harris* (New York: Doubleday, 1985) provide friendly treatments of their subjects. Charles Messenger's *Bomber Harris and the Strategic Bombing Offensive* (London: St. Martin's, 1984) is also useful.

Three important sources are available in English on German air defenses. The Air Historical Branch's *Rise and Fall of the German Air Force* (originally published in 1948; reprinted, New York: St. Martin's, 1983) still has considerable value. Williamson Murray's *Strategy for Defeat: The Luftwaffe, 1933–1945* (Maxwell AFB: Air University Press, 1983) is a valuable work, recommended for its extensive use of German sources. Richard J. Overy's *Goering: The Iron Man* (London: Routledge and Kegan Paul, 1984) provides many valuable insights, not only into the personality of its principal subject, but also the development of German air defense policy.

Among German sources, one should consult Gebhard Aders, *History of the German Night Fighter Force 1917–1945* (London, 1979), originally published in Stuttgart in 1978 as *Geschichte der deutschen Nachtjagd 1917–1945*; Wilhelm Deist, *The Wehrmacht and Nazi Rearmament* (London: Macmillan, 1981); Wilhelm Deist, *"Die Aufrüstung der Wehrmacht,"* in *Das Deutsche Reich und der Zweite Weltkrieg*, vol 1 (Stuttgart: *Deutsche Verlags-Anstalt, 1979*); Klaus Maier, *"Totaler Krieg und operativer Luftkrieg,"* in *Das Deutsche Reich und der Zweite Weltkrieg*, vol 2 (1979); and Horst Boog, *"Der angloamerikanische strategische Luftkrieg über Europa und die deutsche Luftverteidigung,"* in *Das Deutsche Reich und der Zweite Weltkrieg*, vol 6 (1990).

Works on the German war economy include two of Alan Milward's books: *War, Economy, and Society, 1939–1945* (Berkeley: University of California Press, 1977) and *The German Economy at War* (London: Athlone Press, 1965). Also important are Burton H. Klein, *Germany's Economic Preparations for War* (Cambridge: Harvard University Press, 1959), and Berenice Carroll, *Design for Total War, Arms and Economics in the Third Reich* (The Hague: Mouton, 1968). Different interpretations of the first two and one-half years of the war are found in Richard J. Overy's works: "Hitler's War and the German Economy: A Reinterpretation," *Economic History Review*, 2d ser 35 (1982); *Goering: The Iron Man* (London: Routledge and Kegan Paul, 1984); "Mobilization for Total War in Germany, 1939–1941," *English Historical Review*, vol 103 (Jul 1988); and *"Blitzkriegswirtschaft? Finanzpolitik, Lebensstandard und Arbeitseinsatz in Deutschland, 1939–1943," Vierteljahrschefte für Zeitgeschichte*, vol 36 (1988). One should also consult Hans Erich Volkmann, *"Die NS-Wirtschaft in Vorbereitung des Krieges,"* in *Das Deutsche Reich und der Zweite Weltkrieg*, vol 1 (Stuttgart, 1979); Bernard Kroener, *"Die Personnen Ressourcen des Dritten Reiches im Spannungsfeld zwischen Wehrmacht, Bürokratie, und Kriegswirtschaft, 1939–1942"*; and Rolf-Dieter Müller, *"Die Mobilisierung der Deutschen*

Wirtschaft für Hitlers Kriegführung," in *Das Deutsche Reich und der Zweite Weltkrieg*, vol 5, pt 1.

The most recent assessment of the impact of strategic bombing on the German economy is Alfred Carl Mierzejewski, *The Collapse of the German War Economy, 1944–1945: Allied Air Power and the German National Railway* (Chapel Hill: University of North Carolina Press, 1985). In addition, volume 5, part 2 of *Das Deutsche Reich und der Zweite Weltkrieg* will provide recent German analysis and interpretation when it appears in print.

Two early works on the effects of bombing also should be consulted: Fred Charles Ikle, *The Social Impact of Bomb Destruction* (Norman: University of Oklahoma Press, 1958), and Irving L. Janis, *Air War and Emotional Stress: Psychological Studies of Bombing and Civilian Defense* (Westport, Conn.: Greenwood, 1951). See also Earl Beck, *Under the Bombs: The German Home Front* (Lexington: University Press of Kentucky, 1986), and Olaf Groehler, "The Strategic Air War and Its Impact on the German Civilian Population," in Horst Boog, ed., *The Conduct of the Air War in the Second World War: An International Comparison* (New York: Berg, 1992). A recent treatment of the ethical dimension of direct attacks on the civil population is provided in Stephen A. Garrett, *Ethics and Airpower in World War II: The British Bombing of German Cities* (New York: St. Martin's, 1993).

For British aircraft production and the British war economy, see M. M. Postan, *British War Production* (London: HMSO, 1952). On the problems involved in designing, producing, and modifying aircraft, see M. M. Postan, D. Hay, and J. D. Scott, *Design and Development of Weapons* (London: HMSO, 1964). On specific operations, four books by Martin Middlebrook should be consulted, in large measure because of the interviews with aircrews by the author. They are *The Battle of Hamburg* (New York: Scribner's, 1981); *The Nuremberg Raid* (New York: William Morrow, 1974); *The Peenemünde Raid* (London: Allen Lane, 1982); and *The Berlin Raids* (London: Viking, 1988). Other useful accounts of operations include John Sweetman's *Operation CHASTISE* (London: Jane's, 1983); R. Barker, *The Thousand Plan* (London: Chatto and Windus, 1965); and Alexander McKee, *Dresden 1945: The Devil's Tinderbox* (London: Souvenir Press, 1982).

3

The American Strategic Air Offensive Against Germany in World War II

Stephen L. McFarland and Wesley Phillips Newton

On December 7, 1941, the U.S. Army Air Forces (AAF)[1] possessed a strategic doctrine for fighting the air war against the Axis powers, but too few weapons to employ it fully. A few years earlier, future leaders of the wartime AAF at the Air Corps Tactical School, Maxwell Field, Alabama, postulated that strategic bombing by self-defending bombers would become the decisive force in war, capable of defeating an enemy nation, independent of naval and ground forces. Some air leaders believed the shattering of an enemy's industrial infrastructure might obviate prolonged surface campaigns. Could self-defending bombers fight their way to enemy industries? Could the Norden bombsight aim bombs with sufficient accuracy to destroy crucial industries? Could the destruction of these industries shorten or end the war? Despite the convictions of Air Corps Tactical School personnel, only combat could answer these questions.[2]

The American doctrine of high-altitude (above 20,000 feet) daylight precision strategic bombing[3] evolved from the earlier work of Giulio Douhet, Italian prophet of air power and author of *Command of the Air*, and Hugh Trenchard, father of Britain's Royal Air Force (RAF). William "Billy" Mitchell, pioneer American advocate of air power, who also believed bombing was an independent weapon that could win wars, expanded and popularized the concept.[4] The Air Corps Tactical School doctrine was unique, however, in making industry, not urban population centers, the targets of strategic bombing. Though such targets would be more difficult to destroy and would exact more casualties from the AAF, American air leaders spoke as one in opposing attacks on civilians. Lt. Gen. Ira C. Eaker, who commanded American air forces attacking Germany during the war, summed up this feeling: "We must never allow the record of this war to convict us of throwing the strategic bomber at the man in the street."[5]

Background of the Strategic Bombardment Campaign

Early in 1941, American and British leaders developed a cooperative strategy called ABC–1 for use against the Axis. They identified Germany, the most powerful Axis nation, as the main focus for Allied action and the target of a strategic air offensive intended to pave the way for a land offensive. In July 1941, President Franklin D. Roosevelt asked the British and his own armed services for their weapons production requirements for a Victory Program. Contained in an annex to the U.S. Army's contribution to this program was the Air War Plans Division Plan 1 (AWPD–1). As a product of the new planning agency of the Air Staff, AWPD–1 was both an estimate of the Army's aerial requirements for victory and the AAF blueprint for daylight precision strategic bombing. With ABC–1's priorities in mind, AWPD–1 identified and ranked German targets and target systems.

Completed in eleven days in August, AWPD–1 reflected prewar Air Corps Tactical School thinking and "inferred the possibility of German defeat through air war by itself." Based on experience gained in the Battle of Britain, it showed some awareness of the importance of fighter escort for the strategic bombers, but it concluded that the technology required for long-range fighters was unachievable. Haywood Hansell, a major figure in the creation of the plan, later called this omission one of its "greatest faults."[6]

As the Axis powers swept from one victory to another in early 1942, U.S. Navy leaders grew increasingly antipathetic to the army's efforts at planning for a major offensive in Europe and sought a reallocation of men and materiel to the Pacific Theater. If any threat appeared with overriding urgency, they argued, it was the Japanese advance. The Commanding General of the Army Air Forces, Henry H. "Hap" Arnold,[7] fought to prevent such a diversion, insisting that the AAF's planned bombardment campaign against Germany not be weakened through such a major diversion. He had the support of British Prime Minister Winston S. Churchill and President Franklin D. Roosevelt, whose commitment to defeating Germany first remained unwavering.[8]

Indeed, in late December 1941, Roosevelt and Churchill affirmed ABC–1's Germany-first priority at the Arcadia Conference in Washington, D.C. The AWPD–1 concept for the buildup of an AAF heavy bomber force in the United Kingdom, one that could deliver air blows against the Reich, received the conference's endorsement. Preparations for this aerial buildup and appointment of its key commanders occurred in January 1942. In a few months this effort became part of the BOLERO plan fashioned earlier in Washington, D.C., by Brig. Gen. Dwight D. Eisenhower of the War Plans Division, which proposed a more rapid, direct assault against the German enemy, in 1942 if possible. BOLERO, however, violated the step-by-step strategy envisioned by ABC–1, which allowed time for the heavy bomber force in England to grow and mount a strategic bombardment offensive *before* mounting an invasion.[9]

In any event, the shortage of aircrews, airplanes, and British bases; U.S. Navy competition for priority and resources in the Pacific; British reluctance to commit to a cross-channel invasion before 1944; and events in North Africa, the Soviet Union, and the Atlantic hindered the establishment of an American heavy bomber force in England from the outset. The ever-impatient Arnold, who pressed constantly for the buildup, put two of his closest associates in charge of the buildup and future operations: Maj. Gen. Carl A. Spaatz and Brig. Gen. Ira C. Eaker.[10] Eaker went to England in February 1942 as Commanding General, VIII Bomber Command to pave the way, while Spaatz, as commander designate of AAF in Great Britain, remained stateside to organize the Eighth Air Force.

The four-engine Boeing B–17 Flying Fortress and Consolidated B–24 Liberator heavy bombers were to be the instruments of this strategic bombardment doctrine. Their prototypes first flew in 1935 and 1939, respectively. During the war, the AAF continuously modified these bombers to improve their performance, and though the *Luftwaffe* might challenge them forcefully during daylight precision bombing missions, they continued to appear in ever-increasing numbers. American industry ultimately produced 30,865 B–17s and B–24s during the World War II and augmented this heavy bomber force with 39,604 P–38, P–47, and P–51 fighters. In England in the fall of 1942, however, few bombers were available to the Eighth Air Force when it began limited daylight bombing operations.[11]

Early Operations

The original Eighth Air Force heavy bombardment group in England, the B–17-equipped 97th, launched the first, albeit minuscule, heavy bomber raid against the continent on August 17, 1942. One of the twelve bombers on this mission carried VIII Bomber Command's leader, General Eaker. "Each plane's load of bombs," he reported, "made a long mushroom-like pall of smoke, sand and dirt, which slowly rose into the air" in the limited operation against a marshaling yard at Rouen-Sotteville, France, within range of RAF Spitfire escort. It was a beginning, but no real test of American daylight strategic bombing doctrine.[12]

In succeeding months, not only did the slow buildup in England limit the

Arnold sent Brig. Gen. Ira Eaker to England in February 1942 as Commanding General, VIII Bomber Command (*left*). The Consolidated B–24 Liberator heavy bombers (*top*) and four-engine Boeing B–17 Flying Fortress (*bottom*) were to be the instruments of this strategic bombardment doctrine.

effects of the American bombing effort, but the Allied decision for Operation TORCH (the invasion of North Africa in November 1942) led to a sizable diversion of Eighth Air Force's men and machines. Most of the Eighth's small collection of fighters departed for North Africa with bombardment groups that had acquired some battle experience, such as the 97th. But other B–17 groups soon arrived in England — the 91st, 92d, 303d, 305th, and 306th — with the 44th and 93d B–24 units. By Christmas 1942, all had flown in combat, and problems such as frontal vulnerability because of the physical limitations of hand-operated nose guns became apparent. The harsh realities of daytime aerial combat also quickly became

American industry produced 30,865 B–17 and B–24 bombers in addition to 39,604 P–38, P–47, and P–51 fighters in its gigantic lines of production (*left*).

On this mission to Stuttgart, Germany (*right*), the 306th Bombardment Group B–17s encountered heavy, intense flak and strong enemy fighter opposition during the seven-hour flight.

obvious. The war diary of the 306th for October 9, 1942, noted: "intense heavy flak was encountered over the target [the Lille industrial area], FW 190s attacked in force. Tail gunner Sgt. Arthur E. Chapman . . . had one hand shot off and was critically wounded in the chest." The strain of initial combat, of attempting to improve tactical formations under fire, and of missing faces at muster took their toll on morale. The coming twelve months would severely test the stamina of those involved in the air war over Europe.[13]

Army Air Forces heavy bomber missions flown from English bases from August 1942 to June 1943 were small in terms of bomb tonnage (30,658 tons) and in terms of significance of the targets attacked. During the roughly sixty operational missions scattered throughout this 9½-month period, indecisive missions against submarine installations, shipyards, and ports consumed a large share of the effort. Demands of the Combined Chiefs of Staff (the American and the British chiefs of staff) and the war in the Atlantic against the submarine threat delayed strikes against the target American air leaders judged to be the first priority — the *Luftwaffe*. Although nighttime RAF area assaults had been punishing German cities for some time, the first Eighth Air Force strike inside

Germany did not occur until January 27, 1943.

The reasons for the disappointing initial performance of American strategic bombing forces were not hard to decipher: poor weather, diversion of men and materiel to North Africa and to the Pacific, and the slow production of airplanes and crews stateside. From mid-October 1942 until the spring of 1943, six heavy bomb groups (four composed of B–17s and two of B–24s) constituted the entire strategic bombardment force of Eighth Air Force. Still, the British, the AAF's air staff, and the VIII Bomber Command's General Eaker urged an expanded daylight strategic bombing campaign against Germany.[14]

The AAF's assessment of its strategic bombardment in these months was illusory. Bombing accuracy appeared surprisingly good, and that perception helped allay and even reverse, to some extent, earlier British criticism of American daylight bombardment operations. Eaker, who succeeded Spaatz as commander of the Eighth Air Force in December 1942, claimed in February 1943 that the bomber had shown the ability to reach the target and do the job.[15] In fact, he had not yet tested the doctrine of daylight precision bombardment with a large bomber force against enemy defenses because of the slow buildup of American forces and because the Germans were just beginning to shift priorities to defending the day skies over Germany.

Nevertheless, these initial activities reinforced the correctness of the strategic bombardment doctrine in the minds of AAF leaders and planners. Although Arnold had developed misgivings about the self-defending aspect, commanders such as Spaatz and Eaker still believed in late 1942 that the self-defending daylight bomber could carry the attack into Germany's heartland and win the war without long-range escort fighters — and without a land–sea invasion of the continent. Such initial, heady optimism might have been consistent with prewar doctrine, but it was hardly justified at the beginning of 1943. American bombers had not yet flown over any part of Germany, and German priorities still ranked defense against the American daylight bombing campaign fourth after requirements of the Eastern Front, the RAF night bombing campaign, and the Mediterranean Theater.

Short of bombers and aircrews in the winter of 1942–1943, Eighth Air Force also faced fluctuating objectives for accomplishing its campaign of strategic bombardment. On July 21, 1942, Maj. Gen. Dwight D. Eisenhower directed Eighth Air Force to begin "the maximum degree of operations with a view to obtaining and maintaining domination of the air over Western France by April 1, 1943," in preparation for an invasion. Eighth Air Force commander Spaatz, before leaving to head AAF operations in North Africa, therefore identified aircraft factories, marshaling yards, and submarine installations, in that order, as the bombing targets to win Eisenhower's "domination of the air." On October 20, 1942, however, Eisenhower reordered these priorities and instructed Spaatz to concentrate attacks on submarine bases, aircraft factories, and transportation, in that order. The British air staff, the Enemy Objectives

Unit of the Economic Warfare Division of the American Embassy in London, and the Joint Strategic Survey Committee added their opinions to what was becoming a nightmare of conflicting and ill-defined objectives for strategic bombardment.[16]

As if these were not objectives enough, Sir Arthur Harris, commander of the RAF's Bomber Command, wanted the Eighth to join the RAF in a massive night campaign to destroy Germany's cities, claiming he could knock Germany out of the war by 1943. Leaders of the American and British navies meanwhile demanded an all-out bombing effort against the reinforced submarine pens housing the U-boats that were threatening Allied convoys in the Atlantic Ocean. AWPD–1 had identified 154 German targets in five target systems for attack by the Eighth Air Force including, in priority order, the *Luftwaffe* (embracing air bases, aircraft production, aluminum, and magnesium), electrical power, transportation centers, oil, and morale. A successor plan, AWPD–42, complicated matters with 177 targets within six target systems. It added submarine construction yards and synthetic rubber production while it eliminated morale as a target.[17]

Following a press conference on the combined actions of the American and British bomber forces, Air Vice Marshal Richard Peck, Portal's principal public information deputy, and Maj. Gen. Carl Spaatz await the return of bombers from a mission.

In the face of British demands for increased military efforts in the Mediterranean and the U.S. Navy's insistence on a higher priority for offensive military operations in the Pacific Theater (supported by the U.S. Army's General Douglas MacArthur and his airman, General George C. Kenney), Arnold and General George C. Marshall, U.S. Army Chief of Staff, faced the daunting task of establishing a strategy for winning the war in Europe. How were they to prepare a detailed plan for defeating Germany, one that for Arnold would guarantee a strategic bombardment campaign and for Marshall would win the Combined Chiefs of Staff's support for a direct cross-channel offensive against Germany?

The diverse points of view were presented to Allied leaders at the Casablanca Conference in French Morocco in January 1943. Arnold, Spaatz, and Eaker defended before the Combined Chiefs of Staff the AAF's strategy for defeating Germany with daytime strategic bombardment. The AAF's position was to employ American air power to destroy only Germany's essential industries rather than attempt to damage portions of all its industries. The RAF's Harris decried this strategy as a panacea. There were, he argued, no critical industries whose destruction could knock Germany out of the war or greatly affect German military capabilities. Harris remained convinced that Allied air forces should launch nighttime general area attacks designed to achieve such widespread urban destruction that all industrial activity would cease.[18]

Harris's strategy had the advantage of the RAF lessons learned in more than two years of nighttime area bombing in Germany. The AAF, on the other hand, had yet to send any bombers over Germany proper and advocated an untried and untested strategy of daytime precision bombing.[19] According to Churchill, Eaker's brilliant briefing paper, "The Case for Day Bombing," and the U.S. Navy's pressure for a greater effort in the Pacific convinced him to support the American air strategy and maintain the European emphasis of Allied war plans, bombing Germany by day and night.[20]

The Combined Chiefs of Staff instructed the two Allied air forces to follow their respective strategic doctrines and attack Germany around the clock under the strategic direction of British Chief of Air Staff Sir Charles Portal. "Your primary object will be the progressive destruction and dislocation of the German military, industrial and economic system, and the undermining of the morale of the German people to a point where their capacity for armed resistance is fatally weakened." Whatever Harris's contentions, the Combined Chiefs clearly concluded that Germany held certain critical industries that must be attacked by day: submarine, aircraft, transportation, oil, and nebulous "other targets," in that order of priority. The Combined Chiefs nonetheless allowed Harris to interpret the directive as an order for a general area offensive aimed at "the undermining of the morale of the German people," while Eighth Air Force conducted a selective strategic offensive against target systems critical to the German war economy.[21] At Casablanca, the Americans agreed to delay the cross-channel invasion, thus giving Eighth Air Force time to carry out its daylight bombing offensive. This offensive would force the *Luftwaffe* into a battle of attrition for air superiority, a battle viewed as essential to achieving victory in the war.[22]

Among American military leaders, translating the principles of the Casablanca Directive into operational instructions began even before the conference. Anticipating approval of daylight strategic bombardment, in December 1942 Arnold established a board of business, government, and military officials to study the German economy and make a detailed analysis of Germany's vulnerability to aerial attack. On March 8, 1943, this Committee of Operations Analysts submitted a report identifying six target systems, the

The strategy for winning the war was hammered out with the British at the Casablanca Conference in January 1943. American leaders present were (*left* to *right* seated) General George C. Marshall, President Roosevelt, and Admiral E. J. King and (standing *left* to *right*) Harry Hopkins, Lt. Gen. H. H. Arnold, Lt. Gen. Brehon Somervell, and Averell Harriman.

destruction and continued neutralization of which would gravely impair and might paralyze the Western Axis war effort. In order of priority the systems were submarines, the aircraft industry, ball bearing production, oil, synthetic rubber, and military transport. Eighth Air Force and RAF staff personnel then determined which specific targets in these systems Eighth Air Force would attack and with how much force. The resulting list of seventy-six specific targets and the operational plan to destroy them formed the American portion of what was now called the Combined Bomber Offensive.

The Combined Chiefs of Staff approved the Combined Bomber Offensive plan for strategic bombardment in late April 1943 at the Trident Conference in Washington, D.C. They stopped the diversion of Eighth Air Force units to North Africa and the Mediterranean, ordered Eighth Air Force to begin its role in the Combined Bomber Offensive on June 10, 1943, under the code name POINTBLANK, and called on the RAF's Bomber Command to complement the Eighth Air Force effort by attacking "surrounding industrial areas at night." At the Quadrant Conference at Quebec, Canada, in August 1943, the Combined Chiefs assigned POINTBLANK the highest strategic priority as a prerequisite to an invasion of the European continent.[23] With POINTBLANK, Arnold, Spaatz, and Eaker had seemingly won the opportunity to demonstrate the power of strategic bombardment.

The final POINTBLANK plan contained two major points. First, whatever the beliefs of American air commanders, the strategic bombing campaign was not expected to defeat Germany independently; rather, it was to weaken Germany sufficiently to permit initiation of final, combined operations on the Continent (the inauguration of which would be determined by the success of POINTBLANK). Second, the winning of air superiority by the defeat of German fighter strength became an intermediate objective, second to none. The primary industries selected for attack remained submarine, aircraft, ball bearing, and oil. Secondary industries were synthetic rubber and military transport. Because the intermediate objective could only be achieved and these industrial targets destroyed effectively and quickly by precise daytime bombing, completion of POINTBLANK would become the primary responsibility of Eighth Air Force, although the intent of the Combined Chiefs of Staff was that it be a *combined* Anglo-American bomber offensive.

Three serious miscalculations, however, plagued the ambitious POINTBLANK plan. First, without adequate intelligence, American planners based their analysis of German industry on evaluations predicated on equivalent American industry. Not until after the war would they learn that the German and American economies differed dramatically. Second, many aircrews in unescorted bombers would be shot down before AAF leaders abandoned their belief in the self-defending bomber. And third, ULTRA, the top secret Allied signals intelligence project that deciphered high-level German military communications, would provide little information on the impact of Allied

German signal troops at the front transmit vital messages to their commands using the code machine Enigma to protect the security of their transmissions. The allies acquired this device to develop ULTRA.

bombing of economic targets.[24] Aerial reconnaissance, moreover, could view only imperfectly the working interior of Germany's industry. The Allies would assess the achievement of POINTBLANK bombing objectives with imprecise intelligence, and these assessments would lead to mistakes in the conduct of the campaign.

The Crisis in Strategic Bombardment

To perform its assignment, Eighth Air Force in mid-1943 operated with less strength than was required to prosecute the POINTBLANK objectives. Eaker's communications with Washington underscored the need for more airplanes and crews. To Arnold, his typical closing remark was "Please rush replacement aircraft and combat crews as fast as you can." Trying to fight a global war, Arnold and the Air Staff did what they could. By May 1943, insufficient maritime shipping, organizational problems, and aircraft diversions to other theaters restricted Eighth Air Force to twelve heavy bombardment groups, including some not yet combat-ready. The Eighth could field about 250 heavy bombers per mission, on average, while Harris's Bomber Command now flew nighttime raids composed of 500 to 1,000 bombers against German cities.

Three more heavy bomb groups would became operational in June, two in July, two in August, two in September, one in November, and four in December. Counting the nine groups operational in the Mediterranean, the AAF had thirty-five heavy bombardment groups available for the POINTBLANK offensive by January 1, 1944. To execute POINTBLANK, however, the Combined Chiefs of Staff at Trident projected that the American strategic air force would require fifty-one groups by that date.[25]

With the resources at hand, Eighth Air Force continued its largely experimental daytime bombardment operations while the Combined Chiefs of Staff struggled to refine strategy and objectives. These missions served to provide additional information for Combined Bomber Offensive and POINTBLANK plans, including changes in tactics to counter evolving, increasingly effective German aerial and antiaircraft artillery defenses. For example, Brig. Gen. Laurence S. Kuter, commanding general of the 1st Bombardment Wing, and Col. Curtis E. LeMay, commanding the 305th Bombardment Group, arranged squadrons and groups into aerial combat boxes of eighteen to twenty-one heavy bombers each, with two or three combat boxes forming wings and so stacked in flight as to create overlapping fields of defensive fire. This tactical formation became the standard in Eighth Air Force.[26]

The expanding AAF offensive against the *Luftwaffe* in mid-1943 came at

Col. Curtis E. LeMay, commanding the 305th Bombardment Group, greets Brig. Gen. Haywood S. Hansell upon his return from a mission over enemy territory.

a critical point in the war. Germany had increased the priority on defense against American daylight attacks, and the approach had begun to bear fruit. American bomber losses rose as more *Luftwaffe* fighter units appeared on the Western Front (at the expense of units on the Eastern and Mediterranean Fronts). German industry mobilized for increased aircraft production, planning to reach 2,000 single-engine fighters per month by the end of the year. The new *Luftwaffe* Chief of Air Staff shifted priorities from offense to defense, and ULTRA revealed that Hitler had ordered all groups on the Eastern Front to send two of their best pilots to flesh out the defenses of the Reich.

The improved aerial defenses and first efforts at dispersing German industry, now also under way, combined to make high-altitude daylight bombing of the German aircraft industry even more difficult. Dispersal changed German production facilities from a few well-known, large complexes, mass-producing complete aircraft, to hundreds of smaller, unidentified sites producing individual aircraft components assembled elsewhere, in additional dispersed sites.[27] Except for attacks against ball bearing production, the Combined Bomber Offensive target systems would have to await the defeat of the *Luftwaffe* before the power of daytime strategic bombardment could be brought to bear and seriously affect Germany's industrial ability to wage war.

Though targeted for defeat, the *Luftwaffe* and its supporting aeronautical industries and research laboratories had become a complex, interwoven system. If *Luftwaffe* fighters in the sky over Europe presented the AAF with a vexing, tactical problem, the *Luftwaffe* infrastructure on the ground presented a profound strategic problem consisting of supporting airfields, aircraft depots, training centers, final assembly plants, testing facilities, component manufacturing industries (including engines, ball bearings, propellers, air frames, instruments, and similar components), basic raw-material industries (fuel, rubber, and aluminum), and, at the strategic extreme, the urban areas that housed these industries and their workers. Where in this integrated industrial sector of the German war economy were the choke points that if bombed would defeat the *Luftwaffe* and win for the Allies air superiority? Poor weather, insufficient forces, too little time, the dearth of long-range fighter escorts for the bombers, and the growing power of the *Luftwaffe* further complicated the question.

Until mid-August 1943, the Eighth Air Force aimed its offensive against the *Luftwaffe* at the operational end of the target system. It sent 41 percent of heavy bomber sorties against air targets, many in France, that included airfields, depots, and final assembly plants. Eighth Air Force also targeted related industries, including the Huels synthetic rubber plant, the bombing of which caused a two-month shortage of this material. Allied failure to follow up this successful attack, an all too common fault, would prove one of the significant missed opportunities of the war.[28]

The Blitz Week attacks of July 24–30, 1943, were the largest yet by Eighth

Air Force; more than 1,000 individual sorties were sent against some fifteen targets inside Germany. Eighth Air Force, however, was still not concentrating on the critical targets listed in the POINTBLANK plan for the Combined Bomber Offensive (the week's attacks were spread among submarine yards and bases, airfields, and rubber, aircraft, and various other industries). Without sufficient fighter escort, the B–17s and B–24s could not fly into Germany without heavy losses, even when the longer range P–47s began carrying drop-tanks on July 28, 1943. Eighth Air Force required two weeks to repair and reequip after the staggering loss of nearly 100 of the 1,000 bombers involved in the Blitz Week attacks. Worse, the bombardment caused no serious interruption in German production and unquestionably accelerated the dispersal of German industry. Strategic bombing achievements were limited that summer, at the cost of nearly 5 of every 100 bombers that Eighth Air Force launched.[29]

Throughout the summer of 1943, Arnold demanded that Eighth Air Force demonstrate the power of strategic bombardment by striking Germany often and hard. In the same messages, he often wanted the Eighth simultaneously to conserve its bombers to ensure an air force large enough to achieve the POINTBLANK objectives. Arnold complained frequently that the Eighth's organization was too cumbersome, that Eaker was too lenient with combat and service commanders, and that Eaker did not provide him with adequate information to allow him to brief politicians and the public. Arnold especially criticized Eaker's tendency to send only 300 heavy bombers on missions after the AAF chief had sent him more than 1,000 B–17s alone. His repeated messages, in the chief's own words, were intended to "build a fire under General Eaker." He told Eaker: "I am willing to do everything possible to build up your forces but you must play your part . . . a definite change seems to be in order but you have to be tough to handle the situation." Eaker's responses to Arnold were nearly always the same — all problems stemmed from too few airplanes and crews; please send more.[30]

From the outset, Arnold was not entirely fair. For example, he seemed to forget that TORCH had siphoned off much of Eighth Air Force's maintenance personnel. Nevertheless, Arnold earnestly tried to meet most of Eaker's requests, in 1943 ordering to England nearly every new bomber unit then in training. When Eaker complained that the B–17 was vulnerable to attack from the front, Arnold and his chief of air staff, Maj. Gen. Barney Giles, went to the airplane manufacturers and told them no more bombers were to be manufactured without nose turrets, even if it meant losing aircraft to the resulting delays. Giles said it was "a hell of a lot better to lose one hundred airplanes on the production line than to lose one hundred airplanes from the Eighth Air Force with ten hand-picked officers and men in each airplane." He wrote to Eaker, "So much for that — the turrets are coming."[31]

By August 1943 the *Luftwaffe* and Eighth Air Force had established a

General officers planning a bombing for an Eighth Air Force target are (*left* to *right*, top row) Ira C. Eaker, Frank O'D. Hunter, and Robert C. Candee and (*left* to *right*, bottom row), Carl A. Spaatz, Dwight D. Eisenhower, and Walter H. Frank.

pattern of confrontation. The Germans marshaled an early-warning radar net work, aerial spotters, and concentrations of antiaircraft artillery around cities and industrial targets, all organized in seven aerial defensive zones and each under a single controller. Overhead, the *Luftwaffe* used air-to-air bombing (time-fuzed bombs dropped on American bomber formations), frontal assaults on bombers, rocket projectiles, fighter-borne cannon firing from beyond the range of the bombers' defensive fire, and coordinated mass attacks — all in attempts to defeat the Eighth Air Force. Eighth Air Force relied on concentrated bomber formations flying at high altitude (many of the bombers were now mounting nose turrets) that attacked specific strategic targets to execute POINTBLANK. And these targets had to be in Germany if serious damage was to be done to the *Luftwaffe*, regardless that heavy bomber losses were inherent in such missions.

In July and August 1943 the AAF made its largest effort yet at strategic bombing. Looking for what Bomber Command's Harris liked to call a panacea — a critical choke point in the German war economy — Eighth Air Force targeted the ball bearing plants at Schweinfurt and the Messerschmitt aircraft factory complex at Regensburg. One month earlier, heavy bombers based in the Mediterranean Theater had struck the oil refineries at Ploesti and the aircraft factories at Wiener-Neustadt. Planners hoped that the destruction of these targets would immediately curtail *Luftwaffe* expansion plans, cause wholesale industrial shutdowns for want of antifriction bearings, and eliminate the only major source of natural petroleum for the German war machine. All of the targets, located well inside the Third Reich, would test, for the first time, the doctrine of high-altitude daylight strategic precision bombing and would serve as the penultimate test for the unescorted, self-defending bomber.

The aircrews showed great bravery and determination in these operations, arguably the best-known strategic bombing missions in United States aviation history, but the results were decidedly mixed. Ploesti temporarily lost 40 percent of its production capacity, although surplus capacity and quick repairs prevented any serious disruption of supply. The attacks on Schweinfurt, Regensburg, and Wiener-Neustadt forced a reorganization of the ball bearing and aircraft industries (including dispersal and a scramble to buy foreign bearings). Both sides won Pyrrhic victories. Factory buildings were leveled, but they could and would be rebuilt in a matter of weeks unless they were bombed repeatedly. The cost to the attacking bomber forces was appalling. The losses, 116 bombers of the 557 that attacked their targets, made vital follow-up attacks impossible. Victory over the German Air Force would not be quick. Increasingly, it would turn on the war of attrition in the skies over occupied Europe.[32]

Some American airmen surmised that "the standard bomber, such as the B–17, cannot possibly cope with such a situation [unescorted, deep penetration]," and the only solution was to acquire long-range escort fighters that could accompany bombers on daytime missions, or else follow the British example and convert to nighttime area attacks. In Washington, AAF Chief of Staff Barney Giles pushed for more fuel capacity in the new P–38s and P–51s while, in England, Eaker pushed for drop-tanks for his P–47s; however, both men

In the first mass attack at low level by heavy bombers, the B–24 Liberators participated in a mission against Ploesti from a base in North Africa.

NORWAY
SWEDEN
GREAT BRITAIN
DENMARK
Baltic Sea
North Sea
IRELAND
Hamburg
Alconbury
NETHERLANDS
Essen
Berlin
POLAND
London
Cologne
BELGIUM
GERMANY
Leipzig
Dresden
Frankfurt
Paris
SLOVAKIA
Stuttgart
Vienna
Budapest
Munich
600 miles from Foggia
FRANCE
HUNGARY
AUSTRIA
SWITZERLAND
ROMANIA
600 miles from Alconbury
Milan
YUGOSLAVIA
ITALY
SPAIN
CORSICA
Rome
Foggia
Naples
MAJORCA
SARDINIA
SICILY
Mediterranean Sea
Eighth and Fifteenth Air Force
Heavy Bomber Radii of Action
In the Combined Bomber Offensive
January 1943–May 1945
0 75 150
miles

The best known raids of the war were those against the Ploesti oil refineries in Romania, in which aircrews showed unflinching bravery, flying through dense smoke, heavy flak, and at low level. As a result, Ploesti lost 40 percent of its production capacity, as storage tanks lie desolate at the Astro Romano oil refinery.

realized POINTBLANK could not wait for these developments. The VIII Bomber Command commander, Brig. Gen. Frederick L. Anderson, remained upbeat, if doctrinaire. "The high-level, precision bomber . . .," he told the AirStaff, "is the weapon which is breaking Germany," and "the VIII Bomber Command is destroying and will continue to destroy the economic resources of Germany to such an extent that I personally believe no invasion of the Continent or Germany proper will ever have to take place." Giles wrote back tartly: "our losses on those... raids were a bit hard to take. We can prevent heavy losses of our bombers by the use of escort fighters. You should use every possible opportunity to push the fighter commander into extending his effort."[33]

During September 1943, bad weather limited Eighth Air Force to only two missions against German targets. During the second week in October the skies over Germany cleared and allowed the Eighth to launch another intensive effort to accomplish Combined Bomber Offensive objectives. Losses again were heavy on these missions into Germany, which became generally known as Black Week. At the end of that week, on October 14, 1943, 229 B–17s returned to Schweinfurt to attack the ball bearing plants, and the *Luftwaffe* was waiting. Eighth Air Force pressed the attack, and the resultant damage temporarily reduced Germany's ball bearing production capacity by 50 percent. It was a near-perfect example of the doctrine of daylight precision strategic bombing, for American air power had choked a critical point in the German war economy, at least until workers could repair the damage. On the other hand, Eighth Air Force counted 60 of the 229 bombers missing in action, an awesome 26 percent loss rate, plus dozens of bombers heavily damaged, some beyond repair. Bomber aircrews that returned claimed to have destroyed in the air an absurd number of enemy fighters: 702 (twice as many fighters as the *Luftwaffe* had available in Germany at the time).

Again, bomber losses precluded follow-up missions. During Black Week, Eighth Air Force lost 148 bombers, nearly 13 percent of all its attacking forces, and nearly 1,500 experienced crewmen. Whatever the tenets of strategic bombardment, the *Luftwaffe*, using the superiority of its fighters over American self-defending bombers, had fought the American air effort over Germany to a standstill. Without fighter escorts, Eighth Air Force could not gain air superiority over Europe, and POINTBLANK's objectives would remain beyond reach.[34]

General Eaker initially viewed the second Schweinfurt aerial battle "pretty much as the last final struggles of a monster in his death throes . . . we now have our teeth in the Hun Air Force's neck." Arnold at first supported his commander in England, writing Eaker that "the cornered wolf fights hardest." Such bravado hid deeper sensitivities to the American bomber losses over Schweinfurt. Eaker objected to Arnold's subsequent press conference in Washington in which the chief raised certain doubts about the way the mission had been run. Eaker defended his operational procedures, claiming "we have not been lazy, cowardly or ineffective."[35]

The reality of inflated combat claims and actual combat losses in the second attack on Schweinfurt soon set in. Arnold expressed increasing doubts about the outcome, telling Eaker he hoped "the losses are worth the results."

Extremely heavy opposition faced the bombers over Schweinfurt, suffering some of the greatest losses of the war. *Top photo*: B-17s on the run to Schweinfurt; *center*: on return assaults, the Fortresses attacked remaining buildings to prevent rebuilding of the ball bearing industry; *bottom*: ruins of Schweinfurt's Kugelfischer works, Germany's largest individual bearing producer, are examined by American observers after the war.

The Messerschmitt aircraft factory at Regensburg was a perpetual target to eliminate the source of latest types of German fighters. At the top is shown a heavy concentration of bombs being released from B–24 Liberators of the Fifteenth Air Force on the Obertraubling Airdrome at Regensburg, a base for jet-propelled aircraft. Below are seen B–17s of the Eighth Air Force attacking the aircraft factory that produced the Bf 109G.

After the war Arnold wrote: "Could we keep it up? To this day, I don't know." For the Air Staff, the strategic bombing missions of mid-October 1943 sounded the death knell for unescorted, deep penetration raids into Germany during daylight hours. Eaker later admitted to Arnold that "there are no definite indications yet that the *Luftwaffe* is on the verge of collapse." Neither Eaker nor Arnold knew that the *Luftwaffe* in the west was now stronger than ever, or that German fighter production continued to grow. Although Portal in early December reported that POINTBLANK was three months behind schedule, he convinced the Combined Chiefs of Staff that POINTBLANK and American day bombing should continue because they were the only means at hand for defeating the *Luftwaffe* in time for the invasion of France, now scheduled for 1944. Portal, too, could not know that the actual and planned German fighter production figures for the first six months of 1944 would exceed Allied intelligence estimates by 250 and 325, respectively.[36]

On the other side of the English Channel, the Germans too were facing serious problems. To defend Germany itself from daylight raids, the *Luftwaffe* had to forego defending large areas of occupied Europe. ULTRA revealed to Allied commanders that the *Luftwaffe* had instructed factories to deliver fighters directly to operational units, bypassing reserve pools and depots, and had ordered damaged or inoperative aircraft cannibalized to overcome parts shortages. Moreover, valuable test pilots had been detailed to fly in defense of aircraft factories. The *Luftwaffe* had based its successful aerial tactics on heavily armed and armored fighters firing cannon and rockets beyond the range of American bomber defensive fire. Now, *Reichsmarschall* Hermann Goering ordered the development of fighters that would make all-out attacks on Allied bombers at close range "without regard to losses." Losses at the time were bad enough; in September and October 1943 the *Luftwaffe* recorded 560 of its fighters destroyed.[37] Swift American long-range escort fighters in the future would make of the heavily armored, slow-moving German fighters, as Eaker saw it, "duck soup."[38]

The Winning of Air Superiority

While weather seriously limited bombardment operations during the last two months of 1943, the AAF was preparing for the time when the skies cleared. Although he had never been an avid proponent of escort fighters, Eaker now urged British authorities to greater efforts in the production of range-extending paper drop-tanks for the P–47s and P–38s, claiming that a one-month delay in the Ministry of Aircraft Production manufacture of drop-tanks had cost Eighth Air Force twenty bombers over Schweinfurt. Arnold asked Portal to transfer to the Eighth Air Force the P–51s scheduled for the RAF in late 1943, charging RAF Fighter Command with "not making use of their full capabilities" to defeat the *Luftwaffe*. He also pressed the British to assume complete responsibility for fighter aircraft aid to the Soviet Union by sending the Soviets less useful, short-range Spitfires, thus freeing American production to equip Eighth Air Force. Portal irritated Arnold by not agreeing to these requests. On October 30, 1943, with North Africa no longer drawing away equipment from Europe, Arnold ordered all shipments of long-range P–38 and P–51 fighters from the United States to Eighth Air Force for the last two months of the year.[39]

The bad weather, if improvidential in Arnold's view, held a hidden benefit. It sharply curtailed bombardment efforts in the POINTBLANK campaign and made time available to reequip the Eighth Air Force. And it delayed the inevitable confrontation with the *Luftwaffe* before these and other American measures could bear fruit. In early 1944, the *Luftwaffe* would use the same leaders, tactics, and weapons that allowed it to inflict so much damage on Eighth Air Force in the summer and fall of 1943. Eighth Air Force, on the other hand, would be an entirely different protagonist in 1944.

The poor wintertime flying weather over Northern Europe caused Arnold at the Quebec Conference in August 1943 to question Portal. Should the Allies keep all of the strategic bombers at bases in England? Portal suggested that supplementary bombardment operations might soon be launched from more sunny Italy, which the Allies were scheduled soon to invade. Portal made his point: well over half of Germany's fighter production took place in adjacent

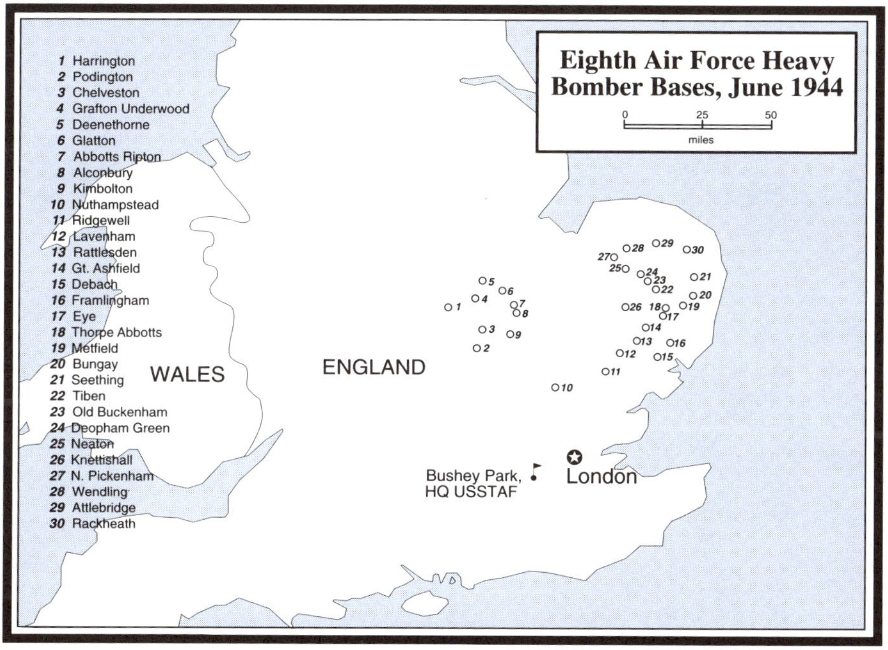

Austria — beyond the flying radius of heavy bombers based in England. In October 1943 Arnold offered the Combined Chiefs a blueprint for revamping the AAF's setup in the Mediterranean. Twelfth Air Force would be retained, but as a tactical air force. He proposed creating a new strategic air force, the Fifteenth. Both the Twelfth and the Fifteenth would be based in Italy. The Combined Chiefs of Staff and Eisenhower concurred. Arnold's subsequent diversion to Italy of fifteen heavy bomber groups, originally intended for England, did not please Eaker.[40]

Arnold also boosted his idea for an overall Anglo-American air commander and championed a revised bombardment strategy that gave "overriding first priority . . . to the early defeat of the *Luftwaffe* with emphasis upon short term effect on the sources of German air strength." Army Chief of Staff General Marshall told him that the time was not right for an overall air commander. For defeat of the *Luftwaffe*, Allied planners developed an operations plan in November 1943, code named ARGUMENT, which called for an all-out Allied bombing assault on German single- and twin-engine fighter, synthetic rubber, and ball bearing production, to begin whenever forecasts promised three or more continuous days of clear weather. So pressing was the need to reduce enemy fighter output that the assault was to be made without long-range fighter escorts if such were still unavailable. This decision came despite Arnold's and apparently most members of the Air Staff's new-found conviction that long-range escort fighters were vital. If they had shed the self-defending bomber concept, they still were willing, in apparent desperation, to dispatch more "self-

Generals Arnold and Marshall.

defending bombers on deep penetration daylight missions." Lacking sufficient numbers of heavy bombers to meet ARGUMENT's goals, however, planners deleted synthetic rubber from the air plan and reduced the emphasis on ball bearing production. By so doing they admittedly lowered their expectations for strategic bombing, from eliminating "Germany's capacity to wage war" to crippling "her war potential." Whatever the expectations, at the beginning of 1944, ARGUMENT remained on hold because of the constant overcast above Hitler's *Festung Europa*.[41]

For aircrews, coping with cloud cover was constantly vexing. Not only did one face problems of limited visibility, moisture, cold, and icing when aloft, but it was quite impossible to see anything from 20,000 feet through a carpet of absolutely thick cloud cover. In answer to Arnold's carping about Eighth Air Force's lack of performance, Eaker rejoined that not since mid-October 1943 had his crews accomplished any visual bombing, and in the entire Eighth Air Force he had only ten aircraft equipped with serviceable H2X sets. Even so, these H2X bombing radars, the AAF's version of the RAF's H2S terrain-identifying radar used to find and bomb *cities*, lacked the resolution needed for precision bombing of strategic targets through clouds.[42]

Arnold's irritation with Eaker and the Eighth Air Force's performance grew throughout the summer and fall of 1943, and Arnold continued to criticize Eaker's service organization and subordinate commanders. He complained that Eaker was not launching enough sorties against critical, and therefore more dangerous, targets. He also objected to the low priority Eaker still assigned to

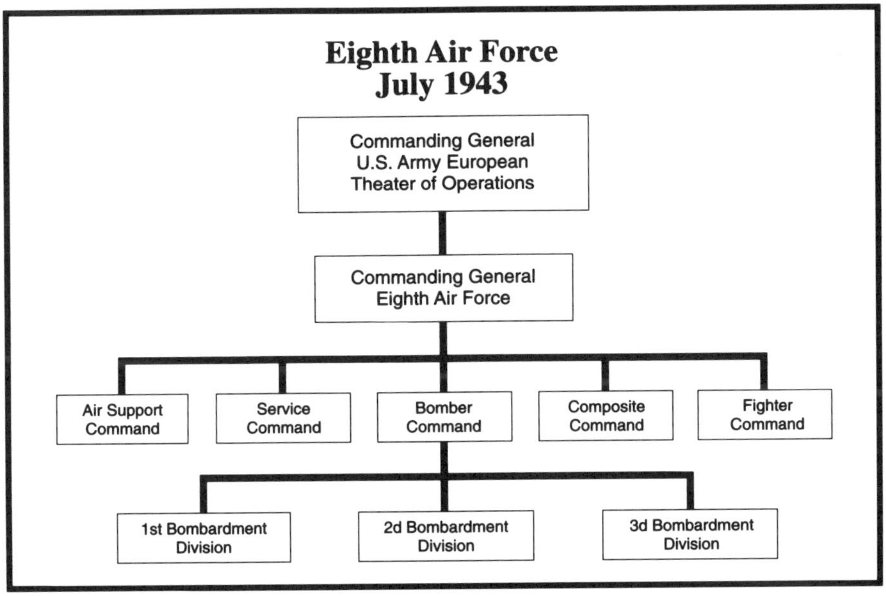

escort fighters, and he chastised him for assigning a higher priority to armor and self-sealing fuel tanks for transport aircraft. Although Arnold had taken pains in October to send Eaker all available long-range P–51 escort fighters, Eaker, to simplify maintenance, turned them over to the tactical air force in England, the Ninth. Pressured by his own fighter commander, Eaker later reassigned them to VIII Fighter Command as long-range escorts for his bombers.[43]

The Arnold–Eaker collision was classic: Arnold set policy and provided the men and equipment and expected results, regardless of cost. Eaker too wanted results, but he was unwilling to see his air force decimated in the process, especially not on missions when precision bombing could not be accomplished because of winter cloud cover. Arnold's impatience with Eaker and the need to control the operations of two strategic and two tactical air forces deployed in Europe (the Eighth and Fifteenth, and the Ninth and Twelfth, respectively) led him to change the European AAF organizational structure. The Combined Chiefs of Staff approved Arnold's reorganization in early December 1943, and Arnold notified Eaker of the change at midmonth.[44]

The change produced one of the major and most persistent controversies concerning the AAF in World War II. Arnold reassigned Lt. Gen. Carl A. Spaatz from the Mediterranean Theater to command the newly created United States Strategic Air Forces (USSTAF) headquartered in England, with operational control over the Eighth and Fifteenth and administrative control over the (tactical) Ninth. (Operational control of the Ninth Air Force was exercised by Supreme Headquarters Allied Expeditionary Force/Allied Expeditionary Air Force.) The original commander of Fifteenth Air Force, Lt. Gen. James "Jimmy" Doolittle, moved from Italy to England and assumed command of the Eighth Air Force. Eaker, now a lieutenant general, went to Italy to command the new Mediterranean Allied Air Forces, a headquarters with administrative control over the Fifteenth Air Force but direct control of Twelfth Air Force and other Allied units operating in the area.

Determined to shake up things to produce a winning team, Arnold had convinced Spaatz, Eisenhower, and Marshall that the change in command was in the best interest of the cause. Spaatz apparently made the suggestion that Doolittle be Eaker's replacement to direct Eighth Air Force and that Frederick Anderson, former VIII Bomber Command commander, be Spaatz's deputy for operations.

Although Arnold wanted a senior American commander in the Mediterranean to counterbalance the British, Eaker was most unhappy about being reassigned to a scene of lesser action, especially since the Eighth Air Force was reequipped and ready to return again to the skies over Europe in numbers sufficient to challenge the *Luftwaffe* for air superiority. He protested to all his superiors and threatened to make his displeasure public. In the end, however, he saluted and accepted the new assignment. As air power historian W. A. Jacobs phrased it, Eaker "suffered the common fate of those commanders sent

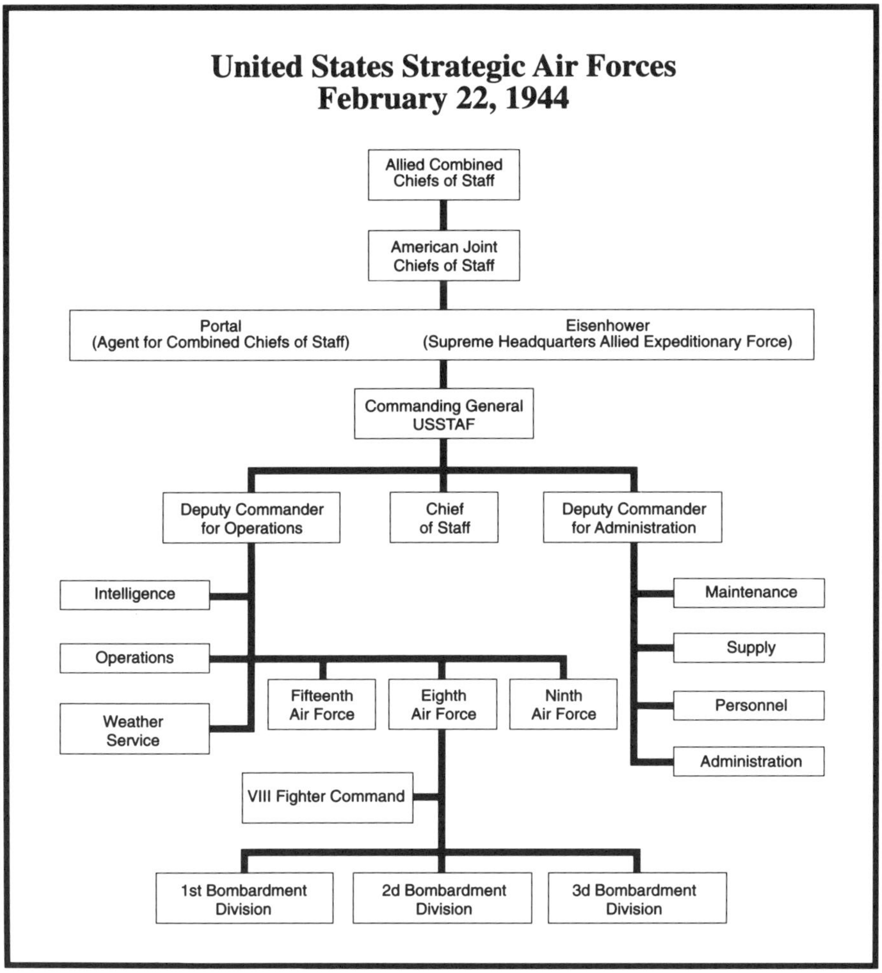

into battle too soon with too little."[45]

Besides increased numbers of heavy bombers, the new Eighth Air Force leadership would have one advantage Eaker never had: great numbers of efficient, long-range escort fighters. The delay in acquiring fighters that combined speed, maneuverability, and range stemmed at least as much from AAF doctrine as it did from the slow evolution of technology. The AAF, to be sure, entered the war with fighter planes barely able to compete with the enemy's best. But the self-defending bomber tenet of strategic bombardment doctrine hardly motivated American air leaders to develop a long-range escort fighter for the bombers' protection. Nearly all military aviators of the day believed that any fighter with a combat radius sufficient to accompany heavy bombers for a thousand miles would be too slow and lack the maneuverability needed to compete on even terms with short-range, highly maneuverable, fast fighter interceptors.[46]

The first American fighter to assume this role was Lockheed's twin engine P–38, first test-flown in 1939. Powered by liquid-cooled in-line Allison engines and with distinctive twin booms, gondola, and tricycle landing gear, the P–38 Lightning evolved from a powerful high-altitude interceptor into a long-range bomber escort. When the crisis of the self-defending bomber emerged in the summer of 1943, the P–38's unrefueled ferry range (point to point) of 900 miles drew immediate attention. With two large 300-gallon drop-tanks, its range could be extended to about 1,600 miles. The second American fighter to meet this need was the Republic P–47 Thunderbolt, a single radial engine aircraft first test-flown in 1941. Though originally lacking effective range, by late 1943 it could carry two large drop-tanks and achieve a combat radius (out and back) of 650 miles.

Another single-engine fighter would complement and eventually replace, for the most part, both the P–38 and the P–47 in the long-range escort role. The P–38 suffered periodic problems with its Allison engines, while the rugged P–47 proved more effective as a dive bomber and close support aircraft. The single in-line Merlin–engined North American P–51 Mustang had the best combination of speed, agility, and range of them all. With an unrefueled ferry range of 1,300 miles — a range extended to 2,000 miles with two 110-gallon drop-

The P–38 Lightning evolved from a powerful high-altitude interceptor to a long-range bomber escort.

Two large drop-tanks on the P–47 extended the essential range for protecting bombers.

tanks — later models could reach and return from nearly every corner of Germany. Built originally for the RAF with an Allison engine but reequipped with a Rolls-Royce Merlin engine, the Mustang was later ordered for the AAF, reportedly at Arnold's insistence. The first AAF P–51 was test flown in May 1942, and increasing numbers of improved Mustangs appeared in England in late 1943, in time for the great battles for air superiority in early 1944.[47]

Under the direction of Maj. Gen. William E. Kepner, named VIII Fighter Command commander in July 1943, and General Giles, Chief of Air Staff, manufacturers added greater internal fuel capacity to increase the range of these fighters. At the operational end, mechanics added discardable paper fuel tanks. Problems with the procurement of these tanks resulted when the AAF failed to request them in adequate numbers or assign them sufficient priority and because Eighth Air Force overestimated the capacity of British industry to produce them. Each American day fighter eventually carried drop-tanks, including the Mustang when it first entered combat with the Ninth Air Force's 354th Fighter Group in December 1943. At that time, however, the appearance of an American long-range fighter escort appeared to be a case of too little and nearly too late to seize the air superiority required for OVERLORD, the planned cross-Channel invasion of France. It would take increasing numbers of aircraft and innovative tactics in the American air war against Germany for POINTBLANK to succeed in time.

Jimmy Doolittle had been on the job only a few weeks in January 1944 when he made an important decision — he "freed" the escort fighters, reversing a previous practice that called for them to remain within one hundred feet of the heavy bombers *until* German fighters attacked the formation. Doolittle's North

African experience provided a precedent. Praising his P–38 escort fighters in that theater, he wrote Arnold that "these ships have done a magnificent job as indicated by our low bomber losses. The P–38s save our bombers by breaking up the enemy fighter formations before they can get set to attack the bombers." Now in England, Doolittle discovered a sign on the wall in his fighter commander's office left over from an earlier commander: "THE FIRST DUTY OF THE EIGHTH AIR FORCE FIGHTERS IS TO BRING THE BOMBERS BACK ALIVE." Doolittle ordered General Kepner of VIII Fighter Command to replace it with one that read: "THE FIRST DUTY OF THE EIGHTH AIR FORCE FIGHTERS IS TO DESTROY GERMAN FIGHTERS." Doolittle informed Kepner, "We'll still provide a reasonable escort for the bombers, but the bulk of the fighters will go out hunting for the Jerries. Flush them out in the air and beat them up on the ground on the way home. Your first priority is to take the offensive."[48]

Sources disagree about when this policy was first implemented, but on the VIII Fighter Command mission of January 24, 1944, fighters provided continuous area coverage rather than close support. Each fighter group designated one of its three squadrons to be a "bouncing squadron," one that could be "ordered off at any time the group leader thought . . . necessary to investigate or attack enemy aircraft operating at a visual distance from the bomber route. In other words, this squadron did not have the responsibility of remaining with the bomber at any cost." Once ordered out, the bouncing squadron had the "freedom of action of being able to roam about and smack down any Huns that may be forming up for future attacks well outside the usual range of our fighters' aggressive movements, which had hitherto been controlled by the requirement to remain with bombers." For January 29th and 30th, official fighter claims soared to record heights: forty-five aerial victory credits on the 29th, fifty on the 30th.[49]

Respecting Doolittle's admonition that fighters should beat up the enemy on the ground on the way home, on February 5, 1944, 20th Fighter Group P–38s combined escort duties with strafing attacks against German aircraft on the ground. After leaving the bombers, 78th Fighter Group Thunderbolts sprayed airfields, locomotives, a tug boat, and a flak tower. On February 8, the Mustangs of the 354th Group took the bombers to Frankfurt and then, in the words of the group historian, were "permitted to wander about when and where they would, and to destroy enemy aircraft in the air and on the ground."[50]

On the eve of launching ARGUMENT, the weather-dependent bombardment raids against Germany's aeronautical infrastructure, Spaatz questioned the validity of a policy that freed escort fighters; perhaps he was nervous about the impending air battle and any actions that could make predicted bomber losses even greater. This policy, Doolittle replied, was made necessary by the *Luftwaffe*'s tactic of ignoring the escorts to concentrate on the bombers.[51] If, in early 1944, AAF fighter pilots continued pursuing and engaging *Luftwaffe* fighters, they would gain experience and, in time, a significant advantage in aerial combat. The new policy, Doolittle claimed, had thus far proved a marked success. Convinced, Spaatz authorized Kepner to cross the chain of command and deal directly with Maj. Gen. Elwood R. Quesada, commander of IX Fighter Command, or even with individual Ninth Air Force wing commanders to borrow any additional long-range fighters he needed for ARGUMENT.[52]

Spaatz needed every advantage. A report by the AAF Board Project flatly stated that the AAF must intensify its efforts even if it meant "attriting U.S. forces at a greater rate than replacements are available." Arnold had underlined the urgency of defeating the *Luftwaffe* in a New Year's message to his field commanders, exhorting them to "destroy the enemy air forces wherever you find them, in the air, on the ground and in the factories." Spaatz now vowed a war of attrition, basing his calculations on the cold logic of a "production-wastage differential." The defeat of the *Luftwaffe* depended on its losses exceeding the number of aircraft produced, through attacks on *Luftwaffe* production sources *and* on its fighters aloft. Doolittle decided to conserve the strength of his Eighth Air Force on noncritical operations "to avoid having the hammerhead force depleted between critical operations." At Arnold's bidding, he also raised the number of combat sorties required of his flyers before they could be considered for rotation back to the United States. Heavy bomber crews would only be eligible for rotation after completing thirty missions, not rotated automatically after twenty-five as they had been before. Fighter pilots would only become *eligible* for rotation after 200 hours of combat flying, and not rotate *automatically* after 200 hours of combat flying as before.[53]

On February 19, 1944, with the prediction of good flying weather, a resolute Fred Anderson, USSTAF Deputy for Operations, pressed Spaatz to unleash ARGUMENT. Meteorologists had made an unusual forecast for the

Maj. Gen. Elwood Quesada, commander of the IX Fighter Command.

Maj. Gen. Frederick L. Anderson, USSTAF Deputy for Operations.

depths of a European winter — about a week of essentially clear skies in late February. By this time Eighth Air Force had accumulated enough heavy bombers to send a thousand at a time into Germany. And Eighth and Ninth Air Forces together possessed enough long-range fighters to escort these bombers on all legs of daylight bombing missions, both into Germany *and* back. Spaatz agreed; he authorized ARGUMENT to begin in the last week of February, even without long-range fighter escorts if necessary, because of the perceived urgent need to reduce the strength of the *Luftwaffe*. The stage was set for what would become known as Big Week.[54]

Because their own meteorologists on the night of February 19, 1944, forecast bad weather for the next day, General Doolittle and Ninth Air Force commander Maj. Gen. Lewis Brereton repeatedly urged Spaatz to cancel this first ARGUMENT mission. Anderson fought all night to save the mission, even sending for his toothbrush so he could stay by the telephone and in contact with Spaatz. After checking the latest weather reports early in the morning, Spaatz notified Doolittle in England and Eaker in Italy: ARGUMENT would begin as scheduled.[55]

On the night of February 19/20, meanwhile, 730 heavy bombers of the RAF Bomber Command began ARGUMENT by hitting Leipzig with the normal area bombing. Losses totaled seventy-eight RAF bombers, a rate of nearly 11 percent, including 569 crewmen for the night's work.[56] While the surviving British bombers returned to their bases, First Lieutenant William Lawley of the American 305th Bombardment Group was preparing for his ARGUMENT operation. On Sunday morning, February 20, 1944,

> we were awakened the usual time for a mission — about 3:00. The weather people had called the shot right; the weather was not a factor that morning. We took off just after daybreak. We were crowded all over England.[57]

The primary targets for Eighth Air Force's Mission 226 were the aircraft industries in and around Leipzig, Gotha, and Brunswick that produced Bf 109s and 110s, Ju 88s and 188s, and FW 190s.[58] If weather over the primary targets prevented visual bombing, the bombers were to make the first AAF strike on Berlin. Although the RAF bombardment the night before had not harmed the aircraft factories, the effort to disrupt the vast RAF bomber stream had served to tire the flak crews for the daylight battle to come. After crossing the English Channel, the American bombers picked up 332 P–47 and 36 P–38 fighter escorts plus 16 RAF Mustang and Spitfire squadrons.

> But we always lost our escort for the same reason the German fighters themselves seldom came in right over the target — they didn't want to be hit by the flak. About the time we hit the IP [initial point], everything went calm except for the flak. Over the target I picked up one flak shell, an 88-mm, that did not explode; it went right through the left horizontal stabilizer and cut the control cable. Suddenly I was very nose heavy. At bombs away, my bombs did not go. I tried to salvo [emergency immediate bomb release], but that didn't work either. The racks had moisture on them and were frozen.

Maj. Gen. Lewis H. Brereton, Commanding General, Ninth Air Force, with Generals Spaatz and Eisenhower, discusses plans during an inspection tour of medium bomber stations in England.

> In the meantime the group had moved on out, because when they released their bombs their planes jumped a hundred feet or so. I managed to get back in formation one time, but I knew I couldn't keep running like that because I'd never get home — I was pulling too much power.
>
> The fighters came at us front and rear. I saw those pretty little things winking at us — the tracers. On the first pass it was like the airplane was hit all over. A shell came through the right windshield and hit my co-pilot in the face and exploded. He put his hands over his face and fell forward. He hit the control stick and knocked it out of my hands and forced the plane into a steep spiral. Almost immediately we went from 28,000 feet to 12,000 feet. The left windshield was covered with blood. My window was clear and by looking out I realized I was descending and the aircraft was in a turn and I was able to correct it. The group reported we had spun in.

But the copilot had played a martyr's role. His death slump had sent the plane spinning downward as though it were in its dying fall. Once the plane went into a spin, Lawley later reported:

> The fighters did not bother me again. By the time the aircraft had leveled out, I had already rung the alarm bell to get out and then got word that two guys in the back were too critically wounded to bail out. The bombardier came up and said, "Can we make this thing fly?" I said, "I don't know. We'll try."

Not only did Lawley make it fly, he brought it in safely with a belly landing at a Canadian base in England.

The Big Week Allied offensive of February 19/20 through 25/26 generated some 6,200 bombing sorties that dropped more than 19,000 tons of bombs on eighteen German airframe and two ball bearing manufacturing centers. Combined Allied losses during Big Week were 370 heavy bombers and 38 fighters; a loss rate of 8 percent for RAF night bombers, 6 percent for American day bombers, and less than 1 percent for Allied fighters. Aircrew casualties exceeded 5,000. Allied bombers and fighters claimed nearly 600 German aircraft destroyed in aerial combat, although the *Luftwaffe* admitted to losing only 230. Loss statistics, however, fail to reveal the significance of Big Week. Aerial combat among AAF fighter escorts and *Luftwaffe* day fighters was critical to the developing battle of attrition. Whereas a flood of new fighter aircraft allowed VIII Fighter Command to end Big Week with 90 percent more P–51s than the number with which it began, the *Luftwaffe* lost over one-third of its authorized strength, including irreplaceable veteran pilots and air commanders.[59]

The Big Week bombardment was the most concentrated, intense effort of World War II by either side to destroy one sector of a nation's economy and war production. Nevertheless, the AAF had not achieved air superiority. Quantities of *Luftwaffe* fighters still rose to oppose Allied bombers. Their performance might vary, but they could still menace the OVERLORD landing. The flaw in ARGUMENT now became clear: the campaign concentrated on bombing the aeronautical infrastructure, with attrition of the airborne *Luftwaffe* viewed only as a side effect, yet it was aerial combat that directly and adversely reduced German fighter strength. American fighters had to shoot down German fighters and, as Doolittle had admonished, "beat them up on the ground."[60]

Arnold, too, had come to believe that *Luftwaffe* fighters had to be lured into battle and destroyed while aloft. Allied air leaders now concluded that American bombers had to attack a prestige target in daylight, one that would expose the bombers to the maximum number of German fighters, a location the *Luftwaffe* would have to defend. Using this Verdun-type reasoning for initiating a battle to impair an enemy force's will and capacity to fight on, planners selected the German capital, Berlin, which the RAF had been bombing periodically at night since November 1943 in the vain hope of shattering enemy resistance. Fred Anderson maintained that avoiding aerial wastage was the "major German preoccupation" and that USSTAF's "major determination" was to promote that wastage. "We've got to go to Berlin with three bomb divi-

A P–47 (above *left*) and a P–51 (above *right*).

Capt. Robert K. Morgan, pilot of the famed B–17, *Memphis Belle*, the first aircraft in VIII Bomber Command to complete a twenty-five mission tour, accepts well wishes from Lt. Gen. Jacob L. Devers, U.S. European Theater Commander. Following rest and recuperation in the United States, Morgan joined the Twentieth Air Force in the Pacific, where he flew on the first bombing mission to Tokyo.

sions," he declared, "then they'd [*Luftwaffe*] come up" to fight. Missions to Berlin would expose the bombers over Germany for nearly five hours. However, the real target was not Berlin itself: it was the city's airspace. Spaatz had planned on attacking the German capital since December 1943 when he decided the quickest way to force combat between the opposing aerial forces was to send American bombers deep into Germany. Missions to Berlin were withheld, however, until a sufficient number of long-range escort fighters was available to win a battle of attrition.[61]

Again, Fred Anderson championed the new campaign. On March 2, 1944, he called Spaatz to complain about Doolittle's delays in attacking Berlin. Spaatz and Doolittle had decided to cancel the mission scheduled on March 4, but Anderson talked Spaatz into giving the word to go, arguing that "the minutes are ticking away" and the time for completing POINTBLANK was "getting critically short." He convinced Spaatz to cancel passes for Eighth Air Force crews so that Doolittle's command would be ready to take advantage of every opportunity to fly — passes that Spaatz had already approved because bad weather was forecast. Bad weather did indeed interfere with the March 4th mission, and Doolittle tried to cancel the mission scheduled for the ninth on the basis of weather, but Anderson prevailed, and the bombers went to Berlin.[62]

Fewer bombers went to Berlin than had flown in Big Week, but the Americans launched more fighters than ever before. True to American bombing doctrine, these bombers had as their targets Berlin's industrial complexes, but aerial combat was the heart of the battle. Anderson clearly understood what was happening. "It didn't matter if Berlin was overcast," he observed. "The resulting air battle would result in attrition, which makes it more important than any destruction on the ground. Going to [any other target] to find clearer skies won't achieve the same result."[63] Anderson was correct about the *Luftwaffe*'s reaction. The German air force had tapped bomber and transport units for pilots to fly the day fighters, pressed pilot trainees into early service, ordered night fighters to fly day missions, and brought in units from Italy and the Eastern Front to defend the Nazi capital. By March 1944, the *Luftwaffe* would have 72 percent of its operational fighters in central Germany within reach of Berlin.[64]

Two decisions, one American and one German, would go far in determining the outcome of the Battle of Berlin. First, Spaatz departed from past procedure: he ordered his forces to plan their routes to encounter and attract the *Luftwaffe*, not to avoid it. Second, back on December 27, 1943, *Reichsmarschall* Hermann Goering had ordered *Luftwaffe* fighters to avoid American fighters and go only for the bombers. Until then, German fighters had different aerial tactics for use against each. General Adolf Galland, chief of German fighters, believed that Goering's order permitted American fighters to roam the skies at will and attack the Germans under the most advantageous circumstances possible.

Late in March 1944, Goering relented somewhat. He allowed one-third of

Adolf Hitler with his top air commander, Hermann Goering.

each interceptor force to attack Allied fighters in an effort to draw escorts away from the bombers, but by then it was too late. The Germans subsequently tried battle formations of assault *Sturmgruppe* led by cover groups flying straight to the bombers. The idea of battle formations, said Galland, was not "the decisive error; the fatal mistake was not to lay down beforehand which fighters were to go for your fighters and which were to attack your bombers." In his postwar memoir, *The First and the Last*, Galland described the deterioration of his once-confident force: "Our losses rose unrelentingly. Forced onto the defensive, our units forgot how to conduct a dogfight. Now it had come to banking and diving away. Naturally, any cohesion of the unit was lost, and singly our fighters were finished off by the enemy who outnumbered us greatly." Banking and diving away constituted suicide, Galland said, because American fighters could nearly always out-dive the German fighters.[65]

The great daylight battles over Berlin of early March 1944 illustrated Galland's central point. Weather, or the excuse of it, thwarted Spaatz's first efforts to overrule Doolittle's caution and cause the *Luftwaffe* additional attrition. Tactical commanders on March 3 ordered the bombers to turn back when the weather appeared insurmountable, although a flight of P–38s became the first American formation to appear over the German capital. Most bombers hit targets of opportunity on March 4, with weather the reason for the change

General Adolf Galland, chief of German fighters.

in targets, even though a combat wing of B–17s, with its P–51 escort, managed to reach Berlin and drop its bombs. The weather cleared on March 6, and the order came for a maximum effort.

This mission to Berlin, and the one of March 8, produced two of the fiercest air battles of the war as Eighth Air Force armadas filled the sky over the German capital. The Americans and Germans, as usual, made extravagant claims. Nevertheless, on March 6 a record 69 bombers went down. The *Luftwaffe* posted losses of nearly 120 fighters for these two days. The most damaging loss to either side was that of experienced German pilots. The AAF readily replaced lost men and airplanes in 1944, while its forces steadily increased in size and strength. *Luftwaffe* crew losses were virtually irreplaceable, especially when quality was considered, because the training system was increasingly starved for aviation fuel and flightworthy aircraft. When another air armada returned to Berlin on March 9, 1944, it met practically no fighter opposition. Bombs dropped by radar through clouds found the city and added to the flames, smoke, and rubble of previously hit residential areas and occasional industries. This strategic bombing, however, was of minor significance; air superiority in daylight had passed to the Americans.[66]

The Diversion from Strategic Bombardment

In mid-March 1944 the AAF in Europe appeared on the verge of the kind of strategic bombardment capability that Douhet, Trenchard, Mitchell, Arnold, and others had foreseen. The heavy bombers, the crews, the opportunity, and the will were there. So too were the competing needs of the ground forces for interdiction strikes in preparation for OVERLORD and the need for preemptive strikes against German offensive aerial capabilities. Both needs would put great demands on Allied strategic bombers and create diversions from the strategic bombardment campaign. The first distraction was a new German weapon designed to rain explosives on England.

Back in 1943, the Allies became aware of a secret German effort to complete two new weapons: the V–2 ballistic rocket and the V–1 air-breathing, guided missile. The first visual confirmation of the V–2 came on June 23, 1943, and that of the V–1, on November 13, 1943. Fearing that Adolf Hitler might launch chemical or biological warfare against England using these weapon systems, Allied leaders responded with Operation CROSSBOW.[67] In late August 1943 the Combined Chiefs of Staff ordered Eighth Air Force and the RAF to bomb any targets involved in the research, experimentation, manufacture, launching, or transport of these new weapons with an "over-riding priority."[68]

Eighth Air Force was to bear most of the burden of Operation CROSSBOW, in spite of tests in the United States indicating that low-level fighter-bomber attacks would be the most effective method to use against such missile installations. The first major Eighth Air Force attack was launched against the Watten "large" site on August 27, 1943. From August 1943 to March 1945 the Allies dispatched 68,913 sorties (65 percent by the AAF) and dropped 136,789 tons of bombs (6.8 percent of total Allied wartime bomb tonnage) against CROSSBOW targets at a cost of 498 aircraft and 1,950 crewmen (80 percent American) lost. The entire strategic defensive delayed Germany's V-weapons offensive by three or four months — to the night of June 12/13, 1944, after OVERLORD had begun — but the attacks did not materially affect the total number of V-weapons launched.[69]

Bomber commanders generally opposed the CROSSBOW effort, believing that heavy bombers could achieve little in attacks against these targets and that Allied strategic bomber forces should not be diverted from offensive strikes against Germany's war-making capabilities. From late 1943 to May 1944, when Eighth Air Force was locked in a fight to the death with the *Luftwaffe*, CROSSBOW diverted 6,100 sorties from POINTBLANK. In the summer of 1944, CROSSBOW claimed one-fourth of all bombing missions launched. Hitler told confidants that V-weapons constituted a better defensive than offensive weapon because for every Allied bomb dropped on a CROSSBOW target in France, one less bomb was dropped on Germany.[70] The new German weapons nevertheless represented a serious threat to England, especially to the Allied air, sea, and

Dilution of Bombing Effort in World War II: Army Air Forces Bombing by Target Category

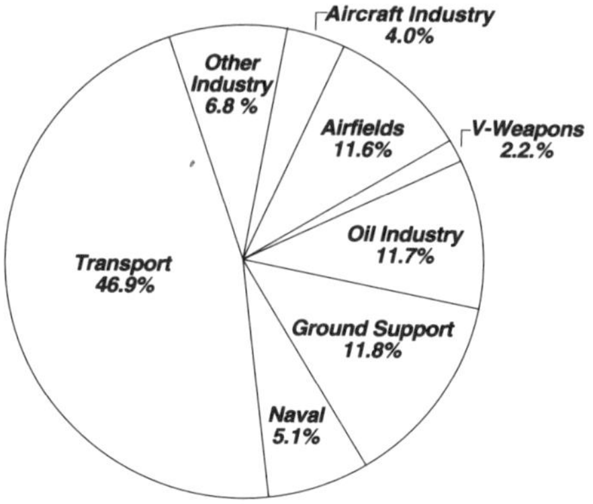

land forces concentrated there in preparation for the invasion of France. Wars are fought in the present, not the future, and the Combined Chiefs of Staff had therefore little choice but to order the CROSSBOW strategic defensive.

The second major diversion of the strategic bombing effort revolved around OVERLORD, the pending cross-channel invasion of France. While the fury of the Big Week and Berlin air battles slackened, and as attrition in the air took its toll on the *Luftwaffe*, an internal battle was building among the Allies. In part it turned on the command structure for the air component of OVERLORD and in part on what would be the most effective use of strategic bombers, other than their POINTBLANK objectives, in support of OVERLORD. British Air Chief Marshal Trafford Leigh-Mallory, commander of the Allied Expeditionary Air Force, the tactical air command for OVERLORD, officially unveiled the transportation plan on March 3, 1944. It called for strategic bombers to assail key transportation focal points (marshaling yards, repair facilities, and certain bridges) in France, Belgium, and western Germany to delay the movement of German reinforcements into the beachhead on D-Day and immediately thereafter. This plan, authored by Professor Solly Zuckerman, a British scientific consultant to Leigh-Mallory's headquarters, also called for continued POINTBLANK and CROSSBOW attacks. It assigned a majority of Allied air power, however, to strikes against the previously defined focal points in the German-controlled European transportation net for the ninety-day period preceding before OVERLORD. A direct interdiction of the Normandy landing area would commence twenty-four days before the invasion.[71]

Carl Spaatz and the Economic Objectives Unit, responsible for USSTAF targeting, presented a plan on March 5, 1944, making oil the key target for strategic bombers. Appealing for support, Spaatz told Arnold that this proposal constituted "the most far-reaching use of strategic air power that has been attempted in this war." Spaatz had a lingering conviction that OVERLORD was risky at best and that the strategic bombers, especially his, could still bring the Germans to their knees, if only the weather and his superiors would cooperate. He accepted OVERLORD, but he thought air superiority had to be maintained to the moment of invasion if the endeavor was to succeed. Spaatz judged oil the weak link in the German wartime economy, and he believed any attack on oil would force the *Luftwaffe* actively to continue the war of attrition begun over Berlin in early March. Leigh-Mallory's insistence that air superiority over the invasion area could not be achieved until the beginning of OVERLORD irked Spaatz. The air chief marshal's announcement that he also expected to take charge of all strategic air forces on March 1, when POINTBLANK would become subordinated to OVERLORD, infuriated Spaatz. Although Spaatz expected and approved, as did Arnold, of Eisenhower's assuming control of the strategic air forces at a propitious time, Spaatz remained dead set against Leigh-Mallory's taking charge of his bombers.[72]

The showdown between the oil plan and the transportation plan came at a meeting on March 25, 1944, with Eisenhower to have the final say (Arnold had informed Spaatz that the determination was strictly up to Eisenhower). Eisenhower's British deputy, Air Chief Marshal Arthur Tedder, favored the transportation plan as a means for concentrating all Allied air elements in a common cause. Portal favored the same plan, but he wanted the railroad aspect confined to France. Harris, who at first declared against the transportation plan, now shifted toward it, but only as a subordinate part of his city-busting concept. Churchill, who had come to distrust Leigh-Mallory and opposed his taking charge of the heavy bombers, made it known that he favored oil. Spaatz's deputy, Fred Anderson, disparaged the transportation plan.

At the March 25 meeting, Spaatz argued that his oil plan would lead to further attrition of the *Luftwaffe* by forcing fighter aircraft to defend these key facilities and by denying them fuel. Expert testimony from the other side contended it would be several months before attacks on oil had any appreciable effect on German fuel supplies, too late for OVERLORD. On March 26 Eisenhower decided in favor of the transportation plan.[73]

Leigh-Mallory's victory was not complete, for Spaatz won the battle over the command of Allied air forces in OVERLORD. Eaker found Spaatz to be "jubilant and overjoyed. The strategic British and American air forces were not to be put under Leigh-Mallory." There would be under Eisenhower "three coordinate air commanders: Leigh-Mallory, tactical; Spaatz and Harris, strategic." Tedder was to head the coordinating staff agency. Eaker observed that Spaatz was not too disappointed about transportation's winning out since "all had firmly agreed that the German Air Force was to be the all-consuming priority."[74]

The transportation plan did not prevent William Kepner's VIII Fighter Command from conducting far-ranging fighter-bomber operations across Germany in April and May 1944. These attacks shifted from the earlier "unorganized air guerilla warfare," as Kepner later characterized it, into a systematic campaign that divided Germany into zones for roaming bands of P–38s, P–47s, and P–51s. This wide-ranging aerial campaign was in essence a strategic aerial offensive, using bullets instead of bombs to destroy enemy airfields, locomotives, barges, radio stations, and the like. Germany found it impossible to cope with these widespread aerial attacks in April and May; they "created chaos at German airfields and at ground control operations, cut enemy lines of communications, and had an indeterminable impact on German morale."[75]

This campaign was, however, tangential to the transportation plan. On April 17, 1944, three days after the strategic air forces came officially under the operational control of Eisenhower, Tedder issued a directive intended to implement Eisenhower's March 26 decision. Weighted heavily toward POINTBLANK objectives, it gave the transportation plan secondary status in USSTAF operations, with no mention of oil. It also appeared to free Harris to perform his cherished city-busting. Tedder seemingly drafted his directive to soothe Churchill's objections to the transportation plan. The Prime Minister's opposition continued until early May when Roosevelt stifled the oil advocates' objections with his view that target selection should be a militarily decision.[76]

During April and May 1944, the air assault on transportation gathered momentum. By day, Eighth Air Force heavy bombers and fighters, Ninth Air Force medium B–26s and light A–20s, and Ninth Air Force fighter-bombers joined Typhoons and other ground support aircraft from the British Second Tactical Air Force of the Allied Expeditionary Air Force to drop tons of explosives on marshaling yards, bridges, and supply and maintenance facilities and fire thousands of bullets and air-to-ground rockets at locomotives, boxcars, and now, even passenger coaches. Historians disagree about the results, but one thing seems clear — adding bridges across the rivers Meuse and Seine as an addendum to the plan insured a satisfactory measure of success.[77]

While concentrating on transportation targets, the AAF nevertheless launched occasional attacks against sensitive German targets to draw out and continue the attrition of the *Luftwaffe*. On May 21, 1944, a strategic fighter mission over Germany not only wreaked devastation on parked aircraft, locomotives, barges, locks, motor transport, oil refineries, and radio and radar installations, with paltry *Luftwaffe* fighter opposition, but it obtained important intelligence. All German aircraft seen on the ground were east of Hamburg, some 500 miles from the projected Normandy landing areas, too far to interfere with invasion forces assembling in English ports. Both the attrition from the recent great air battles and Allied fighter assaults on *Luftwaffe* bases had achieved the most significant accomplishment in Europe of Allied aerial forces since the Battle of Britain, pushing German air power back beyond the effective range of the OVERLORD beaches.[78]

When paratroops came down in the predawn hours of June 6, 1944, and assault forces clung to the beaches in the first perilous hours of D-Day, "there were no great air battles." It was the denouement of the battle for air superiority that USSTAF and the *Luftwaffe* had fought in the first quarter of 1944. While Allied heavy bombers flew thousands of sorties in support of the invasion and the push inland, Eighth Air Force suspended its strategic bombing campaign until June 14. Despite the lull, the *Luftwaffe*'s well-founded fear of the heavy bomber attacks continued to keep the major part of German fighter strength at home. The Allies now possessed air superiority, and the small *Luftwaffe* effort to stop or stall OVERLORD failed miserably. Most of the eighteen *Luftwaffe* fighter groups belatedly sent to France to help contain the landings, an ULTRA intercept revealed, quickly returned to Germany in tatters. General Galland tried to reorganize the German fighter arm to counter Allied ground operations in France, but he was unsuccessful against overwhelming odds. Hitler's insistence in August 1944 on committing Galland's strategic fighter reserve against the oncoming Allies was to no avail, and it wasted resources badly needed for the strategic defense of the Reich.[79]

Operation FRANTIC — the AAF effort to bomb eastern Germany by shuttling between bases in England, Italy, and the Soviet Union — occurred simultaneously with the Allied drive across France in the summer of 1944. FRANTIC was part of a general exchange between the Soviets and Western Allies involving air power, including Lend Lease supplies to the Soviets. Stalin earlier called for AAF units to aid his beleaguered armies. American forces, operating from bases in Western Europe, proposed to shuttle-bomb targets, particularly oil, in Eastern Europe, and then land in Soviet territory. But the effort met with little Soviet support, despite agreements made at the Tehran Conference late in 1943. The Soviets appeared more interested in the tactical uses of air power, as befitted their approach to air war. If the Soviets were obstinate, the Americans were overoptimistic, hoping that shuttle raids with Soviet support would set a precedent for such operations in the East, if the Soviet Union declared war on Japan.

Grudgingly, Moscow approved three bases in the Ukraine for AAF operations; the first shuttle mission from an English base occurred on June 22, 1944. The most successful FRANTIC operation originating in England was directed against previously untouched oil refineries in Poland three weeks later. Alarmed, the *Luftwaffe* retaliated the next day with a surprise attack against American bombers that had landed at the poorly defended Ukrainian airfield near Poltava. In a single stroke, the Germans destroyed forty-three B–17s, one of the greatest losses of Allied aircraft in World War II.

Although several other shuttle missions occurred during the summer and early fall, the Soviets became increasingly cool to these operations, whose results seemed meager in terms of the number of planes involved and the logistical support that they demanded. At best, FRANTIC suggested Allied unity and proved to German leaders that strategic bombers could strike the Third Reich from all sides; at worst, it portended the Cold War. FRANTIC ended in September 1944 amidst growing tension between the United States and the Soviet Union.[80]

The Final Strategic Bombardment Campaign

Operation FRANTIC was secondary to the main AAF activities in the Allied thrust across France. British pressure occasionally forced diversion to CROSSBOW targets, but ground forces had priority claim to the strategic bombers, as they had at St-Lo on July 24–25, 1944. Eisenhower allowed Spaatz and Harris some freedom of action, and Spaatz could now attempt an oil campaign. A restricted version of this campaign in May was one aspect of Spaatz's Verdun-based strategy to lure *Luftwaffe* fighters into a battle of attrition. ULTRA intercepts confirmed the potential of his approach to undermine Germany's ability to continue the war after three attacks in May. Albert Speer, Hitler's minister of armaments and war production, stated afterward that "a new era" in the air war had begun, one that would mean the end of armaments production.[81]

While Eighth Air Force was deeply involved in OVERLORD, Fifteenth Air Force began renewed attacks from the Mediterranean area against German oil facilities on June 12, 1944, with a strike against Ploesti. Fifteenth Air Force struck repeatedly at this target and at oil plants in southern Germany, Austria, and Hungary. During the last week in June 1944, Eighth and Fifteenth Air Forces dropped 20,000 tons of bombs on oil targets in five missions. Immediately, ULTRA revealed a *Luftwaffe* order that units previously sent to France to challenge OVERLORD be returned to Germany to defend the oil plants. Speer wrote Hitler that the raids in May had cut aviation fuel production by half and that continuation of the raids had reduced production to 1,218 tons per day at the end of June, compared with 5,645 tons per day at the beginning of May. *Luftwaffe* consumption during May and June was running at 19,500 tons per month, nearly four times the production rate. Gasoline engines were consuming fuel at twice the available production capacity, and diesel fuel consumption exceeded its production by a factor of 2½. Unless Germany took emergency measures, Speer warned bluntly, existing stocks of fuel would be depleted by August 1944: "*From this time onward there will be an unbridgeable deficit which must lead to tragic consequences* [emphasis in original]."[82]

Speer, with Hitler's approval, responded with an extensive program of passive and active defenses. He ordered millions of cubic meters of concrete to reinforce the plants, moved more antiaircraft guns to protect the plants at the "expense of the protection of German cities," began the construction of underground refineries, and built dummy plants to confuse Allied bombardiers. He reduced fuel allocations for the home front and restricted the *Luftwaffe* only to flights by aircraft involved in actual operations against the Allies. A construction army of 350,000 workers repaired damage, but Allied bombing increased.[83]

In September 1944, Spaatz officially made the German oil industry the number one priority of his strategic offensive. Only weather impeded the effort. The German chemical industry felt the ripple effect as feeder stocks dried up. Only four synthetic plants in all of Germany, prime targets for the offensive, could produce the high-octane gasoline needed for *Luftwaffe* fighters. The ULTRA network reported *Luftwaffe* units unable to fly for want of fuel and *Wehrmacht* ground units, increasingly exposed to Allied aerial attacks, wondering what had become of the *Luftwaffe*. In November, the Allies dropped more than 37,000 tons of bombs on oil targets, mostly by using H2X radar-assisted bombing through cloud cover. However accurate or inaccurate the bombing may have been, by year's end German fuel production met only 55 percent of consumption. The rest had come from dwindling stockpiles. The AAF seemed finally to have found a choke point.[84]

Allied airmen could welcome any shortages in German petroleum stocks in light of a somewhat revived *Luftwaffe* day fighter threat first detected in midsummer of 1944. In July Doolittle expressed concern about the enemy's rocket and turbojet aircraft that were beginning to appear in the sky over Germany. He requested some of America's still experimental turbojet fighters for combat testing in Europe. Arnold at first reacted with scorn to Doolittle's apprehension, but in a few months he too voiced alarm over German Me 262 jets. Only occasionally, however, did the *Luftwaffe* demonstrate that it was still a force worthy of the AAF's attention. On September 27 the Germans inflicted heavy losses on an American heavy bomber force when the 445th Bombardment Group became separated from the main bomber stream. The group historian reported, "Out of no-where, more than a hundred German fighters tore through the formation from behind, blazing away with their cannon." Only four of the group's thirty-seven B–24s returned to base.[85]

Galland resumed the buildup of a fighter reserve for the Big Blows program against American bomber fleets. He hoped to turn the bombers away from oil targets toward fighter production, now widely dispersed, and thereby retain enough fuel to hold back the Soviets until Germany could conclude a separate peace with the other Allies. The potential for great harm continued to worry AAF planners and commanders, but weather, diversions, and attrition prevented any of Galland's Big Blows until it was too late.[86]

September 1944 was in some respects another pivotal month in the strategic aspect of the air war over Europe. It saw the strategic bombers removed from the control of General Eisenhower, with Eighth and Fifteenth Air Forces returning to USSTAF in the regular AAF chain of command and Bomber Command reverting to the RAF normal command chain that ended at the Air Ministry. Arnold, a longtime advocate of Eisenhower's direction of the heavy bombers, at the second high-level conference at Quebec late in September "flopped over." As he later informed a disturbed Spaatz, who strongly favored Eisenhower's continued control, the change gave him "the advantage of having you as my representative determine the targets and objectives for the Strategic Air Force on a co-equal status with Portal... [giving] us a position in the scheme of things that we have never had before." Eisenhower, assured of assistance when the tactical situation called for it, accepted the change, as did Spaatz, the good team player.[87]

In the same communication to Spaatz, Arnold expressed his disappointment that the Allies had bogged down in the West and the "German Air Force has made a strong comeback." For his part, Spaatz continued to press the campaign against oil, but the onset of bad weather and Allied disagreements over priorities complicated the targeting process. Air Chief Marshal Tedder was aware that the Combined Chiefs of Staff were restive with the direction of strategic bombing and contemplated applying it more directly to support the stalled ground campaign in the West. He advanced the idea that attacking transportation best complemented the oil campaign. In October Tedder convinced Spaatz and Eisenhower that a transportation campaign focused on the German rail system would weaken the enemy at the point where his resistance was stiffest, please the Combined Chiefs, and allow Spaatz to continue to hammer at oil. Because the oil targets required precision bombing, American bombers targeted such facilities when visual bombing was possible. When weather interfered (which it did much of the time), they now used radar-assisted

In December 1944, it was concluded that widespread attacks against the rail network would hasten the breakdown of the transportation system. The Ulm rail yards were half demolished by the Eighth Air Force, thus denying supplies to the Nazis fighting the Seventh Army.

bombing to attack railroad marshaling yards. Sir Arthur Harris continued to criticize these panaceas and persisted in bombing German cities, despite the evident desire of the British air staff to phase out such area bombardment.[88]

On December 13, 1944, Spaatz reported to Arnold "increasing evidence that the attacks on rail communications and industrial areas in Germany are having a cumulate effect . . . the breaking point may be closer at hand than some of us are willing to admit." To hasten this breaking point, Spaatz and the British planned "widespread attacks directed against rail communications in an attempt to secure a widespread breakdown of the transportation system." Bad weather delayed the effort for a time. Moreover, Hitler's counterattack in the Ardennes, launched on December 16, forced USSTAF to divert most of its bombers and fighters in an all-out effort to check the German advance, delaying the broader transportation campaign until 1945, although heavy bombers of the Fifteenth Air Force continued the oil campaign from Italy.[89]

Resistance by Allied ground troops, particularly American forces who bore the brunt of the offensive, disrupted the German timetable in the Battle of the Bulge and eventually doomed it to failure. Inclement weather, counted on by the Germans, initially stunted Allied air support of the harassed ground defenders. When the weather eventually cleared, Eighth Air Force heavy bombers joined with medium bombers and fighter-bombers in bringing air power to bear to control the German breakthrough. Beginning with a massive raid on December 23, 1944, and continuing for the next twelve days, Fortresses and Liberators concentrated on targets that would likely have an immediate effect on German Bulge operations in an area behind the front, west of the Rhine. When bad weather returned, the bombers resorted to the use of radar. By December 26, "the tremendous effort of the tactical and strategic air forces . . . began to show its effect upon the enemy's effort to continue the offensive."

This flak-peppered Liberator crash-landed in England after a strenuous mission.

Photographs and other intelligence revealed "cratering and cutting of the main highways and railroads, the destruction . . . of road and rail bridges, the blocking of chokepoints and narrow passes and the heaping of huge piles of rubble upon the narrow streets of innumerable villages through which the enemy's movement had . . . been canalized."[90] Strategic bomber efforts to down selected bridges spanning the Rhine were mostly unsuccessful. Nevertheless, by the second week in January 1945, the Germans were in retreat.

This last German offensive on the Western Front, including the last major *Luftwaffe* offensive strike of the war against Allied airfields in Belgium and Holland on January 1, 1945, produced, among the Allies, the view that the war in Europe might well continue for an extended period. On January 12, 1945, Spaatz issued a new bombing directive reflecting this pessimism. He continued the priority strikes against oil and transportation but reinstated the priority of the *Luftwaffe* and jet aircraft production as targets, assuming a protracted war would allow the *Luftwaffe* to offer a real threat to Allied air superiority.[91]

In February 1945, the USSTAF again endeavored to destroy Germany's will and to wage war through strategic bombing. While continuing the bombing of industrial targets, Spaatz ordered a concentrated blow against German transportation targets. American heavy bombers struck deliberately chosen targets close to small German towns as yet untouched by the air war. Called CLARION, the bombing results were mixed, although they further weakened an already crumbling network of roads, railroads, and canals. German flak and occasional Me 262 jets still bedeviled Allied aerial operations, but the cumulation of American and British bombs secured the collapse of the German economy by the end of February.[92]

February 1945 also saw several Allied strategic night–day missions attain a climax that still evokes heated controversy. In the summer of 1944, Anglo-American air planners, under pressure to retaliate against the German V–1 cruise missile attacks on England, developed a plan code-named THUNDERCLAP that specifically targeted Berlin for a massive one-shot night and day assault by strategic bombers. Clearly the idea bothered planners, who worked hard to rationalize the attack. For the British, such an effort would be a continuation of their area bombing campaign, with civilian morale as the principal target. For the Americans, it would be precision bombing against military and industrial targets, but because of their willingness to use radar bombing if necessary, that, in effect, meant area bombing. The underlying theme of THUNDERCLAP was to produce a definitive event to push over a teetering Third Reich. The Bulge offensive delayed the plan, but Allied success there and the Soviet advance into the eastern edge of the Reich brought the plan to the forefront of strategic operations. Planners added Dresden and several other eastern German cities to the target list on the rationale that the resultant disruption of surface transport would impede the movement of German reinforcements from other fronts to the east.

B–17 Flying Fortresses dropped high-explosive bombs on railway centers in Dresden.

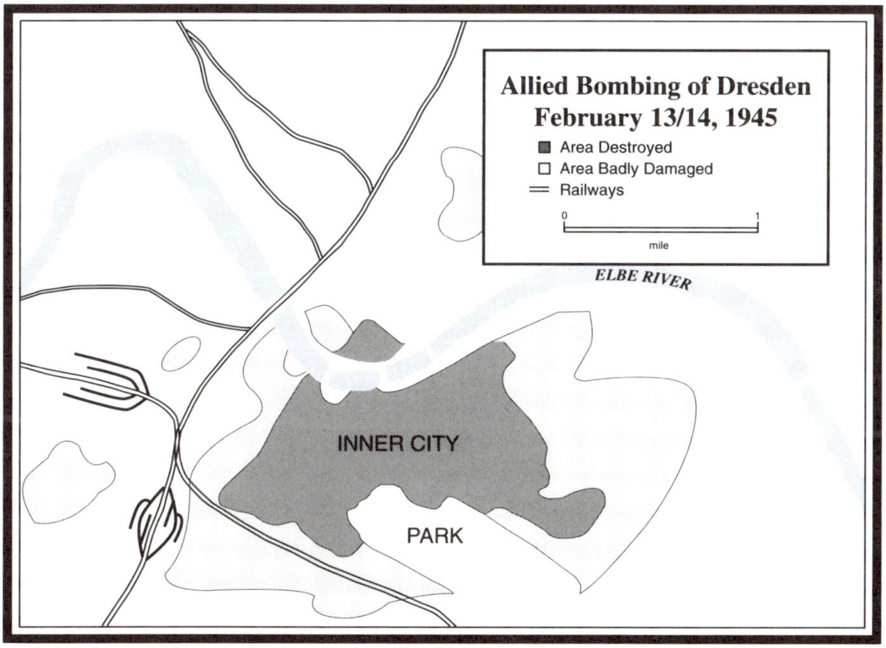

Arnold did not favor the indiscriminate bombing of civilians, and Doolittle, Eaker, and Anderson opposed the idea. At first, Spaatz seemed firmly opposed to making civilians a direct target, but in the end he accepted it for expediency's sake while trying to finesse his reasoning. The upshot was a THUNDERCLAP mission to Berlin by Doolittle's Eighth Air Force on February 3, 1945. The number of civilian casualties is still the subject of disagreement among historians, but Germany did not collapse as a result.[93]

A few days later in Dresden, on the night of February 13/14, RAF formations on a THUNDERCLAP mission with a lethal mixture of incendiaries and high-explosive bombs produced a roaring firestorm, magnified the next day as Eighth Air Force bombers dropped their loads into the smoke and flames. Both air forces repeated the attacks on Dresden the following night and day. If, other than a railroad marshaling yard, few vital targets existed in the ancient city, some 35,000 people died nonetheless. Still, these violent aerial assaults did not bring Germany to its knees.[94] Only after Hitler killed himself on April 29, 1945, and only after Allied armies swarmed over the bomb- and shell-ravaged land, did the war end.

Evaluation

The AAF went to England in 1942 to launch a type of war for which practically no precedent existed. What little was known came from the limited experiences of England and Germany, each of which in 1940–1941 tried daylight strategic precision bombing, more or less against each other's war economies, and each had given up the effort as beyond its capability. The lesson the AAF leaders drew from these British and German failures was not that the doctrine of strategic bombardment was incorrect, but that the two warring powers had been unwilling to commit the resources necessary for the effort and had not employed the weapon correctly.[95]

The AAF pushed ahead and, because this was a new type of war that had to be tried and tested in combat, inevitably made mistakes of its own. By April 16, 1945, when Spaatz officially declared an end to strategic bombardment, AAF strategic bomber forces in Europe had launched 501,536 bomber sorties and dropped 1,005,091 tons of bombs, at a cost of 67,646 personnel killed and 8,325 bombers lost. In the process, they did much to destroy Germany's "sustaining resources" while keeping Germany's armed forces in a constant state of defense, part of a larger strategic plan for the land invasion and occupation of Germany.[96]

In conducting the strategic bombardment of Germany, Allied air leaders had to make a series of command decisions, most based on theory or inadequate intelligence and usually in the heat of battle. For the British and Americans alike, air offensive doctrine had its roots in the thinking of Douhet,

Trenchard, and Mitchell and ended with the decisions of Arnold, Eaker, Spaatz, Doolittle, Anderson, Portal, Tedder, and Harris. The American commanders had the support of heavy bomber and fighter designers and manufacturers in supplementing and changing the doctrine of daylight precision bombing and in developing the technology to make it possible.

Six days after the invasion at Normandy, Generals Arnold and Omar Bradley, Commanding General of the Twelfth Army Group, walk along a French beachhead in their inspection tour of progress in France.

Aided by the development of the fast, self-defending bomber — the B–17 Flying Fortress — the Air Corps Tactical School in the 1930s influenced such key figures as Arnold and a whole generation of Army air leaders to accept a doctrine of bomber invincibility and superiority. In 1941, the staff of the Air War Plans Division, championed by Marshall and Arnold and with approval from the Roosevelt administration and Congress, asked for the massive forces necessary to make this doctrine a reality. From 1939 to 1945, consistent support allowed 25 percent to 35 percent of the total American war production to be devoted to the construction of 324,750 aircraft for the global war effort. A substantial part of this air fleet eventually overwhelmed Germany's 111,077 wartime aircraft. (The latter figure included aircraft returned to German factories for major repairs, and thus counted twice.)[97] Moreover, Germany built its aircraft too late to change the fortunes of war, because by the middle of 1944 the *Luftwaffe* had too few trained pilots and too little fuel to fly them.

Political maneuver and persuasion gave the wartime strategy of daylight precision strategic bombardment a chance. At Casablanca in January 1943, Generals Eaker, Andrews, Arnold, and Spaatz helped win the Combined Chiefs of Staff support for the continuation of daylight precision bombing, even though the AAF had not yet tested it against Germany proper. The Combined Chiefs also had the foresight to support both the RAF's and the AAF's plans for the air war. Rather than chancing that only night area bombing *might* work, or that only day precision bombing *might* work, they selected the Combined Bomber Offensive as the surest means to victory. Realizing also the need for air superiority, they assigned the highest and most immediate priority to defeat of the *Luftwaffe*. As a follow-up, Portal decided to continue with POINTBLANK in December 1943 even though not only had the *Luftwaffe* proved the shortcomings of the American doctrine of unescorted daylight precision bombing in the summer and fall of 1943, the campaign was also three months behind schedule. According to the authors of the official British history of the strategic air war, Portal's action was "a courageous decision and, perhaps, one of the most important of the war" because POINTBLANK was the only method available to the Allies for defeating the *Luftwaffe* in time for OVERLORD.[98]

Other war-winning command decisions were more tactical, revolving around the specific measures to be used for defeating the *Luftwaffe*, weakening Germany sufficiently to permit the initiation of combined operations on the Continent, and eliminating Germany's economic capacity to make war. The appearance of the AAF's long-range escort fighters in late 1943 reversed the *Luftwaffe*'s successes against massed strategic daytime bomber attacks. The success of day bombers in Europe owed largely to Generals Giles's and Kepner's pushing of American engineers and manufacturers to do what the British, Germans, and many Americans had assumed was impossible. The fact that a sufficient number of these long-range fighters was available in the crucial battles for air superiority in early 1944 was due to Arnold's decision to give priority to his European strategic air forces over the air forces in other theaters of war for *all* P–38s and P–51s during the critical months of late 1943 and early 1944. Arnold also decided late in 1943 on a controversial change of command that affected the field leadership of the European strategic bombardment offensive. No one can now argue with the results of his change.

One of the commanders affected by this realignment, Jimmy Doolittle, freed the fighters on escort duty. This proved significant in an early 1944 general policy change that allowed the USSTAF to win air superiority in time for the invasion of France. Air superiority could be judged "the greatest single achievement of the air attack on Germany."[99] The other change that most contributed to the winning of air superiority was the Spaatz and Fred Anderson decision for a Verdun-type strategy of attrition: exposing American bombers as bait over Germany whenever possible and choosing targets that would force *Luftwaffe* fighters to do battle and be destroyed.

The final critical command decisions in the American campaign of strategic bombardment were those of Spaatz, with the backing of Tedder, first to strike the German oil industry and then, when Eisenhower released the strategic

Maj. Gen. William E. Kepner, Commander of VIII Fighter Command (*left*), and Maj. Gen. Barney Giles, Arnold's Chief of Air Staff (here as a brigadier general, on the *right*), both pushed American engineers and manufacturers to come up with long-range escort fighters.

bomber forces from their task of supporting Allied land armies, to strike the German transportation network. Regardless that reduced German fuel supplies kept its armed forces from regaining the offensive initiative, the United States Strategic Bombing Survey judged "the attack on transportation [to be] the decisive blow that completely disorganized the German economy."[100]

The Strategic Bombing Survey was an attempt by a group of civilian and military specialists to measure the results of the strategic bombardment campaign. It stemmed from the desires of certain AAF figures — principally Generals Arnold, Kuter, Spaatz, Anderson, and Maj. Gen. Muir S. Fairchild — to evaluate what air power had accomplished in the Pacific and the European wars and what lessons might apply to a hoped-for independent air force in the postwar era. Initial planning had begun at least by 1942; became more definite in 1943, as shaped by the Committee of Operations Analysts and the Air Staff; and coalesced by late 1944, when business executive Franklin D'Olier accepted Arnold's offer to head a committee of specialists to measure the effects of the strategic bombing on Germany. The survey, later expanded to include the war in Japan, had many problems, not the least of which was D'Olier's decision to refrain from any exact definition of strategic bombardment; it nonetheless produced a creditable product that scholars continue to cite. Its most significant conclusion, in the words of David MacIsaac's study of the work, was that "only *repeated* and *sustained* attack could assure the permanent dislocation of

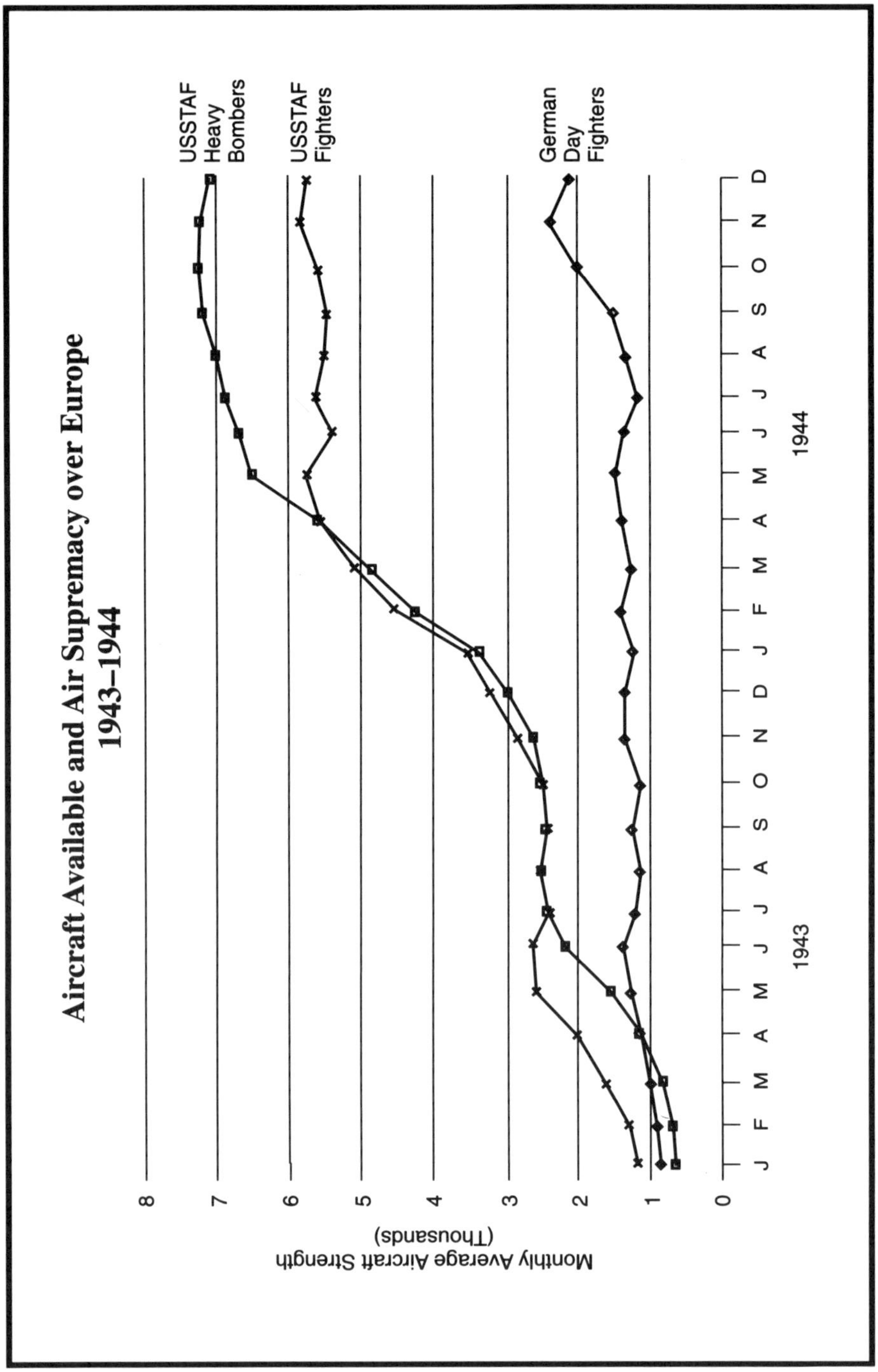
Aircraft Available and Air Supremacy over Europe
1943–1944
USSTAF Heavy Bombers
USSTAF Fighters
German Day Fighters
Monthly Average Aircraft Strength
(Thousands)
0
1
2
3
4
5
6
7
8
J F M A M J J A S O N D
1943
J F M A M J J A S O N D
1944

manufacturing effort."[101]

Altogether, these command decisions insured significant air contributions to Allied victory, but at a major sacrifice from the men and planes of the U.S. AAF in Europe. As that victory demonstrated, in the words of the Strategic Bombing Survey, "even a first-class military power — rugged and resilient as Germany was — cannot live long under full-scale and free exploitation of air weapons over the heart of its territory." The Arnold, Spaatz, and Doolittle decisions affecting the air war made possible the successes attained in the daylight strategic bombing campaign.[102]

Ultimately, USSTAF fighters and bomber defensive gunners achieved air superiority through attrition in the air, a success not achieved through the bombing of German aircraft in their factories. According to the Strategic Bombing Survey, the single greatest cause for the *Luftwaffe*'s defeat was the decline in quality of German pilots, a result of attrition. Allied bombing brought oil shortages that affected the training of new pilots directly and adversely, to be sure, but a decline in *Luftwaffe* fighter crew quality was apparent even by March 1944, two months before Eighth Air Force's first attack on German oil.[103] It was bombers in the air over Germany that brought defending enemy aircraft into the sky and into an aerial combat that, in terms of sheer numbers, they could not win. Bombers also inflicted damage on the ground: Germany lost about 13,000 aircraft from planned production as a result of the bombing of the German aircraft industry in 1944, which constrained the *Luftwaffe* order of battle to an increase of only 700 aircraft in that year. The near total absence of the *Luftwaffe* over the Normandy beachhead on D-Day, its general ineffectiveness in the crucial next few days as the Allies moved inland, and its subsequent failure to prevent Allied air harassment of the *Wehrmacht* in France proved the importance of strategic bombing and the winning of air superiority. Air superiority made the daylight strategic bombing campaign possible; ironically, given the virtual abandonment of day bombing by the British early in the war, American bomber losses by day proved smaller than RAF bomber losses at night.[104]

The strategic bombing offensive also forced diversion of German war resources to defend against the Second Front of the Allied air invasion. Besides the personnel assigned to the *Luftwaffe*, 2 million laborers manned the industries that supplied it with equipment, 2 million German soldiers and civilians were committed to ground antiaircraft defense, and 2½ million more laborers had to clean the debris, repair bomb-damaged industries, and build dispersal plants. All the while, the role of the *Luftwaffe* changed dramatically. From a force in 1942 that devoted 60 percent of its manpower and airplanes to offensive warfare in support of the *Werhmacht*, by 1944 the *Luftwaffe* was devoting 82 percent of its manpower and airplanes to an aerial defense of the homeland. And Germany's commitment of 40 percent of war production to a defensive air arm kept its industry from producing the other weapons needed to win the war.[105]

American bombing forced German industry to disperse, thereby reducing the losses a single plant would experience if struck. By so doing, Germany lost the advantages of scale and mass production, so important to American industry and for its prodigious wartime output. After the war, Professor Willy Messerschmitt confided that forced dispersal cost Germany 50 percent of its potential aircraft production. In the words of the Strategic Bombing Survey: "It may well be that more aircraft were lost out of production because of dispersal than because of direct bombing." But the Allied bombing gave to Germany only the choice between destroyed industry or dispersed industry. "In the end," however, "dispersal defeated itself." The effort placed heavy strains on an already overburdened German transportation system. The intensification of Allied attacks on German railways in late 1944 cut off the dispersed plants and, according to German plant managers, eventually caused the greatest single reduction in German wartime production. The transportation and oil campaigns of 1944 and 1945, responsible for nearly 40 percent of the total American bomb tonnage dropped in the European war, produced the collapse of the German economy that had begun in early 1945. During the last sixteen months of the war, under the impact of nearly 90 percent of all the bombs dropped on Germany by Allied air forces in World War II, Germany lost 20 percent of its industrial war production. The mortal blows to oil and transportation complemented one another, preventing plant managers from switching to oil-driven road transport when rail access was cut.[106]

Despite the accomplishments of strategic bombing, mistakes were made. The lives of many airmen could have been saved if AAF leaders had recognized earlier that, no matter what doctrine said, heavy bombers needed fighter escorts to keep losses down and such escorts could be quickly developed. High-altitude daylight strategic bombing suffered from poor accuracy (only about 20 percent of all bombs hit within 1,000 feet of their aiming points), insufficient follow-up bombings, and inadequately powered bombs (American 500-lb bombs could not destroy German machine tools).

Poor target choices numbered among the biggest mistakes. Based on hindsight, the Strategic Bombing Survey discovered that aircraft engine and propeller production rather than airframe assembly would have made a better bombing target. Notwithstanding Albert Speer's postwar contention that a relentless campaign against the ball bearing industry could have shortened the war by a year, ball bearings probably should never have been targeted. Ball bearing production facilities proved difficult to destroy, bearings could be imported, and German industry showed great flexibility in overcoming shortages of these devices. The synthetic oil industry, on which Germany depended for fuel, chemicals, explosives, and rubber, should have been struck much earlier and harder. Participants in the air war and experts alike are in almost unanimous agreement here.

Seeking to ascertain the validity of his command decisions during the war, General Spaatz (*right*), along with Lt. Gen. Vandenberg, interviewed POW Hermann Goering (*center*), chief of the German Air Force.

In retrospect, the most significant missed opportunity was the failure to launch concerted attacks on Germany's four tetraethyl lead and ethylene dibromide production plants and its electric power plants. Destruction of the lead-additive facilities would have grounded the *Luftwaffe* immediately. Without these knock-reducing agents, high-performance World War II aircraft engines would have lost 40 percent of their power. Likewise, the destruction of forty-five electric generator plants would have reduced the capacity of Germany's already understrength power network by 40 percent. However, Ira Eaker's postwar conclusion probably has some validity that if the AAF had concentrated exclusively on one target system, its bomber losses would have been unacceptably high.[107]

The American daylight strategic bombing campaign took so long to produce results for three reasons. First, the slow buildup of the American air forces in the early years of the campaign limited the force that could be brought to bear on Germany. Nearly 84 percent of all the tons of bombs dropped by American strategic forces on Germany arrived in the last twelve months of the war; not until September 1944 did American forces drop more bombs on targets inside German territory than on those outside. Second, diverting these bomb tonnages from strategic targets, and failure to concentrate on one or two target categories, reduced the bombing effect. Finally, the reserve capacity of the German economy delayed results. Despite heavy bombing, German industrial output continued to rise until late 1944. As the world's largest manufacturer of machine tools, Germany was able to absorb considerable bomb damage with no serious, immediate losses in production. Many plants were able to meet production requirements with single shifts, switching to double shifts only in late 1944 to compensate for the damage produced by American bombs. Hitler, for reasons of his own, did not order increased industrial production until after the military reverses before the gates of Moscow in early 1942 and, despite continued setbacks, never ordered a general economic mobilization.[108]

Such excessive industrial capacity resisted the best efforts of American precision strategic bombing, but by 1945 Germany was nevertheless defeated. Allied bomb tonnages accumulated rapidly in late 1944 and early 1945, defeating the best efforts of the Germans to maintain their economic production. The combination of aerial bombardment and advancing armies ground down the German war machine. American daylight strategic bombing forces had been essential to obtain victory.

NOTES

1. On June 20, 1941, the War Department reconstituted the Army Air Corps as the Army Air Forces (AAF). Though it remained a combatant arm of the U.S. Army, this action provided for a separate Air Staff and marked the beginning of what would become a separate air force. A War Department reorganization of March 9, 1942, created three autonomous Army commands: Army Ground Forces, Services of Supply (later, Army Service Forces), and the AAF. "This reorganization dissolved the Office of the Chief of the Air Corps and the Air Force Combat Command, and merged all elements of the air arm into the Army Air Forces," commanded by Lt. Gen. Henry H. Arnold. *The Army Almanac* (Washington, D.C.: GPO, 1950), pp 213–214; Charles A. Ravenstein, *The Organization and Lineage of the United States Air Force* (Washington, D.C.: GPO, 1986), pp 8–9.

2. Robert F. Futrell, *Ideas, Concepts, Doctrine: A History of Basic Thought in the United States Army Air Force, 1907–1964*, 2 vols (Maxwell AFB, Ala.: Air University, 1971), vol 1, pp 74–77, 92–94; Thomas H. Greer, *The Development of Air Doctrine in the Army Air Corps, 1917–1941* (Maxwell AFB, Ala.: Air University, 1955); Wesley Frank Craven and James Lea Cate, eds., *The Army Air Forces in World War II*, 7 vols (Chicago: University of Chicago Press, 1948–1958; Washington, D.C.: Office of Air Force History, 1983), vol 1, pp 51–53, 68–71, 597–599; Ronald Schaffer, *Wings of Judgment: American Bombing in World War II* (New York: Oxford University Press, 1985), pp 20–34; Craven and Cate, eds., *Army Air Forces in World War II*, vol 6, pp 203–208; Haywood S. Hansell, Jr., *The Strategic Air War against Germany and Japan* (Washington, D.C.: Office of Air Force History, 1986), pp 7–19.

3. "Daylight precision bombing" was a misleading phrase that promoted and justified the AAF's strategy for winning the war in Europe. It was certainly daylight and it was certainly bombing, but it was hardly precise. A postwar study of American attacks on German oil targets, in many ways the model for precision bombardment, revealed that of every hundred bombs dropped on these targets, one hit oil pipelines, two hit production areas, two failed to explode, three hit decoy plants, eight landed in open terrain inside the target area, and eighty-four missed the target area entirely — yielding about a 3 percent success rate. United States Strategic Bombing Survey (USSBS), *Oil Division Final Report*, Rpt no. 109 (Washington, D.C.: GPO, 1947), fig 7.

4. Giulio Douhet, *The Command of the Air*, trans. Dino Ferrari (New York: Coward-McCann, 1942; Washington, D.C.: Office of Air Force History, 1983); Andrew Boyle, *Trenchard: Man of Vision* (London: William Collins, 1962); Alfred F. Hurley, *Billy Mitchell: Crusader for Air Power* (New York: Franklin Watts, 1964). Richard Overy argues that Mitchell's immoderate advocacy of bombers winning wars was a political exaggeration designed to win an independent status for the air arm. R. J. Overy, *The Air War, 1939–1945* (New York: Stein & Day, 1980), p 16. Still, Mitchell warned of the prerequisite of air superiority. Hurley, *Billy Mitchell*, p 83.

5. Lt Gen Ira Eaker and Arthur G. B. Metcalf, "Conversations with Albert Speer," *Air Force Magazine*, vol 60 (Apr 1977), p 57.

6. Maurice Matloff and Edwin M. Shell, *The War Department: Strategic Planning for Coalition Warfare, 1941–1942* (Washington, D.C.: GPO, 1959), p 9; Louis Morton, "Germany First: The Basic Concept of Allied Strategy in World War II," in Kent Roberts Greenfield, ed., *Command Decisions* (Washington, D.C.: GPO, 1961), pp 3–38; Craven and Cate, eds., *Army Air Forces in World War II*, vol 1, pp 136–150; Air War Plans Division, AWPD–1, Aug 1941, pt 3, file 145.82–1, U.S. Air Force Historical Research Agency, Maxwell AFB, Ala. (USAFHRA); Craven and Cate, eds., *Army Air Forces in World War II*, vol 2, pp 368–369; Hurley, *Billy Mitchell*, p 83; Hansell to Arnold, Aug 11, 1941, file 145.95 (WP–III–A2), book 3, USAFHRA; W. A. Jacobs, "Strategic Bombing and American National Strategy, 1941–1943," *Military Affairs*, vol 50 (Jul 1986), pp 134–135; Craven and Cate,

eds., *Army Air Forces in World War II*, vol 2, pp 290–296; Haywood S. Hansell, Jr., *The Air Plan That Defeated Hitler* (Atlanta: Higgins-McArthur/Longino & Porter, 1972), pp 80–85, 88, 91, 101–110.

AWPD–2 was an interim plan of October 1941 for short-term allocations of equipment to the Soviets, British, and AAF. Craven and Cate, eds., *Army Air Forces in World War II*, vol 1, pp 133–134. AWPD–4 was an inflated AWPD–1 stressing an immediate buildup of aircraft in England for an air offensive against Germany that was prepared a week after Pearl Harbor. It was not accepted. Craven and Cate, eds., *Army Air Forces in World War II*, vol 1, p 246. AWPD–42 was a new estimate in 1942 of the aerial needs of the U.S. armed services and the Allies. U.S. Navy objections to a proposed crimp in its shipbuilding program, compromises over resources available for 1943, and plans for giving the Army control over all four-engine bombers led to a reduction in the number of airplanes to be procured, to a restoration of much of the Navy's shipbuilding program, and to Navy control over some land-based four-engine bombers. The earlier AWPD–1, according to the postwar bombing survey, more closely identified "the optimum target system for the destruction of Germany's industrial life" than did AWP–42 or the actual bombing campaign as waged. Air War Plans Division, AWPD–1, Aug 1941; Air War Plans Division, AWPD–42, Sep 1942, file 145.82–42, USAFHRA; Roosevelt to Marshall, Aug 24, 1942, file 145.96–96, USAFHRA; Futrell, *Ideas, Concepts, Doctrine,* vol 1, pp 115, 143.

7. In June 1941 Arnold's official title was Chief of the United States Army Air Forces and Deputy Chief of Staff for Air. In March 1942 he became the commanding general of the AAF. See note 1 above. See also Mark S. Watson, *Chief of Staff: Pre-War Plans and Preparations* (Washington, D.C.: GPO, 1950), pp 289–293.

8. Jacobs, "Strategic Bombing and American National Strategy," pp 133–136; Allan R. Millett and Peter Maslowski, *For the Common Defense: A Military History of the United States* (New York: Free Press, 1984), pp 404–407, 414–423, 436.

9. Craven and Cate, eds., *Army Air Forces in World War II*, vol 1, pp 558–559, 563–564.

10. Arnold to Eaker, Jan 31, 1942, file 145.95 (WP–III–A2), USAFHRA; Thomas Coffey, *HAP: The Story of the U.S. Air Force and the Man Who Built It* (New York: Viking Press, 1982), pp 308–333; Schaffer, *Wings of Judgment*, pp 10–17; James Parton, *"Air Force Spoken Here": General Ira Eaker and the Command of the Air* (Bethesda, Md.: Adler & Adler, 1986), pp 131–135, 139; DeWitt S. Copp, *A Few Great Captains: The Men and Developments That Shaped the Development of the U.S. Air Force* (Garden City, N.Y.: Doubleday, 1980), pp 443–444.

11. Michael S. Sherry, *The Rise of American Air Power: The Creation of Armageddon* (New Haven, Conn.: Yale University Press, 1987), p 97; Jean H. DuBuque and Robert F. Gleckner, *The Development of the Heavy Bomber, 1918–1944* (USAF Hist Study 6, Maxwell AFB, Ala.: Air University Press, 1951), pp 73–82, 105–120, 147–49, 166–167, 171–172; Hansell to Chief of Air Staff, Dec 20, 1941, file 145.95 (WP–III–A2), USAFHRA.

12. Roger A. Freeman, *Mighty Eighth War Diary* (London: Jane's, 1981), pp 9–10; Eaker to Spaatz, Aug 19, 1942, Eaker Papers, Manuscript Div, Library of Congress, Washington, D.C. (LC).

13. Craven and Cate, eds., *Army Air Forces in World War II*, vol 1, pp 641–642; Leo J. Meyer, "The Decision to Invade North Africa (TORCH) (1942)," in Greenfield, ed., *Command Decisions*, pp 129–153; Spaatz to CG, ETOUSA, Nov 9, 1942, and Eaker to Stratemeyer, n.d. [October 1942], file 145.95 (WP–III–A2), Great Britain no. 2, USAFHRA; Gp hist, 92d Bom Gp (H), Mar–May 1942, file GP–92–HI, USAFHRA; Hansell to Chief of Air Staff, Dec 26, 1941, file 145.95 (WP–III–A2), USAFHRA; Gp hist, 306th Bom Gp (H), Mar–Dec 1942, file GP–306–HI, USAFHRA.

14. Millet and Maslowski, *For the Common Defense*, pp 416–418; Freeman, *Mighty Eighth*, pp 9–64; Churchill to Roosevelt, Sep 16, 1942, as cited in Charles Webster and Noble Frankland, *The Strategic Air Offensive against Germany*, 3 vols (London: HMSO, 1961), vol 1, pp 355–356; Arnold to Eisenhower, Jul 30, 1942, file Bomber Command to Great Britain (hereafter BC to GB) no. 231, box 49, Arnold Papers, Manuscript Div, LC; Stratemeyer to Spaatz, Oct 7, 1942, and Stratemeyer to Asst Chief of Air Staff, Nov 21, 1942, file 145.95 (WP–III–A2) Great Britain no. 2, USAFHRA; Eaker to Arnold, Feb 26, 1942, file BC to GB no. 226, box 48, Arnold Papers. A heavy bombardment group at this time consisted of 4 squadrons of 12 aircraft each, or 48 aircraft per group — a total of 288 heavy bombers in Eighth Air Force's 6 groups. By war's end the number of aircraft in a squadron had increased to 18, with 72 aircraft in each group.

15. Craven and Cate, eds., *Army Air Forces in World War II*, vol 2, pp 226–229; Eaker to Andrews, Feb 27, 1943, file BC to GB no. 226, box 48, Arnold Papers.

16. Special rsch hist, National Security Council (NSC), "ULTRA: History of U.S. Strategic Air

Force Europe vs. German Air Force," Jun 1945, RG 457, National Archives, Washington, D.C. (NA), p 22; Craven and Cate, eds., *Army Air Forces in World War II*, vol 2, pp 215–216, 237; Hist, 8th AF, vol 1, file 520.01–1, USAFHRA; Parton, *"Air Force Spoken Here,"* pp 248–249; W. W. Rostow, *Pre-Invasion Bombing Strategy: Eisenhower's Decision of Mar 25, 1944* (Austin: University of Texas Press, 1981), pp 15–23.

17. Webster and Frankland, *Strategic Air Offensive*, vol 2, pp 5–7, 47–52; rpt, Committee of Operations Analysts (COA), Mar 8, 1943, files 118.04 and 118.01, USAFHRA.

18. Rpt, COA, Mar 8, 1943; Webster and Frankland, *Strategic Air Offensive*, vol 2, p 5.

19. The first Eighth Air Force mission against a target in Germany was the January 27, 1943, effort to Wilhelmshaven.

20. Hist, 8th AF, Aug 17–May 1, 1943, vol 2, pt 1, file 520.01, USAFHRA.

21. Combined Chiefs of Staff (CCS) 166/1/D, Jan 21, 1943, file 119.151–1, USAFHRA; Webster and Frankland, *Strategic Air Offensive*, vol 2, p 14, citing ltr, Harris to Air Ministry, Mar 6, 1943.

22. Though no evidence exists to indicate whether the Allies ever used this information, at the same time the AAF pressed for daylight bombing at Casablanca, ULTRA revealed that *Luftwaffe* leaders had concluded the only way they could be forced into ruinous attrition or forced to allow unopposed bombing was for the Allies to bomb Germany during the day. Special rsch hist, NSC, "ULTRA," pp 44–45.

23. CCS 217, May 14, 1943, file 119.04–6, USAFHRA; CCS 309, Aug 1943, file 119.151–3, USAFHRA.

24. Special rsch hist, NSC, "ULTRA," pp 1–4.

25. Eaker to Stratemeyer, Jan 2, 1943, file 168.491, vol 1, USAFHRA; CCS 224/1, May 25, 1943, in Webster and Frankland, *Strategic Air Offensive*, vol 4, p 273–283; memo, Mar 3, 1943, file 145.95 (WP–III–A2), USAFHRA.

26. Craven and Cate, eds., *Army Air Forces in World War II*, vol 2, p 266.

27. Special rsch hist, NSC, "ULTRA," pp 60, 71, 74.

28. USSBS, *Huels Synthetic Rubber Plant*, Rpt no. 128 (Washington, D.C.: GPO, 1947), p 1.

29. Craven and Cate, eds., *Army Air Forces in World War II*, vol 2, p 682; Freeman, *Mighty Eighth*; Bomber Narrative of Operations, file no. 519.332, USAFHRA. All mission summations and sortie and loss figures come from Craven and Cate, eds., *Army Air Forces in World War II*, vol 2, appendices.

30. Giles to Eaker, Jun 11, 1943, Arnold to Eaker, Aug 1, 1943, Arnold to Eaker, Apr 10, 1943, Eaker to Arnold, Feb 15, 1943, and Eaker to Stratemeyer, Feb 19, 1943, file 168.491, vol 1, USAFHRA; Eaker to Stratemeyer, Jul 2, 1943, file 168.491, vol 2, USAFHRA; Coffey, *HAP*, p 314; Arnold to Eaker, Jun 15, 1943, file no. 227, box 48, Arnold Papers.

31. Deputy Chief of Air Staff to Assistant Chief of Air Staff (AC/AS), Plans, Aug 14, 1943, file 145.95 (WP–III–A2), book 2, USAFHRA; Giles to Eaker, Jul 30, 1943, file 168.491, vol 1, USAFHRA.

32. ULTRA revealed at this time that attrition was beginning to tell on the *Luftwaffe*, as urgent demands for more fighter production to keep up with losses made plain. Moreover, *Luftwaffe* headquarters ordered its group and squadron commanders not to fly in operations against American bomber formations because of the high rate of casualties among irreplaceable personnel.

33. Atkinson to Kuter, Jun 19, 1943, file 145.95 (WP–III–A2), book 2, USAFHRA; Eaker to Giles, May 13, 1943, Giles to Eaker, Jul 30, 1943, Eaker to Portal, Aug 1, 1943, Arnold to Eaker, Aug 1, 1943, Anderson to Stratemeyer, Jul 21, 1943, Anderson to Giles, Aug 4, 1943, and Giles to Anderson, Aug 30, 1943, file 168.491, vol 1, USAFHRA.

34. Craven and Cate, eds., *Army Air Forces in World War II*, vol 2, pp 704–705. Parton (*"Air Force Spoken Here,"* pp 327–328) contends that the conclusions drawn by Arthur B. Ferguson and reported in Craven and Cate (*Army Air Forces in World War II*, vol 2, pp 704–705) were incorrect and suggests that deep penetrations without fighter escort were possible, that Second Schweinfurt was not a disaster, that losses could be absorbed, and that Eighth Air Force had not lost air superiority because it had not yet achieved air superiority. Kenneth P Werrell ("The Strategic Bombing of Germany in World War II: Costs and Accomplishments," *Journal of American History*, vol 73 [Dec 1986], p 705) maintains that the AAF's peacetime theory of strategic bombing had "failed in wartime practice." The USSBS concluded: "As a result of these battles, the Eighth AF temporarily lost air superiority over the target areas of Germany. The necessity for full fighter escort all the way to the target became painfully obvious. All penetration beyond the range of fighters as then equipped ceased until the fighter range extension program could be completed." USSBS, *Aircraft Division Industry Report*, Rpt

no. 4 (Washington, D.C.: GPO, 1947), pp 67–68.

35. Eaker to Arnold, Oct 15, 1943, file 145.95 (WP–III–A2), book 2, USAFHRA; Arnold to Eaker, Oct 15, 1943, folder 4, box 16, Eaker Papers.

36. H. H. Arnold, *Global Mission* (New York: Harper, 1949), p 495; Eaker to Arnold, Oct 15, 1943, file 145.95 (WP–III–A2), book 2, USAFHRA; Eaker to Arnold, Oct 22, 1943, file BC to GB no. 228, Arnold Papers; Arnold to Eaker, Oct 15, 1943, folder 4, box 16, Eaker Papers; Eaker to Arnold, Oct 19 and Oct 22, 1943, file 168.491, vol 1, USAFHRA; Eaker to Arnold, Oct 21, 1943, folder 15, box 16, Eaker Papers; Arnold to Eaker, Oct 16, 1943, folder BC to GB no. 227, box 48, Arnold Papers; CCS 403/1, Dec 3, 1943, in Craven and Cate, eds., *Army Air Forces in World War II*, vol 2, p 715; USSBS, *Aircraft Division Industry Report*, p 74, exhibit 2.

37. According to Eighth Air Force claims, the *Luftwaffe* had lost 1,139 fighters during those months.

38. Special rsch hist, NSC, "ULTRA," pp 103, 107, 112–114; Williamson Murray, *Luftwaffe* (Baltimore: Nautical & Aviation Publication, 1985), p 215; Eaker to Arnold, Oct 15, 1943, in memo, Sec Air Staff to AC/AS, Plans, Oct 23, 1943, file 145.95 (WP–III–A2), book 2, USAFHRA.

39. Memo, Sec Air Staff to AC/AS, Oct 23, 1943; Arnold, *Global Mission*, p 495; Eaker to Arnold, Oct 15, 1943, folder Oct 1943, box 16, Eaker Papers; Eaker to Portal, Oct 14, 1943, file 168.491, vol 1, USAFHRA; Arnold to Portal, and Portal to Arnold, Sep 25, 1943, Oct 14, 1943, Oct 15, 1943, Oct 24, 1943, and Oct 31, 1943, file 145.95 (WP–III–A2), book 2, USAFHRA; Giles to Eaker, Dec 16, 1943, file 168.491, vol 1, USAFHRA; Arnold to Portal, Oct 31, 1943, file 145.95 (WP–III–A2), book 2, USAFHRA.

40. Kuter to Arnold, Aug 30, 1943, file 145.95 (WP–III–A2), book 2, USAFHRA; Hansell, *Strategic Air War*, pp 91–92; Arnold to CCS, Nov 1, 1943, file Strategy (and Command), box 39, Arnold Papers; Arnold to Eaker, Oct 22, 1943, file 168.491, vol 1, USAFHRA.

41. Arnold to CCS, Nov 1, 1943, and Marshall to Arnold, Nov 5, 1943, file Strategy (and Command), box 39, Arnold Papers; ARGUMENT folder, n.d., subject file, box 135, Carl Spaatz Papers, Manuscript Div, LC; Plans for ARGUMENT, Dec 1943, file 520.4231B, USAFHRA. The plan for an overall Allied air commander was dropped for good at the SEXTANT Conference because both the British and American camps feared the position would be filled by the other side.

42. Eaker to Arnold, Feb 26, 1943, file BC to GB no. 226, box 48, Arnold Papers; Portal through Welch for Arnold, Dec 14, 1943, file 145.95 (WP–III–A3), book 2, USAFHRA; USSBS, *A Study on the Bombing Accuracy of the USAAF Heavy and Medium Bombers in the ETO*, Rpt no. 63 (Washington, D.C.: GPO, 1947), p 1.

43. "Minutes of General and Specific Staff Meeting," Mar 29, 1943, file 520.141–1, USAFHRA; Kepner to Giles, n.d. and Dec 27, 1943, cited in Craven and Cate, eds., *Army Air Forces in World War II*, vol 3, pp 11, 811; Eaker to Comdrs, 9th AF and VIII Ftr Comd, and to Chief of Staff, 8th AF, Oct 31, 1943, file 520.161–1, USAFHRA.

44. Arnold to Eaker, Jun 15, 1943, file no. 227, box 48, Arnold Papers; Eaker to Arnold, Jul 20, 1943, file 168.491, vol 1, USAFHRA; Coffey, *HAP*, p 314; Arnold to Giles, Jan 5, 1944, file 168.491, vol 5, USAFHRA; intvw, James C. Hasdorff with Lt Gen Barney M. Giles, Nov 20–21, 1974, K239.0512–814, USAFHRA; Eaker to Arnold, Jul 20, 1943, file 168.491, vol 1, USAFHRA; minutes, 8th AF Comdrs Mtg, Jun 7, 1943, file 520.141–1, USAFHRA; Murray, *Luftwaffe*, p 168; Arnold/Marshall to Eaker, Sep 21, 1943, file 520.1622, USAFHRA; Eaker to Kepner and Brereton, Oct 31, 1943, file 519.1612, USAFHRA.

45. Giles intvw, Nov 20–21, 1974; Benjamin Franklin Cooling, ed., *Case Studies in the Achievement of Air Superiority* (Washington, D.C.: Center for Air Force History, 1994), p 285.

Those who, in recent works, have cited Arnold as the person most responsible for the change are Parton, *"Air Force Spoken Here,"* chaps 18, 19, and Richard G. Davis, *Carl A. Spaatz and the Air War in Europe* (Washington, D.C.: Center for Air Force History, 1993), pp 271–280. The essential difference between the interpretation of the change by these historians is that Parton, who was Eaker's aide in the war, feels that Arnold was often irrational and unpredictable and could have prevented Eaker's stunned reaction to the change with a personal letter at the time he cabled Eaker of the change (p 336). Davis, though not totally in agreement with Arnold's method of removing Eaker from command of Eighth Air Force, seems to agree with Arnold's motives. Arnold was clearly less than satisfied with Eaker's performance. On December 27, 1943, for example, Arnold's chief of staff, Barney Giles, wrote Spaatz to advise him that "General Arnold has not been satisfied with the efforts made to date." Giles to Spaatz, Dec 27, 1943, file 168.491, vol 2, USAFHRA.

46. The development of a long-range escort is covered in Benjamin S. Kelsey, *The Dragon's*

Teeth? (Washington, D.C.: Smithsonian Institution Press, 1982), and in Stephen L. McFarland, "The Evolution of the American Strategic Fighter in Europe, 1942–1944," *Journal of Strategic Studies*, vol 10 (1987), pp 189–208. See also I. B. Holley, Jr, "Of Saber Charges, Escort Fighters, and Spacecraft: The Search for a Doctrine," *Air University Review*, vol 34 (Sep–Oct 1983), pp 6–10.

47. Bernard Boyland, "The Development of the Long-Range Escort Fighter," 1955, file 101–136, pp 47–53, 118–183, USAFHRA; Kelsey, *Dragon's Teeth?* p 65; Ray Wagner, *American Combat Planes* (Garden City, N.Y.: Doubleday, 1968), pp 212–233.

48. Doolittle to Stratemeyer, Jan 31, 1943, file 168.491, vol 1, USAFHRA; Jimmy Doolittle with Beirne Lay, Jr., "Daylight Precision Bombing," in *Impact* (New York: James Parton, 1980), vol 6, p xv; Doolittle to McFarland and Newton, Oct 28, 1985.

49. Craven and Cate, eds., *Army Air Forces in World War II*, vol 3, p 48; Hist, 8th AF headquarters, Feb 1944, file 520.01, USAFHRA; narrative of Ops, VIII Ftr Comd, 1943–1944, esp Jan 24, 29, and 30, 1944, file 168.6005–55, USAFHRA; Wesley P. Newton and Calvin F. Senning, *USAF Credits for the Destruction of Enemy Aircraft, World War II* (USAF Hist Study 85, Maxwell AFB, Ala.: Office of Air Force History, 1978); Daily Bul no. 32, Feb 1, 1944, file GP–352–HI(FTR), USAFHRA.

50. Gp Hists, 354th, 20th, and 78th Ftr Gps, Feb 1944, files GP–354–HI(FTR), GP–20–HI(FTR), and GP–78–HI(FTR), USAFHRA.

51. 8th AF, "History of Eighth Air Force," 1942–1945, vol 1, pp 16–17, file 520.01, USAFHRA; Spaatz to Arnold, Apr 10, 1944, file 519.161–4, USAFHRA. Doolittle's reference to the *Luftwaffe* tactic of ignoring escort fighters was based on Hermann Goering's order in December 1943 for the *Luftwaffe* to limit attacks on Allied bombers and to avoid the fighters. This was a fatal mistake, according to *Luftwaffe* General of Fighters Adolf Galland in a interrogation at the end of the war, In fact, the *Luftwaffe* had always tended to concentrate on the bombers. Boyland, "Development of the Long-Range Escort Fighter," p 203; interrogation by the RAF of Adolf Galland, *General-leutnant Jagdflieger*, "The Birth, Life, and Death of the German Day Fighter Arm," ADI (K) 273/1945, copy in file 168.6005–82, USAFHRA.

52. Kepner to all units, Feb 14, 1944, file 524.03, USAFHRA; 8th AF HQ Hist, Feb 1944; conference memo, Feb 19, 1944, subject file, box 14, Spaatz Papers.

53. Arnold to Spaatz, Dec 27, 1943, subject file, box 14, Spaatz Papers; Arnold to 8th AF, Dec 27, 1943, file 168.491, vol 5, USAFHRA; Spaatz to Arnold, Jan 21, 1944, and memo, Feb 19, 1944, subject file, box 14, Spaatz Papers; minutes, comdrs mtg, Feb 8, 1944, file 520.141–1, USAFHRA; Arnold to Doolittle, Feb 11, 1944, file 519.245–1, USAFHRA; Doolittle to Arnold, Mar 4, 1944, file 168.6007–2, USAFHRA.

54. Plans for ARGUMENT, Nov 2–29, 1943, file 508.401, USAFHRA; Plans for ARGUMENT, Dec 1943, file 529.4231B, USAFHRA.

55. Intvw, Col C. G. Williamson, Jun 14, 1944, subject file, Big Week folder, box 136, Spaatz Papers; Anderson Diary, entry for Feb 19, 1944, box 316, Spaatz Papers.

56. Webster and Frankland, *Strategic Air Offensive*, vol 2, p 205; memos, Feb 1944, subject file, Big Week folder, box 169, Spaatz Papers.

57. The first-person narrative account that follows is from intvw, Wesley P. Newton and Stephen L. McFarland with Col William R. Lawley, Jr., USAF (Ret), Mar 21, April 5, and May 8, 1985. First Lieutenant Lawley was awarded the Congressional Medal of Honor for his combat action on Feb 20, 1944, the first day of Big Week. The copilot was Lt. Paul Murphy.

58. John F. Ramsey, The War against the *Luftwaffe*: AAF Counter-Air Operations, Apr 1943–Jun 1944 (USAF Hist Study 110, Maxwell AFB, Ala.: Air University Press, Aug 1945), pp 116, 124–126, 140–142.

59. USSBS, *The Defeat of the German Air Force*, Rpt no. 59 (Washington, D.C.: GPO, 1947), table 5–5, pp 61–66; Freeman, *Mighty Eighth*, pp 183–188; Kenn C. Rust, *Eighth Air Force Story* (Temple City, Calif.: Historical Aviation Album, 1978), p 32; USSBS, *Aircraft Division Industry Report*, pp 60–66, 69; Ramsey, "War against the *Luftwaffe*," pp 140–42; USSBS, *Overall Report (European War)*, Rpt no. 2 (Washington, D.C.: GPO, 1945), p 2; Craven and Cate, eds., *Army Air Forces in World War II*, vol 3, pp 43–46; "Material Behind the 'Big Week,'" Feb 20–25, 1944, file 519.04–1, USAFHRA; John to Art [employees of the USAF Historical Office in England researching captured German documents], Jun 28, 1949, file K110.8–22, USAFHRA; narrative of Ops, VIII Ftr Comd, Feb 20–25, 1944, file 168.6005–55, USAFHRA; "The Contribution of Air Power to the Defeat of Germany," 1945, based on "Basic Returns of the German Quartermaster Staff," file 519.601C, vol 4, USAFHRA.

60. Doolittle with Lay, in *Impact*, vol 6, p xv.

61. Arnold to Spaatz, Jan 4, 1944, rpt file, box 248, Arnold Papers; memo, Dec 12, 1943, subject file, box 145, Spaatz Papers; memo, Jan 4, 1944, rpt file, box 248, Arnold Papers; Anderson Diary, entries for Feb 27, 29, 1944. Spaatz's interest in Berlin came at the same time that the Soviet Union officially requested an American bombing of Berlin. "Planning and Organization for Large Scale Raids on Berlin," Jan 12, 1944, File 519.3171–1, USAFHRA.

62. Anderson Diary, entries for Mar 2–9, 1944. Spaatz had already found it necessary to write Doolittle a "Dear Jimmy" letter criticizing him for being too irresolute when it came to weather. Spaatz to Doolittle, Jan 28, 1944, file 519.1612, USAFHRA. Spaatz's diary for Feb 9, 1944, carried the following: "Today is to go on record as completely wasted. Good weather at bases, good weather over target, and Doolittle sent *no* bombers." Spaatz personal diaries, entry for Feb 9, 1944, box 14, file Feb 1944, Spaatz Papers.

63. Anderson Diary, entry for Mar 2, 1944.

64. Intel Summary, Mar 16/31, 1944, file 519.3073, USAFHRA; United Kingdom, Air Ministry, "The Rise and Fall of the German Air Force (1933 to 1945)," file 512.042–248, USAFHRA; special rsch hist, NSC, "ULTRA," p 155; intvw, Stephen L. McFarland and Wesley P. Newton with Lt Gen Günther Rall, Nov 12, 1985.

65. Spaatz to Arnold, Mar 11, 1944, subject file, box 14, Spaatz Papers; ADI (K) 273/1945, pt 3, p 9; "Allied Air Power, Comments and Opinions by High-Ranking German Officers and Industrialists," May 16, 1945, file 519.04–5, USAFHRA; Adolph Galland, *The First and the Last*, trans. Mervyn Savill (New York: Henry Holt, 1959), p 250; ADI (K) 273/1945, pt 3, pp 24, 35.

66. Rall intvw, Nov 12, 1985; narrative of Ops, VIII Ftr Comd, Mar 3, 4, 6, 8, 9, 1944, file 168.6005–55, USAFHRA; Intel Summary, Mar 1–15, 1944, file 519.3070, USAFHRA; intvw, Bruce C. Hopper and Lt Charles A. Foster with Maj Gen William E. Kepner, Jul 15, 1944, file 524.0581, USAFHRA; Anderson Diary, entry for Mar 6, 1944; USSBS, *Overall Report*, p 9; *General-leutnant* Josef Schmid, "The Struggle for Air Supremacy over the Reich, Jan 1, 1944–March 31, 1944" [claims based on the war diary of the *Luftwaffe's* I Fighter Corps], 1954, files K113.107–158 through –160, USAFHRA; "The Contribution of Air Power to the Defeat of Germany" [based on "Basic Returns of the German Quartermaster Corps"], 1945, vol 4, file 519.601C, USAFHRA. The story of the winning of air superiority by the AAF in 1944 in time for OVERLORD is told in Stephen L. McFarland and Wesley Phillips Newton, *To Command the Sky: The Battle for Air Superiority over Germany, 1942–1944* (Washington, D.C.: Smithsonian Institution Press, 1991).

67. F. H. Hinsley *et al.*, *British Intelligence in the Second World War: Its Influence on Strategy and Operations*, 5 vols (New York: Cambridge University Press, 1979–1990), vol 3, pt 1, pp 369–412; Stephen L. McFarland, "Preparing for What Never Came: Chemical and Biological Warfare in World War II," *Defense Analysis*, vol 2 (1986), pp 107–121.

68. Craven and Cate, eds., *Army Air Forces in World War II*, vol 2, pp 85, 92. The Allied air forces did not attack poison gas plants for fear any coincidental release of gas might encourage the Germans to claim the Allies had initiated gas warfare. Plants involved in the production of raw materials for the gas plants were targeted. USSBS, *Powder, Explosives, Special Rocket and Jet Propellants, War Gases and Smoke Acid*, Rpt no. 111 (Washington, D.C.: GPO, 1947), p 54. Operation CROSSBOW began officially on November 15, 1943.

69. Craven and Cate, eds., *Army Air Forces in World War II*, vol 2, pp 97–99, 103, and vol 3, p 535; USSBS, *V-weapons (CROSSBOW) Campaign*, Rpt no. 60 (Washington, D.C.: GPO, 1947), p 26; "The Contribution of Air Power to the Defeat of Germany," Jan 1944–May 8, 1945, file 519.601C, app J, USAFHRA.

70. Craven and Cates, eds., *Army Air Forces in World War II*, vol 3, pp 529, 533; USSBS, *V-weapons*, pp 27, 31, exhibit 9; David Irving, *The Rise and Fall of the Luftwaffe* (Boston: Little, Brown, 1973), p 268.

71. Webster and Frankland, *Strategic Air Offensive*, vol 3, pp 18–22; Craven and Cate, eds., *Army Air Forces in World War II*, vol 3, pp 72–77; Davis, *Carl A. Spaatz*, pp 300.

72. Spaatz to Arnold, Mar 6, 1944, with "Plan for Completion of the Combined Bomber Offensive," Mar 5, 1944, file BC to GB no. 226, box 48, Arnold Papers; Davis, *Carl A. Spaatz*, pp 331–339.

73. Webster and Frankland, *Strategic Air Offensive*, vol 3, pp 28–30; Craven and Cate, eds., *Army Air Forces in World War II*, vol 3, pp 76–79; Rostow, *Pre-Invasion Bombing Strategy*, pp 3–6, 44–51, app A, "Final Minutes of the March 25 Meeting," pp 88–98; Hansell, *Strategic Air War*, pp 104–105; Davis, *Carl A. Spaatz*, pp 331–339; Anderson Diary, entry for Mar 25, 1944; Spaatz to Arnold, Mar

24, 1944, and Eaker to Arnold, Mar 1944, file 168.491, vol 5, USAFHRA.

74. Eaker to Arnold, Apr 8, 1944, file 168.491, USAFHRA; Davis, *Carl A. Spaatz*, pp 352–353.

75. Kepner intvw, Jul 14, 1944; McFarland, "Evolution of the American Strategic Fighter," pp 20–26.

76. Portal to Spaatz, Apr 13, 1944, file 519.1612–3, USAFHRA; Webster and Frankland, *Strategic Air Offensive*, vol 3, pp 36–37, and copy of Apr 17 directive in vol 4, app 8; Winston S. Churchill, *The Second World War*, 6 vols (Boston: Houghton Mifflin, 1950), vol 5, pp 528–530 [contains the exchange of correspondence with Eisenhower and Roosevelt concerning the bombing of the railroad centers]; Davis, *Carl A. Spaatz*, pp 400–403.

77. Craven and Cate, eds., *Army Air Forces in World War II*, vol 3, pp 138–153; Webster and Frankland, *Strategic Air Offensive*, vol 3, p 39. Webster and Frankland (*Strategic Air Offensive*, vol 3, pp 5, 39) indicate the railroad attacks alone, with Bomber Command leading the way, were a success. John E. Fagg (in Craven and Cate, eds., *Army Air Forces in World War II*, vol 3, pp 156–57), Rostow (*Pre-Invasion Bombing Strategy*, pp 59–65), and Hansell (*Strategic Air War*, p 107) contend that adding the bridges capped the pre-OVERLORD transportation performance, which was lagging beforehand.

78. Arnold to Spaatz, Apr 24, 1944, file 168.491, vol 5, USAFHRA; McFarland, "Evolution of the American Strategic Fighter."

79. "Historical Study of Air Support by the Eighth Air Force for the Land Invasion of Europe, Jun 2–16, 1944," file 168.6005–72, USAFHRA; "Air Operations over the Invasion Front in Jun 1944," 8th *Abteilung*, Aug 27, 1944, trans. British Air Ministry, Air Historical Branch, Jun 24, 1947, file 512.621, 7/32, USAFHRA; Freeman, *Mighty Eighth*, pp 248–267; special rsch hist, NSC, "ULTRA," p 196; Galland, *First and the Last*, pp 273–293.

80. Von D. Hardesty, *Red Phoenix: The Rise of Soviet Air Power, 1941–1945* (Washington, D.C.: Smithsonian Institution Press, 1982), pp 199–203; Kuter to Loutzenheiser, Sep 7, 1943, file 145.95 (WP–III–A3), book 2, USAFHRA; Richard C. Lukas, *Eagles East: The Army Air Forces and the Soviet Union, 1941–1945* (Tallahassee: Florida State University Press, 1970), pp 192–201; Giles to Eaker, Sep 7, 1944, file 168.491, vol 5, USAFHRA.

81. Hansell, *Strategic Air War*, p 129; Craven and Cate, eds., *Army Air Forces in World War II*, vol 3, pp 287–308; Davis, *Carl A. Spaatz*, pp 384–400; intvw, Bruce C. Hopper and Lt. Charles A. Foster with Carl A. Spaatz, May 20, 1945, file 519.1612, USAFHRA; special rsch hist, NSC, "ULTRA," pp 177–179; Albert Speer, *Inside the Third Reich* (New York: Macmillan, 1970), p 346.

82. Speer to Hitler, Jun 30, 1944, file 137.1–3, USAFHRA; Davis, *Carl A. Spaatz*, pp 442–443.

83. Craven and Cate, eds., *Army Air Forces in World War II*, vol 3, pp 298, 287; special rsch hist, NSC, "ULTRA," pp 205, 212–13, 216–17; Speer to Hitler, Jul 29, 1944, file 137.1–3, USAFHRA; Davis, *Carl A. Spaatz*, p 443–445.

84. Speer to Hitler, Oct 5, 1944, and Jan 19, 1945, file 137.1–3, USAFHRA; special rsch hist, NSC, "ULTRA," pp 25, 234, 291, 344, 270.

85. Henry L. Stimson, "Notes on a Conference with General Doolittle," Jul 20, 1944, and Arnold to Doolittle, Jul 26, 1944, file BC to GB no. 226, box 48, Arnold Papers; Arnold to Spaatz, Sep 21, 1944, file 168.491, USAFHRA; Gp Hist, 445th Bom Gp, Sep 1944, file GP–445–HI(BOM), USAFHRA.

86. ADI (K) 273/1945, pt 3, p 59; Davis, *Carl A. Spaatz*, pp 518–519; Spaatz to Arnold, Nov 5, 1944, Arnold to Spaatz, Nov 30, 1944, and Dec 30, 1944, file 168.491, vol 2, USAFHRA.

87. Arnold to Spaatz, Sep 29, 1944, and Eisenhower to Arnold, Oct 26, 1944, file BC to GB no. 227, box 48, Arnold Papers; Spaatz to Arnold, Sep 30, 1944, file 168.491, vol 2, USAFHRA.

88. Arnold to Spaatz, Sep 29, 1944, file BC to GB no. 227, box 48, Arnold Papers; directive no. 2 for the control of Strategic Air Forces, Oct 28, 1944, file 168.491, vol 2, USAFHRA; David R. Mets, *Master of Airpower: General Carl A. Spaatz* (Novato, Calif.: Presidio Press, 1988), pp 259–261; Webster and Frankland, *Strategic Air Offensive*, vol 3, pp 63–74, 80–81.

89. Spaatz to Arnold, Dec 13, 1944, file 168.491, vol 2, USAFHRA; Craven and Cate, eds., *Army Air Forces in World War II*, vol 3, pp 672–711.

90. Craven and Cate, eds., *Army Air Forces in World War II*, vol 3, p 695.

91. Mets, *Master of Airpower*, pp 261–266; Craven and Cate, eds., *Army Air Forces in World War II*, vol 3, p 665; USSTAF, "Results of Dec 1944 and January 1945 Operations," Feb 1945, file 519.551–6, USAFHRA; Davis, *Carl A. Spaatz*, pp 542–543; Spaatz to Arnold, Jan 16, 1945, incl directive no. 3 for the Strategic Air Forces, Jan 12, 1945, file 168.491, vol 2, USAFHRA.

92. Davis, *Carl A. Spaatz*, pp 548–549, 553, 556, 570–573.

93. Craven and Cate, eds., *Army Air Forces in World War II*, vol 3, p 638; Spaatz to Arnold, Mar 24, 1945, file 168.491, vol 5, USAFHRA; Craven and Cate, eds., vol 3, p 535; Schaffer, *Wings of Judgment*, pp 62, 88–89; Davis, *Carl A. Spaatz*, pp 432–439, 543–553. Richard G. Davis has probably the best account of the evolution of THUNDERCLAP and — in contrast to Craven and Cate, Mets, and Sherry — argues for a low casualty figure of some 1,500 on the basis of the mainly visual nature of the bombing and the low mix of incendiaries. He feels that the February 3 Berlin raid, more than the one in Dresden, saw the Americans depart from their policy of not deliberately and directly bombing civilians. Davis, "Operation THUNDERCLAP: The U.S. Army Air Forces and the Bombing of Berlin," *Journal of Strategic Studies*, vol 14, no 1 (Mar 1991), pp 90–111.

94. "Prospect for Ending War by Air Attack against German Morale," Mar 5, 1944, file 519.318–1, USAFHRA; David MacIsaac, *Strategic Bombing in World War Two: The Story of the United States Strategic Bombing Survey* (New York: Garland, 1976), pp 78–79; Spaatz to Arnold, Jan 7, 1945, in Kuter memo to Arnold, Jan 15, 1945, file 519.612, USAFHRA; Schaffer, *Wings of Judgment*, pp 96, 98–103; Anderson Diary, entry for Jan 31, 1945; Davis, *Carl A. Spaatz*, pp 556–565; Sherry, *Rise of American Air Power*, pp 259–263. Sherry (p 262) characterizes the Dresden raids, and by implication the February 3 raid on Berlin, as "less the product of conscious callousness than of casual destructiveness." He and David Mets (*Master of Airpower*, p 275) stress Dresden as the most significant departure from established policy.

95. Davis, *Carl A. Spaatz*, p 53, 56; USSBS, *The United States Strategic Bombing Survey: Summary Report (European War)*, Rpt no. 1 (Washington, D.C.: GPO, 1945), p 2.

96. USSBS, *Statistical Appendix*, table 1, p 1; USSBS, *Summary Report*, pp 2–3.

97. Werrell, "Strategic Bombing of Germany in World War II," p 708; USSBS, *Summary Report*, pp 6–7; USSBS, *Overall Report*, pp 11–16.

98. Webster and Frankland, *Strategic Air Offensive*, vol 2, p 52.

99. USSBS, *Overall Report*, p 11.

100. USSBS, *Summary Report*, p 12.

101. MacIsaac, *Strategic Bombing in World War Two*, pp 1–3, 26–28, 154–167.

102. USSBS, *Overall Report*, p 22; USSBS, *Defeat of the German Air Force*, p 3.

103. Fuel shortages throughout the war led the *Luftwaffe* to give its pilots one-half the training American pilots received through June 1944 and one-third thereafter. USSBS, *Overall Report*, chart 10.

104. USSBS, *Defeat of the German Air Force*, p 34; USSBS, *Statistical Appendix*, p 1, table 1.

105. Special rsch hist, NSC, "ULTRA," pp 100–101; "American Bombardment and the German Air Force," n.d., file 519.04–7, USAFHRA; USSBS, *Overall Report*, p 11.

106. Interrogation, of Willy Messerschmitt, Jun 25, 1945, file 519.607A–82, USAFHRA; USSBS, *Aircraft Division Industry Report*, pp 23, 7; USSBS, *Messerschmitt A.G., Augsburg, Germany, Part B*, Rpt no. 11 (Washington, D.C.: GPO, 1947), vol 3, p 13; USSBS, *Defeat of the German Air Force*, p 32; USSBS, *Erla Maschinenwerke GmbH, Heiterblick, Germany*, Rpt no. 7 (Washington, D.C.: GPO, 1947), pp 2–4; USSBS, *Overall Report*, p 10; Werrell, "Strategic Bombing of Germany in World War II," p 711; USSBS, *A Brief Study of the Effects of Area Bombing on Berlin, Augsburg, Bochum, Leipzig, Hagen, Dortmund, Oberhausen, Schweinfurt, and Bremen*, Rpt no. 39 (Washington, D.C.: GPO, 1947), p 35; USSBS, *Junkers Aircraft and Aero Engine Works Dessau, Germany*, Rpt no. 6 (Washington, D.C.: GPO, 1947), p 5; USSBS, *Gothaer Waggonfabrik, AG Gotha, Germany*, Rpt no. 9 (Washington, D.C.: GPO, 1947), p 11.

107. German Air Force Staff, Plans Division, "The Anglo-American Air War against Germany," Apr 25, 1944, file 137.317–3, USAFHRA; Hansell, *Strategic Air War*, pp 34–35; USSBS, *Messerschmitt AG, Augsburg, Germany, Part A*, vol 2, p 26; USSBS, *Summary Report*, pp 6, 14–15; USSBS, *Overall Report*, pp 25, 29, 40, 45, 82–84, 108; Eaker and Metcalf, "Conversations with Albert Speer," pp 54–55; USSBS, *Aircraft Division Industry Report*, p 8.

108. USSBS, *Summary Report*, p 2; USSBS, *Overall Report*, pp 85–86.

Bibliographic Essay

The starting point for any study of American strategic bombardment in Europe during World War II must be the documents collection of the U.S. Air Force Historical Research Agency (USAFHRA) at Maxwell AFB, Alabama. Immense and, most importantly, organized and catalogued, the collection includes some official and personal papers of major participants in the American air war, captured German documents, unit histories ranging from USSTAF's to those of the squadrons and detachments, and the working papers of the Air Staff. These are but some of the USAFHRA's treasures. The unit histories represent personal records of dramatic events and personalities largely untapped by scholars. Microfilm copies of the USAFHRA's collection are available at the United States Air Force History Support Office, Bolling AFB, Washington, D.C.

The papers of the major decision-makers identified in this study are an additional primary source for understanding the AAF strategic bombardment campaign. The Manuscript Division of the Library of Congress maintains the papers of Arnold, Spaatz, and Eaker; important supplements reside at USAFHRA. The Arnold Operational Letters at Maxwell AFB, for example, include correspondences of Arnold, the Air Staff, and various commanders. Doolittle has a few papers in the Manuscript Division, but the bulk is being deposited at the University of Texas in Dallas. Frederick Anderson's papers are at the Hoover Institute, Stanford University, but his war diary and some of his correspondence can be found in the Manuscript Division as part of the Spaatz collection. The USAFHRA houses the Kepner papers, including VIII Fighter Command field orders, operations reports, and interviews.

The official history, *The Army Air Forces in World War II* 7 vols (Chicago: University of Chicago Press, 1948–1958; Washington, D.C., Office of Air Force History, 1983), eds Wesley Frank Craven and James Lea Cate, has held its own against all the monographs and articles researched and written since its original appearance more than three decades ago. The weaknesses of *The Army Air Forces in World War II* are generally those related to its official nature and its publication so soon after the war. The authors were too uncritical of the actions of leading American figures and too critical of British officials. They knew nothing or could reveal nothing of ULTRA, and they relied occasionally on their own reminiscences of the war unsupported by documentation, as in the case of the "stand-down" from unescorted flights deep into Germany after the second attack on Schweinfurt.

The official British record of the Combined Bomber Offensive is Sir Charles Webster and Nobel Frankland's four-volume *Strategic Air Offensive against Germany* (London: HMSO, 1961). It is basic for the British effort, but it is too pro-Portal and pro-Tedder and too hard on Harris. It treats the role of the U.S. AAF fairly.

Biographies of the major characters in the American strategic bombing effort are available, though their quality is uneven. At the top rests Richard G. Davis's *Carl A. Spaatz and the Air War in Europe* (Washington, D.C.: Office of Air Force History, 1993). Based on an exhaustive analysis of the Spaatz papers, it is the best single volume work on Anglo-American relations in the cause of tactical and strategic operations. David Mets's biography, *Master of Airpower: General Carl A. Spaatz* (Novato, California: Presidio Press, 1988), covers Spaatz's life and, although insightful, lacks a critical balance due to editorial decisions. James Parton's *"Air Force Spoken Here": General Ira Eaker and the Command of the Air* (Bethesda, Md.: Adler & Adler, 1986), is based on extensive research and a long, personal acquaintance with Eaker, and hence tends to be overly sympathetic to the subject. Thomas M. Coffey's *HAP: The Story of the U.S. Air Force and the Man Who Built It, General Henry H. "Hap" Arnold* (New York: Viking, 1982), is more popular and useful, mostly for its account of his career development, his family relationships, and his health. Somewhat more useful than Coffey's work is the popular biography of Doolittle by Lowell Thomas and Edward Jablonski, *Doolittle: A Biography* (Garden City, N.Y.: Doubleday, 1976). It tells of Doolittle's "freeing" of the fighters, among its other contributions. DeWitt Copp's *A Few Great Captains: The Men and Events That Shaped the Development of U.S. Air Power* (Garden City, N.Y.: Doubleday, 1980) is detailed, but would have been more useful if fully documented. Paul F. Henry has a useful short biography of Kepner entitled "William E. Kepner: All the Way to Berlin," in John L. Frisbee, ed., *Makers of the United States Air Force* (Washington, D.C.: Office of Air Force History, 1987).

Autobiographies of Arnold, Doolittle, and Hansell are also available. Arnold's memoir, *Global Mission* (New York: Harper & Row, 1949), leaves out a lot, but it is the best available book about the commander of the AAF until Richard G. Davis completes his biography of the AAF commander. Doolittle, with the aid of Carroll V. Glines, produced a memoir, *I Could Never Be So Lucky Again* (New York: Bantam Books, 1991), containing some fascinating anecdotes and revealing insights, but much is glossed over including his role in the winning of air superiority in 1944. The semiautobiographical works of Haywood S. Hansell, Jr., including *The Air Plan That Defeated Hitler* (Atlanta: Higgens/McArthur/Longino & Porter, 1972) and *The Strategic Air War Against Germany and Japan* (Washington, D.C.: Office of Air Force History, 1986), are especially valuable for first-hand accounts of the evolution of the air war plans.

The consensus among scholars of the air war seems to be that R. J. Overy's *The Air War, 1939–1945* (New York: Stein & Day, 1981) is the best short, general history. As yet it has no close rival. A recent study of the evolution of U.S. air might is Michael S. Sherry's *The Rise of American Air Power: The Creation of Armageddon* (New Haven, Conn.: Yale University Press, 1987). The subtitle reveals the driving theme of a book whose climax is in the watershed years of World War II. The variety of Sherry's sources for a work with sociological, psychological, philosophical, and doomsday ramifications is enviable. He dissects the role of such figures as Arnold with harsh but convincing logic. Serviceable is the popular history, *Fortress without a Roof: The Allied Bombing of the Third Reich* (New York: St. Martin's Press, 1982), by Wilbur H. Morrison. W. A. Jacobs has a provocative article on "Strategic Bombing and American National Strategy, 1941–1943" in *Military Affairs*, vol 50 (Jul 1986), pp 133–139. A fine semischolarly book on one of the great air battles is Martin Middlebrook's *The Schweinfurt-Regensburg Mission* (New York: Charles Scribner's Sons, 1983).

Lee Kennett's *A History of Strategic Bombardment* (New York: Charles Scribner's Sons, 1982) places the subject in its general historical context with operational, moral, and philosophical themes. A significant study of the moral and morale aspects of strategic bombardment, especially by the AAF (though it strains to prove its point) is Ronald Schaffer's *Wings of Judgment: American Bombing in World War II* (New York: Oxford University Press, 1985). A more balanced postrevisionist evaluation of this issue is Conrad C. Crane's *Bombs, Cities, and Civilians: American Airpower Strategy in World War II* (Lawrence: University Press of Kansas, 1993). Also of interest is Richard G. Davis, "German Rail Yards and Cities: U.S. Bombing Policy 1944–1945," *Air Power History*, vol 42 (summer 1995), pp 46–63. An article that places the strategic bombing of Germany into the perspective of the Western concept of a "just war" is Louis A. Manzo, "Morality in War Fighting and Strategic Bombing in World War II," *Air Power History*, vol 39 (fall 1992), pp 35–50.

Probably the best book on the *Luftwaffe* in English, based on the extensive use of both American and German sources, is Williamson Murray's *Luftwaffe* (Baltimore: Nautical & Aviation Publication, 1985). The development of the long-range escort and strategic fighters is well served by a participant in their development, Benjamin S. Kelsey, in his *The Dragon's Teeth*? (Washington, D.C.: Smithsonian Institution Press, 1982), and by Stephen L. McFarland in "The Evolution of the American Strategic Fighter, 1942–1944," *Journal of Strategic Studies*, vol 10 (Jun 1987), pp 190–208. McFarland and Wesley Phillips Newton have a scholarly study of the AAF's winning of air superiority in Europe, *To Command the Sky: The Battle for Air Superiority over Germany, 1942–1944* (Washington, D.C.: Smithsonian Institution Press, 1991). Walt W. Rostow in *Pre-Invasion Bombing Strategy: General Eisenhower's Decision of March 25, 1944* (Austin: University of Texas Press, 1981), defends the case for oil. Operation FRANTIC is the subject of a prize-winning scholarly study by Richard C. Lukas, *Eagles East: The Army Air Forces and the Soviet Union, 1941–1945* (Tallahassee: Florida State University Press, 1970).

The effect of ULTRA on the air war is scrutinized in in the special research history prepared by the National Security Council, "ULTRA: History of the U.S. Strategic Air Forces vs. German Air Force," a copy of which can be found in Record Group 457 of the National Archives in Washington, D.C.; in Diane T. Putney, ed., *ULTRA and the Army Air Forces in World War II: An Interview with Associate Justice of the U.S. Supreme Court Lewis F. Powell, Jr.* (Washington, D.C.: Office of Air Force History, 1987); and in F. H. Hinsley *et al.*, *British Intelligence in the Second World War*, 5 vols (London: HMSO, 1978–1990).

Scholars are just now beginning to utilize another source of generally untapped primary documentation. Though first-hand accounts must be carefully weighed against certain inherent weaknesses, oral histories of some of the thousands of men and women who served in the AAF during the war await systematic development, analysis, and usage.

Printed documents on the AAF's participation in strategic bombardment are surprisingly rare. The only outstanding collection of printed documents comprises the 216 volumes of the United States Strategic Bombing Survey. The work of the survey group is definitively told by David MacIsaac in *Strategic Bombing in World War Two: The Story of the United States Strategic Bombing Survey* (New York: Garland, 1976).

Recently, historians have tried to reevaluate the effect of strategic bombing on Germany in World War II. R. J. Overy, in "Hitler's War and the German Economy: A Reinterpretation," *Economic History Review*, vol 35 (May 1982), pp 272–291, believes the Germany economy failed to supply German forces because the "larger war arrived before preparations for it were complete." Alfred C. Mierzejewski, *The Collapse of the German War Economy, 1944–1945: Allied Air Power and the German National Railway* (Chapel Hill: University of North Carolina Press, 1988), argues that the somewhat tardy attacks on transportation, specifically rail marshaling yards, eliminated the one link, coal, that united the German war economy. He fails, however, to consider the direct impact of the oil bombings on the German military. Henry D. Lytton, "Bombing Policy in the Rome and Pre-Normandy Invasion Aerial Campaigns of World War II: Bridge-Bombing Strategy Vindicated — and Railyard-Bombing Strategy Invalidated," *Military Affairs*, vol 47 (Apr 1983), pp 53–58, asserts that transportation was indeed the key, but that its destruction was best accomplished by bombing bridges. Richard G. Davis, "Bombing Strategy Shifts, 1944–45," *Air Power History*, vol 36 (winter 1989), pp 33–45, provides a more balanced analysis, recognizing the economic impact of the coal/transportation effort and the military impact of the oil campaign. *Air Power History* also has carried the remarks of scholars and historical participants on joint RAF–USAAF concerns in World War II at a seminar held in England in 1990. One of the panel's proceedings appeared as "Joint RAF/USAF Seminar, Part II: The Strategic Air Offensive in Europe," *Air Power History*, vol 38 (fall 1991)], pp 39–48. Kenneth P. Werrell has produced a useful comparison of "Friendly Rivals: 8th and 15th Air Forces in World War II," *Air Power History*, vol 38 (summer 1991), pp 36–42.

Two recent books have examined the strategic bombing campaign from the inside out. Stephen L. McFarland, *America's Pursuit of Precision Bombing, 1910–1945* (Washington, D.C.: Smithsonian Institution Press, 1995), examines the problem of bombing accuracy and the impact it had on the technology, tactics, targets, and effectiveness of the effort. Charles W. McArthur, *Operations Analysis in the U.S. Army Eighth Air Force in World War II* (Providence, R.I.: American Mathematical Society, 1990), concentrates on the means Eighth Air Force used to determine bomb sizes and types, fuze settings, target selection, and bombing procedures.

For descriptions of the missions, units, order of battle, and bombing statistics, five sources are readily available. Roger A. Freeman, *Mighty Eighth War Diary* (London: Jane's, 1981), is the essential record for the day-to-day actions of the Eighth Air Force. It also provides anecdotes to spice up the otherwise bare statistics. Report no. 2a of the United States Strategic Bombing Survey, *Statistical Appendix to Overall Report* (*European War*) (Washington, D.C.: GPO, 1945), is perhaps the most authoritative source for statistical information. For a bare-bones chronology, see *U.S. Army Air Forces in World War II: Combat Chronology, 1941–1945* (Washington, D.C.: Center for Air Force History, 1991). Maurer Maurer, in *Combat Squadrons of the Air Force: World War II* (Maxwell AFB, Ala: USAF Historical Division, 1969), supplies important dates, base assignments, commanders, insignias, campaigns, and aircraft for all AAF wartime combat squadrons. *Air Force Combat Units of World War II* (Washington, D.C.: Office of Air Force History, 1983) does the same for AAF organizations from the combat group through the numbered air forces and USSTAF.

4

Strategic Bombing in the Pacific 1942–1945

Alvin D. Coox

During the 1920s and 1930s, military leaders on both sides of the Pacific Ocean could agree, at least implicitly, that in some unspecified future, hostilities between the United States and Japan were a distinct possibility. Indeed, the U.S. Navy based most of its fleet in the Pacific because of "the unspoken conviction of every senior naval officer that the next war would be with Japan, the only maritime power likely on any near or remote contingency to challenge the United States."[1] In Japan as early as 1921, a retired Army lieutenant general, Kōjirō Satō, published a widely read treatise, *If Japan and America Fight*. In it he claimed that America posed a "threat" to Japan's existence comparable to the threat posed by the Mongolian forces of Kublai Khan in the 13th century and those of Tsarist Russia in 1904.[2]

Other hypothetical national enemies, however, demanded consideration in American and Japanese contingency planning and diffused the strategic priorities. Thus, in the 1930s, the attention of the American government in general, and of the U.S. Army in particular, was increasingly riveted on the Nazi Third Reich and Europe, not on Japan and the Pacific. This focus of attention would underlie a decision by U.S. leaders, even after the Japanese attack on Pearl Harbor, to pursue a "Germany-first" military strategy in World War II.[3]

The Japanese and American Dimension

For its part, the Imperial Japanese Army (IJA) envisaged fighting enemies that differed from those confronting its sister service, the Imperial Japanese Navy (IJN). The IJA's traditional foes were nearby continental land powers, primarily the Soviet Union and China. The IJN's hypothetical enemies, after the fateful sundering of the Anglo-Japanese alliance in the early 1920s, were two first-class but distant naval and maritime powers: Britain and the United States.[4] Insofar as military aviation was concerned, the identities of the national foes and the geographical difficulties of the anticipated theaters of operation directly affected doctrine, training, and materiel requirements.

In the absence of a separate air force organization at the command level, the Imperial Japanese Army Air Force (IJAAF) was subordinated to land force commanders. Consequently, the IJAAF was obsessed with providing tactical ground support — flying artillery — against massive enemy armies on narrow fronts, often in cold weather and on the Asian continent. But IJAAF heavy bombers, in range and payload resembling medium bombers of other major powers, needed to fly only slightly more than 300 miles from Manchuria to reach targets at Chita in Siberia. "We were in trouble," admitted an IJAAF general, as soon as the undeclared war with China began in 1937 and most certainly when the war with the United States began in 1941. One IJA War College instructor of the day derided IJAAF heavy bombers as nonexistent or useless ornaments, far inferior to their long-range Imperial Japanese Navy Air Force (IJNAF) counterparts.[5]

The IJNAF, though also subordinated to IJN surface commands, possessed a somewhat broader conception of the use of air power than did the IJA. Assigned long-range, overwater missions, IJNAF aircraft had a striking reach that permitted them to hit targets deep within a foe such as China. The newest twin-engine Mitsubishi Type 96 Nell attack bomber, for example, possessed a point-to-point maximum range of more than 1,425 miles. Nevertheless, as the history of World War II would demonstrate, the IJNAF fought mainly in terms of striking enemy warships and in support of amphibious landings and ground forces already ashore.

Some IJA officers, of course, wanted true heavy bombers capable of attacking strategic targets. During the decade after World War I, when the Soviet Union succeeded Russia and that country no longer appeared to be a powerful military factor, and when relations with the United States deteriorated after the Washington naval-limitation accords in 1922, the IJA began to consider the Philippines as a possible theater of operations in a future war. As early as 1927–1928, at military direction, the Mitsubishi firm commenced design work on a large four-engine bomber, the Type 92, or Ki–20, that could strike at such targets as Corregidor from bases in Formosa. Since Japanese aeronautical technology was still primitive, Mitsubishi turned to the German firm of Junkers for production rights to a passenger model. It took three years to manufacture the first military prototype of the Ki–20, which was delivered in 1931 and of which only six were produced.

The Mitsubishi Type 92 was a superheavy bomber, weighing in at 25.4 tons. It had a range of 1,550 miles and a bombload of 2 tons, but its maximum speed was only 110 mph at a time when even the Japanese Type 91 fighters were already flying at 185 mph. Line units had no practical use for the top-secret Type 92 bomber, and it never saw combat. Its swan song occurred during the Imperial review of January 1940, when three of the huge planes flew in formation over Tokyo and dazzled civilian onlookers. Already obsolescent, the Type 92 struck IJA critics as a "lumbering monster." The plane warrants mention, however, as a rare but unsuccessful Japanese approach to meeting the still-nebulous needs of strategic aerial bombardment.[6]

During a visit to the United States in 1937, one IJAAF officer was enthralled by the sight of a four-engine B–17 bomber at Langley Field, Virginia. Proceeding to Washington, D.C., the Japanese officer sought to arrange the purchase of B–17 aircraft from Boeing; his request was denied.[7] In an effort to create Japanese heavy-bomber combat capability, other IJAAF buyers turned to a foreign producer that would sell these aircraft: Italy. A number of twin-engine, twin-tail Fiat BR–20s (designated I–100s by the Japanese) were imported, but they proved unpopular with IJAAF airmen and performed with questionable effectiveness in action against the Russians in the Mongolian border war at Nomonhan in 1939.[8]

Before the outbreak of war in the Pacific Theater, one further effort to develop a long-range aircraft originated in a commercial quarter. In 1939–1940 Asahi Press sponsored a nonstop airplane flight between Tokyo and New York — a distance of more than 9,300 miles — at a minimum cruising speed of 186 mph. Since the IJAAF lacked organic heavy aircraft to attempt even an appreciable portion of the projected flight, IJA Air Staff officers beat the system by working with Tokyo University and the Tachikawa Aviation Company to design and manufacture the Type A–26, or Ki–77, experimental long-range research airplane — without consulting either the hidebound IJA General Staff or the War Ministry. A twin-engine, all-metal aircraft with a loaded weight of about eight tons evolved, but Ki–77 development was overtaken by the more pressing priorities in the theater. Only two were built by 1942–1943. One, intended to fly to Germany, made it to Singapore in 1943, but it was lost over the Indian Ocean. The second survived the war.[9]

The IJAAF experience with the Mitsubishi Type 92/Ki–20 and Tachikawa Type A–26/Ki–77 long-range aircraft illustrated the outlook of both Japanese air services toward bombardment aviation: tentative design approaches, minuscule industrial infrastructure, piecemeal development, and worst of all, an utter lack of vision for strategic employment of long-range bombers in modern warfare. On the other side of the Pacific, however, U.S. Army airmen managed these matters better.

Quite apart from the superior material power and technological prowess of the United States, American airmen in the 1930s had a better understanding of strategic bombardment, its functions, and a doctrine for its employment. The Americans emphasized high-altitude strategic precision bombing directed at military targets and could, therefore, castigate the Axis powers for indiscriminate air strikes against civilian populations in China and Europe. It was but a short step from venting moral condemnation of Japanese outrages to demanding military deterrence or even retaliation on behalf of a brutalized China. Indeed, some administration officials in Washington revealed unusually aggressive private thoughts. A year before the Pearl Harbor attack, Secretary of State Cordell Hull told the sinophile Treasury Secretary, Henry Morgenthau: "What we have to do, Henry, is to get five hundred American planes to start from the Aleutian Islands and fly over Japan just once.... That will teach them a lesson." Even more remarkable, Hull mused further: "If we could only find some way to have them drop some bombs on Tokyo." For his part, Morgenthau talked with Nationalist Chinese officials about the possibility of transferring American bombers to the Chinese Air Force for strikes against the Japanese capital and other cities, but that unsound idea fell through.[10]

The U.S. Army Air Corps had for some time considered the technical requirements for a long-range heavy bomber. In 1933, thinking in terms of a 2,000-lb bombload and a maximum point-to-point range of perhaps 5,000 miles at a speed of 200 mph, it requested proposals from aircraft manufacturers. In due course, the four-engine B–17 (delivered in 1937) and, in later years, the B–24 and the very heavy B–29 bombers of World War II fame were conceived.[11] Bombers in these categories could be employed over great distances to strike enemies anywhere in the world.

Clearly, the Japanese numbered among the major hypothetical foes against which these American bombers might be employed. After visiting Japan in 1924, Army Brig. Gen. William "Billy" Mitchell prepared a report detailing the vulnerability of Japanese cities to aerial assault.[12] As 1941 approached, however, man-made technical tools of strategic bombardment had not yet caught up with the theoretical objective of wreaking decisive havoc on America's potential transpacific antagonist. U.S. Navy leaders also believed that the next war would be with Japan, although the belief held that time worked on the side of the United States. If an actual Japanese attack was not impossible, it was not feasible. According to Billy Mitchell, President Franklin D. Roosevelt for years clung to the fallacious "idea that a war in the Far East would be impracticable and that an attack upon us by Japan is inconceivable." To Mitchell, that was "Navy thinking."[13] Presumable experts shared the President's optimism until the day of Pearl Harbor.

An underestimation of Japanese military capabilities supported this notion. When Japan invaded China in 1937, Adm. Harry E. Yarnell in Shanghai

Explosion of the USS *Shaw* in a view across Pearl Harbor, as a result of the Japanese attack on December 7, 1941.

informed higher headquarters that Japanese "aviation is distinctly inferior to ours. A few of our well trained squadrons would drive their planes out of the air and their ships out of the [Yangtze] river in a short time."[14] A former director of the Office of Naval Intelligence, Capt. W. D. Puleston, advised that the Japanese, though vigorous in their efforts to develop naval aviation, were "usually a phase behind." He judged that the IJN was unable to match U.S. aircraft carriers in the number of planes carried onboard, and IJN personnel could not launch or retrieve aircraft as rapidly as could the Americans.[15]

This American conceit persisted until the bitter end. *Aviation Week* in September 1941 advised its readers: "Isolated from her Axis fellow aggressors . . . her air force of low offensive strength . . . Japan, if engaged in a great air war, would crumble like a house of cards."[16] In mid-November 1941, well-known American military critic Maj. George Fielding Eliot judged that Japan would be in no condition to wage hostilities against a coalition of major foes. The IJA, he declared, was "sadly out of date," and Japanese air power was "almost nonexistent."[17]

Feelings of racial superiority buttressed the military assessments. Ignorance and contempt shaped an ethnic stereotype that depicted the Japanese samurai as small, scrawny, and slow-witted, encumbered of an overbite and protruding front teeth, whose poor eyesight made him a wretched shot by day and half blind at night. As for Japanese aviators, naval author Fletcher Pratt declared: "Every observer concurs in the opinion that [they] are daring but incompetent."[18] In the words of a French military critic, "The Japanese seem to lack something physical as pilots."[19] An American practitioner of strategic air power, General Curtis E. LeMay, whose B–29 bombers ravaged Japan in 1945, summed up his prewar understanding of the Japanese with characteristic candor:

> Well, first of all I certainly wasn't an expert on [them]. Prior to the war we had practically a non-existent intelligence system. So I personally consider that I knew nothing about the Japanese except that they were pretty tough fighters and that they did consider a defeated enemy even worse than a dog, and treated them as such. I had respect for them as an enemy, but not much respect for them as a people.[20]

General LeMay may be regarded as representative of most of his prewar compatriots.

The Pacific War: From Defense to Offense

The first strategic plan of the Air War Plans Division (AWPD–1), an air requirement annex to a War Department document submitted to the Joint Army-Navy Board in August 1941, called for a strategic defensive stance in the Far East, regardless of Japan's decision to declare or not declare war on America. U.S. strategic aviation would only be employed to protect the Western Hemisphere and the Pacific possessions and interests of the United States. Only after the Third Reich had been defeated in Europe would a strategic offensive commence against Japan.[21] Throughout the course of the protracted, but feckless, diplomatic conversations with the Japanese in the autumn of 1941, the U.S. Chiefs of Staff repeatedly recommended that President Roosevelt avoid any political action that might provoke a Japanese attack before the United States could reinforce the Philippines and the Western Pacific. This program could not be completed before February or March 1942.[22]

The stunning success of the Japanese aircraft carrier attack on Pearl Harbor on December 7, 1941, and, in the next few months, the swift Japanese conquest of the Western Pacific and Southeast Asia left the AWPD–1 assumptions in doubt and the high strategy of the United States and its new allies in disarray. With America's enforced involvement in the war, public expressions of anger and resentment, and a blazing desire for revenge among civilians and military alike, were predominantly directed against Japan. At the same time, considerable concern existed in the United States that the Japanese might return to bomb or invade the nearly defenseless Hawaiian Islands, or even attack U.S. defense industries on the West Coast.

It was a terribly bleak period for the Allies. U.S. Army Air Forces (AAF) Commander General Henry H. "Hap" Arnold observed that exposed outposts desperately needed reinforcements and commanding officers everywhere were clamoring for aircraft, including precious heavy and light bombers. With the U.S. Navy's battle fleet largely immobilized at Pearl Harbor and the Royal

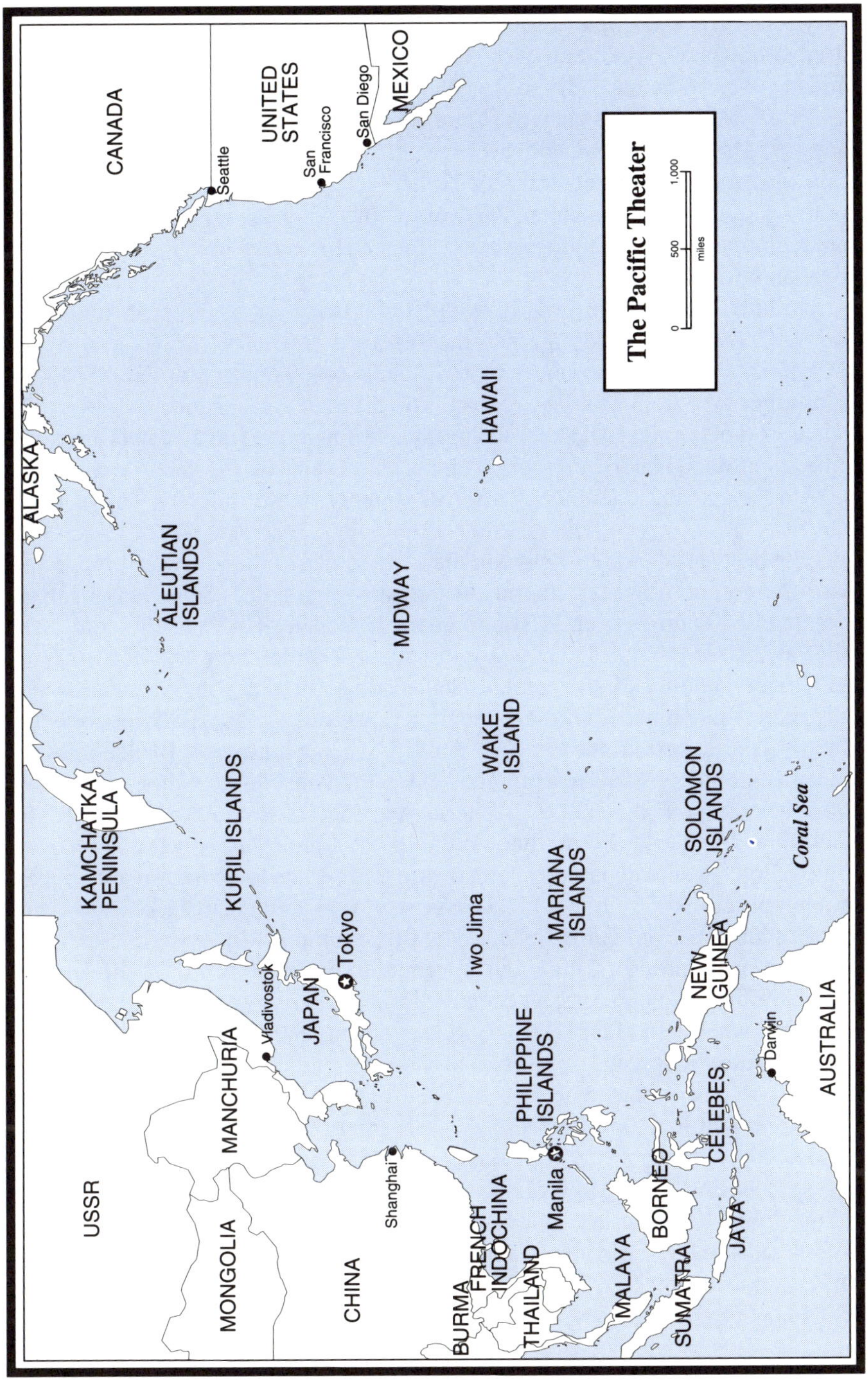
The Pacific Theater
0
500
1,000
miles
CANADA
UNITED STATES
MEXICO
Seattle
San Francisco
San Diego
ALASKA
ALEUTIAN ISLANDS
HAWAII
MIDWAY
WAKE ISLAND
KAMCHATKA PENINSULA
KURIL ISLANDS
USSR
MANCHURIA
Vladivostok
JAPAN
Tokyo
Iwo Jima
MARIANA ISLANDS
SOLOMON ISLANDS
Coral Sea
NEW GUINEA
Darwin
AUSTRALIA
CELEBES
BORNEO
JAVA
SUMATRA
MALAYA
THAILAND
FRENCH INDOCHINA
BURMA
CHINA
MONGOLIA
Shanghai
PHILIPPINE ISLANDS
Manila

Navy's only two capital ships in the Far East sunk off Malaya by land-basedaircraft, how could one even think of striking at Japan? As British Prime Minister Winston Churchill put it: "We had lost the command of every ocean except the Atlantic . . . Japan was supreme and we everywhere [were] weak and naked."[23] President Roosevelt, however, insisted that the American military "find ways and means of carrying home to Japan proper, in the form of a bombing raid, the real meaning of war."[24] In Congress, among many others demanding similar retribution, Senator Lister Hill called for gutting the heart of Japan with fire.[25]

Striking Japan by air directly from North America was out of the question during the war, considering that the distance from San Francisco to the Philippines via Oahu, Hawaii, is approximately 6,900 miles, plus another 1,400 miles tagged on from Manila to Japan. The distance from Seattle to Tokyo via Hawaii is 6,600 miles. An approach to Japan through the Kuril Islands attracted some attention in American military circles, but even over the great-circle route, via Alaska and the Aleutians, it is still roughly 4,900 miles to Tokyo from Seattle.[26]

The best aerial road to Tokyo in the early stages of the war was through the maritime region of Soviet Siberia. At the narrowest point of the Sea of Japan, it is merely 200 miles from Russia to Otaru-Hakodate; Tokyo itself is only 700 miles away. No point in Japan was more than 875 miles from Soviet soil. With the Soviet Union and the United States allied in the struggle against the European Axis powers, it did not seem preposterous to expect Soviet cooperation, overt or covert, in the Far East. Indeed, when a high-level British mission led by Churchill visited Washington, D.C., for consultations little more than two weeks after Pearl Harbor, General Arnold told Royal Air Force (RAF) Chief of Staff Charles Portal that, on the basis of preliminary negotiations, the United States would apparently soon obtain permission to operate against Japan from air bases in the vicinity of Vladivostok. Portal doubted that likelihood, and his skepticism proved correct. A more realistic jump-off locale might be found in unoccupied zones in China, but information available to the Americans on the feasibility of such a project remained scanty.[27]

In the weeks after Pearl Harbor, Roosevelt emphatically and repeatedly told senior military and naval advisers that the fight had to be taken to the enemy, that a bombing raid had to be launched against Japan proper "as soon as humanly possible to bolster the morale of America and her Allies." When the subject came up again in late January 1942, Arnold reviewed the various strategic basing possibilities — China, Russia, and the Aleutian Islands. Advised that the distances involved in a strike from Alaska, in particular, were excessive, Roosevelt directed that the China option be explored.[28] A few weeks earlier, on December 31, 1941, General Arnold sent a memorandum to the Air War Plans Division that typified the feeling of frustration in Washington:

1. The job given [Maj. Gen. George H.] Brett [formerly Chief of the Air Corps, now

> assuming command of the AAF in Australia] was to determine the way to bomb Japan from China with heavy bombers. He was not given the job to determine ways and means for *not* doing it. The attached is a cable [from Brett, dated Dec 26, 1941] full of "nots." I want to find out how to do, not how not to do it.
> 2. Study this and recommend what additional cable we should send to [another AAF officer] now in China. The theme song must be "How can we bomb Japan from China?"[29]

The British military staff, especially Portal, advised American military leaders that air operations against the Japanese homeland should be assigned to the U.S. Navy. Aircraft carriers, in the British view, could get in close and surprise the Japanese much the same as the IJN aircraft carriers had sneaked in on Oahu and the British carriers had jumped the Italian fleet at Taranto. Such raids might even pressure Japanese naval task forces to return to home waters. Because the combat radius of existing carrier planes did not exceed 300 miles, the Americans remained unconvinced by Portal's arguments. Arnold termed it suicidal to deploy aircraft carriers within reach of enemy land-based airplanes. The AAF chief privately harbored another reason to distrust the British rationale:

> I always thought that Portal mixed wishful thinking in with his reasoning concerning the Pacific aerial strategy. I thought he was afraid if our Air Force planned to use heavy bombers against Japan it would cut down the number he would receive.[30]

If none of the early schemes for dispatching land-based U.S. bombers against Japan from Russia, China, or Alaska proved realistic, they bred a bold and innovative solution — a modification of the British suggestion. In January 1942, the operations staff of the U.S. Navy commander Admiral Ernest J. King devised a novel plan to use Army medium bombers from aircraft carriers for a retaliatory strike against Honshu Island. The bombers would not try to return to the carriers but would proceed to destinations in unoccupied China. Thus was born the Doolittle Raid, named for the bomber unit commander Lt. Col. James H. "Jimmy" Doolittle, who led sixteen twin-engine B–25s from the carrier *Hornet* on April 18, 1942, in the harrowing low-level, daylight attack on Tokyo and three other cities.[31] Though ineffective militarily — only about 50 Japanese died, some 400 suffered wounds, and about 200 houses burned — the raid boosted American morale and assumed a great symbolic significance. Japanese propaganda, to be sure, played down the raid as much as possible, but for the first time, a degree of uneasiness pervaded the Japanese civilian population.[32]

Long comfortable in their insular world, certain Japanese military leaders now comprehended the nature of the menace from the air and began to devote attention to reinforcing fighter and antiaircraft artillery defenses of the homeland, though this attention would wane during the hiatus in bombing that followed. Because the Japanese armed forces failed to protect the Imperial capital from air invaders sweeping in from the western Pacific, IJN leaders, especially Admiral Isoroku Yamamoto, Commander-in-Chief of the Combined Fleet, now felt compelled to neutralize the threat emanating from Midway and

Doolittle Raid. A B–25 on the deck of the *Hornet* is shown as it prepares to take off on the historic attack against Japan on April 18, 1942 (*top left*). Despite rough seas, the first B–25 Army bomber is given the go-ahead signal on the flight deck of the carrier (*bottom left*) and takes off (*bottom right*). According to plan, the crew bailed out in China after the raid. Col. "Jimmy" Doolittle and his crew pose for this photo with the Chinese who helped (*top right*).

the Aleutian Islands, a decision that would have far-reaching implications for the future course of the war.

In the United States, the Doolittle air strike served precisely as Roosevelt and his high command had hoped. The attack repaid the Japanese, at least in part, for Pearl Harbor, and it challenged the myth of Tokyo's inviolability. Vice Adm. William F. Halsey termed Doolittle's flight "one of the most courageous deeds in all military history,"[33] though subsequent discussions between Arnold and the President made it apparent further bombing attacks against Japan "were for the moment impossible, since we had neither the bases nor the required types of planes."[34] No more Doolittle-type, one-way raids would be attempted and the Japanese homeland would be spared aerial attack for another two years, until a strategic bombing offensive could be mounted from islands captured in the western Pacific.

The absence of air attacks against the Japanese home islands between 1942 and 1944 did not signify AAF willingness to let the U.S. Navy shoulder further burdens alone. In late February 1942, even while the Doolittle raid was being prepared, Air War Plans and A–3 staff officers in Washington advised General Arnold:

> Air action should never be arbitrarily restricted to the limited operating capabilities of naval surface forces. The restrictive nature of such a policy or doctrine is readily apparent in the present use of our military and naval forces in the Far East. If this policy were invoked in this theater today it would cripple air initiative and striking power by limiting air operations to the capabilities of surface naval forces which have been seriously restricted and intermittent in character.[35]

AAF planners did not lack enthusiasm or imagination in conjuring up ways of hitting Japan again, as often and as soon as possible. In May 1942, for example, Headquarters Seventh Air Force forwarded to General Arnold a memorandum prepared by military intelligence in Hawaii on the potential effect upon Japanese morale of an air attack on the Imperial holy shrine at Ise, not far from the ancient capital of Kyoto.[36]

Unfortunately, the round-trip flying range necessary to reach and bomb Japan remained the bugbear. In December 1943, Maj. Frank Short in U.S. Army Ordnance submitted a most ingenious proposal to reduce that range. At the core of his carefully researched and eloquently written seventeen-page plan was a series of six man-made floating islands, each designed as concentric triangles consisting of two "battle islands" and two "stepping stones" located at maximum intervals of 700 miles between Midway and Japan. Two more "supply islands" would be placed between Hawaii and the continental United States. The four feeder islands would constitute a continuous chain to supply the two battle islands in staging an unrelenting aerial assault of 1,000-plane raids against targets in Japan, a strategic bombing campaign similar to that being conducted against Germany from bases in England.[37]

Examined by several headquarters, Short's proposal received short shrift. Army engineers, in particular, discerned serious problems: bunching 1,700 friendly fighters and 930 bombers in a restricted (and underestimated) parking area; the immobility and vulnerability of the floating islands to aerial and submarine attack; the enormous problems of supply and maintenance for tremendous forces concentrated in so small a zone; and the unrealistic financial requirements for constructing the floating islands. In early April 1944, Headquarters AAF returned Short's report to him without action.[38] In any case, the ultimate American strategic bombardment campaign would not draw upon Short's revolutionary notion, but on establishing bases on real islands and using them as stepping stones on the long road to Tokyo.

The United States might have expended enormous amounts of energy and effort in the Doolittle Raid, but critics could still call it "a grandstand play [that] appealed to President Roosevelt."[39] The U.S. Navy had earlier sent aircraft carriers to raids against Japanese targets in the Marshall Islands on February 1, 1942, Wake Island on February 24, Marcus Island on March 4, and

New Guinea on March 10. Though these effects, like those of the Doolittle Raid, proved beneficial to American morale and offered experience in real combat, a well-placed naval officer admitted that the Japanese did not mind these first U.S. Navy air strikes "any more than a dog minds a flea."[40]

One approach, stubbornly argued by many highly placed U.S. Navy staff planners at the outset of the war (with undoubted support from much of the public), called for jettisoning the whole Europe-first strategy, refocusing attention on the Asia-Pacific Theater, and devoting the main national effort to an early counteroffensive against Japan. U.S. Navy leaders, their forces mauled at Pearl Harbor, would consequently press for a more aggressive campaign in the Pacific. Rebuffed, they sidestepped objections of the "Europe-firsters" by asking only for "a limited 'active' defense," a term that really meant a limited but escalating offensive. In this sense, the Navy sought "in combination, though not in cooperation, with . . . [General Douglas] MacArthur to change the strategic position of the war against Japan," thereby posing a direct challenge to AAF and Army roles already accepted in Allied contingency planning.[41]

Particularly distasteful to the overextended, underequipped AAF was the Navy's desire for a U.S. Army strategic air force committed to the Navy. The AAF concept for the use of strategic air power was no more popular within the Navy than was the AAF espousal of the Europe-first policy. Despite some continuing discord, the Joint Chiefs of Staff (JCS) accepted, and the Anglo-American Combined Chiefs of Staff (CCS) endorsed, the principle of a strategic defensive against Japan, which would eventually be replaced by maximum strategic offensive action in the Pacific after Nazi Germany had been conquered. In the meantime, Japan would be denied access to vital raw materials.[42]

General Douglas MacArthur

Concerning aerial warfare, the CCS agreed that "the first essential is to gain general air superiority at the earliest possible moment, through the employment of concentrated air power." Piecemeal commitment of

Admiral Chester W. Nimitz

the few available aircraft would be avoided. The main objectives of air offensive operations in Eu rope were visualized in some detail, but the Pacific Theater received less definitive attention. In the eyes of the CCS, the key to Allied victory remained Germany's defeat. In the early phase of the war, Soviet involvement against Japan was not endorsed, for "if Russia took on another adversary it might jeopardize its efforts against Germany." Only "the minimum of force necessary for the safeguarding of vital interests in other theaters," the CCS concluded, "should be diverted from operations" against the Third Reich. In short, the Atlantic and European regions were deemed decisive, and Allied strategy was shaped accordingly.[43]

In the Pacific, at the very time U.S. forces surrendered on Corregidor in May 1942, Japanese reversals began with a draw at the battle of the Coral Sea and abandonment of major landings in New Guinea, followed by a decisive defeat at the battle of Midway in June. In August, American troops landed on Guadalcanal and Tulagi islands, marking the beginning of America's counteroffensive in the Pacific region. After fierce fighting, Japanese forces withdrew from Guadalcanal in February 1943. Soon afterward they lost their Aleutian Island foothold at Attu in May and on Kiska in July, the same month U.S. forces landed on New Georgia. A sequence of widespread Allied offensive operations followed in rapid succession: at Lae and Salamaua in New Guinea in September; on Bougainville in the Solomons, and on Makin and Tarawa in the Gilberts in November; and on New Britain in December 1943.

Meanwhile, Allied commanders in the Far Eastern theaters accumulated at least as many men and as much materiel as their counterparts did in the Germany-first European Theater. The bulk of the U.S. Navy, as always, remained deployed in the Pacific, and by the end of 1942, the U.S. Army also had 460,000 men in the Pacific, compared with 380,000 committed against the European Axis.[44]

By the autumn of 1943, a two-pronged Pacific counteroffensive was under way, considerably earlier than the Japanese had expected. The U.S. Navy pursued the limited active defense for which it had clamored, and had ostensibly been denied, in the form of a Central Pacific thrust directed from Admiral Chester W. Nimitz's headquarters at Pearl Harbor.

From his headquarters at Lennon's Hotel in Brisbane, Australia, General MacArthur also directed a separate, Southwest Pacific thrust.

> Because of its material superiority, the United States could afford such expensive — and occasionally dangerous — luxuries as divided command and the lack of an overall strategy in its war against Japan . . . for the United States, the record of the Pacific War is not so much a story of how the services forgot their differences but rather of the ingenuity displayed by service leaders in devising courses of action which allowed them to get on with the war without having to settle those differences.[45]

For its part, the AAF was particularly unhappy that the Navy's "production schedule was not in consonance with the agreed [Allied] joint strategy, and . . . competed with and jeopardized the buildup of forces for the main effort." Brig. Gen. Haywood S. Hansell, for one, deplored the Navy buildup of aircraft carriers; their "multiple aircraft complements and combat crews . . . enjoyed equal priority in resources with the Army Air Forces, which were committed to the top priority strategic effort against Axis Europe."[46] In short, the AAF believed the Navy's circumvention of U.S. grand strategy impeded its own buildup of forces that conformed with the agreed-upon joint strategy. AAF leaders would often voice this complaint as events unfolded in the Pacific.

The Biggest and Best: The B–29

From the first tentative Anglo-American military thinking about a strategy for conducting a war in the Far East until the last Allied wartime summit conference at Potsdam in July 1945 (aptly code-named TERMINAL), the JCS developed a mix of war-winning measures that included blockade, aerial bombardment, and plans for an invasion of Japan. As early as the Casablanca Conference in January 1943, General Arnold spoke of new, very long-range (VLR) bombers able to strike Japanese targets from land bases. Pressed by the British to identify the specific bases, Arnold referred only to the Nanchang region in China and maritime Siberia.[47]

Clearly, Arnold had in mind the B–29 Superfortress employed in the Pacific Theater. Whereas the Japanese no more than investigated the manufacture of superheavy bombers such as the Type 92, the Americans already had launched a sustained program to develop a state-of-the-art VLR aircraft in response to the Kilner-Lindbergh board's recommendations submitted back in the summer of 1939. The prospective bomber was to be superior to the four-engine B–17 in terms of range and bombload, but the Western Hemisphere and the Atlantic Ocean, not the Pacific, were its original theater of operation. To justify its original mission, the Army Air Corps proposed it be used for "defensive bombardment," not offensive strategic attack against an enemy homeland. Boeing, Consolidated-Vultee, Douglas, and Lockheed all bid on the project, and in June 1940 the first two companies won modest contracts to initiate design studies based on Army Air Corps specifications. Before the year was out, Boeing and Consolidated-Vultee were further funded to build three prototypes each, the XB–29s and the XB–32s. The latter design eventually lost out to Boeing's version, and only fifteen B–32 production models made it to the Pacific Theater before the end of the war.[48]

Everything about Boeing's XB–29 was of novel and gargantuan proportions, from its three-story-high tail fin and four 2,200-hp Wright Cyclone R–3350 turbo-supercharged engines to its radar navigation system and pressurized crew cells. Although the Air Staff kept adding design refinements, Boeing strove to cut one full year from the five customarily needed to carry military aircraft from conception to operational employment. In June 1941, almost two years and 8,000 drawings after the draftsmen and engineers had begun their tasks, Boeing received a contract to deliver fourteen B–29s by January 1943 — an order that soon rose to 250. A month after the Japanese attack on Pearl Harbor, the AAF ordered 500 Superfortresses, and this number soared to 1,664 even before the first prototype took to the sky in September 1942.

The daring B–29 procurement became known as the $3-billion-gamble. The Superfortress became the first military aircraft ordered into industry-wide production straight off the drawing board. And this, before the first prototype had been built, much less flown.[49] These demands strained Boeing's sizable manufacturing facilities. In addition to using its main plant at Seattle for B–29 production, the company converted a Navy plant at Renton, also in Washington State, and started a new factory at Wichita, Kansas, for that purpose. The Glenn L. Martin Company and Bell Aircraft, invited to join the Superfortress project, built B–29s at Marietta, Georgia, and Omaha, Nebraska, under Boeing's direction. General Arnold told President Roosevelt that Boeing "had really gone to town," and he applauded the American aircraft industry's "will to do."[50]

On September 11, 1942, Edmund T. Allen, Boeing's chief test pilot, first flew the prototype XB–29 for seventy-five minutes at Seattle. Col. Donald Putt, the AAF project officer, took to the air in the giant bomber next day. The XB–29, however, was infested with technical bugs, particularly in its Cyclone engines, and disaster struck about five months after the initial flights. On February 18, 1943, with Allen at the controls, the second prototype, engine flaming, crashed and exploded at Boeing Field in Seattle. As it struck a meat packing plant, Allen, his ten crewmen, and a number of workers in the factory all perished. The AAF and a watchdog committee of the Senate, headed by Harry S. Truman, conducted a rigorous investigation of the stunning accident.

Edmund T. Allen, Boeing's chief test pilot at the controls of the prototype XB–29 that first flew for seventy-five minutes at Seattle, Washington.

The culprit apparently was the magnesium used in the Cyclone engines. Tightened direction of the whole program was initiated at the oversight level, and the AAF promptly established a B–29 Special Project staff to exercise direct control of all testing, production, and crew training activities.[51] If the XB–29 crash impeded progress on the Superfortress project (and some feared the great gamble lost), the AAF staff's basic faith in the VLR bomber program remained unshaken. Fortunately for the cause of the Allies in the war with Japan, this setback and the ensuing delay proved relatively brief.

Western sources generally have assumed that the first Japanese inkling of the B–29 occurred after the Superfortress appeared in combat. Actually, IJA intelligence knew before the war that the United States was designing both the XB–29 and the XB–32, though it lacked details. Sometime between the spring and summer of 1943, the IJA High Command learned from various sources of intelligence culled from abroad, including aviation journals, about the XB–29 crash in Seattle.[52] This news shook the IJA Air Staff, since only two years had elapsed from the time the design first appeared in intelligence briefs.

Despite the failure of one test flight, the American VLR bomber obviously had progressed well beyond the design stage, was nearing mass production, and soon would appear in combat. Little doubt remained among the Japanese that such a plane had the capability of striking their homeland from distant land bases and that the enemy had just such a plan in mind.

After the test flight crash, IJA intelligence acquired bits and pieces of evidence about U.S intentions from public releases, obtained primarily in Buenos Aires and Washington, D.C.[53] The IJA Air Staff, which had been working on ways to cope with the B–17, by the summer of 1943 sensed that the difference between the Superfortress and the Flying Fortress was one between a parent and a child. The B–29 was twice the size of the B–17 and had a maximum takeoff weight of 70 tons. The Superfortress could carry half a railroad freight car's load in bombs, 10 tons, and hauling that maximum bombload, it could fly a one-way, nonstop route of 3,400 nautical miles.[54]

At the end of 1942, the IJA High Command had established a B–17 Countermeasures Committee, and in the summer of 1943 this committee was converted into a body that would study defenses against the B–29. Chaired by the Vice Minister of War, the group included section chiefs and experts from the IJA General Staff, IJAAF Central Headquarters, War Ministry, and Ordnance Bureau. One of its first tasks was to collect and analyze hard facts about the B–29: What were its characteristics? When would it go into serial manufacture? What would be the rate of production? When, where, and in what numbers would it enter combat?[55] Drawing on limited information in mid-1943, Japanese military analysts estimated that the B–29 would enter mass production around September or October of 1943, some six months behind schedule, a result of the crash of the prototype in February. Starting at about forty planes per month, the rate of manufacture could be expected to rise steadily from the fifth month, climbing to an output of 75 to 100 and then perhaps to 150 per month soon afterward.

The IJA surmised that the total B–29 inventory would reach approximately 200 units between March and the end of May 1944, 400 by the end of June, and at least 1,000 to 1,100 by the close of 1944. Some IJAAF analysts believed that the B–29s would go into action as soon as the inventory numbered 200. If that total was attained by March 1944, it was thus not unreasonable to expect raids as early as April. Most of the IJA Air Staff, however, did not think that the B–29s could conduct real strategic bombardment against the home islands until May or June 1944, when their total number had reached a critical mass of 400.[56]

No consensus as yet existed among the IJA General Staff regarding the geographical direction from which the expected B–29 units might strike at the homeland. The Military Intelligence Bureau's 6th (Euro-American) Section thought in terms of raids originating from a site such as Wake Island in the Pacific, well within the predicted reach of the Superfortress. Some staff officers remained skeptical about the reality of this threat; thus Wake was not included within the zone regarded as indispensable to the defense of Japan. The 7th (China) Section intelligence analysts judged bases on the Chinese mainland the more likely source of B–29 attacks. In late December 1943, the IJA General Staff Operations Bureau conducted a week-long map exercise, code-named KO-GO, or TIGER, to test the hypothetical axes of forecasted B–29 bombardment. A possible enemy base on Wake Island was not excluded from consideration; nevertheless, the exercise concluded that the IJA must launch a ground campaign to eliminate potential enemy bomber bases in China. The campaign was originally intended to open the corridor from North to South China, but its focus subsequently changed to mainly southwest China, with objectives such as Kweilin and Liuchow. The IJA High Command issued the appropriate order on January 24, 1944, calling for an offensive beginning in May against base areas in Hunan and Kwangsi.[57]

As for the technical data on the Superfortress aircraft needed by the B–29 Countermeasures Committee, IJAAF Air Intelligence directed a lieutenant specializing on fuselages to play the part of the chief body designer of the Boeing Company and draw up blueprints for the giant four-engine bomber. Another lieutenant, also an expert, was to do the same thing for the engines to be used in the new American VLR bomber. Picking the brains of other aeronautical engineers and comparing the results of their own studies with materials gleaned from abroad, these officers prepared a useful and realistic preliminary analysis by the end of 1943. After much more concrete intelligence became available in early 1944, the two officers developed highly detailed specifications for the B–29. Their analysis stood up well against the authentic Boeing design materials revealed after the war.[58]

Despite all of the effort invested by the Japanese military in studying the threat posed by the Superfortress, a field-grade IJAAF staff officer lamented that the army's concern, though evident, did not sink in very deeply at that time. "It never reached the point," he confided, "where we devoted special effort to air defenses. The truth is that we just continued to confront the B–29 with the same old thinking that we had applied to air defense before the B–29 appeared on the scene."[59] Certainly, this curious failure in *preparing to meet the threat*, and the overall Japanese aerial defensive weakness, would be fully demonstrated and exploited, once U.S. strategic bombardment of the homeland began in earnest.

While the Japanese in 1943–1944 analyzed the B–29's characteristics and capabilities and expected to bear the brunt of its fury (without knowing the direction of its attack), American military leaders debated the optimum theater in which to base the Superfortress. Only at the end of 1943 did the JCS decide against sending B–29s to England for use against targets in Europe, since the vast distances of the Pacific-Asian region required a long-range bomber. Various commanders, however, held different views about the best ways to employ the weapon. General MacArthur's chief of the Fifth Air Force, Maj. Gen. George C. Kenney, for example, wanted B–29s in quantity for use against enemy targets in the Southwest Pacific. The U.S. Navy also had its own idea of using B–29s for long-range scouting and antisubmarine missions.[60]

Headquarters AAF in Washington, D.C., had a more prophetic vision of the B–29s' optimal employment and control. With their initial introduction into combat in 1944, they would be used to strike directly against Japan proper, and not employed against peripheral targets. They would also be committed to combat under a unified command, for they would be operating in regions behind advancing forces belonging to either MacArthur or Nimitz. In view of the unsynchronized spheres of Army and Navy operations, General Arnold could see no alternative but to retain in his own hands central control of all B–29 operations, though he did not want to do so.

"In the end," Arnold declared, "while everybody wondered why I kept personal command of the . . . B–29s, there was nothing else I could do, with no unity of command in the Pacific. I could find no one there who wanted unity of command, seemingly, unless he himself was made Supreme Commander."[61]

Thus, on April 4, 1944, the Twentieth Air Force was born, a very heavy bombardment (VHB) force functioning directly under the JCS, with Arnold in command and headquartered in Washington. Only the responsibilities of logistic support and some administration would be delegated to the commander of the local theater. Under Arnold's command, Twentieth Air Force was directed to "achieve the earliest possible progressive dislocation of the Japanese military, industrial, and economic systems and . . . undermine the morale of the Japanese people to a point where their capacity and will to wage war was decisively weakened." B–29 support of Pacific military operations existed only as a secondary mission.[62]

The Joint Intelligence Committee in Washington, D.C., meanwhile, carefully considered potential bases from which B–29s could reach the most important enemy targets. Assuming a combat radius of 1,300 or 1,600 miles, the bombers could operate from the Aleutian Islands off Alaska; Calcutta, India; Chengtu, China; Broome or Darwin, Australia; Port Moresby, New Guinea; or the Marianas Archipelago in the Pacific Ocean. Saipan Island in the Marianas was clearly the ideal pick from the standpoint of logistics and defense, but it remained in Japanese hands. The second best was the Chinese city of Chengtu, though its choice entailed numerous unsatisfactory features:

1. The Chinese Army would have to be equipped and supplied to secure the airfields area from expected Japanese air and, perhaps, ground attack.
2. In this remote location, the B–29s would require great stocks of materiel and high-octane gasoline flown-in from India, *and* they would have to act as their own transports.
3. Because Maj. Gen. Claire Chennault's Fourteenth Air Force would need to protect the Chengtu base against the Japanese, additional airlift would have to augment and support his forces. That would demand far more tonnage than was now being hauled across the "Hump" of the Himalayan Mountains.
4. With the Japanese in control of northern Burma, it was dubious whether the Ledo Road and the pipeline to Kunming would be completed in time to ease the burden on the already overtaxed air transport system between India and China.[63]

Only the best and most experienced personnel were selected for the B–29 units. Brig. Gen. Kenneth B. Wolfe (*left*), chief of B–29 development and flight testing, became the first commander of the XX Bomber Command of B–29s and moved it to the China-Burma-India Theater. First Lieutenant Leonard F. Harman, shown in front of a Martin B–10 during the Alaska flight in 1934 and who figured significantly in testing bombers in the mid-1930s, commanded the 58th Bombardment Wing in late 1943.

In short, the security and logistical requirements of basing B–29s in Chengtu, inaccessible by water or surface transportation and vulnerable to enemy counteraction, appeared daunting. Nevertheless, that city's geographical location in Szechwan Province, about 275 statute miles northwest of the Nationalist Chinese wartime capital of Chungking, was in its favor. The closest Japanese ground forces were located some 400 miles distant, and, barring a total collapse of the Chinese regime, the Japanese were too far away to threaten it except by aerial strikes. (Even the Japanese ground offensive in May 1944 made scant progress up the Yangtze River, although it did cut China in two, north to south.) Finally, strong political reasons existed for bombing Japan from unoccupied China, most notably the desire to shore up Chiang Kai-shek's shaky regime psychologically and keep it in the war.

The AAF also had its own reasons to commence strategic bombardment expeditiously. With a core force of B–29s already in being, precious time would be lost if one waited for island bases to be conquered in the Pacific region, on the eastern approaches to Japan. Also, President Roosevelt was insistent that the new and costly bombers strike immediately at Japan. Finally, the new aircraft could be tested in combat; flight and ground personnel could be trained in operations and maintenance; and logistical support could be assessed in the field.[64]

The AAF activated the first B–29 field headquarters — XX Bomber Command — on November 20, 1943, and moved it to the China-Burma-India (CBI) Theater under Brig. Gen. Kenneth B. Wolfe in early 1944. In March 1944, Brig. Gen. LaVerne G. "Blondie" Saunders took command of the B–29 force placed under Wolfe — the 58th Bombardment Wing (Very Heavy), previously headed by Col. Leonard F. Harman. The AAF prided itself that "the best and most experienced personnel available . . . within the continental United States were placed in the B–29 organizations."[65] Saunders's prior, impressive record with the 11th Bombardment Group on Guadalcanal bespoke that pride.

Because of the geographical and logistical environment in which it was obliged to work, XX Bomber Command had an unusual operating structure. Combat units were based in India, in the Calcutta area, but they staged to advanced bases in China. Fuel had to be conveyed to the forward bases by B–29 tanker planes traversing the mountainous and desolate Hump route, over the immensely high Santsung Range in the Himalayas, and in wretched weather. Repair and maintenance facilities in China were substandard, and enormous preparations had to be mounted for every major bombing strike. Consequently, the frequency and the scale of the B–29 raids would be sharply curtailed. During the first nine months of the B–29 China operations, AAF planners expected only 14 percent of the total number of B–29 flights to be devoted to strikes against the foe.[66]

The B–29s would assuredly have attractive strategic targets, for the industrial heartland of Japan and the heavy industry of Manchuria had never been subjected to high-altitude precision bombardment. On the other hand, XX Bomber Command B–29s could only reach Korea, Manchuria, Formosa, western Honshu, and northern Kyushu. Important targets in the Tokyo-Yokohama and Nagoya-Osaka-Kure districts of Honshu lay beyond their effective range, causing an obvious and significant limitation on projected B–29 operations.[67] AAF bombardment doctrine would be employed: high-altitude precision bombing, in formation, by daylight. A B–29 wing contained four combat groups, each group consisting of three squadrons. The basic bombing unit, the squadron, numbered nine to twelve airplanes exclusive of reserves. The principal bomb-sighting action was conducted by the lead aircraft. To allow daylight returns and landings, the bombers had to take off from the forward bases before daybreak; formations assembled on the way to the target.[68]

Despite the many mechanical difficulties yet to be worked out (such as chronic engine overheating), XX Bomber Command possessed a remarkable strategic weapon in the B–29. Manned by a crew of eleven, the VHB aircraft mounted an advanced, full-circle airborne radar, had a cabin pressurization system, and used the powerful new Cyclone radial engines. The Superfortress

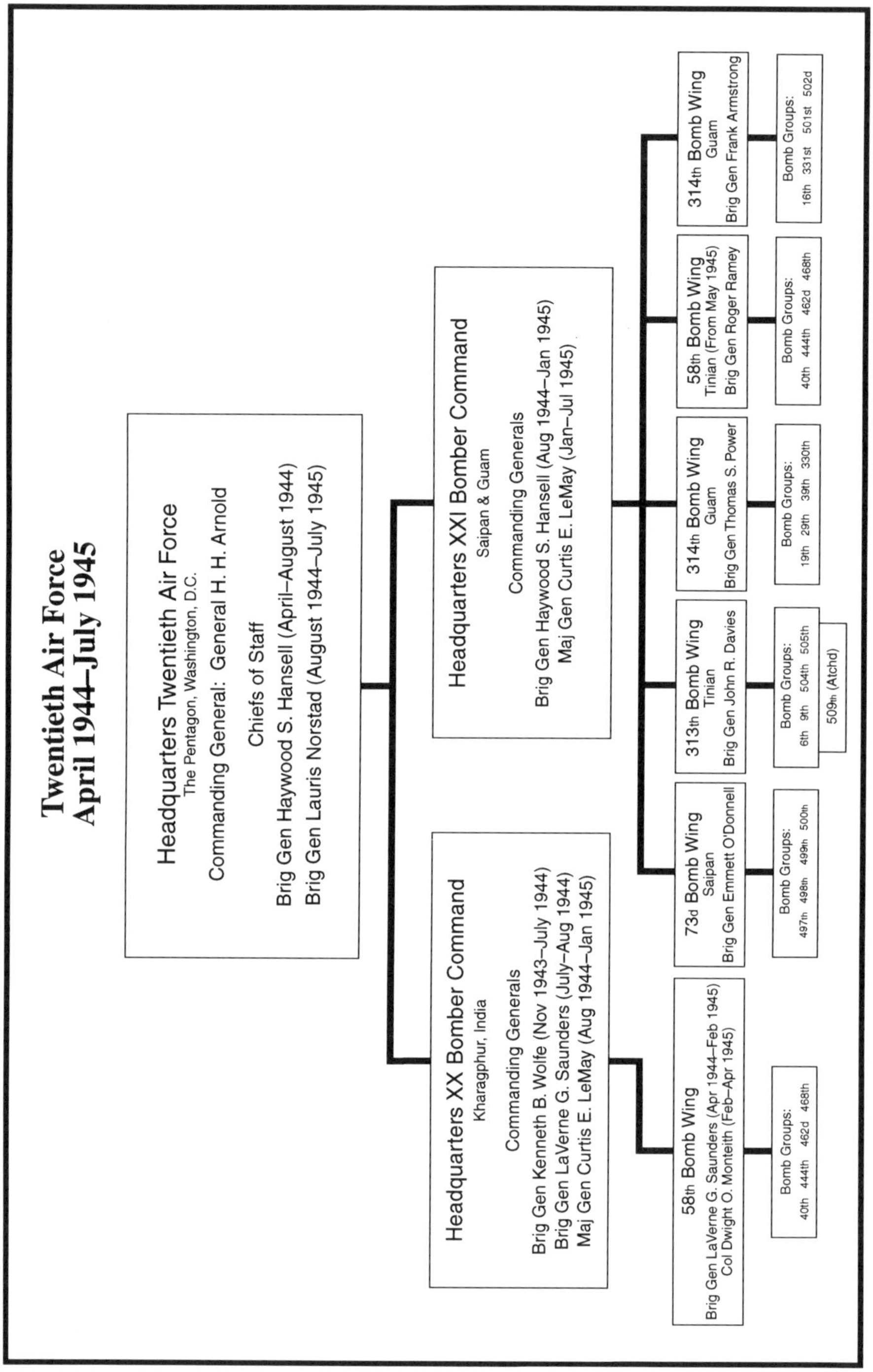
Twentieth Air Force
April 1944–July 1945
Headquarters Twentieth Air Force
The Pentagon, Washington, D.C.
Commanding General: General H. H. Arnold
Chiefs of Staff
Brig Gen Haywood S. Hansell (April–August 1944)
Brig Gen Lauris Norstad (August 1944–July 1945)
Headquarters XX Bomber Command
Kharagphur, India
Commanding Generals
Brig Gen Kenneth B. Wolfe (Nov 1943–July 1944)
Brig Gen LaVerne G. Saunders (July–Aug 1944)
Maj Gen Curtis E. LeMay (Aug 1944–Jan 1945)
58th Bomb Wing
Brig Gen LaVerne G. Saunders (Apr 1944–Feb 1945)
Col Dwight O. Monteith (Feb–Apr 1945)
Bomb Groups:
40th 444th 462d 468th
Headquarters XXI Bomber Command
Saipan & Guam
Commanding Generals
Brig Gen Haywood S. Hansell (Aug 1944–Jan 1945)
Maj Gen Curtis E. LeMay (Jan–Jul 1945)
73d Bomb Wing
Saipan
Brig Gen Emmett O'Donnell
Bomb Groups:
497th 498th 499th 500th
313th Bomb Wing
Tinian
Brig Gen John R. Davies
Bomb Groups:
6th 9th 504th 505th
509th (Atchd)
314th Bomb Wing
Guam
Brig Gen Thomas S. Power
Bomb Groups:
19th 29th 39th 330th
58th Bomb Wing
Tinian (From May 1945)
Brig Gen Roger Ramey
Bomb Groups:
40th 444th 462d 468th
314th Bomb Wing
Guam
Brig Gen Frank Armstrong
Bomb Groups:
16th 331st 501st 502d

also deployed remote-controlled automatic gun turrets — the central fire-control system directed five turrets. A Norden bombsight was mounted in the nose. With a tare weight of about 75,000 pounds, the B–29 could attain a top speed of approximately 350 mph at 25,000 feet, had a tactical radius of action exceeding 1,500 miles, and could carry a maximum payload of 10 tons of bombs.[69]

China-based B–29 strategic bombardment operations against Japan did not proceed rapidly, not even as rapidly as IJA intelligence predicted. The tasks of building heavy-duty runways and other giant facilities in India and China proved arduous, labor intensive, and time-consuming. Exasperated by the delays, in October 1943 President Roosevelt snapped at Army Chief of Staff, General George C. Marshall:

> I am still pretty thoroughly disgusted with the India-China matters. The last straw was the report from Arnold that he could not get the B–29s operating out of China until March or April next year. Everything seems to go wrong. But the worst thing is that we are falling down on our promises every single time. We have not fulfilled one of them yet.

He even questioned the rationale for the proposed employment of the AAF's newest weapon of war: "I do not see why it is necessary to use B–29s. We have

General George C. Marshall leaves the conference building after a meeting of the JCS in Berlin after World War II.

several other types of bombing planes."[70]

Presidential complaints receive prompt replies. The American investment in all strategic bombardment aircraft now approached the commitment to the Manhattan District — a fact doubtless known to AAF Commander General Arnold, who prepared the apologia for the President at Marshall's direction. Not only military careers but the future of an independent air force turned on the demonstrated efficacy of strategic aerial bombardment. "Because our present B–24s and B–17s do not have sufficient range," Arnold explained, "the B–29 is the only bombardment type aircraft with which we can reach Japan from bases presently available to us." Arnold recounted the technical troubles affecting B–29 engine production and the logistical problems encountered in India and China. Consequently, he observed, "the present plan for using B–29s [from China] against Japan proper is still being studied within the Air Staff."[71]

The B–29s, to be sure, were not coming off the assembly lines in large numbers, at least not in units that were fully combat-ready. The first production models had been delivered by Boeing's Wichita plant in July 1943, and 14 became available by the end of August. Six months later, however, only 97 B–29s had been delivered, of which a mere 16 were operational; the balance required as many as fifty-four modifications each to be cleared. At this rate, the AAF could not meet Arnold's target of 150 battle-ready B–29s (and 300 crews) in March 1944. To expedite factory production and delivery, the AAF froze the B–29's design in October of 1943.[72]

In training at Salina, Kansas, in 1943, the new 58th Bombardment Wing never had more than a dozen aircraft available for flying use during the entire year. Arnold was horrified when he visited Salina on March 9, 1944, and found not one B–29 ready to leave for the CBI Theater. By Herculean effort, the first 11 of the 150 Superfortresses scheduled for India were prepared by March 26. Colonel Harman led the initial flight on the 11,500-mile route from Kansas to the Calcutta area via Newfoundland, Morocco, Egypt, and Karachi in the east of India. Taking off from Salina in daily flights of nine or ten aircraft, 130 B–29s had made it to India by May 8, with seven lost en route.[73]

Operation MATTERHORN

The U.S. bomber offensive from India and China was appropriately code-named after an Alpine counterpart to the Himalayas, MATTERHORN. After some disagreement among the planners regarding the best Chinese site, the airfield complex at Chengtu in Szechwan province was chosen over the more advanced, but more exposed, base at Kweilin, home of Chennault's Fourteenth Air Force. AAF target analysts picked the following objectives for XX Bomber Command's first strikes: iron and steel works, aircraft factories, oil storage plants and refineries, urban industrial zones, transportation, shipbuilding facilities, and

B–29 Radius of Action
Operating from Chengtu
June 1944–January 1945
Enemy-Held Territory, May 1944
0
250
500
miles
USSR
MONGOLIA
MANCHURIA
Vladivostok
Anshan
Mukden
Peking
Port Arthur
KOREA
JAPAN
CHINA
Shanghai
Chengtu
Chungking
NEPAL
BENGAL
INDIA
BURMA
Formosa
1,500 miles from Chengtu
THAILAND (Siam)
FRENCH INDOCHINA
SARAWAK
MALAYA
BORNEO
Singapore
PAPUA NEW GUINEA
CELEBES
SUMATRA
JAVA
Port Moresby

B–29 Superfortresses of the 444th Bombardment Group, XX Bomber Command, on an airfield in China, December 1944.

docks in Manchuria, Korea, Japan, and Thailand. To a considerable extent, target selection was affected by range constraints and weather considerations.[74]

B–29s from Chengtu flew an initial combat strike on June 5, 1944, a thousand-mile, high-altitude daylight foray against railway yards and shops at Bangkok, Thailand, in what was regarded as a shakedown mission. Of ninety-eight bombers that made it into the air, one in five encountered mechanical problems which, coupled with bad weather, forced pilots to turn back, crash-land, ditch in the Bay of Bengal, or land at other bases. Seventy-seven B–29s bombed Bangkok against feeble flak and fighter opposition. Photoreconnaissance later showed bombing results to have been, at best, mediocre: eighteen bombs hit the main target area, the remainder missed by as much as two miles. Five B–29s were lost to mechanical/operational causes.[75]

Arnold, through the JCS, kept the pressure on Wolfe and XX Bomber Command to strike at Japan itself, not only to impress the enemy with the power of the Superfortress but also to assist the Fourteenth Air Force in east China and the U.S. Navy's vital invasion operation at Saipan. Wolfe, desperately short of gasoline, somehow got sixty-eight of seventy-five of Blondie Saunders's B–29s airborne on June 15, 1944, for the first raid against a major target in Japan — the coke ovens of the great iron and steel works at Yawata in northern Kyushu. The only concession that Wolfe extracted from Twentieth Air Force headquarters was permission to fly at night, at lower altitude than usual, and in a loose formation on the 3,200-mile round-trip flight. Of sixty-three B–29s, each hauling two tons of bombs, that kept going toward Yawata, only forty-seven located the target. Bombing results again were unimpressive; a distant powerhouse was hit, but the coke ovens went unscathed.[76]

Japanese radar on Chejudo Island detected the approaching B–29s at 11:30 P.M. on June 15, 1944, and warned Western Army headquarters at Fukuoka. The army's 4th Air Group (19th Air Wing, Western Army) possessed thirty-five available fighters, most importantly the eight Kawasaki Ki–45 Tōryū (Dragon Killer) twin-engine night fighters, known to the Americans as Nick. Led by two captains, the Tōryū interceptors scrambled at 12:52 A.M. on June 16, and at 1:11 A.M. they reached their objective. More than thirty B–29s and B–24s were reported coming from the direction of China, and the first aerial encounter occurred at 1:30 A.M. against B–29s arriving singly every two or three minutes. As often happened, U.S. and Japanese accounts of the air fighting do not resemble one another. According to the 19th Air Wing, seven American bombers were shot down: four as confirmed and three as probables. One charred hulk found by the Japanese was identified as the new B–29. The Americans admitted losing seven bombers on this mission, but only one to enemy action on the return flight to China. The Japanese senior staff officer of the Western Antiaircraft Group subsequently was reprimanded by the Western Army, and on June 24 the commander of the 131st Antiaircraft Regiment (covering Yawata) was transferred to Manchuria.

On June 16, 1944, the 19th Air Wing conducted an investigation to determine why the minimum objective of shooting down 50 percent of the raiding planes had not been achieved. The main conclusions were that the speed and firepower of Japanese fighters were woefully insufficient, a lack of night-radar equipment prevented catching enemy planes out of reach of searchlights, an insufficient number of airfields limited the concentration of defending fighters, and night fighter strength was scant. With respect to the last point, investigators found that most of the actual combat on June 15/16 was conducted by the eight Tōryū night fighters; inexplicably, not many of the other available fighters saw action.[77]

American as well as Japanese commanders answered for their unimpressive performance in the first bombing of Japanese targets in June. Wolfe and XX Bomber Command received new mission orders from the Twentieth Air Force almost immediately, this time including a raid against the giant coke ovens of the Showa steel plant at Anshan in Manchuria, 1,600 miles from Chengtu. When Wolfe pleaded a crippling lack of fuel after the Yawata mission, Arnold relieved him of command (the dismissal eased by promotion to major general) and on July 6 assigned XX Bomber Command to General Saunders, who took over a B–29 inventory that had been reinforced in driblets, but worn down through heavy use (in particular by ferrying fuel from India) and by abominable weather.

On the night of July 7/8, 1944, XX Bomber Command launched small harassing raids by seventeen B–29s against various targets on Kyushu. The Japanese air warning system in China, greatly improved by now, transmitted considerable details beforehand. One wave of U.S. aircraft bombed blindly through the clouds over Nagasaki and Sasebo while a second overflew northern Kyushu but did not penetrate to Kokura-Yawata. Japanese air defenses admitted no losses and made no claims about intercepting the enemy.[78]

When the Anshan mission was finally launched on July 29, 1944, Saunders fell somewhat short of committing the hundred planes that Arnold requested. Nineteen of twenty-four bombers, airborne too late for the Anshan raid, struck at alternate targets. Of seventy-one B–29s that headed for Manchuria, eleven could not make it to Anshan. Nevertheless, the daylight raid was judged a relative success. Sixty Superfortresses flying in formation hit Anshan from 25,000 feet, dropping 160 tons on target. This time considerable damage was achieved: the Showa steel works were hit by ninety-five bombs. One-fifth of the coke ovens were damaged, but of these, half could be repaired within two or three days. Although the blast furnaces were unhurt, hits on the attached facilities made three of nine furnaces inoperative for about a month. Six months were required to restore another furnace to operation. Consequently, the Japanese feared that Anshan's projected annual steel production of 1.25 million tons would be reduced by 300,000 tons. The bombers also inflicted some damage on railways and communication facilities. Casualties among workers in the steel plant numbered about 100 killed and 170 wounded; in addition, 40 were killed and 50 wounded among the general population. Fighter opposition was not very serious, and the effectiveness of the antiaircraft barrages was not proportional to the number of rounds fired by the ground batteries. Over China, Japanese fighters did down one B–29 which crashed in friendly Chinese territory. In all, five B–29s were lost in the operation.[79]

The highest Japanese claims against the B–29s were three downed and two damaged; the most modest report spoke of two enemy planes observed trailing smoke and lagging behind their formations. Only nine Nakajima Ki–44 Type 2 single-seat fighters (Shōki or Tojo) were on-station at Anshan on July 29, 1944. Claims were made, however, that the 70-mm antiaircraft guns defending the town had damaged two B–29s. Appraising the actions at Anshan, the IJA commander found the air defense system confused, inaccurate, and ineffective, and he discerned no sense of urgency in southern Manchuria in general. The defending fighters were too few in number, needed a lot of time to climb to the bombers' altitude, and could not catch up with the B–29s even when pursuing them for a distance of sixty miles. In short, the interceptors were unable to cope effectively with the Superfortress at altitude. Though acquainted with some of these problems, the Kwantung Army chief of staff, after inspecting the situation at Anshan, judged that laxity had been present in the air defense measures and that the defending fighter unit did not conduct its mission thoroughly; he personally reprimanded the local 15th Air Group commander.[80]

Reflecting the improved capabilities of Saunders's command and the excess number of B–29s available, Twentieth Air Force authorized a small night raid against Nagasaki on Kyushu Island on August 10, 1944 — the same day that another B–29 force struck Palembang from an RAF base in Ceylon. Of twenty-nine aircraft that took off from Chengtu, twenty-four bombed Nagasaki through cloud cover, with poor results. One B–29, suffering mechanical problems, came down in friendly hands in China. Enemy resistance at Nagasaki was feeble, but a B–29 reported shooting down one Japanese fighter in flames, the first official claim by a Superfortress.[81] Japanese sources record almost no air-to-air engagements on that date, although B–29s tried in vain to attack Nagasaki, Kokura-Yawata, and Shimane through rain and clouds.[82]

The Americans were still not of one mind concerning the optimum targets and tactics for the B–29s. General Chennault, for example, repeatedly asked that Superfortress be used against the Hankow waterfront; when his recommendations were rejected, he asked that the B–29s be either thrown into a counterair campaign or pulled out of China.[83] In XX Bomber Command, some staff officers wanted to use the B–29 in radar-guided night raids; others pressed for precision daylight bombardment. As for targets, General Saunders preferred trying to knock out Japanese iron and steel before turning to the aircraft industry.

In the seventh bombardment mission against Japan on August 20, 1944, a force of seventy-five B–29s took off from Chengtu. Of these, sixty-one dropped ninety-six tons of bombs on Yawata that afternoon and night; six other Superfortresses struck secondary targets. Japanese resistance at Yawata was the most severe encountered to date. Heavy flak made a direct hit and brought down one B–29; eight other bombers were damaged. "Superfort" gunners claimed great results in air-to-air combat: seventeen fighters shot down, thirteen probables, and twelve damaged. Japanese Tōryū interceptors, however, destroyed one B–29 by combined gunfire and air bombing, and another two in spectacular ramming attacks — the first encountered to date and deemed to have been intentional. That night, ten of thirteen B–29s again hit Yawata with fifteen tons of bombs, without loss. The Japanese reported the return of B–29s to Kyushu shortly after midnight, though they caused little damage. But the commander of the IJAAF's 52d Air Group and another pilot died in crashes; both were flying Nakajima Ki–48 Hayate fighters (Gale or Frank), the best Japanese interceptor to see combat in the last phase of the war. The Americans judged their own losses as heavy on August 20. In addition to the four bombers destroyed over Yawata, ten B–29s went down from operational causes. Ninety-five crewmen were killed or missing, although, as was later learned, one crew bailed out over Siberia and was interned by the Soviets. Once more, the material results of the strikes at Yawata appeared particularly unimpressive.[84]

General LeMay, in training with B–29 aircraft at a flying field in Nebraska, learned he would succeed Saunders after the Yawata raid. He arrived in India on August 29, 1944, to take over XX Bomber Command. As he had with Eighth Air Force in the European Theater, LeMay determined to fly at least one combat mission from the start, stating: "I won't know what's going on until I

Brig. Gen. Curtis E. LeMay would receive another star and take command of the XX Bomber Command in August 1944.

do." The second raid on Anshan gave the general his chance to learn firsthand the quality and tactics of the "reputedly excellent" Japanese fighter defense deployed at this high-priority Manchurian target. Under his command on September 8, 1944, 108 B–29s took off and 95 struck Anshan that afternoon, mainly hitting the coke works. Intelligence later estimated the damage inflicted to be moderate. As for Japanese fighter resistance, LeMay was unimpressed: "they never mounted a decent attack," he declared, probably because they misjudged the speed of the U.S. bombers and "turned the wrong way." When Saunders asked LeMay about the performance of the Japanese interceptor pilots, it is said he replied: "Stinkin' .... My first impression is that they won't be as tough as the Germans." Ironically, the general's plane had been holed by flak at 25,000 feet. In all, four B–29s were lost to causes other than air combat. The American gunners claimed eight enemy fighters shot down, nine probables, and ten damaged.[85]

Although the B–29s missed Anshan's blast furnaces, according to the Japanese more than fifty 500-lb bombs smashed the coke ovens. Another twenty or more bombs hit a portion of the special steel factory. In Anshan city, nine projectiles struck the railway line, some 500 structures housing workers collapsed, and 267 people were killed. The Japanese considered the defense of Anshan a failure in view of the severe damage to ground facilities and the casualties inflicted on personnel. After inspecting the scene next day, the Kwantung Army chief of staff voiced regret that the defenders not only had been unable to protect the area, but also had caused so few losses to the enemy.

Great clouds of smoke rising from the Showa steel works in these two photos attest to the accuracy of the B–29 Superforts during their raid on Anshan, Manchuria.

In the days following the Anshan raid, Kwantung Army headquarters sought to pinpoint problems and devise countermeasures. Japanese fighters lacked the performance and Japanese pilots lacked the training to cope with the ultrahigh-altitude tactics and speed of the B–29, which could operate between

The best interceptor to see combat in the last phases of the war was the Frank (*above*). Bombers in the Japanese inventory included the Betty (*center*) and Sally (*bottom*). (Americans identified Japanese fighters with males' names and bombers with females'.)

26,000 and 30,000 feet. Since the forty-one Tojo fighters committed to defend Anshan mounted 13-mm guns ineffective against the bombers, the Japanese decided to shift to ramming tactics. The 20-mm guns of the three Nick fighters employed at Anshan held greater promise, and more Nicks with heavier-caliber armament were needed quickly.

In Chengtu, before the ninth B–29 mission was launched in late September, LeMay worked to improve maintenance and modify tactical doctrine, stressing daylight precision raids. A rigorous new bomber training program introduced his battle-tested, twelve-plane combat box in place of the four-plane diamond formation. These innovations were scarcely under way when the B–29s were ordered to go after Anshan for a third time. On September 26, 1944, LeMay launched 109 of his 117 bombers. Of 88 aircraft that made it to the Anshan complex, 73 were obliged to bomb the coke ovens by radar because of unfavorable weather. Their results were very poor, but no B–29s were lost despite vigorous Japanese fighter activity.

That night, Japanese bombers struck the Chengtu area for the second time, hitting five B–29s and damaging two severely. LeMay requested night fighters and antiaircraft artillery to defend the forward airfields, but Japanese night raids, which continued sporadically until December, remained small in scale, producing relatively minor damage.[86]

Material effects obtained by the B–29 campaign against the Japanese from Chengtu also left much to be desired. LeMay struggled with the same obstacles that beset his predecessors, Wolfe and Saunders, obstacles that ranged from the seemingly unending mechanical bugs in the B–29 to the wretched weather and high mountains between India and Szechwan. LeMay later wrote:

> About four missions a month was the best we could do out of China, and sometimes we didn't even manage that. It was all due to the bad logistics, wherein our greatest share of flying effort had to be expended in bringing gas up to those Chengtu strips . . . but when it came to bombing Japan, it was bombing Japan from Chengtu or no place. So gasoline had to be hauled over the Himalayas.

He concluded: "There can be no sustained and intensive effort by any bombers [that] have to feed their own fuel to themselves."[87]

Apart from the scantiness of missions, another problem became so vexatious as to cause "deep concern" to General Arnold: the low amount of bomb tonnage carried by the China-based B–29s. "Our concept of the B–29," Arnold complained to LeMay in September 1944, "was an airplane that would carry

The grainy photo of the Tojo fighter was captured during the invasion of Kwajalein, and it served as one of the top identification photos of the new aircraft.

very heavy bomb loads for very long range. We have attained some of the distances but we have not as yet obtained the bomb loads.... It is my desire that you give the bomb load problem a great deal of thought." Arnold also told General Hansell, Twentieth Air Force Chief of Staff, that the B–29s had "not carried any more bombs and in most cases considerably less than the B–24s and B–17s carry. One of the greatest factors in the defeat of Japan will be the air effort. Consequently every bomb that is added to each airplane that takes off for Japan will directly affect the length of the war."[88]

While LeMay succeeded in bettering B–29 sortie rates, increasing bombloads, and restructuring XX Bomber Command, a marked increase in supplies also became available to his forward units. The AAF also shifted the choice of strategic targets, as recommended by the Committee of Operations Analysts (COA), from coke ovens in Manchuria to the Japanese aircraft industry. Additionally, the AAF now approved B–29 missions intended to support directly Allied military operations in the Pacific Theater (code-named PAC-AID).[89]

The American strategic comeback in the western and southwestern Pacific region in late 1944, to be sure, presented Twentieth Air Force and the U.S. Navy with new challenges and opportunities for the concerted use of air power. In particular, MacArthur's planned advance from New Guinea to the Philippines focused American attention on the geographical zone lying between those islands and on Formosa and the China coast. XX Bomber Command still fielded only one VHB wing at Chengtu, China (in the far-off CBI), and its *raison d'être* was overtaken by the new geostrategic objectives and target systems (such as Japan's aircraft industry) evolving elsewhere. Still, the command demonstrated considerable flexibility and clout in responding to the requirements of PAC-AID and of strategic bombardment as stipulated by higher headquarters.

Japanese air facilities on Formosa received particular attention, first by swarms of U.S. Navy carrier-based fighters from Task Force 38 on October 10–11, 1944, then by B–29s from Chengtu. Of 130 bombers that sortied on October 14, 104 struck the assembly plant and air base at Okayama (Kanzan), Formosa, with excellent results, at a cost of twelve emergency landings, one crash, and one person missing. Follow-up raids were conducted on the 16th and 17th, completing the battering of Okayama, Tainan, and Takao, Formosa. As the official AAF history records it, PAC-AID "brought little aid to Pacific forces but accomplished a minor strategic job with admirable thoroughness."[90] Thereafter, as XX Bomber Command received more B–29s, it devoted most of the remaining five China-based missions to the important Omura aircraft factory on Japan's southernmost island, Kyushu.

IJA intelligence in China, meanwhile, well aware of the B–29 buildup in CBI, wondered whether XX Bomber Command was merely being reinforced or whether a second wing was about to be created. The IJN's C-Team had

B–29s over China, on a raid against Japan.

broken Chinese military codes, allowing interception of communications involving U.S. bases in China and the movement of American planes. Predicting B–29 raids from China, accordingly, was "rather easy."[91]

Of 103 B–29s that staged into Chengtu, China, in late October 1944, 78 took off for Omura, Japan, and 59 bombed it on October 25, 1944, at a cost of two aircraft. Aluminum-fabrication works sustained considerable damage.[92] That day over northern Kyushu — at Nagasaki, Sasebo, and Omuta, as well as at Chejudo — the IJAAF claimed to have downed one bomber and damaged twelve. Intercepting the bombers during their departure and return over China, IJAAF fighters claimed to have shot down one and damaged another. When an air raid alert sounded again on Kyushu in the morning of October 26, the IJNAF sent up all available fighters over Omura, Nagasaki, and Sasebo. At Omura, near noon, in thick clouds, four Zeros detected a lone B–29 at an altitude of about 20,000 feet, chased it for thirty-five minutes past Sasebo, and claimed a hit but not a kill. The Japanese surmised that this single plane had flown in to survey the results of the raid the day before.[93]

The next B–29 raids against Omura were costly and ineffective. Though ninety-six bombers were airborne on November 11, 1944, only twenty-nine got to Omura in weather so poor that even radar bombing was a problem. Five B–29s were lost to operational causes.[94] That morning, the Japanese reported, a total of about eighty U.S. bombers raided Omura, Nagasaki, Fukuoka, and Chejudo. Cloud cover at about 10,000 feet cloaked western and northern Kyushu, and although more than seventy-five IJAAF and IJNAF fighters scrambled to intercept them, the B–29s dropped their bombs from above the clouds and departed without being intercepted by aircraft or shot at by antiaircraft guns. The naval arsenal at Omura again was targeted by B–29s, and twenty-four bombs fell in clusters around it, burning down four workers' dormitories. The naval airfield was back in operation within four or five hours. Special efforts were made to prevent the Americans from confirming any effects of their raid, since it was apparent that they had to bomb by radar.

On November 21, 1944, 109 B–29s took off from Chengtu, and 61 again bombed Omura by radar in vile weather, with apparently poor results. They encountered aggressive fighter opposition and inaccurate antiaircraft fire in the process. Despite unusually high claims of forty-six Japanese interceptors shot down (including nineteen probables), the American raiding force admitted the loss of five bombers to enemy action and a sixth to a crash on takeoff. The IJNAF fighters intercepted the raiders and claimed to have thrown them into confusion. A Japanese pilot shot up one B–29 and then brought it down by ramming, at the cost of his life. Once again, the naval arsenal at Omura incurred considerable damage. Other B–29s hit Saga, Omuta, and Kumamoto. Intercepting the bombers on their way back to China, Japanese fighter pilots claimed good results. For the day, the Japanese estimated between five and twelve B–29s shot down and nineteen damaged. Though not all Japanese antiaircraft batteries opened fire, coast artillery guns were active between Omura and Sasebo, claiming to have brought down one B–29.[95]

XX Bomber Command had every intention of attacking Omura again in early December, but Mukden in southern Manchuria was substituted as the target on December 7, 1944. From Chengtu, 108 B–29s took off, and 91 reached the Mukden area rather easily in fine but very cold weather. Although ten hit a rail yard nine miles from the objective, eighty others released 262 tons of bombs on primary targets, mauling the arsenal and doing some damage to the main factory. Japanese interceptors were active, making 247 passes and unintentionally colliding with two bombers. They destroyed two fighters and a B–29. In another incident, a Japanese pilot deliberately rammed a bomber; both planes went down. Finally, an air-launched phosphorus bomb hit, remained embedded, and continued to burn in the wing of a B–29, but the bomber returned safely to base.[96]

The rammings reported by the Americans were apparently conducted by Japanese pilots of the Manchukuoan Air Force, whose aircraft (otherwise unidentified) were too poor to accomplish much except by means of special attack, or *kamikaze*, tactics. In a subsequent XX Bomber Command raid against Mukden on December 21, 1944, one lieutenant successfully rammed and destroyed a B–29 at the cost of his own life, and a major who got in very close was downed by his intended victim. Many of the 104th Air Group's planes started out carrying phosphorus bombs, but most jettisoned them because they interfered with climbing. The first squadron reported no effective hits by phosphorous bombs and blamed the poor results on insufficient training in the bombs' use. Noticing that the B–29s on this mission dropped their bombs blindly and too soon, the Japanese surmised that the two leading planes were lost in the attack by the 104th Group. Ground targets at Mukden suffered almost no damage.

PAC-AID objectives supporting the December 1944 invasion of Luzon in the Philippines, meanwhile, caused XX Bomber Command B–29s to be directed again against Japanese targets on nearby Formosa through which (American military leaders believed), the *kamikaze* attacks against Luzon were being staged. U.S. Navy Task Force 38 teamed up with XX Bomber Command to attack Formosan bases in January 1945. After emergency replenishment of fuel supplies, 46 B–29s headed for the island on January 9. The 39 that bombed Kiirun harbor by radar faced no opposition, but the result of their effort was undetermined. On January 14, 83 B–29s (all fully modified for the first time) took off again for air installations on Formosa, where 54 planes hit Kagi and 13 struck Taichu airfield, in good weather and very effectively. Two days later, 79 of 92 airborne B–29s dropped 397 tons of bombs on Shinchiku, Formosa, again without opposition, adding to the damage U.S. Navy carrier-borne planes had already inflicted upon buildings and parked aircraft. The January 16, 1945, raid against Shinchiku marked the end of XX Bomber Command's involvement in PAC-AID operations; indeed, it would be the last B–29 combat mission flown from the Chengtu Valley in China.[97]

After the China-based B–29 strikes drew to a close, XX Bomber Command flew missions from Kharagpur in India throughout Southeast Asia until March 30, 1945. Kharagpur at least possessed the advantages of ample fuel and bombs, and the command flew twenty missions from there in two months — compared with twenty-nine missions in the preceding seven months — all directed against secondary targets: Saigon, Cam Ranh Bay, Phnom Penh, Penang, Kuala Lumpur, an anchorage near Bangkok, ports on the Burma coast, and familiar targets at Bangkok, Rangoon, and Singapore (though primary objectives at the last-named were placed increasingly off-limits by the British). The last combat mission from India — the forty-ninth — occurred on March 29/30, 1945, when twenty-nine B–29s from Kharagpur hit an oil storage area at Singapore. Using tactics that LeMay was introducing against the Japanese homeland, these bombers came in singly at night, in-trail, and very low, at 5,000 to 7,000 feet.[98]

The impending demise of XX Bomber Command could be foretold when General LeMay departed for Guam on January 18, 1945, to take command of XXI Bomber Command. His experienced successor, Brig. Gen. Roger M. Ramey, directed the last quasi-tactical B–29 bombing missions for Lord Mountbatten before heading for the Marianas himself in late April as the 58th Bombardment Wing Commander. It remained for Brig. Gen. Joseph Smith to close up shop at Kharagpur and proceed to Okinawa with the last air echelon in July 1945. XX Bomber Command was inactivated on July 18, when the

Eighth Air Force, Pacific (redeployed from Europe), took over direction of the new B–29 wings on Okinawa.[99] "On the eve of victory over Japan," historians of the AAF wrote, "the organization, once Arnold's pride but now stripped of its combat units, died quietly like an old man who had outlived his usefulness and his friends."[100]

No American commander, of course, expected to deal Japan a knockout blow through strategic aerial bombardment from western China. As official AAF historians admitted, "the strategic results of VHB operations from Chengtu were not a decisive factor in the Japanese surrender." When Operation MATTERHORN ended, "one might have found it difficult to round up a decent showing of mourners for the interment of that plan."[101] General LeMay called the logistical situation at Chengtu "utterly impossible." In a classic of understatement, General Arnold termed the operations from China, "not simple." Others called Operation MATTERHORN fantastically uneconomical and barely workable. In fact, in September 1944 LeMay declined the Twentieth Air Force's implied offer of more B–29 units. The "whole operating scheme," he declared, "was basically unsound and justified only by the lack of other bases." Writing later in more picturesque terms, LeMay concluded:

> I've never been able to shake off the idea that General Arnold himself never believed that this would work. It didn't work. No one could have made it work. It was founded on an utterly absurd logistics basis. Nevertheless, our entire Nation howled like a pack of wolves for an attack on the Japanese homeland. The high command yielded. The instrument wasn't ready, the people weren't ready, nothing was ready. Folks were given an impossible task to perform. They tried to be good soldiers and do their duty. The whole thing was stacked against them. Under these trying conditions I think that they did one whale of a job with what they had.[102]

U.S. Strategic Bombing Survey (USSBS) analysts of the B–29 campaign against Japan put the best possible face on the performance of XX Bomber Command. "Despite the tremendous obstacles imposed by theater logistics, extreme operating ranges, a new type of combat weapon, and the limited force available . . . ", they asserted in words reminiscent of those that accompany a military decoration, XX Bomber Command "succeeded in carrying the war home to Japan by destroying or seriously damaging many vital installations which were within its striking radius." Between June 1944 and January 1945, China-based B–29s operating at high altitude dropped 5,200 tons of bombs on Japanese heavy industry, two aircraft plants, transportation facilities, oil storage sites, and naval installations — with admittedly mediocre results. In the process, though, XX Bomber Command "provided thorough combat service testing of the B–29 and produced invaluable data upon which many important improvements and modifications were made to production line aircraft, and from which sound cruise control charts and other operational data were developed." LeMay, however, assessed the campaign for what it was when he first went to Kharagpur, calling it a "tiny B–29 war . . . being waged from there."[103]

Critics of the MATTERHORN strategic bombing campaign, citing its lack of success, have observed:

1. the logistical effort invested in the B–29s based in China and India would have been better used in support of Chennault's Fourteenth Air Force's tactical bombing efforts in China;
2. the training and combat experience that the B–29s received by attacking targets in Manchuria and Japan's inner zone could have been obtained by hitting targets in the outer zone from bases much easier to supply in Australia and the Southwest Pacific; and
3. the strategic bombing missions by the CBI-based bombers were not directed against optimal targets and, instead, should have been concentrated, with U.S. Navy submarines, on inland waters, tactical missions "for search, low-level attacks and mining in accelerating the destruction of Japanese shipping, or on attacking oil and metal plants in the southern area."[104]

Supporters of MATTERHORN have rejoined:

1. the war in the Pacific was not won on the continent of Asia, but on the home islands in Japan, which was precisely what AAF headquarters intended from the outset; and
2. the combat units of XX Bomber Command showed how effective they had become after they left Chengtu and Kharagpur for the Marianas, where operating conditions were much more favorable for them.[105]

These rather limp responses, to a great extent, beg the MATTERHORN questions raised by its critics.

The Final Aerial Assault Under Way

Bombing Japan from western China was never the AAF's first choice; it was an interim measure invoked until the preferred base area in the Marianas became available. Even while XX Bomber Command was setting up shop in the CBI Theater and flying combat missions, planning proceeded for the establishment of XXI Bomber Command in the Marianas chain, which lay within Admiral Nimitz's Pacific Ocean Areas Command. In Washington, D.C., General Arnold retained operational control over Twentieth Air Force B–29 forces, and in April 1944 he decided to limit the MATTERHORN force to the 58th Bombardment Wing and divert the 73d Bombardment Wing, when ready, directly to the Marianas.

The Mariana Islands possessed many advantages lacking in western China: more direct access to friendly maritime and air supply, insulation from enemy ground attack, relative immunity from hostile air interference, customarily good weather conditions, and a radius of action that exposed the heart of metropolitan Japan to strategic bombardment. Its disadvantages, including global communication and logistical problems (though not on the order of Chengtu-Kharagpur and the Hump), included vast distances from the U.S. mainland (about 5,000 miles from California) and from targets in Japan (it was still more than 1,200 miles to Tokyo). The Marianas also required huge air bases and

B–29 Radius of Action
Operating from the Marianas
November 1944–August 14, 1945

related facilities to accommodate a much larger force of B–29s, where each bombardment wing (with 12,000 men and 180 planes) would be deployed on a single field.

After many vicissitudes in the development of strategic plans for the Pacific Theater in general and B–29 employment in particular, the JCS agreed to move up the date of Nimitz's assault on the all-important Marianas, bypassing Truk and Ponape in the process. By mid-June 1944, just as Operation MATTERHORN got under way in China, U.S. forces invaded Saipan in the Mariana archipelago. Although a temporary runway was finished on Saipan by August 6, building the Marianas aerial infrastructure — two big airfields and an air depot on Guam, two airfields on Tinian, and one airfield on Saipan — proceeded much more slowly than expected. Severe enemy resistance on Saipan, which continued into July, set back the plans for Guam and Tinian by thirty days. Moreover, Admiral Nimitz assigned construction priority to U.S. Navy facilities on Guam that were to be his forward headquarters as well as a staging location for the proposed invasion of Formosa. That priority delayed completion of the B–29 bases by approximately one hundred days. On Saipan itself, hard coral formations just below the earth's surface, tropical rains all summer long, and intermittent enemy air raids introduced further difficulties.[106]

When General Hansell (who had transferred from Chief of Staff of the Twentieth Air Force to take over XXI Bomber Command on August 29, 1944) landed his first B–29 on Saipan on October 12, the final paving of Isley Field remained incomplete. Aviation engineers worked around the clock to extend and widen one runway and to build a second one and dispersal areas with hardstands. But the full panoply of B–29 facilities at Isley Field would not become fully available until April 1945, six months later. Still, Brig. Gen. Emmett O'Donnell, commander of the 73d Bombardment Wing, had all four of his bombardment groups (497th, 498th, 499th, and 500th) crowded along the single runway by the first week of November 1944, in time for XXI Bomber Command to launch its first B–29 mission from Saipan against Tokyo on November 24.

When Saipan fell to the American invaders, the Japanese military surmised that B–29 long-range bombers would be based there, probably by the end of September or the beginning of October 1944, and be targeted against the Kanto plain (on which Tokyo and Yokohama lay) and central Honshu. IJNAF aircraft succeeded in photographing Isley Field for the first time on September 23, 1944. Cloud cover blocked most of the main runway, but, in conjunction with collateral information, it was now apparent that Isley was ready to receive B–29s. Radio intercepts in late October and the first small B–29 raids on Truk suggested that a considerable number of Superfortresses, perhaps one group of four squadrons, might have already reached Saipan. On November 6, 1944, two Japanese scout planes staged through Iwo Jima in good weather and took excellent photographs of the facilities on both Saipan and Tinian. About forty

Brig. Gen. Emmett O'Donnell commanded the 73d Bombardment Wing.

B–29s were sighted at Isley; the South Field (about 2,500 yards by 165 yards) was judged to have the capability of supporting one hundred Superfortresses. Other 2,350-yard strips were under construction or newly built on Saipan and Tinian and were thought suitable for accommodating B–29s in December. On the 17th, IJNAF scout aircraft confirmed the presence of numerous B–29s on Guam.[107]

The Japanese launched raids via Iwo Jima against Saipan and its growing nests of B–29s. Two of these small-scale but effective attacks in early November 1944 wiped out four B–29s, crippled six, and damaged another twenty-two. XXI Bomber Command responded by dispersing some Superfortresses from Isley to Guam and devising plans to team up with the Navy to preempt the staging strips on Iwo.[108] IJAAF records note that seven heavy bombers and five scout planes attacked Saipan in the middle of the night on November 6/7, 1944, reportedly hitting more than eleven aircraft while losing one of their own. (IJNAF materials are much less optimistic about the effects of this raid.) A few weeks later, on November 27, three Type 4 IJAAF bombers strafed and bombed Isley Field South at very low altitude, reportedly causing fires at two locations and burning more than twelve planes while losing none in the process.

The IJN Air Staff, dissatisfied with the IJAAF's earlier performance, trained their own fighters for this mission and, on the morning of November 27, 1944, dispatched twelve Zeros and two Saiuns against Saipan via Iwo Jima. Though B–29s had raided Tokyo in force this day, there was still potentially happy hunting for the Japanese in the Marianas, and the IJNAF planes went in before noon. Although the Japanese could not ascertain the results of the attack directly, intercepts of U.S. message traffic indicated that Isley Field had been

The Japanese loss of the Mariana Islands to the Allies spelled doom to the career of Tojo Hideki, Premier of Japan.

thrown into confusion by the raiders. The IJNAF planes had come in low, strafed numerous parked B–29s, destroyed four, and severely damaged six.[109] The Americans con ceded that two twin-engine intruders in the early morning had struck at low altitude, caught the B–29s loading bombs, and destroyed one and damaged eleven. Later that same day, at noon, ten to fifteen single-engine fighters got through the radar screen by flying low and, while the 73d Bombardment Group was off to Tokyo, shot up B–29s on Saipan, wrecking three Superfortresses and damaging two.

The Japanese would mount more counterraids in the next weeks, the last on January 2, 1945. Between November and January, more than 80 Japanese planes attacked Saipan and Tinian at a cost of about 37, but in the process they destroyed 11 or 12 Superfortresses, crippled 8, damaged 35, and inflicted more than 245 casualties. The Americans admitted that the constant threat of counter raids exerted "a significant and adverse effect on morale." In due course, U.S. thinking focused on the need to seize Iwo Jima, the intruders' vital staging base, though efforts would first be made to suppress the bases on the island by air and sea attack.[110]

On November 5, 1944, goaded in part by a Japanese bombing and strafing attack against Isley Field on the night of the 2d/3d (which, though unsuccessful, presaged more counterraids),[111] Hansell and O'Donnell sent their green

Saipan established: Members of an engineer aviation battalion lay mat in preparation for the landing of equipment on the beaches of Saipan (*top*). They rigged garbage chutes over the cliffs through which they sent trash into the sea (*center*). A B–29 flies over Saipan, a new base for the Superfortresses (*bottom*).

squadrons against airfields on Iwo Jima, 725 miles from Saipan. Again, the bombing resulted in minimum damage to Japanese installations; a dozen airborne enemy fighters did not interfere. The Japanese admitted that one IJNAF Tenzan (Jill) suffered moderate damage on the ground. Another B–29 so-called training mission on November 8, also directed against Iwo, accomplished nothing. In the process, one of eight Japanese interceptors managed to hit a B–29 with a phosphorus bomb, causing minor damage.

On November 11, 1944, an untried, third B–29 group attacked Truk, achieving no better than fair effects. This was the final shakedown raid. Three of the four B–29 groups had seen action against the enemy and had emerged essentially unscathed. For the rest of the war, XXI Bomber Command would often use battered Truk as a practice field for other B–29 groups as they arrived — thirty-two runs in all. Now the AAF wanted the Superforts to begin striking Japan itself in the final aerial assault in the Pacific Theater. Heartland Honshu would become the first strategic-bombardment target for the Saipan-based B–29s. The overriding AAF doctrine still stressed high-altitude precision strategic bombing by large formations in daylight. General Arnold would finally be able to apply his basic principle to the fullest: "the main job of the Air Force is bombardment."[112]

In Washington, D.C., the COA had recently submitted updated target recommendations for use by the new XXI Bomber Command. The JCS considered two alternate assumptions: that an air and sea blockade alone could bring Japan to its knees or that an invasion of the home islands would be necessary. The JCS opted for the second alternative, and the Twentieth Air Force chose its targets accordingly, after fine-tuning by the Joint Target Group. The first priority was the Japanese aircraft industry followed by urban industrial areas and intensified attacks against shipping, to include the laying of sea mines by B–29s.

The aviation industry was considered a particularly lucrative target system because it was highly concentrated at several factories located in and around the major cities of Tokyo, Nagoya, and Osaka. Second in importance only to Mitsubishi's engine plant at Nagoya was the Nakajima factory at Musashino-Tama (in Tokyo's suburbs), which turned out nearly 30 percent of all Japanese combat aircraft engines. As a target, Tokyo had added advantages: it was the Imperial capital, Japan's largest city and the nerve center for civil and military administration, preeminent in industry, communications, commerce, and transportation. A symbolic site (for both adversaries), Tokyo had not been struck from the air for two and a half years, not since Doolittle's daring raid in April 1942.[113] Moreover, based on the experience of XX Bomber Command, Japanese aerial defenses over the home islands appeared feeble.

At the end of 1944, Japanese aerial defenses were indeed weak. Even against solitary F–13 (the high-flying photoreconnaissance version of the B–29) scout airplanes over Japan, IJAAF and IJNAF interceptors were impotent, and heavy antiaircraft artillery was in woefully short supply. For example, in broad daylight on the afternoon of November 1, 1944, a clear fall day, Eastern Army headquarters in Japan learned that a four-engine aircraft had been detected flying into the airspace of the homeland. The 10th Air Division alerted its interceptor units. Dozens of IJAAF fighters scrambled, trying to catch up with the intruder, but it got away "because it was flying at an altitude in excess of thirty-three thousand feet." On November 7 the Japanese thought they had deployed a perfect defensive setup against intruders, and as many as 300 fighters scrambled to attack a lone B–29 being tailed by a Japanese reconnaissance plane. The B–29 soared to almost 40,000 feet and escaped to the south. Meanwhile, reports came in to the 10th Air Division, responsible for defending the greater Tokyo region, that three formations of six B–29s each were overflying the entire Boso area. The 10th Air Division Commander, Kihachirō Yoshida, issued a stirring exhortation to his whole command, but the operation had to be canceled when division officials learned that the ground observers had confused friendly airplanes with enemy ones. On numerous occasions thereafter, prior to the B–29s' first Tokyo raid of November 24, Japanese fighters swarmed up to intercept single B–29 scouts, but "we simply weren't able to nab the enemy."[114]

Japanese interceptor pilots were well aware of the problems they faced contending with the B–29. Lt. Col. Shizuma Matsumura, 10th Air Division Staff officer, lamented that friendly fighters and antiaircraft guns could not reach B–29s flying at more than 31,000 feet. Japanese planes could only climb to about 28,000 feet, and the available antiaircraft artillery could reach only about 23,000 feet. It was frustrating and humiliating, said Matsumura, to try to defend the Imperial capital "in front of an audience" under such circumstances: "Watching our numerous planes being made fools of publicly, some citizens spoke ill of us — 'What in blazes are the air defense units doing up there?'"[115]

In late 1944, General Yoshida searched for answers. First, he converted fighter crew training to high-altitude defense against B–29s rather than to low-altitude attack against aircraft carriers. To deal with the high-altitude American bombers, Japanese interceptors, flown by only the most experienced pilots, would climb above 33,000 feet, although it took them about forty-five minutes to reach that height after takeoff. Unless the pilot flew precisely a level course in the thin air at that height he would immediately lose altitude if he dipped a wing even slightly. Moreover, serious problems with the oxygen equipment had to be corrected.

Given the shortage of true night fighters, nighttime flying training was difficult, and Yoshida endeavored to improve the pilots' coordination with searchlight batteries. He set up beacon lights at key points around Tokyo so that the aviators could fix their position during blackouts. Additionally, Yoshida had pilots work on tactics to break through cloud layers and perfect their skill in dropping phosphorus shells on enemy bombers.

Yoshida devoted special attention to early warning by picket ships and ground lookouts, and he increased searchlight coverage on land and at sea. He intended to engage enemy bombers as far from vital metropolitan locations as possible. To do this, he had to improve on the sixty-minute lead time ordinarily available after a warning from the Hachijojima lookouts, when the enemy was 125 to 150 miles south of the watch post. On November 12, 1944, he organized a special intelligence unit to work in concert with naval pickets deployed about 200 miles offshore to detect B–29s on their departure from the Marianas and then to track their approach. Teams of cryptanalysts and English-language specialists were assembled to work on intercepting and decoding U.S. radio traffic emanating from Saipan and Tinian. The special intelligence units became quite adept in plotting B–29 raids before takeoff, but the naval picket project made little progress because it lacked proper materiel.

As early as October 1944, General Yoshida convened all of his unit commanders for map maneuvers against the expected B–29 raids. Empirical evidence for the testing of tactics stemmed from the knowledge of German experience in the Allied bombing of Hamburg and Berlin and of XX Bomber Command raids against northern Kyushu. Yoshida's staff expected that XXI Bomber Command would come in high (26,000 to about 30,000 feet) in daylight, flying in waves, with formations of more than several planes each. Thereafter, in practice, Yoshida tried to compensate for materiel deficiencies by committing his air division en masse against enemy raiders, no matter how few. When the first efforts to intercept B–29s failed dismally, Yoshida began to rely on *kamikaze* — ramming — tactics. Also, to enhance climbing capability, the Japanese fighters were lightened by divesting them of bulletproof steel plating, bombing equipment, and even some armament. Radio equipment, gun sights, and lightweight oxygen devices without fixed tanks were retained.

On November 7, 1944, Yoshida ordered each of his fighter groups to form special attack units consisting of four planes each. At this early stage of *kamikaze* thinking, pilots were allowed to return to base if they could not engage the foe. Bailing out was also acceptable if it could be done after ramming an enemy plane. Even when a pilot tried to down a bomber by gunfire, the preferred method, he should "attack with the spirit of a rammer." By all accounts, every unit commander accepted Yoshida's directive readily and without resentment. One group commander remembered that the preponderance of pilots were keen to fly *kamikaze* missions, while a few merely said yes, and nobody said no. The group commander formed his special attack unit only from the category of individuals very eager, taking into consideration family circumstances and squadron needs; squadron commanders were not consulted.[116] By late 1944, however, it was a case of too little, too late for Japanese aerial defenses.

Whatever progress the 10th Air Division had made would be put to the test in XXI Bomber Command's first big raid against the Tokyo area on November 24, 1944. Late that morning, the watch post on Ogasawara notified Eastern Army headquarters that a large force of B–29s was moving north toward Honshu. As reports came in from Izu, 10th Division Commander Yoshida saw a full-scale raid in the making, and he scrambled all of his operational interceptors in the Tokyo region.

A few days earlier, on November 11, 1944, Twentieth Air Force had transmitted the first target directive to Hansell. It called for an early precision-bombardment strike against Tokyo (SAN ANTONIO I), while XX Bomber Command in China continued raids against Omura and Okayama. Ten to twelve unescorted squadrons composed of nine to eleven B–29s from XXI Bomber Command were to conduct visual, daylight bombing from 30,000 feet, each carrying more than 8,000 gallons of fuel and 5,000 pounds of bombs (30 percent incendiaries and 70 percent 500-lb general-purpose bombs).

F–13As from Hansell's 3d Photographic Reconnaissance Squadron safely flew seventeen single-plane missions over Japan before the Tokyo raid. Half of these were to acquire target information; the rest, to scout weather — all infuriated the hapless Japanese defenders. To enhance the chances of success, prestrike ferret electronic eavesdropping missions also would be flown, as would a feint to Nagoya launched the same day by F–13s dropping chaff. But bad weather over Saipan repeatedly delayed SAN ANTONIO I, and the skies did not clear until the 24th.[117]

O'Donnell led the first of the four groups' 111 B–29s (carrying 278 tons of bombs) from Saipan at 6:15 A.M. on November 24, 1944. Seventeen of the bombers dropped out; six others suffered mechanical failures. A high-altitude tail wind (from the jet stream) greatly accelerated their speed over Tokyo, and an undercast almost blotted out the target. Flying north over the Izu peninsula and then east from the vicinity of Mount Fuji, twenty-four B–29s headed off to bomb the Nakajima aircraft factory at Musashino near Mitaka; fifty-nine others hit secondary targets — docks and urban areas — by radar, and then retired over the sea via Kashimanada. The 10th Air Division deplored the raiders' swift passage through the Kanto area; they moved so rapidly that the fighters could not mount the persistent attacks as planned. Only a few of the best Japanese pilots even managed to reach the altitude necessary to close with the B–29s, and they generally could conduct no more than one firing pass and lost altitude whenever they turned to attack again. The enemy was long gone by the time they recovered. The B–29s' speed and altitude also foiled the antiaircraft batteries because the view aloft was obscured by heavy clouds over their positions.

The 47th Air Group Commander at Narimasu, Captain Noboru Okuda, recollected that, without assembling in the air, his fighters went straight after the enemy in flights of two or three, even singly. Okuda saw the B–29s operating in threes, totaling nine per echelon that flew over Tokyo in swift succession. The Japanese fighter pilots attacked the nearest target ahead of them, mostly one-on-one. Okuda flew his Nakajima Ki–44 Type 2 fighter against a B–29 at about 34,000 feet. The best approach, he knew, was to fly toward the Superfortress at the same altitude and from the front. The B–29s' forward firepower was much weaker than it was in the rear, and sighting and shooting was easy for the approaching interceptor, up to the point of running into the bomber. After making his firing pass, Okuda flew under the belly of the Superfortress and disengaged directly. "The B–29 was firing fiercely," he said, "and I felt as if I were traversing red fireworks."[118]

Altogether, the 10th Division claimed to have damaged nine B–29s and shot down five, although only one hulk of an aircraft was detected on the ground. A *kamikaze*-flying noncommissioned officer was responsible for one of the confirmed kills achieved by ramming. Six IJAAF fighters were lost. The B–29s dropped many bombs in the Nakajima factory area, causing a considerable number of casualties but inflicting only slight damage on the facilities. General Yoshida, after reviewing the results of the fighting, ordered each group to increase its *kamikaze* flight to eight planes. In his critique, Yoshida pointed out that the fighting had degenerated into solo engagements, which detracted from overall unit effectiveness. True, this stemmed from the enormous difference in the capabilities of the opposing aircraft, but there was no use even thinking about obtaining new, high-altitude interceptors. Realistically, the best that could be done was to intensify the interceptor training program.

In view of the flight path taken by the B–29s in striking the Kanto region, the Eastern Army's antiaircraft artillery command saw the need to reinforce the antiaircraft positions on the west side of Tokyo. Against such high-flying raiders, it was deemed imperative to go over to heavier caliber guns. Since the Superfortresses's speed, increased by a lively tail wind, caused them to pass rapidly through the antiaircraft batteries' effective arc of fire, special emphasis would be required to emplace the 120-mm guns in particular.[119]

Far to the south on Saipan on November 24, 1944, the surviving bombers straggled for 3½ hours onto the single operating runway, which became so congested that two groups had to proceed to Guam. Japanese fighters had proved less dangerous than feared; their passes were uncoordinated and uneven. Of the 125 Japanese interceptors estimated to have sortied, Superfortress gunners claimed to have shot down 25 (only 7 confirmed) and damaged 9. One bomber, rammed in the tail by a damaged Tony, crashed at sea and lost the entire crew. Antiaircraft fire was no more than moderately active and not particularly accurate. Bomber losses were not judged to be excessive, certainly not as severe as those experienced at the hands of the *Luftwaffe* in Europe. Two B–29s were lost and eight damaged, three by friendly fire. One man died, four suffered injuries, and eleven were missing. The bombing results, though somewhat more effective than thought at the time, were unimpressive in terms of damage to buildings and machinery. However, Hansell's bombers had managed to strike a well-defended target zone, unescorted, and in poor weather. Psychologically, XXI Bomber Command's Tokyo raid showed the Japanese that B–29s could hit Japan's heartland directly, not merely along the fringes reached by XX Bomber Command, and with relative impunity.[120] The nature of the war in the Pacific Theater had changed.

XXI Bomber Command replicated the B–29 raid of November 24 in many ways during the next three months of operations from the Marianas. Possessing only the 73d Bombardment Wing until February 1945, the command relied on an average inventory of about 125 combat bombers. With the addition of the 313th Bombardment Wing in early February and two groups of the 314th

Bombardment Wing toward the end of that month, the B–29 inventory in the Marianas more than tripled, to 385. But foul weather over Japan posed a serious problem during the winter months from December 1944 through February 1945, allowing only eighteen missions in ninety days.

Strategic bombardment remained the same throughout these three months: high-altitude, daylight, precision attacks directed primarily against the Japanese aircraft industry, especially engine plants, targets that drew sixteen of the twenty-two main missions before March 9, 1945. For Hansell's command it was a period of "experimentation and adjustment" intended to determine and capitalize on the Superfortress's operational capabilities and limitations in a combat environment, meanwhile contending with a still-imperfect support infrastructure of airfield facilities, maintenance and supplies, and auxiliary services (weather forecasting, air-sea rescue, communications, and long-range reconnaissance).[121] Miserable weather over Japan, often termed a more serious foe than the Japanese, exacerbated the B–29s' overall loss rate (which rose significantly to 5.7 percent of airborne bombers in January 1945). Until a friendly intermediate base became available at Iwo Jima, many crews, unable to make an emergency ground landing for repairs or fuel, ditched their aircraft at sea.

During this period, Japanese interceptors were the most aggressive, though the AAF did not deem the aerial resistance to be nearly as serious as the type encountered in Europe. Because of the limited number of B–29s available at this stage, however, the strategic bombardment campaign exhibited repetitive features which enabled the defenders to concentrate their fighters over predictably critical target areas. In numbers and persistence, though not particularly in quality, firing passes and ramming posed "a serious although temporary problem" for XXI Bomber Command. The B–29's own gunnery system directed by its airborne fire-control apparatus proved highly dependable in action, and it chalked up an impressive hit or shoot-down score against enemy interceptors that did manage to close.[122]

High-altitude flying at 25,000 feet and above could not entirely protect the B–29s from interception, as even the first raid on Tokyo in late November 1944 had shown. In the course of several subsequent missions, an estimated 200 to 300 Japanese airplanes conducted more than 500 individual passes, and no long-range American fighters yet were available to provide escort. In November 1944 aerial attacks against the bombers totaled fewer than 200; in December, the figure was close to 2,000; in January 1945 it became nearly 3,000; but in February, it was less than 1,500. Daily fighter opposition peaked during a raid against Musashino on January 27, when 984 Japanese aerial passes were made against 56 B–29s, bringing down 9. Antiaircraft guns fired much but accomplished little during this initial phase of high-level operations. Between November and February, XXI Bomber Command acknowledged only a single loss to antiaircraft fire.

To hit COA-selected targets, XXI Bomber Command flew twelve major missions between November 29, 1944, and January 19, 1945. The Musashino works and the Tokyo area accounted for four of them: on November 27 with eighty-one B–29s airborne; on December 3 with eighty-six; on December 27 with seventy-two; and on January 9 again with seventy-two. The bombers went after targets at Nagoya five times: on December 13 with ninety B–29s airborne; on December 18 with eighty-nine; on December 22 with seventy-eight; on January 3 with ninety-seven; and on January 14 with seventy-three. The three other major raids occurred on the night of November 29/30 with some thirty B–29s using night radar for the first time against Tokyo; a tactical strike by sixty-two Superforts against Iwo Jima on December 8; and a raid against the Kawasaki aircraft plant at Akashi, west of Kobe, on January 19 by eighty B–29s. The missions against Nagoya on December 22, 1944, and January 3, 1945, dropped incendiaries, the latter strike conducted for the Twentieth Air Force as a test of high-altitude incendiary area bombing.

Put simply, the bombing results of ten or eleven of these twelve raids did not meet expectations. Evaluations of the high-altitude precision bombing's effectiveness ranged from nil to poor. The night-radar raid of November 29 was unsuccessful; the incendiary attack of December 22 produced negligible results; and the experimental strike of January 3 proved inconclusive. A high number of aircraft aborts (caused by bad weather and mechanical failure) and an inability to bomb primary targets en masse contributed to the generally ineffectual results. As indicated in Table 4–1, the percentage of planes that bombed compared with those that took off underscored the problem.[123] Because formation flying at high altitude demanded heavy loads of fuel, the bombloads carried by the Marianas-based B–29s also were relatively small. Table 4–2 presents the monthly tonnage of bombs dropped by the XXI Bomber Command in high-altitude precision attacks against aircraft industry targets. Nagoya bore the unenviable distinction of being bombarded by 152 tons in January 1945.[124]

If American commanders were dissatisfied with the B–29 bombing results, their Japanese air defense counterparts remained frustrated by their inability to stop the raiders. During the B–29 night raid on Tokyo on November 29/30, 1944, General Yoshida ascended to the rooftop of his 10th Air Division com command post at Eastern Army headquarters to view the bombing. From this vantage point, he confided in his diary:

> I felt a keen sense of responsibility as I watched the fires still burning in the areas of Kanda and Nihonbashi . . . but, against enemy bombers which flew above the clouds at an altitude of twenty-seven thousand feet, in the rain, there was no way to catch and attack them using planes that could not climb above those clouds at night. Our antiaircraft guns also fired more than three hundred rounds without result.
>
> This was the third raid by enemy bombers in a week, and their fighting spirit was very high. I suppose they have complete confidence in their planes' excellent performance, and they possess fine attack techniques such as radar flying above the clouds, radar bombing, and technical proficiency in high-altitude bombardment.[125]

TABLE 4–1
Planes Taking off on Missions versus Planes Bombing
December 1944–January 1945

	No. of Planes		
Date	Taking off	Bombing	Percent
Dec 3	86	59	68.6
Dec 18	89	63	70.8
Dec 22	78	48	61.5
Dec 27	72	39	54.2
Jan 3	97	57	58.8
Jan 9	72	18	25.0
Jan 14	73	40	54.8
Jan 19	80	62	77.5
Total	647	386	59.7

Source: Craven and Cate, eds., *Army Air Forces in World War II*, vol. 5, pp. 560–566.

TABLE 4–2
Monthly Tonnage of Bombs Dropped by XXI Bomber Command in High-Altitude Precision Attacks
November 1944–January 1945

Target	November 1944	December 1944	January 1945
Aircraft Industry	112	247	42
Oil Facilities	—	12	9
Steel Manufacturing	—	3	6
Urban Areas:			
Tokyo	392	102	205
Yokohama	4	4	8

Source: USSBS, *Effects of Air Attack on Tokyo etc.*, p. 6.

The overall air defense commander, Prince Naruhiko Higashikuni, was similarly distressed when the bad weather, as he saw it, prevented Japanese fighters and antiaircraft guns from engaging the B–29s on November 29/30, 1944. He, too, felt a sense of profound responsibility for "allowing the enemy to operate at will over the capital city." But, he reasoned, if countermeasures were studied quickly and diligently, might not the enemy attacks be contained?

One way might be to step up counterraids against the B–29 bases in the Marianas. Another might be to modify Japanese interceptors, permitting them to fight at altitudes in excess of 30,000 feet. Still another method might be to develop or improve radar capabilities to direct antiaircraft fire against invisible targets at extreme altitudes. Higashikuni therefore ordered his staff to study countermeasures on an urgent basis.

Against an estimated seventy B–29s in good weather on December 3, 1944, Yoshida's 10th Air Division deployed its units well in advance at ultrahigh altitude. That day the interceptors claimed to have knocked down twenty-seven B–29s, more than twice as many Superfortresses as the Americans admitted. Japanese airfield battalion personnel did capture three U.S airmen who had parachuted from a B–29 downed over Chiba. In all, the Japanese conceded the loss of six aircraft. Maj. Teruhiko Kobayashi, 244th Air Group Commander, was one of those shot down; he survived the crash and using a reserve Kawasaki Hien (Tony) fighter, took off and tried in vain to catch up again. He later recounted the experience of closing with a B–29 formation:

> They concentrated gunfire against me from all directions, with the B–29s behind and on the flanks of the lead formation lending overwhelmingly superior fire support. It was like flying into a bird net made of small mesh.[126]

Japanese antiaircraft batteries, firing 80-mm and 120-mm guns, claimed to have downed two more B–29s. Though a large number of bombs dropped on and around Nakajima's Musashino works, the damage was slight.[127]

In another raid on December 13, 1944, the Japanese air defenders seem to have done somewhat better against the B–29s. Though Superfortresses, flying singly or in small flights, consistently eluded Japanese interception, defense headquarters tracked them almost minute-by-minute. On the 13th, the 10th Air Division deployed its interceptors over Izu but moved them westward when it became apparent that U.S. bombers were going to hit Nagoya for the first time. There, the 11th Air Division of the Central Army prepared to do battle over Nagoya or over the Han-Shin (Osaka-Kobe) area. Two IJAAF air groups tried to close with the B–29s but had trouble climbing to high altitude. The interceptors' attacks were deemed to have been delivered sporadically. The Superforts mainly bombed the area of the Mitsubishi Heavy Industries factory, which sustained considerable damage. The 11th Division claimed to have shot down two B–29s and damaged eight; antiaircraft batteries claimed no hits. Against a smaller force of B–29s that headed for the Kanto district that day, 10th Air Division patrols endeavored to engage but reported their inability to catch up with the bombers. According to the diary of one of the two IJA commanders defending the skies over Nagoya on December 13:

> About seventy or eighty B–29s invaded the Nagoya area and attacked the vicinity of the Mitsubishi plant. We intercepted and damaged two bombers. These results were unsatisfactory, but the lessons should be applied to future fighting. Regrettably, there was a lot of trouble with our machine cannon, the result of insufficient practice firing and gunnery training. We removed wing guns and protective shields.[128]

Summarizing his experiences contending with the B–29s during 1944, General Yoshida recalled that his 10th Air Division had clashed 40 times with the raiders, sending 4,021 interceptor sorties against 638 bombers. Nine percent of the intruding B–29s were claimed shot down at a known cost of 10 percent of the defending fighters — an unacceptable ratio. Indeed, Yoshida resorted to what he called "deliberately unreasonable means" (the special attack, or *kamikaze*, tactics), a choice that brought tears to the general's eyes. He blamed these desperate straits largely on the backwardness of Japanese science and technology. If only Japan possessed fighters whose normal service ceiling was at least 36,000 feet, a fivefold increase in effectiveness and the thwarting of the enemy's air campaign could have been accomplished with half the effort and half the cost. The situation confronting Imperial Japan, as Yoshida saw it, was "truly deplorable"; at year's end he wrote:

> Everything is too late now. There is no alternative but to continue to force the unreasonable [special attacks]. One consolation is that the effectiveness of enemy bombing has been slight and the raids have affected neither our [national] combat power nor the spirit of the populace . . . but aggressively wiping out the enemy raiders is the best air defense measure; so I hereby resolve to bend every effort to achieve certain victory in decisive combat by expending further energy on innovation and study, by trying harder, and by making up for material insufficiency by enhancing spiritual strength.[129]

The laments and efforts at improvement by General Yoshida and Prince Higashikuni bore some fruit by late 1944. Although spiritual strength in the absence of modern interceptors would prove insufficient to stop the B–29 onslaught, the damage inflicted on B–29s was relatively severe on several occasions.[130]

On January 9, 1945, the IJA 10th Air Division deployed all units to intercept approximately thirty B–29s detected over Shizuoka heading for Tokyo in formations of three to eight planes, though some bombers approached singly. In the course of the ensuing battles, the Japanese claimed to have shot down or damaged four times as many B–29s as the Americans admitted. Six of the bombers confirmed destroyed were brought down by *kamikaze* pilots, three of whom bailed out safely. In addition to the IJAAF fighters lost in *kamikaze* action, eight were crippled and nine less severely hit. The IJNAF sent up fifty-one fighters, which reportedly downed one B–29 and damaged more than ten. Two Raidens (Jacks) and one Zero-sen (Zeke) sustained minor damage in the combat. Damage at the Musashino plant was slight; many bombers missed the target because of strong winds. The B–29 force also dropped bombs at Yokohama, Numazu, and Fujieda, while twenty other Superforts went after Nagoya and twenty more attacked Mie and Wakayama prefectures.

In his diary entry for January 9, 1945, General Yoshida commended the gallant performance of his *kamikaze* airmen and penned a poignant eulogy for the three pilots who died ramming B–29s. It heartened him this time, he wrote,

Commonly used as *kamikaze* were the Jack (*top*) and the Zeke (*bottom*) fighters.

While the *kamikaze* struck the B–29 Superfortresses, their most intensive concentration took place against U.S. naval vessels, especially carriers. Two dramatic photos show how antiaircraft fire, peppering the skies, shot down both *kamikazes* before they were able to reach their objectives. The one above is the USS *Sangamon*, and the one below, the USS *Lexington*.

to convey a feeling of reassurance to the populace of Tokyo who could see the rammings in broad daylight and deduce that the air defense of the capital was "going strong." Word of the fighting was transmitted to the Emperor, in whose name a third Imperial citation was awarded to the 10th Air Division.[131]

Whatever satisfaction the Japanese air defense command gleaned from the performance on January 9, 1945, it was dispelled ten days later. On the 19th, ten B–29s moving east over Hamamatsu convinced the 10th Air Division that a big raid was heading for the Kanto plain. General Yoshida sent up all of his interceptor units to deal with the supposed threat. He lost one fighter for his trouble. At the same time, the 11th Air Division tried to shuffle its forces to cover both Nagoya and Osaka-Kobe. But the main body of eighty B–29s flew west via the Han-Shin area and, sighting visually, hammered Kawasaki's main fuselage and engine construction factory at Akashi, where no less than fifty structures were razed. The sixty-two Superfortresses that struck Akashi dropped 154 tons of high-explosive projectiles. Their unusual success came against Japan's third largest manufacturer of combat aircraft.[132] In this engagement, only seven IJNAF Raidens (Jacks) seem to have engaged the B–29s which, according to Japanese sources, for the first time fully accomplished their doctrinal objective, precision bombardment from high altitude.[133]

On the American side, however, ironies abounded in the conduct and consequences of the Akashi air strike. U.S. photo interpreters grossly underestimated the effectiveness of the raid, actually the B–29s' best performance to date. The day after the raid, on January 20, 1945, General Arnold relieved Hansell as commanding general of XXI Bomber Command and replaced him with General LeMay.

XXI Bomber Command under New Management

In Washington, D.C., dissatisfaction with XXI Bomber Command had been brewing for weeks. Impatient and disturbed with results to date, Arnold had sent his Chief of Staff, Brig. Gen. Lauris Norstad, to Guam in early January 1945, about two weeks before the productive raid against Akashi, first to meet with Hansell and then with LeMay and advise them of the impending changes in leadership he had decided upon. Hansell declined to stay as LeMay's deputy and instead accepted a backwater post with AAF Training Command in the United States.

General LeMay left Kharagpur, India, on January 18, 1945, for the Marianas to assume what has been called "the most coveted operational command job in the AAF." As LeMay viewed the assignment, the Chief of the AAF "had to keep juggling missions and plans and people until the B–29s did well. General Arnold was absolutely determined to get results out of this weapons system."[134]

Before relinquishing command, Hansell reported at length to Arnold. Though convinced the performance of his 73d Bombardment Wing compared favorably with that of the 58th Bombardment Wing in the CBI, he acknowledged four main problems that were already receiving corrective attention: convincing unit members of the merits of daylight high-altitude precision bombardment versus radar night bombing; improving the deplorable level of bombing accuracy; reducing the aircraft abort ratio; and reducing the bomber-ditching rate and enhancing air-sea rescue operations. Hansell submitted that his major weakness had been to push the command too hard in the absence of sufficient depot facilities and maintenance support. Too many aborts and ditchings had resulted.

Widely respected in AAF circles for his brilliant planning background and field experience, Hansell subsequently drew little criticism for his tour with XXI Bomber Command. The worst that could be said of him was that he tended to be inflexible in outlook and reluctant to innovate. Most particularly, he clung to the dogma of high-altitude precision bombing when it was becoming apparent that high-altitude winds over Japan precluded the effective use of this concept with existing technology, and when Arnold and Twentieth Air Force leaders in Washington were espousing experimentation with incendiary area bombardment.[135]

Hard driver though he was, LeMay could not turn the B–29 situation around overnight. Helped by the infusion of more airplanes and better maintenance, he confronted the same problems that Hansell had, including delayed completion of bases, questionable Superfortress serviceability and reliability, and miserable weather and high winds over Japan during the winter. Inevitably, most of XXI Bomber Command's missions under LeMay between January and March 1945 resembled Hansell's efforts, though the numbers of bombers dispatched were usually greater, and the cost sometimes worse. LeMay's XXI Bomber Command sustained especially grievous losses in air raids on January 27, February 4, and February 10, 1945. In the costly January 27 raid, 56 B–29s went against Musashino in Tokyo; Japanese interceptors made 984 passes, and 9 Superforts went down. The United States regarded this as a failed mission, "completely spoiled" by clouds and fearsome winds.[136]

If XXI Bomber Command was disappointed by the ineffectiveness of the January 27 raid as a precision bombing strike, the Japanese were well aware of the results on the ground in Tokyo. Incendiaries and high-explosive projectiles came crashing down on eleven central wards in the capital. Thwarted by jammed air-raid warning sirens, 1,453 people were killed, missing, or injured, the highest number to date. In addition 1,241 houses were totally or partially destroyed and 4,400 persons became homeless. Elementary schools, temples, and other undamaged buildings were pressed into service for the dispossessed, and neighborhood associations and city authorities furnished meals, bedding, blankets, and emergency medical supplies. Roads, railways, high-voltage wires,

The rubble in Tokyo's Ginza area and the devastation of the most populated section of the city made the Japanese well aware of the results of B–29 bombing raids.

and water pipes were particularly damaged.

The 10th Air Division, having had ample lead time to cope with the raid, scrambled all available fighters. The communique subsequently issued by the Japanese Imperial General Headquarters (IGHQ) claimed twenty-two B–29s had been shot down by IJA planes and many more had been damaged. The IJN also sent up fifty interceptors and reported downing six more bombers, making a grand total of twenty-eight claimed destroyed (against nine actuals). The first Superfortress formations, hit hard by fighters, were disorganized when they finally retired toward the sea to the east. During the strike, the Japanese lost twelve IJAAF and three IJNAF interceptors, a ratio of 1 to 2 according to their own claims, but nearly 2 to 1 in actuality.[137]

On February 4, 1945, XXI Bomber Command again tested high-altitude incendiary bombing techniques directed against Kobe. A total of 129 B–29s attacked, including 38 from the new 313th Bombardment Wing. Of these, 69 (flying at 24,500 to 27,000 feet) struck the target zone with about 154 tons of incendiaries and 14 tons of high explosives, severely mauling local war production factories in the Kobe district. Two hundred Japanese fighters put up a stiff resistance; they succeeded in shooting down only one B–29 and damaging thirty-five, apart from another bomber that burned after landing on Saipan.[138] Incendiary bombs caused substantial damage at Kobe, and production at the shipyards was sharply curtailed. IJAAF interceptor units claimed to have downed six Superforts and damaged thirty. If five of the reportedly destroyed bombers were transferred to the damaged category, the Japanese claim would match the American admission. Three IJAAF fighters were lost.[139]

In a February 10, 1945, raid against Nakajima's Ota aircraft assembly plant, eighty-four B–29s bombed the target area with generally unimpressive results. The few incendiaries dropped seemed to have inflicted only moderate damage. No less than twelve Superfortresses were lost and twenty-nine damaged.[140] According to the Japanese, the Ota factory sustained great damage from the bombing. Yoshida's 10th Air Division, reinforced by the remnants of two battered groups returned from the Philippines in December, deployed airborne patrols and claimed fifteen B–29s (close to the American admission) shot down, at a cost of seven planes. An independent antiaircraft gun battalion, shifted to Ota, saw action against the B–29s for the first time and reported shooting down three and damaging eleven of the bombers. The Japanese air defense headquarters later surmised that the heavy losses inflicted on the Superfortresses on this day contributed to XXI Bomber Command's subsequent decision to abandon daylight bombardment in favor of night raiding.[141]

The most impressive B–29 performance to date occurred on February 25, 1945, when 172 B–29s (including elements of the new 314th Bombardment Wing) lent strategic support to the recent U.S. landings on Iwo Jima by pounding Tokyo for two hours in daylight, scattering 454 tons of fire bombs by radar through very heavy cloud cover from altitudes of 23,000 to 31,000 feet. The bombers met no fighter opposition and flak was light. Though only one square mile of the city was thought to have been burned out and the bombing could hardly be termed precise, this was the definitive test of an incendiary attack that Twentieth Air Force had been craving.[142]

The damage and the shock inflicted on the Japanese capital by the fire raid of February 25, 1945, were far greater than American intelligence analysts deduced at the time, or even immediately after the war. Initially, an estimated 600 U.S. Navy carrier planes, hotly contested by 123 IJAAF fighters, had struck airfields, factories, and transportation facilities on the northern and eastern Kanto plain on the morning of the same day. Then, in worsening weather that turned to heavy snow by afternoon, B–29s in seven echelons struck Tokyo from above the clouds with high-explosive bombs and, most particularly, incendiaries, the city's first experience with the latter. Fire after fire broke out, burning more than 10 percent of Tokyo's houses (190,000 of 1,540,000 structures). The firefighters could not contend with the conflagration. Nakajima's Ota aircraft factory, however, did not sustain much damage. Yoshida discerned that emotions were finally mounting among the populace as the result of the repeated air raids, and public criticism appeared to the effect that Japanese air defenses were undependable.[143]

If the February 25, 1945, raid was the most effective B–29 strike so far, the least impressive precision bombing attempt (the biggest to date) directed against Musashino followed on March 4, 1945, constituting the eighth failure there. Heavy cloud cover forced 159 of the 192 aircraft to divert their attacks to secondary targets at Tokyo, and another 17 were required to try objectives of last resort.[144] Still, the diverted Superforts dropped high explosives and many incendiary bombs from above the clouds on a broad expanse of Tokyo and caused considerable damage.[145]

Months before, in September 1944, Japanese production of aircraft and naval ordnance had attained its highest level; the output of IJA ordnance peaked in February 1945. By the beginning of March, however, production of military supplies had declined 20 percent below its zenith. This suggests that factors other than strategic air bombardment contributed significantly to restrict Japanese military production before XXI Bomber Command's raids really got going, factors including interdicted shipping; shortages of steel, aluminum, and other raw materials; restricted land transportation; lack of artisans; and worker absenteeism. By the end of 1944, the decline in aircraft output resulted largely from a lack of engines, caused in turn by an insufficient supply of ferroalloys and, less importantly, a shortage of trained workers.[146]

In one respect, XXI Bomber Command's early raids did play some part in degrading Japanese production. In the aircraft industry, for instance, the relocation of plants and machinery did not really begin until Hansell's attacks of January 1945. Cumulative figures for the dispersal rate of aircraft manufacturing facilities were 1.5 percent for the period between September and November 1944; 4 percent for December 1944; 19 percent for January 1945; and 37 percent for February 1945.[147]

Confronting a weakened though still defiant enemy, American leaders dreaded the prospect of an invasion and a protracted war on the Japanese home islands. They wished an early, climactic, and less expensive termination of hostilities. Since strategic air bombardment was capable of striking the geographical innards of homeland Japan, they focused on alternatives that would best exploit the B–29s' tremendous offensive potential and perhaps obtain a decision after so many months of relative frustration.

In the Marianas, General LeMay deplored the ineffectiveness of the B–29 precision bombardment campaign. Intensive lead-crew training was progressing, to be sure, "but another month of indifferent operations went by," LeMay grumbled. "When I summed it all up, I realized that we had not accomplished very much during those six or seven weeks." Radar bombing still posed problems at ultrahigh altitude in fierce jet stream winds. Visual bombing opportunities, moreover, occurred no more frequently than seven days a month over the Japanese home islands; the average window of opportunity was closer to three or four days per month.

At the beginning of March 1945, LeMay felt the pressures of time and urgency ("General Arnold needed results," he said) to avoid a ground invasion of Japan and, in the bargain, insure the survival of the strategic air command concept. In search of an answer, he devised a daring new form of aerial bombardment: fly at low altitude at night against enemy targets, below effective heavy antiaircraft fire and above light antiaircraft fire, and exploit the apparent lack of Japanese 20-mm and 40-mm antiaircraft batteries. Of the genesis of the plan, LeMay later judged it "a combination of several people's ideas," but "the low altitude part . . . was my own thinking [impelled by] the trouble . . . in Washington in getting a proper role for air power in the war."[148]

The primary ordnance chosen to exploit these tactics was the 6.2-lb M–69 fire bomb, which had been developed by a team of National Defense Research Council scientists in 1942. Three inches in diameter and twenty inches long, the M–69 was unlike any other incendiary. After crashing through the roof of a home or factory and coming to rest, a delay fuze actuated, detonating a TNT ejection charge that spewed magnesium particles through gasoline gel contained in a sock. The resulting "explosion blew burning gel out of the tail of the casing and — like a miniature cannon — shot it as far as 100 feet. If the gel struck a combustible surface and was not extinguished, it started an intense and persistent fire."[149]

Altering B–29 operations from 28,000 or 30,000 feet in daytime to 5,000 or 8,000 feet at night ran counter to decades of precision-bombardment doctrine. Accepting intelligence estimates of inadequate midlevel antiaircraft defense was also risky. The potential advantages, however, were many and enticing. Clusters of M–69 incendiaries, mixed with high-explosive bombs to drive the defenders under cover, would be devastating in an area attack against the congested tinderbox construction and the shadow industry of small contractors thought to characterize urban Japan.[150] One B–29 was considered able to incinerate approximately sixteen acres of ground construction, easily whipping up firestorms. Radar-directed bombing also would be easier at low altitude, and improvements in bombing accuracy could be anticipated, since bombers could come in under the weather and start telltale fires. Also, the strain on B–29 engines would be less severe at low altitudes, and defensive armament aboard the bombers would be largely unnecessary at night. Guns, ammunition, and gunners could be dispensed with, saving precious weight for larger bombloads. Without extra gas in the bomb bays, each B–29 might carry at least 6 tons of incendiaries. LeMay's XXI Bomber Command was about to fulfill Senator Lister Hill's most fervent wartime wish of gutting the heart of Japan with fire.

Though LeMay would have preferred 500 Superforts per nighttime raid, the minimum number he intended to employ for saturation bombing purposes, drawn from three air wings, was 300. By early March 1945, XXI Bomber Command had 380 or 385 B–29s available, with another 600 due from the United States in the near future. LeMay did not believe that Japanese night fighters amounted to a serious threat, though he mused, "we could be wrong as hell." U.S. intelligence reported a perceptible diminishment in the overall Japanese fighter opposition; the total number of enemy aircraft passes against B–29s over targets had fallen from nearly 3,000 in January to fewer than 1,500 in February 1945. Night interception remained ineffective, and no evidence was seen that Japanese fighters were equipped with airborne radar. B–29 crews reported that most of the enemy firing passes at night occurred when the Superfortresses were caught in the glare of searchlights or were illuminated by ground blazes.[151]

While XXI Bomber Command in the Marianas attempted to assess the extent of Japan's nighttime aerial defenses, Japanese defense officials in Tokyo faced a number of other harsh realities. Priority demands of frontline overseas theaters stretched thin the allocation of available air power. As B–29 raids on the homeland intensified in February 1945, the Japanese defense command also began to be tormented by U.S. Navy carrier strikes involving hundreds of fighters per sweep. To meet them, only about 375 interceptors (a bit less than 20 percent of the combined aerial inventory of the IJAAF and IJNAF) were assigned between January and March 1945 to defend all of the home islands. The Japanese fighters available to engage in combat usually numbered between 20 and 50 planes per air defense sector. Thus, quite apart from aircraft shortages, accelerating attrition, and the deteriorating quality of airplanes and pilots, hardly more than one hundred fighters could be brought to bear in any one action. But for harried Japanese defense officials, geography — the limited depth of an insular country — became the gravest shortcoming of all. *Every strategically crucial Japanese installation fronted on the long Pacific littoral.*

IJAAF counsels divided after mid-1944 over the optimal way to employ Japan's limited interceptor strength. Should the fighters be withheld from short-term combat and saved to fight in the anticipated enemy ground invasion? Or should they be committed to defend the home islands against strategic bombers from the outset? But what if they were frittered away, and the invasion was not repelled? What if the fighting dragged on? To these unanswered questions, General LeMay's bombers and Admiral Nimitz's carrier aircraft fighters added new complications in early 1945. If Japanese interceptors were committed to defend the homeland before the invasion, should they engage bombers only (against whose limited-size formations the percentage of interceptor attrition might be relatively low)? Or should they battle it out with swarming U.S. Navy carrier-based fighters, which would chew them up (a situation that would surely worsen if and when Iwo Jima became a base for AAF fighters)?

The IJAAF decision in early 1945 to conserve interceptors for use primarily against bombers distressed General Yoshida, who felt strongly that fighters were designed to fight, not be held on a leash. With the loss of their *raison d'être*, morale among interceptor pilots was bound to plummet. In mid-February 1945, Yoshida sent his chief of staff to complain bitterly at IJAAF headquarters, but to no effect. Soon afterward, however, Japan's front-line, capital defense command was reshuffled. Promoted to lieutenant general effective March 1, Yoshida was sent to command a new air division in China. His chief of staff was reassigned to another air division in Korea.

Lt. Gen. Kanetoshi Kondō, an experienced fighter pilot, took over the 10th Air Division and ordered his fighter units to engage only targets of opportunity instead of attempting to intercept all intruders and protect all strategic sites. Because of the previous emphasis on interceptor versus bomber tactics and reports of the ineffectiveness of Japanese fighters pitted against their U.S. Navy counterparts, Kondō now ordered that training in antibomber combat be subordinated to antifighter training. When B–29s launched large-scale raids, however, defending interceptors would conduct concentrated counterattacks. Commanders of air groups were now delegated to engage U.S. raiders on a case-by-case basis. By terminating unitwide alerts, widespread patrolling, and blanket air cover, Kondō introduced a more flexible system of command and control that reduced fuel consumption and useless effort. His measures, however, could produce only limited effects so late in the war.[152] Weaknesses in Japanese fighter defense, perceived by General LeMay, would soon be exploited to the fullest by the Americans.

In the Marianas, LeMay's proposed, untried tactics of strategic bombardment were debated at XXI Bomber Command headquarters. Not only were the tactics radical, they were of a make-or-break nature for the general himself. LeMay recalled a feeling of "anxiety I'd not wish to experience again." To spare Arnold possible embarrassment should the new tactics fail, and because he had his own negative intelligence assessment of Japanese air defenses, LeMay provided Twentieth Air Force with a minimal, one-day forewarning of the impending fire raid. He would bend every effort, he assured his superiors, to insure that each forthcoming B–29 attack would be dispatched at minimum intervals and with maximum power.[153] Next day, in the late afternoon and early evening of March 9, 1945, in a sequence of takeoffs lasting nearly three hours, one B–29 wing from Guam and two wings from Saipan and Tinian launched a total of 334 Superforts on Mission 40.

Shortly after midnight, in the first hour of March 10, 1945, 279 B–29s were winging over Tokyo. Cloud cover was not daunting that evening, and visibility, if not unlimited, was quite good. Lead squadron pathfinders, like RAF pathfinders over Germany, marked the aiming points, here with napalm-filled projectiles that quickly initiated unmistakable fires. The raiders then flew in between 4,900 and 9,200 feet, attacking a heavily populated rectangle some four by three miles on opposite sides, the rectangle adjoining an industrial zone deemed the most important in the entire capital.

As the B–29s spread out and dropped their bombs, fierce winds of increasing velocity fanned the fires which merged into a huge sea of flame. Smoke billowed skyward and obstructed visibility; searing updrafts knocked some of the B–29s 2,000 feet above their initial altitude. On the way back to base, tail gunners reported seeing the glow of the fires 150 miles away from Tokyo. In all, the Superfortresses loosed 1,665 tons of incendiaries. Poststrike photography showed an area of almost sixteen square miles burned out in the heart of the capital — 63 percent of the commercial district, 18 percent of the industrial area, and the entire core of the residential zone.

Brig. Gen. Thomas S. Power, commander of the new 314th Bombardment Wing on Guam and leader of the Tokyo raid, circled high above his bombers for about two hours, observing the scene closely and taking photographs at LeMay's behest:

> I watched block after block go up in flames until the holocaust had spread into a seething, swirling ocean of fire, engulfing the city below for miles in every direction. True, there is no room for emotions in war. But the destruction I witnessed that night over Tokyo was so overwhelming that it left a tremendous and lasting impression with me.[154]

The Americans judged the Japanese aerial defense of Tokyo on March 9/10 as "only moderately effective." Interceptors put up "nil" resistance to the first formations of bombers; later Superforts encountered "weak opposition." Throughout the three-hour raid, B–29 crewman reported seventy-six sightings and merely forty passes by enemy aircraft, each usually made when searchlights caught a bomber in their rays. Not one B–29 was lost to fighter attack. Japanese antiaircraft batteries proved much more effective, though varying in accuracy and intensity. Automatic-cannon units typically fired too low; heavy guns, too high. The severity of the gunfire, generally accurate and moderately intense at first, decreased appreciably in both respects as the raid progressed and the fires spread; the last elements encountered no antiaircraft fire. Flak hit a total of forty-two B–29s, of which fourteen went down, but five crews were saved at sea. The loss-to-sortie ratio came to 4.2 percent, which XXI Bomber Command deemed an acceptable price for the catastrophe wreaked on Tokyo — "not much like [the loss of B–17s at] Regensburg," LeMay remembered.[155]

Japanese radar had picked up a few unidentified enemy aircraft south of the Boso peninsula late on March 9, 1945, and the Japanese air defense command issued an alert at 10:30 P.M. for the Kanto area. Since the enemy planes did not seem bound for any specific target area, no particularly vigorous countermeasures were initiated. When the intruders disappeared from view, the Japanese concluded that there was nothing to worry about for the time being.

The night was clear and dry above the Kanto plain, but very strong north winds, "too strong for a man to stand up to," were interfering with the effectiveness of Japanese ground radar. Word arrived at midnight that a watch post on southwest Boso had heard something that sounded suspiciously like B–29s approaching. While this intelligence was being assessed, at 12:08 A.M. Tokyo reported an incendiary bombing attack. Swarms of B–29s flying north over the bay continuously, singly or in small formations, appeared at unheard-of altitudes from 6,000 to almost 10,000 feet, some coming in as low as 3,000 feet. When the 10th Air Division realized the scope and intensity of the raid, it scrambled all available night fighters at about 12:15 A.M. A field-grade IJAAF staff officer observed the raid from the ground:

> I was able to see the B–29s' actions clearly, since they were operating at medium altitude and the fires made things bright . . . . In view of the B–29s' moderate altitude, our interceptors as well as our antiaircraft guns were able to oppose them effectively. Even our machine guns fired at B–29s which came in particularly low. I saw two or three bombers go down before my eyes, so I imagine that we must have achieved considerable results that night.[156]

Altogether, the Japanese claimed to have shot down fourteen or fifteen B–29s and to have damaged fifty, claims almost identical to the U.S. admission of plane casualties to flak alone.[157]

At 2:37 A.M. the air raid sirens stopped in Tokyo, and at 3:20 the all-clear sounded, but not until after 5 A.M. on March 10 did the fires began to burn themselves out. Firefighting units from the capital and nearby prefectures struggled all night against the inferno but, crippled by high winds and a lack of water, had lost control of the situation in the first thirty minutes of the raid. Even fire-resistant buildings ignited in the inferno. In Kanda, blazes were so intense that tin sheeting from roofs flew about like snowflakes. Ninety-five fire engines were consumed by flames, five large fire stations were destroyed, and 125 firemen were lost, in addition to more than 500 casualties among auxiliary firemen. The voracious fires left little rubble; glass and even concrete melted.

After scrutinizing the aerial photographs, LeMay concluded that "if it hadn't been for that big river [the Sumida] curving through the metropolitan area, a lot more of the city would have gone."[158] In fact, the fires that started in the teeming Asakusa area jumped the Sumida and gutted Shitamachi, the residential and commercial heart of downtown Tokyo. Fanned by winds that may have reached a velocity of 45 mph, the flames quickly combined with those in other wards, flinging firebrands and sparks, trapping and asphyxiating or scorching panic-stricken residents who tried to reach the presumable sanctuary of firebreaks or waterways. The shallow canals proved treacherous, for in some cases their waters heated to the scalding point. A police bureau official spoke excitedly of the damage when interrogated after the war: "The condition was so terrible, I could not well describe it. After a raid I was supposed to investigate, but I didn't go [on March 10] because I did not like to see the terrible sights."

Metropolitan fire agency records listed 276,791 dwellings burned — 18 percent to 25 percent of the prewar number of structures in the capital. That night between 72,489 and 83,793 people died, and between 24,230 and 40,918 were injured, the higher figures obtained from metropolitan police bureau data. More than 1 million residents (from a February 1945 population of almost 5 million) lost their homes. Twenty-five days were needed to clear all the piles of dead and parts of bodies from the charred ruins. Many corpses, melted together, could not be separated or even differentiated as to sex.[159]

There can be no doubt that the fire raid of March 9/10 contributed significantly toward sapping Japanese resolve at all levels. The bombings could not be hidden from the people or dismissed as enemy propaganda; they constituted what has aptly been called the propaganda of the dead. Emperor Hirohito insisted on inspecting the ravaged Kōtō district on March 18. There is reason to believe that what the shaken, grim-faced monarch saw at firsthand intensified his determination to bring the war to an end as soon as possible.[160]

Neither the Japanese military nor the general public had been prepared for the scale of destruction on the ground or the tactics of the B–29s in the sky. The complete failure of the Japanese air defense system to cope with the area bombardment caused serious concern and helplessness at the top levels of government. Fears arose that the Superforts could isolate parts of Japan from the rest of the country. Higher headquarters worried lest another air attack finish off the capital, strangle the nerve center of the administration and the armed forces, and ruin the morale of the populace.[161] Even today, Japanese publicists call the fire bombing of Tokyo a "scorched-earth strategy."[162] Indeed, the Japanese attack on Pearl Harbor paled before this land-based B–29 assault on Tokyo in March 1945. The shock effects of aerial bombardment on Japanese morale would assume great importance during the last month and a half of hostilities.[163]

For the American airmen, if this predawn raid against Tokyo served as just retribution for Pearl Harbor, it also became the benchmark against which the destructiveness of all future raids would be measured. Subsequent analysts called it "by far the most effective air attack against any Japanese city," causing "a greater degree of death and destruction than that produced by any other single mission in any theater during World War II." Though the magnitude of the devastation could not be known with certainty for the first days afterward, Arnold wired LeMay an enthusiastic: "Congratulations. This mission shows your crews have got the guts for anything."[164]

This incendiary blitz drastically changed the course of the strategic air bombardment campaign against Japan. LeMay would maintain unrelenting pressure on the enemy and on his own crews, for he was convinced it was possible to smash all four of Japan's main industrial cities within the following ten days. He wanted to attack Nagoya the day after the Tokyo raid, but he had to settle for a day's delay. Employing tactics that resembled those against Tokyo (with the exception of issuing some machine-gun ammunition for tail gunners), XXI Bomber Command got off 313 B–29s on the night of March 11/12 for Mission 41, of which 280 to 285 made it to the Nagoya target area. From altitudes of 6,800 to 7,300 feet, and for almost three hours, the Superforts dropped 1,793 tons of incendiaries — 125 tons more than were dropped on Tokyo. Hundreds of fires erupted, but no sea of flame appeared this time because Nagoya had fine firefighters, sufficient water, good firebreaks, and minimal wind. Moreover, the Americans had made mistakes in planning and execution; much of the B–29 force bombed short of the targets at Nagoya. As a consequence, the burned-out sectors were widely dispersed, little more than two square miles in total were gutted, and the aircraft factories were not destroyed. Still, the B–29 force was not hurt badly either: one plane was lost when it ditched after taking off and twenty others suffered damage — eighteen from antiaircraft gunfire and only two from fighters.[165]

The following evening, March 13/14, XXI Bomber Command staged another successful fire raid, this time against Osaka, Japan's second largest city. The Superforts' belly turrets were beefed up with machine-gun ammunition, but in other respects the Osaka strike was a reprise of Tokyo four days earlier. Of 301 B–29s that took off from the Marianas, 274 hit built-up Osaka, where cloud cover forced radar bombing. Dropping 1,733 tons of incendiaries in a dense and even pattern at 7,000 feet, in three hours the bombardiers ravaged more than eight square miles of the dense commercial and industrial core of the city. Japanese interceptor resistance, rated as feeble, mounted forty passes but no hits. Apart from a B–29 lost on takeoff, only one Superfort went down; thirteen received flak damage. After this raid, it became gratifyingly apparent to the Air Staff in Washington that "the Tokyo raid had not been a fluke."[166] According to the firefighting authorities in Osaka, 3,988 people were killed, 8,463 hurt, and 678 reported missing. Those who lost their houses numbered more than half a million.

General Shōzō Kawabe, the 15th Area Army Commander at Osaka, had surmised from the appearance of a solitary enemy airplane late on March 13 that the B–29s might hit Osaka as they had Tokyo and Nagoya. He therefore broadcast a radio warning to the populace that night. When enemy formations were reported approaching Japan, the general rushed to his command post, though he was unsure where the bombers would head. Soon enough it became clear that Osaka was their objective. Kawabe penned in his diary:

> A total of about one hundred planes came in flights of one to three and dropped incendiaries for three hours till 2:40 A.M. From an altitude of about 2,000 meters [6,670 feet] they struck boldly, spreading skillfully from the center of Osaka to the port zone. Despite the desperate efforts made by everybody, including reserves, the fires were enormous and we lost most of the old city. It was most deplorable. Army facilities, the

> division commander's official residence, and many, many individual houses were burned. The firemen did their best but the blazes were still out of control when the evening [of March 14] came.... The military reacted faster than the municipal authorities, and food and emergency assistance were dispensed quickly. The number of razed houses amounted to 100,000–150,000.... I am therefore concerned about the effects that this will exert. It was good fortune within misfortune, however, that the number of casualties was slight. As for our interceptor units, the air division accomplished nothing, though the antiaircraft gun group claimed eleven bombers shot down and twenty-four damaged.[167]

Three days after the Osaka raid, on March 16/17, 1945, American B–29s went after the great port of Kobe, Japan's number six city, located across the bay from Osaka. Kobe was a promising target for it had been previously hit by only 174 tons of bombs (other than strays) in February, but the main limiting factor on XXI Bomber Command now was the depletion of incendiary bomb stocks caused by the intensifying demands of the fire blitz campaign. Even so, the tonnage dropped on Kobe from about 6,800 feet by a massive force of 370 B–29s was the most concentrated and heaviest to date — 2,355 tons in a little more than two hours.

Statistically, the results in Kobe were not overwhelming: an area consisting of a little less than three square miles (20 percent of the urban area) was reported burned out. Still, the conflagrations razed half of Kobe's business zone and the adjacent industrial district. The Kawasaki shipyard, which built large submarines, was also severely damaged. In all, approximately 500 industrial structures were destroyed and 162 damaged. About 66,000 houses burned, leaving more than 242,000 people dispossessed. Japanese authorities had begun moving many civilians nonessential to the war effort out of the major metropolitan areas and into the countryside, and casualties in this raid numbered only 2,669 killed or missing and 11,289 injured. Though General Kawabe's air defense command put up many interceptors (the B–29 crews counted 314) which made 93 separate passes, the fighters proved ineffective. Of the three U.S. bombers that went down that night, none had been struck by an interceptor. The Japanese again grossly underestimated the raiding force, counting only 60 in the force of 370 B–29s that struck Kobe on March 16/17.[168]

With his stocks of fire bombs all but exhausted, LeMay managed to launch a fifth big raid within the ten days he had originally set for XXI Bomber Command — this one a 313-plane return to Nagoya on the night of March 18/19 (Mission 44). One modification was made in the bombload: every third aircraft carried two 500-lb high-explosive projectiles targeted against the motorized fire brigades that would respond. Two hundred and ninety B–29s made it to Nagoya where, relying mainly on radar, they unloaded from 1,858 to 1,863 tons of bombs from 4,000 to 7,500 feet for 2¾ hours. This attack burned out three square miles of the city, producing a total of five square miles razed in the two raids. During the second strike, the B–29 bombardiers commented on problems with blinding searchlights, radar interference, and dummy fires.[169]

Japanese officials estimated the raiding force at 160, closer to the American figure than previous tallies had been. Activity by IJAAF interceptors was "very low-key," accounting for their lack of success against the B–29s. Although the 11th Air Division and the antiaircraft units submitted large claims for this and other night raids on the Nagoya-Osaka-Kobe region, there was no denying that all these urban areas had been devastated.[170]

On March 27, 1945, two bombardment wings (now without incendiaries) struck the IJA airfield, training center, and arsenal at Tachiarai; the IJN air base at Oita; and the aircraft plant at Omura — all on Kyushu. Of the 165 Superforts that took off, 151 hit their primary targets from medium altitude with good results, especially at Tachiarai, encountering scant resistance in the process. That same night, a third wing dropped aerial mines in Shimonoseki Strait to block the Inland Sea. Following a small twelve-bomber strike against the Mitsubishi plant at Nagoya on March 30/31, XXI Bomber Command struck hard against Kyushu for the second time. Its 147 B–29s obliterated the machine works at Tachiarai and tore up the Omura airfield. Fifteen Superforts were hit by fighters but none were lost.

The five large-scale U.S. bombing missions in March benefited from increased bomb tonnage, operations at lower altitudes, and stepped-up schedules of attack. XXI Bomber Command, however, depleted its supply of fire bombs after the second raid on Nagoya, producing a crimp in LeMay's further plans for nearly four weeks. "Apparently the Nimitz crowd," LeMay carped, "thought that we were uttering an empty boast when we told them [on Guam] that we'd be out of incendiaries shortly . . . . Folks hadn't believed us and folks hadn't supplied us . . . we couldn't do another hour of incendiary work. No ammunition for the job."[171]

XXI Bomber Command had decisively demonstrated its revolutionary tactics in the course of the fire blitz against Japan's urban areas. Striking singly from an average height of only 7,000 feet, 380 B–29s flew 1,595 sorties and dropped 9,635 tons of incendiaries, mainly by radar. In four of Japan's primary cities, thirty-two square miles of industry, commerce, and housing were burned out. The sortie rate reached 75 percent of all those flown in the previous fourteen weeks, and the tonnage dropped was threefold greater. Despite the strain (some of the men had participated in all five missions), crew morale remained high and the number of planes reaching their primary targets rose from 58 percent to 92 percent. The cost in crews was 0.9 percent, the lowest rate to date.[172]

A vital consequence of the fire blitz was the American decision to strike two parallel target systems. When visual bombing proved feasible, daylight high-explosive strikes were sent against precision targets, but at medium to high altitudes only — from 12,000 to 20,000 feet. "Precision targets" meant important industrial objectives, such as aircraft and ordnance works, oil facilities, marshaling yards, and arsenals. When visual bombing was not possible, the

B–29s were used to lay mines in Tokyo Bay to intensify the campaign against shipping.

B–29s were to use their new area tactics employing radar and incendiaries against important urban industrial complexes.

Another consequence of significance was that General LeMay increasingly began to speak of an airpower solution, or "victory through air power" (borrowing DeSeversky's famous terminology). He argued that strategic air bombardment alone, with sufficient logistical support, could compel Japan's capitulation in the short term with no need of an Allied ground invasion of the home islands which General Douglas MacArthur so earnestly anticipated taking over. As LeMay said at the time of the first Tokyo raid on March 9/10: "If this raid works the way I think it will, we can shorten this war." Later he declared: "The destruction and demoralization in Japan was being rapidly accelerated. Had increased like cube root."[173] Indeed, in a private message to Norstad in late April 1945, LeMay explained his reasoning:

> I consider that for the first time strategic air bombardment faces a situation in which its strength is proportionate to the magnitude of its task. I feel that the destruction of Japan's ability to wage war lies within the capability of this command, provided its maximum capacity is exerted unstintingly during the next months, which is considered to be the critical period.

Certain members of General Arnold's headquarters staff personally shared LeMay's bold, minority view.[174]

If the seizure of the Marianas projected VLR bomber power within range of Japan, the conquest of Iwo Jima and Okinawa in early 1945 further supported and intensified the B–29 offensive against Japan. Though costly and controversial, the taking of Iwo Jima eliminated the Japanese counterraids against the Marianas, provided a haven for Superfortresses that became crippled or short of fuel (thereby cutting operational losses), and supported friendly fighters that could escort the B–29s or conduct offensive sweeps against Japanese targets. On April 7, 1945, each of the six AAF P–51 Mustang squadrons based on Iwo Jima dispatched aircraft to protect Marianas-based Superforts bound for Honshu.

Between April 7 and the close of June 1945, Seventh Air Force fighters flew 426 escort sorties in daylight. By war's end, scarcely more than four months later, VII Fighter Command had flown 1,700 sorties in support of the B–29s. The Japanese conceded that the swift, maneuverable P–51s ruined their already slim interceptor capability; not their night fighters, bombers, nor armed scout planes could cope successfully with escorted B–29s. American sources estimated that the AAF fighter escorts reduced enemy interception by as much as 70 percent. Moreover, the presence of friendly fighters boosted the confidence, morale, and effectiveness of B–29 crews, who often lauded the spunky "P-Five-Ones," their "little brothers."[175] Additionally, Iwo Jima saved many an American flyer, for some 2,400 B–29s (with 25,000 crewmen aboard) made emergency landings on the island by war's end.[176]

XXI Bomber Command, which had helped soften up Iwo Jima in Hansell's day, was now called on, in concert with U.S. Navy carrier-borne planes, to support the invasion of Okinawa — code-named Operation ICEBERG. Admiral Nimitz was to serve as theater commander, and for this invasion he exercised operational control over the B–29s. Directed against tactical objectives and away from strategic targets, it was difficult for the Twentieth Air Force to demur because the acquisition of Okinawa would provide sites for VHB planes to bomb Japan, less than 500 miles away, as a prelude to the ultimate ground invasion of the Japanese homeland. Though willing to assist when B–29 intervention could be decisive, LeMay chafed in the assignment, and he hoped that the diversion of effort would be of minimum duration.

On April 1, 1945, U.S. forces landed on Okinawa. Initial enemy ground and air resistance was light, though *kamikaze* planes did achieve a number of hits on warships and shipping. Eager to get back to strategic bombardment, LeMay sent a strike against Nakajima at Musashino on the same day. It was a false lull for LeMay; operations at Okinawa soon reached a crisis. Starting on April 6, the Japanese launched hundreds of *kamikaze* and conventional aircraft from Kyushu in a fierce counteroffensive aimed at targets on and around Okinawa. Even a small IJN task force led by the mighty superbattleship *Yamato* tried in vain to break through to the Ryukyus.

The lair of the *kamikazes* appeared to be the airfield complex at Kanoya on southern Kyushu, against which LeMay sent fifty-three B–29s from two wings on April 8, 1945. Cloud cover forced the planes to go after Kagoshima instead. Admiral Nimitz still urgently needed help, and on April 17, XXI Bomber Command resumed missions in earnest against the airfields on Kyushu. Six enemy airfields were attacked by 134 B–29s, with undetermined results. The B–29 attacks, though greatly appreciated by the U.S. Navy, could not suppress the *kamikaze* threat. Until almost the end of June, a total of 1,900 Japanese suicide aircraft, plus many conventional planes, hammered U.S. ships off Okinawa. Usually striking in massive waves, but sometimes singly, these aircraft came mainly from Kyushu, although some 250 originated in Formosa.

The Twentieth Air Force kept its promise to aid Nimitz by dispatching formations of more than one hundred B–29s from all three wings of XXI Bomber Command during the peak of the crisis in April. Flying at 13,000 to 21,000 feet and using mainly general-purpose bombs throughout the later missions (often fuzed to detonate after impact to delay repairs), the Superforts in 93 attacks against airfields on Kyushu and Shikoku launched 2,104 sorties during the 25 days between April 17 and May 11.[177] These operations constituted 75 percent of the B–29 bombing and mining campaign in that period. Airfields at Kanoya and Kanoya East were struck 22 times, and those at Oita, Kokubu, Miyazaki, Miyakonojo, and Tachiarai were attacked 41 times.

Apart from the cost in aircraft and crews lost (which the U.S. Air Staff regarded as light), the question can properly be asked whether this use of B–29s was either appropriate or productive. Putting the best face on the use of the Superfort as a tactical bomber, supporters noted that the size and frequency of *kamikaze* attacks diminished substantially in the face of the airfield interdiction offensive, and this use of Superforts caused the Japanese to disperse their aircraft and rely on camouflage and dummies, which reduced the effectiveness of their air force. Skeptics abounded in the Twentieth Air Force and XXI Bomber Command. LeMay, for one, was convinced there was nothing to bomb:

> But we were ordered to go over and keep dropping more bombs on those beat-up airdromes . . . . All we were doing at last was plowing the fields . . . . I complained all the way up to General Arnold. The Navy sent a hint in return, declaring that if we didn't stay on the job, they would pull off Okinawa and leave our doughboys there . . . . No matter how we socked away at those airdromes, we could not reduce the *kamikaze* threat to zero. In some proportion it was always there.[178]

According to an estimate prepared by XXI Bomber Command, 95 percent of 1,405 Japanese *kamikaze* combat sorties between April 17 and May 11 were launched on the very day that some of their airdromes came under attack by B–29s. Nevertheless, the battle situation gradually improved at Okinawa. To counter the *kamikaze* threat further, in addition to U.S. Navy carrier-borne airplanes, VII Fighter Command committed hundreds of P–51 fighters based on Iwo Jima and, later, those based in the Ryukyus to sweeps against enemy airfields in Japan. On May 11, 1945, Admiral Nimitz released XXI Bomber Command from its ICEBERG support mission, enabling LeMay to resume his fire blitz against selected urban targets on Honshu.[179]

The *kamikaze* threat, according to the official Japanese military histories, came from the IJA Sixth Air Force and the IJN Fifth Fleet, which conducted the constant attacks from central and southern Kyushu against U.S. airfields on Okinawa and shipping off that island. Miyakonojo, Chiran, and Mansei were the main bases for the air strikes, but B–29 raids against those particular installations were few, Chiran being hit only once. In addition, all the airstrips on Kyushu were dispersed and thoroughly concealed, "so the losses incurred there were comparatively scant although the B–29s raided on a daily basis."[180] Except for the additional criticism that important *kamikaze* bases were not identified as targets by the Americans, the Japanese assessment of the ineffectiveness of the Superforts' tactical campaign against Kyushu bears an uncanny resemblance to LeMay's own doubts.

Temporarily without incendiaries, XXI Bomber Command was unable to mount a fire raid against Japan between March 19 and April 13, 1945.[181] The command nonetheless could experiment with variations in the bombardment of homeland Japan, even during the Okinawa-support phase. Exhorted by the Twentieth Air Force to try nighttime, high-altitude precision attacks, LeMay first sent 251 B–29s against the Mitsubishi works at Nagoya on March 24/25, 1945. The 1,533 tons dropped were concentrated mostly against one aircraft industry objective, but the results were rated poor. A mere 60 tons fell in the main target zone. LeMay next attempted single-wing attacks with differing mixes of bombs and flares. On March 30/31, 12 of 14 Superforts made it over Nagoya. They dropped 50 tons but failed to hit the Mitsubishi works. Another bombardment wing sent a big force against Nakajima-Musashino on April 1/2. Although 115 B–29s dropped 1,019 tons, a skimpy 4 tons impacted the target area, with trifling effect.

LeMay then tried a new combination: sending one bombardment wing against each of three targets (aircraft engine and assembly plants) on a single day. Thus, on the evening of April 3/4, 1945, 61 B–29s dropped 503 tons on Tachikawa, 48 loosed 181 tons on Mitsubishi-Shizuoka, and another 48 bombers dropped 275 tons on Nakajima-Koizumi. Once more, the results were negligible: Shizuoka was not hit, Tachikawa suffered medium damage, and the effect on Koizumi was light. LeMay knew now that the technology for high-altitude nighttime precision bombing was unavailable. For the moment, however, XXI Bomber Command went on pounding Kyushu by night, sometimes striking aircraft industry targets by day.[182]

Events changed for the better when the B–29s, accompanied for the first time by P–51 fighters, tackled LeMay's two priority targets simultaneously by daylight on April 7, 1945. In midmorning, against Nakajima-Musashino, 101 Superforts from one wing carrying new 2,000-lb projectiles battered the target zone with 490 tons in good weather. The P–51 escorts from Iwo Jima deserve much credit for the bombers' success that day. The IJN Air Command scrambled 110 interceptors to deal with the B–29s in the morning, but for its trouble it lost nine Gekkō (Irving), Ginga (Frances), and Suisei (Judy) night fighters to an estimated thirty Mustangs, against only two Superforts claimed shot down. The IJNAF pilots reported that the Japanese night fighters were practically helpless in aerial contests with the P–51s that accompanied the B–29s. The IJAAF, similarly surprised by the appearance of Mustangs over the Tokyo area, lost eleven fighters.

In the afternoon of April 7, two B–29 wings launched a high-altitude precision attack against the Mitsubishi plant at Nagoya, the number two target on XXI Bomber Command's hit list. Dropping 614 tons of high-explosive bombs from approximately 20,500 feet in weather as good as it had been at Musashino, 154 Superforts finally tore apart the Mitsubishi engine works, destroying or damaging 90 percent of the building area and essentially putting the facilities out of action for the remainder of the war. Japanese intelligence reported the B–29s heading toward the Nagoya area that afternoon, and the 11th Air Division ordered one group and one squadron of interceptors to patrol over Ise and two more groups to cover Nagoya city. Few IJNAF fighters were available in the area because most of their strength had been transferred south to Kyushu in support of the Okinawa campaign. As at Musashino, the Japanese interceptor units at Ise were surprised to encounter agile P–51s, which quickly got above them and mauled two of eight newly received Type 4 fighters, including the group commander's aircraft.

One B–29 formation feinted toward the Han-Shin sector, while the main force went on to the factory district at Nagoya, without the Mustangs that fought at Ise. From this sequence of operation, Japanese defenders concluded that the Americans used P–51s only to clear the way for the bombers that followed. A total of sixteen B–29s were reportedly shot down on the 7th, plus forty damaged. Japanese records typically stress claims of success against the bombers while saying relatively little about the destruction of targets on the ground. Of the two raids on April 7, U.S. analysts assessed the damage-effect ratings as "heavy" at Musashino and "very heavy" at Nagoya.[183]

Five days after the powerful attacks against the Nakajima and Mitsubishi plants, 119 B–29s of the 73d Bombardment Wing returned to knock out Musashino on the morning of April 14, 1945. Armed with 1-ton bombs and escorted this time by 102 P–51s, 94 Superforts dropped another 490 tons of high-explosive bombs and essentially finished off the machinery at the Nakajima works. XXI Bomber Command's number one target was finally eliminated, but it had taken eleven B–29 attacks (plus a U.S. Navy carrier strike) to do the job. Refusing to be distracted by a smaller B–29 feint to the north against Kōriyama on April 12, the defenders sent up an estimated one hundred fighters to intercept the last major raid against Musashino and damaged thirty-six bombers (by American admission). The IJN Air Command put in a modest claim that seventy-two fighters had shot down one bomber at the cost of six IJNAF interceptors.[184]

XXI Bomber Command's support of the Okinawa campaign was still going on when new stocks of incendiaries allowed General LeMay to mount more low-altitude area attacks — Missions 67 and 69 — in the greater Tokyo region. For more than 3½ hours, between 10:40 P.M. on April 13 and 2 A.M. the next morning, 327 Superforts dropped between 2,124 and 2,139 tons of incendiaries (two-thirds by radar) on what the Americans called the arsenal district of northwest Tokyo. Coaxed by brisk winds, the blazes generated an enormous firestorm which unnerved the firefighters and razed an estimated 11.4 square miles — most of the area that had survived the March 9/10 bombing. In the heart of the capital, the Empress Dowager's residence and sacrosanct Meiji Shrine were hit, and flames jumped the moat to burn down part of the Imperial Palace. The latter had served as a useful firebreak for the Japanese and as a point of reference for B–29s, though the Americans had placed the palace area off-limits to direct bombing.[185] Humiliated and discouraged, the war minister, the IJA Chief of Staff, and various army commanders went to the palace to apologize. Instead of recriminating the military, the emperor commended it for the good fight it had waged.

The next evening on April 14/15, while Tokyo reeled from the previous assault, another force of 303 B–29s (109 over Tokyo itself) dropped 1,930 tons of incendiaries, all by radar from low altitudes. Nearly an identically sized area, 11.1 square miles, was razed — over half of it in western Tokyo, more than a third in Kawasaki, and the rest in Yokohama. Flak and fighter resistance were deemed intense and the searchlights blinding. IJAAF and IJNAF pilots and antiaircraft gun crews submitted wild claims of seventy B–29s shot down and more than fifty damaged. Nevertheless, municipal authorities had to admit the conflagration on the ground was gigantic and the effectiveness of firefighting efforts, almost nil. In these two fire raids combined, Tokyo and Yokohama lost more than 217,000 structures, and Kawasaki, nearly 32,000, but, because of the movement of civilians from the cities to the countryside, casualties were much fewer than they had been during the first onslaught of March 9/10.[186] The high command could conclude that the enemy's strategic bombardment campaign was deliberately designed to break the will of the people, who had been led to believe that the homeland was invulnerable to attack from the sky or invasion from the sea. By now, however, aircraft production and pilot skills had dwindled, oil was not reaching Japan from Southeast Asia, and there seemed no alternative but to conserve existing aircraft for the last decisive battle for the Empire.[187]

Elsewhere, while XXI Bomber Command's ICEBERG support activities drew to a close, B–29s still attacked strategic industrial targets. Thus, on the morning of April 24, 1945, 101 Superforts visually bombed Hitachi's Tachikawa aircraft engine plant at Yamato, north of Tokyo, dropping about 474 tons of standard projectiles from only 2,000 feet. The B–29s inflicted tremendous damage, losing four of their own; sixty-eight received hits (this against a U.S. claim of thirty-eight enemy fighters destroyed, including twenty-four probables). Then, on the morning of May 5, 148 B–29s, including bombers from the newly arrived 58th Bombardment Wing, went in at 20,000 feet against the Hiro naval aircraft plant at Kure to drop 578 tons of heavy projectiles. The Superforts pulverized their target. Five days later, more than 200 B–29s attacked oil storage and refinery facilities in the Yamaguchi-Hiroshima district, the first but far from the last such raid against the fuel sector of the economy. The next morning, on May 11, about sixty B–29s struck the Han-Shin area via Osaka Bay. Their target was Kawanishi's Konan airframe factory. The IJN Air Command scrambled a meager twenty-five fighters, all that could be mustered over the Kobe-Amagasaki sector, which could not stop the Superforts from smashing the Kawanishi plant.[188]

In Europe a few days earlier, on May 8, 1945, with its forces overwhelmed on land and in the air, Nazi Germany capitulated to the Allies. Afterward, an impatient General Arnold toured the Pacific, where he could not obtain a firm estimated date for Japan's collapse from any American theater commander. The sole exception was XXI Bomber Command's LeMay, who confidently declared that destruction of some sixty Japanese urban areas would force the country to its knees by October 1. The planned invasion might yet be obviated. "Personally," said Arnold, "I was convinced it could be done. I did not believe Japan could stand the punishment from the air that Germany had taken." Planning for the ground landings should continue, Arnold told General Marshall, but, in the meantime, priority should be devoted to intensified air attacks designed to effect the "complete destruction of Japan proper."[189]

LeMay proceeded to launch a month-long series of area fire bombings to complete what he had begun in March. At the same time, precision targets would be hit as appropriate, per the two-track system devised earlier. The new round of incendiary raids began with a maximum-effort 529-plane daylight attack (Mission 174) against northern Nagoya on May 14, 1945, between 7:00 and 8:30 A.M. This was the first fire raid in which the bombing was conducted visually. Flying between 12,000 and 20,500 feet, 472 B–29s dropped 2,515 tons of bombs, which burned out approximately 3.2 square miles of the city and destroyed 20,000 structures. Billowing smoke diminished the bombing accuracy of later formations. Ten B–29s were lost this day — only one to fighters and another to flak in combat. U.S. gunners claimed to have downed 48 fighters (30 probables) and damaged 16.[190]

On the night of May 16/17, 522 B–29s struck Nagoya again on Mission 176. Hitting the main target, 457 bombers came in at 6,600 to 8,300 feet for one hour, but smoke and thermal currents compelled half of them to bomb from 15,000 feet, expending much more time in the process. The B–29s dropped 3,600 tons of incendiaries to burn out about 3.8 square miles that included dock areas, industrial facilities, and a famous shrine. Three bombers were lost, although none to enemy action. By now, some 30 percent of Nagoya's built-up area had been leveled, more than 113,000 structures destroyed, nearly 4,000 people killed, and more than 427,000 made homeless.[191] The raids of May 14 and 16/17 deleted Nagoya as a prime target for aerial bombardment. Only Tokyo sustained as many air attacks.

Much of what remained of Tokyo was obliterated in two subsequent fire raids. During Mission 181 on the night of May 23/24, 1945, the western part of the capital, with its conglomeration of industrial and residential structures, came under attack again by 520 B–29s (from an airborne force of 562, including 44 pathfinders). In poor weather, and relying almost entirely on radar, the planes dropped 3,646 tons of incendiaries in two hours from heights of 7,800 to 15,000 feet. Searchlights and smoke proved troublesome, and antiaircraft fire was heavy; Japanese fighters, though active, accomplished little. Sixty-nine Superforts were damaged and seventeen lost (thirteen to enemy action).[192]

In the second half of the mission that night, Tokyo was struck once again for 2½ hours after 1:30 A.M. on May 24. Approaching from the direction of Suruga and Sagami bays, 250 B–29s flew in, mostly one at a time, dropping 76,518 incendiary bombs. Strong winds drove the flames into a huge cauldron of fire, destroying an estimated 64,487 structures and leaving 224,601 people homeless. Casualties were relatively light on the ground: 762 people were killed and 4,130 injured. Elsewhere, small numbers of B–29s bombed Kawasaki, Yokohama, Shizuoka, Hamamatsu, and southern Saitama.[193] Although 5.3 square miles of Tokyo were razed in the course of the raids on May 23/24, the AAF rated Mission 181 as the least effective of the six major attacks on the

Japanese capital.

Consequently, in Mission 183 on the night of May 25/26, 1945, the B–29s went after Tokyo again. Of 502 airborne Superforts, 464 attacked the western section of the city, slightly north of the district that had been hit two nights earlier, this time closer to the Imperial Palace. The latest targets lay within the governmental, financial, and commercial areas, in addition to the usual clusters of houses and factories. A combination of light clouds and smoke necessitated radar bombing. The planes loosed 3,251 to 3,262 tons of fire bombs (three-fourths by radar) from heights of 7,900 to 22,000 feet. This time, almost seventeen square miles of the capital burned, the largest area destroyed thus far in any raid on Tokyo.[194]

According to Japanese officials, most of northwest Tokyo was destroyed. The cone of destruction centered on the four downtown districts and extended to the city's fifteen wards beyond. High winds once again spurred the fires, rapidly razing 156,430 structures and dispossessing between 560,000 and 620,000 inhabitants. An estimated 3,242 people died and 13,706 were injured. Flames jumped the Imperial outer moat like "bounding tigers," setting much of the monarch's palace afire. The Omiya detached palace also burned. Fires in the city were not brought under control until 5:15 A.M. In her diary, a nineteen-year-old high school girl remembered that traumatic night:

> There was a gigantic explosion nearby and the earth shook. In fear of being cooked alive in the trembling covered shelter where we had taken refuge, my mother and I dashed breathlessly toward the open fields of the distant Tama River. Though it was May, I had my father's coat over my quilted hood, but the coat caught fire and I flung it off. As I ran, I kept my eyes on the sky. It was like a fireworks display as the incendiaries exploded. Blazing petroleum jelly, firebrands, and sparks flew everywhere. People were aflame, rolling and writhing in agony, screaming piteously for help but beyond all mortal assistance.
>
> The road was blocked by the falling bombs, so we had to veer toward Senzoku Pond. By now we had run 10 miles in a gale. We collapsed on the ground near the pond, utterly exhausted. A bamboo grove was on fire, the bamboo knuckles exploding with frightening shots. But we could not budge physically. Finally, at dawn, we began to trudge homeward. When we reached high ground, we could see that our entire neighborhood had been razed. I saw roofs flying in the air and a huge flaming telephone pole being spun by a cyclone.
>
> When we eventually got to the site of my home, it was a charred ruin, still burning and smoking. We had lost all our worldly goods. A nearby bomb shelter had taken a direct hit and everybody within had been killed. One man saved a baby whose mother had a leg blown off. Today, whenever I hear the siren of a fire truck or see logs crackling in a fireplace, my heart pounds and I relive those grisly days of childhood terror.[195]

Since the core of the Imperial Palace itself burned in this raid, a crestfallen War Minister Korechika Anami asked to be dismissed. In reply, however, the Emperor conveyed the following exhortation: "Though we well understand the pain of the War Minister, the nation is currently facing its supreme crisis. The War Minister should therefore remain in his post and do his best to support us."[196] For the rest of the war, the Emperor and Empress lived in quarters improvised in the underground library on the palace grounds.

While admitting that Missions 181 and 183 devastated the greatest area of Tokyo, postwar USSBS analysts judged the attacks of both May 23/24 and 25/26 to have been less effective than preceding ones because the two strikes were centered on less heavily populated parts of the capital.[197] Such an evaluation can only be true in terms of the relative numbers of casualties; certainly it cannot be true in terms of the overall psychological effects on the populace. Japanese plant managers and personnel involved with air raid protection had reached a point where they "seemed to be resigned to the fact that there were no effective countermeasures against a full-scale B–29 attack."[198]

As civilian morale plummeted, myths began to take root and circulate. Some individuals reassured themselves that certain locales would be immune to air attack if American prisoners or industrial plants that the United States wanted to spare were located in the area. Certain cities were rumored to be safe if they contained friends or relations of prominent Americans. Kyoto would remain off-limits because it supposedly held the burial site of General MacArthur's mother. Other areas rumored to be immune to bombing were those in

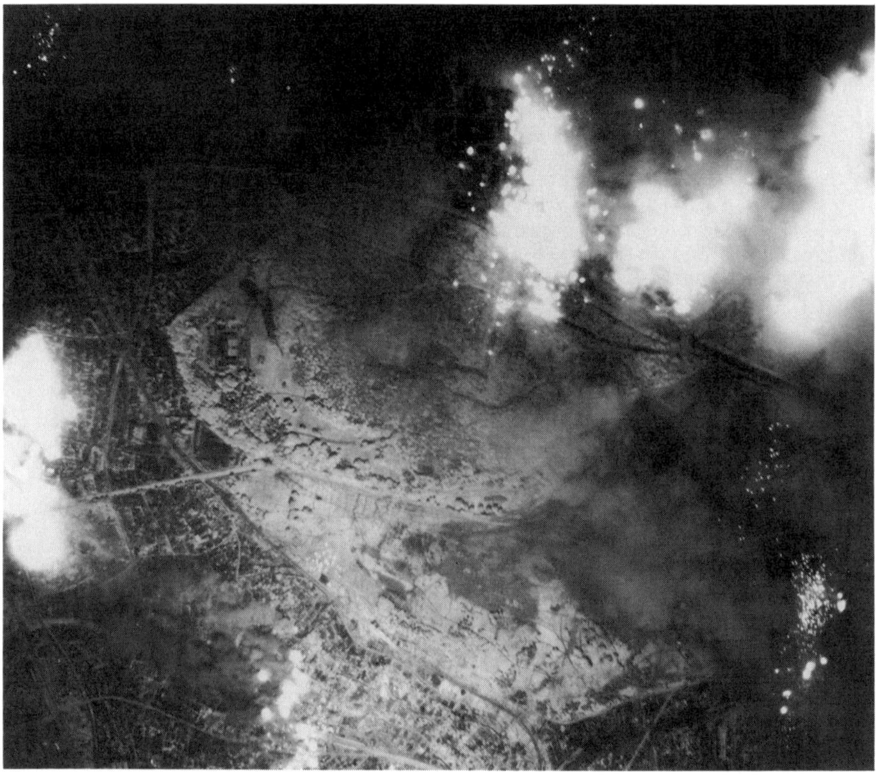

Night raid of May 25/26, 1945 on the Tokyo Palace area.

which large numbers of Christians were located, such as Nagasaki, or scenic spots the Americans liked, such as Miyajima, outside Hiroshima.[199] Psychologists conversant with the relationship between warfare and emotional stress note that extended aerial bombardment induces "compulsive-like avoidances, rituals, and superstitious practices."[200] This analysis well describes the mental state to which much of the Japanese urban populace had been reduced by the B–29 fire raids in the spring of 1945.

In the United States, the COA and later the Joint Target Group had long grappled with questions concerning the potential effectiveness of area bombardment in compelling Japan to consider surrender. By late May 1945, more than half of the urban area of Tokyo had been obliterated, quite apart from the grievous destruction already visited on Japan's other major cities. From the standpoint of military logic, wasn't it time for Japan to lay down its arms? To expedite that process, Joseph C. Grew, Acting Secretary of State and former U.S. Ambassador to Japan, believed President Truman should issue a public statement reassuring the Japanese that unconditional surrender would not necessarily mean elimination of the Emperor. On May 29, Grew convened a high-level meeting to consider the idea. Though everyone agreed to it in principle, U.S. military leaders (Secretary of War Henry L. Stimson, Secretary of the Navy James V. Forrestal, and Army Chief of Staff General George C. Marshall) deemed the issuance of so mollifying a statement at that moment inadvisable "for certain military reasons not divulged."[201] Consequently, no overt, concerted effort was made to combine the duress of aerial bombardment with the suasion of diplomacy. Under the circumstances, XXI Bomber Command continued its unrelenting area fire blitz campaign against Japanese urban targets, and the great port city of Yokohama was next on the incendiary agenda.

The enormous damage and casualties that Tokyo sustained in the course of six major area raids between November 1944 and June 1945 have obscured the decisiveness of the one blow dealt to Yokohama, only twenty miles away. With a population exceeding one million in February 1944, Yokohama was Japan's fifth largest city, although its population density was one-fifth that of the capital's. Yokohama's fine harbor, the country's largest, served as a port of entry for Tokyo and the whole Kanto plain, and it handled approximately one-fourth of Japanese foreign trade before the war. Yokohama maintained the nation's fourth largest shipbuilding yards and had excellent marine repair facilities. The city's sizable industrial output included automotive works, oil refineries, and plants engaged in chemical manufacturing and aluminum processing. Between November 1944 and April 1945, B–29s dropped only 28 tons of bombs on Yokohama. The highly industrialized harbor area of neighboring Kawasaki was hit hard by 1,110 tons of bombs on April 15, but XXI Bomber Command did not designate Yokohama a primary target before the end of May.[202]

The two nighttime raids against Tokyo at the end of May 1945 cost XXI Bomber Command 43 B–29s lost and 169 damaged. To fire-bomb Yokohama, LeMay and his staff devised another, possibly safer, mix of tactics. This time the B–29s would fly in during daylight in carefully timed formations at high altitude and with fighter escorts from Iwo Jima, the Mustangs' first deployment on a fire-blitz mission. At 9:30 A.M. on May 29, 517 Superforts escorted by 101 P–51s went after the heart of Yokahama, where they were met by some 150 Zeros. In good weather the B–29s dumped 2,570 tons of incendiary bombs, first visually and then by radar when the usual torrents of smoke began to mask the targets. This raid knocked out twenty priority targets and gutted most of downtown Yokohama. To the nearly 7 square miles destroyed on this day, if the 1½ square miles burned in the spillover from the Tokyo-Kawasaki raid of April 15 plus some other damage are added, a total of 8.9 square miles of the city lay ravaged and more than 89,000 of 224,000 buildings were razed. The loss of warehouses, docks, piers, wharves, factories, and shipping congregated at Yokohama was rendered worse by the wretched inadequacy of the firefighting equipment.[203] Casualties included at least 4,832 killed and 17,967 injured. Some 399,187, or 39 percent of the 1944 population, were left homeless.[204] Mustangs claimed 49 Japanese aircraft (including 23 probables) shot down and nine damaged, at the minimal cost of 3 P–51s. Even so, 5 B–29s were lost and 175 damaged.

When General George C. Kenney, Far East Air Forces Commander, entered the city three months later, he observed Yokohama to be

> . . . practically a shell of what had once been a thriving city of a million people. The fire bombs of the B–29 attacks had burned out whole blocks, and piles of rubbish had overflowed into the streets . . . . Yokohama had a job on its hands that would take years to accomplish. [Apart from the Bund and Yamashita Park] everywhere else the eye met scenes of desolation and destruction. The fury of modern warfare was an appalling thing to think about.[205]

On Japan's main island of Honshu, two of the three largest and most important urban-industrial centers were now burned out: Kei-Hin (Tokyo-Yokohama-Kawasaki) and the region centering on Nagoya. Among the main Japanese cities, U.S. Joint Target Group photo interpreters judged physical damage to the roof areas alone to be 40 percent in Tokyo, 57 percent in Yokohama, 35 percent in Kawasaki, and 40 percent in Nagoya.[206] For XXI Bomber Command's targeteers, only the third urban-industrial zone known as Han-Shin (Osaka-Kobe) was still worth working over; it had had a respite of more than 1½ months since the great fire raid of mid-March.

LeMay therefore launched his B–29s on three daylight fire raids against Osaka on June 1, 7, and 15, 1945, and one against Kobe on June 5. In the first of the new Osaka attacks, 458 of 511 airborne Superforts came in at 18,000 to 28,500 feet and dropped 2,788 tons of bombs, all but 81 tons being incendiaries. About 3½ square miles of the city were razed, including a small portion of Amagasaki directly to the north, and 66,817 structures were destroyed or

B–29s of the 500th Bombardment Group, 73d Bombardment Wing, dropped fire bombs on Japanese installations of Yokohama, Japan, on May 29, 1945 (*top photo*), causing heavy incendiary damage to large sections of the urban area (*upper right*), as viewed from the air (*upper left*).

severely damaged (most were residential; 1,629 were factories). Total casualties amounted to 13,995, of which 3,960 were killed or missing; an additional 218,508 people were left homeless. Originally, 148 P–51s were to have flown escort for the U.S. bombers, but 27 of their number crashed in the frightfully bad weather and only 27 reached Osaka. Japanese air defenses were vigorous.

Though the Americans claimed to have shot down 24 enemy planes, the B–29 force lost 12 aircraft.[207] The defenders estimated that more than 400 U.S. bombers constituted the formations that began the strike against northwest Osaka from 8 A.M. on June 1. Pilots from the IJA 11th Air Division reported shooting down 24 B–29s and damaging 30. A Japanese fighter group commander called the June 1 raid "utterly devastating":

> The bombers flew north to Osaka in a majestic column of echelons that reminded one of an air show. They came in at 5,000 to 6,000 meters [16,400 to 19,680 feet], their bombs causing huge fires and smoke that soared into the sky and apparently transformed the clouds into a mountainous cumulo-nimbus mass that hung over the key target areas. Our antiaircraft guns fired well. I was deeply affected when I saw the huge hulk of a B–29 plummet to earth.

The local IJA area army commander bestowed a unit citation on the 3d Antiaircraft Artillery Division for its performance on June 1, 1945.[208]

In the same way that Tokyo had been blotted out, quadrant by quadrant, XXI Bomber Command pounded Osaka for a second and third time. On June 7, at 10 A.M., 409 of 458 B–29s struck northern and northeastern Osaka and nearby Amagasaki at altitudes ranging from 17,900 to 23,150 feet. In heavy undercast, the Superforts, using radar, dropped 1,795 tons of incendiaries and 797 tons of high-explosive bombs intended for the important Osaka army arsenal, the only area attack to include such a large proportion of high-explosive tonnage. An escort of 138 P–51s was dispatched, but it was mainly the cloak of clouds that kept the losses of B–29s down to two; flak damaged another eleven of the bombers. The June 7 raid destroyed or damaged 56,576 buildings, of which 1,022 were classified as industrial. Among civilian casualties numbering at least 7,182, an estimated 1,728 were killed or missing. The dispossessed numbered 195,682.[209]

The B–29s returned to what was left of Osaka a week later. From 8:45 to 10:45 A.M. on June 15 at an average altitude of 21,500 feet, 444 of 516 Superforts unleashed about 3,157 tons of incendiary bombs, June's biggest incendiary bombload. Though the targets were scattered, nearly two square miles were burned out in eastern, southern, and northwestern Osaka, and a bit more than one-half square mile burned in the industrial core of Amagasaki. Of 53,327 structures destroyed or badly damaged at Osaka, 1,479 were categorized as industrial. Those killed or missing numbered 481, among a total of at least 1,852 casualties. Another 173,923 people, at minimum, lost their homes. Poor weather on the 15th turned back the P–51 escorts, but the same adverse conditions also thwarted 27 IJNAF fighters that tried to intercept the bombers. Two B–29s were lost to operational causes; enemy action damaged only one.[210]

Between the first and second fire raids on Osaka, B–29s mounted a daylight attack lasting eighty-five minutes on June 5 against Kobe, across the bay, extending to Nishinomiya and Ashiya. Flying at lower heights of 13,650 to 18,000 feet, 473 Superforts dropped 71 tons of high explosives and 3,006 tons of incendiaries that burned out 4.35 square miles of Kobe's urban area and destroyed or damaged 52,327 structures including 692 factories, warehouses, and other industrial buildings. The raid was particularly effective in knocking out important plants not hit in three smaller and nonconsecutive attacks between February and May. Japanese interceptor units contested this raid, which was unescorted. Superfort crews counted 647 passes by enemy fighters, which resulted in the loss of 9 bombers (plus 2 to operational causes) and damage to no less than 176. The IJA 11th Air Division claimed 30 B–29s shot down and another 50 damaged.

Once more, the Japanese military's claims of alleged successes against the B–29s tended to gloss over, though they could not deny, the obvious holocaust on the ground. There, without assistance from stricken Osaka just twenty miles away, Kobe's firefighting and other emergency services broke down. In this raid, an estimated 3,614 people were killed, 10,064 injured, and 179,980 rendered homeless. In the words of Kobe's police chief, the populace now "felt acutely the incompetence and helplessness of the civilian air defense, and the air defense fighting spirit collapsed completely, replaced by fear; they hoped only for the safety of their persons and the security of their own homes and fortunes." Among civilians of Kobe, the chief continued, the "fighting spirit completely vanished and an anti-war, anti-military trend appeared." Conscripted laborers deserted in droves, absenteeism increased in offices and factories, and shortages, especially of food, led to sizable black-market activities.[211]

To the doleful statistics on physical damage inflicted upon Japan's major cities in terms of total roof area alone, by mid-June 1945 one could add the figures for compact Kobe (56 percent) and sprawling Osaka (35 percent). After June 5, XXI Bomber Command "correctly dismissed [Kobe] as a profitable target either for incendiary attacks or for precision high-explosive operations."[212] No large Japanese city, except Kyoto, the fourth largest in the country, was worth an all-out fire blitz any longer. Aerial bombardment operations never were deliberately launched against the old capital and cultural center, which American political leaders had placed off-limits despite Arnold's own preference.[213] Twice American fighters strafed factories in Kyoto's suburbs, destroying no structures but killing one person and wounding thirteen. Single U.S. bombers also hit Kyoto, once in the month of January, once in April, and once in June 1945, either by accident or navigational error. Some buildings were destroyed and 81 people were killed, 388 injured, and 1,855 dispossessed.[214] The relatively minuscule nature of these statistics underscores Kyoto's unique immunity from air attack in World War II, despite the superheated atmosphere of the time.

Immediately following the successful conclusion of XXI Bomber Command's fierce offensive against the Han-Shin area in June 1945, Acting Secretary of State Grew again raised the question of a presidential statement conveying a clearer and more sensitive explanation of the meaning of unconditional surrender, designed to induce the Japanese to capitulate. As a result of having been subjected to the torments of strategic air bombardment, Grew believed, elements existed in Japan "who clearly realize that they have everything to lose and nothing to gain by continuing the war." From the U.S. standpoint, too, this seemed to be the ideal time to seek a merciful end to the war while still dealing from strength. President Truman carefully examined Grew's latest proposal (drafted two days after the third fire raid on Osaka). Though the President said that he liked the idea, he decided to defer considering it further until the Big Three (Britain, Russia, and the United States) met at Potsdam in July.[215] Consequently, the war continued. LeMay, meanwhile, studied the next persuasive actions XXI Bomber Command could take to force Japan's surrender without an invasion. At the end of June, despite losses, LeMay possessed an awesome inventory of 878 B–29s.[216]

Since the President declined to use diplomacy in June to complement air bombardment in bringing about Japan's surrender, American military leaders proceeded with traditional planning — plans in which ground armies would contest ground armies, and air and naval power would play strictly supporting roles. Other possibilities, however, were not ignored. In the absence of General Arnold, who was on an inspection tour of the Pacific in mid-June 1945, LeMay was summoned to Washington to brief the JCS about the AAF concept for hastening the end of the war in the Pacific. Under the agenda being followed, LeMay explained, no important military targets would be left to bomb in Japan by early September 1945. Though XXI Bomber Command would next attack Japan's transportation sector, "We couldn't see much of a war going on after that time." LeMay argued forcefully that conventional air power could end the war by October, without the need for an invasion of the home islands. His staff officers explained the details of the proposed urban and transportation bombing operations, intelligence, and logistics, but none could persuade the JCS that a surface assault and occupation of the Japanese homeland would not be necessary. Army Chief of Staff Marshall, it is said, slept through much of the briefing.[217]

Later, the AAF would be criticized in some quarters for offering the JCS no better strategy than "piling up more rubble than the enemy could tolerate," but a target specialist from XXI Bomber Command recalled the reaction of the men on Guam when General Arnold told them in June that the B–29 campaign "would enable our infantrymen to walk ashore on Japan with their rifles slung": "Even most airmen were skeptical; [men from] other branches of the service expressed their doubts [much] more vividly."[218] General Marshall, for one, certainly entertained some blunt reservations:

> We had to assume that a force of 2.5 million Japanese would fight to the death, fight as they did on all those islands we attacked. We figured that in their homeland they would fight even harder. We felt this despite what generals with cigars in their mouths had to say about bombing the Japanese into submission. We had killed 100,000 Japanese in one raid in one night, but it didn't mean a thing insofar as actually beating the Japanese.[219]

Back at XXI Bomber Command headquarters, cigar-smoking LeMay and his staff reviewed the accomplishments and costs to date of his fire-blitz offensive against Japan's main cities. Between March 9 and June 15, 1945, B–29s had flown 6,960 sorties, dropped 41,592 tons of incendiaries, and gutted 102 square miles of the most densely populated and most important Japanese industrial zones. In the nine most recent maximum-effort raids alone, the Superforts carried 27,493 tons of incendiaries. In the course of the 17 major raids between March and June, 136 B–29s were lost to all causes, equivalent to 2.1 percent of those dispatched on combat missions.[220]

The fifty or sixty smaller remaining urban areas in Japan now assumed importance to LeMay's target analysts, partly because they possessed industrial value and partly because they were simply vulnerable in terms of flammability and congestion. The next phase of aerial bombardment called for just enough bomb tonnage to be dropped on every target to obliterate it on the first strike, and more B–29s were available than were needed for this assignment. The two-track method of strategic bombardment would be refined and its application intensified. Separate forces of B–29s were to attack separate targets, adapting to weather conditions on a day-by-day basis. Thus, when visibility was satisfactory, the Superforts would conduct daylight precision strikes against priority strategic objectives, still mainly in the aviation industry sector, employing squadron formations of nine to eleven planes that would come in at medium altitude with the formation leader handling the primary sighting task. In times of poor weather, night radar bombing would be employed for area incendiary attacks.

XXI Bomber Command planned to concentrate 70 percent to 85 percent of its bombing offensive on radar-guided incendiary fire bombing. Such a level of effort was deemed sufficient to knock out all surviving key industries within the three-month period that LeMay considered necessary to defeat Japan, or at least to paralyze the country before the first surface landing took place. LeMay's approach, known as the Empire Plan, was flexible but complicated. Conventional fire raids, of course, demanded much prior attention to detail, but this plan needed particularly careful and constant photo interpretation; force requirements, munitions, tonnage, and bombing route information; thoughtfully selected aiming points; and accurately computed area densities and requirements for radar bombing.[221]

Raids involving industrial targeting under the Empire Plan got under way even before XXI Bomber Command's maximum-effort fire blitz against the Han-Shin region was finished in mid-June 1945. In raids on industrial targets through July, B–29s dropped 7,045 tons of bombs on many airframe plants and related aviation factories. Eight other precision raids between June and August attacked five arsenals with another 5,270 tons of bombs. The Superforts also went after two light-metal processing plants, dropping 1,020 tons, and a heavy electrical equipment factory, with 806 tons.[222] Some 22 percent of the overall

B–29 effort involved the daylight precision attacks. In June 1945 when the weather proved better than expected, 25 percent of the high-explosive bombardment campaign was directed against 20 important industrial targets. A representative example of how XXI Bomber Command distributed its forces to achieve maximum results in good weather can be drawn from the almost exclusively visual operations conducted against Japan on just one day: June 26.[223] (Table 4–3)

TABLE 4–3
B–29 Precision Raids on June 26, 1945

Target (Location)	Number of aircraft	Average bombing altitude (feet)	Roof area destroyed or severely damaged (square feet)
Sumitomo metal (Osaka)	68	22,500	252,900
Army arsenal (Osaka)	112	23,000	Negligible[a]
Kawasaki aircraft (Akashi)	36	21,000	984,080
Chigusa arsenal (Nagoya)	38	17,500	556,900
Atsuta arsenal & Nihon vehicle (Nagoya)	31	17,500	465,950 & 282,230
Mitsubishi aircraft (Kagamigahara)	77	16,000	227,000
Aichi aircraft (Eitoku)	64	11,000	173,700
Sumitomo duralumin	31	21,000	551,130
Kawasaki aircraft (Kagamigahara, follow-up)	32	22,000	357,800
Utsube oil refinery (2d naval fuel depot, Yokkaichi)	34	15,500	645,750
Total	523	—	4,497,400

Source: USSBS, *Strategic Air*, p 16.

[a] Bombing achieved by radar drops.

As Japanese aerial defenses disintegrated, LeMay took maximum advantage of the situation. The raid against the oil refinery at Yokkaichi on June 26, 1945, marked the long-awaited appearance in the Marianas of an entire wing of

B–29s (the 315th) stripped of defensive armament except for tail guns, carrying enhanced bombloads, and equipped with the improved APQ–7 (Eagle) radar-bombing system. In the first of fifteen attacks launched against ten oil refineries and related installations through mid-August, these specialized bombers (known as B–29Bs) struck unopposed for 1½ hours, beginning at 10:35 P.M. on June 26 in nearly zero visibility. Thirty planes bombed by radar, dropping a total of 223 tons of high explosives (i.e., 891 500-pounders) to good effect against the target. From late June 1945 until the end of the war in August, 1,095 of 1,200 aircraft dispatched by the 315th Bombardment Wing unloaded 9,084 high-explosive projectiles on primary target oil facilities. The stripped B–29Bs increased their bombloads from 7 to 10 tons per plane in the process. Altogether, an estimated 7 percent of the total bomb tonnage dropped on Japan was directed at seventeen oil-industry targets. During its seven weeks of combat operation, the 315th Bombardment Wing lost only four B–29Bs (0.25 percent of its airborne effort). Another sixty-six aircraft were damaged, only one by enemy fighters.

The B–29B raids crippled much of Japan's oil storage-tank capacity, although the importance to the war effort of the heavy damage inflicted on storage and refining units proved far less than it otherwise might have been. USSBS analysts later conceded that the true effectiveness of the B–29B campaign against the Japanese petroleum system could not be accurately determined because "the industry was already defunct when attacked . . . nearly all production had ceased on account of lack of crude oil." Still, in terms of bombs causing damage, the aerial offensive against oil installations in Japan was more effective than it had been in Europe because of the compactness of Japanese plants, their lighter construction, inadequate defensive measures, and the higher average American high-explosive tonnage and heavier bombs employed against these Japanese targets.[224]

Operation STARVATION

As early as August 1944, XX Bomber Command in the CBI had demonstrated in dramatic fashion the effectiveness of sea mines dropped by parachute from low-flying bombers in enemy shipping lanes. In oil-rich Sumatra, B–29s shut down the Moesi River link to Palembang when eight B–29s dropped sixteen 1,000-lb mines from 500 to 1,000 feet. After-action intelligence impressed General Arnold in Washington, D.C.: within a few days these mines sank three Japanese ships and damaged four others, forcing closure of the long river channel to fuel shipments for nearly a month. Thus far in the war, Arnold was aware that heavy bombers using precision bombing from high altitudes had probably never sunk an underway ship at sea or on a navigable river. Now it appeared that B–29s might carve for themselves an important niche in maritime operations. Indeed they would.

Always open to innovative applications of air power, General LeMay embraced this novel use of Superforts to sow heavy sea mines, as did Generals George E. Stratemeyer, Millard F. Harmon, and Chennault. Generals Norstad and Kenney, however, remained dubious and unenthused about diverting payload tonnage from strategic bombing operations to the mining of navigable waters. Enthusiasts in the small and influential Mine Warfare Section of the Chief of Naval Operations, however, worked with receptive AAF counterparts on the Air Staff to extend the B–29 lessons of Sumatra to the mining of Japanese home waters. This planning effort in Washington and the Marianas became the genesis of Operation STARVATION. Largely devised by Naval Commander Ellis A. Johnson, General LeMay's Director of Mining Operations, STARVATION commenced at the end of March 1945.[225]

The strategic objectives of this campaign were designed to choke the import of food and raw materials into Japan, impede maritime movement and supply of Japanese armed forces, and halt, if possible, shipping between China and Japan on the Sea of Japan. These objectives were to be secured by blockading the Inland Sea and industrial and commercial ports of Tokyo and Nagoya with air-dropped sea mines, by singling out and mining choke points such as Shimonoseki Strait between the islands of Honshu and Kyushu (through which 80 percent of Japan's merchant shipping ordinarily sailed), and by mining ports in Korea and on the northern shores of Japan to interdict the sea traffic between those two destinations.

Employing radar for nighttime mine laying from altitudes between 5,000 and 8,000 feet, the 313th Bombardment Wing flew 1,427 mine-laying sorties that were classified as successful (of 4,723 such sorties conducted by all Allied air forces in the Pacific), and between late March and August 1945 the wing dropped 12,135 sea mines (of 21,389 dropped by all Allied aircraft) in the targeted areas.[226] In sum, B–29s flew 34 percent of the sorties but laid 63 percent of all mines in the maritime offensive against Japan. Contrary to the fears of Generals Kenney and Norstad, mine-laying activities consumed only 5.7 percent of XXI Bomber Command's overall airborne effort. Even the limited number of these single-purpose sorties had a devastating effect on Japan's economy. Though an average of only forty B–29 aircraft participated in mine-laying activities, Superforts sank or damaged nearly two-thirds of all Japanese vessels (606 of 961) that were knocked out by mines. In terms of ship tonnage destroyed, of the 2 million tons of Japanese shipping lost to mines, B–29s accounted for more than a half — 1.25 million tons. The number of Superforts lost on all of these sorties was fifteen, less than 1 percent of the planes involved in mine drops.[227]

XXI Bomber Command and U.S. Navy personnel involved in the campaign observed that B–29 air-dropped sea mines sank or damaged "more shipping than any other agent including submarine or direct air attack by both Army and

Naval forces" in the Pacific. Employing equally vivid language, USSBS analysts concluded that Japanese leaders at war's end found themselves "ringed with ports polluted with aerial mines."[228] Japanese sources likewise concede that Operation STARVATION achieved its purpose and lived up to its name. One IJN captain and mine expert said that maritime mining crippled Japanese shipping to such a degree that, in the fall of 1945, the country "eventually starved." The president of Japan's biggest shipping line, Nihon Yūsen Kaisha, declared that the relative proportion of all shipping lost in combat was one part to submarines, six to aerial bombs, and *twelve* to mines. The IJN captain added emphatically: "I think you probably could have shortened the war by beginning [the air-dropping of mines] earlier."[229]

Postwar analysis of the state of Japan's food supply in 1945 confirmed these figures and impressions. Substantial imports, about 20 percent on a caloric basis, had long been needed to meet the nation's requirements for even a minimal diet. With insufficient numbers of Japanese minesweepers to cope with the problem, the sea mines effectively blockaded the country and worsened the situation most gravely, preventing the waterborne import of foodstuffs from mainland Asia and elsewhere. The Japanese government assigned foodstuffs a shipping priority second only to coal in early 1945. With continuing deterioration, in April 1945 the shipping of industrial raw materials was sacrificed in favor of the shipping of salt and food, especially soybeans. Three months later, as aerial mine laying all but precluded any shipping, especially on the Inland Sea, desperate Japanese authorities cut already skimpy food rations 10 percent nationwide.

From a wealth of statistics on the nutritional situation in Japan, data on the percentage of total calories provided by wartime imports appear in Table 4–4. In overall terms, the caloric value of the diet supplied by imports dwindled to 5 percent of prewar imports by 1945. Only soybean importation held up at 71 percent, thanks to its shipping priority. Medical experts agree that the dwindling supply of food commodities resulted in an insufficient and steadily decreasing nutrient intake during the war. Emergency stocks of a key staple, rice, met only a small percentage of annual consumption needs. Finally, the maritime blockade also restricted Japan's access to nitrogen, phosphorus, and potassium fertilizers so essential to its intensive terraced agriculture.[230] Whatever the effect of atomic warfare, the measured body blow that Operation STARVATION delivered to the sustenance of Japan's entire populace unquestionably contributed to the decision of Japanese leaders to surrender unconditionally.

TABLE 4–4
Percent of Required Calories Imported by Japan

Source of Calories	Percent of Imported Calories, by Year 1941	1945
Rice staple	22	3
Wheat	1	0
Other grains and beans	52	35
Sugar	82	56

Source: USSBS (Medical Division), *The Effects of Bombing on Health and Medical Services in Japan*, pp. 30–36.

The Last B–29 Raids

XXI Bomber Command's effective sortie rate and delivery of bomb tonnage increased appreciably during the last few months of the war. Sorties numbered 5,243 in June 1945, 6,168 in July (by which time 979 B–29s were on hand), and 3,145 in the first half of August (when the Superfort inventory reached 1,000). In June, the B–29s dropped 37,542 tons of bombs; in July, 43,091 tons; and in the first two weeks of August, 21,873 tons. To complement XXI Bomber Command, the first forty-two Superforts of the Eighth Air Force would arrive on Okinawa in August. This air force, moved from Europe, was to have an inventory of 600 B–29s in November, when the invasion of Kyushu, code-named Operation OLYMPIC, was scheduled to occur. By then, the Eighth Air Force and XXI Bomber Command would be able to fly 3,500 and 6,500 sorties per month, respectively, and drop a combined total of 70,000 tons of bombs in November if the war continued that long. With B–29s flying around the clock, AAF plans called for an incredible maximum of *115,000 tons of bombs dropped per month* on Japan thereafter.[231]

Meanwhile, about 70 percent of the ongoing strategic air offensive was allocated to the fire bombing, mostly by radar at night, of built-up industrial locations. From the middle of June to the middle of August 1945, B–29s struck individually from altitudes between 8,000 and 14,000 feet and razed another 76 square miles of Japan's secondary urban areas. In July alone, forty-two Japanese cities were burned in radar-directed night area attacks. By mid-August, a total of fifty-two urban regions had been substantially destroyed and six others partially ruined. In fifteen cases, the destruction of built-up areas exceeded 70 percent; examples are Toyama, 99.5 percent; Numazu, 89.5 percent; Fukui, 84.8 percent; Tsu, 81 percent; Hachioji, 80 percent; and Hitachi, 78.2 percent. In another fourteen instances, the destruction ranged from 50

The Air War Against Japan:
Major Cities Burned Out
March 15–August 7, 1945
0
125
250
miles
Hokkaido
Sea of Japan
JAPAN
Honshu
Toyama
Tokyo
Kawasaki
Yokohama
Nagoya
Osaka
Kobe
Hiroshima
Inland Sea
Tokuyama
Yawata
Shikoku
Sasebo
Nagasaki
Kyushu
Kagoshima
Pacific Ocean
Okinawa
Iwo Jima

percent to 70 percent. As LeMay had directed, all were obliterated on the first try. Only three of the fifty-eight targeted cities had to be attacked a second time, and this because of operational conditions, mainly adverse weather over the objective.[232]

The Japanese reaction to the scope, scale, and intensity of this aerial bombardment mixed consternation with despair, bitterness with anger, and exhaustion with lassitude. Admiral Teijirō Toyoda, Minister of Munitions from April 1945 until the end of the war, told a USSBS interrogator: "When you . . . bombed the small cities in which so much of our industry had been scattered in an attempt to avoid destruction, our production was dealt a fatal blow."[233] Maj. Gen. Masakazu Amano, a section chief on the IJA General Staff after February 1945, described the psychological and material dimensions of the final B–29 raids:

> The continued incendiary bombings over all parts of Japan, with their devastating effect on vulnerable towns and cities and the virtual impossibility of repairing the damage, were generally responsible for the subsequent uneasiness and lowering of morale among the people. More and more of the people began to doubt the ability of our armed forces to win such a protracted war. The restlessness of the people was heightened and rather accelerated when the United States changed its tactics [in June 1945] and began to make indiscriminate and inhuman attacks on obscure and militarily valueless villages and towns. These attacks on the whole . . . strengthened the people's enmity toward the United States.[234]

American analysts who investigated the effects of aerial bombardment on the Japanese war economy addressed the theme of General Amano's bitter remarks with considerable candor:

> Although an effort was made to direct [secondary urban] attacks toward targets the destruction of which would do damage to industrial production, the preponderant purpose appears to have been to secure the heaviest possible morale and shock effect by widespread attack upon the Japanese civilian population. To this end, the practice was adopted, in July, of broadcasting in advance the names of towns marked for destruction. Certain of the cities attacked had virtually no industrial importance. Others were significant only as transportation centers.[235]

Japanese accounts of one particular mass raid by B–29s against five secondary targets on the night of August 1/2, 1945, confirmed that American leaflets and radio broadcasts provided advance warning — identifying all of the towns to be attacked — which shook the populace.[236]

By early July 1945 the home islands were entirely at the mercy of XXI Bomber Command. The Japanese military had not provided antiaircraft defenses against the bombers in medium-sized and small urban areas. Moreover, the IJAAF and IJNAF no longer could offer much aerial resistance because of the shortage of interceptors and aviation fuel, and the policy of husbanding remaining aircraft to meet the expected ground invasion of the homeland. Consequently, B–29s now roamed much of the country at will as Japanese leaders essentially abandoned all of the secondary cities to their fate in the saturation raids.[237] Some idea of the power, pervasiveness, and unrelenting tempo of the area bombing of secondary urban and precision bombing of strategic targets is contained in Table 4–5 that follows, detailing the B–29 raids of July and August of 1945.[238]

Having first destroyed Japan's major cities and then its second-tier urban centers, XXI Bomber Command determined in July 1945 to knock out transportation; "[we were] going into the railroad business too," as LeMay put it. The command intended to paralyze the entire transportation network in the home islands before the first Allied surface landings. Admittedly, little damage had yet been done to Japan's main railway trunk system, though attacks had impeded intraurban and interurban movement of supplies, with adverse effects on industrial production, dispersal, and repair.[239]

During the month, the U.S. Navy and XXI Bomber Command attacked and virtually eliminated all transport at Aomori on northern Honshu, a choke point in the rail transportation system connecting central Japan and Hokkaido. American carrier-based planes first conducted low-level attacks against the Aomori–Hakodate rail ferries in mid-July, claiming to have sunk or damaged all 12 of the ferries plus 17 steel ships and 149 smaller vessels. It was an impressive toll, for U.S. naval aviation devoted only about 2 percent (142 tons) of its overall bombing effort to strikes against Japanese transportation.[240] Two B–29 groups from the 58th Bombardment Wing, staging through Iwo Jima, subsequently put 61 Superforts over Aomori on the night of July 28/29. These planes carried 2,082 canisters containing 79,116 new M–74 incendiary projectiles packed with magnesium-gasoline paste, black and magnesium powder, and white phosphorus. Striking for almost 1½ hours from 10,000 to 16,000 feet, the B–29s released 547 tons of the new weapon over the target zone. Flak resistance was minor; only one searchlight tried to illumine the raiders. This raid burned out 88 percent of the built-up portion of Aomori, destroyed 11,330 dwellings, caused 1,016 casualties, and crippled rail and ferry service to Hakodate. Anchored in the harbor, 27 cargo or transport ships burned (25 beyond repair), and 102 of 105 fishing boats were hit (of which 38 sank). Ashore, 61 railway cars were knocked out, and fire consumed numerous warehouses, mills, and railroad structures.[241]

Before more railway attacks could be mounted, the war ended. Postwar U.S. critics have asserted that the shift in attention to Japan's transportation was a case of too little, too late. The vulnerable railways always begged for attention; they were small-scale, seriously overloaded, already short of coal, and narrowly channeled through numerous bottlenecks. U.S. strategic planning, however, remained preoccupied with the presumable need to launch a ground invasion of Japan. Hence, the American command decision to assign railways top target priority was not taken until July, and the campaign against them, not slated to commence in earnest until mid-August, long after Operation STARVATION virtually closed Japan to shipping. In review, USSBS experts concluded:

TABLE 4–5
B–29 Raids of July–August 1945

Date	Target	Number of Bombers	Comments[a]
Jul 3–4	Numerous, on Honshu	470+	
Jul 8–10	Sendai, Sakai, Wakamatsu, and Gifu	497	Other B–29s struck Utsube oil refineries at Yokkaichi and sowed mines in Shimonoseki Strait, Niigata Harbor, etc.
Jul 12–13	Utsunomiya, Tsuruga, Ichinomiya, and Uwa Jima; also Kawasaki petroleum complex	506	IB and HE ordnance, radar-directed drops
Jul 14–15	Kudamatsu (Nippon Oil Co.)	—	Radar-directed
Jul 16–17	Numazu, Kuwana, Hiratsuka, and Oita	471	IBs
Jul 18–20	Fukui, Hitachi, Choshi, and Okazaki	547	4,000-ton IBs and HEs, visual and radar-directed
Jul 21–23	Ube	77	B–29B radar strike against synthetic oil plant
Jul 24	Osaka, Nagoya	599	Visual and radar attacks against four aircraft factories, a textile mill, and Osaka arsenal
Jul 25	Kawasaki	76	B–29B radar strike against oil facilities
Jul 26	Omuta, Matsuyama, and Tokuyama	305	IBs
Jul 28	Tsu, Aomori, Ogaki, and Ujiyamada	562	Conflagrations in broad areas, including Shimotsu oil refinery
Aug 1–2	Nagaoka, Toyama, Mito, Hachioji, and Kawasaki	766	IBs and HEs, raids noted in text
Aug 6	Numerous, throughout Japan	604	Same day the atomic bomb dropped on Hiroshima
Aug 8	Musashino/Tokyo	412	Daylight attack on Nakajima aircraft plant
Aug 9	Amagasaki and Tokyo	165	Same day the atomic bomb dropped on Nagasaki
Aug 14	Numerous, throughout Japan	833	Largest and last set of raids against industrial and urban areas

Source: USSBS, *Air Campaigns*, pp. 52–53; Craven and Cate, eds., *Army Air Forces in World War II*, vol. 5, pp. 674–675.

[a] IB=incendiary bomb; HE=high explosive

> Railroad attack would have carried with it an almost immediate threat of starvation, not only for the major urban concentration but for the entire deficit food areas, such as western Honshu . . . . [In short] the evidence available indicates that a concentration of air attacks exclusively on railroads and urban areas, at least from March 1945 on, would in all probability have led to an earlier surrender and would at the same time have more successfully reduced enemy military capabilities to oppose a landing.[242]

Japanese sources largely agree with this assessment. Naoki Hoshino, a ranking government economic planner, said that if food and coal from Hokkaido and Kyushu had been cut off through destruction of the main rail lines, the populace indeed would have starved, and continued resistance would have become impossible. Moreover, an IGHQ logistics officer later revealed that members of the Japanese Army General Staff were astonished when U.S. bombers continued to assault factories and urban areas instead of knocking out the all-important railway system. Had the latter approach been adopted, he was convinced, Japan's capitulation would have come earlier.[243]

Near the end of the war, in early August 1945, the AAF possessed a mighty inventory of some 3,700 B–29s. Existing bases in the Marianas could accommodate scarcely 1,000. Some 600 or 700 more were to be based in Okinawa by November, before the execution of Operation OLYMPIC. To handle such numbers efficiently, General Arnold and the JCS reorganized the AAF in the Pacific along the lines of the Allied experience in the European Theater. The JCS directed, on July 2, 1945, that an overall Headquarters U.S. Army Strategic Air Forces in the Pacific (USASTAF) be established on Guam to direct the Twentieth Air Force and the Eighth Air Force arriving from England. General Carl Spaatz was named USASTAF commander, Lt. Gen. Barney M. Giles, his deputy, and Maj. Gen. Curtis E. LeMay, Chief of Staff.

LeMay's XXI Bomber Command was redesignated the Twentieth Air Force (still headquartered on Guam) and placed under the command of Lt. Gen. Nathan F. Twining. The B–29 force consisted of five VHB wings (twenty groups), including the highly secret 509th Composite Group (Atomic). On Okinawa in mid-July 1945, Lt. Gen. James H. Doolittle took command of the Eighth Air Force, which absorbed the headquarters of the old CBI-based XX Bomber Command and was about to be reequipped with B–29s. Plans called for Doolittle's command to match Twining's Twentieth Air Force in size and composition by February 1946, but only elements of one bombardment wing actually reached Okinawa by war's end. What some have called Arnold's "unique experiment in air command" (the old Washington-based, Guam-operated Twentieth Air Force) ended in the summer of 1945. LeMay, it is said, had his doubts about the viability, timing, and military necessity of the vast structural realignment of the Pacific-based strategic bombardment units.[244]

In the meantime, the B–29 had been chosen in 1943 to carry and deliver the top-secret atomic bomb. In early 1944, the required modification of some fifteen Superfortresses began. A battle-tested bomber pilot, Col. Paul W. Tibbets, Jr., commanded the 509th Composite Group activated in December

Following the daring attack on the Japanese home islands in 1942, Doolittle assumed command of bombardment units in the Mediterranean and Europe. He appears above in a B–25 as Commanding General, Twelfth Air Force. In the middle of July 1945 he took command of the Eighth Air Force, which absorbed the headquarters of the old CBI-based XX Bomber Command.

1944 (and its special predecessor unit formed in mid-1944). Maj. Charles W. Sweeney commanded the 393d Bombardment Squadron under Tibbets. Training commenced at remote Wendover Field, Utah, drawing on a bombardment squadron from a training base in Nebraska. In January 1945 members of the group flew ten B–29s to Batista Field in Cuba for training in visual and radar bombing from high altitudes. After receiving modified aircraft in May, the group set up shop in the Pacific, at North Field on Tinian. Navigation and combat flight training against targets as far away as Truk and Marcus ensued. Then came conventional pinpoint strikes against targets in Japan, simulating the tactics to be used in eventual nuclear attacks. Forty-nine dummy bombs, configured like the actual atomic devices, were dropped on twenty-eight Japanese cities. Only Colonel Tibbets, however, knew the secret of the mission; his subordinates thought that some sort of special, albeit conventional, projectile — a gimmick — was involved.

Few Japanese cities had escaped B–29 fire raids, so AAF target analysts who were studying the use of the special weapon against a sizable urban industrial center, one offering psychological and experimental advantages, had to choose a location that had been kept off the primary target list for hitherto-

Col. Paul W. Tibbets, Jr., commanded the 509th Composite Group that dropped the atomic bomb.

unrevealed reasons. In Washington, D.C., General Arnold designated Kyoto and Hiroshima, followed by Niigata and Kokura, for the 509th Composite Group's exclusive attention. When Secretary of War Stimson insisted that Kyoto remain excluded from bombing because of its significance as a national treasure, Arnold unenthusiastically substituted Nagasaki.

Near Alamogordo, New Mexico, Manhattan District technicians successfully tested the first atomic bomb, with its plutonium core, on July 16, 1945. Six days later, during the Big Three meeting in a bombed-out Berlin suburb called Potsdam, President Truman convened his "chief advisors in the little White House at Babelsberg to make the final decision about the use of the bomb." Among the military present, only AAF Commander Arnold opposed using the atomic bomb, believing with LeMay that Japan could be coerced to surrender by B–29s dropping conventional ordnance. "None of the other military men — especially General Marshall — concurred with Arnold."[245] On July 26, Allied leaders in Potsdam issued a demand for Japan's unconditional surrender, though the A-bomb itself was never mentioned. President Truman had already authorized Spaatz, on his order, to dispatch the 509th Composite Group against the selected Japanese targets on or about August 3, weather permitting, with additional bombs to be delivered as available. Tibbets's unit had been pronounced fit for action, and one special weapon was made ready for operational use between July 31 and August 3. Since Prime Minister Kantaro Suzuki was wrongly reported to have rejected the Potsdam Declaration, President Truman on his way back from Europe determined to drop the A-bomb on Japan.[246]

Though LeMay and a handful of senior AAF officers in the field had been told that atomic ordnance was involved, they did not actually comprehend what the bomb would do. Tactically, however, it was known that only one bomber would be used (to lull the defenders into ignoring a presumable scout plane); that aiming would be conducted visually; and that the attacking B–29, after banking sharply, would have to get away from the target very fast. The seventh largest urban area in Japan, Hiroshima, remained first on the AAF's proposed atomic target list. The city was the main administrative and commercial center in southwest Honshu, a port of embarkation for troops and supplies, and the location of a considerable number of wartime industrial plants.

At 8:15 A.M. on August 6, 1945, the *Enola Gay*, piloted by Colonel Tibbets, dropped on Hiroshima the first atomic bomb, a uranium 235, gun-type fission device. Weather conditions were eminently suitable for visual attack, thus saving Kokura and Nagasaki from destruction, at least for the moment. From an altitude of 31,600 feet and at a ground speed of 328 mph, Little Boy, a weapon likened to an elongated trash can with fins, fell toward earth, timed to explode at around 2,000 feet with the energy equivalent of approximately 15,000 tons of TNT. The bomb, which would come to be known to the Japanese as *Pika-don* — "flashboom" — fell almost at the exact heart of the nearly flat and symmetrical urban grid, razing approximately 4.4 out of 7 square miles of the central city. The blast and flames whipped up a firestorm that devoured oxygen and generated fierce winds. Approximately 68,000 of the city's 90,000 buildings were wrecked or burned, many by secondary, conventional fires. A telltale mushroom cloud soared skyward, the top of the pillar reaching a height of 27,000 feet within eight minutes of the detonation, visible to U.S. airmen 390 miles away. In the city, black rain fell heavily, in large drops — oily, sticky, and contaminated by radioactive ash.

The attack came forty-five minutes after the all-clear had sounded from an earlier alert. Taken by surprise, generally indifferent to the appearance of an intruding plane or two, and depending on the gods to continue to safeguard the community, Hiroshima's residents had not sought shelter. Many were caught in the open, exposed to the flash. Casualty estimates varied; early data suggested that some 60,000 to 80,000 people were killed and as many or more were injured. The proportion of deaths from acute injury totaled 74 percent on the bombing day and 89 percent within two weeks. The Japanese Defense Agency's official postwar history, drawing on Hiroshima Police Bureau materials as of late 1945, raises the total casualty toll to 129,588, greater than

Hiroshima: At 8:15 a.m. on August 6, 1945, the *Enola Gay* (as it appeared on Tinian Island above), piloted by Colonel Tibbets, dropped the first atomic bomb on Hiroshima — a uranium-235 gun-type fission device. A telltale mushroom cloud soared skyward (*right*), the top of the pillar reaching a height of 27,000 feet in eight minutes after the detonation, visible to U.S. airmen 390 miles away. In the city, black rain fell heavily, in large drops, oily, sticky, and contaminated by radioactive ash.

half the city's wartime population of 245,000;[247] namely, 78,150 killed, 13,983 missing and presumed dead, 9,428 seriously injured, and 27,997 moderately injured. The Japanese Ministry of Health and Welfare continues to tabulate the number of people who have died from the effects of atomic radiation. A summary report on Hiroshima released in 1990 spoke of a total of 107,905 deaths. It is believed in Japan that the grand total of casualties in Hiroshima now exceeds 200,000. American intelligence in 1945 knew of no prisoner-of-war camp at Hiroshima, but it was later learned that twenty-three U.S. captives died there, three at the hands of irate Japanese. An additional 1,000 Japanese-American residents of the city also are thought to have been killed by the atomic bomb.[248]

Sixteen hours after the first A-bomb shattered Hiroshima, an announcement was issued in President Truman's name, identifying the nature of the new weapon and warning Japan that unprecedented destruction lay in store unless it gave up promptly. In Tokyo, however, the complexities of the traditional consensual process and the intransigent attitude of IJA diehards prevented a quick resolution, though the country was by now in extremis. For their part, the Allies had no intention of allowing military operations to be slowed by the Japanese haggling over terms of capitulation. Even in what would become the last week of the war, the B–29s conducted another fifteen missions against the home islands, employing more than 800 of the 1,000 bombers based in the Marianas. Compounding Japan's woes, the Soviet Union terminated the Neutrality Pact it had with Japan since 1941, and powerful Russian forces invaded Manchuria, Korea, and the Kurils on August 8, 1945, a grim corrective to Japanese leaders' wishful thinking that the Soviets could be induced to serve in a broker's capacity and intercede on their behalf.

In early August it was known in Allied capitals that Japanese moderates quietly sought a graceful way to end the war. Still, in the absence of any forthright and positive reply to the Truman warning, American planners wondered if as many as five atomic bombs would have to be unleashed on Japan to force surrender.[249] Japanese military authorities, many of whom knew better, publicly continued to cover up or play down the nature and importance of the atomic bomb, if only to sustain the morale of the general population outside the stricken Hiroshima area. IGHQ communiques merely mentioned "considerable damage" inflicted on Hiroshima by a small number of B–29s. There were only elliptic references to the A-bomb embedded in wording about "an apparently new type of bomb," the details of which were under investigation. Without further explanation, the Japanese people were told to wear protective garb covering the entire body and to seek shelter when even a few enemy planes appeared.[250]

IJA Field Marshal Shunroku Hata, in command at Hiroshima, typified the army diehards. He told War Minister Anami that the A-bomb "had hardly any effect on the ground one foot below the surface." Unaware that little remained standing above the ground, Anami asked the marshal, who was on his way to visit the Emperor, to relay this reassurance and have the monarch "understand that the atomic bomb is not such a dreadful weapon."[251] Hata's disingenuous assessment, of course, was as dangerous as it was ridiculous; even one of his enlisted soldiers wondered where Hiroshima had gone when he reentered the wasteland next day. Hiroshima Castle, site of Second General Army headquarters and barracks — slightly less than one kilometer north of ground zero — had disappeared completely; bloated corpses floated in its moat.[252]

Hiroshima: The two aerial views make the city appear as a ruthlessly trampled-on toy.

Hiroshima: The two ground views show the macabre destruction. Only the heavy walls of the vault prevented the walls from collapsing in the Koa Fire Insurance Company (*left*).

Unknown to all except a select few, America at this stage possessed only enough fissionable ingredients to assemble one more combat weapon, this time the roundish shaped Fat Man projectile, the reputedly more efficient implosion-type device with a plutonium core. If, as it seemed, the Hiroshima raid would not bring about Japan's capitulation, retention of the initiative was indispensable, and President Truman and the JCS, as they viewed the situation, could not long put off a second nuclear strike. Weather forecasts predicted worsening conditions over Kyushu during the typhoon season now at hand. For the B–29 command on Guam, the originally considered strike date of August 11 appeared increasingly unlikely for the prime target of Kokura, or for the secondary target, Nagasaki.

On the morning of August 9, 1945, Major Sweeney, piloting *Bock's Car*, tried three times to sight visually on the preferred Kokura target, but the weather interfered, so he headed for Nagasaki. That city differed topographically from Hiroshima, its irregular urban area lying within a valley basin bisected by a mountain spur that ran to the congested shoreline rimming a fine harbor. The peninsular location rendered access to the city difficult, but there was considerable industrial buildup, and Nagasaki had become essentially a Mitsubishi town, with shipyards, electric-equipment production, steel factories, and an arms plant, all run by the conglomerate firm. Having been struck previously by only five small-scale bombing raids, Nagasaki presented a relatively pristine target.[253]

On the advice of his bombardier and weapons officer, and taking his fuel situation into consideration, Sweeney decided to resort to radar bombing at Nagasaki if a visual run — as required by orders — proved infeasible. Thus, when *Bock's Car* loosed Fat Man from 28,900 feet, set to detonate at about 1,500 feet (which it did at 11:02 A.M.), the approach was by instrument until

A nuclear weapon of the type detonated over Nagasaki. The bomb is 60 inches in diameter and 128 inches long. Fat Man weighed about 10,000 pounds and had a yield equivalent to approximately 20,000 tons of high explosives.

almost the last moment, when a break in the clouds suddenly appeared over the city. The turbulence caused by the detonation was severe, but Sweeney's Superfort and the two observation planes were able to maneuver swiftly away from the blast: "It was as if the B–29s were being beaten by a telephone pole."[254] On the ground, the populace was caught by surprise when the bomb fell. Most were at work, at home, or in the streets when Fat Man exploded. Nagasaki prefecture authorities subsequently issued a report of the episode, based on survivors' accounts:

> When the atomic bomb exploded, an intense flash was observed first, as though a large amount of magnesium had been ignited, and the scene grew hazy with white smoke. At the same time at the center of the explosion, and a short while later in other areas, a tremendous roaring sound was heard and a crushing blast wave and intense heat were felt. The people of Nagasaki, even those who lived on the outer edge of the blast, all felt as though they had sustained a direct hit, and the whole city suffered damage such as would have resulted from direct hits everywhere by ordinary bombs. The zero area, where the damage was most severe, was almost completely wiped out and for a short while after the explosion no reports came out of that area.

Belying Marshal Hata's account, the stunned prefectural chroniclers concluded: "If such a great amount of damage could be wreaked by a near miss, then the power of the atomic bomb is unbelievably great."[255]

The USSBS estimated that 35,000 to 40,000 people died at Nagasaki and that at least 5,000 more were unaccounted for in addition to a somewhat larger total number of those injured. The latest Japanese estimates of the dead in the Nagasaki bombing or its immediate aftermath are on the order of 60,000 to 70,000 from perhaps 270,000 inhabitants at Nagasaki. Postwar deaths by the census date of 1950 were put at 30,000. The latest official Japanese register of total dead now exceeds 100,000[256] The configuration of the Urakami Valley, behind steep hills and at some distance from the central city, constrained the cone of damage. Only one firestorm occurred, and the wind shift limited the blazes. Only in that sense were the effects on Nagasaki "as a whole less shattering than at Hiroshima." With an energy equivalent reliably estimated at 21,000 tons (6,000 tons greater than the energy produced by Little Boy), Fat Man caused a far larger scale of devastation, although the area razed (1.5 to 1.8 square miles, or 44 percent of the built-up area) was smaller, a result of the point of impact and the uneven terrain. Blast effects and primary fire damage were especially evident at Nagasaki. Of nonresidential structures examined after the war, nearly 60 percent of their usable floor space was destroyed or severely damaged. Only 12 percent of these structures escaped damage; the rest as far away as 16,000 feet from ground zero incurred minor degrees of damage. The postraid effects on production in the main industrial targets at Nagasaki were estimated as follows: full production of steel would be set back by one year; electric power would be severely reduced for two months although full capacity would return in six months; and arms production would be restored to two-thirds of capacity in fifteen months.[257]

TABLE 4–6

Comparative Effectiveness of Atomic versus Conventional Bombing by B–29s in 1945

Parameter[a]	Aug 6: Hiroshima	Aug 9: Nagasaki	Mar 9/10: Tokyo	Average of 93 Urban Attacks
Bombload	1 bomb	1 bomb	1,667 tons	1,129 tons
Population density (per square mile)	35,000	65,000	130,000	Unknown
Area destroyed (in square miles)	4.7	1.8	15.8	1.8
KIA and MIA (in thousands)	70–80	35–40	83	1.85
WIA (in thousands)	70	40	102	1.83
Mortality rate (thousands per square mile destroyed)	15	20	5.3	1.0
Casualty rate (thousands per square mile destroyed)	32	43	11.8	2.0

Source: USSBS, *Atomic Bombs*, p. 33.

[a] KIA=killed in action; MIA=missing in action; WIA=wounded in action

The entire concept of strategic air bombardment would undergo a radical rethinking when the atomic weapon entered the AAF inventory of munitions. USSBS analysts came up with a number of startling comparative figures, as depicted in Table 4–6 above.[258] Nuclear weapons inflicted unprecedented mortality and casualty rates (higher at Nagasaki than at Hiroshima), exceeding those at Tokyo by 300 to 400 percent even though the population densities at the two cities hit by the atomic bomb were a fourth to half those of the capital, which an expert on physical damage said had burned more ferociously than any of the other cities.

To match the physical damage wrought by the atomic bombs, some 2,100 tons of conventional munitions (1,200 tons incendiary bombs and 400 tons high explosives) would have had to be dropped by at least 210 B–29s at Hiroshima, and 1,200 tons (225 tons incendiary bombs and 675 tons high explosives) by 120 Superforts at Nagasaki. Even these figures are understated, for the experience at Nagasaki in particular was unrepresentative and imperfect. This is so because even though Fat Man was so much more potent than Little Boy as a weapon, the sheltered nature of the Urakami Valley reduced Fat Man's effectiveness by an estimated 80 percent. Under optimal conditions, truly equivalent conventional bombing against Nagasaki would have required approximately 2,700 tons (2,200 tons incendiary bombs and high explosives to inflict comparable physical damage and 500 tons of fragmentation bombs to inflict comparable casualties) delivered by some 270 B–29s.[259] One need hardly add that only a single nuclear-armed Superfort, flying unopposed above Japan and entirely unescorted in the case of the Nagasaki raid, was employed against each of the target cities.[260]

In Tokyo five days later, on August 14, 1945, Emperor Hirohito and the senior civil and military leaders accepted the Potsdam Declaration and agreed to surrender. Word of the decision was announced publicly next day.

In Retrospect

In recent years, a number of American critics have castigated the B–29 strategic bombardment campaign against Japan, specifically area bombing whether by conventional fire blitz or by nuclear means.[261] They often cite the comment of a certain U.S. air commander who said that the target of the Superforts became "the Japanese mind."[262] To the postwar affliction of conscience, General LeMay offered perhaps the best rejoinder:

> Contrary to suppositions and cartoons and editorials of our enemies, I do not beam and gloat where human casualties are concerned . . . . No matter how you slice it, you're going to kill an awful lot of civilians. Thousands and thousands. But, if you don't destroy the Japanese industry, we're going to have to invade Japan. And how many Americans will be killed in an invasion of Japan? Five hundred thousand seems to be the lowest estimate. Some say a million . . . . Do you want to kill Japanese, or would you rather have Americans killed?[263]

As for the strategic bombardment policy employed against Japan, USSBS analysts observed:

> The vulnerability of the Japanese to air attacks was never a primary consideration of basic Allied strategy. The theater Air Command, however, while selecting urban targets primarily on the basis of their economic value, anticipated that, apart from the economic results of those raids, the impact of mass bombing on the people would seriously undermine the enemy's ability to continue the war.[264]

From a strictly professional standpoint, General Hansell prepared a reasoned critique of LeMay's application of strategic bombardment which laid waste to sixty-six Japanese cities. Long a proponent of precision bombing, he wondered whether the urban area raids had been necessary to force Japan's capitulation:

> It seems to me, in retrospect, that . . . the urban incendiary attacks, which were more devastating by far than the two atomic attacks, could almost certainly have been avoided or their quantity greatly reduced if primary reliance upon selective [precision] bombing had been pursued, even if the end of the war were slightly postponed.[265]

Considering the employment of atomic weapons, LeMay summed up the views of an entire U.S. wartime military generation: "... when informed that we were about to be given a piece of ordnance which would far surpass in accomplishment any bomb ever dropped before by any nation, we all said: '*Swell.*'" And, as if replying to Hansell, he continued: "I think we would have won the war anyway, merely by sticking to our incendiary tactics. But we were given the [atomic] bombs and told to go ahead and drop them." LeMay remained unshakably convinced (as did Truman and other American leaders) that "if a nuclear weapon shortened the war by only a week, probably it saved more lives than were taken by that single glare of heat and radiation." In this context, Japanese of wartime age understood that military expediency, not moral issues, was the deciding factor.[266]

As Hansell has pointed out, Spaatz and LeMay "opposed dropping the atom bombs *if the invasion of Japan was postponed or abandoned*."[267] USSBS experts also concluded in their summary report that the aerial assault on Japan would have been sufficient to end the war without employing atomic weapons:

> ... it seems clear that, even without the atomic bombing attacks, air supremacy over Japan could have exerted sufficient pressure to bring about unconditional surrender and obviate the need for invasion. Based on detailed investigation of all the facts, and supported by the testimony of surviving Japanese leaders involved, it is the Survey's opinion that certainly prior to December 31, 1945, and in all probability prior to November 1, 1945, Japan would have surrendered even if the atomic bombs had not been dropped, even if Russia had not entered the war, and even if no invasion had been planned or contemplated.[268]

A few more observations on the use of nuclear weapons need to be introduced that bear on the history of the strategic air campaign against Japan:

(1) All responsible U.S. military and civilian leaders knew that Japan had been beaten by 1944, and certainly by the summer of 1945. As the Japanese authorities viewed it at the time, capitulation to the United Nations* turned on the problems of unconditional surrender, the integrity of the national polity, and retention of the Imperial institution. The unwillingness or inability of Japan's leaders to treat for peace soon enough brought about the U.S. decision to use the atomic bomb against Nagasaki (the so-called unnecessary bomb).[269]

(2) Hansell has suggested that an excessive feeling of haste colored American military decisions, not least of which was the rush to drop the A-bomb immediately after the plutonium version was successfully tested.[270] The AAF, it must be conceded, neither originated nor challenged the decision to employ nuclear weapons against Japan. Said Spaatz: "... I had no difficulty in that. I ordered the drop. I first had verbal orders but I insisted on written orders. That was purely a political decision, [not] a military decision. The military man carries

*The United Nations Conference on International Organization was held in San Francisco, California, from April 25, 1945, to June 26, 1945, when the charter was signed.

Aboard the USS *Missouri*, on September 2, 1945, General Douglas MacArthur dictated the surrender terms to the Japanese (*above*). MacArthur and Lt. Gen. Richard K. Sutherland observe as Japanese Foreign Minister Mamoru Shigemitsu signs the surrender terms (*below*).

out the orders of his political bosses. So that doesn't bother me at all."[271]

(3) Although racism has been raised and magnified unduly as a major consideration driving U.S. military and state policy against Japan in World War II, we cannot ignore entirely the anger and outrage of the American people against the Japanese, or their craving for revenge for Pearl Harbor, Bataan, and Corregidor. Concerning strategic aerial bombardment in this regard, General Spaatz further observed:

> The Germans never bombed us . . . . We had not the same urge, or the same feeling, as far as bombing Germany is concerned, as we had for the Japs who first attacked at Pearl Harbor . . . . We didn't hear any complaints from the American people about mass bombing of Japan; as a matter of fact, I think they felt the more we did the better. That was our feeling toward the Japanese at that time.[272]

Japanese advocates of peace, strengthened by the Emperor's determination to save his country from national suicide, ultimately won out over the hawks in high places. Arguments still rage in Japan over which issue most impelled the country to capitulate unconditionally on August 15, 1945. Was it the Soviet invasion of Manchuria and Korea, the two nuclear bombs, the B–29 incendiary raids, or the sea-mining campaign? Most observers on both sides of the Pacific would agree that no single element or military service can be assigned preeminent credit for the U.S. victory. Of the atomic bombs, for example, it has been said:

> Even in the target cities [they] did not uniformly destroy the Japanese fighting spirit . . . . The bombs were tremendous personal catastrophes to the survivors, but neither time nor understanding of the revolutionary threat of the atomic bomb permitted them to see . . . a final blow to Japan's prospects for victory or negotiated peace . . . . A quip was current in high Government circles of this time that the atomic bomb was the real *Kamikaze*, since it saved Japan from further useless slaughter and destruction.[273]

As long as the Japanese knew about and expected a ground invasion — with the exalting hope of martyrdom, or at least of securing better surrender terms — they hoarded thousands of special attack aircraft for an all-out final effort, and they bore up under the devastating incendiary raids. The advent of the nuclear weapon served a critical purpose, allowing the Japanese to escape from the hammerlock imposed by the hardnosed, unrepentant military. Gen. David A. Burchinal, who piloted lead B–29s in raids against Japan and became an operations planner under LeMay, summed it up neatly: "The A-bomb certainly gave [the more peaceloving Japanese] an excuse to come out of the closet [and] throw up their hands. . . . "[274]

The cascade of conventional and atomic bombing produced effects unforgettable for the Japanese who lived through it all, and for the Allied occupiers who first landed and confronted the utter devastation. In his diary, MacArthur's Army pilot wrote of his arrival in Japan on August 30, 1945:

> My first impression of the people is that they are to be pitied. The children . . . were crying from hunger . . . . [In Yokohama] the devastation is almost unimaginable and was caused by one fire bomb raid we made on the city in May. For mile after mile everything has burned, and there are no houses standing . . . . Now the people have

> collected . . . strips of scorched, buckled [tin] and constructed small lean-tos atop the rubble and are living there by the thousands.[275]

A British observer, returning to the ashes and debris of Japan at war's end, found the Imperial capital he knew before the war "bore no resemblance whatever to the Tokyo I saw on the first morning of my return. To understand what has happened it is useless to think in terms of 'destruction' as we came to know it in Europe; you have to give a new meaning to the word."[276]

Operation STARVATION contributed mightily to Japan's final collapse, and the role of the B–29 in maritime operations remains generally overlooked. This campaign not only halted virtually all shipping into and out of Japan, it also established for the very heavy bombers an important niche in maritime operations, a niche all but forgotten today — though shipping would be just as severely affected were low-flying bombers used to mine sea lanes today, in the Persian Gulf or elsewhere. (Navies of the world would do well to maintain a respectable number of minesweepers.) In 1945, one can only imagine the effects STARVATION might have had if that campaign had been combined with a simultaneous B–29 campaign against Japan's railway system. Fortunately for the Japanese, or perhaps unfortunately for them in terms of extending the conflict, American airmen, in general, were disinclined to innovate, preferring instead to rely on the hallowed doctrine of high-altitude, precision strategic bombardment.

From Chengtu, to Kharagpur, to Saipan, Guam, and Tinian, the "B-san" ("Mr. B–29"), in concert with Allied land and sea offensives through the Southwest Pacific and westward across the Central Pacific areas, helped pave the way for victory in the Pacific Theater. The Superforts released 147,000 of the 160,800 tons of bombs — more than 90 percent of the total tonnage — dropped by all military aircraft on the Japanese home islands. Some 90 percent of the total U.S. bomb tonnage fell on Japan during the last five months of the war. The B–29s, as President Roosevelt had wished and General Arnold sought to ensure, became the only major Allied combat force capable after 1943 of consistently and forcefully striking Japan's home islands. Indeed, just as General Arnold forecast, they had much to do with the first American infantrymen walking ashore on Japan with their rifles slung. In this sense, the Superforts and the men who served them must be judged a primary though not the sole element contributing to Japan's final collapse in 1945.[277]

NOTES

1. Samuel Eliot Morison, *History of United States Naval Operations in World War II*, 15 vols (hereafter Morison, *U.S. Naval Operations in World War II*) (Boston, 1943), vol 3, p 27.
2. Kōjirō Satō, *If Japan and America Fight*, trans. Jihei Hashiguchi (Tokyo, 1921), pp 3–4.
3. See, for example, Haywood S. Hansell, Jr., *Strategic Air War against Japan* (hereafter Hansell, *Air War*) (Maxwell AFB, Ala., 1980), chap 2.
4. Saburō Hayashi in collaboration with Alvin D. Coox, *Kōgun: The Japanese Army in the Pacific War* (Quantico, Va., 1959), pp 1–3.
5. Maj Gen Kawashima Toranosuke, in *Nihon gunyōki shashin sōshu* (Tokyo, 1970), p 59. The AWC instructor was Fujimuro Ryōsuke.
6. Lt Col Rai Fumio, in *Nihon gunyōki*, p 60; see also p 212.
7. Kawashima, in *Nihon gunyōki*, p 57.
8. Alvin D. Coox, *Nomonhan: Japan against Russia, 1939* (Stanford, Calif., 1985), vol 1, pp 273–276.
9. See note 7 above. See also R. J. Francillon, *Japanese Aircraft of the Pacific War* (New York, 1970), pp 262–264.
10. John Morton Blum, *From the Morgenthau Diaries*, 3 vols (Boston, 1965), vol 2, pp 366–368.
11. Wesley F. Craven and James L. Cate, eds., *The Army Air Forces in World War II*, 7 vols (Chicago, 1948), vol 1, chap 2.
12. Lee Kennett, *A History of Strategic Bombing* (New York, 1982), p 164.
13. Cited in Emile Gauvreau, *The Wild Blue Yonder: Sons of the Prophet Carry On* (New York, 1944), p 171.
14. Ltr, H. E. Yarnell, Shanghai, to Adm A. J. Hepburn, Commander-in-Chief, U.S. Fleet, San Pedro, Calif, Sep 27, 1937, Yarnell Papers, Library of Congress, Washington, D.C.
15. Ernest R. May, ed., *Knowing One's Enemies: Intelligence Assessment Before the Two World Wars* (Princeton, N.J., 1984), p 477.
16. Cited in Clive Howard and Joe Whitley, *One Damned Island after Another* (Chapel Hill, N.C., 1946), pp 22–23.
17. Gauvreau, *Wild Blue Yonder*, p 221.
18. Fletcher Pratt, *Sea Power and Today's War* (New York, 1939), pp 178–179, cited by David Kahn in May, ed., *Knowing One's Enemies*, pp 476–477.
19. E. Guyot in *Vu*, Nov 25, 1936, p 1432, cited in Max Werner (pseud.), *Military Strength of the Powers*, trans. Edward Fitzgerald (New York, 1939), p 286.
20. Alfred F. Hurley and Robert C. Ehrhart, eds., *Air Power and Warfare*, Proceedings of the 8th Military History Symposium, USAF Academy, Oct 18–20, 1978 (Washington, D.C., 1979), p 200.
21. Hansell, *Air War*, pp 5–6.
22. Grace Person Hayes, *The History of the Joint Chiefs of Staff in World War II: The War against Japan* (hereafter Hayes, *JCS History*) (Annapolis, Md., 1982), pp 31–32.
23. Winston S. Churchill, *The Grand Alliance* (New York, 1950), pp 518, 522.
24. H. H. Arnold, *Global Mission* (New York, 1949), p 298.
25. Joe C. Dixon, ed., *The American Military and the Far East*, Proceedings of the Ninth Military History Symposium, USAF Academy, October 1980 (Washington, D.C., [1981]), p 33.
26. Gauvreau, *Wild Blue Yonder*, pp 207–209.
27. Arnold, *Global Mission*, p 276.
28. Carroll V. Glines, *Doolittle's Tokyo Raiders* (Princeton, N.J., 1964), pp 14–19, 24–31.
29. Gen Arnold (H.H.A.) to Air War Plans Div, Dec 31, 1941, U.S. Air Force History Support

Office, Bolling AFB (AFHSO). The AAF officer referred to in the second paragraph was Lt. Col. E. H. Alexander, then in China with Brig. Gen. John Magruder.

30. Arnold, *Global Mission*, pp 276–277.

31. A recent account by the author, drawing upon Japanese-language sources, will be found in chapter 8, "Air War against Japan," B. F. Cooling, ed., *Case Studies in the Achievement of Air Superiority* (Washington, D.C., 1994), pp 383–452. For other American treatments of the raid, see Vern Haugland, *The AAF against Japan* (New York, 1948), p 78; Craven and Cate, eds., *Army Air Forces in World War II*, vol 1, pp 440–444; Glines, *Doolittle's Tokyo Raiders*, chap 6; Morison, *U.S. Naval Operations in World War II*, vol 3, pp 387–398; Comdr Walter Karig and Lt Welborn Kelley, *Battle Report: Pearl Harbor to Coral Sea* (New York, 1944), pp 297–298; Ronald H. Spector, *Eagle against the Sun: The American War with Japan* (New York, 1985), pp 153–155; and E. B. Potter, *Bull Halsey* (Annapolis, Md., 1985), pp 55–65.

32. "Chronicle of the War: Apr 18, 1942," *The Japan Year Book, 1942–1943* (Tokyo, 1943), p 330; *The Orient Year Book, 1942* (Tokyo, 1942), p 417.

33. William F. Halsey and J. Bryan III, *Admiral Halsey's Story* (New York, 1947), p 104.

34. Arnold, *Global Mission*, p 307.

35. AAF/A–WPD (OAA) to Chief of the AAF, subj: Combined Comments of Air War Plans and A–3, Feb 26, 1942, AFHSO.

36. Asst Chief of Staff for Military Intelligence, HQ Hawaiian Dept, to G–2 Hawaiian AF, subj: The Effect of an Attack on Ise Shrine on Japanese Morale, May 22, 1942, indorsed forward by HQ 7th AF, AFHSO.

37. Maj Frank Short, USA, Office of the Chief of Ordnance, "A Shorter Road to Tokyo: A Method Is Proposed for Bringing FULL AIR POWER against Japan," Dec 22, 1943, AFHSO.

38. The engineers' review is dated January 26, 1944; the ATC memorandum, January 27, 1944. See also ltr, Lt Col Short to CG/AAF, attn: Deputy Chief of Staff, [no subj.], May 31, 1945, AFHSO. The Operational Plans Division close-out memorandum to AF/AS Intelligence and AC/AS Plans is undated ("This office has no interest in regard to subject").

39. John B. Lundstrom, *The First Team: Pacific Naval Air Combat from Pearl Harbor to Midway* (Annapolis, Md., 1984), p 137.

40. Morison, *U.S. Naval Operations in World War II*, vol 3, p 268.

41. W. A. Jacobs, "Strategic Bombing and American National Strategy, 1941–1943," *Military Affairs*, vol 50, no 3 (Jul 1986), p 133.

42. This discussion of Allied strategy is based on Hayes, *JCS History*, pp 40–42, and Hansell, *Air War*, pp 10–14.

43. Hayes, *JCS History*, pp 51–52.

44. Spector, *Eagle*, pp 221, 298.

45. *Ibid.*, p 560.

46. Hansell, *Air War*, p 14.

47. Hayes, *JCS History*, p 299; Hansell, *Air War*, pp 18–19.

48. Keith Wheeler *et al.*, *Bombers over Japan* (Alexandria, Va., 1982), pp 21–22; Craven and Cate, eds., *Army Air Forces in World War II*, vol 5, pp 6, 10, 332.

49. Wheeler *et al.*, *Bombers*, p 23; Craven and Cate, eds., *Army Air Forces in World War II*, vol 5, pp 7–9; see also Arnold, *Global Mission*, p 245.

50. Arnold, *Global Mission*, p 284; Wheeler *et al.*, *Bombers*, pp 23–24.

51. Wheeler *et al.*, *Bombers*, pp 26–27; Craven and Cate, eds., *Army Air Forces in World War II*, vol 5, pp 6–7.

52. Hideo Tsuji, "Hondo bōkū sakusen (6)," *Kaikō*, Aug 1987, p 5; the official Japanese Military History series in 102 volumes (*Senshi Sōsho*), Bōeichō Bōei Kenshūsho Senshi Shitsu (hereafter, BBSS), *Hondo bōkū sakusen*, pp 194, 201. Japanese readers track individual volumes in the BBSS series by title, not by number, a practice employed hereafter in this study.

53. Tsuji, "Hondo bōkū," p 5.

54. Marcelle Size Knaack, *Post–World War II Bombers, 1945–1973* (Washington, D.C., 1988), p 494.

55. BBSS, *Hondo bōkū*, pp 199–200; Tsuji, "Hondo bōkū," pp 6–7.

56. IJAAF Maj Ashihara Takeyoshi, cited in BBSS, *Hondo bōkū*, pp 197–199; IJA Maj Gen Sawada Jo'ichirō, diary extract, Dec 26, 1943, cited by Tsuji, "Hondo bōkū," p 7.

57. Hayashi, *Kōgun*, pp 87–89, 287; BBSS, *Hondo bōkū*, pp 199, 227; Tsuji, "Hondo bōkū," pp 5–6.

58. Tsuji, "Hondo bōkū," p 7, citing Ashihara Takeyoshi; BBSS, *Hondo bōkū*, pp 197–198.

59. Tanaka Kōji, cited by Tsuji, "Hondo bōkū," p 7.

60. Hansell, *Air War*, pp 25–26, 30; Wheeler *et al.*, *Bombers*, p 28; Stanley L. Falk, "General Kenney, The Indirect Approach, and the B–29s," *Aerospace Historian*, vol 27, no 3 (fall/Sep 1981), pp 147–155; Craven and Cate, eds., *Army Air Forces in World War II*, vol 5, pp 10–13.

61. Arnold, *Global Mission*, pp 347–348; Hansell, *Air War*, pp 19, 26; intvw, Col Bill Peck with Gen Curtis E. LeMay, Mar 11, 1965, K239.0512–785, AFHSO, pp 5–6.

62. Maurer Maurer, ed., *Air Force Combat Units of World War II* (Washington D.C., 1983), pp 470–471; United States Strategic Bombing Survey (USSBS), Military Analysis Division, Army and Army Air Section, *The Strategic Air Operation of Very Heavy Bombardment in the War against Japan (Twentieth Air Force: Final Report)* (hereafter USSBS, *Strategic Air*) (Washington, D.C., 1946), pp 2–3; Hansell, *Air War*, pp 27–28.

63. Joint Intelligence Committee (JIC) 152/2, "Optimum Use, Timing, and Deployment of V.L.R. Bombers in the War against Japan," Jan 18, 1944, AFHSO.

64. USSBS, *Strategic Air*, p 3; Hansell, *Air War*, pp 19–20.

65. Maurer, ed., *Combat Units*, pp 399-400, 452-453; USSBS, *Strategic Air*, pp 3, 7.

66. JIC 152/2, "Optimum Use," Jan 18, 1944, AFHSO.

67. USSBS, *Strategic Air*, pp 7-8, 27.

68. *Ibid.*, pp 4, 8.

69. *Ibid.*, p 4; Wheeler *et al.*, *Bombers*, pp 30–33.

70. Memo, The White House, Washington (initialled F.D.R.), for Gen Marshall, Oct 15, 1943, AFHSO.

71. Draft reply, addressed "Dear Mr. President," hand-carried, Oct 18, 1943, AFHSO.

72. Wheeler *et al.*, *Bombers*, pp 34–35.

73. *Ibid.*, pp 37–39, 52–53. See also Craven and Cate, eds., *Army Air Forces in World War II*, vol 5, pp 52–57.

74. USSBS, *Strategic Air*, p 8; Wheeler *et al.*, *Bombers*, pp 29, 34; Hansell, *Air War*, pp 20–21; Craven and Cate, eds., *Army Air Forces in World War II*, vol 5, pp 92–95.

75. USSBS, *Strategic Air*, pp 56–57; Craven and Cate, eds., *Army Air Forces in World War II*, vol 5, pp 95–98.

76. USSBS, *Strategic Air*, p 7; Craven and Cate, eds., *Army Air Forces in World War II*, vol. 5, pp 99–103; Wheeler *et al.*, *Bombers*, pp 57–59.

77. BBSS, *Hondo bōkū*, pp 204, 290–302; "Tōryū to Gekkō," *Maru*, Apr 1958, pp 61–67; *Nihon rikugun sentōkitai*, p 344; *Nihon no sentōki* (Kōjinsha, 1973), p 119. For the U.S. counterpoint account, see Craven and Cate, eds., *Army Air Forces in World War II*, vol 5, pp 100–101.

78. BBSS, *Manshū hōmen kōkū sakusen*, p 533.

79. Craven and Cate, eds., *Army Air Forces in World War II*, vol 5, pp 105–107; Wheeler *et al.*, *Bombers*, pp 59–60.

80. BBSS, *Manshū hōmen*, pp 533–535, 545.

81. Craven and Cate, eds., *Army Air Forces in World War II*, vol 5, pp 71–73, 107–112.

82. BBSS, *Hondo bōkū*, p 344.

83. CAKX 6082 (Aug 10, 1944), Chennault to WD, and CM–OUT–79786 (11 Aug 1944), WD to Chennault, both cited in Craven and Cate, eds., *Army Air Forces in World War II*, vol 5, pp 112–113.

84. For American accounts of the raids of August 20, see Craven and Cate, eds., *Army Air Forces in World War II*, vol 5, pp 113–115. For the Japanese side, see BBSS, *Hondo bōkū*, pp 344–356; "Tōryū to Gekkō," *Maru*, Apr 1985, p 78; and Terasaki Takaji, "Yasen 'Gekkō' kichi-shirei no kaisō," *Ibid.*, pp 80–83.

85. Curtis E. LeMay and MacKinlay Kantor, *Mission with LeMay: My Story*, (New York, 1965), pp 328–332; Thomas M. Coffey, *Iron Eagle: The Turbulent Life of General Curtis LeMay* (New York, 1986), pp 113–115; Craven and Cate, eds., *Army Air Forces in World War II*, vol 5, pp 115–116.

86. LeMay, *Mission*, pp 328, 332; Craven and Cate, eds., *Army Air Forces in World War II*, vol 5, pp 116–18; see also BBSS, *Manshū hōmen*, pp 545–547.

87. LeMay, *Mission*, p 332.

88. Msgs, Arnold to LeMay and Hansell, both Sep 22, 1944, cited by Coffey, *Iron Eagle*, pp 119–120.

89. Craven and Cate, eds., *Army Air Forces in World War II*, vol 5, pp 118–137, *passim*.

90. *Ibid.*, pp 135–139; USSBS, *Strategic Air*, p 8; BBSS, *Hondo hōmen kaigun sakusen*, p 298.

91. Intvw, author with H. Agawa, March 8, 1989; Nakamuda Kenichi, *Jōhō shikan no kaisō* (Tokyo, 1985), p 249. What was called the Z-Team conducted the main analysis at Sasebo.

92. Craven and Cate, eds., *Army Air Forces in World War II*, vol 5, pp 139–140; BBSS, *Hondo bōkū*, p 364.

93. BBSS, *Hondo bōkū*, pp 365–366. Minor differences appear in the account in BBSS, *Kaigun sakusen*, p 298. When the weather improved, the China Expeditionary Army's air force attacked the Chengtu airstrips on the night of October 26, 1944, claiming fifteen B–29s burned and forty-two damaged, and three smaller aircraft, presumably fighters, destroyed. BBSS, *Hondo bōkū*, p 365.

94. Craven and Cate, eds., *Army Air Forces in World War II*, vol 5, p 140.

95. BBSS, *Hondo bōkū*, p 367; BBSS, *Kaigun sakusen*, pp 299–300; BBSS, *Manshū hōmen*, pp 556–559, 562–564. No details are given as to the fate of the U.S. airmen who bailed out.

96. Craven and Cate, eds., *Army Air Forces in World War II*, vol 5, pp 141–142.

97. LeMay, *Mission*, p 339; Craven and Cate, eds., *Army Air Forces in World War II*, vol 5, pp 148–150.

98. Craven and Cate, eds., *Army Air Forces in World War II*, vol 5, pp 157–165.

99. USSBS, *Strategic Air*, p 9; LeMay, *Mission*, pp 338–339; Craven and Cate, eds., *Army Air Forces in World War II*, vol 5, pp 165–168.

100. Craven and Cate, eds., *Army Air Forces in World War II*, vol 5, p 165.

101. *Ibid.*, p 150. When XX Bomber Command "said goodbye to the Chengtu Valley . . . no tears were shed then — or ever will be shed, I reckon." LeMay, *Mission*, p 339.

102. LeMay, *Mission*, p 322; Arnold, *Global Mission*, p 480; Craven and Cate, eds., *Army Air Forces in World War II*, vol 5, pp 4, 150.

103. USSBS (Over-all Economic Effects Division), *The Effects of Strategic Bombing on Japan's War Economy* (hereafter USSBS, *Effects on War Economy*) (Washington, D.C., GPO, 1946), p 37; USSBS, *Strategic Air*, pp 8–9; Craven and Cate, eds., *Army Air Forces in World War II*, vol 5, pp 170–174; LeMay, *Mission*, p 323.

104. USSBS, *Summary Report (Pacific War)* (Washington, D.C.: GPO, 1946), p 9. See also the section "Operation STARVATION" in this chapter.

105. Craven and Cate, eds., *Army Air Forces in World War II*, vol 5, pp 174–175.

106. *Ibid.*, pp 512–517.

107. BBSS, *Hondo bōkū*, pp 400–402; BBSS, *Kaigun sakusen*, pp 306–307.

108. For details of these raids, see BBSS, *Kaigun sakusen*, p 302; BBSS, *Hondo bōkū*, p 403; and Craven and Cate, eds., *Army Air Forces in World War II*, vol 5, pp 560, 581.

109. BBSS, *Hondo bōkū*, pp 403, 414. Again the IJN records are not in accord with IJA materials. All IJA raiders made it back to base. For the IJN account, see BBSS, *Kaigun sakusen*, pp 306–307.

110. USSBS, *Strategic Air*, p 23; Craven and Cate, eds., *Army Air Forces in World War II*, vol 5, pp 581–583. Six IJN Type 1 attackers went against Saipan on December 6/7 and caused fires. Two planes were lost in the process. Eight IJA heavy bombers struck Isley Field on December 6/7. They reported destroying ten bombers while losing six of their own planes. Several Gingas and heavy bombers participated in the attacks of December 25–26, but no significant results were reported. BBSS, *Kaigun sakusen*, pp 307–308; BBSS, *Hondo bōkū*, p 420; see also Iwamoto Tetsuzō, *Zero-sen gekitsuiō* (Tokyo, 1986), p 271.

111. Details on the counterraid of November 2/3 are provided in BBSS, *Kaigun sakusen*, p 302. See also BBSS, *Hondo bōkū*, p 403; Craven and Cate, eds., *Army Air Forces in World War II*, vol 5, p 581.

112. Arnold, *Global Mission*, p 290; USSBS, *Strategic Air*, p 9; Craven and Cate, eds., *Army Air Forces in World War II*, vol 5, pp 548–551; BBSS, *Kaigun sakusen*, p 302. The Japanese reported fourteen B–29s over Truk on October 28th, 1944, and eight on the 30th. BBSS, *Hondo bōkū*, p 401.

113. USSBS (Civilian Defense Division), *Field Report Covering Air Raid Protection and Allied Subjects: Tokyo, Japan* (hereafter USSBS, *Air Raid Protection, Tokyo*) (Washington, D.C.: GPO, 1947), pp 1–4; USSBS (Urban Areas Division), *Effects of Air Attack on Urban Complex Tokyo-Kawasaki-Yokohama* (hereafter USSBS, *Effects of Air Attack on Tokyo etc.*) (Washington, D.C.: GPO, 1947), pp 3–5; USSBS (Physical Damage Division), *Effects of Incendiary Bomb Attacks on Japan: A Report on Eight Cities* (Washington, D.C.: GPO, 1947), pp 68–87; Craven and Cate, eds., *Army Air Forces in World War II*, vol 5, pp 551–555.

114. BBSS, *Hondo bōkū*, pp 402–405; Yamamoto Shigeo *et al.*, *B–29 tai rikugun sentōtai: rikugun bōkū sentotai no kiroku* (Tokyo, 1985), pp 74–75.

115. BBSS, *Hondo bōkū*, pp 405–406.

116. *Ibid.*, pp 374–402, 405–409; statements, Maj Kodama (53d Air Group), Lt Col Ashihara Takeichi (HQ IJAAF), Lt Col Matsumura Shizuma (10th Air Division), and Capt Okuda Noboru (47th Air Group), BBSS, Tokyo.

117. Craven and Cate, eds., *Army Air Forces in World War II*, vol 5, pp 553–558.

118. BBSS, *Hondo bōkū*, p 414.

119. *Ibid.* The IJN record of the November 24 raid will be found in BBSS, *Kaigun sakusen*, pp 308–309.

120. USSBS, *Strategic Air*, p 9; Craven and Cate, eds., *Army Air Forces in World War II*, vol 5, pp 558–560.

121. USSBS, *Strategic Air*, pp 9–12.

122. *Ibid.*, pp 11, 19–20.

123. Craven and Cate, eds., *Army Air Forces in World War II*, vol 5, pp 560–566.

124. USSBS, *Strategic Air*, pp 11–12; USSBS, *Effects of Air Attack on Tokyo etc.*, p 6; USSBS (Urban Areas Division), *The Effects of Air Attack on Japanese Urban Economy: Summary Report* (hereafter USSBS, *Effects on Urban*) (Washington, D.C.: GPO, 1947), pp 5–6.

125. BBSS, *Hondo bōkū*, pp 415–417.

126. SS, *Hondo bōkū*, pp 428–429.

127. Yamamoto *et al.*, *B–29 tai rikugun sentōtai*, pp 188–190.

128. Maj Furukawa Haruyoshi, 56th Air Gp Comdr, in BBSS, *Hondo bōkū*, pp.420–422. For IJN interception efforts, see BBSS, *Kaigun sakusen*, pp 310–311.

129. BBSS, *Hondo bōkū*, pp 415–417, 428–429.

130. In three such actions, six bombers were shot down and six damaged at Musashino/Tokyo on December 3, 1944; four were downed and thirty-one damaged at Nagoya on December 13; and six were downed at Musashino/Tokyo on January 9, 1945. Craven and Cate, eds., *Army Air Forces in World War II*, vol 5, pp 561–565, 574.

131. BBSS, *Hondo bōkū*, pp 434–436; BBSS, *Kaigun sakusen*, pp 313–314.

132. In 1944, the banner year for the Japanese military aviation industry, Kawasaki Aircraft Industries built 17 percent of IJA combat airframes and 12 percent of combat engines for Tony and Nick fighters and the Lily light bomber. USSBS (Aircraft Division), *The Japanese Aircraft Industry* (hereafter USSBS, *Aircraft Industry*) (Washington, D.C.: GPO, 1947), pp 16, 111.

133. BBSS, *Kaigun sakusen*, p 314; BBSS, *Hondo bōkū*, pp 437–438.

134. LeMay, *Mission*, pp 338–339; Craven and Cate, eds., *Army Air Forces in World War II*, vol 5, pp 566–567; Hansell, *Air War*, pp 46–49.

135. Air Marshal Sir Robert Saundby, *Air Bombardment: The Story of Its Development* (New York, 1961), p 200; Craven and Cate, eds., *Army Air Forces in World War II*, vol 5, pp 567–568, 574–576.

136. Craven and Cate, eds., *Army Air Forces in World War II*, vol 5, p 568.

137. Rpt, Tokyo Gov Nishio Juzo, in BBSS, *Hondo bōkū*, pp 439–442; BBSS, *Kaigun sakusen*, pp 314–315.

138. Craven and Cate, eds., *Army Air Forces in World War II*, vol 5, pp 568–570.

139. BBSS, *Hondo bōkū*, p 442, *Kaigun sakusen*, pp 315–316.

140. Craven and Cate, eds., *Army Air Forces in World War II*, vol 5, pp 570–571.

141. BBSS, *Hondo bōkū*, pp 442–443; BBSS, *Kaigun sakusen*, p 316.

142. USSBS, *Effects of Incendiary Bomb Attacks*, p 90; Craven and Cate, eds., *Army Air Forces in World War II*, vol 5, pp 572–573; Richard H. Kohn and Joseph P. Harahan, eds., *Strategic Air Warfare: An Interview with Generals Curtis E. LeMay, Leon W. Johnson, David A. Burchinal, and Jack J. Catton* (Washington, D.C., 1988). pp 60–61.

143. BBSS, *Hondo bōkū*, pp 482–485. For the IJN account, see BBSS, *Kaigun sakusen*, pp 317, 323–324.

144. Craven and Cate, eds., *Army Air Forces in World War II*, vol 5, p 573.

145. BBSS, *Hondo bōkū*, p 487.

146. USSBS (Military Supplies Division), *Japanese War Production Industries* (Washington, D.C.: GPO, 1946), pp 1, 6.

147. *Ibid.*, p 12. See also Craven and Cate, eds., *Army Air Forces in World War II*, vol 5, pp 573–574.

148. LeMay intvw, Mar 11, 1965.

149. E. Bartlett Kerr, *Flames over Tokyo: The U.S. Army Air Forces' Incendiary Campaign against Japan, 1944–1945* (New York, 1991), p 14. This volume contains a thorough description of

the development and employment of the M–69 fire bomb.

150. For a discussion of XX Bomber Command's incendiary raid of December 1944 against Hankow, viewed through the eyes of a biographer of Claire Chennault, see Martha Byrd, *Chennault: Giving Wings to the Tiger* (Tuscaloosa, Ala., 1987), pp 265–266. For LeMay's own comments, see his *Mission*, p 351.

151. LeMay, *Mission*, pp 344–352; USSBS, *Strategic Air*, pp 12–13, 19–20; USSBS, *Effects of Incendiary Bomb Attacks*, p 90; Craven and Cate, eds., *Army Air Forces in World War II*, vol 5, pp 612–615; Arnold, *Global Mission*, pp 596–597; USSBS, *Aircraft Industry*, pp 23, 29–30, 106. One USSBS study found that "home" industry had received greatly exaggerated attention in Allied circles during the war. USSBS, *Effects of Air Attack on Tokyo etc.*, pp 29–30.

152. Intvws, author with M. Ikuta, H. Toga, M. Hattori, and I. Kōno, 1990–1991; see also BBSS, *Hondo bōkū*, pp 316, 480–482, 486; Watanabe Yoji, *Nihon hondo bōkūsen* (Tokyo, 1982), pp 155–157, 248–251; *Homeland Air Defense Operation Record*, Japanese Monogr no. 157 (JM 157), HQ USAFFE/Eighth US Army (Rear) (Tokyo, 1952), pp 36, 38, 61, 63–68, 75–78; rsch memo, Alvin D. Coox, "The B–29 Bombing Campaign against Japan: The Japanese Dimension," May 1982, pp 20–22; Alvin D. Coox, "The Rise and Fall of the Imperial Japanese Air Forces," *Aerospace Historian*, vol 27, no 2 (Jun 1980), p 82.

153. LeMay, *Mission*, pp 10, 348; Kohn and Harahan, eds., *Strategic Air Warfare*, pp 63–64; USSBS, *Strategic Air*, p 13; Craven and Cate, eds., *Army Air Forces in World War II*, vol 5, pp 613–614.

154. Gen Thomas S. Power with Albert A. Arnhym, *Design for Survival* (New York, 1965), p 11.

155. USSBS, *Effects of Incendiary Bomb Attacks*, pp 90–116; USSBS, *Effects of Air Attack on Tokyo, etc.*, pp 2, 6, 7; USSBS, *Aircraft Industry*, p 30; LeMay, *Mission*, pp 353, 369; Craven and Cate, eds., *Army Air Forces in World War II*, vol 5, pp 615–616. In the 73d Bombardment Wing, one bomber was lost and twenty-three damaged; in the 314th Bombardment Wing, nine planes sustained minor damage; in the 314th Bombardment Wing, eight bombers were lost and ten damaged.

156. Lt Col Ashihara Takeichi (Rsch Sec, HQ IJAAF), in BBSS, *Hondo bōkū*, p 488.

157. Far to the north of Tokyo on the night of March 9/10, 1945, at the southern edge of the Zaō mountain range in the direction of Yamagata, three B–29s, flying singly at intervals of perhaps an hour apart, crashed into snowy Mt. Fubō, each at about the same altitude. Villagers eventually recovered thirty-four American corpses on the mountain. In 1961 the local Japanese erected a monument in memory of the American B–29 crewmen who died there. Sanematsu Yuzuru, *Tokiori no fude no ato* (Tokyo, 1987), pp 31–36.

158. LeMay, *Mission*, p 353.

159. Kōno intvw, 1991, and intvw, author with H. Suzuki, Oct 3, 1951; see also BBSS, *Hondo bōkū*, pp 487–489, 500; BBSS, *Kaigun sakusen*, pp 342–343. Supplementary information, with sometimes differing statistics, is from USSBS (Morale Division), *The Effects of Strategic Bombing on Japanese Morale* (hereafter USSBS, *Effects on Morale*) (Washington, D.C.: GPO, 1947), pp 36–37; USSBS, *Effects of Incendiary Bomb Attacks*, pp 94, 102; USSBS, *Effects of Air Attack on Tokyo etc.*, pp 3–4, 7–8; USSBS, *Air Raid Protection, Tokyo*, pp 58–63, 79, 83, 173, 183–184; and Craven and Cate, eds., *Army Air Forces in World War II*, vol 5, pp 616–617. For accounts of the scene on the ground, see Martin Caidin, *A Torch to the Enemy: The Fire Raid on Tokyo* (New York, 1960), pp 131, 137, 150–151; Masuo Katō, *The Lost War: A Japanese Reporter's Inside Story* (New York, 1946), pp 207–216; Robert Guillain, *I Saw Tokyo Burning*, trans. William Bryon (Garden City, N.J., 1981), pp 181–188; and Fusako Sasaki, "Tokyo Fire Raids: The Japanese View," *History of the Second World War*, 8 vols (London: Purnell & Sons, 1969–1969), vol 6, no 11, pp 2542–2543.

160. I propound this view in *Japan: The Final Agony* (New York, 1970). For the remark about the pervasiveness of air attack, see USSBS, *Effects on Morale*, p 33.

161. JM 157, p 74; BBSS, *Hondo bōkū*, p 503; Ikuta and Toga intvws; USSBS, *Effects of Air Attack on Tokyo etc.*, p vi.

162. "Record of the Shōwa Era," NHK/TV, Tokyo, Mar 1, 1989, television documentary.

163. USSBS (Over-all Economic Effects Division), *The Effects of Strategic Bombing on Japan's War Economy: Appendix ABC* (hereafter USSBS, *War Economy: Appendix ABC*) (Washington, D.C.: GPO, Dec 1946), p 38.

164. USSBS, *Effects of Air Attack on Tokyo etc.*, p 6; USSBS, *Effects of Incendiary Bomb Attacks*, pp 115–116; Arnold quoted in LeMay, *Mission*, p 353.

165. USSBS, *Effects of Air Attack on Tokyo etc.*, pp 6, 7; USSBS, *Strategic Air*, p 12; LeMay, *Mission*, pp 353–354; Craven and Cate, eds., *Army Air Forces in World War II*, vol 5, pp 618–619.

The Japanese estimated the number of B–29s that hit Nagoya on March 11/12 at 130. BBSS, *Hondo bōkū*, p 490.

166. USSBS (Urban Areas Division), *Effects of Air Attack on Osaka-Kobe-Kyoto* (Washington, D.C.: GPO, 1947), pp 24–25, 32; USSBS, *Effects of Air Attack on Tokyo etc.*, pp 2, 6, 7; LeMay, *Mission*, p 354; Craven and Cate, eds., *Army Air Forces in World War II*, vol 5, pp 619–621.

167. Kawabe in BBSS, *Hondo bōkū*, pp 490–493. See also Craven and Cate, eds., *Army Air Forces in World War II*, vol 5, p 620; USSBS, *Effects on Morale*, p 223; USSBS, *Effects of Air Attack on Osaka etc.*, pp 26, 32.

168. BBSS, *Hondo bōkū*, p 490; Craven and Cate, eds., *Army Air Forces in World War II*, vol 5, pp 621–622; USSBS, *Effects of Air Attack on Osaka etc.*, pp 156–159.

169. LeMay, *Mission*, pp 354, 368; Craven and Cate, eds., *Army Air Forces in World War II*, vol 5, pp 623–624; USSBS, *Effects of Incendiary Bomb Attacks*, p 231.

170. BBSS, *Kaigun sakusen*, p 343; BBSS, *Hondo bōkū*, pp 490, 492.

171. LeMay, *Mission*, pp 355, 368–370; USSBS, *Strategic Air*, pp 14, 29.

172. USSBS, *Strategic Air*, p 12; Craven and Cate, eds., *Army Air Forces in World War II*, vol 5, pp 623–624.

173. LeMay, *Mission*, p 369; USSBS, *Strategic Air*, pp 12–13. For Joint Target Group recommendations in detail, see Craven and Cate, eds., *Army Air Forces in World War II*, vol 5, pp 623–625.

174. TC FN–25–11, LeMay to Norstad, Apr 25, 1945, cited in Craven and Cate, eds., *Army Air Forces in World War II*, vol 5, pp 626–627; memo, Combs for Norstad, Apr 10, 1945, cited in *ibid.*, p 625. See also Kohn and Harahan, eds., *Strategic Air Warfare*, p 65.

175. USSBS, *The Seventh and Eleventh Air Forces in the War against Japan* (Washington, D.C.: GPO, 1947), p 23; Roger A. Freeman, *Mustang at War* (Garden City, N.Y., 1974), p 148; Craven and Cate, eds., *Army Air Forces in World War II*, vol 5, p 597; USSBS, *Strategic Air*, pp 20, 23; Wheeler *et al.*, *Bombers*, p 174; Kevin Herbert, *Maximum Effort: The B–29s against Japan* (Manhattan, Kans., 1983), pp 53–54. See also Ikuta and Kōno intvws and the intvw, author with R. Nomura, 1990–1991; Watanabe, *Nihon*, p 233.

176. Arnold, *Global Mission*, p 568; Craven and Cate, eds., *Army Air Forces in World War II*, vol 5, pp 597–598.

177. On May 10, the B–29s also struck fuel reserves in the Tokuyama region.

178. LeMay, *Mission*, pp 369–370.

179. For the Operation ICEBERG phase, see LeMay, *Mission*, pp 370–372; Kohn and Harahan, eds., *Strategic Air Warfare*, pp 48–51; USSBS, *Strategic Air*, p 17; and Craven and Cate, eds., *Army Air Forces in World War II*, vol 5, pp 627–635.

180. BBSS, *Hondo bōkū*, pp 546–548.

181. LeMay, *Mission*, pp 369–370; Kohn and Harahan, eds, *Strategic Air Warfare*, pp 58–59; Craven and Cate, eds., *Army Air Forces in World War II*, vol 5, p 636.

182. USSBS, *Aircraft Industry*, pp 111–114; Craven and Cate, eds., *Army Air Forces in World War II*, vol 5, pp 646–647; BBSS, *Hondo bōkū*, pp 491–492, 541–542; BBSS, *Kaigun sakusen*, pp 367–368.

183. USSBS, *Aircraft Industry*, pp 110–115; Craven and Cate, eds., *Army Air Forces in World War II*, vol 5, pp 647–648; BBSS, *Hondo bōkū*, pp 542–544; BBSS, *Kaigun sakusen*, p 368. Former AAF Capt. Jerry Yellin, pilot of a P–51 on the April 7 escort mission, recorded a vivid memoir of his wartime experiences. Jerry Yellin with Kevin Brass, "Of War and Weddings," *San Diego Magazine*, vol 41, no 2 (Dec 1988), pp 122–125, 192–214.

184. USSBS, *Aircraft Industry*, p 114; Craven and Cate, eds., *Army Air Forces in World War II*, vol 5, p 648; BBSS, *Hondo bōkū*, p 544; BBSS, *Kaigun sakusen*, p 369. The IJAAF made no particular claims this day.

185. For a reason that would soon assume vital political importance to both belligerents: ". . . the Emperor of Japan is not at present a liability and may later become an asset." Radg 160916Z, CINCPOA Advon to CTF 58, CTF 93, Mar 16, 1945, and TC G–19–5–H, CG 20th AF to CG XXI BC, Mar 19, 1945, both cited in Craven and Cate, eds., *Army Air Forces in World War II*, vol 5, p 638.

186. USSBS, *Effects of Incendiary Bomb Attacks*, p 117; Craven and Cate, eds., *Army Air Forces in World War II*, vol 5, pp 635–636.

187. BBSS, *Hondo bōkū*, pp 544–545; BBSS, *Kaigun sakusen*, p 369.

188. Craven and Cate, eds., *Army Air Forces in World War II*, vol 5, pp 649–650; BBSS, *Hondo bōkū*, pp 546–548; BBSS, *Kaigun sakusen*, p 370. For details on the B–29 campaign against oil that

began on May 10, see USSBS, *War Economy: Appendix ABC*, p 38; and Craven and Cate, eds., *Army Air Forces in World War II*, vol 5, pp 658–660.

189. Arnold, *Global Mission*, pp 564, 566–567. Arnold's tour took place in June 1945. See also Craven and Cate, eds., *Army Air Forces in World War II*, vol 5, pp 636, 650.

190. USSBS (Urban Areas Division), *The Effects of Air Attack on the City of Nagoya* (hereafter USSBS, *Effects of Air Attack on Nagoya*) (Washington, D.C.: GPO, 1947), p 8; USSBS, *Effects of Incendiary Bomb Attacks*, p 231; Craven and Cate, eds., *Army Air Forces in World War II*, vol 5, pp 636–637; BBSS, *Hondo bōkū*, p 552; BBSS, *Kaigun sakusen*, p 374.

191. USSBS, *Effects of Air Attack on Nagoya*, pp 1–2, 4; USSBS, *Effects of Incendiary Bomb Attacks*, pp 231–232; Craven and Cate, eds., *Army Air Forces in World War II*, vol 5, pp 637–638; BBSS, *Hondo bōkū*, p 552; BBSS, *Kaigun sakusen*, p 374.

192. USSBS, *Effects of Incendiary Bomb Attacks*, p 117; Craven and Cate, eds., *Army Air Forces in World War II*, vol 5, p 638.

193. BBSS, *Hondo bōkū*, pp 552–553; BBSS, *Kaigun sakusen*, pp 374–375. See also USSBS, *Effects of Incendiary Bomb Attacks*, p 117; USSBS, *Air Raid Protection, Tokyo*, p 174.

194. Craven and Cate, eds., *Army Air Forces in World War II*, vol 5, pp 638–639; USSBS, *Effects of Air Attack on Tokyo etc.*, p 7.

195. Sasaki, "Tokyo Fire Raids," p 2543. See also Guillain, *Tokyo Burning*, pp 210–211; Kato, *Lost War*, pp 216–217.

196. BBSS, *Hondo bōkū*, pp 553–554; BBSS, *Kaigun sakusen*, pp 375–376. See also USSBS, *Air Raid Protection, Tokyo*, pp 174, 184–186. On May 25 and 26, the Japanese also were subjected to fighter sweeps by P–51 Mustangs, guided by B–29s, aimed mainly at airfields.

197. USSBS, *Effects of Air Attack on Tokyo, etc*, p 7; USSBS, *Air Raid Protection, Tokyo*, pp 3–4.

198. Rpt, "Mitsubishi Oil Company, Kawasaki Refinery," in USSBS, *Air Raid Protection, Tokyo*, p 190.

199. USSBS, *Effects on Morale*, pp 36, 91.

200. Irving L. Janis, *Air War and Emotional Stress: Psychological Studies of Bombing and Civilian Defense*, RAND Series (New York, 1951), pp 166–167.

201. Joseph C. Grew, *Turbulent Era: A Diplomatic Record of Forty Years, 1904–1945*, ed. Walter Johnson, 2 vols (London: Hammond, Hammond & Co., 1953), vol 2, pp 1421–1434.

202. USSBS, *Effects of Air Attack on Tokyo etc.*, pp 3–7; Craven and Cate, eds., *Army Air Forces in World War II*, vol 5, p 639.

203. USSBS (Civilian Defense Division), *Final Report Covering Air-Raid Protection and Allied Subjects in Japan* (Washington, D.C.: GPO, 1947), pp 13, 114, 127–129.

204. USSBS, *Effects of Incendiary Bomb Attacks*, p 117; USSBS, *Effects of Air Attack on Tokyo etc.*, pp 4, 7–8; Craven and Cate, eds., *Army Air Forces in World War II*, vol 5, pp 639–640. The USSBS reports consider Yokohama separately from its neighbor, Kawasaki, thereby creating an artificial geographical distinction that has generated some confusion on the part of analysts who might have wished to assess the effects of strategic air bombardment in a more holistic fashion.

205. George C. Kenney, *General Kenney Reports: A Personal History of the Pacific War* (New York, 1949; reprint, Washington, D.C., 1987), pp 575–576.

206. USSBS, *Effects of Air Attack on Osaka etc.*, p 30; Craven and Cate, eds., *Army Air Forces in World War II*, vol 5, p 643.

207. USSBS, *Effects of Air Attack on Osaka etc.*, pp 24–27, 29–30, 32–33, 36–37. See also LeMay, *Mission*, p 374. In a rare glitch, Craven and Cate confuse the results of the Osaka raid of June 1 with those of the March 13th raid. Craven and Cate, eds., *Army Air Forces in World War II*, vol 5, p 641.

208. Maj Furukawa Haruyoshi, 56th Air Gp, in BBSS, *Hondo bōkū*, pp 555–556; BBSS, *Kaigun sakusen*, pp 375–376.

209. USSBS, *Effects of Air Attack on Osaka etc.*, pp 24–26, 31–33, 36–37; Craven and Cate, eds., *Army Air Forces in World War II*, vol 5, pp 641–642; BBSS, *Kaigun sakusen*, p 376.

210. USSBS, *Effects of Air Attack on Osaka etc.*, pp 25, 31–33, 36–37; Craven and Cate, eds., *Army Air Forces in World War II*, vol 5, p 642. Seven IJNAF fighters were destroyed on the ground at Naruo. BBSS, *Kaigun sakusen*, p 377.

211. USSBS, *Effects of Air Attack on Osaka etc.*, pp 148–170; Craven and Cate, eds., *Army Air Forces in World War II*, vol 5, p 641, BBSS, *Hondo bōkū*, p 555; BBSS, *Kaigun sakusen*, p 377.

212. USSBS, *Effects of Air Attack on Osaka etc.*, pp 30, 168, 170.

213. LeMay, *Mission*, p 389; Craven and Cate, eds., *Army Air Forces in World War II*, vol 5, p

710.

214. USSBS, *Effects of Air Attack on Osaka etc.*, pp 240–243, 246.

215. Grew, *Turbulent Era*, vol 2, pp 1435–1436, 1437.

216. Hansell, *Air War*, p 86.

217. Kohn and Harahan, eds., *Strategic Air Warfare*, pp 63–65; Coffey, *Iron Eagle*, p 175; Michael S. Sherry, *The Rise of American Air Power: The Creation of Armageddon* (New Haven, Conn., 1987), pp 306–308.

218. Robert Nathans, "Making the Fires That Beat Japan," in Horatio Bond, ed., *Fire and the Air War* (Boston, 1946; reprint, Manhattan, Kans., 1974), p 136. See also LeMay in Kohn and Harahan, eds, *Strategic Air Warfare*, p 65.

219. Intvw, with Marshall, 1959, as quoted in Edward S. Shapiro, "Atomic Diplomacy, Chapter Two," *Intercollegiate Review*, vol 23, no 2 (spring 1988), p 34.

220. USSBS, *Strategic Air*, p 14.

221. USSBS (Military Analysis Division), *Air Campaigns of the Pacific War* (Washington, D.C.: GPO, 1947), p 52; USSBS, *Strategic Air*, pp 15–16; Craven and Cate, eds., *Army Air Forces in World War II*, vol 5, pp 650–651; Kohn and Harahan, eds., *Strategic Air Warfare*, p 67.

222. USSBS, *Effects on War Economy*, p 38.

223. USSBS, *Strategic Air*, p 16.

224. USSBS (Oil and Chemical Division), *Oil in Japan's War* (Washington, D.C.: GPO, Feb 1946), p 115 and overview at pp 91–133. See also USSBS, *Strategic Air*, p 18; LeMay, *Mission*, pp 376–377.

225. Ltr, Rear Adm Kenneth L. Veth, USN (Ret), to author, Aug 8, 1985, subj: Mining in World War II; USSBS (Naval Analysis Division), *The Offensive Mine Laying Campaign against Japan* (hereafter USSBS, *Mine Laying*) (Washington, D.C.: Naval Analysis Division, 1946; Washington, D.C.: HQ Naval Materiel Command, 1969), pp 13, 108–109.

226. According to Japanese sources, a large though unspecified number of the parachuted mines fell on land.

227. Ellis A. Johnson with David A. Katcher, *Mines against Japan* (Silver Spring, Md., 1947; reprint, Washington, D.C., 1973), pp 16–17. See also USSBS, *Summary Report (Pacific War)* (Washington, D.C.: GPO, 1946), pp 17–18; USSBS, *Mine Laying*, pp 1, 66–67; Craven and Cate, eds., *Army Air Forces in World War II*, vol 5, pp 662–674.

228. USSBS, *Mine Laying*, pp 1, 15–16. See also Veth ltr, Aug 8, 1985; Kohn and Harahan, eds, *Strategic Air Warfare*, p 72; Hansell, *Air War*, pp 42–43, 76, 77; USSBS (Transportation Division), *The War against Japanese Transportation, 1941–1945* (hereafter USSBS, *War against Transportation*) (Washington, D.C.: GPO, 1947), pp 42–43; Frederick M. Sallagar, "Lessons from an Aerial Mining Campaign," RAND Rpt R-1322-PR (Santa Monica, Calif., 1974); LeMay, *Mission*, p 386.

229. Interrogation, IJN Capt Tamura Kyugo, Oct 22, 1945, NAV no. 26 (USSBS no. 103), in USSBS/Pacific (Naval Analysis Division), *Interrogations of Japanese Officials*, vol 2, p 116. Excerpts from many such interrogations will be found in USSBS, *Mine Laying*, pt 2, pp 37–63. Nihon Yūsen Kaisha (NYK) shipping president Terai Hisanobu, cited in USSBS, *Mine Laying*, p 3. For a detailed IJN account of the U.S. mine laying, its effects, and Japanese minesweeping operations, see BBSS, *Kaigun sakusen*, pp 383–386. IJA attempts to counter the B–29 mining effort are described in BBSS, *Hondo bōkū*, pp 557–558.

230. USSBS (Medical Division), *The Effects of Bombing on Health and Medical Services in Japan* (Washington, D.C.: GPO, 1947), pp 30–36. Lt. Gen. Kan Haruji singled out the lack of salt in the homeland by April 1945 and the lack of rice by June 1945. A–2 questionnaire responses, Mar 4, 1946, GHQ/SCAP, MIS/GS, ATIS (Mar 20, 1946), AFHSO.

231. Hansell, *Air War*, pp 86–87; see also USSBS, *Summary Report*, p 17; *General Marshall's Report: The Winning of the War in Europe and the Pacific* (War Department, 1945), p 84.

232. USSBS, *Strategic Air*, pp 16–17; USSBS, *War Economy: Appendix ABC*, p 38. Japanese data on the losses sustained by secondary urban areas in August 1945 will be found in Mizutani Kōichi and Oda Sanjō, *Nihon Rettō kūshū sensaishi* (Tokyo, 1975), pp 413–414 (Maebashi, Takasaki, Imaji, Ube, Nishinomiya, and esp Toyokawa).

233. Interrogation, USSBS no. 10, as cited in USSBS, *Mine Laying*, pp 47–48.

234. Interrogation excerpt, in *Mission Accomplished: Interrogations of Japanese Industrial, Military, and Civil Leaders of World War II* (Washington, D.C., 1946), p 74. See also Col Satō Hiroo, a War Ministry section chief after Apr 1945, in A–2 questionnaire responses, Mar 4, 1946, GHQ/SCAP, MIS/GS, ATIS (n.d.), AFHSO.

235. USSBS, *Effects on War Economy*, p 38. For a rosy assessment of the value of radio broadcasting and of leaflets in particular in undermining the Japanese will to resist, see USSBS, *Effects on Morale*, pp 127–135. See also LeMay, *Mission*, pp 375–376.

236. BBSS, *Kaigun sakusen*, p 396. The Japanese Army also reported the leaflet drops of July 29 and 31. BBSS, *Hondo bōkū*, p 610.

237. BBSS, *Hondo bōkū*, pp 556, 576.

238. USSBS, *Air Campaigns*, table, pp 52–53. See also Craven and Cate, eds., *Army Air Forces in World War II*, vol 5, table, pp 674–675.

239. LeMay, *Mission*, p 386; USSBS, *Air Campaigns*, p 52.

240. USSBS, *Summary Report*, p 17; USSBS, *Effects on War Economy*, pp 44, 47. Little information concerning the USN carrier strikes on the Aomori area is to be found in USSBS, *Effects of Incendiary Bomb Attacks*, p 154.

241. USSBS, *Effects of Incendiary Bomb Attacks*, pp 121–122, 125–126, 129, 137–146.

242. Hansell, *Air War*, pp 77–78; Craven and Cate, eds., *Army Air Forces in World War II*, vol 5, p 700; USSBS, *Effects on War Economy*, pp 44, 63–65. For detailed tables depicting Japanese rail capacities, see USSBS, *Effects on War Economy*, pp 174–175. The entire issue of identifying and knocking out Japanese land and sea transportation is addressed exhaustively in USSBS, *War against Transportation*, esp chap 1.

243. Interrogations, Hoshino and Lt Col Iwakeshi [*sic*: Iwakoshi?], cited in USSBS, *Effects on War Economy*, p 64 n. 7.

244. LeMay, *Mission*, pp 386–387; *General Marshall's Report*, p 87; USSBS, *Strategic Air*, p 4; Hansell, *Air War*, pp 86, 109–113; Craven and Cate, eds., *Army Air Forces in World War II*, vol 5, chap 22.

245. Margaret Truman, *Harry S. Truman* (New York, 1973), p 273. Truman asserts that Arnold had a change of mind and now opposed using the bomb, which was not the case. LeMay and other AAF commanders had convinced him at least by June 1945 that conventional bombing would force Japan's capitulation before the scheduled OLYMPIC invasion in November. But Arnold had other, more parochial reasons for opposing its use. It is known that he worried about the atomic bomb's effects on the public perception of air power; would people overlook thereafter what had been achieved through strategic bombing of the Axis powers?

246. USSBS, *Strategic Air*, p 18; Craven and Cate, eds., *Army Air Forces in World War II*, vol 5, pp 703–715; LeMay, *Mission*, pp 379–380, 384–386.

247. Evacuation of the civilian population in the face of air raids had reduced the peacetime population from 340,000. BBSS, *Hondo bōkū*, p 642.

248. *Ibid.*; *Yomiuri Shimbun* (Tokyo, May 16, 1990), citing Ministry of Health and Welfare data *ca.* 1985–1988; intvw, author with Dr Yamamoto Osamu, Research Institute for Nuclear Medicine and Biology, Hiroshima University, Apr 1, 1984; Naomi Shohno, *The Legacy of Hiroshima: Its Past, Our Future*, trans. Tomoko Nakamura (Tokyo, 1986), pp 14–15, 24–25, 63–64, 78–80; Mizutani and Oda, *Nihon Rettō*, p 425. American sources include USSBS, *Summary Report*, pp 22–24; Craven and Cate, eds., *Army Air Forces in World War II*, vol 5, pp 716–717; USSBS (Chairman's Office), *The Effects of Atomic Bombs on Hiroshima and Nagasaki* (hereafter USSBS, *Atomic Bombs*) (Washington, D.C.: GPO, 1946), p 8; LeMay, *Mission*, p 385; and Bond, ed., *Fire and the Air War*, p 137. For reminiscences by an American military government officer at Hiroshima, see John D. Montgomery, "The Lesson of Hiroshima," *Harvard Magazine*, vol 87, no 6 (Jul–Aug 1985), pp 40–48.

249. LeMay, *Mission*, p 385.

250. Mizutani and Oda, *Nihon Rettō*, p 425; Shohno, *Legacy*, p 17; Craven and Cate, eds., *Army Air Forces in World War II*, vol 5, p 718.

251. Intvw, author with S. Hata, 1962; testimony, Lt Col Inaba Masao, Military Affairs Bureau staff officer, War Ministry, Oct 13, 1949, in *Statements of Japanese Officials in World War II* (GHQ/FEC, MIS/GS, Hist Div), vol 1, p 585. On this episode of August 14 more complete details appear in my translation of Hayashi, *Kōgun*, p 216 n. 4. For what little the Emperor was actually told on the afternoon of August 6, see Senshi Kenkyūkai, *Genbaku no ochita hi* (Tokyo, 1972), p 321.

252. Soldier Mizuki Toshiyuki from Hiroshima provides vivid recollections in Shohno, *Legacy*, pp 56–57.

253. USSBS, *Atomic Bombs*, p 9; Craven and Cate, eds., *Army Air Forces in World War II*, vol 5, pp 718–720.

254. Craven and Cate, eds., *Army Air Forces in World War II*, vol 5, p 720.

255. Cited in USSBS, *Atomic Bombs*, p 3.

256. *Ibid.*, pp 3–4, 15–20; Craven and Cate, eds., *Army Air Forces in World War II*, vol 5, pp 722–723, 725; *Rafu Shimpo*, Aug 10, 1993; Mizutani and Oda, *Nihon Rettō*, p 428; Shohno, *Legacy*, pp 19–20, 22–24, 27–30, 63–65, 80. According to Shohno, 1,670 U.S., British, and Dutch prisoners had been working in the Nagasaki shipyards. A Japanese soldier revealed in 1976 that at least fifty Allied prisoners of war were killed in the Nagasaki bombing and that he had led fifteen survivors out of the Uragami camp. *Japan Times Weekly*, Aug 7, 1976, p 11.

257. USSBS, *Atomic Bombs*, pp 9–15; USSBS, *Summary Report*, p 24; USSBS, *Strategic Air*, pp 18–19; Craven and Cate, eds., *Army Air Forces in World War II*, vol 5, pp 720, 724–725; Shohno, *Legacy*, pp 19, 54–57.

258. USSBS, *Atomic Bombs*, p 33; Maj Forrest J. Sanborn, AUS (Physical Damage Division, USSBS), "Fire Protection Lessons of the Japanese Attacks," in Bond, ed., *Fire and the Air War*, p 169.

259. USSBS, *Atomic Bombs*, p 33.

260. See Arnold, *Global Mission*, p 371.

261. Among the American historians who have passionately raised the moral issue, the writings of the following are particularly relevant: Schaffer, *Wings of Judgement*, and Sherry, *The Rise of American Air Power*, both of whom have been cited earlier in this study; and John M. Dower, *War without Mercy: Race and Power in the Pacific War* (New York, 1986).

262. USSBS, *Effects on Morale*, p 33.

263. LeMay, *Mission*, pp 352–353. As Winston Churchill, echoing Harry Truman, viewed it: "We [would] have sacrificed a million American and a quarter of a million British lives in the desperate battles and massacres of an invasion of Japan." "The Supreme Triumph" (Aug 16, 1945), in *Victory: War Speeches by the Right Honorable Winston S. Churchill*, comp. Charles Eade (London, 1946), p 229. This contention has been called into question by recent U.S. scholarship. The author, however, basing his conclusions on close study of Japanese records of the day and on interviews with Japanese respondents, remains convinced that Operations OLYMPIC and CORONET indeed would have incurred casualties at the upper range of the wartime estimates of personnel losses because of the ferocity and unyielding resistance of the Japanese defending the homeland and the unresolved problem of the *kamikazes*.

264. USSBS, *Effects on Urban*, p 5.

265. Hansell, *Air War*, pp 90–91.

266. LeMay, *Mission*, pp 380–381. For the enlightened views of a senior American scientist, see Vannevar Bush, *Pieces of the Action* (New York, 1970), pp 62–63.

267. Hansell, *Air War*, p 90.

268. USSBS, *Summary Report*, p 26.

269. I draw the notion of "unnecessary" from conversations with Harvard Professor Edwin O. Reischauer, a Japanologist who was an Army G–2 officer during the period under study.

270. Hansell, *Air War*, p 90.

271. Intvw, Alfred Goldberg with Gen Carl A. Spaatz, May 19, 1965, K239.0512–755, AFHSO.

272. *Ibid.*

273. USSBS, *Atomic Bombs*, pp 22–23.

274. Kohn and Harahan, eds, *Strategic Air Warfare*, p 72.

275. Weldon E. (Dusty) Rhoades, *Flying MacArthur to Victory* (College Station, Tex., 1987), p 448. See also *General Kenney Reports*, pp 575–576, and Montgomery, "Lesson of Hiroshima," p 42, for graphic impressions of the ravaged Kei-Hin sector.

276. John Morris, *The Phoenix Cup* (London, 1947), p 18.

277. General Nathan Twining offers decidedly stronger views in a letter dated March 1, 1963 (AFHSO). See also Craven and Cate, eds., *Army Air Forces in World War II*, vol 5, p 749. For the most thorough technical appraisal of the entire B–29 campaign, see USSBS, *Strategic Air*, pp 19–32. For Japanese impressions of the importance of the B–29 campaign in bringing down Japan, I interviewed R. Sejima in 1979; H. Toga in 1991; H. Agawa in 1989; Y. Sanematsu in 1989; and T. Kase in 1991, among others.

Bibliographic Essay

The strategic air war against Japan has yielded a relatively smaller quantity of literature than its counterpart in the European Theater. Still, the coverage accorded to the B–29s far surpasses the attention paid to fighter forces. Among the reasons for the thinner treatment of strategic air issues in the Asia-Pacific Theater was the Allied emphasis on Europe from the outset of American participation in World War II and the geographical immunity of the Japanese homeland to serious aerial assault until the eventual appearance in 1944 of true, very long-range bomber aviation — the B–29.

Remarkably hardy has been the engagingly written, authoritatively researched, and carefully thought-out series edited by Wesley F. Craven and James L. Cate, the seven-volume official history, *The Army Air Forces in World War II* (Chicago, 1948–1958; reprint, Office of Air Force History, 1983). Most relevant to the present study are volume 1, *Plans and Early Operations, January 1939 to August 1942* (1948) and volume 5, *The Pacific: MATTERHORN to Nagasaki, June 1944 to August 1945* (1953). The Japanese dimension is quite thin and, of course, numerous memoirs have appeared and documents have been declassified in the intervening years, but the intrinsic work has withstood the passage of time without serious erosion.

The U.S. doctrine of air power is soundly examined by Robert F. Futrell in *Ideas, Concepts, Doctrine: A History of Basic Thinking in the United States Air Force, 1907–1964* (Maxwell AFB, Ala.: Air University, 1974). Carrying the topic to the wartime period, and to the European Theater most particularly, is Thomas A. Fabyanic in "A Critique of United States Air War Planning, 1941–1944" (unpublished doctoral dissertation, Saint Louis University, 1973). Stephen L. McFarland treats *America's Pursuit of Precision Bombing 1910–1945* in the Smithsonian History of Aviation Series (Washington, D.C., 1995).

Grace Person Hayes authoritatively presents *The History of the Joint Chiefs of Staff in World War II: The War against Japan* (Annapolis, Md., 1982). Of less value despite the enticing title is James C. Gaston's *Planning the Air War: Four Men and Nine Days in 1941: An Inside Narrative* (Washington, D.C., 1982).

Refreshingly knowledgeable insight into strategic planning as well as important field operations will be found in an indispensable series of studies and reminiscences by Maj. Gen. Haywood S. Hansell, Jr.: *Offensive Air Operations against Japan* (U.S. Air Force Air University, Jan 27, 1953); *Strategic Air War against Japan* (Maxwell AFB, Ala., 1980); and *The Strategic Air War: A Memoir* (Office of Air Force History, Washington, D.C., 1986).

For U.S. end-of-war strategic thinking, the OLYMPIC/CORONET/DOWNFALL operational plans have been declassified, including the air annexes: GHQ U.S. Army Forces in the Pacific (USAFP), "G–2 Estimate of the Enemy Situation with Respect to an Operation against Southern Kyushu in November 1945" (Apr 25, 1945); "DOWNFALL: Strategic Plan for Operations in the Japanese Archipelago" (May 28, 1945); and a series of Staff Studies for operations directed against Kyushu (OLYMPIC) and the Kanto Plain on Honshu (CORONET).

Early but still useful general accounts are Vern Haugland, *The AAF against Japan* (New York, 1948), and James F. Sunderman, ed., *World War II in the Air — the Pacific* (New York, 1962). Numerous other histories that accord the Pacific Theater less than central billing include Basil Collier, *A History of Air Power* (New York, 1974), and Lee Kennett, *A History of Strategic Bombing* (New York, 1982). The AAF dimension is treated briefly by Carroll V. Glines, Jr., in *The Compact History of the United States Air Force* (New York, 1963); Alfred Goldberg, ed., in *A History of the United States Air Force, 1907–1957* (New York, 1957); and Anthony Verrier in *The Bomber Offensive* (London, 1968). Of the more recent works, Ronald H. Spector's *Eagle against the Sun: The American War with Japan* (New York, 1985) offers far better treatment of U.S. ground and naval operations than it does of air operations, and it studies the Eagle more thoroughly than the Sun. The latter remark describes almost every book in English about the campaigns in the Pacific War. For that matter, one would have expected much more on the AAF role in the *Biennial Report of General George C. Marshall, the Chief of Staff of the*

United States Army, July 1, 1943 to June 30, 1945 to the Secretary of War (Washington, D.C., 1946).

Several memoirs by high-ranking AAF officers greatly reinforce General Hansell's materials: H. H. Arnold's *Global Mission* (New York, 1949), though it does not provide as much coverage as we might like; Curtis E. LeMay's *Mission with LeMay: My Story* (Garden City, N.Y., 1965), whose collaborator MacKinlay Kantor is undoubtedly responsible for the exasperatingly conversational and impressionistic style, but which remains well worth reading; and George C. Kenney, *General Kenney Reports: A Personal History of the Pacific War* (New York, 1949; reprint, Office of Air Force History, 1987), whose account has relatively little to say about the B–29s that he never managed to obtain or command. Also from the Southwest Pacific Theater we have Weldon E. (Dusty) Rhoades, *Flying MacArthur to Victory* (College Station, Tex., 1987), but it is of limited value to the study of strategic bombardment itself. Complementing the published reminiscences are Office of Air Force History transcripts of interviews with Generals Curtis E. LeMay, Carl A. Spaatz, and James H. Doolittle. Also published by Air Force History is *Strategic Air Warfare: An Interview with Generals Curtis E. LeMay, Leon W. Johnson, David A. Burchinal, and Jack C. Catton*, eds. Richard H. Kohn and Joseph P. Harahan (Washington, D.C., 1988). Biographies of AAF commanders are few. An example is Thomas M. Coffey's *Iron Eagle: The Turbulent Life of General Curtis LeMay* (New York, 1986), which draws on a wide range of research materials and many personal interviews including more than one hundred hours of conversation with LeMay himself. Colonel Tibbets' own career is depicted in Paul W. Tibbets with Clair Stebbins and Harry Franken, *The Tibbets Story* (New York, 1978).

Editor Maurer Maurer provides authentic data on all AAF groups, wings, divisions, commands, and air forces that fought in the Asia-Pacific Theater in his comprehensive compendium on organization lineage: *Air Force Combat Units of Word War II* (Washington, D.C., USAF Historical Division, 1961). Keith Wheeler edited a well-illustrated volume specifically addressing the history of B–29 operations in *Bombers over Japan* (Alexandria, Va., 1982). A useful title in the Ballantine Weapons Book Series is Carl Berger, *B–29: The Superfortress* (New York, 1970). The best B–29 crewman history is Kevin Herbert, *Maximum Effort: The B–29's against Japan* (Manhattan, Kans., 1983). E. Bartlett Kerr provides a popular account of the fire-bombing campaign in *Flames over Tokyo: The U.S. Army Air Forces' Incendiary Campaign against Japan 1944–1945* (New York, 1991). Kenneth P. Werrell has written *Blankets of Fire: U.S. Bombers over Japan during World War II* (Washington, D.C., 1996).

Among specific B–29 operations against Japan, the incendiary raid on Tokyo on March 9/10, 1945, has drawn considerable attention. Apart from a number of articles on the raid, two journalists' memoirs record the recollections of eyewitnesses: one is the Frenchman Robert Guillain's *I Saw Tokyo Burning*, trans. William Byron (Garden City, N.Y., 1984); the other is by the Japanese Masuo Kato, *The Lost War: A Japanese Reporter's Inside Story* (New York, 1946). Reminiscences by Japanese survivors appear in Haruko Taya Cook and Theodore F. Cook, *Japan at War: An Oral History* (New York, 1992).

The largest compilation of literature on the air bombardment of Japan naturally focuses on the atomic attacks. A well-illustrated chronicle is provided by Donald M. Goldstein, Katherine V. Dillon, and J. Michael Wenger, *Rain of Ruin: A Photographic History of Hiroshima and Nagasaki* (Washington, D.C., 1995). In recent years, a number of American historians have raised the moral issue involved in both the conventional incendiary raids and the Hiroshima-Nagasaki A-bombings. The best documented and most challenging books are Michael S. Sherry, *The Rise of American Air Power: The Creation of Armageddon* (New Haven, Conn., 1987), and Ronald Schaffer, *Wings of Judgement: American Bombing in World War II* (New York, 1985). Theological and philosophical approaches to what has been called nuclear morality are typified by Malham M. Wakin, ed., *War, Morality, and the Military Profession* (Boulder, Colo., 1986), and by Richard A. Wasserstrom, *War and Morality* (Belmont, Calif., 1970). The moral element is also addressed by Conrad C. Crane, in *Bombs, Cities, and Civilians: American Airpower Strategy in World War II* (Lawrence, Kans., 1993). The air campaign against Japan recieves a chapter's attention in Robert A. Pape, *Bombing to Win: Air Power and Coercion in War* (Ithaca, N.Y., 1996). Barton Bernstein discusses "The Perils and Politics of Surrender: Ending the War and Avoiding the Third Atomic Bomb," in *Pacific Historical Review*, vol 46 (Feb 1977), pp 1–27. He is also the editor of *The Atomic Bomb: The Crucial Issues* (Boston, 1976). John W. Dower explores an ugly racial element in his uneven, controversial *War without Mercy: Race and Power in the Pacific War* (New York, 1986). Paul Fussell introduces a knowledgeable personal dimension in *Thank God for the Atom Bomb and Other Essays* (New York, 1988).

The best and most comprehensive studies of the effects of strategic air bombardment on the people of Japan were conducted by the postwar U.S. Strategic Bombing Survey. These civilian and military experts had the unmatched advantage of being on the scene in Japan shortly after hostilities ended in 1945. Despite difficulties in obtaining records (many of which had been destroyed) and in translating and interpreting information, the USSBS teams performed conscientious and intelligent work that has not received the attention it eminently deserves. The dozens of monographs consulted in detail for the purposes of the present study cover such germane topics as air raid protection; health and medicine; consequences of the atomic bombs; economic analyses of the effects of bombing on the aircraft industry, electric power, military supplies, oil and chemicals, shipping and land transportation; and the extent of destruction wrought at such urban centers as Tokyo, Kawasaki, Yokohama, Nagoya, Osaka, and Kobe.

Several specialized studies complement the USSBS field work. They include Ellis A. Johnson with David A. Katcher, *Mines against Japan* (Naval Ordnance Laboratory, Silver Spring, Md., 1947; reprint, Washington, D.C., 1973); and a USSBS monograph, *The Offensive Mine Laying Campaign against Japan* (Naval Analysis Division, 1946; reprint, HQ Naval Materiel Command, Washington, D.C., 1969). A remarkable collection of essays and papers was edited by Horatio Bond, Chief Engineer of the National Fire Protection Association, under the title of *Fire and the Air War* (Boston, 1946). The authors discuss not only wartime fires set by incendiaries and atomic bombs, but also the hitherto-secret activities of American fire protection engineers who assisted in the planning of the destruction of Japanese (and German) cities and industrial plants.

Other sources in English on the Japanese side are less than plentiful. In the 1950s, Japanese Army and Navy consultants prepared a series of original monographs for use by the American military, which were translated, edited, annotated, and disseminated through the Japanese Research Division of Headquarters U.S. Army Forces Far East/Eighth U.S. Army (Rear) in Tokyo, where the present author served as military historian and editor. Particularly relevant to the present study are *Air Defense of the Homeland*, Japanese Monograph (JM) 23 (1956); *Homeland Air Defense Operations Record*, JM 157 (1952); *Homeland Operations Record*, JM 17 (n.d.); and *Outline of Preparations prior to Termination of War and Activities Connected with the Cessation of Hostilities*, JM 119 (1952).

The writer has prepared a number of works that draw heavily on Japanese-language sources. They include "The Rise and Fall of the Imperial Japanese Air Forces," *Aerospace Historian*, vol 27, no 2 (Jun 1980), pp 74–86; "The B–29 Bombing Campaign against Japan: The Japanese Dimension," a research memorandum developed in 1982 exclusively from Japanese materials for use in the previously cited *Bombers over Japan*, ed Keith Wheeler; and *Japan: The Final Agony* (New York, 1970).

From the translated literature on Hiroshima, mention should be made of Naomi

Shohno, *The Legacy of Hiroshima: Its Past, Our Future*, trans. Tomoko Nakamura (Tokyo, 1986); Dr. Arata Osada, comp., *Children of Hiroshima*, translated from the 1951 edition (Cambridge, Mass., 1982); and Kyoko I. Selden and Mark Selden, trans. and eds., *The Atomic Bomb: Voices from Hiroshima and Nagasaki* (Armonk, N.Y., 1990).

A few postwar recollections by former Japanese military and naval officers are available in English translation. Colonel Saburo Hayashi of the Japanese Army wrote a concise but well-researched military history in 1951, translated as *Kōgun: The Japanese Army in the Pacific War*, in collaboration with the present author (Quantico, Va., 1959). Japanese airmen's accounts are found in *Samurai!* by Saburo Sakai with Martin Caidin and Fred Saito (New York, 1958) and in *Zero!* by Masatake Okumiya and Jiro Horikoshi with Martin Caidin (New York, 1956). Robert Mikesh treats the last days of Japanese military aviation in *Broken Wings of the Samurai: The Destruction of the Japanese Air Force* (Annapolis, Md., 1993).

The paucity of English translations reflects considerations other than the availability of materials in Japanese, for the literature on the air war — conventional and nuclear — is enormous in that language. Every major city and many small ones have published histories of the raids they experienced. Yokohama City alone prepared six volumes averaging close to 600 pages each (*Yokohama no kūshū to sensai*) in the 1970s. But the largest and most important Japanese source for serious students of the air war is the 102-volume official military history series (*Senshi Sōsho*) written by the historians of the Japan Defense Agency (Bōeichō Bōeikenskūsho Senshi Shitsu [BBSS]). The most pertinent BBSS volumes, published between 1968 and 1979 but never translated, treat such topics as the Okinawa campaign, Iwo Jima and the Marianas, homeland air defense, naval aviation, protection of Manchuria and Korea, air base construction and operation, and aerial ordnance. Imperial Japanese Army Air Force and Imperial Japanese Navy Air Force volumes were much used in preparing the present study, as were Japanese Defense Agency National Institute for Defense Studies (Boeikenshusho) research monographs (Kenkyū Shiryō) on topics dealing with homeland air defense measures.

Secondary Japanese-language sources that have been most useful in studying the air war include books by S. Kubota, S. Matsumura, A. Nishijima, T. Iwamoto, I. Iwai, T. Tagata, N. Orihara, Y. Sanematsu, M. Ikuta, H. Tsuji, T. Terasaki, K. Nakamuda, S. Yamamoto, Y. Watanabe, K. Mizutani, and S. Oda.

In the course of conducting field research during residence in and annual trips to Japan since 1951, the author interviewed former officers of the Imperial Army and Navy, serving and retired officers of the present-day Japanese Self-Defense Forces, and Japanese historians, writers, and civilian survivors of the B–29 raids. Special acknowledgment is made of research assistance rendered over many years by Y. Imaoka, M. Hattori, S. Itonaga, S. Kōno, H. Toga, K. Sawada, I. Kōno, I. Hata, M. Ikuta, H. Suzuki, T. Nakamura, H. Agawa, T. Kase, R. Sasaki, T. Takao, and I. Mutsu. Early respondents include S. Hata (1962), T. Ohmae (1956), T. Hattori (1956), R. Nomura (1956), S. Hayashi (1956), M. Yano (1956), S. Tomioka (1963), and R. Kusaka (1961). Special thanks are owed H. Takahashi. Finally, valuable information on air-laid sea mines was supplied to the author by Rear Adm. Kenneth L. Veth, USN (Ret.), and on U.S. air war planning by Professor Ronald Schaffer.

5

U.S. Strategic Bombardment Doctrine Since 1945

Steven L. Rearden

In the immediate aftermath of World War II, the dominant voice shaping U.S. strategic military doctrine came from the Army Air Forces (AAF), whose senior and most influential officers in large part had risen through the ranks of bomber commands. Most of them firmly believed that the war validated the thesis advanced and refined by Giulio Douhet, William "Billy" Mitchell, and others in the 1920s and 1930s: bombardment from the air, concentrated on targets of vital economic and military importance, dictated the outcome of modern wars. Even though the air campaigns against Germany and Japan failed to accomplish all of their aims, Allied air power, as the U.S. Strategic Bombing Survey found, "was decisive in the war in Western Europe,"[1] while in the Far East, area bombing of Japan's major cities, followed by atomic bombing of Hiroshima and Nagasaki, appeared to have brought about unconditional capitulation. Altogether, the evidence strongly suggested that air power, especially strategic aerial bombardment, had contributed significantly to defeat of the Axis. By the end of World War II, strategic bombing was no longer a theory; it had become as much a part of modern warfare as combat on land or at sea.[2]

Emergence of the Strategic Air Force

As a direct outgrowth of the wartime experience, thinking in the AAF tended more strongly than ever to regard land-based strategic air power as having replaced sea power as the country's first line of defense. It followed, in the postwar makeup of the armed forces, that the AAF and its strategic component would claim at least parity, if not priority, in the allocation of resources. In practical terms, this meant a coequal, independent air arm of no fewer than seventy groups, organized and operated under a unified command structure.[3] Although AAF enthusiasts probably would have asserted such claims in any case, they did so after the war with the added confidence that strategic air power now embraced nuclear weapons that guaranteed it a central, if not preeminent, place in postwar American defense strategy. "With the atomic bomb," as one historian has put it, "airpower could be said to have come of age."[4] Spokesmen such as General Carl A. Spaatz, who succeeded General Henry H. "Hap" Arnold as the postwar AAF Commanding General and became the first Chief of Staff when the U.S. Air Force became a separate service in 1947, reveled in the opportunities. "Air Power is not only our first line of defense," argued Spaatz, "it is the only instrument using the third dimensional medium, the air; it is the only weapon which has the speed, flexibility, and versatility to cope with the cataclysmic forces yet to be released in the Atomic Age."[5]

As Spaatz's remarks indicated, AAF leaders entered the postwar era expecting the combination of air power and atomic weapons to affect significantly the way future wars would be fought, giving the United States a net advantage for the foreseeable future. At the time, the United States was the sole country in the world in a position to exploit this combination. Not only did it have a monopoly on nuclear weapons, it was also the only major power evidently willing to continue investing the energy and resources in maintaining a strategic air force of any substantial size and capability. The only other countries that could possibly do so — Great Britain and the Soviet Union — were either too disillusioned by their recent wartime experience with strategic bombing to pay it any further heed for the time being (as was the case in Britain) or were apparently unconvinced (as Soviet leaders of the day professed) that nuclear air warfare represented the wave of the future. They were also preoccupied with massive reconstruction efforts that had first claim on resources.[6]

For most American air planners and thinkers, the most provocative question raised by the atomic bomb's advent was no longer *whether*, but *how* it would affect the conduct of future wars. Within some quarters of the scientific and intellectual community, a consensus was emerging that the unprecedented destructiveness of nuclear weapons, and the absence of any effective defense against them, would either lead to their being outlawed or perforce create totally new rules governing conflicts between nations. But the American monopoly could not last forever; once other countries acquired the knowledge and capacity to build their own A-bombs, a war involving their use would spell disaster for all concerned. As Bernard Brodie appraised the situation, the atomic bomb came close to being the "absolute weapon"; therefore it compelled a rethinking of military doctrine based on a recognition that any use of such devices would be impractical and that their only conceivable function was deterrence. "Thus far," Brodie argued, "the chief purpose of our military establishment has been to win wars. From now on its chief purpose must be to avert them. It can have almost no other useful purpose."[7]

Most members of the armed forces readily concurred with Brodie that the atomic bomb's potency set it apart from other weapons. Its potential as a deterrent was obvious, especially after two demonstrations during the CROSSROADS tests in the summer of 1946. In reporting their findings on the CROSSROADS operation, team observers representing the Joint Chiefs of Staff (JCS) concluded that atomic weapons were so destructive that, if "used in numbers," they could "nullify any nation's military effort" and "demolish its social and

On September 19, 1951, in the presence of General Hoyt S. Vandenberg, Representative Overton Brooks, and Secretary of the Air Force Thomas K. Finletter, President Truman signed into law the Air Force Organization Act of 1951 "to provide for the organization of the Air Force and the Department of the Air Force."

economic structures and prevent their reestablishment for long periods of time."[8] Nonetheless, as awesome and intimidating as these weapons appeared, they had certain limitations. First and foremost was a shortage of bombs. Though the exact size of the U.S. nuclear stockpile was a closely guarded secret, atomic weapons were known to be few because they were difficult and expensive to produce. In fact, as late as mid-1948 the United States had stockpiled no more than fifty Nagasaki-type nuclear cores, each with an estimated yield of 20 kilotons. Whether there would be enough of these weapons in the foreseeable future to have the radical impact that Brodie and others imagined remained a matter of conjecture.[9]

All the same, the very existence of the bomb raised awkward questions about whether a large postwar air force was really necessary, as planners during World War II had projected. With the possibility of so much destructive power packed aboard a single aircraft, it was entirely conceivable that a relatively small bomber force would suffice, thereby making the seventy-group program redundant. To address this issue, the AAF hastily convened a study panel, headed by Spaatz, which reported in October 1945. Drafted with the collaboration of Lt. Gen. Hoyt S. Vandenberg and Maj. Gen. Lauris Norstad (both ardent supporters of the strategic bombing concept), the Spaatz report roundly endorsed the full exploitation of atomic technology but saw no reason why the existence of nuclear weapons should alter current plans concerning the size, composition, organization, or deployment of the postwar Air Force. The board

Lt. Gen. Hoyt S. Vandenberg (*left*) and Maj. Gen. Lauris Norstad (*right*), strong supporters of the strategic bombing concept, endorsed full exploitation of atomic technology but saw no reason why nuclear weapons should alter plans for the size, composition, organization, or deployment of the postwar Air Force.

reasoned that while atomic weapons were indeed an important breakthrough, the restricted availability of fissionable material would probably confine their use exclusively to high-priority targets for some time to come. In other words, the advent of nuclear weapons, though potentially of enormous significance, did not immediately signal a new era in strategic air warfare requiring drastic doctrinal changes. "The atomic bomb," the board concluded, "has not altered our basic concept of the strategic air offensive but has given us an additional weapon."[10]

So long as the stockpile of atomic bombs remained small, then, strategic planners were reluctant to place too high a premium on them, even though they fully appreciated these weapons' immense deterrent value and military potential. They therefore assumed that, in any future conflict, strategic bombing would still involve the use of conventional munitions augmented with nuclear weapons, depending on the availability of atomic bombs and the existence of suitable targets. In fact, not until late 1947 to early 1948, at a series of meetings held by the U.S. Air Force Aircraft and Weapons Board, did senior Air Force planners begin to pay close attention to the problems of developing the full range of equipment and facilities required for a comprehensive program of atomic warfare.[11] Until then, most Air Force leaders probably subscribed to Brig. Gen. Alfred R. Maxwell's assessment that the atomic bomb fell into the category of an "experimental weapon."[12] According to Air Force historian Robert Futrell, making the transition from the use of conventional to atomic weapons required, among other things, a laborious mental adjustment: "Although Air Force thinkers never underestimated the destructive capabilities of atomic weapons, they apparently required time to grasp the potential gamut of effects these weapons held for air operations and the modification of air doctrines that could be accepted when they were employed."[13]

One reason the Air Force may have been slow to explore the intricacies and new requirements of air-atomic warfare was the service's own internal inability to agree on a suitable set of doctrinal guidelines. Though the fundamental tenets of strategic air doctrine were well-established, opinions sometimes differed over methods of application, especially as they affected unification and interservice relations. Among the more sensitive issues were the degree of autonomy air operations should have from ground and sea action; the extent to which air defenses could disrupt a strategic offensive; and the proper balance between mobilization potential and existing forces. By 1949, with no consensus apparent, the Air University all but suspended its program of publishing doctrinal manuals.[14]

A further impediment to the development of air-atomic doctrine was President Harry S. Truman's policies on atomic energy. Following public and congressional preferences and his own inclinations, he vested custody and control of nuclear weapons with a civilian-run organization, the Atomic Energy Commission (AEC), established by the Atomic Energy Act of 1946. Though the AEC was supposed to collaborate and consult regularly with the armed forces, enforcement of its custody-and-control mandate often prevented military personnel from gaining needed information or unhindered experience with actual weapons. Matters culminated in the summer of 1948, amid the Berlin blockade crisis, at which time Secretary of Defense James Forrestal urged Truman to transfer custody of atomic weapons to the armed forces. Citing the overriding importance of "political considerations," Truman refused and told the services and the AEC to concentrate on perfecting emergency transfer arrangements instead. While the situation gradually changed and the military obtained more direct access, it was not until after the outbreak of the Korean War, two years later, that Truman modified his stand to allow the military to take custody of weapons, limited initially to nonnuclear components.[15]

In fact, throughout his presidency, Truman resisted firm commitments regarding the use or allocation of atomic weapons. The first and only chief executive to have ordered the dropping of the bomb, he hoped never again to be called on to repeat that decision. At one point, in 1948, maintaining that international control might still be possible, he even went so far as to urge the JCS to develop alternative emergency war plans that relied solely on the use of conventional munitions.[16] Though the project was accorded a low priority and never seriously pursued owing to the enormous expense of the forces it would involve, the fact that it arose in the first place served notice of a need for clearer weapons-use policy. But because of the president's attitude, no one in the Air Force was eager to press the matter, especially after Truman's ruling on the custody question. The result was a generalized statement of policy written to Air Force specifications and endorsed in September 1948 by the National Security Council (NSC). The document, NSC 30, sanctioned the inclusion of

Secretary of Defense James Forrestal urged President Truman to transfer custody of atomic weapons to the armed forces.

bomb in planning exercises, but any decision on its actual use remained with the President.[17]

A similar air of uncertainty and vacillation surrounded the early development of the principal aerial striking arm — the Strategic Air Command (SAC). Originally established in December 1944 as the Continental Air Force, it was redesignated the Strategic Air Command on March 21, 1946. In effect, SAC became the postwar extension of the wartime aerial organizations that embodied the doctrine of strategic bombardment. Along with the Tactical Air Command and Air Defense Command, SAC was one of three major combat commands around which the postwar Air Force developed.[18] SAC however, remained the first and only command organized and equipped for air-atomic warfare until the early 1950s. As such, and at the insistence of the Army and the Navy, SAC occupied a special place in the overall postwar defense organization — becoming a "specified command" — and like the Twentieth Air Force in World War II, SAC reported directly to the JCS with the Air Force acting as their executive agent.[19]

After the Air Force became a separate service in 1947, SAC's assigned mission remained one of conducting "long-range offensive operations in any part of the world either independently or in cooperation with land and Naval forces."[20] But for the first few years of its existence, affected by budget cutbacks, a shortage of technicians, and the posthaste demobilization that followed World War II, SAC barely posed a credible threat to any would-be aggressor.[21] The nucleus of its strike force consisted initially of seven bombardment groups, only one of which, the 509th, was capable of sustained combat operations. Based at Roswell Field, New Mexico, near the Sandia nuclear storage facility, the 509th was the only bombardment group at that time flying Silverplate B–29s, each specially modified to carry a single atomic bomb weighing around 10,000 pounds. The rest flew conventionally configured B–29s or obsolete B–17s, the latter reserved for reconnaissance missions. With an unrefueled range of around 3,250 miles for the B–29 and about 1,700 miles for the B–17, neither was capable of attacking intercontinental targets, and since few suitable overseas bases were available to SAC at the time, its ability to respond in a serious emergency would have been sorely tested.[22]

Not until late 1948 did SAC organize its first squadrons of air refueling tankers (modified B–29s designated KB–29Ms), making it possible for the bombers to fly greater distances with in-flight refueling. Around the same time, SAC took delivery of its first operational B–36s, which the Air Force advertised as having an 8,000-mile range. The plane's numerous and well-publicized defects, however, raised serious questions about its reliability and effectiveness. "As a deterrent to aggression in its early years," concluded one senior Air Force officer, "SAC was far more symbol than reality."[23]

A further problem that SAC faced was the unknown military capability of the most likely potential enemy — the Soviet Union — whose menacing behav-

General Lyman L. Lemnitzer, Army Chief of Staff, Admiral Arleigh Burke, Chief of Naval Operations, Mr. Neil McElroy, Secretary of Defense, General Earle E. Partridge, North American Air Defense Command (NORAD) Commander, and General Thomas D. White, Air Force Chief of Staff, examine the site of the new underground NORAD Command Combat Operations Center at Colorado Springs.

ior and expansionist tendencies after World War II caused increasing concern in Washington, D.C. By 1947, American foreign policy had become fixed on halting the spread of Soviet power and influence under a strategy of "containment." According to George F. Kennan, the Foreign Service officer credited with the idea, containment called for the "adroit and vigilant application of counterforce at a series of constantly shifting geographical and political points, corresponding to the shifts and manoeuvres of Soviet policy."[24] Kennan himself downplayed the role of military power in the application of containment; he thought that two well-trained divisions of Marines were probably all the military force the United States would need. In facing the realities of the task, however, senior decision-makers generally agreed that such a policy, if applied vigorously and consistently, carried with it a risk of military confrontation and therefore the need for a higher degree of peacetime military preparedness than the United States had known in the past.[25]

For Air Force planners, the prospect of a possible war with the Soviet Union posed enormous logistical challenges. Among the most formidable problems were the USSR's immense size and unfamiliar terrain, which made targeting especially difficult. Existing maps of the 1940s were often old and untrustworthy; some dated from Czarist days. Often the only available information on potential targets came from captured German World War II records, including photographs that were so outdated as to make synthetic radar target predictions exceedingly difficult, if not impossible.[26] In the absence of solid intelligence, when strategic planning finally began in earnest in the late 1940s, SAC tended to concentrate targeting on some seventy urban-industrial centers because they were recognized as crucial to the Soviet war-making economy and because they were the most readily identifiable and the easiest targets to find and hit.[27] Similar targeting procedures were employed with devastating effect in World War II by the British in the area bombing of Germany and, toward the end of war, by the United States in the area bombing of Japan. With the availability of nuclear weapons, the results promised to be greatly magnified, leading some targeting specialists to echo Brodie in speculating that atomic air power could conceivably "kill a nation."[28]

Preliminary work on such a plan, code-named HARROW, began in February 1948 in a special Air Force committee known as the Ganey Panel. With a readiness date of January 1, 1951, the concept incorporated a three-pronged approach embodying the "Spaatz concept" of operations. The first and presumably decisive phase would be an all-out atomic attack to begin as soon as possible after the outbreak of hostilities. Consisting of a paralyzing blow delivered within forty-eight hours, mainly against the Soviet Union's transportation and petroleum production systems, it would incapacitate or at least slow the Soviet war machine while the United States and its allies mobilized. The subsequent phases of the air campaign were to proceed at the fastest rate possible over the next six months.[29]

In defense of their counter-economy (or countervalue, as it came to be called) targeting system, Air Force planners stressed the crippling effects such attacks would have on the Soviet Union's war-making capacity, but a 1949 study by a six-member ad hoc JCS committee chaired by the Air Force's Lt. Gen. Hubert R. Harmon claimed the impact would be less than the Air Force had estimated and the attack would by no means be fatal to the Soviet war effort, even if it proved 100 percent successful.[30] Subsequently, in analyzing the ability of SAC to execute the planned offensive, the Weapons Systems Evaluation Group (WSEG), a technical advisory body to the JCS, concluded that while 70 percent to 85 percent of the bombers would reach their targets, only 50 percent to 70 percent would return, and permanently crippling damage would be confined to one-half to two-thirds of the Soviet industrial facilities in the targeted areas.[31]

Air Force leaders consistently maintained that few, if any, of these problems need have persisted were the Air Force given sufficient resources. Among the Air Force's most ardent advocates were the members of the White House–appointed Air Policy Commission, headed by Thomas K. Finletter. In January 1948, the commission issued a well-publicized report urging a defense policy relying primarily on land-based strategic air power.[32] Despite strong support for such a program in Congress, administration policy set forth by Secretary of Defense Forrestal favored "balanced forces," which he defined as an effective mix of military capabilities on land, at sea, and in the air. As actually applied, this formula tended to produce a roughly equal, three-way division of defense funds among the U.S. Army, the U.S. Navy, and the U.S. Air Force. Since President Truman insisted on holding military expenditures to a bare minimum and gave priority (despite his concern over the Soviet threat) to reducing the national debt and balancing the budget, competition was naturally fierce among the three services for the available funds.[33]

In the give-and-take of these postwar budget battles, Air Force leaders believed their service was being short-changed and denied essential equipment and facilities, particularly since it was being required to underpin much of American foreign policy. The premier case in point was the Berlin blockade crisis of 1948–1949, which saw the Air Force provide not only the bulk of the Allied airlift that kept Berlin supplied, but also the show of force (nonatomic B–29s deployed to Britain and West Germany) to indicate to the Soviets the seriousness of Allied resolve. Yet as successful as the entire operation may have appeared to the public, it also revealed Air Force limitations. In fact, with most of its transport planes committed to the airlift, the Air Force would have been hard-pressed to provide SAC with adequate logistical support had the situation worsened to the point that strategic air operations became necessary.[34] For those who had gone through the air campaigns of World War II, such a situation was especially frustrating: it ran counter to what they considered to be the hard-learned lessons of the war, and it underscored the obvious need to rethink the postwar allocation of defense resources. "It is plain," argued one senior officer that

> the situation we face today is far more comparable to that which we faced in respect to Germany in the last war than to that we faced in respect to Japan. Obviously, a balanced force shaped to meet the task before us today, if war should come, is not created by providing a ratio of one dollar for the Army, one for the Navy, and one for the Air Force, or one division for the Army, one ship for the Navy, and one group for the Air Force. . . . In consideration of the potential enemy's present military capabilities, we need a Military Establishment composed of Army, Air Force and Navy forces, but the major portion of this Military Establishment must be Air Force because the enemy's only means of striking at the United States decisively is through the air and our only means of striking at him decisively is through the air. The national security is . . . placed in jeopardy, not by the limit on funds, but by the manner of application.[35]

Similar attitudes, of course, pervaded thinking in the other U.S. services, particularly the Navy, which had its own, long-established ideas on how air power should be developed and applied. In the view of Navy leaders, long-

President Harry S. Truman insisted on holding military expenditures to a bare minimum, reducing the national debt, balancing the budget, and managing a sound strategic posture.

range carrier-based aviation could most effectively provide for the country's future security.[36] Pressing its case, the Navy disavowed any intention of trying to usurp the Air Force's strategic mission but insisted, as part of the settlement reached at the 1948 Key West and Newport conferences between the service chiefs and the Secretary of Defense, that its air arm needed access to nuclear weapons and should not be excluded from planning for strategic air operations.[37] With funds in chronic short supply, whether the country could afford the luxury of two strategic air forces — one land-based and the other sea-based — remained doubtful. This rivalry intensified over time, eventually boiling over in the "revolt of the admirals" and a series of highly publicized congressional investigations in the summer and fall of 1949. The most contentious issues aired during these hearings involved interservice roles and missions, budget shares, and the relative merits of competing weapons systems, such as the Navy's proposed advanced generation of flush-deck supercarriers versus the Air Force's new B–36 intercontinental bomber. Just below the surface, however, lurked a more fundamental and distressing problem: the potential loss of confidence on the part of the American public in its armed forces.

Against this background of incessant interservice bickering, congressional inquiries, tight budgets, and worries about the Soviet Union, a defense posture

The B–36, the largest land-based intercontinental bomber at that time, is inspected by President Truman, waving at photographers.

tilt toward the Air Force and preference for strategic bombardment became virtually inevitable, if only to reassure the public and America's allies that the United States would make optimum use of its superior technology and its most intimidating weapon — the atomic bomb. Despite alleged drawbacks and shortcomings of such a policy, critics had nothing more effective or readily available to offer as an alternative. Moreover, as the results of the 1948 SANDSTONE tests in the Pacific indicated, atomic bombs that achieved higher yields with less fissionable material soon could be mass-produced. By 1949, the era of atomic scarcity was nearing an end, giving advocates of strategic bombing a stronger case than ever and making it virtually certain that future Air Force strategic doctrine would be tailored to waging a nuclear war.[38]

The administration's decision to enhance the Air Force's strategic bombing capabilities relative to the capabilities of the other services was underscored by the discovery, in early September 1949, that the Soviet Union had successfully detonated its first nuclear device, years ahead of most predictions. In the wake of this ominous development, the Truman administration sought to reassure the public that the threat was manageable and that the strategic posture of the United States would remain sound. "As long as America retains (as it can) a tremendous advantage in A-bomb quantity, quality, and deliverability," said JCS Chairman General Omar N. Bradley, "the deterrent effect of the bomb will continue."[39] After a brief but spirited debate among his advisers over the most appropriate response to the Soviet Union's achievement, President Truman announced in January 1950 that the United States would begin work promptly on a potentially even more powerful nuclear weapon utilizing thermonuclear, or hydrogen, design principles.[40]

The Air Force's immediate response to the Soviet A-bomb was a hastily arranged review of existing plans and capabilities, including an intense, but inconclusive, look at continental air defense deficiencies.[41] A more thorough interdepartmental review, growing from the hydrogen bomb decision of early 1950, culminated in a report to the President that recommended an all-around strengthening of American defenses in the face of a Soviet threat likely to worsen. Projecting into the future, the document, NSC 68, estimated that the Soviet Union, by 1954, would have a sufficient number of nuclear weapons and delivery vehicles to seriously damage the United States. Using this "year of maximum danger" as its target, the report urged "a rapid and sustained buildup of the political, economic, and military strength of the free world." Although NSC 68 favored heavy increases in conventional forces, its operating premise held that nuclear weapons and command of the air were crucial to U.S. security and were likely to become increasingly so as the Soviet Union expanded its stockpile of atomic weapons.[42] Despite Truman's initially sympathetic reaction to NSC 68 findings and recommendations, it took the outbreak of the Korean War a few months later (in June 1950) to convince him to lift the lid on military spending.

During the ensuing buildup, a formidable strategic deterrent rated high priority. Though direct combat remained confined to a peninsula in Asia, a constant worry in Western capitals was that Korea might be a diversion for a surprise Soviet invasion of Western Europe. Consequently, strategic forces that could strike directly at the Soviet industrial heartland or retard invading Soviet armies in Europe continued to exercise the strongest claim on air resources.[43] In line with this policy, by 1951 SAC had expanded considerably. As the new year began, SAC comprised a force of some 85,000 personnel and fewer than 1,000 aircraft; by the end of 1951 it had grown to nearly 145,000 personnel and 1,200 aircraft, including 98 B–36s, 340 B–29s, and 219 B–50s, a somewhat longer-distance version of the B–29.[44] Yet in reality the expansion was only beginning. The immediate goal for the Air Force was to establish a total of 95 wings,[45] and by 1952 there was talk in Congress and in the Pentagon of a possible 126-wing or even 150-wing force. A third or more of those wings would be dedicated to SAC.[46]

At the same time, SAC acquired access to what promised to be a substantially larger nuclear arsenal. On top of the nuclear breakthroughs, confirmed in the SANDSTONE tests, and the expansion of nuclear production facilities ordered in 1949, Truman approved the production of more fissionable material (one order was issued in the summer of 1950 and another in January 1952) that would eventually increase plutonium and uranium-235 output by 50 percent and

Secretary of the Air Force Thomas K. Finletter along with senior officers recognized that land-based air was the "keystone of American military power."

150 percent, respectively.[47] One reason was to meet the demands for fissionable material in the H-bomb program, but the increased production was also intended to support new weapon design techniques, such as those promising the development of smaller, more manageable "tactical" bombs and warheads, while not depleting the stockpile of basic nuclear weapons needed for strategic purposes.[48] The increase was indeed impressive, from 299 bombs in stockpile at the outbreak of the Korean War to 22,229 bombs and warheads by mid-1961.[49]

Expanded nuclear weapons development and the Korean War air buildup seemed only to fuel renewed friction among the services. No longer was the debate over funds, now relatively abundant; it concerned the underlying philosophy that would govern the buildup of and determine the overall role for air power in the country's security when money again became scarce. Disagreements were perhaps most intense at the joint planning level and involved the relative priority of ground and sea forces versus the needs of what the Air Force termed the strategic objective of air neutralization, that is, the progressive attrition of the Soviet Union's war-sustaining industries.[50]

When it became apparent that resolution of this issue was unlikely through any acceptable compromise or the development of some joint doctrine, the Air Force Council, composed of the Vice Chief of Staff and deputy chiefs, unilaterally adopted a statement of policy in June 1952 that affirmed, "The war objectives of the United States and her Allies will be imposed upon the USSR by the application of military air force, properly supported by essential land and sea forces."[51] Such air power sentiments were hardly new, but to have them expressed in such frank terms by such a high-level body of one service was unusual. Later that summer, at a meeting in Bar Harbor, Maine, attended by Secretary of the Air Force Thomas K. Finletter, Acting Chief of Staff General Nathan F. Twining, and other senior officers, Air Force leaders went a step further and drew up a lengthy paper that recognized "land-based air . . . as the keystone of American military power."[52] Although the ultimate use of the Bar Harbor memorandum is unclear, its immediate purpose was apparently to provide guidance for Air Force personnel.[53] Beyond that, it may also have been meant to contribute to the preparation of a new series of Air Force doctrinal manuals, the first of which appeared the following year.[54] It is also possible that the paper was meant as a new agenda, not only for the Air Force but also for a future reorientation of basic national security policy. If so, as the ascendant service, the Air Force was in a strong position to make known its views when the opportunity finally presented itself.

The Era of SAC and Massive Retaliation

With the advent of the Eisenhower administration in January 1953, most of the questions and uncertainties surrounding the Air Force's role in national policy soon disappeared. Commander of Allied Forces in Europe in World War II and subsequently Army Chief of Staff, acting JCS Chairman, and NATO Supreme Commander, Dwight D. Eisenhower well understood the importance of strong defense preparedness. Even so, while running for the presidency in 1952, he took the position that the Korean War buildup had become excessive and that, in the interests of preserving a healthy economy, defense expenditures should be reduced immediately by $5 billion and be retailored to meet the actual Soviet threat over the long haul. Once in office, he directed the JCS and the NSC to develop recommendations and he told congressional leaders he intended to "follow a new policy which would continue to give primary consideration to the external threat but would no longer ignore the internal threat" of a weakened economy caused by heavy defense expenditures.[55]

A year earlier, the British Chiefs of Staff conducted a similar review. The resultant white paper, "Defence Policy and Global Strategy," when briefed to U.S. Air Force planners in the summer of 1952, won enthusiastic praise and endorsement.[56] Faced with their own financial difficulties, British leaders sought some measure of relief through reduced defense expenditures, hoping their nascent atomic energy program and air power would help shore up any weakening of their overall defense posture. Although the British realized it was beyond their capacity to match either the United States or the Soviet Union in nuclear power, they felt compelled to acquire atomic and thermonuclear weapons as a hedge against losing further political influence. As one authority on the subject noted, "the basic drive behind the British program was the need to have some sort of capability rather than a specific military concept."[57] Hence, the desire to develop an independent deterrent, which in the 1950s and early 1960s came to be organized around Royal Air Force (RAF) Bomber Command's V-force composed of Valiant, Vulcan, and Victor bombers.[58]

The rationale behind the global strategy paper was a combination of the British Chiefs' increased appreciation for the value of atomic weapons as a deterrent coupled with their firm belief that war with the Soviet Union was not necessarily imminent. Indeed, they thought the Cold War between East and West might last indefinitely. They therefore concluded that the most effective and economical defense posture the West could hope to achieve and maintain for any length of time would be one that relied on the retaliatory threat of long-range air power and atomic weapons.[59]

Combining the British findings with those of its own inquiry, the Eisenhower administration in October 1953 adopted a new statement of basic security policy (NSC 162/2) that called for

1. a strong military posture, with emphasis on the capability of inflicting massive retaliatory damage by offensive striking power,
2. U.S. and allied forces in readiness to move rapidly initially to counter aggression by Soviet bloc forces and to hold vital areas and lines of communication, and
3. a mobilization base and its protection against crippling damage, adequate to insure victory in the event of general war.[60]

General Dwight D. Eisenhower, on temporary leave as President of Columbia University, discusses global strategy with Secretary of Defense James Forrestal.

The Royal Air Force answer for an independent deterrent was organized around Royal Air Force Bomber Command's V-force of Valiant, Vulcan, and Victor bombers. Appearing here are the Vulcan (*right*) and the Victor (*below*).

The new JCS Chairman Admiral Arthur W. Radford publicly described the administration's defense policy as a "New Look" in national security, a choice of words Eisenhower later said he regretted because it seemed to imply "a far more radical change in the composition of our armed forces than was truly the case."[61] A more apt description was probably supplied by Secretary of State John Foster Dulles in January 1954 when he publicly referred to the administration's policy in terms that suggested a doctrine of "massive retaliation."[62] What Dulles had in mind for U.S. strategic forces was twofold. One aspect, to be sure, was to deter a Soviet attack on the United States; the other was to provide a shield against a Soviet conventional offensive against NATO Europe, a concept known as extended deterrence. In its effort to economize, the administration hoped to make up for what it lost in conventional numbers through increased reliance on strategic air power and battlefield nuclear weapons, two areas where the United States continued to enjoy substantial superiority over the Soviets. "This power," Eisenhower insisted, "is for our own defense and to deter aggression. We shall not be aggressors, but we and our allies have and will maintain a massive capability to strike back."[63]

Though it was apparent that the Eisenhower administration would give priority to offensive strategic forces, its commitment to the protection of the country's mobilization base pointed up the need for bolstering continental air defense capabilities as well. This led to a succession of special studies, many done under the supervision of the NSC's Net Evaluation Subcommittee (NESC), a special panel created to monitor Soviet capability to inflict damage on the United States. Programs initiated as a result of these actions in the 1950s were indeed substantial, including extension and completion of the defense early warning radar system (DEW line) authorized under President Truman; deployment of a nationwide air defense missile system and nuclear-armed air defense jet interceptors; construction of alternate command posts for relocation of government and military leaders in case of Soviet attack; and annual tests of the government's ability to relocate and continue functioning through an emergency.[64]

Despite a substantial investment, intelligence estimates in the early 1950s consistently found that improvements in U.S. air defenses failed to keep pace with gains in Soviet offensive capabilities. This reconfirmed what air advocates had long contended about the essentially unstoppable nature of offensive air power and underscored the threat of surprise attack in the nuclear era. In consequence, to gauge accurately the Soviet military threat and provide adequate American defenses against it, President Eisenhower encouraged, approved, and supported fundamental improvements in the technology for acquiring and using intelligence. All the while, and for the next several decades, offensive forces would remain at the center of U.S. strategic policy and defense doctrine.[65]

From its inception, the New Look seemed to confirm once and for all what many airmen had been saying for years about the importance of strategic bom-

Admiral Arthur W. Radford, Chairman of the Joint Chiefs of Staff (*right*), described the "New Look" of the administration's defense policy; John Foster Dulles, Eisenhower's Secretary of State (*below*), preferred to describe it as a doctrine of "massive retaliation."

bardment and the role it should play in national policy.[66] During the eight-year Eisenhower presidency, the Air Force received an annual average share of 47 percent of the defense budget compared with an average share of 31 percent during the Truman years. SAC's budget alone hovered around 18 percent, slightly less than the budget for the entire Department of the Army.[67] There were other incentives as well, perhaps because the New Look finally offered the Air Force the opportunity to demonstrate the full range of its capabilities. As Lt. Gen. Laurence S. Kuter, Commander of the Air University, put it, the role of air power was not only to fight wars if necessary, but "to influence the behavior of other nations by actions short of war in support of national policy."[68]

Because Air Force leaders tended to regard the New Look as essentially a confirmation of the superiority of air power (especially strategic air), there was no urgent reason to reexamine or seriously question its fundamental concepts and theories. As a result, doctrinal manuals issued to Air Force personnel in the 1950s generally described the existing state of the art, with emphasis on strategic bombardment as the most decisive use of air power yet developed.

For planning purposes, nuclear and conventional weapons were treated much alike. The lessons taught in the 1953 manual (and two succeeding volumes issued in 1954 and 1955) were those learned in World War II. This was to concentrate air power on three basic tasks: securing command of the air, a prerequisite to mounting effective strategic operations; attacking the enemy's heartland; and attacking peripheral targets when such attacks would contribute to the success of the offensive. By the time the 1959 version appeared, the impact of new technologies, such as guided missiles and earth-circling satellites, had become apparent, with "aerospace power" replacing "air power" to describe the Air Force's medium of operation.[69]

At the operational level in SAC, the major influence on doctrine was not any particular set of old or new ideas: it was the dominating personality and experience of SAC's commander from 1948 to 1957, General Curtis E. LeMay. An architect of the air campaigns in World War II, LeMay regarded strategic bombardment as the most decisive form of military power in the American arsenal. Tough, dedicated, and exacting, he worked tirelessly to turn SAC into "a real strong and efficient outfit."[70] What eventually made it so was not only LeMay's patience and determination, but also the shared outlook and experience of its personnel, especially those at the senior level. As long as he commanded SAC, LeMay insisted that it be staffed with the "Right People," meaning colleagues whose abilities he admired, whose word he trusted, and with whom he had served in World War II, leaders such as Thomas S. Power (who became his deputy and later succeeded him as commanding general of SAC), Walter "Cam" Sweeney, Emmett "Rosie" O'Donnell, Francis "Grizzy" Griswold, J. B. Montgomery, and Clarence S. "Bill" Irvine (who was his maintenance expert in the war and upon whom he relied initially as commander

Lt. Gen. Laurence S. Kuter, Commander of Air University, felt that one role of air power was to influence behavior of other nations by actions short of war in support of national policy.

of the 509th to give SAC an effective nuclear capability). "I started gathering the old boys in, fast as I could," LeMay recalled.[71] The result was a group of exceedingly able senior officers who not only knew one another well and worked comfortably together, but who also brought to bear a similar mindset on the problems they faced.

LeMay gave SAC a worldwide offensive reach, reinforcing that reach with the command's reputation for excellence in performing its mission. "In my opinion," LeMay once remarked, "SAC's deterrent influence on USSR aggressive intentions can only be maintained by an effective force in being, properly manned, equipped and trained, at the proper time period, and whose combat capability is universally recognized and unquestioned."[72] By 1956, with orders placed for turbojet tankers to air-refuel jet bombers, SAC was well on its way toward realizing LeMay's goal. Perhaps one of the most elite organization in the U.S. armed forces, SAC was virtually an air force within an air force, at that time consisting of more than 200,000 personnel and 55 active bases in the United States and overseas.[73]

The key in the 1950s to SAC's power and effectiveness as a deterrent was its formidable fleet of bombers, operated and maintained by dedicated personnel. Because of the rapid development of jet fighters after World War II, it was only a matter of time before SAC's piston engine B–29s, B–50s, and B–36s became obsolete. The first medium turbojet bomber, the six-engine B–47 Stratojet, with its distinctive sweptback wings and tail surfaces, entered service

An architect of the air campaigns in World War II, Curtis E. LeMay regarded strategic bombardment as the most decisive form of military power in the American arsenal. In 1948 he assumed command of the Strategic Air Command and in 1951 became a full general.

As long as he commanded SAC, LeMay insisted on a staff of the "Right People," such as General Emmett "Rosie" O'Donnell, later to command PACAF.

in October 1951, opening the way for SAC to phase out its B–29s and B–50s. All B–29s were gone by the end of 1954 as were all B–50s by June 1955, making the 1,300 B–47s then in service SAC's principal workhorses. Also, in 1955, the B–52 Stratofortress, an eight-engine jet, began replacing the problem-prone B–36, the last of which were phased out in 1959. Like the B–47, the B–52 employed a sweptback wing design, but it was capable of carrying substantially larger payloads over much longer distances. Early B–52 models (the A through D series) had an unrefueled range of more than 6,000 miles while carrying as many as four gravity-fall thermonuclear bombs. Later models, like the G and H series, would have an unrefueled range of approximately 10,000 miles and could carry as many as eight nuclear bombs; they also incorporated such refinements as Quail decoy missiles and nuclear-tipped Hound Dog air-to-surface cruise missiles for clearing a path through the Soviet Union's increasingly sophisticated air defense system. Moreover, when SAC eventually acquired more than 300 KC–135 jet tankers capable of refueling aircraft at 500 mph at altitudes above 35,000 feet, the B–52s could claim a truly global reach.[74]

Along with improved equipment and facilities, SAC gradually acquired access to better intelligence which allowed it to pinpoint Soviet targets and develop more effective countermeasures to Soviet air defenses. By the early 1950s, SAC reconnaissance planes were flying routine patrols along the Soviet Union's borders to gain information on Soviet air defense radar.[75] Occasionally, they intentionally overflew Soviet territory, apparently to test Soviet air defense reaction times or demonstrate their penetration capability. LeMay recalled that on one occasion in the mid-1950s, he was directed to send some fifty planes, unopposed, flying in clear daylight at 40,000 feet, over the Soviet port of Vladivostok.[76] For future reconnaissance purposes, the United States was also beginning to develop a space satellite program that would eventually revolutionize intelligence-gathering methods.[77] But, for purposes of surprise attack warning, by far the most significant breakthrough of the decade was the extraordinary U–2 high-altitude photoreconnaissance airplane that, equipped with new, sophisticated cameras and film, made its debut in 1955 and began flying mapping missions over the Soviet Union on July 4 the following year. Although developed and controlled by the Central Intelligence Agency, Air Force representatives played a significant role in the U–2 program from the beginning. They sat on the Ad Hoc Requirements Committee that decided the airplanes' missions and were thus in a position to obtain targeting information for SAC war-planning.[78]

With improved intelligence and more nuclear weapons, it became feasible for SAC to develop expanded target lists which took in a wider range of targets than just urban-industrial concentrations. Though gradual, this change was was already in progress by the outbreak of the Korean War. Looking ahead, LeMay began advising his superiors as early as May 1950 that SAC's mission would

The six-engine B–47 Stratojet (*above left*) entered service in 1951, opening the way for SAC to phase out its B–50s (*upper right*) as well as its B–29s (*lower right*) in the mid-1950s.

be radically different by mid-1954, the "year of maximum danger" cited in NSC 68 and most intelligence estimates as the period when the Soviets were expected to have sufficient nuclear weapons and long-range planes to threaten serious damage to the United States. If this was the case, LeMay argued, the first response to a Soviet attack on the United States should be directed against Soviet long-range air, with American attacks on Soviet economic centers following second.[79] The JCS arrived at a similar conclusion. In August 1950 they established three priority tasks for strategic bombing of the Soviet Union: the "blunting," or elimination, of the Soviet Union's air-atomic capabilities; the retardation of advancing Soviet ground forces; and the destruction of key war-supporting facilities in the Soviet urban-industrial complex, concentrating on petroleum production, electric power, and atomic energy plants. Later, the JCS gave each of these targeting categories a code name: BRAVO (for blunting), ROMEO (for retardation), and DELTA (for disruption-destruction of industry).[80]

To implement the JCS targeting scheme, SAC planners estimated that, within days of the outbreak of war, they would have to be able to mobilize at

By 1955, the eight-engine B–52 Stratofortress (*top* photo) began replacing the problem-prone B–36 (*center*) the last of which were phased out in 1959. Meanwhile, in 1955, the extraordinary U–2 high-altitude photo reconnaissance airplane, equipped with new, sophisticated cameras and film, made its debut and began mapping missions over the Soviet Union in the following year (*below*).

least 150 medium bombardment wings or their equivalent.[81] Since budgetary concerns were likely to preclude ever meeting such a goal, LeMay concluded it would be impractical to plan a bombardment campaign calling for sustained operations such as those conducted against Germany and Japan in World War II. Likewise, for strategic purposes he ruled out the use of conventional munitions (except for limited strategic bombing dictated by political considerations, as in Korea). He began thinking instead in terms of "a large atomic attack" that could inflict devastating blows against the Soviet Union outright, before the Soviets could respond in kind.[82] The major initial drawback to such a strategy was the continuing short supply of trained crews, properly equipped aircraft, and atomic bombs. As the assets available to him increased, so too did LeMay's confidence that the planned "Sunday punch" offensive, as it was sometimes called, could be carried out with swift, decisive, war-winning results.[83] Though the Air Force never officially clarified what its spokesmen meant when they talked about the Sunday punch, at least one well-connected observer implied strongly that it would take the form of a preemptive, first-strike attack.[84]

Although the details of SAC's war plans were a tightly guarded secret (and in any case could not be executed without approval of the President), spokesmen for the Air Force by the mid-1950s were beginning to convey the impression that military, or counterforce, targets now took priority, and that countervalue targeting against industry and civilian population centers was virtually a thing of the past. The intention may not have been to make the prospect of nuclear war more acceptable to a skeptical public, but the overall effect was yet another attempt to rebut criticism that such a conflict would be nothing more than the wanton slaughter of innocent noncombatants, a criticism that had dogged strategic bombing since the days of Douhet. In a February 1954 speech to the Chamber of Commerce of Galveston, Texas, Air Force Chief of Staff General Twining readily acknowledged that a "major attack using weapons of modern war, of course, will cause heavy loss of life." But, he added: "Machines and weapons, not people, are the principal targets to be destroyed.... It would be a moral blunder and a military blunder to concentrate our hopes for victory on the piling up of casualties when the opportunity now exists to concentrate with great effect on the enemy's weapons and weapons factories."[85]

Twining's remarks, though doubtless sincere, were nonetheless somewhat misleading. While counterforce targeting steadily gained favor as the 1950s wore on, not until the end of the decade did SAC acquire the technology and intelligence to make an effective counterforce strike appear feasible. Until then, U.S. targeting doctrine had tended to remain fluid; it emphasized the selection of aim points that would cause the maximum amount of damage to the Soviet military-industrial complex in the shortest possible time. As a rule, SAC planners preferred to concentrate on high-priority counterforce targets, but they had to allow also for the limited assets at their disposal. Thus, they strove to select as aiming points important civic, industrial, and governmental centers that if destroyed would yield an extra measure of effect. The goal was not only to incapacitate the Soviet Union's military machine in one swift blow, but at the same time, in keeping with time-honored doctrine, to destroy the enemy's will and ability to continue or resume hostilities.

By March 1954, SAC had identified 1,700 counterforce targets, 409 of which were airfields.[86] In analyzing SAC's "optimum mix" targeting scheme, the WSEG found that planned attacks against military targets such as troop concentrations, while of some immediate value, would not be "lasting or conclusive," whereas without destruction of the Soviet urban-industrial sector "Russia could support *immense* armed forces for at least two years of intensive warfare." Based on these findings, WSEG concluded that for "*some* level of atomic attack" to produce decisive results, SAC's best chance of success was to concentrate its bombardment "against the civic-political structure of the interior of Russia."[87] Of 134 major cities in the Soviet Union, SAC subsequently designated targets in 118 of them for attack. Reporting in February 1955 on the probable effects, WSEG estimated fatalities in the target areas to be in the range of 75 percent to 84 percent of the population.[88] Clearly, if a general war erupted, the suffering inflicted on civilians would be substantial, despite the counterforce orientation of SAC's war plans.

Emerging British doctrine in the 1950s also tended to give precedence to targets (in particular, air bases from which the Soviets could attack the United Kingdom). But with time, financial constraints and the Royal Navy's resistance to a large V-force compelled the RAF to rethink its basic strategy.[89] What eventually emerged was a bombing doctrine, fashioned after the one employed in World War II, that all but ruled out attacks against Soviet military installations because of the limited size and effectiveness of Britain's strategic arsenal. Not until July 1956, when the RAF took delivery of its first Vulcan medium jet bombers, could Britain claim to pose a serious threat to Soviet territory.[90] The following May, Britain tested its first thermonuclear device. Instead of a V-force of 240 front-line planes as originally contemplated, Bomber Command eventually had to settle for a force of around 170. And with a stockpile of only about 60 atomic bombs by 1958, British planners saw no choice but to program the bulk of their resources for attacks against Soviet cities.[91]

The 1950s also witnessed significant changes in Soviet strategic doctrine, which until then had been under the restrictive hold of Stalinist military science. Unlike the United States and Britain, the Soviet Union emerged from World War II with little knowledge of or experience in strategic bombardment. The Soviet air force that had grown up in the interwar period was mainly a tactical air force developed in accordance with Stalin's philosophy that all available power should be concentrated on annihilating the opposing side's military forces in the field. During the war, its chief job had been to support Soviet ground troops, with little attention paid to attacking urban-industrial concentrations or targets deep behind enemy lines. After the war, however, while publicly denigrating the military importance of nuclear weapons, Soviet Premier Josef Stalin established high-priority research and development programs for atomic energy and for equipping a long-range air force, organized in 1946. The earliest Soviet strategic bomber, the Tupolev Tu–4 Bull, which made its maiden appearance in 1948, was virtually a carbon copy of a U.S. B–29 that had accidentally fallen into Soviet hands during World War II. As time passed, the originality and sophistication of Soviet bomber design steadily improved.[92]

After Stalin's death in 1953, Soviet pronouncements on strategy and doctrine tended to acknowledge the contributions strategic bombing had made in World War II and the influence it might have on future wars, especially in conflicts involving the use of nuclear weapons and enemies "on other continents." It followed, as one senior Soviet strategist put it, that "long-range strategic weapons *above all* will be required," not only to annihilate the opponent's armed forces, still the number one priority, but also to destroy "targets deep in his territory" in order to "break up the organization of the country."[93] Among the new strategic bombers unveiled by the Soviets in the early 1950s were two that appeared capable of threatening the continental United States: the Tu–95 Bear, a lumbering four-engine turboprop with a range of 7,800 miles and a bomb capacity of 25,000 pounds, and the Myasishchev Mya–4 Bison, a four-engine jet (sometimes mistakenly compared in performance and capabilities with the U.S. B–52) having an estimated range of 7,000 miles and a payload of 10,000 pounds. By the mid-1950s, based on faulty estimates of production, some U.S. intelligence analysts (including those in the Air Force) projected a bomber gap in which the Soviets would lead the United States two-to-one at the end of the decade, a prediction that proved to be widely off the mark. In fact, only about 200 Mya–4s and Tu–95s were deployed, with many of the latter in later years dedicated to maritime reconnaissance.[94]

Instead of long-range heavy bombers, the Soviets built their strategic air force of the 1950s around medium-range planes. The reasons are still not wholly clear. Possibly it was because medium-range bombers were easier to design and manufacture, but the decision may have been doctrinally driven as well, a logical product of the Soviet military's traditional emphasis on close air support and interdiction. From the Soviet perspective, the most urgent time-sensitive targets in the early and mid-1950s were military bases associated with American strategic air power in NATO Europe and elsewhere around the Soviet Union's borders, targets that did not require delivery systems with intercontinental capabilities. By 1955, U.S. intelligence sources counted approximately 1,300 Soviet medium-range bombers, including some 1,100 obsolete Bulls and about 200 Tu–16 Badgers, a new twin-engine turbojet estimated to have a range of 3,000 miles with a 6,600-lb payload.[95] Subsequently, during the latter half of the 1950s and with the emergence of the ballistic missile age, the Soviet

Among the new strategic bombers unveiled by the Soviets in the early 1950s, capable of threatening the continental United States, were the Mya–4 Bison (*top*) and the Tu–95 Bear (*bottom*).

In 1955, U.S. intelligence sources counted in the Soviet inventory about 200 Tu–16 Badgers, twin-engine trubojet medium bombers.

Union followed a similar pattern. While the Soviets pursued a much publicized (but initially limited) intercontinental ballistic missile (ICBM) program, they accorded priority to the deployment of a large arsenal of more than 600 SS–3, SS–4, and SS–5 medium- and intermediate-range ballistic missiles capable of striking at distances of 700 to 2,500 miles, thus bringing within reach targets such as NATO tactical airfields and nuclear storage bunkers in West Germany and SAC B–47 bases in Great Britain and North Africa.[96]

The emergence of a Soviet strategic air force and, shortly, a ballistic missile force meant that sooner or later the United States would be targeted for direct attack. The most immediate concern, however, was the possible vulnerability of SAC's overseas bases. Studies in the early 1950s by civilian analysts at the RAND Corporation, an Air Force–sponsored think tank, indicated that just a small Soviet strategic air force could inflict heavy losses on U.S. bombers based abroad.[97] As a result, the Air Force, acting on RAND's findings and studies of its own, began in 1955 to recall many of its bombers to dispersed bases in the continental United States and to plan on using overseas bases mainly for presrike and poststrike refueling and arming. The advantages of overseas operations were thus retained, but with a substantially reduced risk to the bomber fleet.[98]

Nonetheless, the threat of another Pearl Harbor–type disaster, this time in the continental United States itself, weighed heavily on American political leaders. It prompted President Eisenhower in March 1954 to appoint a special surprise-attack study panel headed by James R. Killian, president of the Massachusetts Institute of Technology.[99] The committee's top secret report, presented along with a briefing to the NSC in February 1955, depicted a general situation that was bound to worsen unless the United States took vigorous steps over the next decade or so to shore up its strategic posture. "For the first time in history," the report warned, "a striking force could have such power that the first battle could be the final battle, the first punch a knockout." To avoid being the victim of such a blow, the report urged across-the-board improvements in U.S. strategic capabilities, including better early warning and air defense systems, more widely dispersed basing and improved alert practices for SAC's bombers, and, most important of all in the committee's view, increased emphasis on high-technology intelligence gathering and on weapons such as ICBMs and land- and sea-based intermediate-range ballistic missiles (IRBMs) that could respond swiftly and effectively in an emergency with minimal prior warning.[100]

Although the Killian report made no mention of any need for fundamental changes or alterations in basic strategy, its recommended course of action, which stressed the role and importance of ballistic missiles, left no doubt that at some point the Air Force would have to find a more effective operational doctrine, not only for the use of missiles but also for reducing the vulnerability of SAC's bombers to a Soviet missile attack. Like the other services, the Air Force had been developing and testing a variety of guided missiles since World War II, but it had never accorded any of its programs (i.e., long-range ground-to-ground and air-to-ground missiles) the same priority as it had to improving the planes in SAC's bomber fleet.[101] Though the reasons were mainly budgetary and technical, airmen also had a good deal of apprehension over the role that missiles would play. "We were not sure," recalled one senior SAC planner, "whether missiles were going to replace airplanes, or supplement them. . . . I think we wanted to be shown the reliability and the performance capability of the new weapon. Having said that, SAC indeed pushed hard for ICBMs, as well as for IRBMs."[102]

Others, seeing numerous possibilities, wanted to move even faster. Trevor Gardner, Special Assistant (later Assistant Secretary) for Research and Development to Secretary of the Air Force Harold Talbott, was among those who at the outset of the Eisenhower administration saw the need and opportunity for an accelerated missile program. Despite gaps in the intelligence reporting of the day, Gardner suspected that the Soviet Union had pulled ahead of the United States in long-range missile development and would remain there unless the Air Force initiated a stepped-up program committed to the operational deployment of an ICBM no later than 1960.[103] The discovery that high-yield thermonuclear explosives (first tested by the United States in November 1952 and by the Soviet Union in August 1953) could be reduced in size and weight and carried on a missile, as initially reported by the Teapot Committee in February 1954, was further incentive to press ahead quickly, lest the Soviets achieve a similar technological breakthrough.[104]

In December 1955, following an NSC review of the Killian report's recommendations and a Defense Department study on implementing them, President Eisenhower ordered immediate acceleration of research and development work on the ICBM and IRBM programs, giving both equal priority. In doing so, however, he expressed doubts about the military value of these weapons and explained the incentive for his action as "the enormous psychological and political significance of ballistic missiles."[105] A year later, while reviewing the progress of the programs, Eisenhower said he thought "a cold-blooded look at the big missile program was needed" and insisted he wanted only efforts showing significant promise to proceed.[106] On another occasion, he questioned whether much more than a demonstration capability was needed, and he apparently voiced no dissent when Secretary of Defense Charles E. Wilson estimated that, together with other offensive forces, "one hundred and fifty well-targeted missiles might be enough." "We must remember," Eisenhower said, "that we have a great number of bombardment aircraft programmed, and great numbers of tankers that are now being built, and we must consider how to use them."[107]

Eisenhower's major concern that ballistic missiles would become the focal point of intense interservice rivalry and competition because of their high public visibility, potential importance, and numerous applications doubtless contributed to his apprehension. Under the original assignment of service tasks approved by Secretary Wilson in November 1955, the Air Force was to exercise responsibility over long-range land-based ballistic missiles: two first-generation liquid-propellant ICBMs (the Atlas and a backup, the Titan), and an IRBM (the Thor). Meanwhile the Navy and the Army were to work together on a fourth missile, a 1,500-mile liquid-propellant IRBM named Jupiter that could be launched either from land or from a ship at sea.[108] Quarreling over the allocation of resources and competition for production facilities erupted almost immediately between the Air Force and the Army. Additionally, within a year the Navy withdrew from the Jupiter project to concentrate on a new missile, the solid-propellant Polaris, less volatile than the liquid-propellant Jupiter and potentially almost invulnerable because it could be carried aboard submarines and fired from underwater.[109] Apart from the added cost, duplication, and potential waste of so many programs, it was clear that each service had (or was in the process of acquiring) its own strategic arsenal, contrary to the 1948 Key West agreement on service roles and missions that acknowledged the Air Force's primary responsibility for strategic air warfare. To tame the services' free-for-all behavior, Secretary Wilson in November 1956 imposed a range limit of 200 miles on all future Army land-based missiles, and the the Air Force subsequently became the owner of the Jupiter IRBM. No comparable range limitations, however, were imposed on Navy sea-launched ballistic missiles.[110]

In the mid-1950s, Defense Secretary Charles E. Wilson (*right*) outlined for President Eisenhower the missile program. In February 1959, the Titan I ICBM was launched at Cape Canaveral, Florida (*below*).

In addition to a growing list of competitors for the job of strategic bombardment, SAC found itself having to contend with a vocal gallery of critics who wanted to reduce reliance on strategic weapons in order to devote more resources to conventional forces. The argument, raised by some, including General Maxwell D. Taylor, Army Chief of Staff from 1955 to 1959, held that the country was overinvesting in strategic forces that, while effective as a deterrent to general war, were ineffective in coping with the more likely possibility of limited, localized conflicts. As Taylor saw it, what was needed was less emphasis on strategic forces and more military forces capable of dealing with contingencies at a variety of threshholds, from small-scale conventional conflicts to all-out nuclear war, a concept of force-structuring that came to be known as flexible response.[111]

A different line of argument to this issue was advanced by the American proponents of a relatively new concept called finite or minimum deterrence, which boasted the advantages of a comparatively small, secure retaliatory force described earlier by President Eisenhower. Modeling their strategy after RAF Bomber Command's countercities bombing doctrine, minimum deterrence advocates called for tying strategic forces to the destruction of soft countervalue targets (i.e., population and industrial centers) rather than to the destruction of opposing counterforce military targets. For such a purpose the Polaris ballistic missile–firing submarine, a weapons system then still on the drawing board, seemed ideal. Not surprisingly, some of the earliest and most innovative work on minimum deterrence came from the Navy whose Warfare Analysis Group at the Massachusetts Institute of Technology issued the first in-depth report on the subject in 1957.[112] As far as the Air Force was concerned, minimum deterrence failed to persuade, because "it does not confront an enemy with a credible threat of defeat."[113]

Still, as long as SAC assets appeared vulnerable, minimum deterrence and similar concepts purporting to offer cheaper yet more effective countermeasures to rising Soviet strategic power continued to attract followers. Intelligence reports of a bomber gap in 1955 proved to be greatly exaggerated. This miscalculation, however, set the stage for an even larger controversy over a possible "missile gap" that began in late 1957 when the Soviet Union gained precedence over the United States by demonstrating an ICBM capability and then launching the first earth-orbiting space satellite, *Sputnik I*. From all outward appearances, the United States was lagging while the Soviet Union appeared to be surging ahead and capturing a lead in new technologies that might render SAC obsolete and shift the strategic balance permanently in Moscow's favor.

Among those studying the increasingly complex problems of nuclear strategy, none seemed more knowledgeable than Albert J. Wohlstetter, a civilian analyst who had been principal contributor to many of the earlier RAND studies of SAC vulnerability. In 1956 Wohlstetter and a colleague, Fred

General Maxwell D. Taylor, Army Chief of Staff from 1955 to 1959, was appointed by President Kennedy in 1962 as Chairman of the Joint Chiefs of Staff, the first JCS chairman ever to come out of retirement.

Hoffman, produced a report with map exercises showing how a combination of Soviet long-range air and ICBMs could effectively cripple SAC in a series of carefully planned and coordinated raids. "Our SAC," the report found, "presents soft, relatively few, relatively undefended targets." Among possible remedies, Wohlstetter especially favored improvements in early warning and response time and the construction of hardened, protective shelters for SAC's bombers. The Air Force, however, doubted the technological feasibility of his protective shelter idea and was therefore disinclined to pursue it.[114] The federal Civil Defense Administration projected a different response to the threat of Soviet attack and came up with a plan in December 1956 that called for $32 billion in civil defense improvements.[115] Before making a commitment to so expensive an enterprise, Eisenhower wanted the matter studied further; he called on H. Rowan Gaither, Jr., a West Coast attorney who also sat on the boards of the Ford Foundation and the RAND Corporation, to chair an inquiry into the "various active and passive measures to protect the civil population in case of nuclear attack and its aftermath."[116] Shortly after accepting the assignment, Gaither discussed the Civil Defense Administration study with Wohlstetter, whose position was that SAC's vulnerability, rather than civil defense, should be the central issue. Convinced that the threats facing the country went beyond the realm of civil defense, Gaither and his associates began delving into virtually any matter affecting the country's strategic posture, although they apparently did not have access to imagery of the Soviet Union obtained in the super-secret U–2 program.[117]

Early in November 1957 the Gaither Committee tendered its report, *Deterrence and Survival in the Nuclear Age*, to President Eisenhower. As the title suggests, the committee's main concern was not civil defense; it was the state of the U.S.–Soviet nuclear balance and the danger posed by apparent recent progress in Soviet strategic capabilities, especially in ICBM testing. "By 1959," the report warned, "the USSR may be able to launch an attack with its ICBMs carrying megaton warheads, against which SAC will be almost completely vulnerable under present programs." To cope with this and other threats that the Soviet strategic systems posed, the committee recommended a $44 billion program spread over five years — $19 billion to expand and upgrade offensive capabilities and $25 billion for civil defense and related programs. In developing specific recommendations, the committee drew heavily on Wohlstetter's recent suggestions. As part of the offensive buildup, it endorsed such measures as improved readiness and alert procedures, blast shelters providing protection of 100–200 pounds per square inch for planes, increased and accelerated ICBM and IRBM deployment, and an expanded early warning system with capability to detect incoming ICBMs. The report also recommended that SAC bases be provided with an active antimissile defense against ICBMs, and it urged a start be made with off-the-shelf technology. For civil defense, the report called for increased research and development on radars, antimissile defenses, and antisubmarine warfare and for the construction of civilian fallout shelters (but no civilian blast shelters) nationwide.[118] Several committee members told Eisenhower privately, however, if forced to make a choice, they believed it more cost-effective to stress improvements in offensive capabilities rather than defensive measures.[119]

After reviewing the Gaither report's proposals, the JCS acknowledged that weaknesses in offensive capabilities did exist, but they pointed out that virtually every program the committee recommended was either under active study or had been approved and was under way. The major exception was Wohlstetter's protective shelter plan, which the JCS did not favor in view of the Air Force base dispersal program and other measures. Otherwise, timing and the scope of programs tended to be the principal points of contention. The Gaither Committee wanted to accelerate the missile deployment schedule while the administration (having access to U–2 intelligence) preferred a slower pace, largely to avoid heavy immediate outlays.[120]

In fact, by the end of the decade, SAC's overall offensive forces were markedly improved. The accelerated missile program in 1959 deployed the first Atlas D ICBMs. Other improvements, such as equipping an increasing number of B–52 bombers with decoy and attack missiles (Quails and Hound Dogs) and low-level penetration aids, were designed to help improve SAC's ability to overcome Soviet air defenses.[121] By 1960 SAC had redeployed its aircraft from most overseas bases, dispersed them to some sixty-six bases in and around the continental United States and Canada, and brought one-third of its bombers and tankers to fifteen-minute ground alert.[122] As a further precaution, SAC began testing the feasibility of an airborne alert program in 1958, but because of its high cost and extensive demands on personnel and equipment, it remained for the most part an orientation and training program for emergency use only.[123]

Combat aircraft were now being fitted with a new high-frequency radio system called Short Order (later known as Giant Talk) that provided a means of exercising positive control over airborne B–47s and B–52s. The intent of such positive control procedures was to allow SAC to launch its bombers and send them to designated points outside Soviet territory where they would loiter and then return to base unless ordered to do otherwise.[124] The system worked well when tested under ideal peacetime conditions, but in actual emergency it was better suited for a preemptive strike because transmission signals in post-strike situations were susceptible to high-frequency blackout and enemy jamming and because SAC's transmitters were not built to survive an enemy attack or withstand the electromagnetic pulse effects of nearby nuclear blasts.[125]

Although Air Force leaders never openly acknowledged their intentions, many considered SAC's survival and effectiveness to be dependent upon its ability to anticipate and deny the Soviets the opportunity of launching a knockout first blow. Preemption, as far as LeMay was concerned, was always a live option, national policies to the contrary notwithstanding. "I want to make it clear," he was quoted as saying at a 1954 briefing, "that I am not advocating a preventive war; however, I believe that if the U.S. is pushed in the corner far enough we would not hesitate to strike first."[126] Testifying before Congress in 1959, Air Force Chief of Staff General Thomas D. White suggested that national leaders adopt a policy of launch on warning; that is, upon receipt of "tactical and strategic warning" of an impending Soviet attack, SAC's "first priority" would be "to destroy the enemy's capability to destroy us."[127] To what extent Eisenhower shared these views is unclear. Early in his presidency he seemed to see no realistic alternative but to plan for a preemptive launch on warning that allowed for aircraft to be recalled. At one point, while meeting with his military advisers in 1954 he "indicated his intention to launch a strategic air force immediately in case of alert of actual attack."[128] The United States, he said on another occasion in 1957, should "not allow the enemy to strike the first blow."[129] For this reason, he wanted as much advance warning as possible of a Soviet attack in order to optimize the effectiveness of U.S. capabilities. Having already invested heavily in bombers and air refueling tankers, he was not about to discard them until a more viable alternative, presumably secure second-strike retaliatory forces consisting mostly of protected land-based ICBMs and sea-based IRBMs, was perfected. That development, even by the late 1950s, still appeared a way off. Until then, Eisenhower was apparently content to accept the risk that preemption might become necessary, though this was not his only or his preferred option.[130]

Air Force Chief of Staff Thomas D. White, shown with Vice President Richard M. Nixon and Lockheed designer Clarence L. "Kelly" Johnson, suggested that national leaders adopt a launch on warning strategy in case of impending attack by the Soviets.

Along with improving its capacity to respond, SAC was also intent on establishing closer coordination and control over strategic operations, including those contemplated by the British. In November 1957, a delegation from SAC visited Headquarters Bomber Command at High Wycombe to discuss, as Air Vice Marshal Stewart Menaul later described it, "joint operational planning, targeting, and other problems of common interest, with the object of coordinating nuclear strike plans to their mutual advantage." As a result of this and subsequent discussions, the two commands developed a regular exchange of information on a wide range of matters relating to training, tactics, and equip ment for nuclear and conventional bombing that lasted until the abolishment of Bomber Command in 1968. Although the British retained operational control over their forces and the final decision on the selection of their targets, coordination was such, according to Menaul, that "each made the maximum use of the other's knowledge and obtained the maximum coverage of, and concen tration on, priority targets." For the British, this presumably meant continued emphasis on countercity bombing, which worked to SAC's advantage in that it freed additional American assets for counterforce missions.[131]

From SAC's perspective, a more urgent problem was the growing threat to its control over strategic warfare posed by the Navy's Polaris ballistic missile submarine program. In fact, the Air Force thought that in time it would have an antidote to Polaris in short-range air-to-surface attack missiles (SRAMs). These would be loaded aboard a new high-performance Mach 3 bomber, the B–70 Valkyrie, planned as a replacement for the B–52.[132] The likelihood that a substantial number of either the B–70 or a suitable SRAM would be available in the near future was small, certainly not before the first Polaris boat put to sea in 1960. In the opinion of the SAC commander-in-chief (CINCSAC) in Omaha, Nebraska, the only other solution was to establish closer coordination between SAC's forces and all other systems with strategic potential, and with Polaris in particular. In March 1959, CINCSAC General Power raised the issue with the JCS, urging them to support a proposal he wanted to place before the Secretary of Defense "to assign control of the Polaris weapon system to the Strategic Air

The B–70, a high-performance Mach-3 bomber, was planned as a replacement for the B–52. Shown here is the Air Force's first flight of the aircraft in September 1964.

Command in view of its strategic capabilities."[133] The arrangement eventually recommended by the Air Staff in Washington, D.C., was a single United States Strategic Command composed of two subordinate commands, SAC and a new organization called the Strategic Naval Command, that would exercise operational responsibility over the Polaris submarines. Though the latter command would still be part of the Navy, its mission in effect would be to augment and support SAC operations, since the Air Force fully intended to exercise the dominant voice in the overall command structure.[134] Outraged Navy leaders responded by pressing for reductions in the size of SAC and relying more on Polaris for retaliatory protection.[135] Regardless of how vigorously the Navy protested, closer coordination over the increasing number and variety of strategic systems in the U.S. arsenal was clearly needed.

Supported by President Eisenhower's Scientific Advisor George Kistiakowsky, in August 1960 Secretary of Defense Thomas S. Gates, Jr., ordered creation of the Joint Strategic Target Planning Staff (JSTPS) to be located at SAC headquarters near Omaha, Nebraska, where it would have ready acccess to SAC's computers and other essential support.[136] A former Secretary of the Navy, Gates doubted the need for a strategic command; nevertheless, along with other administration officials, he concurred that more effective command and control over strategic forces should be instituted, with an integrated target list keyed to a single operational plan to accomplish this goal. His solution was to consolidate target selection and planning under one organization headed by an Air Force officer with a Navy deputy and having representatives from each major command with nuclear weapons that reported directly to the JCS in Washington.[137]

Approved in December 1960, the first single integrated operational plan (SIOP) for 1961 was virtually a carbon copy of SAC philosophy and operational doctrine. Designed as retaliation to Soviet aggression *or* as a preemptive strike, the plan called for a massive attack at the outset of hostilities. Targeting guidance used in preparing the plan came from an earlier study directed by Lt. Gen. Thomas Hickey, U.S. Army, for the NESC. The Hickey report (NESC 2009) strongly recommended continued adherence to SAC's "optimum mix" philosophy. Of the 1,060 targets, or designated ground zeros, eventually included in this version of the plan, nearly 80 percent were classed "military targets," though the label was somewhat misleading because a substantial number were collocated with urban-industrial centers, in keeping with SAC's practice of trying to maximize the effects of an assault. According to JCS Chairman General Lyman L. Lemnitzer, U.S. Army, civilian casualties would still be "millions in number" due to nuclear fallout and other collateral effects of the bombing even though the brunt of the attack would be against military installations.[138] Herbert F. York, Eisenhower's Deputy Director of Defense for Research and Development and a physicist whose involvement in defense nuclear matters

In 1960, Secretary of Defense Thomas S. Gates, Jr., established a more effective command and control over strategic forces with an integrated target list keyed to a single operational plan.

dated from the Manhattan Project, recalled his first SIOP briefing at SAC somewhat more vividly. "Toward the end," he said, "it seemed that the purpose was simply to strip-mine much of the USSR."[139]

With the completion and adoption of the first SIOP, U.S. strategic doctrine entered a new era of more closely coordinated planning. Massive retaliation may still have been the cornerstone of deterrence theory, but it was gradually giving way to new ways of thinking about the use of weapons, the selection of targets, and the coordination of attacks. "What we would actually do depends on circumstances," Secretary Gates insisted, "but we are adjusting our power to a counterforce theory; or a mixture of a counterforce theory plus attacks on industrial centers and things of that character. We are not basing our requirements on just bombing Russia for retaliation purposes."[140]

The McNamara Era: From Counterforce to Assured Destruction

At the end of the Eisenhower administration, claims of a missile gap notwithstanding, the United States had amassed an unprecedented degree of strategic power, most of it concentrated in SAC. For the most part, SAC's wartime effectiveness hinged on its ability to preempt rather than survive or escape a Soviet first strike. This reflected SAC's continuing reliance on relatively vulnerable manned bombers. With improved warning and alert procedures, a growing missile force, and a new long-range supersonic bomber (the B–70) in the advanced engineering stage, Air Force leaders were increasingly confident that eventually SAC could prevail in almost any circumstance, even after absorbing a surprise attack.[141] Essentially, what had become the basis for U.S. strategic doctrine, as reflected in the SIOP, was the Air Force concept of a war-fighting, war-winning deterrent, designed, as LeMay put it, "to provide a decisive counterforce potential" in the 1960s and beyond.[142]

For the new administration taking office in 1961, however, the prevailing view of what constituted reliable and effective deterrence seemed exceedingly rigid and narrow. As a senator and presidential candidate, John F. Kennedy had been highly critical of the Eisenhower defense program and its reliance on massive retaliation. Kennedy also claimed, although Eisenhower knew otherwise from U–2 photographs, that the United States was lagging behind the Soviet Union in ICBMs and that this missile gap was seriously endangering national security. In Kennedy's view, the need was urgent not only for a strengthened defense effort but also for a broader range of military options, including greater flexibility in the use of nuclear power and improved capabilities for conventional warfare. "We have been driving ourselves into a corner," he contended, "where the only choice is all or nothing at all, world devastation or submission — a choice that necessarily causes us to hesitate on the brink and leaves the initiative in the hands of our enemies."[143]

The task of redeeming Kennedy's campaign pledge fell largely to his Secretary of Defense, Robert S. McNamara, formerly a president of the Ford Motor Company and an AAF statistical control officer in World War II. Confident of his ability to tackle almost any problem, McNamara came to exercise a degree of personal influence over the Pentagon and its policies unprecedented for any Secretary of Defense, before or since. He surrounded himself with a generally able and experienced but somewhat controversial body of likeminded advisers, many drawn from RAND, the academic community, and the Truman administration. To be his White House–based military adviser, Kennedy named General Maxwell D. Taylor, a leading proponent of the flexible-response doctrine and Army Chief of Staff in the Eisenhower administration. In 1962 Kennedy recalled Taylor to active duty and appointed him JCS Chairman. To McNamara and most members of his staff, the first order of business was to get away from the massive retaliation mentality of the 1950s. Translated into administration policy, this meant more emphasis on conventional capabilities and less immediate reliance on the threat of wholesale strategic action in a crisis.

Once in office, McNamara launched a full-scale review of U.S. defense policy to determine how strategic plans and doctrine could be adjusted to reduce dependence on nuclear striking power. As members of the outgoing Eisenhower administration had recommended, he paid special attention to reviewing and refining the SIOP.[144] In 1960 the Air Force had conducted a similar exercise, looking at the possible development of a more controlled counterforce targeting that would avoid attacks on enemy cities, a subject of intermittent, low-level study at Air Force headquarters and at RAND since the mid-1950s. By 1960 proponents of controlled counterforce led by Brig. Gen. Noel F. Parrish of the U.S. Air Force and William Kaufmann of RAND had attracted the attention of Air Force Chief of Staff General White. Impressed by their preliminary findings, White encouraged them to continue with their studies. But important segments of the Air Force, SAC especially, "did not want to be pushed into a change which they thought would weaken deterrence and perhaps make a thermonuclear war impossibly difficult to fight."[145]

For McNamara, the initial appeal of the no-cities strategy was its discrimination and control in the selection of targets, which accorded with the President's desire for greater options in the possible use of force. Shortly after assuming his duties as Secretary of Defense, McNamara asked the JCS to develop a "doctrine" that would permit a controlled, graduated response and negotiating pauses following a nuclear exchange. The JCS reply, submitted in April 1961, doubted whether the requisite capabilities for carrying out such actions existed, and it warned that any attempt under current conditions to implement a doctrine of that kind, or to declare an intent to do so, would be premature and likely weaken the U.S. deterrent posture. The JCS Chairman, General Lemnitzer, acknowledged that his colleagues were split on the matter, and he urged McNamara not to pursue it.[146]

Realizing he was unlikely to receive full cooperation and support from the services, McNamara reassigned the project to a group of civilians working under Alain Enthoven, a leading advocate of systems analysis in problem solving. By the end of the summer, Enthoven and members of his staff had largely completed their task, which by now had come to focus on the preparation of new strategic guidance for a revision of the SIOP. It placed greater emphasis on retaliation, as opposed to preemption, and on the selection of second-strike counterforce targets as part of a series of graduated options to add flexibility not only in the choice of aim points and timing of attacks, but also in the types of warheads and bombs that would be used. Instead of one massive attack as contemplated in the current SIOP, the new plan could be executed in stages.[147] As McNamara explained to the President, the rationale for such an approach was twofold:

> . . . first, to strike back against Soviet bomber bases, missile sites, and other installations associated with long-range nuclear forces, in order to reduce Soviet power and limit the damage that can be done to us by vulnerable Soviet follow-on forces, while, second, holding in protracted reserve forces capable of destroying the Soviet urban society, if necessary, in a controlled and deliberate way.[148]

Service responses to the proposed new plan were apparently mixed, though the dominant view, especially in the Air Force, seems to have been one of apprehension over whether such a complex targeting scheme was technologically and operationally feasible. Some, including the CINCSAC, General Power, an ardent proponent of the counterforce philosophy, deemed it foolhardy to abandon the option of a preemptive strike or expect that SAC could mount a retaliatory attack able to knock out the Soviet Union's entire remaining strategic force.[149] Nonetheless, President Kennedy reportedly liked the plan, and the JCS adopted and approved it. In any case, in January 1962 SIOP–63 became official targeting policy.[150]

With a new military strategy officially approved, though not yet fully translated into operational plans, McNamara set out to convince others of its virtues. On May 5, 1962, he described it in detail to a secret session of the North Atlantic Council in Athens, Greece. A month later in Ann Arbor, Michigan, he expounded publicly:

> The U.S. has come to the conclusion that to the extent feasible, basic military strategy in a possible general nuclear war should be approached in much the same way that more conventional military operations have been regarded in the past. That is to say, principal military objectives, in the event of a nuclear war stemming from a major attack on the Alliance, should be the destruction of the enemy's military forces, not his civilian population.
>
> In other words, we are giving a possible opponent the strongest imaginable incentive to refrain from striking our own cities.[151]

General Curtis E. LeMay, as Air Force Chief of Staff (*center*), and his successor as CINCSAC, General Thomas S. Power, greet President John F. Kennedy upon his arrival at SAC.

Having settled on a new strategic doctrine, McNamara turned his attention to developing a suitable military posture. Initially, this involved the stepping-up of ICBM procurement and, where practicable, greatly increasing the ground and airborne alert status of SAC's B–52s. To pay for these improvements, he ordered reductions in what he considered less promising aspects of the Eisenhower defense program: accelerate the phase-out of the B–47 fleet, suspend the aircraft nuclear propulsion program intended for the development of a nuclear-powered bomber, and curtail work on the high-altitude supersonic B–70 bomber. Looking ahead, he and his staff of analysts concluded it was more sensible to downplay the role of strategic bombers because of their vulnerability and to develop a more survivable, second-strike retaliatory capability built around relatively invulnerable weapons systems — Polaris fleet ballistic missile submarines and two second-generation Air Force ICBMs housed in hardened underground silos. The first of these second-generation ICBMs was Titan II, a large rocket that burned liquid storable propellants, capable of lifting a four-ton warhead; the second was the solid-propellant Minuteman I, also a single-warhead missile carrying a thermonuclear payload.[152]

As the U.S. doctrine of massive retaliation changed during the Kennedy administration to embrace flexible response and assured destruction, it directly affected force levels. SIOP–63, however, contained options for tactics and weapons that were both numerous and broad. It was also the document of record for sizing future strategic requirements. Air Force planners certainly viewed the new SIOP in this light when they testified before Congress early in 1962 in connection with presentation of the administration's defense budget for fiscal year 1963. Air Force leaders made clear that they considered the cutbacks in strategic weapons systems of the previous year injurious and the five-year force projections in the budget inadequate if the service was to meet its strategic obligations and carry out its assigned tasks. "I worry about the trend as established by this year's budget," General LeMay, Air Force Chief of Staff, told one Senate committee. "I do not think you can maintain superiority in this field with that sort of a program." Along with other Air Force spokesmen, LeMay indicated that he would make every effort to secure increases in next year's budget, beginning with a request for restoration of production funds for the RS–70, a reconnaissance/strike version of the B–70.[153]

Behind LeMay's objections was the belief, well-established throughout the Air Force, that as the nation's principal strategic strike force, it had to have the forces and weapons capable of exercising the "power of decision" over the outcome of a general nuclear war. For force-structuring purposes, this meant a blend of manned and unmanned systems fitted to the counterforce concept of operations that SAC had developed and refined over the previous decade. While estimates of force requirements were subject to constant revision, they generally adhered to projections developed in the late 1950s that called for strategic forces capable of attacking 8,400 enemy targets with nuclear weapons by 1963 and an estimated 10,400 targets by the late 1960s. Accordingly, the Air Force calculated that by the end of fiscal year 1970 it would need 3,000 Minuteman, 110 Atlas, and 70 Titan missiles, and 840 B–70 bombers with appropriate tanker support along with heavy bombers of a more advanced design (i.e., nuclear-powered) to continue assuring SAC's set goal of a 90 percent destruction of targets selected in the Sino-Soviet Bloc.[154] Although SAC welcomed the introduction of missiles into its weapons inventory, its leaders still did not believe they possessed sufficient flexibility or accuracy to replace or duplicate the counterforce mission assigned to the manned strategic bomber, the only weapon, General Power declared, that would allow SAC to retain "a credible war-winning capability."[155]

By the fall of 1961, new intelligence derived from the CORONA satellite program confirmed that reports of a missile gap had been exaggerated. Air Force estimates in particular, some crediting the Soviets with an ICBM force of 200 or more missiles by 1961, had proved inflated. What the Soviets actually had, as it turned out, was an operational force of around thirty-five SS–6 ICBMs and possibly a few still experimental SS–7s, all potent but rather primitive and cumbersome rockets, erected above ground, that took hours to arm and fuel for launching.[156]

In confronting these new realities, McNamara straddled two schools, not wanting to ignore the possibility of future large increases in the Soviet strategic force, but not eager to invest any more than he had to in strengthening U.S. capabilities. In presenting the administration's fiscal year 1964 defense budget, he rejected the higher force level recommendations of the Air Force in favor of smaller (but still substantial) increases that accorded with his personal view, revised from a year earlier, that first priority should be "to provide the United States with a secure, protected retaliatory force able to survive any attack within enemy capabilities and capable of striking back and destroying Soviet urban society." "It has become clear to me," he confided to President Kennedy,

> that the Air Force proposals, both for the RS–70 and for the rest of their Strategic Retaliatory Forces are based on the objective of achieving a first-strike capability. What is at issue here is whether our forces should be augmented beyond what I am recommending in an attempt to achieve a capability to start a thermonuclear war in which the resulting damage to ourselves and our Allies could be considered acceptable on some reasonable definition of the term.[157]

From this point on, McNamara shunned controlled counterforce as a strategic objective, though he continued to retain a substantial number of counterforce options in the SIOP, presumably as damage-limiting insurance should deterrence fail. According to McGeorge Bundy, Kennedy's national security adviser, McNamara concluded "that it would be much easier to control strategic procurement if he [McNamara] did not at the same time challenge SAC's targeting doctrines."[158] Even so, McNamara categorically rejected the notion that the United States would ever initiate a strategic nuclear exchange, implying that any counterforce potential the United States might have would be for second-strike retaliatory purposes only. "Because we have a sure second-strike capability," he insisted, "there is no pressure on us whatsoever to preempt. . . . One point I was making in the Ann Arbor speech is that our second-strike capability is so sure that there would be no rational basis on which to launch a preemptive strike."[159]

A number of considerations doubtless played a part in his thinking, but it seems clear that the crucial factor in McNamara's decision to abandon counterforce was his dawning realization that the costs would be enormous, entailing ever-increasing new expenditures. As more weapons were made available, still more targets could be added to the SIOP, which in turn would require more weapons, not to mention more active and passive American defenses. Although Kennedy and McNamara both supported a more vigorous civil defense program, it never caught on either with a Congress that was lukewarm toward the idea or with an indifferent American public.[160] Moreover, as Henry S. Rowen pointed out, McNamara and his advisers came to believe that the task of preplanning nuclear options was hopelessly difficult, and that the more effort they expended on developing nuclear capabilities, the less likely it would be that they would ever succeed in achieving a conventional alternative.[161]

If one were to single out the turning point in McNamara's thinking, it was probably the Cuban missile crisis of October 1962. Despite what he later estimated as an American advantage of sixteen to one in strategic nuclear bombs and warheads, McNamara credited the peaceful outcome to a recognition on both sides that a nuclear exchange would be mutually suicidal. "I didn't believe then, and I don't believe today," he later commented, "that it [strategic nuclear power] did affect the outcome of crises."[162] Rather, in McNamara's view, it was a combination of factors, including not only U.S. nuclear capabilities and Kennedy's threat to use them, but also conventional forces, such as those at sea enforcing the quarantine, that turned the crisis around.[163] From the overall experience, McNamara concluded that efforts aimed at developing and perfecting viable counterforce capabilities were less important than maintaining a retaliatory potential that would make the other side think twice before launching a nuclear attack.[164]

As a substitute goal, in 1963 McNamara introduced the doctrine of assured destruction, which harked back to the minimum deterrence theories favored by the Navy in the late 1950s. As initially defined, assured destruction was "our ability to destroy, after a well-planned and executed Soviet surprise attack on our Strategic Nuclear Force, the Soviet government and military controls, plus a large percentage of their population and economy (e.g., 30 percent of their population, 50 percent of their industrial capacity, and 150 of their cities)."[165] Generally speaking, a strategy of assured destruction was much less costly and easier to implement than a counterforce strategy because the former required not only the maintenance but also the coordination of substantially smaller forces than the latter, two points McNamara would use repeatedly during his remaining tenure in strong arguments to resist further increases in the size of U.S. strategic forces. Such forces, moreover, would be differently constituted as ballistic missiles replaced manned bombers in the strategic weapons inventory.

Militarily, the result was a force structure smaller than one needed for an effective first- or second-strike counterforce strategy, yet probably substantially larger than one necessary for assured destruction.[166] Considerable savings followed, allowing the Defense Department to reduce expenditures on strategic programs by about half, compared with what it had been spending in the late 1950s.[167] In assessing the strategic balance, McNamara believed the number of targetable bombs and warheads, not delivery vehicles, to be the most important issue. Operating on this premise, he persuaded Kennedy's successor, President Lyndon B. Johnson, to accept a leveling off of strategic offensive programs at 41 missile submarines and 1,000 Minuteman launchers. For SAC an important turning point came in 1964 when, for the first time in its history, the number of ICBMs on alert equaled the number of bombers on alert. Thereafter, as the number of missiles entering the inventory increased, reaching their programmed peak in 1969, the ICBM alert force increasingly overshadowed the bomber alert force.[168]

As a concession to the JCS, who favored a larger buildup, McNamara agreed to augment U.S. strategic systems with such refinements as penetration aids for missile warheads, multiple independently targeted reentry vehicles (MIRVs) for some ICBMs and sea-launched ballistic missiles (SLBMs), and a limited antiballistic missile system for use in countering an embryonic Chinese Communist ICBM threat.[169] He also agreed to keep some 600 B–52s on active duty, thus creating a strategic triad composed of land- and sea-based missile systems and land-based bombers. But he canceled the Skybolt air-to-surface missile intended to enhance the B–52's penetration capabilities; he turned down Navy proposals to develop a system for tracking and destroying Soviet SLBMs; and he consistently vetoed Air Force proposals to expedite research and development on a new advanced manned strategic aircraft, later redesignated the B–1, that the Air Force began campaigning for after it lost the B–70.[170] As a result, SAC had no choice but to make do for the foreseeable future with an aging fleet of B–52s and B–58s and a new medium-range plane, the FB–111, a modification of the swing-wing variable-geometry experimental tactical fighter-bomber that had originally been designed for the Tactical Air Command. Lacking the range and payload of the B–70, the FB–111 seemed a poor substitute and was in the eyes of many, especially at SAC, a weapon of dubious value. According to LeMay, the FB–111 was "no goddamn good as a strategic bomber, it wasn't a strategic bomber and that's a fact."[171]

Despite controversy and criticism of his decisions, McNamara continued to caution the need for flexibility and control in the application of military power, especially if it involved planning the use of nuclear weapons. Perhaps

President Lyndon B. Johnson accepted a leveling-off strategic offensive program of submarines and missiles as suggested by his Secretary of Defense, Robert S. McNamara.

his greatest success in this regard was in convincing the European allies, who had initially reacted to his flexible response doctrine with intense skepticism. Basically, the European allies worried that flexible response would be prohibitively expensive to implement because of the need for larger conventional forces, and that it would weaken extended deterrence by removing the threat of immediate SAC retaliation in the event of Soviet aggression.[172] Finally, in 1967, after the United States agreed to dedicate five Polaris submarines to Europe's defense, NATO adopted a new strategic concept paper (MC 14/3) in which the NATO Military Committee called for "a flexible and balanced range of appropriate responses" to meet a Soviet attack.[173]

Yet by the time NATO acted, the issue had become almost academic. In fact, from 1965 on, as he became increasingly preoccupied with the war in Vietnam, McNamara lost interest in trying to refine the strategic concepts he had helped to introduce and advance, especially the precise interaction between flexible response and assured destruction. One result was a widening gap between employment policy, as handed down in the annual guidance for the SIOP (which continued to stress counterforce flexibility and restraint), and operational behavior and procurement policy, as governed by the more limited needs of assured destruction. For McNamara, however, there was no immediate urgency in trying to reconcile these discrepancies as long as the goal of American strategic policy was one of deterrence based on the threat of assured destruction. For him, the ultimate virtue of assured destruction was not only that it served as an eminently effective force-bounding device, but it committed the United States to a strategy so horrendous in its consequences as to make general nuclear war seem virtually unimaginable and therefore, he decided, practically impossible.

SALT and the Challenge of Waning Superiority

By the time McNamara left the Pentagon in 1968, senior military planners, especially in the Air Force, were increasingly concerned that the introduction of the assured destruction concept (mutual assured destruction, or MAD as it came to be called in the 1970s) had eroded the basis of deterrence. Heretofore, deterrence as conceived and incorporated into U.S. strategic doctrine from the early 1950s on had stressed the need for the most modern, up-to-date weapons available, developed around a counterforce potential that would limit damage to the United States by dealing decisively with any aggressor. But in McNamara's view, efforts to develop and maintain such capabilities were, by the 1960s, no longer tenable. For his purposes, deterrence would be adequately served by a second-strike capability sufficient to destroy some percentage of the enemy's industry and population, a proposition that the Air Force saw as seriously flawed. "The Assured Destruction strategy," observed the CINCSAC,

General Bruce K. Holloway, in 1969,

> not only fails to provide high confidence of deterring an all out attack on the United States, it also does not provide options to the National Command Authority for military responses to lower levels of conventional and nuclear aggression. Although we presently have some margin of strength for attacks on military targets to deny the Soviets an unscathed reserve, we cannot assure termination of hostilities in a position of relative advantage.[174]

It seems clear that the principal impetus for change in the doctrine of assured destruction came from within the Air Force, which wanted to break away from its strictures in an effort to revitalize SAC's capacity to function in a counterforce role. The steady and larger-than-expected growth in Soviet strategic capabilities that followed the Cuban missile crisis provided a major inducement. Other considerations included a desire to arrest the decline of SAC that began under McNamara and restore some measure of the command's former prestige and potency. For these reasons, once McNamara was gone, the Air Force launched a vigorous program that stressed SAC's need for more survivable and effective forces and an acceptance by higher authority of a less simplistic view of deterrence than one of assured destruction. The goal became a more effective and secure counterforce war-fighting capability that would allow the United States to prevail in attaining its war aims should deterrence fail.[175]

By 1969, when Richard M. Nixon assumed the presidency, the balance in strategic nuclear power, though still in favor of the United States and its allies, was moving ever closer to numerical parity as the Soviet Union added steadily to its strategic arsenal. Although American strategic forces, because of their size and striking power, remained the cornerstone of Western security, the recent advent of France as a nuclear power, combined with Britain's strategic nuclear forces, presented the overall picture, on paper at least, of a formidable deterrent. Only the United States and Britain, however, coordinated their strategic war plans. Britain in 1963 had placed the operational planning of its strategic nuclear assets, then consisting of about 170 combat planes in Bomber Command's V-force, under NATO, which relied on the JSTPS in Omaha for target allocation.[176] With the end of Bomber Command in the late 1960s, however, British strategic capabilities rested on a flotilla of four *Resolution*-class fleet ballistic missile submarine, all armed with Polaris missiles purchased from the United States.[177]

The French *force de dissuasion*, in contrast, operated independently as a result of President Charles de Gaulle's decision in the 1960s to withdraw all French forces from NATO command and develop a separate strategic nuclear deterrent. As in the United States, French strategic doctrine adhered to the concept of a triad of land- and sea-based missiles and manned aircraft, though it was not until 1971 that France deployed the first in a force of six Polaris-type (*Redoubtable*-class) missile-firing nuclear submarines. Subsequently, France also deployed eighteen silo-based IRBMs, organized in squadrons of nine missiles each. Until then, France's sole means of delivering nuclear weapons consisted of a force of approximately thirty-four Mirage IV medium-range bombers.[178]

By far the most important development of the late 1960s and early 1970s was the increase in the size and effectiveness of Soviet strategic forces compared with U.S. capabilities. In 1965 the United States had 934 ICBM launchers and 464 SLBM launchers compared with a Soviet force of 224 ICBMs and 107 SLBMs. By the end of 1969 the U.S. buildup had peaked at 1,054 ICBMs (1,000 Minuteman launchers and 54 Titan IIs) and 41 fleet ballistic-missile submarines with 656 launchers, while the Soviets continued to expand, deploying three new types of improved ICBMs at a rate of more than 200 annually and launching, on average, 8 ballistic-missile submarines a year, each with 16 launch tubes. Within three years, by 1972, the Soviet Union had a force of more than 1,500 ICBMs, many of which, like the SS–9 and its successor, the SS–18, carried substantially larger payloads — 10 or more warheads in the case of the SS–18 — than their American counterparts did.[179] Also, the Soviets were in the process of beginning to deploy their first new bomber since the 1950s, the supersonic Backfire, which U.S. intelligence estimates assessed as being capable of reaching targets in the continental United States, albeit on a one-way mission. Only in targetable strategic bombs, warheads, and long-range aircraft did the United States enjoy a numerical edge, regardless that McNamara's decision to downplay strategic bombers meant no new planes with intercontinental range had been added to the force since delivery of the last B–52H in 1962. In time, the Soviets developed a MIRV capability of their own, which further enhanced their growing strategic capability.[180]

In the past, such developments would probably have been the tocsin calling the United States to respond with an offsetting buildup in offensive weapons and/or an increase in damage-limiting measures such as civil, air, and ballistic missile defenses. By the end of the 1960s, however, disillusionment and frustration over the inconclusive war in Vietnam had reached the point at which it would have been exceedingly difficult, if not impossible, to mobilize public and congressional support for such a course, as witnessed by the 1969 debate over the deployment of a limited antiballistic missile system. The implementing legislation for the system passed the Senate by a single vote.[181] Though Nixon was determined to preserve a credible strategic posture, he knew he stood little chance of restoring and maintaining a clear strategic superiority. He decided, therefore, that he would make do with what he termed strategic sufficiency, an elastic term that strategic planners in the Pentagon interpreted as including all the elements required to achieve their basic objective of assured destruction.[182]

In actually applying its doctrine of strategic sufficiency, the Nixon administration's early signals generally indicated a more open attitude and a greater willingness to support the Air Force's pleas for near-term modernization

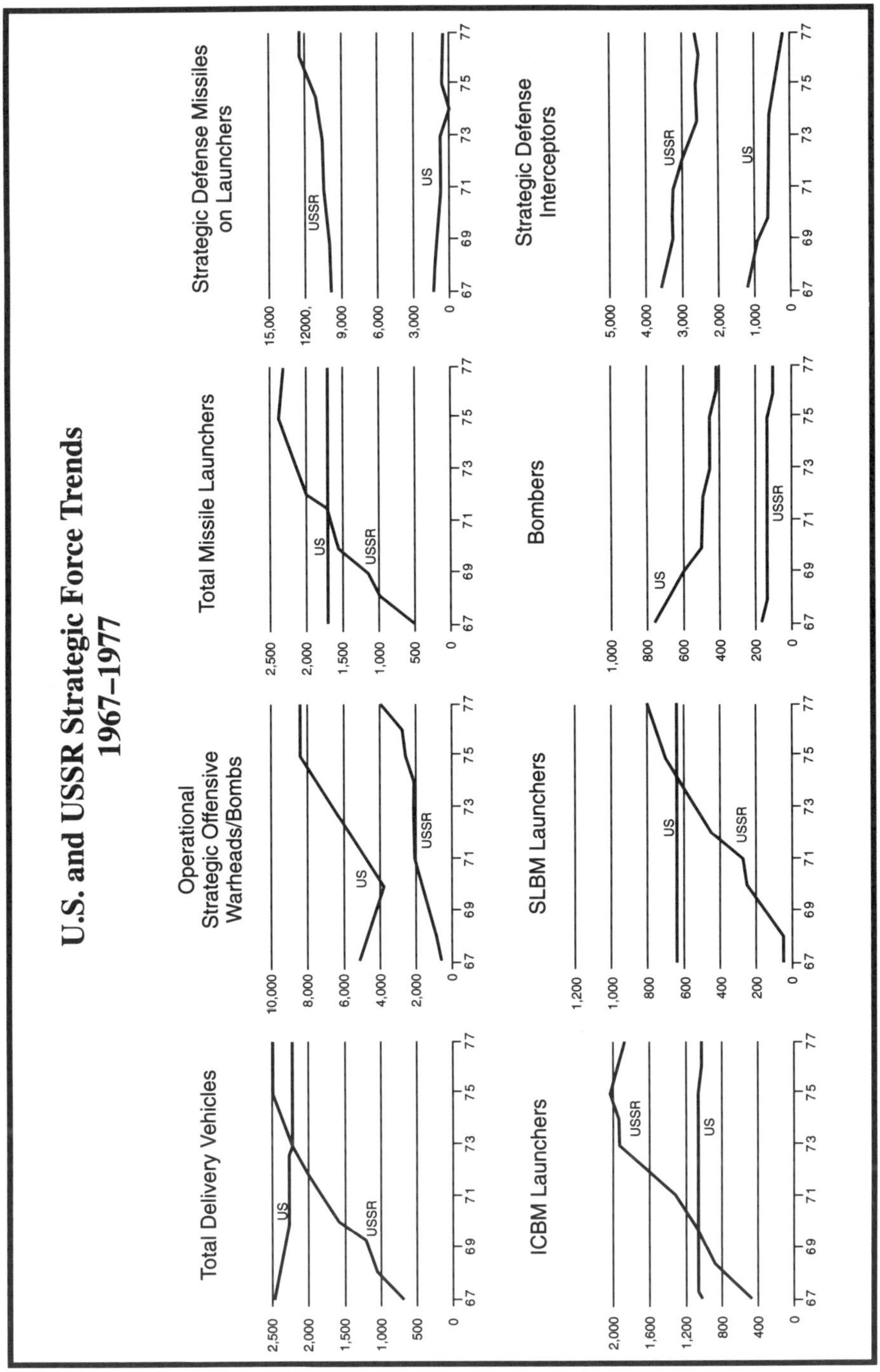
U.S. and USSR Strategic Force Trends
1967–1977
Total Delivery Vehicles
Operational Strategic Offensive Warheads/Bombs
Total Missile Launchers
Strategic Defense Missiles on Launchers
ICBM Launchers
SLBM Launchers
Bombers
Strategic Defense Interceptors
US
USSR
67
69
71
73
75
77

of its strategic component. The immediate result was stepped-up research and development for the advanced manned strategic aircraft (B–1) bomber project and, as previously planned, the phaseout of all Minuteman I ICBMs in favor of a mixed force of 450 improved Minuteman II ICBMs, each carrying a single thermonuclear reentry vehicle, and 550 MIRVed Minuteman III ICBMs, each armed with thermonuclear warheads. But the administration did not appear particularly eager to commit itself to SAC's longer-range modernization goals, which included a more accurate and powerful ICBM preferably housed in highly survivable, hard rock silos, operational deployment of a satellite missile-launch detection and warning system, and a larger advanced airborne command post to accommodate a larger battle staff and advanced data-handling equipment. Meanwhile, as the costs of FB–111 bombers soared out of sight, the Nixon administration scaled back the costly procurement from a planned 253 aircraft to a token force of 76 and accelerated the early retirement of all B–58s and some older B–52s, mainly C and F models built in the 1950s. Only 422 B–52s remained in the active inventory by the end of 1973.[183]

In developing strategic policy, Nixon and his White House advisers (including Henry Kissinger, his influential assistant for national security affairs) also emphasized the need for progress in arms control, which they viewed as offering the best opportunity to curb future growth in Soviet capabilities. Prior to the launching of the Strategic Arms Limitation Talks (SALT) in 1969, however, the record of accomplishments in the arms control field had been meager. In fact, not until 1963, after eight years of negotiations begun by President Eisenhower, did the United States and the Soviet Union sign their first significant arms control measure — the limited nuclear test ban treaty that prohibited weapons tests in the atmosphere, in outer space, and underwater. At that time, Air Force spokesmen, including Generals LeMay and Power, testified against the treaty before the Senate. They argued that, given the status of the respective test programs, such an agreement would leave the Soviets with a potentially stronger and more diversified nuclear arsenal, especially in high-yield weapons able to threaten hardened U.S. targets like missile silos. Their arguments did not prevail, and over the next seven years both sides adhered to the provisions of the treaty, which suggested that other accords might be possible.[184]

Although more ambitious than any previous undertaking, the SALT I negotiations (1969–1972) yielded few surprises, embracing a limited spectrum of weapons — ICBMs, SLBMs, and certain types of defenses against them. Among the most far-reaching controls agreed upon was a treaty of indefinite duration that contained specific restraints on the deployment of antiballistic missile systems. This agreement was a major breakthrough in some respects, but the actual concessions involved were minor since neither side had yet perfected the technologies. In coming to grips with the more complex and sensitive issue of offensive weapons, the best the two sides could do for the

U.S. and USSR ICBMs, 1985

U.S.

	Titan II	Minuteman II	Minuteman III	Peacekeeper (MX)
Number Deployed	26*	450	550	In Development
Warheads	1	1	3	Up to 10
Max Range (KM)	12,000	12,500	14,000	14,000
Launch Mode	Hot	Hot	Hot	Cold

USSR

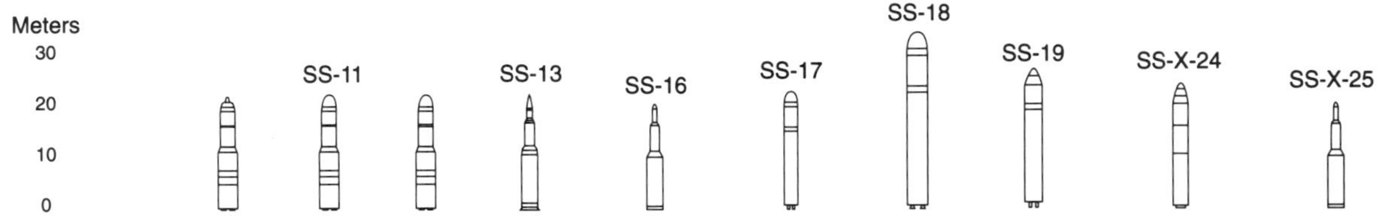

	SS-11 Modification 1	SS-11 Modification 2	SS-11 Modification 3	SS-13 Mod. 2	SS-16	SS-17 Mod. 3	SS-18 Mod. 4	SS-19 Mod. 3	SS-X-24	SS-X-25
Number Deployed	100	420		60	Undetermined	150	308	360	In Development	Nearing Deployment
Warheads	1	1	3MRVs**	1	1	4 MIRVs***	10 MIRVs	6 MIRVs	Up to 10 MIRVs	1
Max Range (KM)	11,000	13,000	10,600	9,400	9,000	10,000	11,000	10,000	10,000	10,500
Launch Mode	Hot	Hot	Hot	Hot	Cold	Cold	Cold	Hot	Cold	Cold

*As of early 1985.
**Multiple reentry vehicle.
***Multiple independently targetable reentry vehicle.

To many in SAC, the FB–111 was of dubious value. To LeMay it was "no goddamn good as a strategic bomber, it wasn't a strategic bomber and that's a fact."

moment, as they met to assess their progress at the 1972 Moscow summit, was a temporary, five-year freeze on the number of ICBM and SLBM launchers, in the hope that future negotiations could devise something more permanent and effective.[185]

Some observers at the time thought the SALT I antiballistic missile treaty reaffirmed the continuing validity of assured destruction because by practically doing away with antimissile defenses, it seemed to reinforce the concept of mutual deterrence based on holding each other's population centers hostage to a nuclear exchange.[186] The substantial size and increasing sophistication of Soviet strategic forces, however, raised questions as to whether the Soviet Union really accepted the notion of mutual assured destruction. The configuration of Soviet strategic forces seemed to suggest the purposeful development of a war-fighting strategy capable of flexibility and selectivity, in line with the traditional counterforce orientation of Soviet military science. Since the Soviets also had never drawn a significant distinction between war-fighting and deterrence, it followed they could be planning to employ their nuclear forces for use in conflicts short of an all-out exchange in the expectation that the United States would not resort to full-scale retaliation for such a provocation. If this was indeed true, the Soviet buildup supported SAC's contention that the United States should have a more responsive strategic doctrine than one of assured destruction, something, as Kissinger later put it, to deal with "contingencies other than all-out nuclear challenge."[187]

The earliest public sign that a reassessment of strategic doctrine was in the offing came from the White House in the first two presidential state of the world messages given to Congress in 1970 and 1971. Although ICBMs accounted for only about 20 percent of the megatonnage in the U.S. strategic arsenal at this time, their high degree of accuracy, low maintenance costs, and relatively reliable command and control made them a valuable component of the triad. Because of the growth in the size, effectiveness, and throw-weight of the opposing Soviet ICBM force, America's missiles were also increasingly vulnerable to a Soviet preemptive first strike. Since the preservation of deterrence remained paramount, it was imperative from the President's perspective to have a broader range of options available should retaliation became necessary. "I must not be — and my successors must not be," Nixon insisted, "limited to the indiscriminate mass destruction of enemy civilians as the sole possible response to challenges."[188] What gradually emerged as the administration's preferred alternative was a refurbished version of the controlled counterforce doctrine, the product of a series of studies and reviews that culminated in National Security Decision Memorandum (NSDM) 242, signed by President Nixon in January 1974.[189]

As explained by Secretary of Defense James R. Schlesinger, who had done much of the earlier conceptual work behind selective targeting while an analyst with RAND, the intent behind the new doctrine was not to produce "radical departures from past practice," but was to provide "a series of measured responses to aggression which bear some relation to the provocation, have prospects of terminating hostilities before general nuclear war breaks out, and leave some possibility for restoring deterrence." The key element of such a strategy consisted of escalation control, obtained by the carefully calibrated use of force and targeting options to further limit any strategic nuclear exchange. Accordingly, selective attacks on critical politico-military targets, including even those in the hard-kill category (missile silos; hardened launch control centers; command, control, and communications bunkers; nuclear storage facilities; and the like) were to be given priority as an alternative to assured destruction's city-busting.[190] However, if escalation could not be controlled, NSDM 242 also sanctioned attacks that would impede Soviet recovery by destroying 70 percent of Soviet industry, a strong indication that the concept of assured destruction had not yet been discarded.[191]

Perhaps the most immediate effect of NSDM 242 was to bring declaratory policy more closely back into line with targeting policy, which had been fairly consistent since the mid-1960s in stressing military counterforce over civilian countervalue targets. Among the changes that Schlesinger subsequently ordered in the SIOP was the requirement for a Strategic Reserve Force (SRF). Composed of air and naval nuclear forces, the SRF was intended to provide a flexible, retargetable mixed force for use after initial hostilities and for assessing damage. Additionally, Schlesinger directed the JSTPS to program for limited nuclear options designed to demonstrate national resolve and inhibit

Soviet military action. These included practically the entire spectrum of involvement, from show of force missions to all phases of nuclear war including recovery and reconstitution periods after an initial exchange.[192]

In addressing the question of what the new doctrine would entail for implementation, Schlesinger promptly disavowed any intention of acquiring capabilities that might give the appearance the United States was seeking to develop a counterforce capacity to launch a disarming first strike against the Soviet Union. Instead of new weapons systems, he opted for such refinements as a higher yield and a more accurate reentry vehicle (the Mark 12a) for Minuteman IIIs; two more powerful and sophisticated thermonuclear bombs (the B–61 and the B–77); and improved accuracy for submarine-launched ballistic missiles in order to increase their counterforce potential.[193] Although he acknowledged that the Air Force needed a new heavy bomber (the B–1) to replace its aging B–52s and an advanced-technology ICBM (the MX) to augment the Minuteman force, he played down the urgency of going into early production of either.[194] This last decision was generally applauded by the political doves in Congress but was a setback for the Air Force, which after the signing of the SALT I accords had hoped to expedite both programs to offset modernization in the Soviet's strategic forces and keep pace with the Navy's Trident missile submarine program. Looking ahead, Secretary of the Air Force John L. McLucas told Congress in 1975 that the rate at which the B–1 and MX were being allowed to progress indicated that they would not become operational until sometime in the 1980s (rather than the late 1970s as the Air Force had envisioned originally).[195]

The next major innovation in strategic doctrine appeared late in the presidency of Jimmy Carter, though studies and analysis were under way almost from the moment he took office. Carter's main interest was arms control, and he hoped to negotiate deep reductions in the offensive arsenals of both sides. When this proved unacceptable to the Soviets, he settled for a limited-duration treaty (SALT II), which he and Soviet General Secretary Leonid I. Brezhnev signed at a 1979 Vienna summit. The main feature of the treaty was the establishment of equal aggregates in certain categories of strategic launchers, with subceilings on those that were MIRVed. Critics assailed the agreement for being filled with loopholes and ambiguities that effectively conceded numerous advantages to the Soviets, especially aggravating the threat posed to U.S. missile silos by the superior throw-weight of Soviet land-based ICBMs. Within months, Carter withdrew the treaty from Senate consideration, partly owing to the opposition it aroused but also in protest over the recent Soviet invasion of Afghanistan. Even so, Carter and his successor Ronald Reagan agreed to abide by its terms until the latter, in 1986, declared the treaty's numerical ceilings outdated and no longer applicable.[196]

How closely Carter was able to coordinate arms control policy on the one hand with arms development policy on the other is uncertain, although it is

President Jimmy Carter covered the globe to achieve arms control. He finally settled for a limited duration treaty, which he and Soviet General Secretary Leonid I. Brezhnev (shown heavily — if only partially — bemedaled) signed at a 1979 Vienna summit.

clear that, in reaching decisions on strategic requirements, Carter and Secretary of Defense Harold Brown (the former Secretary of the Air Force in the Johnson administration) were prone to opt for levels lower than the JCS preferred. A persistent worry among many defense planners, from the time the SALT process began, was that the civilian leadership would withhold needed strategic improvements in order to demonstrate U.S. good faith at the negotiating table. Apart from the restraining influence of arms control, the combination of high inflation and chronic energy shortages in the late 1970s combined to usher in an era of severe austerity throughout the defense establishment. The net result was a further weakening of strategic offensive capabilities, especially in the two land-based legs of the triad. Although the ICBM force held steady at 1,054 launchers (the maximum permitted the United States under the SALT agreements), the number of bombers on active duty declined from 487 (419 B–52s and 68 FB–111s) to 406 (343 B–52s and 63 FB–111s) between the end of 1976 and the beginning of the Reagan administration in 1981.[197] SAC's first priority was to maintain strategic forces roughly on a par with those of the Soviet Union. Considering the circumstances, many in the Air Force began to question seriously whether a credible strategic deterrent posture could be preserved. By the end of the 1970s, the numerical balance in strategic forces shifted markedly in favor of the Soviets.

At the same time, declaratory national policy continued to acquire refinements that increased the need for an effective war-fighting deterrent. Most of the changes (the product of a Cabinet-level review coordinated by Carter's national security assistant, Zbigniew Brzezinski) were minor, but the overall effect was a reworking of policy and strategy, summarized in the 1980 White House directive PD 59 that Brown termed the doctrine of countervailing power.[198] Like the Schlesinger doctrine it superseded, the Brown doctrine stressed deterrence through the application of flexible options so that at a variety of levels of nuclear exchange, "no potential adversary of the United States could ever conclude that the fruits of his aggression would be worth his own costs." In practical terms, this entailed developing "a continuum of options" for the controlled application of nuclear power against "a broad spectrum of targets," though personally Brown remained skeptical that controlled escalation was a "viable concept" once either side crossed the nuclear threshold.[199] Moreover, he considered it impractical, as Schlesinger had advocated, to try to pick out specific countervalue targets that would impede Soviet recovery; instead, should an all-out nuclear exchange prove unavoidable, he preferred retaining a capacity to assure the destruction of at least 200 major Soviet cities.[200]

The most original aspect of the countervailing doctrine was its attempt to incorporate Soviet perspectives into U.S. targeting policy. In addition to the counterforce and countervalue targets already in the SIOP, PD 59 called for the introduction of a new category of counterpolitical targets. These included crucial elements of the Soviet command-and-control infrastructure, which if destroyed, could pose a danger to the survival of the regime. Though similar targets (the Kremlin, for example) had appeared in war plans dating from the late 1940s, the new list was considerably longer and more sophisticated than any previously assembled. Altogether, the 25,000 potential Soviet targets listed in the 1974 SIOP now jumped to more than 40,000. The new SIOP included key communications facilities, military command posts, and governmental and Communist Party headquarters.[201] "A clear U.S. ability to destroy them," Brown contended, "poses a marked challenge to the essence of the Soviet system and thus contributes to deterrence."[202]

To implement the countervailing strategy, Brown said that improvements in the strategic force structure would be needed, though he was never publicly precise about what that entailed. According to prevailing support studies and analyses, the most glaring American weaknesses were in the C^3I system (command, control, communications, and intelligence system) which Brown agreed would have to have more reliable backups. Also, in light of development problems, he questioned the capabilities of the B–1 bomber and recommended that Carter cancel it in favor of concentrating on a less expensive, long-range

Secretary of Defense Harold Brown adopted a policy and strategy that he termed a doctrine of countervailing power.

Defense Secretary Brown questioned the capabilities of the B–1 bomber (above) and recommended that President Carter cancel the program in favor of less expensive long-range missiles.

cruise missile program and a new bomber employing Stealth technology to assist it in evading enemy radar.[203] In the estimate of Lt. Gen. Alton D. Slay, Deputy Air Force Chief of Staff for Research and Development, the remaining level of modernization was "austere but adequate."[204] Proponents of the countervailing strategy, such as Walter B. Slocombe who served as Brown's Deputy Under Secretary for Policy Planning, saw it as an immediately usable concept in many respects but, as Slocombe hastily added in a 1981 article on the subject, more like a decade would be needed to acquire the full range of requisite capabilities.[205]

In Britain and France, official thinking followed a similar line, as new technologies coming to fruition in the 1970s allowed a wider range of strategic targeting options. Heretofore, because of the limited size and flexibility of their strategic forces, British and French planners had not paid much attention to counterforce and counterpolitical targets. They preferred instead to key on Soviet urban-industrial centers, believing the threat of retaliation against these targets would deter Soviet attacks on British and French cities. Now with the advent of microelectronics and multiple-warhead technology in the 1970s, it became possible to broaden target lists to include, as the leading analyst in Britain's Ministry of Defence put it, aspects of "Soviet state power," presumably meaning high-priority politico-military installations similar to those mentioned in PD 59.[206] In Britain, the change in planning began with the development of a new reentry vehicle for Britain's missiles, called Chevaline, which entered operational service in the summer of 1982. Designed initially to penetrate Soviet antiballistic missile defenses around Moscow, Chevaline involved two separately maneuvering clusters of as many as six warheads with yields in the kiloton range, plus decoys.[207] In France, where a similar, multiple-warhead program was in progress at the same time, the shift in targeting doctrine was more gradual, but it generally followed the same pattern as had occurred in Britain.[208]

Despite the use of a somewhat different rhetoric stressing maintenance of a strong deterrence posture and, should deterrence fail, "flexibility in our forces and in our options of response,"[209] the strategic doctrine adopted by the Reagan administration was essentially the same as the previous administration's had been. Attention focused on procuring and improving strategic forces and the C^3I capabilities to support such a doctrine rather than on the development of fresh strategic concepts. Among the measures taken to bolster the country's strategic posture were a revitalization of the bomber program, including not only additional Stealth research but also a revival of one hundred B–1 bombers; the purchase of additional wide-body KC–10 aerial tankers and the reengineering of the KC–135A aerial tanker force; the deployment of four E–4B National Emergency Airborne Command Post aircraft to serve the National Command Authority in time of emergency; and the development of a larger, more accurate reentry vehicle (the D–5 RV) for the Trident missile submarine.[210]

Reagan and his advisers also determined to press on with modernization of the ICBM force, a process that had stalled when differences arose between the administration and Congress over the proper basing mode for the MX, future ICBM procurement policy, and arms control. To help resolve these issues, Reagan appointed a blue-ribbon panel headed by Lt. Gen. Brent Scowcroft (U.S. Air Force, retired), who had been national security adviser to Presidents Nixon and Gerald R. Ford. Reporting in April 1983, the Scowcroft Commission endorsed a limited range of ICBM modernization measures and urged more effort at obtaining arms control agreements to reduce the growth of heavy-MIRVed ICBMS in the United States and the USSR, weapons the commission saw as contributing to crisis instability because of their increased vulnerability and high priority as targets. As a temporary first step, the commission proposed to confine the number of MX missiles deployed in existing Minuteman silos to just 50 (of the 100 requested by the Reagan administration and the deployment goal of 200 proposed by President Carter) and to expedite the engineering effort in the development of a new and presumably less-threatening, single-warhead ICBM known as Midgetman, deployable in a less vulnerable mobile mode.[211] Defense officials responded that such measures fell short of providing the United States with the effective, hard-target kill capacity for which the MX had originally been designed and that waiting for the Midgetman would entail unnecessary risks.[212] Nevertheless, on Capitol Hill the Scowcroft Commission report became the model for a compromise on the MX deployment issue.

Less influential in shaping administration policy was the Scowcroft Commission's recommendation that the United States pay more attention to arms control. Although the Reagan administration probably was never against arms control (as some critics have claimed), many senior Pentagon officials including Secretary of Defense Caspar W. Weinberger, Assistant Secretary of Defense Richard N. Perle, and others were leery of rushing into agreements with the Soviets until the United States had regained some of the strategic strength it had lost in the 1970s. Nor were the Soviets, who suffered a succession of leadership crises until Mikhail Gorbachev took power in 1985, particularly willing or eager to negotiate seriously. As a result, arms control for a time came close to foundering. But with the conclusion in 1987 of the intermediate-range nuclear forces treaty, which phased out all land-based intermediate- and medium-range nuclear weapons, arms control seemed to acquire new momentum.[213]

Interwoven with the Reagan administration's actions on arms control and its efforts to acquire more up-to-date strategic hardware were two further, controversial developments. The first was the administration's decision, reflected in its 1981 Nuclear Weapons Employment and Acquisition Master Plan, to draw up a new SIOP giving closer attention to the requirements in a prolonged

To revitalize strategic deterrence posture, the Reagan administration sought to purchase additional KC–10 tankers for the aerial refueling of bombers, as shown here refueling a B–52 in flight.

Not until Mikhail Gorbachev took power in 1985 were the Soviets willing to negotiate arms control. The Soviet leader appears above with President Ronald Reagan during his arrival ceremony in Washington in 1987.

or protracted nuclear conflict.[214] As a part of that effort, Defense Secretary Weinberger stated that one of his central aims was "a multiplicity of survivable strategic forces."[215] Herman Kahn, head of the Hudson Institute think tank, among others, had advocated a similar approach to strategy and targeting while at RAND in the late 1950s and early 1960s. Thus the idea (then known as war endurance, i.e., the ability to continue operations beyond a first strike or a massive first exchange) was by no means new, although it had fallen from favor when assured destruction became the reigning orthodoxy under McNamara.[216] Critics dismissed the Reagan administration's attempt to revive the war-endurance concept as impracticable, not just for military reasons but also because of the exceedingly costly (and, in some circles, unpopular) civil defense needs associated with it. Nonetheless, it was Weinberger's view that as a long-term goal, credible and effective deterrence should rest on something more than the ability to threaten the annihilation of Soviet cities.[217]

The other major development, announced by Reagan in March 1983, was the Strategic Defense Initiative (SDI), initially a five-year intensive research program aimed at "eliminating the threat posed by strategic nuclear missiles" and making all nuclear weapons someday "impotent and obsolete."[218] Unlike earlier antiballistic missile programs such as Nike-X, Sentinel, and Safeguard, all of which had been Army-run, SDI was conceived and organized as an interservice endeavor.[219] Actually, study of ballistic missile defenses had been continuous, both in the United States and in the Soviet Union after the signing of the 1972 Anti-Ballistic Missile Treaty. This treaty, which prohibited large-scale deployment of such systems, did not proscribe research and development of system components. The intent of SDI was to investigate, within terms of the treaty, whether new directed-energy and space-based technologies could yield more effective systems than those studied in the past.[220] By the time Reagan left office in 1989, testing had yet to yield conclusive answers, and a decision to deploy was in abeyance. Consequently, it was still too early to tell whether SDI was more likely than the doctrine of the past to produce any significant changes leading to a more defensively oriented strategic posture.

In 1984, General Bennie L. Davis, then CINCSAC, suggested a further possibility. Strategic and tactical nuclear weapons might gradually lose some of their military utility and be significantly augmented, if not supplanted, by highly refined conventional munitions (more advanced versions of the smart bombs first introduced late in the Vietnam War) and the Exocet missile, which was used with considerable effect by the Argentine Air Force against the British Royal Navy in the 1982 Falklands conflict. As described by Davis, major conflicts of the future could turn on long-range aircraft launching conventional weapons with unprecedented accuracy and sophistication in standoff attacks against enemy defenses.[221] One drawback to adopting such a doctrine was that, while the United States and its Western allies could feasibly produce the necessary technology, the Soviet Union would not necessarily be

U.S. and USSR Strategic Forces*
1970–1980

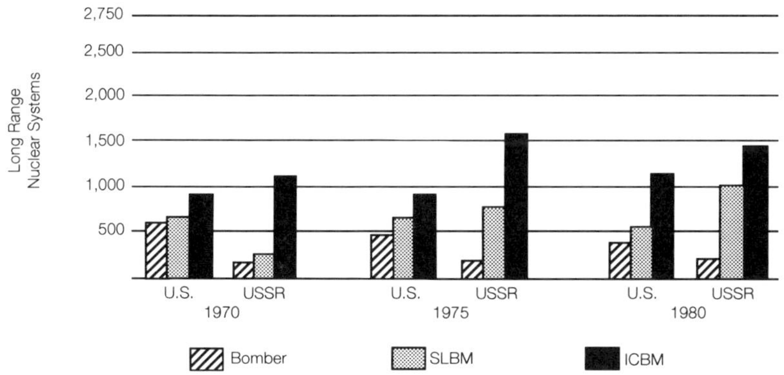

*Includes FB-111 and Backfire bombers.

able to compete on a high-technology aerial battlefield because of the retarded development of Soviet industry in microelectronics and computer systems. Soviet responses, therefore, were likely to be nuclear from the outset in any conflict involving strategic forces. Nonetheless, in countering third-country threats, such as those posed by Iraq in the early 1990s, the conventional alternative had distinct advantages. Accordingly, on January 1, 1987, SAC began training for a dual strategic role, one involving both conventional and nuclear operations.[222] Strategic planning had come full circle.

Conclusion

Since World War II, the most significant changes to have taken place in strategic doctrine, both in the United States and abroad, were those associated with the introduction and spread of nuclear technology: larger and more plentiful weapons; improved delivery systems such as longer-range aircraft and intercontinental ballistic missiles; and more accurate intelligence-gathering and analysis mechanisms to assist in early strategic warning, target identification, and assessment of enemy capabilities. Equally important was the demonstrated ability of a number of countries to develop strategic nuclear capabilities with varying degrees of size and effectiveness. Though the United States initially established an early lead, it did not take long for others, especially the Soviet Union, to perceive the political and military advantages of a strategic arsenal and develop their own capabilities. What began in the early postwar period on both sides as an effort to explore and exploit the military potential of this new and rapidly expanding technological frontier gradually evolved, by the 1980s, into a virtual strategic standoff.

In these circumstances, perceptions of the role of strategic military power changed considerably from the days of Douhet and Mitchell. In the United States, perhaps the most remarkable difference between pre– and post–World War II strategic thinking was the emphasis in the latter period on the concept of deterrence, which in postwar parlance soon became synonymous with prevention of nuclear war. Before World War II and before the advent of nuclear weapons, deterrence was rarely mentioned in connection with U.S. strategic air doctrine. After the war, and especially after the Soviet Union demonstrated a nuclear capability in 1949, it gradually became axiomatic that the first and foremost function of U.S. strategic forces (and those subsequently developed by the British and the French) should be to deter the Soviets from using nuclear weapons. Since that time, the most lively doctrinal debates were those that addressed the question, How should this goal be pursued?

After World War II, American thinking on the problems of strategic air warfare and strategic deterrence followed a fairly consistent course. First and foremost, it emphasized the need for strong offensive forces capable of denying the Soviet Union attainment of its politico-military objectives. Such ideas were a direct outgrowth of the interwar years, when strategic air doctrine first took shape around the organizing concept of offensive operations. By the time World War II had ended, the offensive mentality was firmly set, though it took time for American planners to adjust their thinking to the new realities of the nuclear age and the complexities of possibly having to face the Soviet Union as a future foe. Strategic targeting, which the Air Force among the military services dominated, remained rooted in the Army Air Forces' World War II practice of countervalue/countereconomy bombardment aimed at destroying the most vital parts of the enemy's military-industrial base. Because of the restricted range and effectiveness of available aircraft, sketchy intelligence about target sites and enemy defenses, and the limited size of the U.S. nuclear stockpile, real questions remained as to what impact an air offensive against the Soviet Union would have. Despite its drawbacks and limitations, strategic nuclear bombing held more appeal for senior policymakers and congressmen than did the other two most commonly mentioned options: a costly and politically unpopular conventional buildup, or a questionable maritime strategy centering on the use of carrier-based air power that would, in effect, duplicate the function of land-based air power.

The advent of the massive retaliation doctrine in 1954 seemingly sealed the Air Force's claim to supremacy. It confirmed that the Air Force in general, and SAC in particular, would now provide the country's first line of defense, and that nuclear weapons would have priority in shaping the country's basic defense posture. Both ideas made sense at the time, not only because of the vast preponderance of strategic nuclear power that the United States then enjoyed, but also because SAC's improved fleet of jet bombers could enforce such a policy. From SAC's standpoint, it followed that the deterrent effect of its weapons and war plans was not something that could be separated or made independent from contemplating their actual use. Credible deterrence, in other words, required credible planning, the availability of weapons, and a willingness to use them.

The counterforce targeting doctrine that came to prevail called for an optimum mix of aim points, which gave priority to attacks on military targets that would also yield heavy collateral, or bonus, damage to a broad spectrum of countervalue targets. As long as Soviet strategic forces remained comparatively small, it was entirely conceivable that they, along with much of the rest of the Soviet Union's war-making apparatus, could be effectively neutralized in one sure, swift, Sunday punch, as derived from the Spaatz concept of operations. But by the end of the 1950s, as the vulnerability of the United States to a Soviet surprise missile attack became more apparent, the concept of massive retaliation began to lose credibility.

The ensuing search for a more effective replacement doctrine turned up numerous new ideas but few that could be easily or economically adopted with any assurance of success. Among the United States' allies (initially the British, and later the French), the preferred course, for financial reasons, embraced a posture of minimum deterrence: a rather modest force structure keyed to the destruction of the Soviet Union's most vulnerable targets, i.e., its cities. Though usually drawing strong dissent from the U.S. Air Force, similar proposals emerged from time to time in the late 1950s in the United States. From the Air Force perspective, growth in the Soviet strategic forces meant a growing target list, which in turn underscored the need to continue modernizing and expanding SAC's counterforce capabilities.

The incoming Kennedy administration in 1961, however, saw no point in making further heavy investment in what it regarded as the insatiable demands of massive retaliation. Giving priority to strengthening conventional capabilities, McNamara set about developing a survivable, second-strike retaliatory strategic force that reduced dependence on the more vulnerable weapons systems, particularly manned bombers. Eventually, McNamara seized on the concept known as assured destruction to rationalize his actions, despite his initial inclination to develop a broader range of strategic options along "controlled" counterforce lines. By the time he left the Pentagon in 1968, many such options had in fact been incorporated into the SIOP. As a practical matter, for deterrence purposes, it is doubtful whether McNamara regarded these options as realistic alternatives to threatening large-scale nuclear retaliation that would lay waste to the Soviet Union's cities and industry.

Actually, McNamara never believed such a war would ever occur. By and large, he regarded strategic weapons as possessing a limited, largely deterrent value and and general nuclear war as almost inconceivable. What appealed to him about the assured destruction concept was that it avoided the need for constant and wholesale modernization of the force structure while it emphasized the horrendous consequences of a nuclear exchange. As the size and strength of the Soviet ICBM force climbed after 1964, so too, ironically, did McNamara's confidence that nuclear war was unlikely. "With huge survivable arsenals on both sides," he later wrote, "the likelihood that a strategic attack would be met in kind is very great indeed. Thus, strategic nuclear weapons have lost whatever military utility may once have been attributed to them. Their sole purpose . . . is to deter the other side's first use of its strategic forces."[223] Indeed, fear of likely nuclear suicide may have tempered the Soviet reactions in the 1962 Cuban missile crisis.

The McNamara era also left a U.S. strategic force structure that, in size and composition, remained basically static for the next twenty or more years. After McNamara's departure as Secretary of Defense, SAC urged moving away from the assured destruction concept toward a more effective war-fighting deterrent capability, especially in circumstances that fell short of an all-out nuclear exchange. Although SAC's proposed changes in doctrine generally received a sympathetic hearing and endorsement at higher levels, they were not always translated into practice because of a preoccupation with arms control negotiations in the 1970s and high-level vacillation over strategic doctrine and associated procurement policies. As a result, by the end of the Carter administration, national policy guiding the programming and behavior of U.S. strategic forces tended to equivocate. Though strong on rhetoric resembling that of controlled counterforce in describing highly sophisticated targeting requirements and operations, it was weak in developing a force posture capable of supporting such a strategy. As strategy reviews of the 1970s generally concluded, maintenance of an effective assured destruction threat should be the ultimate concern in any case.

During the early 1980s, this situation began to change as the Reagan administration strengthened U.S. military capabilities, paying special attention to strategic needs including weapons systems and C^3I support. At the same time, in a dramatic move to revitalize research on strategic defenses, President Reagan in 1983 launched the Strategic Defense Initiative, a high-priority program to find a way of rendering nuclear weapons impotent and obsolete. All the same, the role and relative importance of strategic forces, when compared with their conventional counterparts, had diminished since the 1960s. In part, this represented a shift in priorities associated with the flexible response doctrine and a growing acceptance of arms control measures to help regulate the strategic balance. It was also indicative of a general change in thinking along the lines that McNamara had earlier suggested: that parity in strategic power between the United States and the Soviet Union made general nuclear war an increasingly unattractive and remote possibility. As a result, for military planning purposes, attention focused on what appeared to be the more probable threats facing the United States, particularly regional conflicts outside Europe in which conventional forces likely would have the dominant role. The 1991 Gulf War that arrayed United Nations forces against Iraq was one such example. By the early 1990s, with the end of the Cold War and the breakup of the Soviet Union, the need for a large strategic nuclear arsenal appeared open to question. Reflecting the new realities, SAC's 46-year reign as the nation's strategic sword ended on June 1, 1992, when some of its functions and responsibilities passed to a new air combat command and some to a joint Strategic Command. Although strategic weapons remained an element in national security policy, their days of dominance had long since passed.

NOTES

1. United States Strategic Bombing Survey (USSBS), *Summary Report (European War)* (Washington, D.C.: GPO, Sep 30, 1945), p 15.

2. For a balanced assessment of what the wartime air campaigns achieved, see Kenneth P. Werrell, "The Strategic Bombing of Germany in World War II: Cost and Accomplishments," *Journal of American History*, vol 73 (Dec 1986), pp 702–713. See also Ronald Schaffer, *Wings of Judgment: American Bombing in World War II* (New York: Oxford University Press, 1985), pp 20–34, which clearly establishes Douhet's influence on the shaping of American strategic air doctrine during the interwar period.

3. On postwar preparations and emerging Air Force aspirations, see Perry McCoy Smith, *The Air Force Plans for Peace, 1943–1945* (Baltimore, Md.: Johns Hopkins University Press, 1970); and Herman S. Wolk, *Planning and Organizing the Postwar Air Force, 1943–1947* (hereafter Wolk, *Postwar Air Force*) (Washington, D.C.: Office of Air Force History, 1984), pp 45–112.

4. Lawrence Freedman, *The Evolution of Nuclear Strategy* (New York: St. Martin's, 1981), p 22.

5. Address, Gen Carl Spaatz, "Air Power Pre-Eminent," Business Advisory Council, Hot Springs, Va., Oct 27, 1945, Spaatz Papers, box 268, Library of Congress, Washington, D.C. (LC).

6. Much of the adverse postwar British reaction to strategic bombing was directed against the wartime architect of RAF Bomber Command's offensive. For an overview, see Dudley Saward, *Bomber Harris: The Story of Marshal of the Royal Air Force Sir Arthur Harris* (Garden City, N.Y.: Doubleday, 1985), pp 324–330, *passim*. For declared Soviet views, see especially Stalin's denigration of atomic weapons in his September 25, 1946, *Pravda* interview, quoted in Mark E. Miller, *Soviet Strategic Power and Doctrine: The Quest for Superiority* (Miami, Fla.: Advanced International Studies Institute, University of Miami, 1982), p 3.

7. Bernard Brodie, ed., *The Absolute Weapon: Atomic Power and World Order* (New York: Harcourt, Brace, & World, 1946), p 76.

8. JCS Evaluation Board for Operation CROSSROADS, "Final Report," Jun 30, 1947, RG 330 (Records of the Office of the Secretary of Defense), CD 25–1–12, National Archives and Records Administration (NARA).

9. For early stockpile numbers, see David Alan Rosenberg, "U.S. Nuclear Stockpile, 1945 to 1950," *Bulletin of the Atomic Scientists*, vol 38 (May 1982), pp 25–30.

10. Rpt, Spaatz Board, "Effect of the Atomic Bomb on Employment, Size, Organization, and Composition of the Postwar Air Force," Oct 23, 1945, RG 341 (Headquarters, United States Air Force), OPD 384.3 Atomic (Aug 17, 45), sec 1, NARA.

11. Memo, Maj Gen W. E. Kepner, Chief, Special Weapons Gp, for DC/S, Materiel, sub.: Agenda for 2d Mtg USAF Aircraft and Weapons Board, Jan 16, 1948, RG 341, 334 Committees and Reports, NARA.

12. Memo, Maxwell for Lt Gen H. S. Vandenberg, ACAS/Ops, subj: Publicity on the Atomic Bomb, Dec 13, 1945, RG 341, OPD 384.3 Atomic (Aug 17, 1945), sec 1, NARA.

13. Robert F. Futrell, *Ideas, Concepts, Doctrine: A History of Basic Thinking in the United States Air Force, 1907–1964*, 2d printing (Maxwell AFB, Ala.: Air University, 1974), p 122.

14. *Ibid.*, p 185–187.

15. Steven L. Rearden, *History of the Office of the Secretary of Defense: The Formative Years, 1947–1950* (hereafter Rearden, *Formative Years*) (Washington, D.C.: Historical Office, Office of the Secretary of Defense, 1984), pp 425–432; Richard G. Hewlett and Francis Duncan, *A History of the United States Atomic Energy Commission* (hereafter Hewlett and Duncan, *History of the AEC*), 3 vols (Washington, D.C.: U.S. Atomic Energy Commission, 1972), vol 2, pp 154–184.

16. Diary Entry, William D. Leahy, May 6, 1948, LC.

17. Rearden, *Formative Years*, pp 434–436; NSC 30, "U.S. Policy on Atomic Warfare," Sep 10, 1948, U.S. Department of State, *Foreign Relations of the United States*, 1948, vol 1, pp 624–628 (hereafter *FRUS*).

18. Gen Carl Spaatz, "The Future of the Army Air Forces," *Military Review*, vol 26 (Jul 1946), pp 3–4; Wolk, *Postwar Air Force*, pp 118–124.

19. JCS 1259/27 para 4, subj: Unified Command Plan, Dec 11, 1946, and decision on Dec 12, 1946, RG 341, DCS/Ops, Dir of Plans, PO 323.361 (Nov 8, 1943), case 1, sec 4, NARA.

20. J. C. Hopkins and Sheldon A. Goldberg, *The Development of Strategic Air Command, 1946–1986* (hereafter Hopkins and Goldberg, *Development of SAC*) (Offutt AFB, Neb.: Office of the Historian, HQ SAC, 1986), p 2.

21. For an excellent assessment, see Harry R. Borowski, *A Hollow Threat: Strategic Air Power and Containment before Korea* (Westport, Conn.: Greenwood Press, 1982).

22. Group strength is compiled as of November 30, 1946, from memo, Gen George C. Kenney to Gen Carl Spaatz, subj: SAC Strength and Capabilities, Jan 15, 1948, Spaatz Papers, box 257, Command-Strategic folder, LC; Bomber range capabilities are from Lloyd S. Jones, *U.S. Bombers 1928 to 1980s* (Fallbrook, Calif.: Aero Publishers, 3d ed, 1980), pp 48–50, 97. For the status of access to overseas bases, see James F. Schnabel, *The Joint Chiefs of Staff and National Policy, 1945–1947*, vol 1 in *The History of the Joint Chiefs of Staff* (Wilmington, Del.: Michael Glazier, 1979), pp 299–346.

23. Lt Gen Walter E. Todd, "Evolution of Aerospace Power," *Air University Quarterly Review*, vol 12 (winter–spring 1960–1961), p 11.

24. [George F. Kennan], "The Sources of Soviet Conduct," *Foreign Affairs*, vol 25 (Jul 1947), pp 566–582.

25. For an excellent (and largely overlooked) analysis of the debate over the role of military power in containment, see Seyom Brown, *The Faces of Power* (New York, N.Y.: Columbia University Press, 1968), pp 31–52. See also Samuel P. Huntington, *The Common Defense* (New York, N.Y.: Columbia University Press, 1961), chaps 1–5.

26. David Alan Rosenberg, "The Origins of Overkill: Nuclear Weapons and American Strategy, 1945–1960," *International Security*, vol 7 (spring 1983), p 15; Ltr, Lt Gen Curtis LeMay, CG SAC, to Gen Nathan Twining, Vice COS, subj: Recce and Intelligence Capabilities, Apr 26, 1951, Twining Papers, box 54, Reading file, May 1951, LC.

27. For a contemporary analysis of targeting policy, see Col John H. de Russy, "Selecting Target Systems and Targets," *Air University Quarterly Review*, vol 1 (spring 1947), pp 69–78. On the development of early postwar strategic plans, see Steven T. Ross, *American War Plans, 1945–1950* (New York: Garland, 1988), and Alfred F. Hurley and Robert C. Ehrhart, eds., *Air Power and Warfare*, Proceedings of the 8th Military History Symposium, USAF Academy, Oct 18–20, 1978 (Washington, D.C.: Office of Air Force History and U.S. Air Force Academy, 1979), pp 224–231.

28. Futrell, *Ideas, Concepts, Doctrine*, p 122.

29. R. D. Little, *Foundations of an Atomic Air Force and Operation SANDSTONE, 1946–1948* (unclassified version), vol 2 in *The History of Air Force Participation in the Atomic Energy Program, 1943–1953* (Pentagon: AF/CHO, n.d.), pp 255–265.

30. Thomas H. Etzold and John Lewis Gaddis, eds., *Containment: Documents on American Policy and Strategy, 1945–1950* (New York: Columbia University Press, 1978), pp 360–364.

31. WSEG Rpt No 1, JCS 1952/11, Feb 10, 1950, RG 218 (Records of the Joint Chiefs of Staff), CCS 373 (10–23–48), sec 6, Bulky Package. See also David Alan Rosenberg, "American Atomic Strategy and the Hydrogen Bomb Decision," *Journal of American History*, vol 66 (Jun 1979), pp 83–84.

32. U.S. President's Air Policy Commission, *Survival in the Air Age* (Washington, D.C.: GPO, Jan 1, 1948).

33. For a more detailed account, see Rearden, *Formative Years*, chaps 11–13. The best discussion of congressional attitudes is Edward A. Kolodziej, *The Uncommon Defense and Congress: 1945–1963* (Columbus: Ohio State University Press, 1966).

34. Rearden, *Formative Years*, 298–299. The best account of the crisis, focusing on its military aspects, is Daniel F. Harrington, "American Policy in the Berlin Crisis, 1948–1949" (Ph.D. diss., Indiana University, 1979).

35. Memo, Gen S. E. Anderson to Gen William F. McKee, subj: Air Force Policy Line, [Mar 10, 1949], Vandenberg Papers, box 57, Policy Letters folder, LC.

36. For the Navy's viewpoint on this controversy, see Jeffrey G. Barlow, *Revolt of the Admirals: The Fight for Naval Aviation* (Washington, D.C.: Naval Historical Center, 1994); Hurley and Ehrhart, eds., *Air Power and Warfare*, 245–271; Paolo E. Coletta, *The United States Navy and Defense Organization, 1947–1953* (Newark: University of Delaware Press, 1981); and Vincent Davis, *Postwar Defense Policy and the U.S. Navy, 1943–1946* (Chapel Hill: University of North Carolina Press, 1966).

37. Rearden, *Formative Years*, pp 393–402. For the Key West and Newport agreements, see Alice C. Cole *et al.*, *The Department of Defense: Documents on Establishment and Organization, 1944–1978* (hereafter Cole *et al.*, *DoD Documents*) (Washington, D.C.: Historical Office, Office of the Secretary of Defense, 1979), pp 275–293.

38. Hewlett and Duncan, *History of the AEC*, vol 2, pp 161–165. For the development of U.S. policy, see Samuel R. Williamson, Jr., and Steven L. Rearden, *The Origins of U.S. Nuclear Strategy, 1945–1953* (New York: St. Martin's, 1993).

39. Futrell, *Ideas, Concepts, Doctrine*, p 140.

40. Harry S. Truman, *Memoirs: Years of Trial and Hope* (Garden City, N.Y.: Doubleday, 1956), p 309.

41. Memo for the Record, Gen Muir S. Fairchild, subj: Mtg to Review Air Defense Deficiencies, Sep 30, 1949; memo, Fairchild for Vandenberg, subj: Air Defense, Sep 30, 1949; memo, Vandenberg for JCS, subj: Air Defense of the United States, Nov 16, 1949, all in Muir S. Fairchild Papers, box 1, Correspondence Sep 30 and Dec 7, 1949, folder, LC.

42. NSC 68, "A Report to the National Security Council by the Executive Secretary on United States Objectives and Programs for National Security, Apr 14, 1950," with Editor's introduction, in *Naval War College Review*, vol 27 (May–Jun 1975), p 108, *passim*.

43. Memo, Col David A. Burchinal for Asst for Programming, DCS/O, *et al.*, subj: Air Force FY 1953 Program, Nov 23, 1951, Twining Papers, box 117, AF Council 1951–1952 folder, LC.

44. Hopkins and Goldberg, *Development of SAC*, pp 21, 30.

45. In the late 1940s, wings succeeded groups as the primary USAF organizational entity.

46. Doris M. Condit, *History of the Office of the Secretary of Defense: The Test of War, 1950–1953* (Washington, D.C.: Historical Office, Office of the Secretary of Defense, 1988), pp 261–305.

47. Roger M. Anders, ed., *Forging the Atomic Shield: Excerpts from the Office Diary of Gordon E. Dean* (Chapel Hill: University of North Carolina Press, 1987), pp 188–189.

48. Walter S. Poole, *The JCS and National Policy, 1950–1952*, vol 1 in *The History of the Joint Chiefs of Staff* (Washington, D.C.: 1979), pp 142–149; Robert H. Connery and Demetrios Caraley, eds., *National Security and Nuclear Strategy* (New York: Academy of Political Science, 1983), pp 68–72; Memo, LeMay for Vandenberg, subj: Composition of Atomic Weapon Stockpile, Feb 17, 1951, LeMay Papers, box B–197, folder B–9814, LC.

49. DOE Press Release, U.S. Dept of Defense and Dept of Energy, "Summary of Declassified Nuclear Stockpile Information," Jun 1994.

50. Poole, *JCS and National Policy*, p. 173.

51. Memo, Lt Gen Laurence S. Kuter for COS, USAF, subj: Development of a Military Strategic Concept, Jul 1, 1952, Twining Papers, box 56, Reading file, Jul 1952 folder, LC.

52. "The New Phase — A Statement of Air Force Policy," Aug 2, 1952, Twining Papers, box 122, TS folder no. 1, LC.

53. Memo, Lt Gen Laurence S. Kuter, Act Vice COS, subj: Bar Harbor Paper, Aug 3, 1952, Twining Papers, box 122, TS folder no. 1, LC.

54. Ltr, Twining for Gen Benjamin W. Chidlaw, CG ADC, subj: Issuance of Basic Doctrine Manual, May 11, 1953, Vandenberg Papers, box 44, Command Letters (2) folder, LC.

55. Notes, L. A. Minnich, Jr., subj: Legislative Leadership Mtg, Apr 30, 1953, Eisenhower papers, Whitman file, DDE Diary Series, Apr 1953 Diary folder, Dwight D. Eisenhower Library, Abilene Kans. (DDEL). On the background and origins of the administration's reassessment, see Robert J. Watson, *The Joint Chiefs of Staff and National Policy, 1953–1954*, vol 5 in *History of the Joint Chiefs of Staff* (Washington, D.C.: Historical Division, JCS, 1986), pp 1–37; John Lewis Gaddis, *Strategies of Containment* (New York: Oxford University Press, 1982), pp 127–163; Warner R. Schilling, Paul Y. Hammond, and Glenn H. Snyder, *Strategy, Politics, and Defense Budgets* (New York: Columbia University Press, 1962), pp 383–524; and Connery and Caraley, eds., *National Security and Nuclear Strategy*, pp 52–68.

56. COS (52) 362, "Defence Policy and Global Strategy," Jul 15, 1952, in Alan Macmillan and John Baylis, *A Reassessment of the British Global Strategy Paper of 1952* (College Park, Md.: Center

for International and Security Studies at Maryland, 1994), 72–108; see also Air Staff Study, "Defense Policy and Global Strategy," Jul 24, 1952, Tab 7 to memo, Maj Gen Herbert B. Thatcher, Dep Dir, Directorate of Plans, for Twining, subj: Rpt of UK Chiefs of Staff on UK Defense Policy and Global Strategy, Jul 28, 1952, RG 341, DCS/Ops, Dir of Plans, OPD 323.361 sec 6 annex (Aug 16, 50), NARA.

57. Desmond Ball and Jeffrey Richelson, eds., *Strategic Nuclear Targeting* (Ithaca, N.Y.: Cornell University Press, 1986), p 111.

58. On the background and origins of the British decision, see Andrew J. Pierre, *Nuclear Politics: The British Experience with an Independent Strategic Force, 1939–1970* (London: Oxford University Press, 1972), chaps 4, 5; and Ian Clark and Nicholas J. Wheeler, *The British Origins of Nuclear Strategy, 1945–1955* (Oxford: Clarendon Press, 1989), pp 160–182.

59. Cable, Twining to Norstad, subj: JCS Mtg with Slessor, Jul 31, 1952, RG 341, DCS/Ops, Dir of Plans, OPD 323.361 sec 6 annex (Aug 16, 50), NARA.

60. NSC 162/2, "Basic National Security Policy," Oct 30, 1953, *FRUS*, 1952–1954, vol 2, p. 582.

61. Dwight D. Eisenhower, *The White House Years: Mandate for Change, 1953–1956* (Garden City, N.Y.: Doubleday, 1963), p 449.

62. John Foster Dulles, "The Evolution of Foreign Policy," *Department of State Bulletin*, vol 30 (Jan 15, 1954), pp 107–110.

63. Annual Message to the Congress on the State of the Union, Jan 7, 1954, *Public Papers of the Presidents of the United States: Dwight D. Eisenhower, 1954* (Washington, D.C.: GPO, 1960), p 10.

64. Rosenberg, "Origins of Overkill," pp 31–32.

65. Jerome H. Kahan, *Security in the Nuclear Age: Developing U.S. Strategic Arms Policy* (Washington, D.C.: Brookings Institution, 1974), 26–30. For the impact of threat assessments, see Richard K. Betts, "A Nuclear Golden Age? The Balance Before Parity," *International Security*, vol 11 (winter 1986–1987), pp 3–32; see also John Logsdon, ed., *Exploring the Unknown: Selected Documents in the History of the U.S. Civil Space Program* (Washington, D.C.: NASA, SP–4407, 1995), vol 1, pp 217–233.

66. Brig Gen Dale O. Smith, *U.S. Military Doctrine* (Boston: Little, Brown & Co., 1955).

67. Office of the Assistant Secretary of Defense (Comptroller), *National Defense Budget Estimates for FY 1985* (Washington, D.C.: DoD, March 1984), table 6–13, pp 116–117. For SAC budget figures, see Paul R. Schratz, ed., *Evolution of the American Military Establishment Since World War II* (Lexington, Va.: George C. Marshall Research Foundation, 1978), p 63.

68. Lt Gen L. S. Kuter, "No Room for Error," *Air Force Magazine* (Nov 1954), p 30.

69. Robert F. Futrell, "Some Patterns of Air Force Thought," *Air University Review*, vol 15 (Jan–Feb 1964), pp 81–88; AFM 1–1, *Basic Aerospace Doctrine*, Mar 16, 1984, p A-3.

70. Thomas M. Coffey, *Iron Eagle: The Turbulent Life of General Curtis LeMay* (New York: Crown Publishers, 1986), p 280.

71. Gen Curtis E. LeMay with McKinlay Kantor, *Mission with LeMay: My Story* (Garden City, N.Y.: Doubleday, 1965), p 431.

72. Cable, LeMay to Twining, subj: SAC Capabilities, May 15, 1953, LeMay Papers, box B–203, B–27106 folder, LC.

73. Hopkins and Goldberg, *Development of SAC*, p 58.

74. Todd, "Evolution of Aerospace Power," p 17; Hopkins and Goldberg, *Development of SAC*, pp 34, 51, 55–56; Walter J. Boyne, *Boeing B–52: A Documentary History* (London: Jane's, 1981), pp 123–128, 145–148.

75. John Prados, *The Soviet Estimate: U.S. Intelligence Analysis and Russian Military Strength* (New York: Dial Press, 1982), p 29; David Alan Rosenberg, "Toward Armageddon: The Foundations of United States Nuclear Strategy, 1945–1961" (Ph.D. thesis, University of Chicago, 1983), p 158; intvw, John T. Mason with Gen Nathan F. Twining, Aug 17, 1967, Arlington, Va. (Columbia University Oral History Project), pt 3 of 4, 131–132.

76. Coffey, *Iron Eagle*, p 340. Twining mentions a similar episode involving forty-seven U.S. planes, though he provides no date for when it might have occurred. See Twining intvw, Aug 17, 1967, p 134.

77. R. Cargill Hall, "The Eisenhower Administration and the Cold War: Framing American Astronautics to Serve National Security," *Prologue*, vol 27 (spring 1995), pp 59–72.

78. Prados, *Soviet Estimate*, pp 31–35; Anne Karalekas, "History of the Central Intelligence Agency," p. 59, published as U.S. Congress, Senate, Select Committee to Study Governmental Operations with Respect to Intelligence Activities, *Supplementary Detailed Staff Reports on Foreign*

and Military Intelligence, book 4, Sen Rpt 94–755, 94th Cong, 2d sess (Washington, D.C.: GPO, 1976).

79. Memo, Finletter for Vandenberg, [subj: LeMay's Views on SAC's Mission], May 12, 1950, Vandenberg Papers, box 84, Selected Memos folder, LC.

80. JCS 2056/7, Aug 12, 1950, and decision on Aug 15, 1950, cited in Rosenberg, "Toward Armageddon," p 156.

81. Diary Entry, Curtis E. LeMay, Jan 16, 1951, LeMay Papers, box 103, diary no. 3 folder; Memo, LeMay for Vandenberg, subj: Force Requirements of the Strategic Offensive, Jan 20, 1951, LeMay Papers, box B–197, B–9328 folder, LC. LeMay did not specify exactly how many planes this would involve.

82. Memo, LeMay for Vandenberg, subj: SAC Capabilities, Jan 20, 1951, LeMay Papers, box B–197, B–9327 folder, LC.

83. For an example of LeMay's comments, see "Briefing Given to the Representatives of All Services at SAC Headquarters" (hereafter "Briefing at SAC HQ"), Mar 15, 1954, *International Security*, vol 6 (winter 1981–1982), pp 26–28.

84. T. F. Walkowicz, "Counter-Force Strategy," *Air Force Magazine* (Feb 1955), p 29.

85. Address, Gen Nathan F. Twining, Untitled, Chamber of Commerce Banquet, Galveston, Tex., Feb 9, 1954, Twining Papers, box 123, Speeches and Writing file, Feb 1954 folder, LC.

86. "Briefing at SAC HQ," p 18.

87. Memo, no author, subj: Comments on WSEG #10 for Gen Twining, n.d. [*ca.* early 1954], Twining Papers, box 120, Ready file, 1954 folder, LC. Emphasis in the original.

88. "Briefing on WSEG Report No 12," Apr 6, 1955, *International Security*, vol 6 (winter 1981–1982), p 34.

89. N. J. Wheeler, "British Nuclear Weapons and Anglo-American Relations, 1945–54," *International Affairs* (London), vol 62 (winter 1985–1986), pp 78–86.

90. Andrew Brookes, *V-Force: The History of Britain's Airborne Deterrent* (London: Jane's, 1982), p 48.

91. Ball and Richelson, eds., *Strategic Nuclear Targeting*, pp 113–114.

92. Robert A. Kilmarx, *A History of Soviet Air Power* (New York: Praeger, 1962), pp 83–117, 125–130; M. N. Kozhevnikov, *The Command and Staff of the Soviet Army Air Force in the Great Patriotic War, 1941–1945* (Moscow: 1977; Washington, D.C.: GPO for USAF, n.d.); David Holloway, *Stalin and the Bomb* (New Haven, Conn.: Yale University Press, 1994).

93. V. D. Sokolovskii, ed., *Soviet Military Strategy*, trans. Herbert S. Dinerstein, Leon Goure, and Thomas W. Wolfe (Englewood Cliffs, N.J.: Prentice-Hall for RAND Corp., 1963), p 305. Emphasis added.

94. Kahan, *Security in the Nuclear Age*, p 30; Prados, *Soviet Estimate*, pp 38–43; Paul J. Murphy, ed., *The Soviet Air Forces* (Jefferson, N.C.: McFarland, 1984), p 182. Tu–95 and Mya–4 characteristics are from *Jane's All the World's Aircraft, 1973–1974*, pp 485, 489–490.

95. NIE 11–3–55, May 17, 1955, cited in Rosenberg, "Origins of Overkill," p 39; *Jane's All the World's Aircraft, 1973–1974*, pp 488–489.

96. Barton Wright, *Soviet Missiles*, vol 1 in Randall Forsberg, ed., *World Weapon Database*, 2 vols (Lexington, Mass.: DC, Heath, 1986), pp 31, 34.

97. The principal studies in this connection were *Special Staff Report: The Selection of Strategic Air Bases* (RAND R–244–S, Mar 1, 1953) and A. J. Wohlstetter *et al.*, *Selection and Use of Strategic Air Bases* (RAND R–266, Apr 1954).

98. Hopkins and Goldberg, *Development of SAC*, pp 72–73.

99. For the origins and workings of the panel, see James R. Killian, *Sputnik, Scientists, and Eisenhower: A Memoir of the First Special Assistant to the President for Science and Technology* (Cambridge, Mass.: MIT Press, 1977).

100. Technological Capabilities Panel of the Science Advisory Committee, "Meeting the Threat of Surprise Attack," Feb 14, 1955, vol 1, 5, pp 37–46, Eisenhower Papers, White House Office of Special Assistant for National Security file, DDEL.

101. The best account of Air Force missile programs is Jacob Neufeld, *The Development of Ballistic Missiles in the United States Air Force, 1945–1960* (hereafter Neufeld, *Ballistic Missiles*) (Washington, D.C.: OAFH, 1990). See also Edmund Beard, *Developing the ICBM: A Study in Bureaucratic Politics* (New York: Columbia University Press, 1976).

102. Richard H. Kohn and Joseph P. Harahan, eds., *Strategic Air Warfare: An Interview with Generals Curtis E. LeMay, Leon W. Johnson, David A. Burchinal, and Jack J. Catton* (Washington,

D.C.: Office of Air Force History, 1988), p 100.

103. Memo, Gardner for Twining, subj: ICBM Program, Jan 27, 1954, Twining Papers, box 72, SecAF Correspondence 1954, folder 2; Memo, Gardner for Talbott and Twining, subj: ICBM Acceleration Plan, Mar 11, 1954, Twining Papers, box 122, TS file 1952–57, folder 1, LC.

104. Dwight D. Eisenhower, *The White House Years: Waging Peace, 1956–1961* (Garden City, N.Y.: Doubleday, 1965), p 208; Futrell, *Ideas, Concepts, Doctrine*, pp 244–247.

105. Memo, subj: Discussion at 268th Mtg of the NSC on Dec 1, 1955, Dec 2, 1955, Eisenhower Papers 1953–61, Whitman file, NSC Series, box 7, DDEL.

106. Memcon, Col A. J. Goodpaster, subj: Mtg of Nov 8, 1956, on Policy Questions Affecting DoD, Nov 9, 1956, Eisenhower Papers, Whitman file, DDE Diary Series, box 19, Nov 1956 Staff Memos folder, DDEL.

107. Memcon, Goodpaster, subj: Mtg of Dec 19, 1956, on DoD Budget, Dec 20, 1956, DDE Diary Series, box 20, Dec 1956 Staff Memos folder, DDEL.

108. Neufeld, *Ballistic Missiles*, pp 119–147.

109. Robert William Love, Jr., ed., *The Chiefs of Naval Operations* (Annapolis, Md.: Naval Institute Press, 1980), pp 278–279.

110. Memo, SecDef for AFPC, subj: Clarification of Roles and Missions . . ., Nov 26, 1956, in Cole *et al.*, *DoD Documents*, p 311.

111. Maxwell D. Taylor, *The Uncertain Trumpet* (New York: Harper & Row, 1960).

112. Graham Spinardi, "Why the U.S. Navy Went for Hard-Target Counterforce in Trident II (And Why It Didn't Get There Sooner)," *International Security*, vol 15 (fall 1990): p 152.

113. Brig Gen Robert C. Richardson III, "The Fallacy of the Concept of Minimum Deterrence," *Air University Quarterly Review*, vol 12 (spring 1960), p 112.

114. Albert J. Wohlstetter and Fred Hoffman, *Protecting U.S. Power to Strike Back in the 1950s and 1960s* (RAND R–290, Sep 1, 1956), pp 1–4, 11–12, 21; Fred Kaplan, *The Wizards of Armageddon* (New York: Simon & Schuster, 1983), pp 117–124.

115. Harry B. Yoshpe, *Our Missing Shield: The U.S. Civil Defense Program in Historical Perspective* (Washington, D.C.: Federal Emergency Management Agency, 1981), pp 204–205.

116. NSC Action No 1691b (1), quoted in Memo of Discussion, 318th Mtg, NSC, Apr 4, 1957, in *FRUS*, 1955–1957, vol 19, p 463. For the background and origins of the Gaither Committee, see Thomas E. Cronin and Sanford D. Greenberg, eds., *The Presidential Advisory System* (New York: Harper & Row, 1969), pp 185–208.

117. Kaplan, *Wizards of Armageddon*, pp 119–129. The U–2 program was so secret that it was not mentioned in the Killian surprise attack report; for the President, the information was attached as a separate, highly classified annex. Except for those directly involved in the program, few knew of its existence before May 1, 1960. Karl Harr affirms that the U–2 never even appeared as an item in National Security Council deliberations "until it tore its britches." Kenneth W. Thomson, ed., *The Eisenhower Presidency: Eleven Intimate Perspective of Dwight D. Eisehnower*, vol 3 in *Portraits of American Presidents* (Lanham, Md.: University Press of America, 1984), p 97; see also Logsdon, ed., *Exploring the Unknown*, vol 1, pp 217–233.

118. Security Resources Panel of the Science Advisory Committee, *Deterrence and Survival in the Nuclear Age* (Nov 7, 1957; Washington, D.C.: GPO, 1976), p 14, *passim*.

119. Memo, Goodpaster, subj: Conf. with the President on Nov 4, 1957, Nov 6, 1957, Eisenhower Presidential Papers, Whitman file, DDE Diary Series, box 28, Nov 1957 Staff Notes folder, DDEL.

120. JCS 2101/284, Dec 4, 1957, RG 218, Modern Military Division, NARA.

121. Memo, LeMay for White, subj: Vulnerability of SAC Bombers in Early 1960s, Feb 18, 1959; Memo, LeMay for White, subj: B–52 Penetration Aids Modification Program, Feb 20, 1959, both in White Papers, box 25, Air Force Council file, LC.

122. Hopkins and Goldberg, *Development of SAC*, pp 89, 92.

123. Futrell, *Ideas, Concepts, Doctrine*, p 320; Ltr, Gen Thomas S. Power to JCS, subj: Establishment of an Airborne Alert, Jun 23, 1960; Memo, LeMay for Power, subj: Establishment of Airborne Alert, Jul 11, 1960, both in Thomas D. White Papers, box 34, SAC folder, LC.

124. Hopkins and Goldberg, *Development of SAC*, p 92.

125. Bruce G. Blair, *Strategic Command and Control* (Washington, D.C.: Brookings Institution, 1985), p 103.

126. Quoted in David Alan Rosenberg, "'A Smoking Radiating Ruin at the End of Two Hours': Documents on American Plans for Nuclear War with the Soviet Union, 1954–1955," *International Security*, vol 6 (winter 1981–1982), p 27.

127. White testimony, Feb 19, 1959, U.S. Congress, House, Committee on Appropriations, *Hearings: Department of Defense Appropriations for 1960*, 86th Cong, 1st sess (Washington, D.C.: GPO, 1959), pt 1, p 929.

128. Marc Trachtenberg, *History and Strategy* (Princeton, N.J.: Princeton University Press, 1991), p 162.

129. Rosenberg, "Origins of Overkill," p 47.

130. *Ibid.*, pp 62–63.

131. Stewart Menaul, *Countdown: Britain's Strategic Nuclear Forces* (London: Robert Hale, 1980), pp 91, 113.

132. Memo, White for Gerhart, subj: Massive Retaliation Forces, May 29, 1958, Thomas D. White Papers, box 19, COS TS folder no. 3, LC.

133. Ltr, Power to Twining, subj: Command and Control of Polaris, Mar 6, 1959, Thomas D. White Papers, box 27, Command SAC folder, LC.

134. "Staff Study on U.S. Strategic Command," encl to Memo, Maj Gen Hewitt T. Wheless, Dir of Plans, DCS/P&P, for Gen White, subj: Staff Study, Mar 31, 1960, White Papers, box 34, SAC folder, LC.

135. Rosenberg, "Arleigh Albert Burke," pp 302–304; Rosenberg, "Origins of Overkill," p 56.

136. For the organization and internal operation of the JSTPS, see Capt Mark D. Mariska, USA, "The Single Integrated Operational Plan," *Military Review*, vol 52 (Mar 1972), pp 32–39; and Maj Gen Jerome F. O'Malley, USAF, "JSTPS — The Link between Strategy and Execution," *Air University Review*, vol 28 (May–Jun 1977), pp 39–46.

137. Memcon, John S. D. Eisenhower, subj: Mtg on Jul 6, 1960 re Strategic Targets, Jul 6, 1960, Eisenhower Papers, Whitman file, DDE Diary Series, box 51, Jul 1960 Staff Notes folder, DDEL; Memcon, Goodpaster, subj: Mtg on Aug 11, 1960 to Discuss Targeting, Aug 13, 1960, DDE Diary Series, box 51, Aug 1960 Staff Notes folder no. 3, DDEL; Hopkins and Goldberg, *Development of SAC*, p 90.

138. Briefing, JCS Chairman for the President, JCS Single Integrated Operational Plan 1962 (SIOP–62), Sep 13, 1961, encl to JCS 2156/281, RG 218, NARA (also printed in *International Security*, vol 12 [summer 1987], pp 41–51). See also Ball and Richelson, eds., *Strategic Nuclear Targeting*, p 62.

139. Herbert F. York, *Making Weapons, Talking Peace* (New York: Basic Books, 1987), p 185.

140. Testimony, Gates, Jan 13, 1960, U.S. Congress, House, Committee on Appropriations, *Department of Defense Appropriations for 1961*, 86th Cong, 2d sess (Washington, D.C.: GPO, 1960), pt 1, p 26.

141. Gen Curtis E. LeMay, "The Present Pattern," *Air University Quarterly Review*, vol 12 (winter–spring 1960–1961), pp 25–39.

142. *Ibid.*, p 25.

143. Quoted in Alain C. Enthoven and K. Wayne Smith, *How Much Is Enough? Shaping the Defense Program, 1961–1969* (New York: Harper & Row, 1971), p 122.

144. Memcon, Goodpaster, subj: Conf. with the President, Kistiakowsky, *et al.*, Nov 25, 1960, Dec 1, 1960, Eisenhower Papers, DDEL.

145. Richard Fryklund, *100 Million Lives: Survival in a Nuclear War* (New York: Macmillan, 1962), p 28. On the origins and development of the "no-cities" strategy, including the work initiated by Parrish and his study team, see Alfred Goldberg, *A Brief Survey of the Evolution of Ideas about Counterforce* (hereafter Goldberg, *Evolution of Counterforce*) (Santa Monica, Calif.: RAND Corp., RM–5431–PR, Oct 1967; rev. Mar 1981), pp 9, 17–18.

146. Memo, Lemnitzer (for JCS) for McNamara, subj: "Doctrine" on Thermonuclear Attack, Apr 21, 1961, DoD/FOIA Collection, Pentagon.

147. Ball and Richelson, eds., *Strategic Nuclear Targeting*, pp 63–64; Rowen, "Formulating Strategic Doctrine," p 230.

148. App I, Draft Memo, McNamara for Kennedy, subj: Recommended Long-range Nuclear Delivery Forces, 1963–1967, Sep 23, 1961, DoD/FOIA Collection, Pentagon.

149. Thomas S. Power, *Design for Survival* (New York: Coward-McCann, 1966), pp 83, 119–120.

150. Goldberg, *Evolution of Counterforce*, p 25.

151. Robert S. McNamara, "The 'No-Cities' Doctrine," in Robert J. Art and Kenneth N. Waltz, eds., *The Use of Force: International Politics and Foreign Policy*, 2d ed (Lanham, Md.: University Press of America, 1983), p 149.

152. William W. Kaufmann, *The McNamara Strategy* (New York: Harper & Row, 1964), pp

53–55; Futrell, *Ideas, Concepts, Doctrine*, pp 332, 422–423; and Robert S. McNamara, *The Essence of Security: Reflections in Office* (New York: Harper & Row, 1968), pp 71–73. Weapons characteristics from Thomas B. Cochran, William M. Arkin, and Milton M. Hoenig, *Nuclear Weapons Databook*, 5 vols (Cambridge, Mass.: Ballinger, 1984), vol 1, pp 11, 111–112.

153. Futrell, *Ideas, Concepts, Doctrine*, pp 344–349.

154. Memo, LeMay for White, sub.: USAF Tasks and Objective Force Structure (1959–1970), [*ca.* Apr 28, 1959], approved May 5, 1959; memo, LeMay for White, subj: USAF Objective Force Structure (1959–1970), approved May 12, 1959; decision memo, Maj Gen Jacob E. Smart, Assist. V/COS, for D/COS (P&P), and D/COS (M), subj: USAF Tasks and Objective Force Structure (1959–1970), May 15, 1959, all in White Papers, box 25, Air Force Council 1959 folder, LC.

155. Power, *Design for Survival*, p 128. Also see LeMay's comments in Kohn and Harahan, eds, *Strategic Air Warfare*, pp 101–102.

156. Prados, *Soviet Estimate*, pp 111–119.

157. Draft memo, McNamara for the President, subj: Recommended FY 1964–1968 Strategic Retaliatory Forces, Nov 21, 1962, DoD/FOIA Collection, Pentagon.

158. McGeorge Bundy, *Danger and Survival* (New York: Random House, 1988), p 354.

159. McNamara quoted in Steward Alsop, "Our New Strategy," *The Saturday Evening Post*, Dec 1, 1962: p 18.

160. Bundy, *Danger and Survival*, p 355.

161. U.S. Commission on the Organization of the Government for the Conduct of Foreign Policy, *Appendix K: Adequacy of Current Organization: Defense and Arms Control* (Washington, D.C.: GPO, Jun 1975), pp 231–232.

162. Quoted in Michael Charlton, *From Deterrence to Defence: The Inside Story of Strategic Policy* (Cambridge, Mass.: Harvard University Press, 1987), pp 22–23. McNamara added: "Once one had a sufficient force to deter one's opponents from initiating the use of nuclear weapons, 'your' nuclear force or 'our' nuclear force had no impact on actions in a political crisis." *Ibid.*, p 23.

163. *Ibid.*, p 23.

164. Robert S. McNamara, *Blundering into Disaster: Surviving the First Century of the Nuclear Age* (New York: Pantheon Books, 1986), pp 8–9, 44–46.

165. Draft memo, McNamara for the President, subj: Recommended FY 1965–1969 Strategic Retaliatory Forces, Dec 6, 1963, DoD/FOIA Collection, Pentagon.

166. McNamara estimated that for assured destruction purposes, as he outlined them, 200 1-megaton warheads on target would be sufficient. His estimate of the number of survivable launchers this would require was less clear. For a more detailed examination of the force-sizing problem, see U.S. Commission on the Organization of the Government for the Conduct of Foreign Policy, *Appendix K*, pp 139–149.

167. Based on constant dollar comparisons, the reduction described was calculated from *Budget Estimates for FY 1985*, table 6–3, p 80.

168. Hopkins and Goldberg, *Development of SAC*, p 126.

169. *Ibid.*, p 110; Enthoven and Smith, *How Much Is Enough?* pp 251–262; McNamara, *Essence of Security*, pp 63–64.

170. Alton H. Quanbeck and Archie L. Wood, *Modernizing the Strategic Bomber Force* (Washington, D.C.: Brookings Institution, 1976), p 2.

171. LeMay, quoted in Schratz, ed., *Evolution of the American Military Establishment*, p 66. For an account of the F–111, see also U.S. Commission on the Organization of the Government for the Conduct of Foreign Policy, *Appendix K*, pp 118–138.

172. David N. Schwartz, *NATO's Nuclear Dilemmas* (Washington, D.C.: Brookings Institution, 1983), pp 165–173.

173. NATO Communiqué, Dec 14, 1967, quoted in Freedman, *Evolution of Nuclear Strategy*, p 285.

174. *History of the Strategic Air Command FY 1970*, p 13. Unless otherwise indicated, the material in this section is derived from SAC histories and the author's interviews.

175. On the need for being able to "prevail," see Scott D. Sagan, *Moving Targets: National Strategy and National Security* (Princeton, N.J.: Princeton University Press, 1989), pp 70–72.

176. Ball and Richelson, eds., *Strategic Nuclear Targeting*, pp 118–119.

177. The best account of the transition to Polaris is Peter Nailor, *The Nassau Connection: The Organization and Management of the British Polaris Project* (London: HMSO, 1988).

178. Barry M. Blechman and Edward N. Luttwak, eds., *International Security Yearsbook, 1983–84*

(New York: St. Martin's, 1984), 74–76; Ball and Richelson, eds., *Strategic Nuclear Targeting*, p 135.

179. The SS–9 carried a single warhead with a yield of 18–20 megatons. Wright, in Forsberg, ed., *World Weapon Database*, vol 1, p 32, By way of comparison, the largest warhead in the U.S. arsenal at the time was the Titan II W53 with an estimated yield of 9 megatons. Cochran, Arkin, and Hoenig, *Nuclear Weapons Databook*, vol 1, p 39.

180. Second Annual Report to the Congress on U.S. Foreign Policy, Feb 25, 1971, *Public Papers of the Presidents of the United States: Richard Nixon, 1971* (hereafter *Nixon Public Papers, 1971*) (Washington, D.C.: GPO, 1972), p 309; John M. Collins, *American and Soviet Military Trends Since the Cuban Missile Crisis* (Washington, D.C.: Center for Strategic and International Studies, 1978), p 92.

181. For a more detailed analysis of the various restraints operating at the time, see Warner R. Schilling, "U.S. Strategic Concepts in the 1970s," *International Security*, vol 6 (fall 1981), pp 48–79.

182. Richard Nixon, *RN: The Memoirs of Richard Nixon* (New York: Grosset & Dunlap, 1978), p 415; Lt Gen John W. Pauly, USAF, "The Thread of Doctrine," *Air University Review*, vol 27 (May–Jun 1976), p 8.

183. Hopkins and Goldberg, *Development of SAC*, pp 157–159, 171, 183; Schratz, ed., *Evolution of the American Military Establishment Since World War II*, p 67. The actual cost of 76 FB–111s eventually approximated the sum first estimated necessary to procure the 253 originally ordered.

184. Power, *Design for Survival*, pp 76–77; Futrell, *Ideas, Concepts, Doctrine*, pp 369–371.

185. For SALT I agreements, see U.S. Arms Control and Disarmament Agency, *Arms Control and Disarmament Agreements: Texts and Histories of Negotiations* (Washington, D.C.: ACDA, 1982), pp 132–157. The fullest accounts of this phase of the negotiations are Gerard Smith, *Doubletalk: The Story of SALT I* (New York: Doubleday, 1980), and John Newhouse, *Cold Dawn* (New York: Holt, Rinehart & Winston, 1973).

186. See Thomas W. Wolfe, *The SALT Experience* (Cambridge, Mass.: Ballinger, 1979), pp 110–111.

187. Henry Kissinger, *White House Years* (Boston: Little, Brown, 1979), p 216.

188. Second Annual Report to the Congress on U.S. Foreign Policy, Feb 25, 1971, *Nixon Public Papers, 1971*, p 310.

189. Lynn E. Davis, *Limited Nuclear Options: Deterrence and the New American Doctrine*, Adelphi Paper 121 (London: International Institute for Strategic Studies, 1976), pp 1–4.

190. U.S. Department of Defense, *Report of the Secretary of Defense James R. Schlesinger to the Congress on the FY 1975 Defense Budget and FY 1975–1979 Defense Program* (hereafter *SecDef FY 1975 Report*) (Washington, D.C.: GPO, Mar 4, 1974), pp 38, 40, 42, *passim*.

191. U.S. Congress, House, Subcommittee of the Committee on Appropriations, *Department of Defense Appropriations for 1978*, 95th Cong, 1st sess (Washington, D.C.: GPO, 1977), pt 2, p 212.

192. Ball and Richelson, eds., *Strategic Nuclear Targeting*, pp 62–74; Col William M. Charles, Jr., "Rethinking the Unthinkable — Limited Strategic Nuclear Options — Credible or Dangerous?" *Air University Review*, vol 28 (May–Jun 1977), pp 61–71.

193. Maj Gen Edward B. Giller, USAF, "Nuclear Technology in Support of Our Strategic Options," *Air University Review*, vol 28 (Nov–Dec 1976), 27–34; Graham Spinardi, "Why the U.S. Navy Went for Hard-Target Counterforce in Trident II (And Why It Didn't Get There Sooner)," *International Security*, vol 15 (fall 1990), 168–190.

194. *SecDef FY 1975 Report*, pp 42, 51–57; U.S. Department of Defense, *Report of the Secretary of Defense James R. Schlesinger to the Congress on the FY 1976 and Transition Budgets, FY 1977 Authorization Request and FY 1976–1980 Defense Programs* (Washington, D.C.: Department of Defense, Feb 5, 1975), pp I-13–I-17, II-28, II-36–II-38.

195. McLucas statement, Feb 10, 1975, U.S. Congress, Senate, Committee on Armed Services, *Fiscal Year 1976 and July–September 1976 Transition Period Authorization for Military Procurement . . .*, 94th Cong, 1st sess (Washington, D.C.: GPO, 1975), pt 2, p 469.

196. For the history of the SALT II negotiations, see Jimmy Carter, *Keeping Faith* (New York: Bantam, 1982), pp 212–265, and Strobe Talbott, *Endgame: The Inside Story of SALT II* (New York: Harper & Row, 1979). One of the most influential critics of the SALT II treaty was former Deputy Secretary of Defense Paul H. Nitze, who had been a member of the U.S. SALT negotiating team from 1969 to 1974. For Nitze's criticisms, see his testimony in U.S. Congress, Senate, Committee on Foreign Relations, *Hearings: The SALT II Treaty*, 96th Cong, 1st sess (Washington, D.C.: GPO, 1979), pt 1, pp 435–535.

197. Hopkins and Goldberg, *Development of SAC*, pp 203, 230.

198. For the background and origins of PD 59, see Zbigniew Brzezinski, *Power and Principle: Memoirs of the National Security Adviser, 1977–1981* (New York: Farrar, Straus, Giroux, 1983), pp 454–459; Leon Sloss and Marc Dean Millot, "U.S. Nuclear Strategy in Evolution," *Strategic Review*, vol 12 (winter 1984), pp 19–28; and Ball and Richelson, eds., *Strategic Nuclear Targeting*, pp 75–78.

199. U.S. Department of Defense, *Report of the Secretary of Defense Harold Brown to the Congress on the FY 1982 Budget . . .* (hereafter *SecDef FY 1982 Report*) (Washington, D.C.: GPO, Jan 19, 1981), pp 38–43.

200. U.S. Department of Defense, *Annual Report Fiscal Year 1979, Harold Brown, Secretary of Defense* (Washington, D.C.: GPO, Feb 2, 1978), p 55.

201. Desmond Ball, *Targeting for Strategic Deterrence*, Adelphi Paper 185 (London: International Institute for Strategic Studies, 1983), pp 21–23.

202. *SecDef FY 1982 Report*, p 41.

203. Carter, *Keeping Faith*, 81–83; Harold Brown, *Thinking about National Security* (Boulder, Colo.: Westview, 1983), pp 72–74.

204. House Subcommittee of the Committee on Appropriations, *DoD Appropriations 1978*, pp 289.

205. Walter Slocombe, "The Countervailing Strategy," *International Security*, vol 5 (spring 1981), p 27.

206. Ball and Richelson, eds., *Strategic Nuclear Targeting*, p 124.

207. *Ibid.*, pp 122–123.

208. *Ibid.*, pp 127–156.

209. U.S. Department of Defense, *Report of the Secretary of Defense Caspar W. Weinberger to the Congress on the FY 1985 Budget . . .* (hereafter *SecDef FY 1985 Report*) (Washington, D.C.: GPO, Feb 1, 1984), p 29.

210. U.S. Department of Defense, *Report of the Secretary of Defense Caspar W. Weinberger to the Congress on the FY 1984 Budget . . .* (hereafter *SecDef FY 1984 Report*) (Washington, D.C.: GPO, Feb 1, 1983), pp 219–231; *SecDef FY 85 Report*, pp 185–199.

211. *Report of the President's Commission on Strategic Forces* (Washington, D.C.: The Commission, Apr 1983).

212. For a summary of criticisms, see William P. Snyder and James Brown, eds., *Defense Policy in the Reagan Administration* (Washington, D.C.: National Defense University Press, 1988), pp 317–321.

213. Paul H. Nitze with Ann M. Smith and Steven L. Rearden, *From Hiroshima to Glasnost: At the Center of Decision — A Memoir* (New York: Grove, Weidenfeld, 1989), pp 366–462; Strobe Talbott, *Master of the Game: Paul Nitze and the Nuclear Peace* (New York: Knopf, 1988).

214. Sloss and Millot, "U.S. Nuclear Strategy in Evolution," pp 25–26; Ball and Richelson, eds., *Strategic Nuclear Targeting*, pp 79–80.

215. *SecDef FY 1984 Report*, p 54.

216. The classic treatment on the subject of war endurance was Kahn's *On Thermonuclear War* (Princeton, N.J.: Princeton University Press, 1960). Also see Kahn's remarks in attach 2, "Subjects Discussed," in memo, LeMay for COS, USAF, subj: Rpt by the RAND Military Advisory Group, Jul 19, 1960, White Papers, box 36, Air Force Council folder 2, LC.

217. *SecDef FY 84 Report*, pp 54–55.

218. Address to the Nation on Defense and National Security, Mar 23, 1983, *Public Papers of the Presidents of the United States: Ronald Reagan, 1983*, 2 vols (Washington, D.C.: GPO, 1984), vol 1, p 443.

219. For the background of this program, see Donald R. Baucom, *The Origins of SDI, 1944–1983*, (Lawrence: University of Kansas Press, 1992).

220. For a summary of the program's goals, see Lt Gen James A. Abrahamson, USAF, "The Strategic Defense Initiative," *Defense 84*, Aug 1984, pp 3–11.

221. Gen Bennie L. Davis, "Indivisible Airpower," *Air Force Magazine*, Mar 1984, pp 46–50. See also Carl H. Builder, *The Prospects and Implications of Non-Nuclear Means of Strategic Conflict*, Adelphi Paper 200 (London: International Institute for Strategic Studies, 1985).

222. Testimony, Gen John T. Chain, CINCSAC, Feb 3, 1987, U.S. Congress, Senate, Committee on Armed Services, *Hearings: Department of Defense Authorization for Appropriations for Fiscal Years 1988 and 1989*, 100th Cong, 1st sess (Washington, D.C.: GPO, 1987), p 1693.

223. McNamara, *Blundering into Disaster*, p 30.

Bibliographic Essay

Primary Sources

The starting point for research into the origins and development of postwar U.S. strategic air doctrine is the Military Reference Branch of the National Archives and Records Administration in Washington, D.C. Among the collections housed there, the most pertinent for the purposes of this essay were Record Group (RG) 341, Headquarters, United States Air Force; RG 218, Joint Chiefs of Staff; and RG 330 (Office of the Secretary of Defense). All three are immense collections and have, perforce, been consulted selectively. At present, declassification has reached the point where the cutoff date for access to these files is generally the mid-1950s. For additional material, particularly on the Air Force's role, see the massive archives of the U.S. Air Force Historical Research Agency (HRA) at Maxwell Air Force Base, Alabama, including its extensive oral history collection. Nearly all of the HRA's holdings have been microfilmed for the National Archives, though much material remains classified and not readily accessible.

Also useful were the personal papers of the early Air Force Chiefs of Staff. These are held by the Manuscript Division of the Library of Congress and include the papers of Generals Carl Spaatz, Hoyt S. Vandenberg, Nathan F. Twining, Thomas D. White, and Curtis LeMay. Also here are the papers of Vice Chief of Staff General Muir S. Fairchild, which offer some interesting insights into Air Force thinking on postwar priorities. The Spaatz, Vandenberg, and Fairchild papers are nearly all declassified; however, substantial portions of the others are still withheld.

The presidential library system continues to be one of the most important resources of all for understanding the political context in which postwar strategic doctrine evolved. For the crucial period of the 1950s, the Eisenhower papers, including his personal diary, memoranda of conversations, and NSC minutes, at the Dwight D. Eisenhower Library in Abilene, Kansas, are essential. For the period from 1961 on, however, the material thins considerably due to the slow pace of declassification. Neither the John F. Kennedy Library in Boston nor the Lyndon B. Johnson Library in Austin, Texas, has yet released much of significance bearing on the reorientation of doctrine and strategy that occurred in the 1960s.

Pending a fuller opening of the Kennedy and Johnson papers, perhaps the richest source currently available on the post-1961 period is the unindexed collection of miscellaneous documents that the Department of Defense has accumulated as a result of processing Freedom of Information Act requests. At present, these papers are in the custody of the Office of the Secretary of Defense's Office of Public Affairs in the Pentagon. Among the documents in this collection are the draft presidential memoranda on strategic offensive and defense forces prepared by McNamara between 1961 and 1968, setting forth his recommendations for strategic force levels and the rationale behind them.

As slow as declassification may seem for documents in the United States, no other country is nearly as forthcoming. Despite the collapse of the Soviet Union and the opening of hitherto inaccessible Soviet archives, primary source materials on the Soviet atomic energy program and the Soviet Strategic Rocket Forces remain fragmentary. In France a fifty-year rule governs the opening of official archives. In Britain, the rule is thirty years, with the result that the Public Record Office at Kew, outside London, is now making available files extending well into the 1950s, including records of the Air Ministry, Bomber Command, and the all-important Chiefs of Staff Committee and its supporting subcommittees. The latter are, unfortunately, heavily sanitized ("weeded," in British parlance), but they still yield a reasonably clear picture of how British postwar strategic air doctrine emerged and evolved.

Secondary Sources

In contrast to the rather restricted availability of primary sources, secondary literature on strategic doctrine and policy since World War II has proliferated enormously. Among general surveys, the most comprehensive and reliable are Lawrence Freedman, *The Evolution of Nuclear Strategy* (New York: St. Martin's, 1981), and McGeorge Bundy, *Danger and Survival: Choices about the Bomb in the First Fifty Years* (New York: Random House, 1988). Others I have found helpful and informative are Scott D. Sagan, *Moving Targets: National Strategy and National Security* (Princeton, N.J.: Princeton University Press, 1989), a brief but penetrating analysis of the ongoing problems of nuclear war planning; Jerome H. Kahan, *Security in the Nuclear Age: Developing U.S. Strategic Arms Policy* (Washington, D.C.: Brookings Institution, 1975); Bernard Brodie, *Strategy in the Missile Age* (Princeton, N.J.: Princeton University Press, 1959), now somewhat dated, it remains one of the most perceptive books ever written on the subject; John Lewis Gaddis, *Strategies of Containment* (New York: Oxford University Press, 1982), the best overall survey of basic national security policy; and Fred Kaplan, *The Wizards of Armageddon* (New York: Simon & Schuster, 1983), a popular, rather anecdotal account of strategic nuclear theorizing, with emphasis on RAND's influence. Another popular history is Gregg Herken, *Counsels of War* (New York: Knopf, 1985), which takes numerous liberties with the documentary record.

For article-length surveys, I especially recommend Henry S. Rowen, "Formulating Strategic Doctrine," originally published in U.S. Commission on the Organization of the Government for the Conduct of Foreign Policy, *Appendix K: Adequacy of Current Organization: Defense and Arms Control* (Washington, D.C.: GPO, Jun 1975), pp 219–234, and now reprinted in Laurence Martin, ed., *Strategic Thought in the Nuclear Age* (Baltimore, Md.: Johns Hopkins University Press, 1979), pp 131–156, an excellent analysis by a knowledgeable veteran of strategic planning. In addition, see Lawrence Freedman, "The First Two Generations of Nuclear Strategists," in Peter Paret, ed., *Makers of Modern Strategy from Machiavelli to the Nuclear Age* (Princeton, N.J.: Princeton University Press, 1986); Brent Scowcroft, "Understanding the U.S. Strategic Arsenal," in R. James Woolsey, ed., *Nuclear Arms: Ethics, Strategy, Politics* (San Francisco: Institute for Contemporary Studies, 1984); Aaron L. Friedberg, "A History of U.S. Strategic 'Doctrine' — 1945–1980," *Journal of Strategic Studies*, vol 3 (Dec 1980), pp 37–71; and David Alan Rosenberg, "Reality and Responsibility: Power and Process in the Making of United States Nuclear Strategy, 1945–1968," *Journal of Strategic Studies*, vol 9 (Mar 1986), pp 35–52.

The most reliable survey tracing the development of strategic thinking in the U.S. Air Force is Robert F. Futrell, *Ideas, Concepts, Doctrine: A History of Basic Thinking in the United States Air Force, 1907–1984*, 2 vols (Maxwell AFB: Air University Press, 1989), which while weak on interpretation and analysis, contains a wealth of information not readily available elsewhere. For an overview, see Herman Wolk's essay on the Air Force in Paul R. Schratz, ed., *Evolution of the American Military Establishment Since World War II* (Lexington, Va.: George C. Marshall Research Foundation, 1978); David MacIsaac, "Voices from the Central Blue: The Air Power Theorists," in Paret, ed., *Makers of Modern Strategy*; and the chapter on strategic airpower in Alfred Goldberg, ed., *A History of the United States Air Force, 1907–1957* (Princeton, N.J.: Van Nostrand, 1957).

On early strategic planning Walton S. Moody, *Building a Strategic Air Force* (Washington, D.C.: Air Force History and Museums Program, 1996), covers the subject better than anyone. Also see the contributions by John T. Greenwood and David Alan Rosenberg in Alfred F. Hurley and Robert C. Ehrhart, eds., *Air Power and Warfare*, Proceedings of the 8th Military History Symposium, USAF Academy, Oct 18–20, 1978 (Washington D.C.: Office of Air Force History and U.S. Air Force Academy, 1979). Among Rosenberg's numerous other writings on this subject, I strongly recommend his richly detailed "The Origins of Overkill: Nuclear Weapons and American Strategy, 1945–1960," *International Security* vol 7 (spring 1983), pp 3–71; and his "Toward Armageddon: The Evolution of United States Nuclear Strategy, 1945–1961" (Ph.D. thesis, University of Chicago, 1983).

For the early history of SAC, see Harry Borowski, *A Hollow Threat: Strategic Air Power and Containment Before Korea* (Westport, Conn.: Greenwood Press, 1982), a solid, scholarly account of the 1945–1950 period. It is best read in conjunction with Noel Francis Parrish, *Behind the Sheltering Bomb: Military Indecision from Alamogordo to Korea* (New York: Arno Press, 1979); and William S. Borgiasz, "Struggle for Predominance: Evolution and Consolidation of Nuclear Forces in the Strategic Air Command" (Ph.D. dissertation, American University, 1991). Unfortunately, nothing compares to either of these works on SAC for later periods. A few of SAC's early official histories have been declassified, but in the main one must rely on the official chronology by J. C. Hopkins and Sheldon A. Goldberg, *The Development of Strategic Air Command, 1946–1986* (Offutt AFB: Headquarters SAC, 1986).

Two recurring issues that have drawn considerable attention in the strategic doctrine debate are counterforce targeting and the impact of new technologies. For an overview of targeting practices since World War II, see William C. Martel and Paul L. Savage, *Strategic Nuclear War: What the Superpowers Target and Why* (New York: Greenwood Press, 1986). Other useful material may be found in Alfred Goldberg, *A Brief Survey of the Evolution of Ideas about Counterforce* (Santa Monica, Calif.: RAND, RM–5431–PR, Oct 1967; revised, Mar 1981), a summary analysis; T. F. Walkowicz, "Counter-Force Strategy," *Air Force Magazine* (Feb 1955), which helped introduce and popularize the theory; Richard Lee Walker, *Strategic Target Planning: Bridging the Gap between Theory and Practice* (Washington, D.C.: National Defense University Press, 1983), an informed, analytical account focusing on targeting problems of the 1970s and early 1980s; and the essay by Ball on the evolution of the SIOP in Desmond Ball and Jeffrey Richelson, eds., *Strategic Nuclear Targeting* (Ithaca, N.Y.: Cornell University Press, 1986). Peter Pringle and William Arkin, *SIOP: The Secret U.S. Plan for Nuclear War* (New York: Norton, 1983), is as its title suggests a sensationalized account based on less-than-authoritative research.

For the effect of new technologies in shaping doctrine and strategic thinking, see especially Jacob Neufeld, *Ballistic Missiles in the United States Air Force, 1945–1960* (Washington, D.C.: Office of Air Force History, 1990). See also Edmund Beard, *Developing the ICBM: A Study in Bureaucratic Politics* (New York: Columbia University Press, 1976); Harvey M. Sapolsky, *The System Development: Bureaucratic and Programmatic Success in Government* (Cambridge, Mass.: Harvard University Press, 1972); and Desmond Ball, *Politics and Force Levels: The Strategic Missile Program of the Kennedy Administration* (Berkeley: University of California Press, 1980). For the shift of interest to defensive systems in the 1980s, see Donald R. Baucom, *The Origins of SDI* (Lawrence: University of Kansas Press, 1992), and Janne E. Nolan, *Guardians of the Arsenal: The Politics of Nuclear Strategy* (New York: Basic Books, 1989).

Regrettably, memoirs by those who played key parts in shaping postwar U.S. strategic doctrine are few and far between. The only postwar U.S. chief of staff to publish memoirs was Curtis E. LeMay (with McKinley Kantor), *Mission with LeMay: My Story* (Garden City, N.Y.: Doubleday, 1965), parts of which LeMay later repudiated as having been distorted by his editor for publicity purposes. Thomas M. Coffey, *Iron Eagle: The Turbulent Life of General Curtis LeMay* (New York: Crown Publishers, 1986), and Edgar F. Puryear, Jr., *George S. Brown, General, U.S. Air Force: Destined for Stars* (Novato, Calif.: Presidio Press, 1983), are lively popular biographies written from secondary sources and interviews. For the early postwar years see Phillip S. Meilinger, *Hoyt S. Vandenberg: The Life of a General* (Bloomington: Indiana University Press, 1989). Nathan F. Twining,

Neither Liberty nor Safety: A Hard Look at U.S. Military Policy and Strategy (New York: Holt, Rinehart & Winston, 1966), and Thomas S. Power, *Design for Survival* (New York: Coward-McCann, 1965), are both critiques of American defense policy and contain little in the way of personal reminiscences that would shed light on their roles. The same is true of Robert S. McNamara's *The Essence of Security* (New York: Harper and Row, 1968) and *Blundering into Disaster: Surviving the First Century of the Nuclear Age* (New York: Pantheon Books, 1986), though both provide interesting insights into McNamara's thinking. Also useful in understanding how recent high-level policy has developed is Harold Brown, *Thinking about National Security* (Boulder, Colo.: Westview Press, 1983).

For the Soviet perspective, see V. D. Sokolovskii, ed., *Soviet Military Strategy*, trans and annotated by Herbert S. Dinerstein, Leon Goure, and Thomas W. Wolfe (Moscow: Ministry of Defense, 1962; Englewood Cliffs, N.J.: Prentice-Hall, 1963), a compilation of papers by fifteen leading Soviet officers elaborating on the changes in Soviet strategic doctrine following Stalin's death. Among studies in the West on Soviet strategic forces and doctrine, see David Holloway, *Stalin and the Bomb: The Soviet Union and Atomic Energy, 1935–1956* (New Haven, Conn.: Yale University Press, 1994), an excellent study of the early period using many original Soviet sources; Robert A. Kilmarx, *A History of Soviet Air Power* (New York: Praeger, 1962); Paul J. Murphy, ed., *The Soviet Air Forces* (Jefferson, N.C.: McFarland and Co., 1984); Mark E. Miller, *Soviet Strategic Power and Doctrine: The Quest for Superiority* (Miami: Advanced International Studies Institute, University of Miami, 1982); Thomas B. Cochran, Robert S. Norris, and Oleg A. Bukharin, *Making the Russian Bomb: From Stalin to Yeltsin* (Boulder, Colo.: Westview Press, 1995); and Joseph D. Douglass, Jr., and Amoretta M. Hoeber, *Soviet Strategy for Nuclear War* (Stanford, Calif.: Hoover Institution Press, 1979). David S. Yost, "French Nuclear Targeting," in Ball and Richelson, eds, *Strategic Nuclear Targeting*, surveys the development of French doctrine and provides a guide to further reading on the subject. On British policy and doctrine, see Andrew J. Pierre, *Nuclear Politics: The British Experience with an Independent Strategic Force, 1939–1970* (London: Oxford University Press, 1972); Lawrence Freedman, *Britain and Nuclear Weapons* (London: Macmillan, 1980), which focuses on the 1970s; Steward Menaul, *Countdown: Britain's Strategic Nuclear Forces* (London: Robert Hale, 1980); Ian Clark and Nicholas J. Wheeler, *The British Origins of Nuclear Strategy, 1945–1955* (Oxford: Clarendon Press, 1989); Martin S. Navias, *Nuclear Weapons and British Strategic Planning, 1955–1958* (Oxford: Clarendon Press, 1991); and Ian Clark, *Nuclear Diplomacy and the Special Relationship: Britain's Deterrent and America, 1957–1962* (Oxford: Clarendon Press, 1994).

6

Strategic Bombardment Constrained: Korea and Vietnam

Thomas C. Hone

In January 1943, British and American leaders agreed to use their combined long-range bomber forces for "the progressive destruction and dislocation of the German military, industrial and economic system, and the undermining of the morale of the German people to a point where their capacity for armed resistance is fatally weakened."[1] Here was the essence of American strategic bombing doctrine: destroying simultaneously the enemy's industrial basis for sustaining war and the desire of its civilian population to support a war effort. Such persistent bombing was intended to affect German civilian morale directly and adversely, provoking the hardships (reduced supplies of energy and clothing, for example) associated with paralysis of the economy. Strategic bombing was intended to be a war-winning weapon.

Wartime experience suggested that civilian morale could, indeed, be affected. Although the strategic air war against Germany, for various reasons, was not waged as U.S. Army Air Forces (AAF) planners had urged, the U.S. Strategic Bombing Survey conducted after Germany's surrender argued that the bombing of German transportation, especially of railways and bridges in the autumn of 1944, was "the decisive blow that completely disorganized the German economy."[2] Consequently, the survey authors recommended that XXI Bomber Command B–29s then pounding Japanese cities from bases in the Mariana Islands focus their attacks on Japan's railroads and coastal shipping. Their suggestion was not accepted.[3] Instead, Japanese cities continued to be systematically fire bombed, with devastating effects on Japanese civilian morale.[4] The high level of civilian suffering inflicted by the B–29s, however, left a lasting stigma. The image of American bombers terrorizing Asian women and children would resurface in the Korean and Vietnam conflicts, and it would constrain the use of strategic bombing in subsequent practice.

From its wartime experience, the AAF learned a number of important lessons: First, an overwhelming air offensive should be launched as early as possible in a war, before an enemy's defenses are prepared and its industry dispersed. Second, the critical points of support in an enemy's war effort should be found quickly and attacked until they collapse. Third, rapid and accurate damage assessments should be the key for shifting available bombers from one priority target to another. Finally, centralized control of air units (tactical and strategic) should be used to maximize the effect. These lessons, embodied in postwar Air Force doctrine, were not abandoned when nuclear weapons began to be produced in quantity at the end of the 1940s. The primary weapons of the Strategic Air Command (SAC), however, remained conventional, not nuclear, for the first few years after the command was established in 1946. Only after 1949 did the United States have a sufficient number of nuclear bombs to conduct a limited nuclear bombing offensive. Moreover, the destructive potential of early nuclear weapons still required accurate targeting and carefully prepared postattack damage assessments.[5]

Too powerful for pinpoint attacks, nuclear bombs eliminated the distinction between military and civilian targets. That distinction had been blurred during World War II, sometimes deliberately as in the fire bombing of Japanese cities, and sometimes unavoidably as when AAF bombers raided German strategic targets located among civilian housing. Nuclear weapons offered little choice in making such distinctions because their use meant killing civilians. Yet these potent weapons represented the only military alternative available to the repeated striking of heavily defended targets. For example, the allied air war against Germany in 1943 and 1944 had been a campaign of attrition waged by the combined bomber forces of the United States and Great Britain. It was an ordeal that cost the Allies in time and casualties (79,265 American and 79,281 British airmen). The length of time required for aerial bombardment allowed Germany to repair bomb damage, disperse its industry, and build defenses.[6] SAC leaders did not intend to repeat such a campaign, but they would be compelled to do so nonetheless.

The Conflict in Korea

War in Korea began on June 25, 1950, when the North Korean Army invaded South Korea. The United States immediately asked the United Nations (UN) Security Council to condemn the attack, and the council quickly passed a resolution calling on both sides to stop fighting. The next day, the UN commission in Korea reported that North Korea had indeed attacked South Korea first, and the government in the south appealed to the UN for help. On June 27, at UN headquarters in New York, the Security Council adopted (without the Soviet representative, who was boycotting the meeting on orders from Moscow) a resolution calling on its members to furnish military aid to South Korea. With that resolution as justification, President Harry S. Truman ordered U.S. air and naval forces to cover and support the South Korean Army as it retreated in the face of North Korea's attack.

From his headquarters in Tokyo, U.S. Army General Douglas MacArthur exercised overall command of all U.S. (and later, UN) forces in the Far East. Reporting to him was the Far East Air Forces (FEAF) commanded by Lt. Gen. George E. Stratemeyer. The Air Force's largest overseas command, FEAF, possessed 1,172 aircraft on May 31, 1950, dispersed among its three air forces: the Thirteenth (one fighter-bomber wing and a troop carrier squadron among other organizations) stationed in the Philippines; the Fifth (two fighter-bomber wings, an interceptor wing, a light bomber wing, a troop carrier wing, and other elements) stationed in Japan; and the Twentieth (a fighter-interceptor wing and a B–29 group with supporting units that were not a part of SAC) stationed in Okinawa and the Marianas.[7] When Truman ordered MacArthur to cover the retreat of South Korean forces, Stratemeyer shifted his B–29s (the 19th Bombardment Group) from Andersen Air Force Base (AFB), Guam, to Kadena Air Base (AB), Okinawa. On June 30, 1950, the President authorized the U.S. Air Force to attack North Korea.[8] He also ordered U.S. Army units to assist South Korean forces in defending their territory and the U.S. Navy to blockade North Korea's harbors.

On July 3, 1950, Air Force Chief of Staff General Hoyt S. Vandenberg, with the support of the other members of the Joint Chiefs of Staff (JCS), approved the temporary transfer of two B–29 groups (the 22d and 92d) from SAC to FEAF.[9] On July 7, the Security Council authorized the United States to establish a unified UN command in Korea. The next day, President Truman named General MacArthur the UN commander. Meanwhile, at Yokota AB, Japan, General Stratemeyer established FEAF Bomber Command, composed of the one B–29 group he already possessed plus two being sent as reinforcements. General Vandenberg appointed Maj. Gen. Emmett "Rosie" O'Donnell, Jr., as its first commander, and on July 7 O'Donnell carried to Japan a list of possible targets and his own plan to use incendiaries against North Korean cities.[10]

By July 11, 1950, FEAF Bomber Command had three groups of B–29s for a total of eighty-one aircraft. Since June 28, bombers of the Kadena-based 19th Bombardment Group had been flying combat sorties against North Korean lines of communication in South Korea. General Stratemeyer had to decide whether to continue such attacks to shore up the collapsing defense of South Korea or to turn the 20,000-lb bomb capacity of each of his veteran Superfortresses

USSR
MANCHURIA
NORTH KOREA
Sea of Japan
Pyongyang
38°
38°
Seoul
SOUTH KOREA
JAPAN
Kunsan
Tokyo
Pusan
Yellow Sea
Osaka
CHINA
RYUKYUS
East China Sea
Okinawa
Pacific Ocean
FORMOSA
Philippine Sea
Manila
MARIANAS
PHILIPPINES
Guam
The Korean Theater 1950–1953
Mig Alley
0
150
300
miles

Lt. Gen. George E. Stratemeyer, commander of Far East Air Forces, with General Kum Chung Ell, of the Army of the Republic of Korea.

against North Korea. On July 9 his operations deputy, Brig. Gen. Jarred V. Crabb, prepared a memo on FEAF air employment doctrine in which he argued that the B–29s should inflict "the maximum damage to the enemy with the reservation that the situation does not at this time warrant the mass destruction of population centers. . . ." Crabb recommended attacking North Korean "industrial targets contributing to the combat effort of North Korea" while exercising "[r]easonable care . . . to avoid providing a basis for claims of 'illegal' attack against population centers."[11] He also cautioned that the bombers must be kept well clear of the Manchurian border. Finally, he suggested giving Bomber Command the collateral duties of destroying North Korean airfields, attacking North Korea's supply lines from Manchuria, and directly supporting UN ground forces.

Stratemeyer endorsed Crabb's recommendations in his mission directive to FEAF Bomber Command on July 11, 1950. He forbade Bomber Command to "attack urban areas as targets," though it was directed to attack "industrial targets contributing to the combat effort of North Korean forces. . . ."[12] This directive, though Stratemeyer's, was surely influenced by the expressed concerns of the JCS and the President. Dean Acheson, President Truman's Secretary of State, argued in his memoirs that the President intended to keep the war in Korea limited from "the very start of hostilities."[13] Recently declassified documents, however, show as early as June 25 that Truman and his closest advisers considered the use of nuclear weapons and the possibility the United States might fight the Soviet Union.[14]

U.S. Army Chief of Staff General J. Lawton Collins recalled that the JCS opposed using incendiaries against North Korea's cities not only because the United States would probably end up rebuilding what it had destroyed, but also because such attacks might not succeed in reducing the morale of North Korea's civilians.[15] What General Collins did not mention was that the President, Secretary of State, and JCS all feared the campaign in Korea was only the harbinger of a larger conflict with the USSR. The President and his advisers always acted with this potentiality in mind, and the threat of a larger, even nuclear, war dominated their thinking and that of the Air Staff in Washington.[16] Thus, two elements shaped Stratemeyer's decision: fear that incendiary attacks on North Korean cities would lead to negative propaganda images the North Koreans would exploit, and concern in Washington that Korea represented only the first engagement in a larger war, with the crucial area of conflict not Korea, but western Europe.

Without an adequate force structure, Truman and his advisers had good reason for trying to avoid a major war. SAC's B–36 intercontinental bomber had only entered service in November 1948, and the B–50 (an improved version of the B–29), did not enter regular SAC squadrons until 1949.[17] While the President and JCS supported efforts to enhance SAC's strength and increase the Atomic Energy Commission's stock of fission cores for nuclear bombs, funds for sustained growth and modernization were limited until 1950 when the Truman administration decided the time had come for United States to rearm.

Thus, in July 1950 two B–29 SAC groups, only some capable of carrying nuclear weapons, were sent to England to reinforce the sole group already there,[18] and the next month a group of B–29s, modified to carry nuclear bombs, was sent to Andersen AFB to replace the 19th Group.[19] In September, the first operational KB–29P tankers joined a SAC refueling squadron, and the JCS authorized the Air Force to activate 180 B–29s in reserve and modify them to carry nuclear weapons.[20] Additionally, the Air Force's Air Materiel Command was "instructed to give the highest priority" to renewing the B–29s and modifying deployed B–50s for nuclear bombing missions.[21] This mobilization and rearmament effort would add significantly to SAC's strength, but the advantage would not be immediate. For example, in 1951, SAC units could field only 98 long-range B–36s and 12 B–47 medium jet bombers; in 1952, these aircraft still numbered only 154 and 62, respectively.[22] This limited strategic bombing capability was a major reason why the JCS counseled Truman against any action across the Yalu River into Manchuria.

Secretary of State Acheson observed that although "the field and home

Army Chief of Staff General J. Lawton Collins recalled that the JCS opposed using incendiaries against North Korea's cities.

commands were united on political and military policy," this union came through dire necessity.[23] During the first two months of the conflict, the JCS feared that South Korea might be lost to the North, and they did not want American air assets in Japan depleted in an effort to hold back the North Koreans. General MacArthur reportedly asked the JCS for nuclear weapons as early as July 9, 1950, and General Omar N. Bradley, JCS Chairman, considered asking President Truman to relinquish control of such weapons to MacArthur. The chiefs, however, chose instead to husband the nuclear stockpile.[24] In fact, Korea was always considered a secondary theater, and General Stratemeyer's requests for aerial reinforcements, for the most part, were refused. In May 1951, Lt. Gen. Idwal H. Edwards, the Air Force's Deputy Chief of Staff for Operations remarked to Stratemeyer that "the Air Force was caught off balance when the Korean War started and is still off balance today," and "industry is not yet producing sufficient aircraft to permit an increase in your strength."[25]

Strategic and political factors beyond Korea would compel Stratemeyer to use whatever aerial assets he could obtain. But how available air assets in the theater should be used was an issue that divided the services. On July 14, 1950, MacArthur's staff created a General Headquarters (GHQ) Target Group, "a part-time organization" of four officers (two from the Army) to analyze and recommend lists of North Korean targets for the B–29s. Stratemeyer's FEAF already had a Target Section and Target Committee, and their target recommendations did not agree with those of the GHQ Target Group in Tokyo. To resolve this conflict, Stratemeyer recommended to MacArthur that the UN commander organize a Target Selection Committee composed of MacArthur's deputy chief of staff and intelligence deputy, FEAF's vice commander for operations, and a "Navy representative." The existing Target Group would report to the new Target Selection Committee that, in turn, came to depend upon the support of FEAF's own Target Committee, sustained by FEAF's Target Section. MacArthur approved Stratemeyer's proposal on July 22, 1950, thereby returning "the bulk of target selection to FEAF" and making FEAF's Target Committee "the basic theater agency for target selection for the remainder of the war. . . ."[26] This committee focused first on North Korea's limited industry, directing Bomber Command's B–29s to attack North Korean facilities in Hungnam, Wonsan, Pyongyang, and Konan.[27] North Korea's major hydroelectric power plants along the Yalu were deliberately not targeted.

On July 29, 1950, the JCS authorized General Vandenberg to dispatch two more B–29 groups (the 98th and 307th Bombardment Groups) and a strategic reconnaissance squadron from the United States to General O'Donnell's FEAF Bomber Command in Japan.[28] The reinforcements gave O'Donnell a force of two B–29 groups and a reconnaissance squadron at Yokota and three bomber groups at Kadena. The first strategic bombing campaign since World War II was about to begin. On July 30, forty-seven B–29s from the 22d and 92d Bombardment Groups bombed the Chosen nitrogen explosives factory in a single attack, with all the Superfortresses dropping their bombs in just four minutes. Because clouds obscured this North Korean target, the SAC squadrons relied on their AN/APQ–13 bombing radars, justifying SAC's "intensive radar-training programs." Two days later, B–29s, flying and bombing visually from 16,000 feet, attacked Chosen's nitrogen fertilizer company.[29] Hungnam's chemical plants came next, and then Wonsan's railroad shops and oil refinery.

B–29s at Kadena Air Base, Okinawa.

Air Force Chief of Staff General Hoyt S. Vandenberg meets with FEAF Commanding General Stratemeyer to discuss strategy.

With Bomber Command's success in destroying the North Korean Air Force, it could now use its most reliable tactics. For example, airborne strike commanders, preceding attacking bombers in meteorological planes, could specify the method of attack and decide whether the B–29s should strike an alternate target. The Superfortresses usually attacked in V formations consisting of three aircraft, with the trailing bombers dropping their ordnance when the lead aircraft salvoed. When clouds obscured the targets, "a bomber stream of individual aircraft crossed the target at one-minute intervals, bombing individually by radar."[30] By mid-September 1950, only hydroelectric plants and targets close to the Russian and Manchurian borders had not been hit.

In just over a month, FEAF Bomber Command's five B–29 groups had completed their primary mission. By September 15, 1950, the bombers had dropped 30,000 tons of bombs in "approximately four thousand sorties, attacking factories, chemical plants, oil refineries, arsenals, railroad marshalling yards, locomotive foundries, docks, and warehouses."[31] Only four B–29s were lost in these initial operations. Targets were recommended by the JCS, SAC, and FEAF's own Target Section. In accordance with JCS direction, only one hydroelectric plant (at Fusen) was bombed, and radar attacks on Rashin, near the Russian border, were halted after bombs fell too near Soviet territory on August 12.[32] Bomber Command's success paralleled the advance of the UN and South Korean armies. After a successful amphibious landing at Inchon in South Korea on September 15, MacArthur's forces rolled up North Korean forces from the west and south. On October 4, the Security Council approved MacArthur's request to send UN forces across the 38th parallel, the border between North Korea and South Korea. Consequently, on October 19, U.S. forces captured Pyongyang, the capital of North Korea. Five days later, Bomber Command suspended its operations over Korea "because no targets were available on the Korean side of the Yalu."[33] Since they were unneeded, two conventionally armed SAC B–29 groups left FEAF and returned to the United States.

Among the initial targets selected for bombing was the North Korean capital of Pyongyang. Appearing is the wreckage of the railroad yard after bombing.

Unfortunately, MacArthur's decision to conquer and occupy North Korea brought the Peoples Republic of China into the war. Chinese forces first attacked U.S. troops on November 2, 1950. By Thanksgiving, UN and South Korean ground units were very hard-pressed and in retreat. Meanwhile, in Washington, General Bradley worried that the Chinese attacks might precipitate a general war between UN and communist forces. On November 30, President Truman stated publicly that the United States would use "every weapon that we have." He did not, however, favor using nuclear weapons against China because he did not want to kill large numbers of civilians.[34] Nevertheless, that same day, SAC was ordered to dispatch B–29s equipped to carry nuclear weapons to Guam to replace the ones that had been sent in August and withdrawn at the end of October.[35]

On November 5, 1950, General MacArthur ordered FEAF back into North Korea to "destroy every means of communication and every installation, factory, city and village" except "Rashin, the Sui-ho Dam and other electric power plants . . . . This destruction is to start at the Korean-Manchurian border and progress south."[36] The following directive was supplemented by a personal insert from General Stratemeyer to General O'Donnell: "Under present circumstances all such [factories, cities, and villages] have marked military potential and can only be regarded as military installations."[37] At the same time, MacArthur ordered FEAF not to bomb Manchuria or to strike targets "north of a line running from Chongjin to Musan" near North Korea's border with the Soviet Union. Bomber Command's role in this campaign was to "destroy the cities and large towns,"[38] and for the first time in this deteriorating military situation, the use of incendiaries was authorized.[39]

Chinese armies, however, continued to advance southward. On December 5, 1950, they marched into Pyongyang. General MacArthur pressed the JCS for permission to bomb Manchuria and to use Nationalist Chinese troops in Korea, even directly against the communist regime in China. His urgent requests to widen the war in Korea were denied because the JCS did not want to risk an escalation unless a "major military disaster" was in the offing.[40] The British and French governments also wanted the conflict contained, not expanded or deepened. President Truman moved cautiously to hold the UN coalition together. On December 14, the General Assembly of the UN adopted a resolution urging a negotiated end to the fighting in Korea, giving Truman and MacArthur a diplomatic alternative if they could find an opportune moment to employ it. But China, on the march southward, seemed unwilling even to consider that opportunity. Chinese troops captured Seoul, the capital of South Korea, in the first week of January 1951, and their drive south was only halted at the end of that month.

By March 15, 1951, UN troops were again in Seoul, and MacArthur, anxious to strike at the North Koreans and Chinese before any negotiations began, again asked the President for permission to enlarge the war and for authority to use nuclear weapons against Manchurian targets. Truman again refused, although he did have the U.S. delegation to the UN warn their Soviet counterparts that the United States would respond dramatically if Soviet air units, based in either the Soviet Union or Manchuria, attacked UN ground forces fighting in Korea.[41] Both the UN and its communist opponents seemed to have tacitly accepted certain limits on the scope of the war: the United States did not bomb Manchuria, and Soviet forces did not intervene directly in the conflict. In February and March 1951, however, it seemed that this understanding might collapse when the JCS received reports the Soviet Union "had moved three divisions into Manchuria and had positioned other forces for an attack on Japan."[42]

On March 27, 1951, the JCS recommended that "nine nuclear cores be transferred to the Defense Establishment" from the Atomic Energy Commission (AEC), the civilian agency responsible for developing and stockpiling atomic weapons. The chairman of the AEC agreed to make the transfer, but he counseled President Truman against giving General MacArthur the authority to decide when and how to use the nuclear weapons.[43] On April 6, the President signed an order directing the AEC to release the nine bomb cores to General Vandenberg, who ordered them flown to Guam where they could be mated with bomb casings stored at Andersen AFB.[44]

This was a major crisis. On the one hand, MacArthur feared UN forces would fall victim to a surprise Russian assault; on the other, the JCS wanted to preserve nuclear weapons for use against the Soviet Union if the limited war in Korea should suddenly degenerate into a worldwide conflict.[45] General MacArthur, convinced that the war in Korea would expand, now demanded the initiative to widen it as he chose; the President, however, refused to grant him that right. When MacArthur challenged the Commander-in-Chief's decision, Truman fired him on April 11 and appointed General Matthew B. Ridgway to succeed him. After Ridgway received command authority, the JCS authorized him to attack Manchurian and Russian air bases if planes based there assaulted his ground forces as part of a concerted communist offensive.[46] Truman also approved the transfer to Guam of more B–29s to carry atomic bombs, and SAC sent a command and control team to Tokyo to cooperate with FEAF targeteers in case the war should spread rapidly, since control of nuclear-armed aircraft

When General Douglas MacArthur, UN commander in Korea, challenged President Truman's decision to contain the war, the Commander-in-Chief fired him, replacing him with Lt. Gen. Matthew B. Ridgway. MacArthur (*below center*) and Ridgway (*below right*) are shown leaving UN forces headquarters for the front.

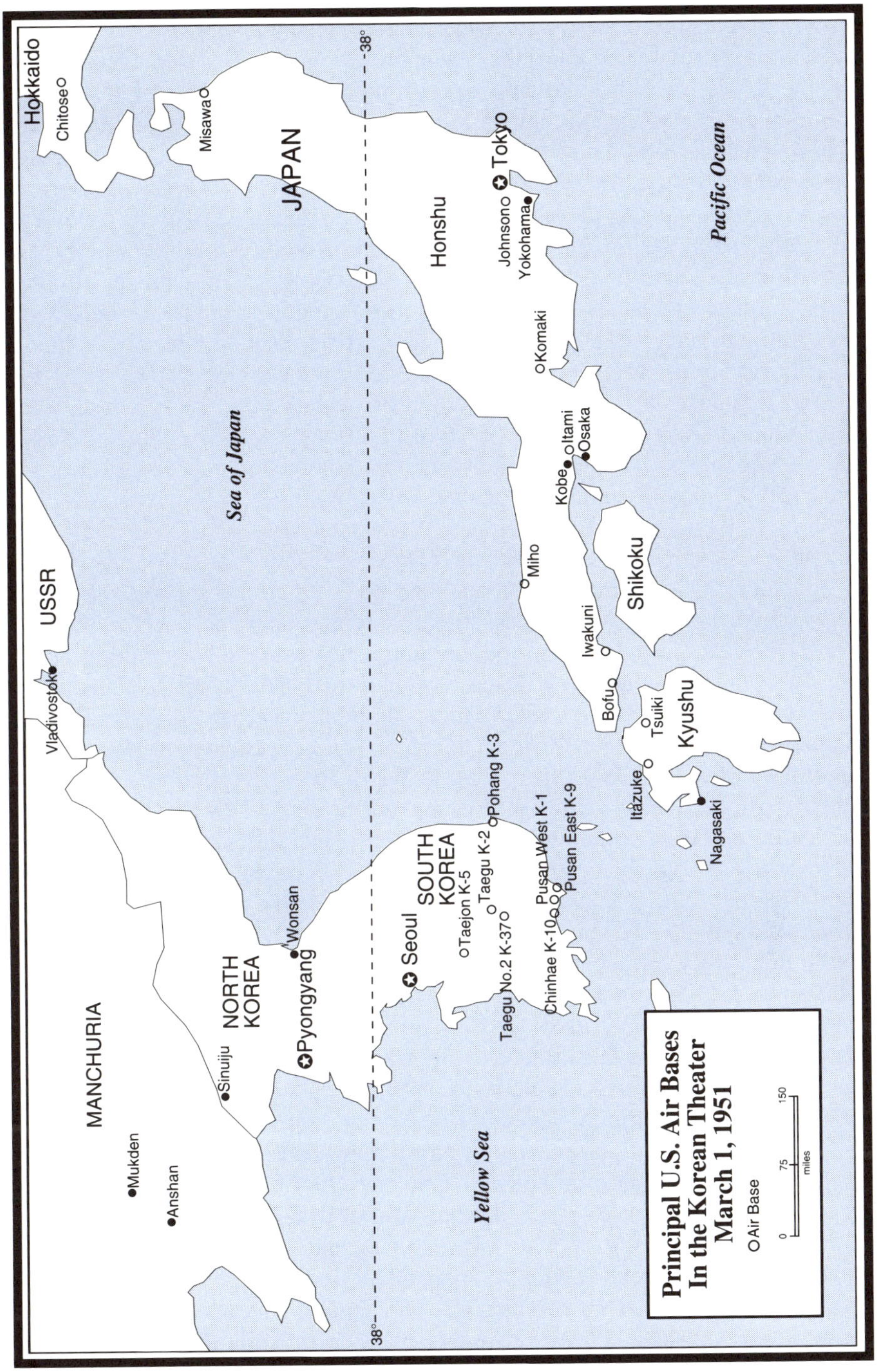
Principal U.S. Air Bases
In the Korean Theater
March 1, 1951
Air Base
0
75
150
miles
Hokkaido
Chitose
Misawa
JAPAN
Tokyo
Johnsono
Yokohama
Honshu
Pacific Ocean
Komaki
Itami
Osaka
Kobe
Miho
Shikoku
Iwakuni
Bofu
Tsuiki
Kyushu
Itazuke
Nagasaki
Sea of Japan
USSR
Vladivostok
MANCHURIA
Mukden
Anshan
Sinuiju
NORTH
KOREA
Pyongyang
Wonsan
SOUTH
KOREA
Seoul
Taejon K-5
Taegu K-2
Pohang K-3
Taegu No.2 K-37
Chinhae K-10
Pusan West K-1
Pusan East K-9
Yellow Sea
38°

was not practicable from SAC headquarters in Omaha, Nebraska.[47]

Several scholars and journalists using newly declassified documents argue that the Korean war was the first case of nuclear diplomacy, and some have even argued that nuclear intimidation was an integral part of U.S. (and hence UN) policy.[48] In 1950 and 1951, when the Korean conflict was anything but stable and controlled, American leaders saw nuclear weapons less as a deterrent than a trump card to be played in the event Japan appeared likely to be lost as a result of enemy action. Facing the small North Korean Air Force, which flew no jet turbine aircraft in the summer of 1950, General Stratemeyer's forces quickly gained and easily held air superiority, but once Chinese air units with their fast and maneuverable MiG–15 began flying missions from Manchuria, FEAF's resources were constantly strained. From December 1950 until the Korean armistice in July 1953, the aircraft available to UN air units were always outnumbered in combat, so from FEAF's perspective, the Korean air war was a conventional campaign of attrition. Although composed of three air forces, FEAF was ill-suited to fight a conventional war and replace extensive losses.

After Chinese forces crossed the Yalu River from Manchuria into Korea in November 1950, Bomber Command's three B–29 groups supported FEAF's interdiction campaign by attacking bridges and railroad tunnels in North Korea. The B–29s also participated in the air superiority struggle by cratering North Korean airfields. Because the Chinese-piloted (and, as it later became known, Soviet-piloted) MiGs had a limited range, communist engineers continually had to repair the North Korean airfields to extend MiG coverage closer to the 38th parallel. In May 1951, Maj. Gen. Otto P. Weyland (who had directed close air support for the Third Army in Europe during World War II) received a promotion and took command of FEAF after General Stratemeyer suffered a heart attack. Weyland's primary concern was to retain air superiority over North Korea. The Communist MiG–15s were a major threat to Bomber Command's B–29s because FEAF's range-limited F–84 and F–86 fighters could not easily protect the slower bombers, and the B–29's defensive armament was ill-suited to warding off the fast, highly maneuverable and heavily armed MiGs. In response, Bomber Command shifted from daylight to nighttime raids. As a result, B–29 night sorties increased from 178 in the January through March 1951 period to 496 during April through June and to 1,032 for October through December.[49]

What made the night attacks on North Korean airfields possible was the short-range navigation, or shoran, system, a method of using radio-frequency transmissions to direct B–29s and other aircraft to targets they could not see. A refinement of the World War II British Gee and Oboe, shoran consisted of radio transmitters placed at widely separated ground installations and a receiver and computers located on the aircraft. The ground stations transmitted continuous radio pulses at a specific frequency. By measuring the time (in micro-

Maj. Gen. Otto P. Weyland, who had directed close air support for Patton's Third Army in Europe, was promoted and took command of FEAF after Stratemeyer suffered a heart attack.

seconds) that it took the different pulses to reach the airplane, a crew using triangulation could determine their distance from a given ground station. After reaching proper altitude, the B–29s would fly along arcs that would take them over their targets. The bombers would keep station on their arcs by staying at a constant distance from one of the ground beacons. When a signal from a second beacon indicated that the aircraft were over their target, they would salvo their bombs. As a FEAF report noted, "Shoran bombing was ideally suited to the Korean theater.... Most of the essential equipment was already in the area... targets were within Shoran operating range; the land area had been surveyed by Japanese surveyors...."[50] Moreover, bombing a runway called for less accuracy than bombing a bridge or a railway line. Shoran-directed attacks by small numbers of bombers became routine for the B–29s of FEAF Bomber Command, and they continued until the fighting ended in July 1953.

General Weyland, however, was unwilling to wage an aerial campaign of attrition simply to preserve UN air superiority and interdict the supplies flowing south down the peninsula to North Korean and Chinese troops.[51] To emerge from the armistice talks, begun on July 10, 1951, with a settlement favorable to the UN, Weyland believed his air forces would have to deny Chinese and North Korean ground and air forces both military and psychological initiatives.[52] Denying them the military initiative was a function of maintaining

Communist MiG–15s (*above*) were a major threat to Bomber Command's B–29s, as neither F–84 (*left*) nor F–86 fighters (*below*) could easily protect the slower bombers.

control of the air and of reducing the flow of supplies, which FEAF had been attempting to do since Chinese troops and aircraft entered the war. Denying them the psychological initiative within the political, strategic, and military constraints established in Washington proved far more difficult.

In January 1952 Brig. Gen. James E. Ferguson, vice commander of Weyland's Fifth Air Force (the tactical command in Korea), asked that his aerial units be permitted to attack North Korea's Yalu River hydroelectric power stations. Most of them had been off limits to FEAF bombers since the beginning of the war but Ferguson believed it was now time to strike them. Coincidentally, Brig. Gen. Jacob E. Smart, Weyland's new operations officer, at that time was trying to find a set of targets so valued by the North Koreans and Chinese that FEAF attacks on them would force the communist governments to conclude an armistice on terms acceptable to the UN.[53] Smart agreed that attacking the hydroelectric power stations (along with Manchurian MiG bases and the Manchurian side of Yalu River bridges) would pay the most political, military, and psychological dividends, and he recommended that Weyland press General Ridgway and the JCS for authority to launch them.[54] Although Weyland did not think it militarily wise to bomb Manchuria, he asked General Ridgway for permission to strike the North Korean power stations, but Ridgway denied his request. In May 1952, General Mark W. Clark replaced Ridgway, and Weyland again raised the issue, this time successfully. The JCS concurred and authorized attacks on all North Korean power stations along the Yalu except the one at Sui-ho, only thirty miles upriver from Antung, home to one of the largest MiG bases in Manchuria.

General Mark W. Clark replaced General Ridgway as UN commander in Korea.

On June 17, 1952, Clark ordered Weyland and Vice Adm. Robert P. Briscoe, Commander Naval Forces Far East, to attack the installations at Hyosen, Fusen, and Chosen.[55] Two days later, the JCS reversed their earlier decision and approved an attack on Sui-ho (at that time the fourth largest hydroelectric station in the world). Navy and Fifth Air Force fighter-bombers opened the attacks

Among the targets to be attacked near Pyongyang were the marshaling yard at Sunchon, north of the fallen capital, and the Kan-ni arsenal, a main source of North Korean ammunition.

During the FEAF bombing assault on North Korea's industry and transportation lines, B–29s (*top photo*) concentrate on the chemical plant at Konan, the largest in Asia. It was left in shambles (*above left*). Bomber Command B–26s (*upper right*) start a run on their target of railroad yards. The marshaling yard (*lower right*) is on the main rail line leading south from Wonsan, an important East Coast port city.

with a strike against Sui-ho on June 23. Because the presence of large numbers of MiGs made daylight raids by Bomber Command's B–29s too risky, the four-engine veterans had since October 1951 been kept out of the region in northwest North Korea known as MiG Alley during daylight. Smaller, faster fighter-bombers, masking their approach to targets by flying at low altitude, could strike and dash off, often before the MiGs could reach them. By June 27, in four days these attacks had shut down 90 percent of North Korea's electric power generation.[56]

General Smart had also recommended to Weyland that air attacks be designed to undermine the "morale of the civilian population actively engaged in the logistic support of the enemy forces."[57] Accordingly, in the first week of July 1952, General Clark and the JCS approved a FEAF Target Committee recommendation that the North Korean capital, Pyongyang, a military as well as political center, be bombed heavily. On July 11, after a warning-leaflet drop, fighter-bombers of the Fifth Air Force, the Seventh Fleet, the 1st Marine Air Wing, and allied units — all shielded from MiG attacks by Fifth Air Force F–86 barrier patrols — worked over targets in and around Pyongyang. That night, Bomber Command B–29s made shoran area strikes.[58] In August, UN aircraft attacked the headquarters of North Korea's Army and, again, Pyongyang. On September 1 Navy carrier planes attacked North Korea's synthetic oil production plant at Aoji, only eight miles from the Soviet Union's border.[59]

Such attacks were meant to have psychological as well as military impact. On September 9, 1952, for example, 175 MiGs challenged Fifth Air Force fighters and fighter-bombers attacking North Korea's military academy, north of Pyongyang. At stake in this incident was the image of control. Strictly military considerations were secondary. The same was true when Bomber Command B–29s struck at Radio Pyongyang in January 1953.[60] In March 1953, the commander of the Fifth Air Force, Lt. Gen. Glen O. Barcus, personally led an attack on Radio Pyongyang, all the while broadcasting to North Koreans the taunt, Where is the communist Air Force? Such attacks were part of General Weyland's effort to force the North Koreans to pay a high price for their intransigence at the armistice talks.[61]

Weyland's "air pressure" campaign was supported by a new President, former Army general and the Supreme Allied Commander in Europe during World War II, Dwight D. Eisenhower, whose Republican administration was committed to ending the Korean conflict as soon as possible. When the armistice talks stalled yet again in April 1953, the JCS suggested that Eisenhower consider nuclear weapons as a means of forcing a settlement.[62] Like his predecessor, he was reluctant to use such weapons in a limited war; instead, he authorized Clark to renew air attacks on Yalu River hydroelectric stations and consider new types of targets. FEAF's Target Committee recommended attacking the dams that held the water essential for North Korea's rice farming, and Clark won JCS approval for this action.[63] General Weyland favored bombing

President Dwight D. Eisenhower, whose Republican administration was committed to ending the Korean conflict as soon as possible, is shown on an inspection tour of bases.

the dams if it would lead to the destruction of North Korean roads, railroads, and bridges downstream; he did not support the bombing of dams as a means of starving the North Korean civilian population.[64]

With General Clark's permission, Weyland ordered Fifth Air Force tactical fighter-bombers to hit the dam at Toksan, twenty miles north of Pyongyang, on May 13 as a test of this new bombing policy. Weakened by the bombing, the dam collapsed, and General Clark later observed that the resulting flood "has been [as] effective as weeks of rail interdiction."[65] Tactical aircraft attacked and holed another dam near Pyongyang on May 15 and 16, and Bomber Command's B–29s unsuccessfully went after a third in a nighttime shoran raid on May 21/22. Thus, the Eisenhower administration had raised the level of conventional destruction in North Korea another notch. The dam-busting not only threatened North Korea's rice crop, it also flooded numerous underground bunkers near Pyongyang.

In all, the evidence does not suggest that President Eisenhower overtly threatened to use nuclear weapons. Some tactical nuclear weapons were deployed to the Far East, though not in large numbers because the European theater had a higher priority. By June 1953, Weyland's 49th Fighter-Bomber Wing, based in Japan, had one squadron of F–84Gs equipped with tactical nuclear weapons.[66] These aircraft were not under SAC's control and their firepower was reserved for a battlefield emergency, such as a desperate, last-ditch effort by the North Koreans and Chinese to gain territory before the signing of an armistice.

The armistice was indeed signed on July 27, 1953. General Weyland remained convinced that FEAF's "air pressure" campaign had been a major factor in persuading the Chinese and North Koreans to accept a settlement.[67] Additionally, evidence exists that the morale of civilian North Koreans was nearly broken by the summer of 1953.[68] Whether true or not, Weyland and the JCS thought the threat of tactical nuclear weapons would deter any violation of the armistice after it was signed.[69] To make clear the strength of U.S. armaments after 2½ years of new production, the JCS authorized SAC to deploy a wing of B–36 bombers to Guam, Okinawa, and Japan at the end of August 1953.[70] They believed such signals of U.S. strength were essential to deterring further fighting in Korea.

Various constraints affected FEAF aerial operations. Bomber Command's initial assault on North Korea's industry and transportation lines in August and September 1950 was the only operation during the war in Korea when strategic bombing was carried out as it had been in World War II. To avoid giving North Korea any propaganda advantage, however, incendiaries were proscribed. When Chinese forces entered the war, General MacArthur's order to regard all villages in communist territory as military targets gave FEAF a free hand to attack civilian morale, but it did not do so. In part this was so because too few aircraft were available for such attacks while simultaneously pursuing air superiority and interdiction. Also, President Truman's desire to maintain unity among UN allies further constrained FEAF's actions. In the event of war with the Soviet Union, the United States would need British bases as staging areas for SAC's B–29s; consequently, official British views strongly influenced U.S. air war policy. Despite this consideration, the first attacks on the Yalu River power stations were not cleared with London beforehand, an oversight that led to a momentary crisis in allied relations.[71]

Even large-scale chaff drops (to cloud enemy radar screens) were forbidden until November 1952. This was to prevent Soviet advisers in North Korea from learning too much about U.S. electronic warfare techniques.[72] No first-line American bombers, not even the B–50, ever appeared in combat over North Korea. B–36s, B–50s, and B–47s were reserved expressly for possible strategic nuclear strikes and reconnaissance missions.

Finally, no theater air commander was present in Korea, contrary to Air Force doctrine. General Stratemeyer had asked General MacArthur for control of all theater air assets in the first days of the war, but U.S. Navy and Marine Corps officers challenged Stratemeyer's request, leaving Air Force–Navy coordination informal and consequently weak. General Weyland, as a result, worked to improve the level of Air Force–Navy cooperation in aerial operations. Shortly after assuming command in 1952, he ordered FEAF's Target Committee to circulate the biweekly targets for his "air pressure" campaign to the Commander, Naval Forces Far East. He also supported the efforts of his Fifth Air Force commander, General Barcus, in establishing a Joint Operations Center in Korea to coordinate tactical strikes by Navy, Air Force, and Marine Corps squadrons.[73]

Army General Mark Clark also favored greater interservice cooperation. In August 1952 he assured General Weyland he would create a joint Far East Command headquarters staff. Though he eventually did so, the close working coordination during the last year of the war between FEAF and Naval Forces Far East "came from the fortunate personalities of the commanders concerned rather than from more stable dictates of command authority and organization."[74] Weyland judged this a serious problem in a stalemated conventional conflict like the one in Korea; he believed the UN command needed to wield its air assets quickly and deftly in combination to gain psychological and political, as well as military, victories.

The Conflict in Vietnam

If, for U.S. forces, the conflict in Korea had a definite beginning, the conflict in Vietnam did not. From the perspective of the Kennedy administration, the regime of South Vietnamese leader Ngo Dinh Diem appeared unable to control the remnants of the anticolonial Viet Minh Army still loyal to Ho Chi

Ho Chi Minh, shown here in 1965, led the Communist Viet Minh forces.

Minh, head of both the North Vietnam government and the ruling communist party. The unrest in South Vietnam therefore represented a temptation the Soviet Union and the Peoples Republic of China could not resist. Tensions between the United States and the Soviet Union were very high in the early 1960s, and Nikita Khrushchev, General Secretary of the communist party in the Soviet Union, proclaimed that his government would not be kept from supporting wars of national liberation by U.S. threats to use nuclear weapons. The Kennedy administration believed Khrushchev was engaged in an indirect strategy to break the bonds of the containment policy that was developed during the Truman presidency. South Vietnam, therefore, was seen as another potential

President Eisenhower and Secretary of State John Foster Dulles greet South Vietnam's leader, Ngo Dinh Diem (*far right*) at National Airport in Washington, DC, in 1957. Diem appeared unable to control the remnants of the anticolonial Viet Minh Army still loyal to Ho Chi Minh.

Korea, a victim of communist expansionism, although the major difference was that this war was more covert than overt.

In South Vietnam, unfortunately, the autocratic Diem government, encumbered with corrupt officials, proved unable to build the kind of popular support necessary to field an Army and a police force able to defeat veteran Viet Minh cadres, many of which had remained in the south when Indochina partitioned at the end of the French-Indochina War in 1954. Had North Vietnam invaded South Vietnam the way North Korea had invaded South Korea in 1950, President Kennedy would have had no problem convincing Congress to support direct U.S. attacks on North Vietnam. But North Vietnam made no overt military move, leaving Kennedy with the problem of developing a suitable strategy for waging a guerrilla war. The President and his advisers believed the South Vietnamese government could fight the war with U.S. military supplies and training, but without popular support, the Diem government was unable to defeat the guerrilla forces (the Viet Cong) arrayed against it. When Diem seemed unable to govern effectively, Kennedy supported a coup that in early November 1963 removed him from the scene entirely. A few weeks later, in late November 1963, the U.S. President too was assassinated. In South Vietnam, the military and political situation continued to deteriorate.

Kennedy's successor, former Vice President Lyndon B. Johnson, inherited a potentially embarrassing foreign crisis. That December, Secretary of Defense Robert S. McNamara briefly visited South Vietnam. On his return he told Johnson that "current trends, unless reversed in the next two–three months, will lead to neutralization at best and more likely to a communist-controlled state."[75] McNamara did not think an increase in direct U.S. military action would be a solution, but he did advise the President to prepare "for more forceful moves if the situation does not show early signs of improvement."[76]

In a January 1964 memo to McNamara, however, JCS Chairman General Maxwell D. Taylor advised the Defense Secretary that the JCS believed the time had come for "bolder actions which may embody greater risks." Convinced that the Viet Cong were "externally directed and supported," Taylor argued "that if the U.S. program succeeds in South Vietnam it will go far toward stabilizing the total Southeast Asia situation."[77] Speaking for the service chiefs, Taylor recommended committing "U.S. forces as necessary in direct actions against North Vietnam."[78] He implied that if the United States failed to act, it might find itself facing communist states throughout Southeast Asia.

Secretary of Defense McNamara, Secretary of State Dean Rusk, and Johnson's National Security Adviser McGeorge Bundy, however, were more cautious. If the conflict in South Vietnam was in fact externally directed and aided, then the real enemy, as in Korea, was probably China, or China and the Soviet Union acting together. These men clearly did not want to risk another war like the one in Korea, but they did not believe President Johnson could avoid aiding South Vietnam. The JCS and Johnson's closest civilian aides and

President Kennedy, in discussion with Ambassador to South Vietnam Henry Cabot Lodge, faced a problem of developing a suitable strategy for waging a guerrilla war.

cabinet officers believed that the struggle for supremacy in South Vietnam held significant implications for the larger competition between the United States and the Soviet Union. They disagreed among themselves, often very strongly, over what level of U.S. military action in South Vietnam and against North Vietnam was appropriate. Indeed, for the next four years, the JCS and their subordinates in the Pacific would press the President to use more force in Vietnam, especially against North Vietnam. Civilians such as McNamara would resist, fearing direct Chinese and Soviet intervention in the fighting. The resulting war strategy was a product of negotiations among Johnson's military and civilian advisers and deputies.[79]

As a first step, the President in February 1964 approved covert U.S. and South Vietnamese military actions against North Vietnam. In April, at Defense Secretary McNamara's request, the JCS compiled a list of ninety-four bombing targets in North Vietnam. In May, the administration drafted a resolution that it could submit to Congress to justify further U.S. military action, including air attacks on North Vietnam.[80] In the meantime, South Vietnamese forces continued to lose territory to the Viet Cong, and the administration found itself in a deadly race with events over which it had only limited direct control. On August 2, U.S. destroyers in the Gulf of Tonkin reported that they had been attacked by North Vietnamese patrol boats. Five days later, Congress passed the Tonkin Gulf Resolution, giving the President the authority he needed to increase direct U.S. participation in the warfare in Vietnam.

Soon afterward, the JCS told McNamara, "U.S. military actions . . . have demonstrated our resolve. . . . Failure to resume and maintain a program of pressure through military actions . . . could signal a lack of resolve."[81] The President, however, rejected the JCS's position, choosing instead to adopt McGeorge Bundy's approach of "graduated and continuing reprisal" for acts of terror committed by Viet Cong in South Vietnam.[82] Materials published in *The Pentagon Papers* show that Johnson wanted the American public to see him as being forced to respond — reluctantly — to the deliberately provocative acts of the Viet Cong and North Vietnamese. On October 30, 1964, Viet Cong units attacked a U.S. air base at Bien Hoa, South Vietnam, and the JCS again urged the President to order major bombing attacks on North Vietnam. Johnson again refused. By January 1965, even Defense Secretary McNamara supported some sort of bombing, if only as a signal that the United States was committed to supporting the South Vietnamese government. On February 7, 1965, Viet Cong guerrillas struck a U.S. Special Forces camp at Pleiku, triggering "a swift, though long-contemplated presidential decision to give an 'appropriate and fitting' response."[83] On February 24, Operation ROLLING THUNDER, a major aerial interdiction campaign, characterized by gradually increasing pressure against North Vietnam, began. It would last for nearly four years.

On the ground, the military balance in the south continued to turn against the Saigon government. Partly as a consequence of the worsening situation, in

April 1965 the Under Secretary of State proposed that the United States withdraw completely. The Commander, Military Assistance Command, Vietnam (MACV), Army General William C. Westmoreland, had already asked that his ground combat strength be nearly tripled, and the President had to decide whether to greatly increase U.S. involvement or pull the forces out. On July 1, 1965, Secretary of Defense McNamara drafted a memorandum for the President. It gave General Westmoreland the reinforcements he needed to take the offensive against the Viet Cong, ordered the U.S. Navy to mine North Vietnam's ports, authorized U.S. Air Force units to attack North Vietnam's rail links to China, and called for U.S. Navy and U.S. Air Force air squadrons to destroy North Vietnam's MiG bases and surface-to-air missile (SAM) sites.[84]

Later that month, McNamara spent a week in South Vietnam. On his return, he altered his July 1, 1965, memo, canceling support for the mining of North Vietnam's harbors and restoring restrictions on the conduct of ROLLING THUNDER. In August, he told the congressional armed services committees that an all-out bombing campaign against the North might bring China into the conflict. Privately, to the JCS, he took the same position, rejecting their recommendation that North Vietnam's strategic oil storage facilities and electric power plants be bombed. The administration's policy of restraint seemed to be justified when, in September 1965, the Chinese Communist government announced it would not intervene directly in the Vietnam conflict except by providing volunteer labor and quantities of foodstuffs.

Air Force leaders were unsure that the conflict would stay limited. Chief of Staff General Curtis E. LeMay in January 1965 had authorized modification of SAC's D and F model B–52s so that they could carry approximately 70,000 pounds of 500- and 750-lb bombs. In February 1965, the JCS had ordered General Thomas S. Power, SAC Commander, to dispatch thirty B–52s to the Eighth Air Force at Andersen AFB, Guam.[85] Intent on meeting single integrated operational plan commitments, Power opposed the modifications and the movement of these aircraft.[86] His objections were overruled, however, and B–52s began flying conventional bombing missions from Andersen against targets in South Vietnam in June 1965.

North Vietnam's government took the February 1965 bombings very seriously. All urban residents not engaged in work essential to North Vietnam's war effort were ordered into the countryside. By the end of 1965, almost 100,000 people had left Hanoi under this edict. Many soon returned, however, when they learned that U.S. aircraft were not attacking areas near Hanoi and Haiphong. As U.S. air attacks intensified in 1966 and 1967, people moved (or were moved) again, which by December 1967 reduced Hanoi's population by one-third.[87] The Hanoi government in 1965 also began dispersing the nation's limited industry and building tunnels and air shelters.[88] Supported by supplies of food and war materiel from China and the Soviet Union and aided by 40,000 Chinese construction workers who repaired bomb damage on the rail links to

CHINA
NORTH VIETNAM
Dien Bien Phu
Hanoi
Haiphong
BURMA
LAOS
Gulf of Tonkin
Vientiane
Udorn
Nakhon Phanom
THAILAND
Hue
Da Nang
SOUTH VIETNAM
Takhli
Korat
Ubon
Don Muang
Dak To
Bangkok
Pleiku
Phu Cat
Tuy Hoa
U-Tapao
CAMBODIA
Nha Trang
Cam Ranh Bay
Phnom Penh
Phan Rang
Bien Hoa
Tan Son Nhut
Saigon
Principal U.S. Air Bases In Southeast Asia 1965–1973
Air Base
0
75
150
miles
South China Sea

SAC commander General Thomas S. Power displays wares to President Kennedy and Defense Secretary Robert S. McNamara (*far left*), as Lt. Gen. Howell M. Estes (*right background*) and others look on.

China, North Vietnam's leaders steadfastly refused to give in to U.S. reprisal and limited aerial attacks.[89]

In effect, North Vietnam stayed one step ahead of the gradually escalating ROLLING THUNDER bombings by dispersing its industry and oil supplies and by building up what eventually became one of the strongest area air-defense systems in the world. In August 1964, for example, North Vietnam had about 1,400 antiaircraft guns of all types but only twenty-two early warning (EW) radars and only four fire control (FC) radars to direct these guns. Its air force was also rudimentary. By the end of March 1965, when ROLLING THUNDER was only a month old, North Vietnam's air defenses mounted thirty-one EW radars and nine FC radars.[90] The North Vietnamese had also more than quadrupled the number of their antiaircraft guns and were building a ground control intercept (GCI) radar and communications network to direct their expanding jet interceptor force.[91] On July 14, 1965, the U.S. Air Force lost its first aircraft to a Russian-made SA–2 SAM, a missile designed to intercept aircraft of all types at medium and high altitudes. By 1966, North Vietnam had deployed nearly a hundred SA–2 launchers, many of them mobile and all with reloads.[92]

North Vietnam's resistance and growing air defenses prompted the JCS to increase their pressure on President Johnson to reduce the number of areas in North Vietnam declared off limits to U.S. bombing. In November 1965 they pushed for the systematic destruction of North Vietnam's oil supplies, and in February 1966, after a 37-day-pause in ROLLING THUNDER produced no diplomatic breakthrough, the JCS again told the President the bombing had to

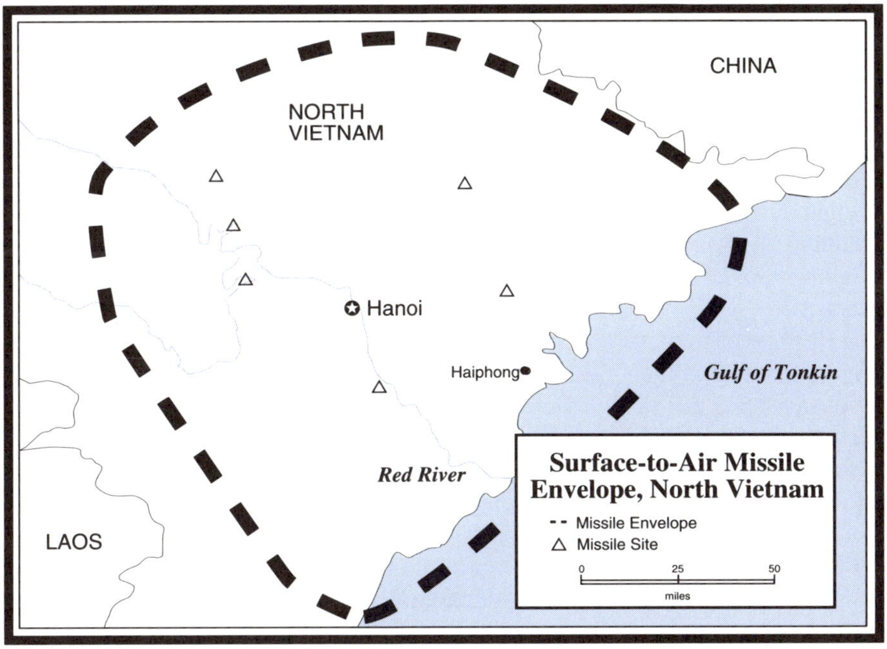

The Soviet SA–2 surface-to-air missile, which shot down the first U.S. Air Force aircraft in July 1965, became a dreaded threat to U.S. air power in Vietnam.

intensify. From his headquarters in Hawaii, Admiral Ulysses S. Grant Sharp, overall Pacific Commander, sided with the JCS. Gradually, over a period of months, Johnson accepted some of their recommendations. In June 1966, for example, North Vietnam's oil storage facilities were bombed for the first time. That December, U.S. tactical aircraft struck targets in Hanoi. In February 1967 President Johnson approved a JCS proposal that Navy aircraft mine North Vietnam's canals. In March, the President approved the basing of B–52s in Thailand. In April, U.S. aircraft were ordered to bomb two power plants in Haiphong and two MiG airfields; in May, Hanoi's main power station was attacked.

ROLLING THUNDER followed a uniform pattern. The initial approach was limited bombing; then, when North Vietnam continued to reject American proposals for a political settlement, the JCS would press for escalation. The President's advisers, meanwhile, would wage a war of memos over which step to take. Reluctantly, the President accepted some escalation, though less than the JCS recommendation. When the new round of bombing failed to bring the North Vietnamese to the conference table, the cycle began anew. President Johnson was so concerned about controlling the escalation of pressure that he personally directed the bombing campaign. Every week the JCS would present an updated list of North Vietnamese targets to Defense Secretary McNamara and Secretary of State Rusk. The two Cabinet officers usually removed some

On hand to greet Vietnamese Chief of State Nguyen Van Thieu for conferences were President Johnson and his Secretary of State, Dean Rusk. Following Thieu are Vietnamese Prime Minister Nguyen Cao Ky, U.S. Chief of Protocol Lloyd, and Mrs. Ky.

suggested targets before reviewing the list with the President at regular Tuesday lunches. Johnson "would make the decision personally as to what targets would be bombed."[93] This unusual review process at the White House level seriously degraded the effectiveness of ROLLING THUNDER.[94] It also created hard feelings within the Department of Defense, particularly between the Secretary and the JCS (and Admiral Sharp). The officers' perception that their professional judgment was not respected by Secretary McNamara gave the debate over ROLLING THUNDER a sharp, personal edge.[95]

The theater command organization responsible for ROLLING THUNDER also was not what the Air Force wanted. In October 1962 the commander of Pacific Air Forces (PACAF) established the 2d Air Division. Headquartered in Saigon to control air operations in Southeast Asia, it became a component of the MACV and also an element of the Thirteenth Air Force, which reported through PACAF to the Commander-in-Chief, Pacific (CINCPAC).[96] In June 1965, the commander of the 2d Air Division was given additional duty as Deputy Commander, MACV.[97] To further complicate the structure, Navy aviation was organized as a separate command that reported first through Seventh Fleet and then through the Navy component commander in the Pacific (CINCPACFLT) who worked for CINCPAC. Since the Thai government did not want a U.S. Air Force field commander in Thailand subordinate to his counterpart in South Vietnam, U.S. Air Force units in Thailand were placed directly under the command of the Thirteenth Air Force, with strike orders coming from the 2d Air Division in Saigon.

As the U.S. air campaign against North Vietnam increased in 1965, Admiral Sharp, CINCPAC, divided North Vietnam into seven districts called route packages, and split command responsibility for each among the area commanders. The 2d Air Division controlled strikes in route packages (RPs) I, V, and VI–A, but MACV selected targets in RP I while final responsibility for selecting targets in RPs V and VI–A remained in the hands of the PACAF commander.[98] The latter, in March 1966, upgraded the 2d Air Division to the Seventh Air Force, whose commander thereafter reported directly to PACAF. General William W. Momyer, Seventh Air Force commander from July 1966 through July 1968, argued that his command should coordinate all air action in South Vietnam and Laos as well as all ROLLING THUNDER strikes against North Vietnam.[99] He was not given this authority, however, and the route package system persisted through the end of ROLLING THUNDER.

When B–52 strikes began in South Vietnam, a SAC Advanced Echelon (ADVON) was assigned to MACV.[100] MACV, however, could only request B–52 missions. SAC retained control over its B–52s and the KC–135 tankers refueling all strategic bombers that flew from Guam and Seventh Air Force tactical aircraft that operated from bases within Southeast Asia. Aerial refueling proved essential for the tactical squadrons of the Seventh Air Force. It allowed heavily loaded F–105 fighter bombers to cover all worthwhile targets in RPs

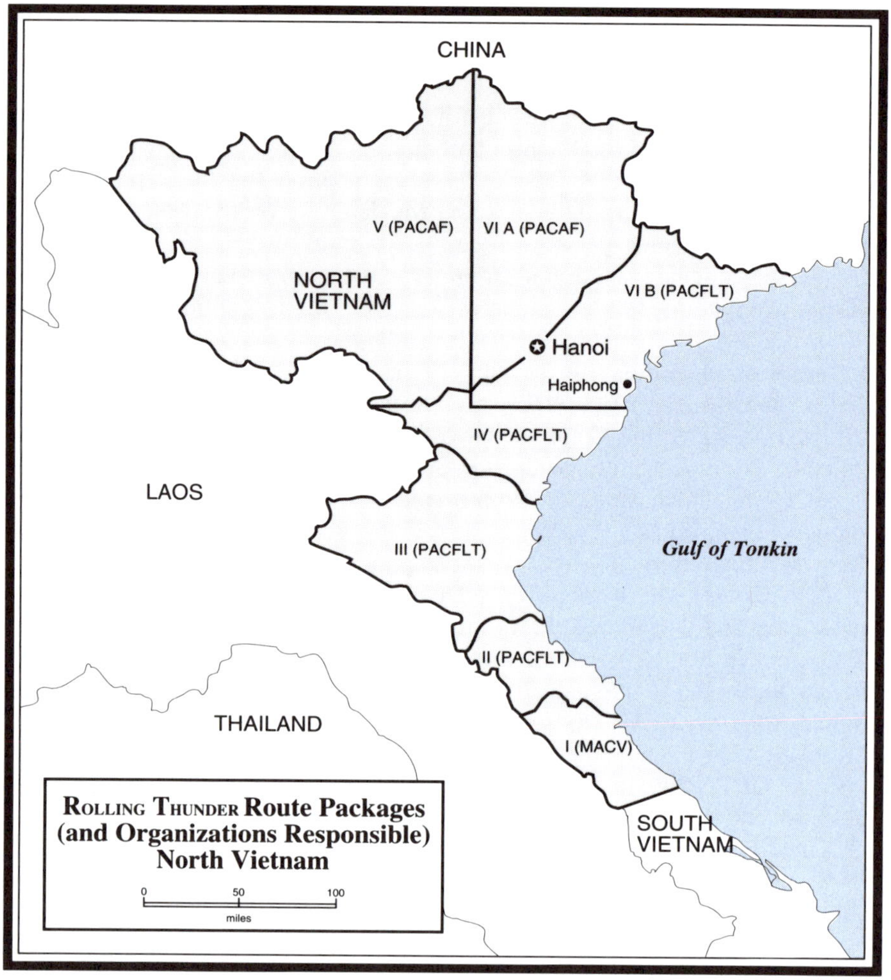

V and VI–A with the exception of those placed off limits by higher political authorities. It also led Seventh Air Force commander Momyer to demand support from SAC's tankers, which SAC headquarters was reluctant to give.[101] The issue between SAC and Seventh Air Force was the scheduling of tanker missions. The former wanted regular, around-the-clock tanker sorties that would produce predictable workloads on planes, aircrew, and ground support personnel (while leaving a reserve of tankers if required for strategic missions). Seventh Air Force, on the other hand, wanted the tankers to surge in support of YOUNG TIGER operations, that is, to support formations of tactical fighter bombers, electronic warfare aircraft, and fighters conducting ROLLING THUNDER missions. Despite SAC's expressed concerns, ROLLING THUNDER missions were given YOUNG TIGER support.

ROLLING THUNDER gradually increased in scale and tempo, reflecting a shift toward the hawks among President Johnson's advisers. By the end of 1967, most of North Vietnam's major electric power and industrial targets had been bombed. U.S. aircraft routinely fought their way into and out of North Vietnam's airspace, attacking transportation links such as the Paul Doumer railroad bridge which carried the rail line from China into Hanoi, and even MiG bases. North Vietnam's government, in response, dispersed its petroleum supplies, evacuated its bureaucracy (but not its highest leadership) from Hanoi, and constructed more air-raid shelters.[102] By mid-1967, nearly three-fourths of North Vietnam's agricultural work force and about two-thirds of its bureaucracy consisted of women, who had replaced men called up for military service. More than a million people labored to repair damages from U.S. bombing attacks. Young and old were organized into paramilitary units to support truck transport and air defense. As in Korea in 1952 and 1953, whole villages, mostly in North Vietnam's southern panhandle, lived underground during the day.[103] North Vietnam's government prepared to become a target of a strategic bombing campaign and, in so doing (and with much aid from the Soviet Union and China),

Aerial refueling turned out to be essential for the fighter squadrons of the Seventh Air Force. A KC–135 Stratotanker refuels a flight of F–105 Thunderchiefs on their way to a strike in North Vietnam.

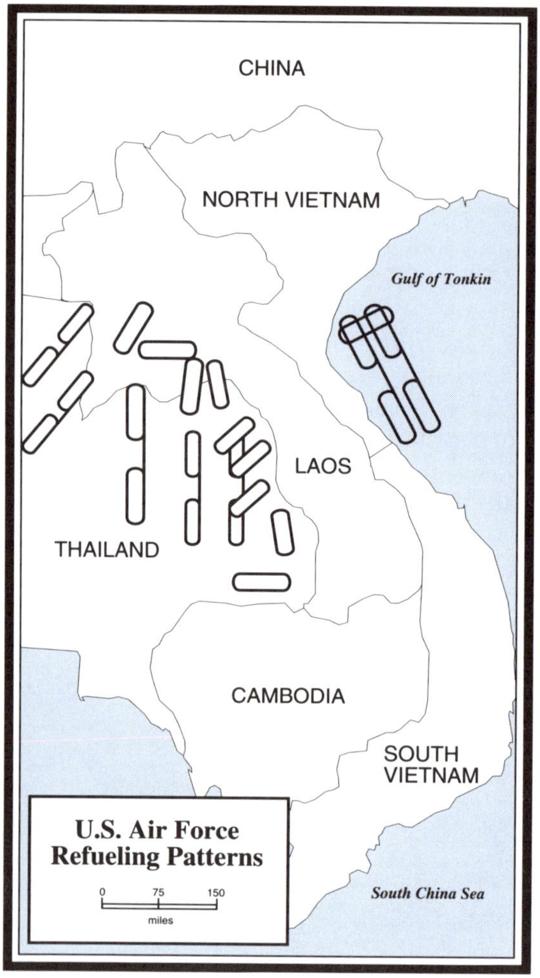

was able to tolerate the damage from ROLLING THUNDER. Additionally, the morale of North Vietnamese civilians did not appear to weaken with the aerial bombardment.[104]

After the Viet Cong launched attacks on South Vietnam's major cities on January 31, 1968 (the start of Vietnam's traditional Tet holiday), the Johnson administration abandoned its strategy of coercive bargaining, applied through the gradually escalating ROLLING THUNDER bombing. General Westmoreland, MACV Commander, subsequently asked for whatever ground reinforcements he could get, and when General Earle G. Wheeler, JCS Chairman, visited South Vietnam in February to evaluate Westmoreland's request, he recommended that 200,000 additional U.S. troops be sent. That same month, the President forced Secretary of Defense McNamara to resign. By this time, McNamara was disenchanted with the policy he had helped fashion in 1964 and 1965. His replacement was Clark Clifford, a Johnson confidant. Clifford, who had accepted the coercive bargaining strategy, turned against it as Secretary of Defense and sought to end the bombing of North Vietnam entirely. When Johnson announced on March 31, 1968, that he would not run again for the presidency, he also revealed he was ordering a halt to ROLLING THUNDER attacks above the 20th parallel (RPs IV through VI). In response, the government of North Vietnam agreed to negotiate, and representatives of North Vietnam and the United States met in Paris that May.

The lengthy, undeclared, and costly war in Southeast Asia cost the Democratic Party the White House. The new President, former Vice President

Richard M. Nixon, was determined to fashion his own strategy for Southeast Asia, but he was equally determined not to withdraw U.S. troops precipitously from South Vietnam. Like Johnson, Nixon feared that the appearance of a U.S. defeat in South Vietnam would lead to the loss of the remaining states in Southeast Asia to international communism. Also, like Johnson, the new President believed in applying military pressure to North Vietnam as a means of reducing the threat posed by Viet Cong and North Vietnamese forces in South Vietnam to the survival of Saigon's government. What the President and his advisers now sought was a means for applying military pressure that worked better than ROLLING THUNDER had.

Their solution was to couple Vietnamization (the incremental substitution of South Vietnamese troops for U.S. soldiers in South Vietnam) with an effort to gradually isolate North Vietnam from both China and the Soviet Union. Nixon's strategy was developed to mollify public opinion in the United States (by reducing U.S. casualties) and to isolate North Vietnam politically. Then, if necessary, an aggressive bombing campaign could be used to apply even more direct pressure against a weakened North Vietnam. Former President Johnson had assumed that the conflict in Southeast Asia, like the war in Korea, was an extension of the conflict between the Soviet Union (and China) and the United States. Nixon shared that assumption. His larger strategy was to change the relationship between the United States and the Soviet Union, and then he would deal with North Vietnam. Henry Kissinger, Nixon's National Security Advisor, was surprised to learn in 1969 that the Johnson

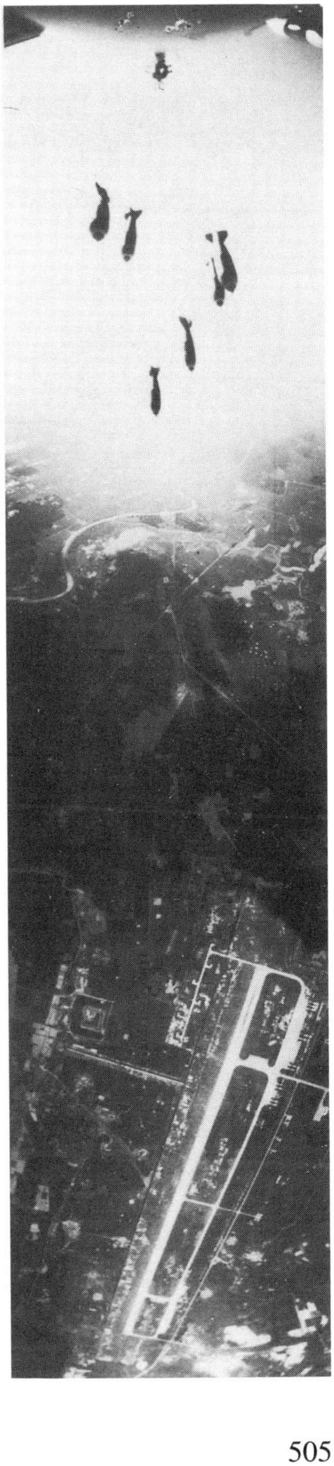

The first bombs to strike previously untouched Phuc Yen airfield in North Vietnam were dropped from an F–105 Thunderchief on October 24, 1967, rendering the field unserviceable.

General Earle G. Wheeler, Chairman of the Joint Chiefs of Staff, at Da Nang Air Base in Vietnam (*top photo*). After seven years as Defense Secretary, Robert S. McNamara (*left* in *center photo*), forced to resign, was succeeded by Clark M. Clifford (*right* in *center photo*). President Johnson reviews troops with General William C. Westmoreland beside him and General Walker to the front at Cam Ranh Bay Air Base in South Vietnam (*bottom*).

administration had no plan to exploit the growing hostility between the USSR and the Peoples Republic of China. Kissinger and Nixon developed such a plan during the first Nixon administration.

The most obvious sign of the change in strategy was the decline in the number of U.S. troops in South Vietnam. At the end of 1968, U.S. forces in South Vietnam numbered 540,000. A year later, they were fewer by al-most 100,000; by the end of 1970, their number had dropped to 280,000; by the end of 1971, it was 140,000. Other signs were present too. In July 1971, for example, Nixon announced he would visit China the following February. The President also worked to organize a summit meeting with Leonid Brezhnev, the General Secretary of the Communist Party of the Soviet Union. Consequently, the North Vietnamese government abruptly withdrew from the Paris peace talks in October 1971. Partly to forestall Nixon's diplomatic initiatives with their Soviet and Chinese sponsors, North Vietnamese leaders planned a major conventional assault against South Vietnam. Nixon authorized U.S. Navy and Air Force air units still in the theater to conduct Operation PROUD DEEP ALPHA, which between December 26 and 30, 1971, consisted of more than a thousand sorties south of the 20th parallel.[105]

Although by far the biggest operation against North Vietnam between the end of ROLLING THUNDER in 1968 and the North Vietnamese offensive in 1972, PROUD DEEP ALPHA would be overshadowed by a series of much smaller raids unauthorized by the Nixon administration. These led to the relief of the Seventh

President Richard M. Nixon on a visit to troops in Vietnam.

Air Force commander, General John D. Lavelle, whose subordinates engaged in deceptive reporting to hide the unauthorized strikes until an enlisted man revealed this practice in a letter to his senator. Lavelle thought that Washington had wanted him to stretch the rules, but when the story broke, he was advised to retire, which he did. Congressional hearings kept his case in the news for months, feeding Americans' distrust of air power in Southeast Asia, a distrust rooted in the covert air operations in Laos and Cambodia.[106]

Nixon, meanwhile, intended his contacts with both Soviet and Chinese leaders to symbolize a decrease in tensions and a changed international situation. The North Vietnamese government was clearly preparing to act before Nixon's diplomatic initiatives affected Chinese and Soviet support. On February 21, 1972, the President departed to the Peoples Republic of China on a state visit, the first ever undertaken by an American president. Four weeks later, on March 30, regular units of the North Vietnamese Army attacked South Vietnam across the demilitarized zone. In early April other North Vietnamese forces launched an offensive in South Vietnam's central highlands, moving toward An Loc, on the road to Saigon.

A few months earlier, in December 1971, Nixon had ordered the JCS to reinforce the Seventh Air Force, and PACAF began transferring F–4 Phantom fighter-bombers to South Vietnam and Thailand at the end of the month. Thirty-seven B–52s also moved to Andersen AFB, Guam, and U-Tapao, Thailand. By mid-February 1972, 84 B–52s were in Southeast Asia: 31 on Guam and 53 at U-Tapao. On April 6, U.S. air units were authorized to resume attacks on North Vietnamese supply lines as far north as the 20th parallel. On April 16, B–52s and other aircraft bombed oil storage facilities near Haiphong. On May 4, the President ordered the JCS to prepare the kind of interdiction campaign first proposed by their predecessors in 1964.[107] Within days, Navy aircraft mined and closed the port of Haiphong. The Soviet Union responded with only mild protests, much to the relief of American military and political leaders. As the commander of the Seventh Fleet recalled: "We always were conscious of the fact that our main enemy was not the North Vietnamese, but the Soviets."[108]

Operation LINEBACKER, the new air campaign throughout North Vietnam, began on May 10, 1972, when 32 F–4s of the 8th Tactical Fighter Wing based in Ubon, Thailand, struck the Paul Doumer bridge in Hanoi, and Navy A–6 attack planes bombed the Hai Duong railroad yard between Hanoi and Haiphong.[109] By the end of May, the rail lines linking China to Hanoi and Haiphong had been cut. By the end of June, the 8th Tactical Fighter Wing wrecked an estimated 106 bridges, cutting North Vietnam's imports of materiel to less than one-fifth of what they had been before LINEBACKER began. One senior Air Force officer observed that LINEBACKER had achieved more "in its first four months of operation than ROLLING THUNDER had in three and a half years."[110]

In 1972, as in 1968, the North Vietnamese had perhaps forty occupied

F–4C aircraft were specially equipped to counter enemy radar-controlled ground defenses.

SAM sites near Hanoi and Haiphong and more as far south as the demilitarized zone, but now, with about 125 MiG–21 and MiG–19 interceptors, their air force was larger and better equipped than it had been during ROLLING THUNDER.[111] To permit U.S. air units to deal more effectively with those defenses, President Nixon relaxed the official rules of engagement and allowed the JCS more freedom in the instructions they gave to the commanders in Vietnam. The buffer zone along North Vietnam's border with China was retained, but in no areas of Hanoi or Haiphong was bombing absolutely prohibited, and areas designated as restricted were only twenty nautical miles in diameter. More importantly, the only targets the JCS had to approve before they could be attacked were those in the buffer zone near China and in the Hanoi/Haiphong restricted areas.[112] Furthermore, the JCS lifted all restrictions on the timing and scheduling of attacks. As General John W. Vogt, Jr., Seventh Air Force Commander, noted, the new policy "permitted us to play the enemy defenses.... This was something we were never able to do in ROLLING THUNDER."[113]

The Air Force and Navy also successfully used newly developed laser-guided bombs. General Vogt observed that "the thing that really made the difference during the early months of LINEBACKER was the precision of our weapons."[114] During the 1972 LINEBACKER interdiction campaign, USAF fighter-bombers mounted two types of laser-target designators to guide "smart" bombs to their targets. The first required that an aircraft circle the objective while keeping its laser designator focused on the target. The second type swiveled, allowing the F–4 to focus on a ground target while still maneuvering to evade antiaircraft fire. Laser-guided bombs directed against ground targets — homing in on the spots of laser light — achieved an accuracy of eight feet circular error average (CEA) and a circular error probable (CEP) of zero feet. General Vogt was amazed and delighted with the incredible accuracy: ". . . we went up and knocked out five bridges on the Northwest Rail Line with a laser strike, and when PACAF ran that through the computers, they determined that where we used 24 total bombs, it would have taken 2,400 bombs to do that by the old conventional method."[115]

Unfortunately, laser-guided ordnance could not be directed effectively against targets during the northeast monsoon season (November through April), and gains made during LINEBACKER would be lost if strike aircraft could not be guided to their targets in the attendant overcast. To compensate, Seventh Air Force finally put an effective long-range navigation (loran) system into operation for North Vietnam. Like the shoran system used in Korea, loran was based on the ability of an aircraft to determine its position by noting the time delays in radio signals broadcast simultaneously by widely separated transmitters. loran's effectiveness, however, was less than shoran's because the distances from loran transmitters in South Vietnam and Thailand to targets in the northern route packages were greater than the distances had been with shoran in Korea.[116]

Much of LINEBACKER's success was also due to improvements in U.S. Air Force and Navy electronic warfare equipment and tactics. During ROLLING THUNDER, U.S. aircraft had to develop increasingly sophisticated tactics, and use much improved electronic gear, to maintain air superiority over North Vietnam. As North Vietnam's radars improved and multiplied, U.S. aircraft added jamming equipment to destroy them. When the North Vietnamese began deploying SA–2 SAMs, Air Force and Navy pilots learned how to evade them. As the MiG threat grew, U.S. Air Force EC–121 aircraft and Navy ships in the Gulf of Tonkin developed means for tracking MiGs and alerting U.S. aircraft to their presence.

After reviewing these developments in 1968, the Air Force's CREDIBLE COMET study group recommended that all tactical aircraft be equipped with effective electronic warfare equipment designed to detect enemy radar signals and transmitters, and jam or confuse enemy electronic sensors.[117] By the start of LINEBACKER, Air Force F–4s carried sensors to tell their two-man crews when enemy FC radar had locked onto their plane, jammers to cloud enemy radar screens with static, and chaff dispensers to release bunches or streams of metallic foil that reflected enemy radar signals to present enemy gunners and missile-launching crews with false radar echoes. WILD WEASEL F–105s were

An F–4 dropping a MK–84 laser-guided bomb over Southeast Asia (*top photo*). By 1972, the North Vietnamese were well equipped, with the Russian MiG–21 in their inventory.

equipped with sensors able to detect the source of ground-based SAM FAN SONG FC radars and with Shrike and Standard antiradiation missiles that could home in on those radars. IRON HAND fighter-bombers whose mission was to attack SAM sites and antiaircraft guns were also available.

Basic principles of aerial electronic warfare, however, had changed little since the technique originated in World War II. Ground defenses depended upon the alert supplied by EW radars. With a sufficient alert, defenders could get their fighters into the air, vectoring them against oncoming bombers with GCI radars. Near a target, attacking aircraft would be "illuminated" by ground-based FC radars directing either missiles or antiaircraft guns. Attackers tried to interfere with or deceive these different radars by clouding EW radar through static from jammers (broadcasting the same signal back at greater power) or by filling them with false targets from chaff.

Jamming told the North Vietnamese that U.S. aircraft were in range of their radars, but when used at the right level of power, it denied enemy radar operators information about the precise location of U.S. planes. EB–66 jammers, for example, would fly racetrack patterns outside the range of North Vietnamese SAMs, continuously broadcasting during a U.S. attack. This tactic was used to make it difficult or impossible for enemy FC radars to gauge accurately the range of U.S. planes by picking out the return of their own radar signals from the signals broadcast by the EB–66s.

The major difference between air defenses over North Korea during 1950 to 1953 and those over North Vietnam from 1965 to 1972 was the presence of SAMs. SA–2s were designed to knock down bombers like the B–52, so the big bombers usually kept away from heavily defended areas of North Vietnam. To compensate for a lack of bombers, the Air Force and Navy developed formations of attacking fighter aircraft to execute their ROLLING THUNDER and LINEBACKER missions. In LINEBACKER, however, the ratio of electronic warfare and escorting aircraft to fighter-bombers was four to one, not counting airborne tankers, surveillance aircraft like the EC–121s, and search and rescue planes and helicopters.[118] Because so much support was needed for the LINEBACKER tactical aircraft strikes, Seventh Air Force could stage only one or two missions per day into RP VI.[119]

During most of the 1972 interdiction campaign, B–52 strikes were staged no farther north than RP I. That October, North Vietnam moved SA–2 missiles into RP I in an effort to drive the B–52s south.[120] The North Vietnamese had reason to react: carrying 108 500-lb bombs, the B–52D was the biggest ordnance delivery system in the theater. On November 20, a MiG–21 interceptor attacked a flight of three B–52s bombing the Ho Chi Minh Trail, which ran south through Laos. This was North Vietnam's attempt to use defensive weapons for air superiority over its own territory, something North Korea had been unable to do twenty years earlier. This attempt, nevertheless, was unsuccessful. The North Vietnamese government, after agreeing to resume negotiations on October 8, now took a harder line when the American election promised to return a Congress eager to withdraw from Vietnam.

At the beginning of December 1972, the U.S. Air Force had the following strategic and tactical aircraft available for missions over North Vietnam: 54 B–52Ds at U Tapao, Thailand; 99 B–52Gs and 53 B–52Ds at Andersen AFB, Guam; 48 F–111s at Takhli, Thailand; 117 F–4s and RF–4s at Udorn, Thailand; 6 F–4C WILD WEASELs, 47 F–4E and F–105G fighter-bombers, and 17 EB–66 jammer aircraft at Korat, Thailand; and 111 F–4s at Ubon. The F–111s were recent arrivals; they had returned to Takhli only that September and were used on single-plane missions against airfields and bridges. In fact, by October 13 they accounted for half of the Air Force's strikes against North Vietnam.[121] Although a B–52G could carry only fifty-one 500-lb bombs and its electronic warfare equipment was less effective than the B–52Ds' against North Vietnamese radars,[122] combined with the available tactical aircraft (Air Force and Navy), B–52 squadrons posed a major threat to North Vietnam. SAC planners had begun selecting targets suitable for the heavy bombers in September 1972, when the fear was that the northeast monsoon would interfere with the LINEBACKER campaign.[123]

Nixon, however, believed he had to do more than threaten their use. He was overwhelmingly reelected that November, and when the North Vietnamese government refused to agree on a cease-fire in South Vietnam, the President informed the JCS on December 15 that he would use the B–52s against Hanoi itself. As Nixon recalls in his memoirs, he told Henry Kissinger, "We'll take the same heat for big blows as for little blows . . . and that means we will have to make the big decision to hit Hanoi and Haiphong with B–52s. Anything less will only make the enemy contemptuous."[124] This decision was consistent with Nixon's efforts to buy time for the government in Saigon by bombing the North. The way in which he made this decision was also consistent with his belief that President Johnson's inability to influence North Vietnam was partly the result of allowing the JCS and civilian advisers to wrangle continually over how to apply military pressure on the North Vietnamese regime.[125] As General Horace M. Wade, Vice Chief of Staff of the Air Force in late 1972, later recalled, "At no time was the JCS asked to comment or to agree or to disagree to what Mr. Kissinger was

An F–105 WILD WEASEL based in Korat, Thailand (*top*), a low-angle front view of a B–52D armed with 500-lb. bombs taking off from U Tapao, Thailand (*center*), and B–52Ds at Andersen AFB, Guam (*below*) are poised for the start of LINEBACKER II Operations on December 18, 1972.

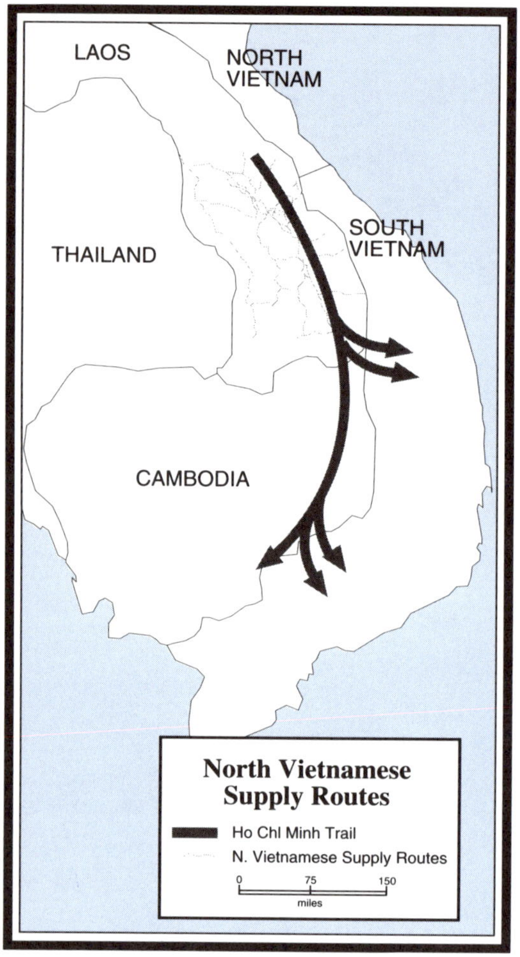

doing."[126] Kissinger and Nixon suspected the JCS might object to the President's plans to hazard B–52s in an attack on Hanoi. The President wanted no interference in his efforts to apply air power to achieve his diplomatic aims.

With the start of what became known as LINEBACKER II on December 18, 1972, B–52 aircrews at Andersen and U Tapao found themselves in a drama familiar to World War II strategic bomber crews. The B–52 crews were ordered to press on, which meant they would fly to the target even if some of their equipment failed: "The loss of two engines enroute or complete loss of the bombing computers, radar system, defensive gunnery, or ECM [electronic countermeasures] capability were [*sic*] not legitimate grounds for abort."[127] The B–52 attacks were scheduled at night; tactical aircraft bombed Hanoi during the day. Both day and night attacks focused on railroad yards, warehouses, communications centers, power plants, MiG airfields, SAM sites, and bridges. If the bombing campaign (planned originally for twelve days) was successful, North Vietnam would be left essentially defenseless to attacks from the air, leaving, Nixon hoped, the North Vietnamese government little choice but to negotiate on his terms.[128]

The LINEBACKER II B–52 strikes were planned and scheduled at SAC headquarters in Omaha. Seventh Air Force and Task Force 77 in the Gulf of Tonkin assembled the flights of supporting aircraft including 8 F–4 chaff spreaders, 8 F–105G or F–4C WILD WEASELS, 10 F–4s flying close escort against MiGs, another 10 Phantoms flying as a distant barrier against MiGs, and 3 EB–66 jammers. The B–52s approached Hanoi from the west, with bombers from Andersen crossing South Vietnam's coastline about halfway between Saigon and Hue. Flying straight on into Cambodia, they turned northwest into Thailand and then northeast over Laos. B–52s based at U Tapao flew northeast over Udorn and then crossed Laos to North Vietnam. Approaching at 32,000 feet, the B–52s initially attacked in three waves at four-hour intervals. The night of the first attack (December 18/19), 129 B–52s attacked in waves of 48, 30, and 51 aircraft flying in three-plane cells. Each wave had the support of 39 chaff spreaders, jammers, WILD WEASELs, and escorts. The bombers were not allowed "to maneuver to avoid SAMs or MiGs from the initial point . . . on the bomb run to the target bomb-release point, a period of about four to five minutes."[129] After dropping their armament, the bomber cells made a hard 100-degree turn away from the target, then withdrew to the west or southwest.

Much careful coordination was required to execute these attacks. The F–4s spreading chaff had to sow their bundles early enough so that the chaff could disperse properly and create alleys down which the bomber cells could safely fly, and the EB–66s had to fly their racetrack ovals close enough to the bombers so that the power levels of their jammers complemented the broadcasts from the jammers carried aboard the B–52s. At least one EB–66 also had to keep its fixed jammer antennas pointed toward the North Vietnamese SAM FC radars at all times. Finally, F–4 escorts and WILD WEASELs had to reach their stations at just the right moment. If they arrived too early, they ran the risks of attracting the SAMs themselves and of running low on fuel; if too late, B–52 losses would be unacceptable. Timing was everything. In a jamming-free setting, for example, F–4s escorting the B–52s could use their own radars to keep station on the bombers. But the bombers' broadcasts also affected the fighters' radars, so escort missions were flown by the clock, with fighter escorts hoping they were flying behind and below the B–52s.[130] Although a complex arrangement, planners could conceive of no better alternative.

A more basic problem, according to Air Force historian Robert Futrell, involved planning the missions from SAC headquarters in Omaha: "In the first three days of this campaign, General John C. Meyer, CINCSAC, used his specified command authority and coordinated targets directly with the JCS. After this, he discussed support arrangements with Admiral Noel Gayler [CINCPAC] and General John Vogt, deputy MACV for air and commander, Seventh Air Force." Vogt then worked with Task Force 77 and the SAC ADVON to arrange support for the B–52 waves. His staff also planned the daylight raids and the F–111 night sorties that complemented the B–52 bombings. This cumbersome system changed on December 21, 1972, when CINCPAC made himself and SAC "jointly responsible for target determination within JCS guidance." After Christmas he "took sole responsibility for air operations over North Vietnam . . . ."[131] This change permitted staffs at B–52 wing-level the freedom to tailor tactics to mission needs.

SAC headquarters apparently considered the first two nights' attacks successful, even though some B–52 crews had discovered that SAM FC radars were able to track their planes successfully during the "post-target turns" away from Hanoi to the west.[132] Risks inherent in allowing Omaha to plan the attacks became clear, however, on the third night of strikes (December 20/21) when North Vietnamese SAM batteries, using a new FC signal frequency, knocked down six of ninety-nine attacking B–52s.[133] By that third night, the SAC-dictated approach and departure tactics of the B–52s were predictable. The North Vietnamese now used their new radar signal to determine the ranges of the bombers and then calculated the B–52s' locations by matching range data against the bombers' jamming signals. When the bomber cells turned away from the their targets, thereby masking some of their own jamming antennas, the North Vietnamese batteries would fire pairs of SA–2s. The missile attacks did not stop the bombers from releasing their bombloads, but a 6 percent loss rate was more than SAC cared to absorb.

To increase their survival rate, the B–52s' approach and departure routes were altered, and the B–52Gs were shifted from attacks against Hanoi to tactical strikes against targets in South Vietnam. The pattern of F–4 chaff drops was also changed, and the number of support aircraft flying with the Thailand-based B–52s on the 21st nearly doubled. SAC targeting experts also identified SAM storage depots, which the B–52s attacked. On December 26, 1972, the eighth day of the bombing, 120 B–52s from Guam and Thailand struck ten different targets near Hanoi in just fifteen minutes while 114 tactical support aircraft helped them saturate Hanoi's SAM defenses.[134] The new tactics worked. For example, on December 28 "Twenty-seven aircraft bombing in the northwest quadrant of Hanoi either crossed directly over one another or flew within five miles or less of other cells on reciprocal tracks." B–29s had used the same basic tactic when North Korea's air defenses stiffened in 1952. One officer argued that Eighth Air Force had "completely revolutionized many modern day bomber tactics" during LINEBACKER II.[135] The last B–52 strikes came in the night on December 29. Only twenty-three SA–2s were fired at the sixty bombers attacking three targets.[136] The North Vietnamese then agreed to resume negotiations.

During around-the-clock bombing that lasted for more than eleven days, B–52s and U.S. Air force fighter-bombers flew 1,364 sorties (724 by the B–52s) against 59 targets. Support sorties numbered 2,066, and more than 42,000 bombs, in excess of 15,000 tons, were dropped on targets in and near Hanoi and Haiphong. One thousand three hundred eighteen North Vietnamese civilians were killed.[137] The North Vietnamese fired 1,241 SAMs, 844 directed at B–52s, which cost SAC 15 B–52s destroyed — a loss rate of about 2 percent.[138]

Conclusion

Strategic bombardment conducted in Korea and Vietnam exhibited vivid contrasts. The strategic bombing campaign in Korea began immediately in 1950, just as soon as U.S. air units gained command of the air. In Vietnam, strategic bombing began gradually and only came into its own as the conflict neared an end for American forces in 1972 after a long, intermittent, and inconclusive aerial interdiction campaign. In Korea in the summer of 1950, FEAF B–29s faced relatively little opposition as they methodically reduced North Korean transportation and industrial centers. In Vietnam, fighter aircraft and bombers flew into the teeth of one of the most sophisticated air defense systems in the world. U.S. strategic bombing in Korea in 1950 closely followed classic doctrine designed to eliminate a state's ability and resolve to wage war: attacks were heavy and sustained until all targets were destroyed. In Vietnam, on the other hand, that doctrine was subordinated to a national policy of coercive bargaining — the application of limited aerial force and the threat of still greater force as forms of political pressure and as signals of strategic intent. When this policy failed entirely to achieve the goals set for it, it was eventually replaced with a strategic bombing campaign similar, if not identical, to the one employed in Korea in August and September 1950.

Nevertheless, strategic bombing in Korea and Southeast Asia was tightly constrained. During the Korean conflict — because the real sources of enemy strength lay in Manchuria and beyond and could not be attacked — General Curtis LeMay, SAC Commander at the time, was moved to declare: "We never did hit a strategic target."[139] In LeMay's view, SAC's real contribution in both Korea and Vietnam was to deter the Soviet Union and China from expanding the war. Yet the damage wrought by U.S. military aviation in Korea and North Vietnam was so great that the consequences appeared strategic, even if the air campaigns themselves were not meant as such. North Korean cities were laid waste, as was North Vietnam's tiny industrial base. Such devastation is strategic, especially in societies having so few strategic resources. Indeed, the Johnson administration warned North Vietnam in 1965 that it would be wiser in the long run to settle than to fight the United States. The North Vietnamese government, however, chose to fight, though a forcibly united Vietnam would not recover economically from the struggle for a decade, and Vietnam, like North Korea, would take a entire generation to repair the air war's human and physical damage.

The bombing campaigns fought in Korea and Vietnam were limited campaigns waged with conventional ordnance. Tactical nuclear weapons, though threatened, were never employed. Instead, U.S. bombers (most of them tactical aircraft) returned again and again to targets which UN and U.S. ground forces could not occupy. Remarkable continuity characterized these tactics. Electronic warfare, incendiaries, and guided munitions were all carried over from World

War II to Korea, and then to Southeast Asia. By the start of LINEBACKER II, the increasingly sophisticated "smart" weapons fired by U.S. tactical aircraft began to achieve the kinds of results that only huge numbers of ordinary bombs, dropped by large aircraft like the B–52, had been able to achieve in the past. This suggested that tactical aircraft equipped with sophisticated weapons could attain the traditional goal of strategic air units: destruction of an enemy's industrial and economic system. Moreover, smart weapons, used after U.S. units gained air superiority, could cripple a smaller nation's economy without killing many civilians. In short, the new weapons and sensors that reached the operating forces in 1972 promised to overturn the accepted notion, based on the experience of World War II and Korea, that strategic bombing entailed high civilian casualties, at least when the enemy was a small nation like Vietnam.

Strategic bombing, however, also may be effective psychologically when huge amounts of ordnance are placed on rather large targets such as factories and railroad yards. LINEBACKER demonstrated the effectiveness of smart weapons used against pinpoint targets, but it took the B–52s to make the psychological point central to strategic bombing. As the Allies understood in 1943, strategic bombing aims to convince people that they, and their forces, cannot win. Yet carrying the war to civilians meant, during World War II, the bombing of civilians and their dwellings directly. LINEBACKER II showed that B–52 bombing could be reasonably accurate and still obtain the desired psychological effect, even if the threat to civilians was mostly implied. LINEBACKER II also showed, however, that a conventional campaign by B–52s against well-defended targets needed a great deal of tactical support, at least with the technology then available to U.S. air units.

Finally, it is clear that U.S. civilian and military leaders tried to use the threat of damage from bombing to intimidate their Korean, Chinese, and Vietnamese opponents. Indeed, the belief is still widespread that bombing is a substitute for actually fighting a ground war with heavy casualties. The Truman, Eisenhower, Johnson, and Nixon administrations all sought to influence the outcome of events by using firepower from the air. Yet neither the North Koreans nor the North Vietnamese had many strategic assets vulnerable to conventional attack by bombers like the B–29 or B–52. In 1953, General Clark recognized this when he ordered North Korea's irrigation dams destroyed. General Weyland, Clark's air deputy, opposed attacking North Korea's rice crop by dam-busting. Clark, on the other hand, saw that the United States had few other alternatives if it chose to use its air advantage to move the North Korean government to the bargaining table.

In Vietnam, the shift from the Johnson administration's coercive bargaining strategy to the Nixon administration's strategy of Vietnamization was significant because it recognized the limited political benefits of conventional air bombardment. Johnson and his advisers hoped that the government of North Vietnam could be persuaded to turn away from its major goal of subverting

South Vietnam. They feared that too harsh an attack on North Vietnam would bring the Peoples' Republic of China into the war, and the Chinese and Soviets played on this fear, hinting they might intervene directly if the United States bombed North Vietnam's irrigation system. What the United States was willing to strike was just not important enough to compel Hanoi to abandon its campaign to capture South Vietnam. The Nixon administration recognized this and accepted the risk of the eventual loss of South Vietnam in exchange for a new political relationship with China and the Soviet Union.

Concerns over a larger war with the Soviet Union or China constrained strategic bombing in Korea and Southeast Asia. Strategic bombing as it evolved after World War II — as a means of quickly winning an intercontinental war — could be applied only in a limited way in limited conflicts. The result in the two cases described here was an extended air campaign of attrition. The considerable experiences gained in Korea and Southeast Asia in strategic bombardment theory and practice would be reflected in the application of air power in DESERT STORM.

NOTES

1. Robert F. Futrell, *Ideas, Concepts, Doctrine: A History of Basic Thinking in the United States Air Force, 1907–1964* (Maxwell AFB, Ala.: Air University Press, 1971), p 76.
2. *Ibid.*, p 81.
3. Michael S. Sherry, *The Rise of American Air Power* (New Haven, Conn.: Yale University Press, 1987), p 308.
4. *Ibid.*, p 279.
5. For an argument that the U.S. Air Force pressed for more nuclear weapons instead of for developing new types of bombs in the late 1940s, see Chuck Hansen, *U.S. Nuclear Weapons: The Secret History* (Arlington, Tex.: Aerofax, 1988). For reasons, see Harry R. Borowski, *A Hollow Threat: Strategic Air Power and Containment Before Korea* (Westport, Conn.: Greenwood Press, 1982).
6. As the United States Strategic Bombing Survey (USSBS) observed, "No indispensable industry was permanently put out of commission by a single attack. Persistent re-attack was necessary." USSBS, *Summary Report* (Maxwell AFB, Ala.: reprint, Air University Press, 1987), p 39. The AAF and RAF casualty figures appear on p 6.
7. Robert F. Futrell, *The United States Air Force in Korea, 1950–1953*, rev ed (Washington, D.C.: Office of Air Force History, 1983), p 58; HQ FEAF, *Korean Air War Summary, Jun 25, 1950–July 27, 1953* (Maxwell AFB, Ala.: U.S. Air Force Historical Research Agency [hereafter AFHRA]), p 2.
8. Acting on his own initiative one day earlier, MacArthur authorized Stratemeyer's air forces to attack targets in North Korea. George E. Stratemeyer, Korean War Diary, Jun 29, 1950, entry, in AFHRA. (This diary is planned for publication by the Air Force History and Museums Program.)
9. General Curtis LeMay recalled that these two units were neither fully manned nor combat-ready for employment in SAC's strategic war plan, and thus available for use in Korea. Hist, SAC, July–Dec 1950, vol 1, pp 15–20; see also, HQ FEAF, "Far East Air Forces Report on the Korean War, Jun 25, 1950–Jul 27, 1953" (hereafter "FEAF Report on the Korean War"), vol 2, AFHRA, p 55.
10. Robert Frank Futrell, *Ideas, Concepts, Doctrine: Basic Thinking in the United States Air Force, 1907–1984*, 2 vols (Maxwell AFB, Ala.: Air University Press, 1989), vol 1, p 295. For comments by General LeMay, see *Strategic Air Warfare*, ed Richard H. Kohn and Joseph P. Harahan (Washington, D.C.: Office of Air Force History, 1988), p 88.
11. Ops Hist, FEAF, Jun 25, 1950–Oct 31, 1950, vol 2, annex 2, p 258, AFHRA.
12. *Ibid.*, p 261.
13. Dean Acheson, *Present at the Creation* (New York: New American Library, 1969), p 540.
14. See R. Dingman, "Atomic Diplomacy during the Korean War," *International Security*, vol 13, no 3 (winter 1988–1989), pp 50–91.
15. Futrell, *Ideas, Concepts, Doctrine*, vol 1, p 295.
16. Mark Trachtenberg, "A 'Wasting Asset': American Strategy and the Shifting Nuclear Balance, 1949–1954," *International Security*, vol 13, no 3 (winter 1988–1989) pp 5–49.
17. Marcelle Size Knaack, *Post–World War II Bombers, 1945–1973* (Washington, D.C.: Office of Air Force History, 1988), pp 25, 169.
18. Futrell, *Ideas, Concepts, Doctrine*, vol 1, p 335. See also Department of the Air Force, *History of Headquarters, USAF, Jul 1, 1950 to Jun 30, 1951* (Department of the Air Force, Jun 1955), p 27 (copy in AFHRA).
19. Futrell, *Ideas, Concepts, Doctrine*, vol 1, p 335.
20. Knaack, *Post–World War II Bombers*, pp 488, 492.
21. *Ibid.*, p 173.
22. J. C. Hopkins and Sheldon A. Goldberg, *The Development of Strategic Air Command, 1946–*

1986 (Offutt AFB, Neb.: HQ SAC, Sep 1, 1986), pp 30, 37. The number of B–47s is misleading, however. The first SAC B–47 wing did not become fully operational for another two years, until early 1953.

23. Acheson, *Present at the Creation*, p 831.

24. Dingman, "Atomic Diplomacy During the Korean War."

25. Ltr, Lt Gen J. H. Edwards, DCS/O, to Lt Gen George E. Stratemeyer, Commanding General, FEAF, subj: Reinforcements, May 18, 1951, Record Group 341, entry 336, "Plans and Project Files, 1942–54," no 381, National Archives, Washington, D.C. (NA).

26. "FEAF Report on the Korean War," vol 2, p 141.

27. "Air War in Korea: XI, Heavyweights over Korea," *Air University Quarterly Review*, vol 7, no 1 (spring 1954), p 101. See also "FEAF Report on the Korean War," vol 2, p 55.

28. "FEAF Report on the Korean War," vol 2, p 56.

29. Futrell, *The United States Air Force in Korea*, p 190.

30. *Ibid.*, p 191.

31. "FEAF Report on the Korean War," vol 2, p 57.

32. Futrell, *The United States Air Force in Korea*, pp 192, 194.

33. "FEAF Report on the Korean War," vol 2,, p 57.

34. Futrell, *Ideas, Concepts, Doctrine*, vol 1, p 298. See also Trachtenberg, "A 'Wasting Asset,'" p 21.

35. Bruce Cumings and Jon Halliday, *Korea: The Unknown War* (New York: Pantheon, 1988), p 123. See also Dingman, "Atomic Diplomacy during the Korean War," p 64.

36. Ops Hist, FEAF, Nov 1, 1950–Feb 28, 1951, vol 2, p 317.

37. *Ibid.*, p 318.

38. *Ibid.*, p 317.

39. "FEAF Report on the Korean War," vol 2, p 58.

40. Dingman, "Atomic Diplomacy during the Korean War," p 67.

41. Futrell, *Ideas, Concepts, Doctrine*, vol 1, p 303.

42. Dingman, "Atomic Diplomacy during the Korean War," p 69.

43. Gordon E. Dean, *Forging the Atomic Shield: Excerpts from the Office Diary of Gordon E. Dean*, ed. Roger M. Anders (Chapel Hill: University of North Carolina Press, 1987), pp 127–130.

44. Dingman, "Atomic Diplomacy during the Korean War," p 73.

45. Futrell, *Ideas, Concepts, Doctrine*, vol 1, p 301.

46. Trachtenberg, "A 'Wasting Asset,'" p 28.

47. Dingman, "Atomic Diplomacy during the Korean War," p 75.

48. See Cumings and Halliday, *Korea: The Unknown War*. Note, too, that Bomber Command B–29s dropped razon (1,000-lb) and tarzon (12,000-lb) guided bombs on North Korean targets beginning in August 1950. When Chinese forces entered the war, tarzon bombs were dropped (not very successfully) on bridges, tunnels, the Hwachon Dam, and one "industrial installation." The tarzon bombs were not used after Mar 1951 because they sometimes detonated too soon after they were dropped. Cumings and Halliday do not seem to understand that the North Koreans might have interpreted the tarzon attacks as rehearsals for nuclear strikes. See "FEAF Report on the Korean War," vol 2, pp 65–66.

49. HQ FEAF, *Korean Air War Summary*, p 22.

50. "FEAF Report on the Korean War," vol 2, p 63. A description of the British Gee and Oboe systems of World War II can be found in Chapter 3 of this volume. See also Office for the Deputy for Communications and Electronics, "Fifth Air Force Communications and Electronics in Korea," AFHRA. For a summary of electronic warfare in Korea, see Dep Ops, Comm–E, HQ FEAF, "FEAF ECM History during Korean Conflict," AFHRA.

51. Dingman, "Atomic Diplomacy during the Korean War," p 78.

52. Eugene E. Emme, *The Impact of Air Power National Security and World Politics* (Princeton, N.J.: Van Nostrand, 1959), p 392.

53. Futrell, *Ideas, Concepts, Doctrine*, vol 1, p 339.

54. Intvw, James C Hasdorff and Arthur W. McCants, Jr., with General Jacob E. Smart, Nov 27–30, 1978, K239.0512–1108, U.S. Air Force History Support Office, Bolling AFB, Washington, D.C. (AFHSO), p 186.

55. The power plants may have played a direct strategic role by supplying electricity to uranium ore processing facilities in Manchuria.

56. "FEAF Report on the Korean War," vol 2, p 143.

57. Futrell, *The United States Air Force in Korea*, p 516.
58. *Ibid.*, p 517.
59. *Ibid.*, p 516. See also Richard P. Hallion, *The Naval Air War in Korea* (Baltimore, Md.: Nautical & Aviation Publishing Co., 1986), p 136.
60. Futrell, *The United States Air Force in Korea*, p 624.
61. Robert F. Futrell, *U.S. Air Force Operations in the Korean Conflict, July 1, 1952–July 27, 1953* (USAF Hist Study 127, Maxwell AFB, Ala.: Air University, USAF Historical Division, 1956), pp 56–57.
62. R. J. Foot, "Nuclear Coercion and the Ending of the Korean Conflict," *International Security*, vol 13, no 3 (winter 1988–1989), pp 92–112.
63. Futrell, *The United States Air Force in Korea*, p 666.
64. *Ibid.*, p 667.
65. *Ibid.*, p 669.
66. Hist, HQ FEAF, FEAF Participation in the Atomic Energy Program, Jul 1, 1953–Jun 30, 1954, pp 1–7, Hist Div, HRC.
67. Emme, *Impact of Air Power*, p 405.
68. Foot, "Nuclear Coercion and the Ending of the Korean Conflict," p 108.
69. Hist, HQ FEAF, FEAF Participation in the Atomic Energy Program, Jul 1, 1953–Jun 30, 1954, p 5.
70. Knaack, *Post–World War II Bombers*, p 35.
71. Futrell, *The United States Air Force in Korea*, p 489.
72. Futrell, *USAF Operations in the Korean Conflict*, p 77.
73. Futrell, *The United States Air Force in Korea*, p 492.
74. *Ibid.*, p 493.
75. Neil Sheehan *et al.*, *The Pentagon Papers as Published by The New York Times* (New York: The New York Times & Bantam Books, 1971), p 271.
76. *Ibid.*, p 274.
77. *Ibid.*
78. *Ibid.*, p 277.
79. Guenter Lewy, *America in Vietnam* (New York: Oxford University Press, 1978), pp 377–378.
80. Sheehan *et al.*, *Pentagon Papers*, pp 286–288.
81. *Ibid.*, p 354.
82. *The Pentagon Papers: The Defense Department History of United States Decisionmaking on Vietnam*, Senator Gravel ed, 5 vols (Boston, Mass.: Beacon Press, 1971), vol 3, p 309.
83. Sheehan *et al.*, *Pentagon Papers*, p 343.
84. *Pentagon Papers*, Gravel ed, vol 4, p 1.
85. *Ibid.*, vol 3, p 319. See also Robert F. Futrell, *The Advisory Years to 1965*, The USAF in Southeast Asia (Washington, D.C.: Office of Air Force History, 1981), pp 263, 266.
86. Intvw, Hugh N. Ahmann with Gen Horace M. Wade, Oct 10–12, 1978, K239.0512–1105, AFHSO, p 395.
87. Edward Doyle, Samuel Lipsman, and Terrence Maitland, *The North: The Communist Struggle for Vietnam* (Boston, Mass.: Boston Publishing Co., 1986), p 70. See also Mark Clodfelter, *The Limits of Airpower* (New York: Free Press, 1989), pp 136–137.
88. Doyle *et al.*, *The North*, pp 72–74.
89. *Ibid.*, p 92.
90. Robert F. Futrell, William H. Greenhalgh, Carl Grubb, Gerard E. Hasselwander, Robert F. Jakob, and Charles A. Ravenstein, *Aces and Aerial Victories: The United States Air Force in Southeast Asia, 1965–1973* (Washington, D.C.: Office of Air Force History, 1976), p 4.
91. William Momyer, *Air Power in Three Wars* (Washington, D.C.: GPO, 1978), p 118. See also Doyle *et al.*, *The North*, p 84.
92. Kenneth P. Werrell, "LINEBACKER II: The Decisive Use of Air Power?" *Air University Review*, vol 38 (Jan–Mar 1987), p 56.
93. Intvw, Dorothy Pierce McSweeny with Gen John P. McConnell, Aug 14–28, 1969, K239.0512–1190, AFHSO, pp 1–20 (copy in Lyndon B. Johnson Library, Austin Tex.).
94. Melvin F. Porter, *LINEBACKER: Overview of the First 120 Days* (Project CHECO, HQ PACAF Dir Ops Anlys, Sep 27, 1973), AFHRA, pp 60–61.
95. General Smart intvw, Nov 27–30, 1978, p 325.
96. Futrell, *Ideas, Concepts, Doctrine*, vol 2, p 280.

97. *Ibid.*, p 281.
98. Jacob Van Staaveren, "USAF Deployment Planning for Southeast Asia, 1966," AFHSO, p 8.
99. William W. Momyer, *Air Power in Three Wars* (Washington, D.C.: Office of Air Force History, 1978; reprint, 1985).
100. Futrell, *Ideas, Concepts, Doctrine*, vol 2, p 282.
101. C. K. Hopkins, "SAC Tanker Operations in the Southeast Asia War," 1979, HQ SAC History Office, pp 57–58.
102. Doyle *et al.*, *The North*, pp 71, 88.
103. *Ibid.*, pp 74, 79, 88.
104. *Ibid.*, p 107.
105. Clodfelter, *Limits of Airpower*, p 151.
106. U.S. Congress, House of Representatives, Investigating Subcommittee of the Committee on Armed Services, *Unauthorized Bombing of Military Targets in North Vietnam*, and Senate, Committee on Armed Services, *Nomination of John D. Lavelle, General Creighton W. Abrams, and Admiral John S. McCain*, both 92d Cong, 2d sess (Washington, D.C., 1972).
107. Clodfelter, *Limits of Airpower*, pp 152, 156–158.
108. William Paden Mack, *The Reminiscences of Vice Admiral William Paden Mack, USN (Ret)*, Oral History Program, 2 vols (U.S. Naval Institute Press, 1979), p 624 (copy in U.S. Navy Operational Archives, Washington Navy Yard). Vice Admiral Mack was Commander of the Seventh Fleet in 1971–1972.
109. John Morrocco, *Rain of Fire: Air War, 1969–1973* (Boston, Mass.: Boston Publishing Co, 1985), pp 131–133.
110. *Ibid.*, p 136.
111. Karl J. Eschmann, *LINEBACKER: The Untold Story of the Air Raids over North Vietnam* (New York: Ballantine Books, 1989), p 27.
112. Paul W. Elder and Peter J. Melly, *Rules of Engagement, November 1969–September 1972* (Project CHECO, HQ PACAF Dir Ops Anlys, Mar 1, 1973), AFHRA, pp 51–55.
113. Porter, *LINEBACKER: Overview*, pp 60–61.
114. *Ibid.*, p 61.
115. *Ibid.*, p 59.
116. *Ibid.*, pp 51–52.
117. Herman S. Wolk, "USAF Plans and Policies: Research and Development for Southeast Asia, 1968," AFHSO, pp 23–24.
118. Eschmann, *LINEBACKER*, pp 30–31.
119. *Ibid.*, p 37.
120. *Ibid.*, pp 64, 83.
121. Clodfelter, *Limits of Airpower*, p 162.
122. Dennis R. Jenkins and Brian Rogers, *Boeing B–52G/H Stratofortress*, Aerofax Datagraph 7 (Arlington, Tex.: Aerofax, 1990), p 28.
123. Eschmann, *LINEBACKER*, p 75.
124. Richard M. Nixon, *The Memoirs of Richard Nixon* (New York: Grosset & Dunlap, 1978), p 733–734. See also Henry Kissinger, *White House Years* (Boston, Mass.: Little, Brown, 1979), pp 1448–1449, 1456–1457.
125. Nixon, *Memoirs* pp 735–736.
126. Wade intvw, Oct 10–12, 1978, p 537.
127. James R. McCarthy and George B. Allison, *LINEBACKER II: A View from the Rock*, USAF Southeast Asia Monograph Series (Maxwell AFB, Ala: Air War College, 1979), vol 6, monogr 8, p 32.
128. Eschmann, *LINEBACKER*, pp 76, 195–201.
129. *Ibid.*, p 76.
130. *Ibid.*, pp 89, 99–100.
131. Futrell, *Ideas, Concepts, Doctrine*, vol 2, p 286.
132. Eschmann, *LINEBACKER*, p 107.
133. *Ibid.*, pp 136–137. See also McCarthy and Allison, *LINEBACKER II*, pp 83–84.
134. Eschmann, *LINEBACKER*, pp 139–140, 162–164.
135. McCarthy and Allison, *LINEBACKER II*, p 158.
136. Eschmann, *LINEBACKER*, p 192.
137. *Ibid.*, p 202. See also, Werrell, "LINEBACKER II: The Decisive Use of Air Power?" p 59.

138. Futrell, *Ideas, Concepts, Doctrine*, vol 2, pp 270, 297.
139. Kohn and Harahan, eds., *Strategic Air Warfare*, p 87.

Bibliographic Essay

Korea

Government Sources

Official records of the Far East Air Forces and the Fifth Air Force are held by the Air Force Historical Research Agency at Maxwell Air Force Base, Alabama. Copies of many of these records are kept on microfilm at the Air Force History Support Office, Bolling Air Force Base, Washington, D.C. Also in Washington, at the National Archives, are the records of Headquarters U.S. Air Force. The Navy's records of the Korean war are located in the Navy Department Operational Archives at the Naval Historical Center, Washington Navy Yard.

The primary official source for the section on Korea was the three-volume study of Air Force operations, originally classified, by Robert F. Futrell, *United States Air Force Operations in the Korean Conflict*. His first volume is subtitled *June 25–November 1, 1950* (USAF Historical Study 71, Maxwell AFB, Ala., 1952); his second, *November 1, 1950–June 30, 1952* (USAF Historical Study 72, Maxwell AFB, Ala., 1955); and his third, *July 1, 1952–July 27, 1953* (USAF Historical Study 127, Maxwell AFB, Ala., 1956). In 1961, Futrell, assisted by L. S. Moseley and Albert F. Simpson, published an unclassified, one-volume summary of the three classified volumes: *The United States Air Force in Korea, 1950–1953* (New York: Duell, Sloan & Pearce, 1961). A revised and improved edition of the same history was reissued by the Office of Air Force History in 1983.

Other primary source materials include the "Far East Air Forces Report on the Korean War, June 25, 1950–July 27, 1953," and the Far East Air Forces operations histories. These documents are located at the Air Force Historical Research Agency. On the Navy side, the "Interim Evaluation Reports" of the Navy's Pacific Fleet discuss naval air operations, particularly those of aircraft carriers. These reports, housed in the Navy's Operational Archives, reveal the vulnerability of aircraft carriers of the day to attacks by Korean war–era jet aircraft. For the historical context of wartime operations, consult Robert Futrell's two-volume set, *Ideas, Concepts, Doctrine: Basic Thinking in the U.S. Air Force* (Maxwell AFB: Air University Press, 1989). Volume 1 is subtitled *1907–1960*, and volume 2, *1961–1984*.

Nongovernment Sources

Much of *International Security*, vol 13, no 3 (winter 1988–1989), is devoted to an assessment of nuclear diplomacy during the Korean war. These articles, coupled with Roger M. Anders, ed., *Forging the Atomic Shield: Excerpts from the Office Diary of Gordon E. Dean* (Chapel Hill: University of North Carolina Press, 1987) (Gordon Dean was Atomic Energy Commission Chairman), provide a glimpse of official policy-making during the war. For Navy operations in Korea, we have Richard P. Hallion's *The Naval Air War in Korea* (Baltimore, Md.: Nautical and Aviation Publishing, 1986). For post–Korea development of peacetime ideas on the utility of less-than-strategic bombing, see Eugene Emme, ed., *The Impact of Air Power* (Princeton, N.J.: Van Nostrand, 1959).

Vietnam

Government Sources

At the time this chapter was written, most documents on the bombing of North Vietnam remained classified; however, the Air Force was beginning to declassify its records of the era. The Air Force has issued a number of titles in its USAF Southeast Asia Monograph Series. Among them are A.J.C. Lavalle, ed., *The Battle for the Skies over North Vietnam* (Washington, D.C.: GPO, 1976); A.J.C. Lavalle, ed., *The Tale of Two Bridges* (Washington, D.C.: GPO, 1976); Drew Middleton *et al.*, *Air War — Vietnam* (New York: Bobbs-Merrill, 1978); and James R. McCarthy and George B. Allison, *LINEBACKER II: A View from the Rock* (Maxwell AFB Ala: Air University, 1979). Also available is Carl Berger, ed., *The United States Air Force in Southeast Asia, 1961–1973: An Illustrated Account* (Washington, D.C.: Office of Air Force History, 1984).

Closer to actual operations are the Project CHECO (Contemporary Historical Examination of Current Operations) reports stored at the Historical Research Agency and at the Air Force History Support Office. Starting in 1962, Project CHECO researchers prepared more than 200 historical reports by 1975. Many are now declassified. Air University publishes a "Research Guide to the Published Project CHECO Reports."

The Pentagon Papers detail policy-making for the war in Vietnam. They cover only the years before 1969, and are incomplete. For example, they do not clearly explain the political pressure placed on the United States by China and the Soviet Union. Two published editions of *The Pentagon Papers* are available. The first, the Defense Department history of U.S. decision-making on Vietnam, was placed originally in the public record by Senator Mike Gravel of Alaska in 1971; it was published commercially by Beacon Press that same year. The second, published by the *New York Times*, also appeared in book form 1971 under the imprints of Bantam Books and Quadrangle Books. Both editions should be supplemented with the oral histories held by the Air Force Historical Research Agency.

Nongovernment Sources

A number of interesting secondary sources are available regarding Vietnam. In 1978, former Seventh Air Force Commander General William W. Momyer's *Air Power in Three Wars* was published by the Government Printing Office in Washington, D.C. In his book, General Momyer argues that the similarities in the uses of air power during World War II, Korea, and Vietnam outweigh their differences. A very different view is contained in James Clay Thompson, *ROLLING THUNDER: Understanding Policy and Program Failure* (Chapel Hill: University of North Carolina Press, 1980). An interesting perspective on ROLLING THUNDER is former Air Force Under Secretary Townsend Hoopes' *The Limits of Intervention* (New York: David McKay, 1969).

ROLLING THUNDER remains a contentious issue. Consider John B. McPherson, "Vietnam: The Perspective of a Former Vice-Director of the Joint Staff," in Alfred F. Hurley and Robert C. Ehrhart, eds., *Air Power and Warfare*, Proceedings of the 8th Military History Symposium, USAF Academy, Oct 18–20, 1978 (Washington, D.C.: Office of Air Force History, 1979); Kenneth P. Werrell, "LINEBACKER II: The Decisive Use of Air Power?" *Air University Review* (Jan–Mar 1987); and Ulysses S. Grant Sharp, *Strategy for Defeat: Vietnam in Retrospect* (San Rafael, Calif.: Presidio Press, 1978) (Admiral Sharp is a former Commander-in-Chief, Pacific). See also Robert A. Pape, Jr., "Coercive Air Power in the Vietnam War," *International Security*, vol 15, no 2 (fall 1990). John B. Nichols and Barrett Tillman discuss Navy operations in *On Yankee Station: Naval Air War over Vietnam* (Annapolis, Md.: Naval Institute Press, 1987). An unusually instructive little book on the electronic warfare of ROLLING THUNDER and LINEBACKER is Larry Davis, *WILD WEASEL: The SAM Suppression Story* (Carrollton, Tex.: Squadron/Signal Publications, 1986). For some perspective on North Vietnam, see Edward Doyle, Samuel Lipsman, and Terrence Maitland, *The North: The Communist Struggle for Vietnam* (Boston, Mass.: Boston Publishing Co., 1986).

Three books cover most of LINEBACKER II. The first, describing the Nixon administration's overall policy, is Mark Clodfelter, *The Limits of Airpower: The American*

Bombing of North Vietnam (New York: Free Press, 1989). The second is Karl J. Eschmann's excellent *LINEBACKER: The Untold Story of the Air Raids over North Vietnam* (New York: Ballantine Books, 1989). The third is the illustrated book by Dennis R. Jenkins and Brian Rogers, *Boeing B–52G/H Stratofortress* (Arlington, Tex.: Aerofax, 1990). The Clodfelter and Eschmann titles are especially recommended.

Among several good overall discussions of the conflict in Southeast Asia are Guenter Lewy's *America in Vietnam* (New York: Oxford University Press, 1978) and William S. Turley's *The Second Indochina War: A Short Political and Military History, 1954–1975* (New York: New American Library, 1987). Two bibliographies are also useful: Richard Dean Burns and Milton Leitenberg, eds., *The Wars in Vietnam, Cambodia, and Laos, 1945–1982: A Bibliographic Guide* (Santa Barbara, Calif.: ABC-Clio Information Services, 1984), and Marvin E. Gettleman, Jane Franklin, Marilyn Young, and H. Bruce Franklin, *Vietnam and America: A Documented History* (New York: Grove Press, 1985).

7

Strategic Bombardment in the Gulf War

Richard G. Davis

From January 17 to February 28, 1991, aircraft of the U.S. Air Force, U.S. Navy, and U.S. Marine Corps, under the control of the United States Central Command (CENTCOM),* as well as contingents of the air forces of eleven other western European and Arab countries,** all under the aegis of the United Nations (UN), systematically attacked and destroyed targets in Iraq and the Iraqi armed forces occupying Kuwait. The air attack was the international community's response to Iraq's occupation on August 2, 1990, of Kuwait, a small oil-rich state at the western end of the Persian Gulf, and its obdurate refusal to abandon its conquest. The air war against Iraq consisted of two separate campaigns distinguished by different sets of targets. In the campaign

---

*CENTCOM was a unified command established and designated by the President under a single commander. It consisted of elements drawn from two or more U.S. military services and corresponded to the World War II–era Theater of Operations. CENTCOM's geographical area of responsibility (AOR) in 1990 covered a broad region comprising much of the Muslim world. Within its purview fell the following nations: Egypt, Sudan, Ethiopia, Kenya, Somalia, Djibouti, the two Yemens, Oman, Saudi Arabia, the United Arab Emirates (UAE), Bahrain, Qatar, Kuwait, Jordan, Iraq, Iraq, Afghanistan, and Pakistan. It also included the Red Sea, the Persian Gulf, the Gulf of Aden, and the Gulf of Oman. U.S. Air Force aircraft that took part in the Gulf War air campaign but flew from bases in Turkey, Spain, and Great Britain, belonged to another unified command, the United States European Command (EUCOM). B–52 bombers that flew from the island of Diego Garcia, a British possession in the Indian Ocean, belonged to the United States Strategic Air Command (SAC), while that island lay in the United States Pacific Command (PACOM) AOR.

**The following countries, all members of the International Coalition taking arms against Iraq, contributed combat aircraft to the War in the Persian Gulf: the United States, the United Kingdom, France, Italy, Canada, Saudi Arabia, the United Arab Emirates, Egypt, Oman, Kuwait, Bahrain, and Qatar.

in the Kuwait Theater of Operations (KTO),* Coalition air forces had three objectives: suppression of Iraqi air defenses in the KTO, preparation of the battlefield for a planned Coalition ground attack (by striking Iraqi ground forces and interdicting Iraqi supply lines), and support of Coalition ground force operations with tactical airlift and aerial firepower.

The second air campaign, the strategic bombardment of Iraq, struck at twelve sets of strategic targets. It sought to disrupt Iraq's air defense system; destroy Iraq's nuclear, biological, and chemical (NBC) weapons and the research, production, and storage facilities associated with them; demolish Iraq's offensive strategic weapons systems (short-range ballistic missiles [SRBMs; Scuds] and bombers); cripple its oil production and electrical industries; impair its war industry; and nullify its communications system. The plan also included strikes on naval forces and port facilities. In addition, U.S. Air Force planners hoped to incapacitate the Saddam Hussein regime.[1] This objective had the readily apparent, but unstated, goal of creating a set of conditions within Iraq conducive to the overthrow of its political leadership. Given the focus of this volume as a whole, this chapter concentrates on the Coalition air operations devoted to the strategic bombardment of Iraq. It does not address directly the large-scale and deadly tactical air operations in the KTO that consumed 75 percent of the total Coalition air effort.

The U.S. Air Force supplied the largest air contingent employed in the Gulf War — approximately 807 aircraft — and conducted the bulk of the strategic air operations. Air Force technological trends and doctrinal thought of the preceding twenty years culminated in this short but precise and destructive strategic bombing campaign. Some technical developments had proceeded openly, such as advances in navigation made possible by signals from satellites of the Global Positioning System (GPS); others, such as electronic combat devices and precision-guided munitions, proceeded in secret; while some, such as stealth flight technology, were hatched and matured out of public sight in the supersecret "black world." During this period the Air Force also began to look at its strategic warfighting doctrine in an effort to adapt to changing circumstances and technology.

---

*The KTO was only a portion of the overall Coalition-Iraqi battlefield. It was defined as the area north of the Saudi Arabian–Iraqi border; south of 31 degrees north latitude; west of the Persian Gulf and the Iranian-Iraqi border, and east of 45 degrees east longitude. This area included Kuwait and southeastern Iraq, including the major Iraqi city of Basra, and stretched approximately from the Iraqi city of As Samawah on the west to the Persian Gulf and from Saudi Arabia to the Iraqi city of An Nasiriyah, in the north. It contained most of the elite Republican Guard, and a large portion of the regular Iraqi armed forces (for a combined total of forty-three understrength divisions). The bulk of the operations of the strategic air campaign against Iraq occurred outside the KTO, while all of the Coalition tactical air operations took place within it.

Post-Vietnam Changes in Technology and Doctrine

Operations LINEBACKER I and II, conducted by U.S. air power over North Vietnam from May to October 1972 and in December 1972, respectively, served as both a harbinger of innovation in strategic bombing and a last hurrah for massed conventional strategic bombardment operations. In LINEBACKER I, launched to counter a massive North Vietnamese ground offensive into the Republic of South Vietnam's northernmost provinces, Air Force fighter-bombers made the first sustained use of precision-guided munitions. They employed electro-optically guided bombs and laser-guided bombs, known generically as guided bomb units (GBUs), to strike key bridges and other pinpoint targets. For example, on May 10, 12, and 13, 1972, precision-guided munitions dropped* seven bridges, including both the infamous Paul Doumer and the Thanh Hoa (the latter dubbed, not without reason, as the bridge that would never go down). Heavy air defenses had frustrated conventional attacks on the two bridges for five years and had taken a heavy toll of attackers. Between April 6 and June 30, 1972, the precision-guided munition–equipped F–4Cs of the 8th Tactical Fighter Wing (TFW), Ubon Royal Thai Air Force Base (AFB), destroyed 106 bridges, including some heretofore off-limits spans near the Chinese border.[2] The dropping of numerous bridges in rapid succession interrupted North Vietnamese logistics and, by denying alternate routes, overtaxed repair capabilities.

The U.S. Air Force was slow to address the doctrinal implications of this new level of bombing accuracy. It did not equip the major portion of its combat aircraft procured between 1972 and 1990 (the F–15C, F–16, and A–10 series) with GBU delivery capability.** The service did upgrade precision-guided munition technology with further developments in its initial delivery system, PAVEWAY I. PAVEWAY II (GBUs–10, –12, and –16) — featuring improved guidance, structural improvements, and folding wings (so that strike aircraft could carry more of them) — became operational in the mid-1970s. PAVEWAY III (GBUs–22, –24, and –27) went into service in the mid-1980s and had improved maneuverability, an autopilot, a laser scanner for target location, and low-level launch capability.[3] By mid-1990 the U.S. Air Force's entire precision-guided munition–capable fleet consisted of only 125 to 135 fighter-bombers:

---

*Such attacks characteristically left bridge spans all or partially severed from their supports and resting in the water, appearing as if they had "dropped" into the water. U.S. Air Force pilots quickly made note of this, and service slang began to refer to bridges as dropped rather than destroyed. It is a more accurate term in that a permanent bridge is not destroyed unless its concrete abutments and piers are demolished.

**In 1990 the F–16 and A–10 could deliver the Maverick air-to-ground missile. A precision-guided munition designed for an antiarmor role, this missile, designated the AGM–65, achieved excellent results in attacks on Iraqi tanks and armored personnel carriers. The Maverick does not have the penetration, weight, nor amount of high explosives required for strategic bombardment. Since 1991 the F–16 when equipped with the appropriate electronic pods has gained a full GBU delivery capability.

64 swing-wing F–111Fs, 56 stealth F–117As, and a handful of F–4Es. A further 24 F–15E Strike Eagles would replace the F–4Es and come on-line as precision-guided munition–capable aircraft by the end of 1990.[4]

While many nations and the other U.S. armed services possessed precision-guided munitions by 1990, the Air Force alone possessed air delivered precision-guided munitions with hard-target penetrating capacity.* By May 1988 the bomb live unit (BLU)–109/B (I–2000) penetrating bomb had completed much of its initial operational tests and evaluations, both in an unguided version (intended for F–16s) and in guided versions for precision-guided munition–capable aircraft.[5] The streamlined 2,000-pound bomb's ballistic and handling characteristics were similar to the standard American Mk–84 blast and fragmentation bomb, which simplified employment to the field. The penetrating bomb's body of high-strength forged steel and its thick bomb walls encased 550 pounds of tritonal explosive. It had a tail-mounted, delayed-action fuze. Striking with a high-kinetic impact at the proper angle, it could penetrate hardened concrete or rubble and other filler.[6] (The I–2000 penetrator also came in PAVEWAY II variants for the F–111F and F–117A, designated GBU–24 and GBU–27, respectively.) When joined to a guided delivery system the BLU–109/I–2000 offered a weapons system of awesome lethality, overwhelming leverage, and surgical precision. Like the Belgians at the fortress of Eben Emael in May 1940, the Iraqis in January 1991 would find hardened concrete woefully insufficient against a foe's ingenuity and unconventionality.

If LINEBACKER I pointed to the future, LINEBACKER II sent a mixed signal to the analyst. LINEBACKER II intended not only to deny valuable materiel and safe areas to the enemy, but also to force him back to the peace table by employing large-scale B–52 strikes on key targets primarily in the Hanoi-Haiphong area. With approximately 200 B–52Ds and Gs available at Andersen AFB, Guam, and U-Tapao Royal Thai AFB, the U.S. Air Force launched 11 days of massed heavy bomber raids of 60 to 129 aircraft fitted with radar bombsights and conventional iron bombs.[7] From December 18 to 29, 1972, American bombers blasted more than thirty-four targets, including marshaling yards, storage and warehouse facilities, fabrication plants, and airfields with 500- and 750-pound bombs. They flew 729 sorties, dropped 15,237 tons of bombs, and lost 15 B–52s, for a loss rate of 2 percent.[8] Because of the enor-

---

*The British Royal Air Force (RAF) possessed the JP–233 airfield denial munition, which dispensed weapons capable of penetrating hardened concrete runways. It was not designed for use against structures. A hardened structure is an individual building or facility sheathed in several feet of specially hardened, steel-reinforced concrete. The structure may also be covered with several feet of rubble and earth. Earth fill, hardened concrete, and sheet steel plate may alternate in covering a single super-hardened facility. A conventional high-explosive bomb will either explode on contact or only dig a slight hole in a hardened structure before exploding. It essentially does no damage other than obliterating aerials and other "soft" protrusions from the target. A penetrating munition burrows through several feet of hardening to explode inside the target.

mous bomblift of the B–52D (up to 50 tons), the amount of high explosives delivered in a raid by 25 B–52s compared favorably with the typical 500-plane B–17 raids of the Eighth Air Force in World War II. The LINEBACKER II missions, with single formations of aircraft occupying more than seventy miles of airspace, marked the end of massed heavy bomber formations conducting strategic bombing. By the end of the 1960s, not even the United States, much less any other power, could afford to build and maintain large numbers of heavy, multiengine bombers. From 624 B–52s and FB–111s in fiscal year 1973,[9] the U.S. Air Force strategic bomber inventory steadily declined to 266 B–1s, B–2s, and B–52s in fiscal 1993.[10]

Another aspect of LINEBACKER II presaged Gulf War air operations. During LINEBACKER II, more combat aircraft supported each raid than the number of attacking bombers committed to it. For example, a raid of thirty B–52s on Haiphong on December 22, 1972, required sixty-five other aircraft for crucial support, including U.S. Navy EB–66s and EA–6Bs that furnished electronic countermeasures, F–105 Iron Hands that suppressed surface-to-air missile (SAM) defenses, F–4 Hunter/Killers that struck antiaircraft artillery (AAA), and numerous other F–4s that provided close escort, counter-MiG combat air patrol, chaff dispensing, and chaff escort.[11] In World War II the Eighth Air Force required one fighter escort for two bombers, but the complexity, layering, and integration of modern electronic/missile/gun/aircraft air defenses reversed that ratio, a consideration that goes far to justify the expense of a modern air defense system.

This mission of thirty bombers exposed a total of ninety-five aircraft to enemy counterair measures. The large number of supporting aircraft required for mass bomber attacks against heavily defended targets militated against future strikes of this kind; it made the attacking force too big to defend adequately. Multifaceted air defenses demanded equally complex responses from attacking aircraft and compelled the attacker to divide his force into aircraft specializing in different roles and to arrange them in packages tailored to meet specific threats within the expected target area. After the war in Southeast Asia, the U.S. Air Force continued to develop the concept of strike packages as part of its overall target planning methodology. Gulf War planners made heavy use of force packaging, but the revolution in precision attack introduced a new twist: smaller packages of strikers went to multiple aiming points in the same area, rather than a single, large package going to just one aiming point.

Making oneself invisible to the foe has been the stuff of legend for thousands of years. More recently it has become the province of technology. From 1940 to the mid-1970s, aerial opponents sought to jam, spoof, or destroy radar with electronic countermeasures: dispensing various configurations of chaff (metallic coated streamers dropped in bundles from aircraft to deceive and mislead enemy radar) and developing antiradiation missiles to home-in on and destroy emitters of electronic signals. Such brute force or active methods proved subject to counter-countermeasures, sometimes interfered with one's own electronic equipment, and were seldom broad enough to blind every specific threat. The American's experience in Vietnam and that of the Israelis in the October 1973 war with ever more complex air defenses, however, stimulated interest in a passive response to the problem: an air vehicle designed to present little or no radar cross section (RCS) from any angle, and one that minimized its heat signature to foil infrared detection. In late 1978 Lockheed Aircraft Corporation and the U.S. Air Force began development of a stealth combat aircraft,[12] the first designed on the basis of its RCS instead of on aerodynamic or flight performance requirements.* The resultant F–117A entered the active Air Force inventory in 1983.

In November 1988, the U.S. Air Force brought the F–117A stealth fighter (known as the Nighthawk) out of the black world and introduced its distinctive appearance, but not its sophisticated technology, to public view.[13] It had first flown in June 1981, and the service took delivery of the last of fifty-nine aircraft in June 1990. Given the aircraft's technical potential against enemy air defenses and its effectiveness as a bombing platform, the Air Force received a bargain, especially since the cost was comparable with conventional (i.e., nonstealthy) alternatives. In 1991 dollars, the total F–117 program cost a relatively modest $8.2 billion overall, with a unit flyaway cost (airframe, engines [installed], electronics, ordnance, and armament) of $52.5 million.[14] For comparison, the F–111F Aardvark had a unit flyaway cost of $10.9 million in 1973 dollars[15] (roughly equal to $45 million in 1991 dollars), but it required a supporting package of electronic jamming aircraft and air-defense suppression aircraft to reach its target. Except for flight refueling, the F–117 needed no other aerial support, thus saving the large costs (fuel, weapons, refueling, and maintenance, not to mention possible loss of aircraft and personnel) associated with escorts. To minimize financial risk and expense and to speed the F–117's development, Lockheed used parts from both the F–18 and F–16 fighters as well as adapted existing attack, computer, and electronics systems. The F–16's fly-by-wire digital flight control system proved particularly beneficial in controlling the dynamically unstable F–117 aircraft, and computer modeling at levels far exceeding those available previously greatly assisted designers in its development.

Stealth technology evolved during World War II, with the introduction of the British Mosquito light bomber, whose plywood construction gave it a low

---

*Lockheed had amassed considerable experience in this field. In the late 1950s and 1960s, designers of the A–12 Oxcart and SR–71 Blackbird strategic reconnaissance aircraft gave considerable attention to reduction of their radar return. In the 1970s Lockheed built the experimental Have Blue stealth technology demonstrator, the first aircraft designed and built where stealth characteristics were paramount. Have Blue, a milestone in its own right, furnished a valuable data base for the subsequent F–117A.

radar return, and with later models of the German U-boat snorkel,* which were relatively small and coated with radar-absorbent material. Stealth technology uses surface shaping to eliminate direct returns, minimize dwell, and produce deceptive returns, and it employs radar-absorbent materials to reduce RCS. (This design feature has no direct relationship to the aircraft's physical cross section.) Not intended to make an aircraft invisible, stealth technology makes an aircraft difficult to detect and virtually impossible to track and engage. Stealth works against all types of radars. A powerful ground search radar may get a weakened return, but less powerful SAM and AAA tracking radars or airborne fighter/interceptor radars will not produce an image suitable for lock-on of their weapons, making them unable to engage stealth aircraft. To reduce the chance of visual sighting, the F–117 attacks only at night. With "eyeball" tracking eliminated, only a very lucky random shot will bring one down. Stealth aircraft, which can operate in areas closed to other attacking planes, can directly threaten a foe's most important targets.[16] Like the British Royal Navy in 1905–1906 when it introduced the first all big-gun battleship, HMS *Dreadnought*, the U.S. Air Force's fielding of the F–117A scored a unique technological triumph of dramatic political, military, and even economic impact. Just as *Dreadnought* rendered obsolete the battle fleets of Britain's enemies, stealth largely nullified the existing air defenses of the Soviet Union and other states. The tremendous Soviet investment in comprehensive and redundant radar warning networks and air defenses now required enormous upgrading with a technology to counter stealth, if such could be developed, at a staggering expense.

Other newly deployed technological devices — from earth-orbiting satellites to secure facsimile (fax) machines — aided air operations in matters both great and small. Space-based systems proved invaluable in the Gulf War. Satellites of the Defense Support System, originally designed to warn of a Soviet ballistic missile attack against the United States, scanned Iraq for the exhaust glow from Scud SRBM launches.[17] The Navstar GPS satellite network revolutionized air and surface navigation. Hand-held receivers and other, more accurate receivers in aircraft, vehicles, ships, and weapons allowed Coalition forces to locate their own positions to within fifty feet instead of to within eight miles, as had been the case with older, ground-based navigation systems. Low-orbiting satellites of the Defense Meteorological Support Program and their civil counterparts operated by the National Oceanographic and Atmospheric Administration provided accurate weather updates in as little as ninety minutes,

*The snorkel was a tube or pair of tubes for air intake and exhaust that extend above the surface of the water above a submerged submarine. The device allowed a submarine to stay below the water's surface to avoid detection and still use its air-breathing diesel engines and conserve or recharge its electric drives. Use of the more powerful diesel engines doubled a submarine's submerged speed. Improved radar, however, permitted the Allies to locate even the snorkel. Hence the Germans countered with radar-absorbent material.

sometimes allowing near real time* retargeting of missions. The Defense Communications Satellite System provided thousands of secure circuits and carried a major percentage of all intertheater and intratheater communications, while intelligence satellites, including the commercial Landsat system, identified enemy surface dispositions for Coalition forces. The secure fax and telephone systems of the U.S. military concealed information from the enemy and allowed unprecedented direct access between lower echelon staffs in both the United States and the theater of operations.[18]

U.S. military war plans and planners, however, had not fully assimilated the import of these changes in aeronautics, munitions, and satellite technology. Although the U.S. armed forces devoted much time to exercising their warfighting capabilities, deployment planning tended to dominate the attention of the staffs of the U.S. Combined Theater Commanders-in-Chief and their subordinate service component commanders. In the U.S. Air Force Air Staff, some general officers reacted against this focus on deployment and against what they felt was too great a departure from theories of independent air power. In particular, they feared that the U.S. Army's AirLand Battle concept,[19] to which the Air Force's Tactical Air Command (TAC) subscribed, would tie air power too closely to ground battle developments and deny air power its necessary freedom of action for interdiction and strategic bombardment. In 1988 Lt. Gen. Michael J. Dugan, Deputy Chief of Staff, Plans and Operations, and Maj. Gen. Charles G. Boyd, Director of Plans, established a Deputy Directorate for Warfighting Concepts within the Air Staff Directorate of Plans. They placed in charge an advocate of independent air operations, Col. John A. Warden III, and encouraged him to refine and disseminate ideas on the independent uses and functions of air power.[20] Within a year the Deputy Directorate acquired the Checkmate Division, which specialized in the analysis of conflicts between Soviet and U.S. Alliances.**

A student of military history, Colonel Warden had read widely and deeply. His book, *The Air Campaign: Planning for Combat*, expressed his ideas on the dominance of air warfare.[21] In keeping with his charter and sheltered by his superiors, Warden encouraged other officers to join him in considering and promoting innovative ideas about the role of air power in modern warfare. Moreover, he encouraged freewheeling discussion and, refreshingly, tolerated

---

*"Near real time" in military parlance accounts for the delay between the occurrence of an event and the receipt of data about it at some other location, introduced by automated data processing and display. "Real time" signifies essentially no delay between the occurrence of an event and receipt of data elsewhere, except for that of transmitting the electromagnetic energy. *The Official Dictionary of Military Terms*, comp Joint Chiefs of Staff (JCS Pub 1) (Cambridge, Mass.: Hemisphere Publishing Corp, 1988), s.vv. "near real time," "real time."

**During the Gulf War, Checkmate, located in the Pentagon subbasement, became the shorthand reference for the entire Deputy Directorate of Warfighting Concepts. The Deputy Directorate funneled most of its communications to the Special Planning Group (the Black Hole) in Riyadh through Checkmate.

dissent.[22] Warden and the like-minded air officers that coalesced about him believed that air planners should first determine the enemy's centers of gravity: those characteristics, capabilities, or locations from which the enemy derives his freedom of action, physical strength, or will to fight. Planners identified these centers of gravity by analyzing the enemy's strength and situation and then by locating the critical vulnerabilities (centers) that, when attacked, would most unbalance him. Centers of gravity might not always equate to specific target systems because critical vulnerabilities might cut across several target systems. The Warden group designated the primary center of gravity in any conflict as the enemy's leadership — its survival, its continued resistance, and its control of its military forces and the other elements of national power. The planner's function did not cease with identification of the centers of gravity. Many considerations such as weather, air defenses, and the overall military situation affected an air campaign. The planners evaluated all these factors as well as the forces at an enemy's disposal to devise the best means for employing U.S. Air Force air assets coherently and cohesively.[23]

The Warden group perceived itself as continuing the tradition of the original Air War Plans Division of the Air Staff of 1941, which devised AWPD–1, the blueprint for American strategic air operations in World War II.* They accepted the time-honored principle of securing air superiority** as the first mission of any air campaign. Next, they accepted the concept of employing force packages as a requirement in mission planning.

In August 1988 Colonel Warden circulated a paper throughout the Air Staff that used graphic representations to codify his targeting and strategic ideas.[24] The model portrayed the relative importance of the strategic targets contained within a nation state arranged in five concentric rings resembling an archery target. Each ring contained specific strategic target sets ranked by the effect their loss would have on an enemy nation. The innermost ring, the bull's-eye, consisted of the nation's leadership and its connections (communications, propaganda media, organs of internal control, etc.) to the population and armed forces. The ring adjacent to the bull's-eye contained the nation's key production centers such as energy, advanced research facilities, and bottleneck industries. The third ring from the center compassed a nation's infrastructure, including its transportation systems. The nation's population constituted the fourth ring. In considering population as a target set, Warden did not propose to inflict physical damage, but rather to conduct psychological operations designed to

---

*During the Gulf War, members of the Deputy Directorate went so far as to put a plaque on the door of the Checkmate office, stating simply, "Air War Plans Division."

***The Official Dictionary of Military Terms*, compiled by the Joint Chiefs of Staff, defines *air superiority* as "that degree of dominance in the airbattle of one force over the other which permits the conduct of operations by the former and its related land, sea, and air forces at a given time and place without prohibitive interference by the opposing force," and *air supremacy* as "that degree of air superiority wherein the opposing air force is incapable of effective interference."

break the morale of the populace or lower its support of the war effort and ruling regime. The fifth, outermost ring contained the most difficult and costly targets to destroy — a nation's military forces in the field. Although the air offensive had to penetrate the hard, outer fifth ring to reach the softer, interior rings, any sustained attack on targets in the fifth ring would consume aerial resources that could be used more effectively elsewhere.

The ordering of the five rings also delineated their relative vulnerability to attack. The outermost ring of fielded military forces consisted of numerous dispersed targets, all of which had the capacity to shoot back (i.e., many targets, costly to destroy), while the fourth ring of population presented an extremely diffuse target, physically and psychologically. Destruction of even a minor portion of a nation's population contravened U.S. national policy and entailed adverse domestic and international political repercussions (i.e., many targets, hard to influence). The third ring, infrastructure, offered a large number of targets of differing degrees of difficulty. Individually they might be vulnerable, but in total, although they could yield significant results, they would also consume much effort (i.e., many targets, good return). The second ring offered a far more vulnerable target. Hitting a few vital spots in key industrial plants could disrupt entire sectors of an enemy's war economy and affect his ability to continue the conflict (i.e., few targets, high return). Finally, an attacker would find enemy leadership the most important and, because of the relatively few number of points to be hit, the most easily targetable objective of all. Theoretically, knocking out a handful of superhardened bunkers could "behead" a nation in a single raid (i.e., very few and difficult targets, very high return). Checkmate adopted this model, which Warden continued to refine in the succeeding months.[25]

The Kuwait Crisis, U.S. Deployment, and War Plans

Kuwait, a small Sunni Islamic Arab state, lies at the northwestern edge of the Persian Gulf. Topographically it is part of the desert covering much of the Arabian Peninsula. Because Kuwait sat astride a pool of 13.5 percent of the world's known supply of petroleum,[26] in 1990 its 2.1 million inhabitants (many of whom were imported laborers) had the highest per capita standard of living in the world. Its armed forces consisted of 20,300 troops, 245 tanks, and 35 combat aircraft. Its much larger and more conservative southern neighbor, Saudi Arabia, a Sunni Arab monarchy, possessed another 25 percent of the world's known petroleum reserves. The Saudi population of 15 million supported an armed force 3½ times larger than Kuwait's. To Kuwait's north and west lay its most powerful neighbor, the Republic of Iraq, a single-party, Arab-socialist state, ruled autocratically by its strongman and president-for-life, Saddam Hussein. Iraq, with a population of 18.8 million, fielded armed forces outnumbering Kuwait's by 50:1 in manpower, 25:1 in tanks, and more than 20:1 in combat aircraft. The Islamic Republic of Iran, with a population in 1990 of 55.6 million and a civil government dominated by conservative Muslim Shiite clerics, lay across the Persian Gulf to the east.

By the summer of 1990, following a bitter, protracted, and ultimately victorious war against Iran, Hussein's regime desperately needed hard currency. At the same time the price of oil, Iraq's only marketable cash commodity, continued to fall. In July, Hussein accused Kuwait of exceeding its Organization of Petroleum Exporting Countries (OPEC) production quota and demanded the Kuwaitis forgive an estimated Iraqi war debt of $10 billion. The Kuwaitis, through pride, miscalculation, or simple refusal to submit to blackmail, refused to accede to any of Hussein's threats. At the OPEC meeting of July 26, 1990, the states of Saudi Arabia and Kuwait and the OPEC majority voted a target oil price of $21 per barrel, 16 percent less than the $25 per barrel advocated by Iraq. Thwarted by OPEC and the other gulf states, and unwilling to cut his grandiose military, social, and nation-building expenses, Hussein dispensed with bluster and resorted to action. He did not calculate that anyone, least of all the world's superpowers, would oppose him. Months earlier, on February 12, 1990, he had informed U.S. diplomat John Kelly, Assistant Secretary of State for Near East and South Asian Affairs, that the Soviets were "finished as a world power."[27]

On July 25, in a remark later repeated in media broadcasts, he told U.S. Ambassador to Iraq, April Glaspie, that the U.S. could not accept ten thousand dead in a single battle, as Iraq had done. The British and the French, mean-

The Republic of Iraq, a single-party Arab-socialist state, was ruled autocratically by its strongman and president for life, Saddam Hussein, who littered the countryside with propaganda symbols supporting his regime.

while, competed with one another until the crisis broke to sell Saddam Hussein advanced weapons such as the Exocet antiship missile, the F–1 fighter to carry it, and other advanced technology. In addition, the Kuwaitis themselves, because of their relatively tightfisted monetary policies and abuse of contract labor, were an unpopular lot in the Arab world.

On August 2, 1990, in a single day of violent blitzkrieg, Iraqi armed forces overran and occupied Kuwait. Within the week Iraq annexed Kuwait as a new province. These moves brought Hussein control of 20 percent of the world's oil reserves and placed his armed forces directly across the border from the Eastern Province of Saudi Arabia, which contained the bulk of that country's petroleum reserves. The prospect of greatly altered international oil production and the changes in distribution arrangements that were likely to follow immediately fixed the world's attention on the Persian Gulf.

At the same time, the Iraqis siezed all foreigners in Kuwait and Iraq and declared them to be hostages for the actions of their governments. This ill-considered, uncivilized action hardened world public opinion against Saddam Hussein. The subsequent dispersal of the thousands of hostages to potential air attack targets throughout Iraq complicated Coalition military planning, but it further lowered the Iraqi's international status. Once the Iraqi invasion forces finished their occupation of Kuwait, they adopted, at least initially, an offensive posture that could only be directed at Saudi Arabia. This helped to further persuade the Saudis to accept U.S. aid.

Before August 1990, the United States had followed a conciliatory policy toward Iraq, hoping as President George H. W. Bush observed, to draw that country "into the family of nations." The U.S. government guaranteed agricultural purchases, which made Iraq one of the principal buyers of American grains, and it granted to Iraq hundreds of export licenses for so-called dual-use manufactured items. The characteristics of these goods, including high-quality machine tools, allowed the recipient country to utilize them for military or civilian purposes.[28] During the Iran-Iraq war in the 1980s, the Reagan administration had shared overhead imagery and communication intelligence with the Iraqis, some in almost real time. Not only did this information aid Iraqi military efforts, it also enabled its military leaders to deduce sensitive U.S. intelligence capabilities and adopt countermeasures, such as emphasizing landline communications by laying fiber optics cables and concealing projects from imaging systems.[29] Sharing of some forms of U.S. intelligence continued until at least May 29, 1990.[30] At the end of the Iran-Iraq War in July 1988, the United States shifted its attention from Iraq to other crises.[31] With finite intelligence resources, U.S. leaders concentrated on other pressing situations, such as the internal crisis within the Soviet Union and events in Eastern Europe. This rational allocation of priorities had unforeseen consequences. The United States' conciliatory policy toward Iraq, compromise of intelligence capabilities, and shortage of up-to-date intelligence on the Persian Gulf area combined to handicap subsequent vital U.S. efforts. From the start of the crisis to the conclusion of the Persian Gulf War, the United States found itself with insufficient intelligence and bomb damage assessment (BDA). In some areas of activity, such as nuclear weapons manufacturing and research capabilities, the United States never realized the extent of Iraqi technical progress until after the cessation of hostilities.

The Americans nonetheless reacted immediately to Hussein's invasion of Kuwait. On August 2, 1990, President Bush froze Iraqi assets in the United States. That day the President also met with the Chairman of the Joint Chiefs of Staff, Army General Colin E. Powell, and Army General H. Norman Schwarzkopf, Commander-in-Chief, CENTCOM. He told them he would consider an Iraqi attack on Saudi Arabia a casus belli.[32] Two days later President Bush and top civilian and military leaders met and agreed to send troops to Saudi Arabia, if the Saudis requested them. They further decided to send to Saudi Arabia a delegation headed by Secretary of Defense Richard B. Cheney, and to include General Schwarzkopf.[33] On August 8, 1990, President Bush publicly stated U.S. objectives in the crisis:

> Immediate, complete, and unconditional withdrawal of all Iraqi forces from Kuwait;
> Restoration of Kuwait's legitimate government;
> Security and stability of Saudi Arabia and the Persian Gulf;
> Safety and protection of the lives of American citizens abroad.[34]

The uncompromising American objectives reassured the nations of the Persian Gulf — and guaranteed a full-scale confrontation with Iraq. On August 6 the U.S. delegation met with the Saudi King, Fahd ibn Al-Aziz, who agreed to accept U.S. troops on Saudi soil. President Bush ordered the deployment to begin at once.[35]

By August 8, 1990, McDonnell-Douglas F–15C Eagle Air Force fighters and the Boeing E–3B Sentry Airborne Warning and Control System (AWACS) aircraft had landed in Saudi Arabia. General Schwarzkopf assigned U.S. Air Force Lt. Gen. Charles A. Horner* to the Saudi capitol, Riyadh, designating him

---

*In addition to his temporary assignment as CENTCOM Forward, General Horner simultaneously held several other positions. As a U.S. Air Force general, he commanded the U.S. Ninth (Tactical) Air Force (a command composed entirely of U.S. Air Force units). When the Ninth Air Force acted as part of CENTCOM, Horner became the Air Component Commander and the Commanding General, Central Command Air Forces (CENTAF), which included all U.S. Air Force units (the Ninth Air Force plus reinforcements), save U.S. Air Force Special Operations Forces units, in the CENTCOM AOR. Schwarzkopf also appointed Horner as the Joint Force Air Component Commander (JFACC), with the duties of planning, coordinating, allocating, and tasking theater-wide air operations (including U.S. Marine Corps and U.S. Navy aviation and all aircraft belonging to Coalition air forces committed to the defense of Saudi Arabia) in accordance with Schwarzkopf's apportionment decisions. Horner's duties as JFACC also made him the Airspace Control Authority and Air Defense Commander, which in practice gave him control over all Coalition and Iraqi airspace because he had the last word in controlling flight paths, preventing excessive airspace congestion, and coordinating flights with air defenses. In order to fly, one had to clear with the JFACC. One way or another, Horner had the dominant air power voice in the in-theater U.S. and Coalition command structures.

General H. Norman Schwarzkopf, Commander-in-Chief, U.S. Central Command (CINCCENTCOM).

CENTCOM Forward. This gave Horner the responsibility of receiving all U.S. forces arriving in Saudi Arabia while Schwarzkopf returned to the United States and the even more complicated task of overseeing the deployment. Although CENTCOM had drafted deployment plans for the Persian Gulf and was in the process of creating a hasty plan for defensive operations, the President wanted more flexibility for a wider range of responses. Therefore, Secretary Cheney ordered Schwarzkopf to plan for an offensive option in case Saddam Hussein engaged in further aggression or began killing hostages.[36] Schwarzkopf's CENTCOM and component staffs, overwhelmed with deployment and defensive planning, lacked the resources to create such plan. On August 8 Schwarzkopf, with General Powell's permission, approached the U.S. Air Force Air Staff in the Pentagon directly, asking for a retaliation plan directed against strategic targets in Iraq and ready for execution by the end of August.[37] He spoke to the U.S. Air Force Vice Chief of Staff, General John M. Loh, who accepted the task and, in turn, asked Colonel Warden and his Deputy Directorate to prepare the air plan.[38]

The air-war plan reflected the ideas of the Warden group. On August 10, Warden briefed the initial version of the plan to Schwarzkopf, who approved it. The next day Warden briefed Powell, who approved the plan and directed Warden to add provisions for attacking and destroying the Iraqi ground forces in Kuwait.[39] The President learned of the plan on August 15. On August 17, Warden returned to CENTCOM headquarters at MacDill AFB, Florida, with a final version and a rudimentary operations order. He called the plan INSTANT THUNDER to emphasize its difference from ROLLING THUNDER, the interminable and constrained bombing campaign over North Vietnam.[40] INSTANT THUNDER had a clear concept of operations: "Conduct powerful and focused attacks on strategic centers of gravity in Iraq over a short period of time (days, not

Lt. Gen. Charles A. Horner was designated CENTCOM Forward.

weeks)." It rested on the following four principles:

Target the Hussein regime, not the Iraqi people
Minimize civilian casualties and collateral damage*
Minimize American and allied losses
Pit U.S. strengths against Iraqi weaknesses.[41]

The plan employed the five rings and within them identified ten target sets. Under government leadership, the primary center of gravity in Iraq, it struck at two target sets: first at Hussein's regime to "incapacitate" it, and second at military and civil command, control, and communications (C^3) to isolate decision-makers and slow the transmission of orders. By targeting Hussein, not his people, the planners would have liked to physically eliminate him, an outcome considered unlikely. Instead, they expected and hoped to create a set of conditions leading to his overthrow. Initially, however, they boldly stated their goal as eliminating Saddam Hussein himself. That objective they modified upon learning U.S. policy forbade the direct targeting of heads of state. In the second ring, key production, INSTANT THUNDER identified electricity as its third target set; oil (internal distribution and storage, not production and export capability) as its fourth set; NBC research, development, and production facilities (including airfields with chemical-capable aircraft) as its fifth target set; and general military research, production, and storage (including mobile SRBM launchers) as its sixth target.

Because the air planners worked with a limited attack force in the third ring, infrastructure, they confined themselves to making Iraqi railroads their

*The U.S. Air Force defined *collateral damage* as "the damage to surrounding resources, either military or non-military, as a result of actions or strikes specifically against enemy forces or military facilities." Air Force Manual 11–1, Air Force Glossary of Standardized Terms, HQ U.S. Air Force, 1989.

Colonel John A. Warden III, in the Checkmate briefing room, discusses the INSTANT THUNDER plan, the code name assigned to the proposed Strategic Air Campaign.

seventh target set. A single-track rail line between Baghdad, the Iraqi capital in the center of the country, and Basra, Iraq's second largest city and lying near the Persian Gulf and Kuwait, carried most of the freight between these two cities and supplied the Iraqi forces in Kuwait. In the fourth ring the plan called for psychological warfare operations against the Iraqi population, foreign workers in Iraq, and the Iraqi soldiers in Kuwait. Against the outermost ring, i.e., Iraq's fielded forces, INSTANT THUNDER made the strategic integrated air defense system its eighth target set. The plan included Iraq's only naval facility, Umm Qasr on the Persian Gulf, and the associated antishipping missile sites as its ninth target set; seven airfields with modern interceptor aircraft constituted the tenth target set. The air planners selected the final two target sets to protect Coalition air and naval forces. Among the ten sets of targets, INSTANT THUNDER required Coalition air forces to bomb 84 targets and fly a total of 5,700 attack sorties in six days. The air plan stressed the use of precision-guided munitions. The planners expected the cumulative impact of this bombing to devastate the Iraqi war effort.[42]

The ordering of the INSTANT THUNDER target sets reflected Checkmate's targeting priorities as of August 17, 1990. The target priorities assigned in this process were the ones desired as of that date, but they were not permanently fixed; the tenth target set did not necessarily have and was not necessarily expected to hold the tenth priority throughout the campaign. Some targets by

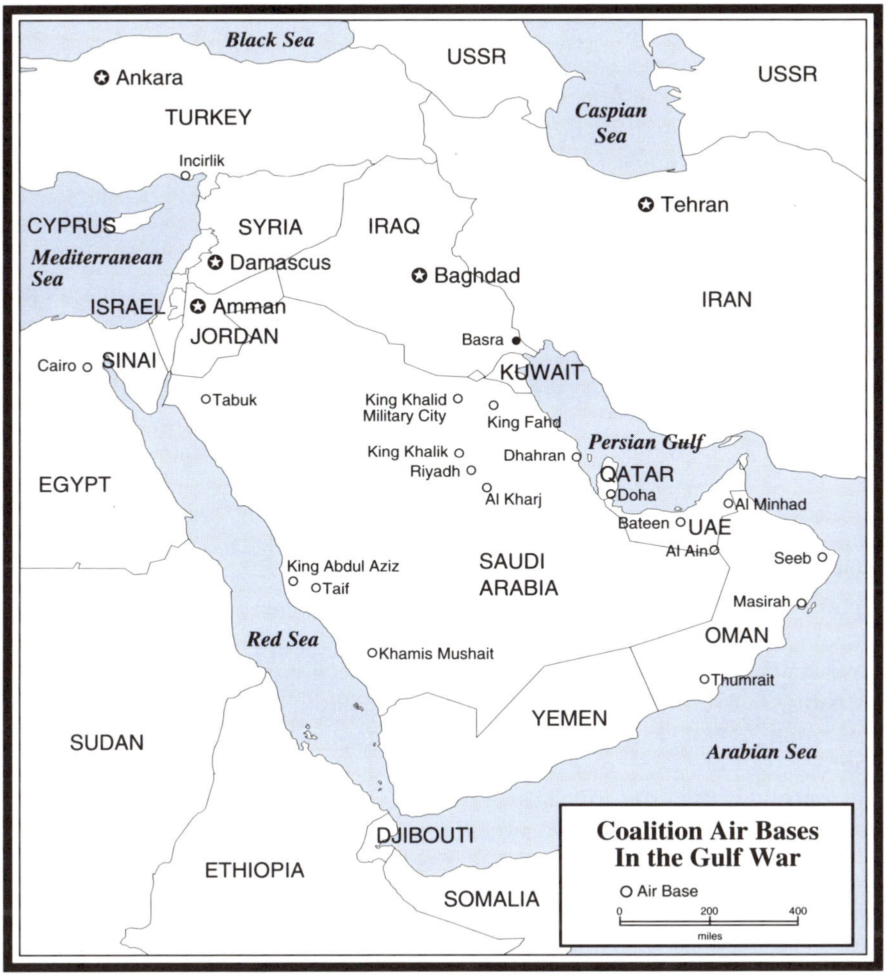
Black Sea
USSR
USSR
Ankara
TURKEY
Caspian Sea
Incirlik
Tehran
CYPRUS
SYRIA
IRAQ
Mediterranean Sea
Damascus
Baghdad
IRAN
ISRAEL
Amman
JORDAN
Basra
Cairo
SINAI
KUWAIT
Tabuk
King Khalid Military City
King Fahd
Persian Gulf
King Khalik
Dhahran
Riyadh
QATAR
EGYPT
Al Kharj
Doha
Al Minhad
Bateen
UAE
Al Ain
King Abdul Aziz
SAUDI ARABIA
Seeb
Taif
Masirah
Red Sea
OMAN
Khamis Mushait
Thumrait
YEMEN
SUDAN
Arabian Sea
DJIBOUTI
ETHIOPIA
SOMALIA
Coalition Air Bases In the Gulf War
Air Base
0
200
400
miles

their nature might fall into more than one ring. For example, an enemy's integrated air defense system (interceptors, SAMs, AAA, and air defense command and control facilities) has large components in the fifth ring, fielded forces, although its command and control system center of gravity was closely associated with the first ring, leadership. Because the overarching operational consideration in any sustained air campaign is obtaining air superiority, that requirement demanded immediate attacks on the opponent's air defense system. Thus, the highest-priority attacks went first to a target system that for the remainder of the conflict may well be attacked only to keep it suppressed — thereafter assuming a very low target priority. In this scheme, few target priorities remain absolutely constant throughout a campaign. Warden conceived the five rings as a guide for air campaign planners, not as a straightjacket for the conduct of actual air operations. Once hostilities began, political and military realities could and did change the order of targeting.

Warden predicted that an aerial attack would effect a change in the political regime, eliminate Iraq's strategic offense and defense capability, and disrupt its internal economy while leaving its ability to export oil intact. It would also enable the nations of the Persian Gulf to deal effectively with Iraq's residual forces.[43] Like air planners before him, Warden set ambitious goals. Still his plan rested on two major problematic assumptions: first, that it was built on a firm foundation of targeting information; and second, that an intense strategic bombing blitz would change the regime of a police state.

By the start of the conflict in 1991, strategic air planners had added two more target sets to INSTANT THUNDER's descendants and modified a third. At Cheney's and Schwarzkopf's direction, they added the Republican Guard in late August 1990. The Republican Guard, a praetorian force politically loyal to Saddam Hussein in the same way the SS had been loyal to Adolf Hitler, served as a prop for the regime by providing the politically reliable military forces necessary to overawe or defeat the regime's internal enemies. Theoretically, weakening it would ease the path to power for the regular army or other plotters who opposed the regime or for Kurdish and Shiite opposition groups. The Republican Guard also formed the most militarily effective portion of the Iraqi ground forces. It had first call on quality personnel, received the most modern weapons available to the Iraqis, and had the most thorough training. During the Iran-Iraq War it performed as a strategic reserve used to seal and drive back Iranian breakthroughs, and at the end of that war it spearheaded the victorious Iraqi counteroffensives that ended the conflict. In August 1990, by then a force grown to two armored, one mechanized, and four infantry divisions and one special forces division, it led the Iraqi invasion of Kuwait and apparently stood posed to invade Saudi Arabia. Immobilizing or severely damaging the Republican Guard with tactical air power would greatly hamper the ability of the Iraqi ground forces to conduct offensive operations, a prime concern to the Coalition's leadership in August 1990. Likewise, air attacks could limit its capacity to counter Coalition ground attacks when and if it became necessary to liberate Kuwait physically. The Republican Guard's political function and military potential made it a prime target for both the strategic and tactical air operations.

By January 1991 the strategic air planners had added fifty-four major highway and road bridges between Baghdad and Basra to the railroads as an eleventh target set. New intelligence, unavailable at the INSTANT THUNDER's inception, revealed that target system's vulnerability to air assault. In addition, by the start of the conflict in-theater planners had modified the Iraqi military poduction and research target set by splitting out Scud ballistic missile research, production, and launchers to form a twelfth target set.[44]

Schwarzkopf fully endorsed INSTANT THUNDER on August 17, 1990, and sent Warden to Riyadh to brief General Horner. Horner rejected the plan's "airpower alone" aspects, but he accepted the plan's target list and selected three key members of Warden's team to compose a small, highly secret CENTAF Special Planning Group. This group completed preparation of an offensive air campaign plan ready for execution by September 15, 1990.[45] In the process, the group became known as the Black Hole when other members of the CENTAF staff observed that resources, personnel, and intelligence all seemed to go into the Special Planning Group, but they never seemed to come out again — activity not unlike that in a collapsed neutron star.* In late August 1990 Horner appointed U.S. Air Force Brig. Gen. Buster C. Glosson to head the Special Planning Group.[46] One of Glosson's first moves was to augment the Special Planning Group to twenty officers. He added one experienced weapons officer from each Air Force combat wing in Saudi Arabia to act as a subject-matter specialist. For the next five months, in response to new intelligence and to the growing number of Coalition aircraft available, the Special Planning Group expanded its target list. The group, too, increased in size with the addition of SAC refueling experts and U.S. Navy and U.S. Marine Corps representatives. RAF planners joined the Special Planning Group on September 19, 1990. Although the Saudis had unofficial knowledge of the plan from September onward, they were not officially informed of it until early December 1990.

If General Horner rejected the form, he kept the substance of INSTANT THUNDER. Although he co-opted the Warden group into a planning process under his own control, he did not fundamentally change its campaign plan. At his prodding, the term INSTANT THUNDER and overt traces of the Warden group disappeared from briefings. By September 2, 1990, INSTANT THUNDER had transmogrified into CENTAF Offensive Campaign — Phase I,[47] although the targets and the philosophy used to identify them remained unchanged. These aspects formed the core of subsequent offensive air planning, which continued

---

*The extreme secrecy met Schwarzkopf's wishes that U.S. offensive planning activities be tightly guarded. For political and diplomatic reasons neither the Saudi nor American government wished to acknowledge, at this stage of the crisis, that the Coalition was planning for offensive operations against Iraq on Saudi soil.

In August 1990 General Horner appointed Brig. Gen. Buster C. Glosson to the Special Planning Group. Here he is conferring with Lt. Col. David Deptula in Riyadh.

to emphasize leadership, electricity, NBC facilities, and the other targets sets derived from the five rings. Glosson himself became a convert to the principles of INSTANT THUNDER, largely owing to the efforts of one of Warden's key planners and a member of the Black Hole, Air Force Lt. Col. David A. Deptula. The strategic air planners, who remained separate from the CENTAF staff in part because the high security classification of their work, also maintained a continuing and close relationship with Checkmate in the Pentagon during the prewar and wartime periods, with mutual exchanges of information. When Powell and Schwarzkopf received the initial version of CENTAF's Offensive Campaign — Phase I on September 13, 1990, they decided that whenever hostilities broke out, even in the event of an Iraqi offensive, they would implement it rather than respond with a defensive air plan. By mid-September, CENTAF's offensive campaign had already become the first part of a larger combined and joint theater campaign prepared by the CENTCOM staff.

Although Warden intended INSTANT THUNDER as a stand-alone warstopper, Schwarzkopf saw it merely as the first phase of an integrated air-ground campaign to liberate Kuwait. As early as August 25, 1990, he had presented a four-phased offensive plan to Cheney and Powell:

> Phase I, "Strategic Air Campaign" against Iraq;
> Phase II, "Kuwait Air Campaign" against Iraqi air forces in Kuwait;
> Phase III, "Ground Combat Power Attrition" [air attacks] to neutralize the Republican Guard and isolate the Kuwait battlefield; and
> Phase IV, "Ground Attack" to eject Iraqi forces from Kuwait.[48]

Air power alone would accomplish the first three phases. Ground and air power working together would execute the last one. By mid-September Horner also assigned Phase II and some Phase III planning to the Special Planning Group under General Glosson. Although detailed planing for offensive ground and joint air-ground operations had begun with the arrival of U.S. Army planners in mid-September 1990, it accelerated after President Bush authorized doubling the number of American forces in the theater, including transferring the U.S. Army's mechanized VII Corps from Germany. For practical purposes, the addition of more aircraft allowed CENTAF to merge the first three phases and execute them simultaneously, instead of sequentially, with situational changes made in the level of effort among them.[49] The initial air assault would include attacks on targets belonging to each of the first three phases. As the ground assault approached, the effort taken from Phase I and devoted to Phase III would grow.

The Republican Guard had figured prominently in the air planning for both Phase I, the strategic air campaign, and Phase III, preparation of the battlefield. (The Guard by November 1990 had withdrawn from forward positions on the Kuwaiti-Saudi border to reserve positions along both sides of the Iraqi-Kuwaiti border, where it assumed its traditional counteroffensive role.) In mid-December 1990, the Special Planning Group and the CENTAF Director of Operations tactical air planners and officers who prepared the air tasking order (ATO) combined to form the CENTAF Directorate of Campaign Plans which was led by General Glosson in Riyadh. Within that directorate, on the eve of the Gulf War in mid-January 1991, direction for bombing the Republican Guard in the field shifted from the strategic planners to the KTO targeteers.*

Meanwhile, U.S. aircraft flowed into the region. The reinforcements swelled the number of aircraft available to deliver precision-guided munitions by almost 150 percent. Altogether, these reinforcements gave CENTAF some 700 Air Force combat aircraft, which included 93 percent of the service's precision delivery–capable aircraft inventory with 63 percent of its stockpile of laser-guided bombs. Offshore, U.S. Navy carriers had 62 A–6Es, all GBU-capable, on their flight decks, although that service faced a serious shortage of

*This was neither a formal change of task nor one called for by doctrine or philosophy. Rather, it acknowledged the fact that the Republican Guard's physical location in the KTO placed them beneath air space controlled by the tactical planners, not the strategic ones.

Lt. Col. David A. Deptula, one of Warden's key planners, briefs General Schwarzkopf in Riyadh on the attack plan.

"smart" munitions. (During the war, the Air Force would drop or launch 90 per cent of all the precision air-to-ground munitions.) Also, by January 15, 1991, the Air Force had deployed more than 200 air-to-air inflight refueling tankers[50] — the single most important aircraft type in the theater. Without flight refueling, the bulk of the missions flown in the strategic air campaign could not have reached their targets and returned. Likewise, no less than 70 percent of the

KC–135 tankers deployed to King Khalid International Airport at Riyadh, from which they regularly operated over Iraqi airspace and ensured the success of the air war.

TABLE 7–1

Major Types of U.S. Military Aircraft in the CENTCOM AOR[a]
September 1, 1990–February 1, 1991

Service	Aircraft	Type	Sep 1	Oct 1	Nov 1	Dec 1	Jan 1	Feb 1
Air Force	F–15C	Fighter	70	72	72	72	96	96
	F–4G	Wild	24	36	36	36	48	49
	F–16	Ftr/Attack	106	120	120	120	168	212
	A–10	Attack	72	96	96	96	120	144
	AC–130	Gunship	0	5	5	4	4	2
	F–117A	**Bomber**	**18**	**18**	**18**	**18**	**36**	**42**
	F–15E	**Ftr/Attack**	**24**	**23**	**24**	**24**	**46**	**48**
	F–111F	**Bomber**	**18**	**32**	**32**	**52**	**64**	**64**
	B–52	Bomber	20	20	20	20	20	36
	RF–4C	Recon	6	6	6	6	6	18
	JSTARS	Srvl & Ctrl[c]	0	0	0	0	2	2
	E–3A	Srvl & Ctrl	6	6	6	6	7	11
	EF–111	EW[d]	10	14	14	14	18	18
	KC–10	Tanker	0	6	6	6	6	30
	KC–135	Tanker	79	93	114	115	164	194
	C–130	Airlift	70	95	96	96	96	149
	EC–130E	ABCCC[e]	6	6	6	6	6	6
Navy	F–14	Fighter	56	76	76	76	76	109
	F/A–18	Ftr/Attack	58	58	58	88	88	89
	A–7E	Ftr/Attack	0	24	24	24	24	24
	A–6E	**Bomber**	**35**	**48**	**48**	**62**	**62**	**96**
	E–2C	Srvl & Ctrl	12	17	17	21	21	29
	EA–6B	EW	12	17	17	21	21	27
	KA–6D	Tanker	8	12	12	16	16	16
	S–3A/B	Recon	23	31	31	31	31	43
Marine Corps	F/A–18	Ftr/Attack	48	48	48	48	72	78
	AV–8B	Ftr/Attack	40	60	60	60	81	84
	OV–10	Srvl & Ctrl	0	8	8	8	8	19
	EA–6	EW	12	12	12	12	12	12
	A–6	**Bomber**	**9**	**10**	**10**	**10**	**20**	**20**
	KC–130	Tanker	6	8	8	8	12	15

Source: Gulf War Air Power Survey (GWAPS) vol 5, *Statistics*, table 9, "U.S. Air Order of Battle by Aircraft and Service," pp 31–32.

[a]Boldface entries indicate aircraft that could could deliver laser-guided bombs using self-designation.
[b]Wild Weasel=aircraft specalizing in attacking enemy air defenses
[c]Srvl & Ctrl=surveillance and control aircraft
[d]EW=early warning aircraft
[e]ABCCC=airborne command, control, and communication

Navy's strike flights* required land-based Air Force tanker support to complete their missions.[51]

The Special Planning Group worked from CENTAF headquarters located at the Royal Saudi Air Force Headquarters Building in downtown Riyadh. Once the conflict started, this group guided strategic air operations. In another section of the headquarters building, officers aided by specialized computer software compiled a daily ATO.** CENTAF transmitted appropriate portions of the ATO to the U.S. Air Force wings, U.S. Navy carrier battle groups, and the air organizations of the Coalition allies. Although the Navy withheld some aircraft from the tasking order for fleet defense flights and the Marine Corps withheld some of its aircraft for organic ground support, as a rule, no allied aircraft flew over Saudi Arabia, Kuwait, or Iraq without an ATO authorization.[52] This order was the crucial enabler of air operations; however, it did not conceive the air operations. That function belonged to the Special Planning Group in Riyadh.

After September 15, 1990, the Special Planning Group constantly updated the strategic air campaign plan, almost always with the expectation that a decision in Baghdad or Washington might mandate execution in a few hours. Additional Coalition forces and new target intelligence accounted for many of the changes. More force meant more known targets could be attacked; new intelligence meant new targets to strike. CENTAF's Offensive Campaign — Phase I had a list of almost 300 targets by January 16, 1991, with an additional 300 potential targets under consideration.[53] The list had obviously become a victim of "target creep," a phenomenon that occurs when every organization (the U.S. Air Force, Navy, and Army, U.S. intelligence agencies, and CENTCOM, among others) having input into the targeting process seeks to insert its own special needs and targets into the overall list. To maximize shock effect, ensure coordination of missions, and strike the most significant

---

*The Navy's F/A–18C had an unrefueled combat radius of action (the distance an aircraft can fly to a target, attack it, and return) of 160 miles, while the A–6E had a 390-mile combat radius. (This means that the F/A–18C could attack targets as far as 160 miles away and then return to base, for a total unrefueled flight of up to 320 miles.) Because a combat radius is specified to include fuel consumed in tactical or evasive maneuvers to and from the target, in carrying weapons, in forming up into formation, and in waiting to land, an aircraft's range far exceeds its actual combat radius. If an F/A–18C flew a shuttle mission (not normally done in the Gulf War), it could take off from Base A, attack a target, and land at Base B as long as its total unrefueled elapsed flight totaled no more than 320 miles. The maximum distance an aircraft can travel in a straight line, without refueling, is its combat range, approximately twice the distance of its combat radius.

**The ATO, known to Vietnam-era U.S. Air Force veterans as the FRAG Order, scheduled and coordinated all daily Coalition air activity over Iraq and Saudi Arabia. It did not cover U.S. Navy Fleet defense sorties. For each individual sortie it supplied call signs, deconfliction of air space, coordination with friendly air defenses, electronic warfare, suppression of enemy air defenses, and combat search and rescue, as well as provided target assignments (sometimes even specific weapons) and escort and tanker rendezvous. A single ATO, when completed, filled a computer printout the size of a telephone book. The ATO controlled all flights over Saudi Arabia and most over Iraq (U.S. Air Force aircraft flying out of Turkey complied with their own ATO, but CENTAF assigned their targets), and in theory one could not fly without specific authorization by the ATO. This made the ATO a powerful tool in the hands of the JFACC General Horner.

Black Hole Special Planning Group at Ridyah in the fall of 1990.

targets such as air defenses, leadership, communications, electricity, and NBC as soon as possible, the Special Planning Group carefully scripted the sorties of the first two days and part of the third day of the air campaign. Horner reasoned that changing circumstances would invalidate any attempt to prepare detailed plans any further in advance.

The Special Planning Group consolidated its targeting recommendations and guidance into the air campaign's key internal planning document — the daily master attack plan.* "It consisted of the sequence of attacks for a twenty-four-hour period and included the time on target, target number, target description, number and type of weapons systems and supporting systems for each attack package." Once the conflict began, and after Glosson approved the daily master attack plan, it became the outline for the officers preparing the daily ATO. As the official Department of Defense (DoD) report on the conflict affirmed, the master plan "drove the process"; it became the basis for all other daily air attack and mission planning.[54]

Colonel Deptula, on loan from the Secretary of the Air Force's staff group and an INSTANT THUNDER author and the chief strategic air planner in the

*The master attack plan reflected targeting strategy, changing priorities of the commander-in-chief and higher authorities, combat and political developments, the latest intelligence from multiple sources, weather, threat, and the availability and suitability of attack assets. It attempted to match the most appropriate weapon and/or delivery system, given current circumstances, with a target. Unlike ATOs, the master attack plan was relatively concise. The first day's (January 17, 1991) master attack plan, the only one to include two nights (January 16/17 and 17/18 and also the day of the 17th), consisted of only twenty-one pages.

Special Planning Group, prepared the daily master attack plan.* That allowed him to maintain, subject to his superiors' approval, much of the strategic philosophy (centers of gravity, the five target rings, the decisiveness of air power, etc.) that underlay INSTANT THUNDER, not only within CENTAF's Offensive Campaign but within the daily execution of the entire strategic bombing campaign against Iraq proper.[55] As he laid out many iterations of plans for operations along the length and breadth of Iraq in the months before the war, he developed a sense for the internal rhythms and the flow of the attack and an understanding of the spatial, temporal, and technological relationships between the targets and the identified Iraqi air defenses. Simultaneous and carefully sequenced attacks in different locations, he realized, could have synergistic effects that would increase the shock and destruction inflicted upon the foe and provide protection for the attackers. The Iraqis would be unable to concentrate their defenses on any one mission. When incorporated into CENTAF's Offensive Campaign — Phase I and executed with precision-guided munitions, force packaging, simultaneity, and stealth technology, it would give air power a preeminent role in the conflict.[56]

Because Colonel Deptula had to spread his forces throughout Iraq, he took issue with Air Force planning and targeting orthodoxy by subtly shifting the emphasis of the desired results of air operations. In the past, traditional mission planning sought to achieve a specified level of destruction, expressed in percentages, for any given target. Target work sheets, photographic imagery and interpretation, and force packaging had revolved around the assignment of sufficient weapons to achieve the desired percentage of destruction. Deptula abandoned this line of thought. He reasoned that overbombing wasted resources — partial destruction might achieve the desired result. Moreover, total destruction often went beyond what was needed or even desired. The advent of guided munitions with their tiny circular error probables reinforced this conviction. By late August 1990 he argued for missions planned in combinations and designed to produce a desired effect on the enemy. Leveling an atomic research facility might have less adverse effects on the enemy's will to fight than would a series of more limited but precise attacks on its leadership's command and control bunkers. Emphasis on effect had wider implications. It encouraged a planner to concentrate not just on discrete targets, but to consider the tactical, operational, and strategic points within a center of gravity,

*Sometimes military planners such as Lt. Col. Deptula occupied important positions with a great deal more leverage on or control over actions than his rank might suggest. Historic examples of other field grade planning officers with seemingly disproportionate influence include Col. Max Hoffman of the Imperial German Army, plans officer for Generals Hindenburg and Ludendorf on the German Eastern Front, 1914–1917; Cmdr. Minoru Genda, Imperial Japanese Navy, who planned the Pearl Harbor attack; and the four U.S. Army Air Forces officers, Lt. Col. Harold L. George, Lt. Col. Kenneth N. Walker, Maj. Haywood S. Hansell, and Capt. Lawrence S. Kuter, who created Air War Plans Division Plan No. 1 (AWPD–1) in eleven days (July 1941). AWPD–1 served as the early blueprint of the AAF's wartime expansion and employment.

to search for interconnections between target systems, and to seek means for achieving the larger political objectives through innovative targeting. One need not target a complex of buildings, for example, when one could target a single office in a specific building.

In addition, Deptula's concept of simultaneity in some instances had the effect of seeming to blur target priorities and make the conduct of the strategic air campaign appear unfocused. For example, to support an attack on a Scud solid-propellant production plant in central Iraq, planned for 0700 hours, one might have to strike airfields in southern and western Iraq at 0645 as well as a communications center in eastern Iraq at 0655. To the uninitiated, these strikes would not necessarily appear related and might even seem haphazard, for they fell in multiple target sets. But the distraction and spreading of Iraqi air defenses decreased the losses of Coalition pilots and aircraft* and increased the chances of a successful attack on the primary target. In such a case, as in most air campaigns, the realities of planning for combat forced compromises in targeting priorities. The Special Planning Group finalized the first day's plans only a week before the conflict.

The overall air plan had to be ready to execute at all times. Moreover, once the air campaign began, Deptula, Glosson, and other air officers feared that Hussein might quit at the first blow, or more likely, that the UN might impose a cease-fire or that some other political decision might end the campaign in as little as a week or at any time thereafter. The possibility of a short campaign caused Black Hole planners to select for attack a broad variety of targets as soon as possible rather than concentrate on destroying completely one target system and then switch to completely destroy another. It seemed better to inflict *some* damage on *many* target systems/centers of gravity than to attack two or three target systems and leave all others untouched. This decision echoed the Warden group's tenet of bombing for effect, not always for destruction, and its belief in the systemic disruption of an enemy's economy and society when multiple centers of gravity are bombed. Consequently, to make the initial and subsequent attacks as broad as possible, Deptula prepared an onslaught that spread his strike forces widely and employed the minimum number of weapons on the maximum number of targets. This placed a premium on smashing highly leveraged targets (centers of gravity whose destruction would have the widest effect on as many target systems as possible) first.

As noted earlier, General Horner reorganized CENTAF air planning in mid-December to streamline and coordinate planning for the looming air offensive. He combined the strategic air Special Planning Group with CENTAF's defensive/tactical air planners to form the Guidance, Apportionment, Tasking

---

*Before and during the conflict, Horner and Glosson stressed to air planners and air wing commanders alike that minimizing Coalition losses came before any other consideration. Intvw, Richard G. Davis, Perry Jamieson, and Diane T. Putney with Maj Gen Buster C. Glosson, Dir Legis Affairs, Air Force Air Staff, Bolling AFB, Washington, D.C., Dec 11, 1991.

Division, in a newly created Directorate of Campaign Plans. From that time onward the Special Planning Group became the Iraqi Target Cell, responsible for the strategic air campaign, and the CENTAF tactical planners became the KTO Cell, responsible for tactical air planning and targeting in the KTO. The Iraqi Target Cell designated its strategic targets individually. It had approximately 600 targets but could only apportion approximately 250–300 sorties a day among them after January 29, 1991. Before then, strategic sorties had numbered more than 600 daily. This method of planning lent itself particularly well to the anticipated campaign against bridges, which overlapped both Iraq proper and the KTO. Thus the Iraqi Target Cell held responsibility for planning all bridge attacks that, historically, had been interdiction targets belonging to tactical air planners.

The Coalition military leadership's concentration of the air effort on Iraqi ground forces and related targets in the KTO meant that during three-fourths of the conflict, the KTO Cell planned the operations of more than four-fifths of the total Coalition air effort. The KTO Cell had large air resources, but it faced a seemingly endless number of potential targets. In theory, each individual artillery piece and armored fighting vehicle in each of the forty-three Iraqi divisions in the theater, not to mention infantry concentrations and the many supply dumps and vehicles in the Iraqi logistics network, could be a separate aiming point. Consequently, KTO tactical air planers, for the most part, adopted an entirely different targeting approach than the one their colleagues in the Iraqi Target Cell used. They did not usually designate individual targets; such an attempt would have overwhelmed planning capacity and the ATO system. Instead, the KTO Cell employed force on force. It sent packages of tactical strike and support aircraft to attack designated kill boxes. These were grids, thirty miles on a side, divided into quadrants and laid out over a standard map of Kuwait and southeastern Iraq. Unless otherwise directed by an airborne controller or other changes in circumstance, the strike aircraft each day would hit all targets of opportunity within its designated kill box. A single quadrant comprised an area almost equal in size to New York City. These devastating and ubiquitous operations accomplished both aerial interdiction of Iraqi supply and destruction of Iraqi military equipment and personnel. Obviously, this planning method lent itself to attacks on the Republican Guard, for which the KTO tactical air planners had responsibility throughout the conflict. The KTO Cell gave the results of its planning to Deptula, head of the Iraqi Target Cell, who incorporated them into the master attack plan.

In mid-December 1990 when General Horner added the CENTAF staff division in charge of preparing the daily ATO and the Airborne Command Element/Current Operations Division to the Directorate of Campaign Planning, he consolidated the training, defensive, and (eventually) offensive ATOs into a single function. This allowed air planners to fly on the AWACS aircraft,

U.S. Air Force Airborne Warning and Control System (AWACS) aircraft.

monitor plan execution, and modify future work in light of actual operational feedback. Horner placed the directorate under the command of General Glosson and further designated Glosson to command the 14th Air Division (Provisional), composed of all U.S. Air Force tactical fighter wings in the theater. Thus the officer in charge of plans also had direct control of the largest single force of Coalition strike aircraft.[57]

The four-month shakedown before the end of 1990 gave Coalition forces the opportunity to fine-tune their training. It also revealed a flaw that would plague the war effort from start to finish. The CENTAF headquarter's intelligence organization (and to a lesser extent, CENTCOM headquarter's own intelligence organization) often failed to supply fully the Special Planning Group and operations officers in U.S. combat wings in the field with timely and accurate target materials.* Neither the Black Hole nor the F–117A-equipped 37th TFW ever established smooth or cordial working arrangements with CENTAF intelligence before the conflict. Officers in flying units complained

*The DoD in 1989 defined *target materials* as "graphic, textual, tabular, or other presentations of target intelligence primarily designed to support operations against designated targets by one or more weapons systems. Target materials are suitable for training, planning executing and evaluating such operations." The Black Hole also had difficulty obtaining, among other items, target folders for specific targets and information about Iraqi air defenses through CENTAF intelligence channels.

that CENTAF intelligence failed to meet their specialized requirements for planning and targeting materials.

For their part, CENTAF intelligence officers complained that the units failed to follow standard procedures and tried to beat the system by establishing work-around links directly with nonofficial contacts in the United States, such as Checkmate. The Special Planning Group (later the Iraqi Target Cell), of course, had a voracious appetite for almost every sort of intelligence concerning Iraq and a need for it immediately to plan the first phase of the air campaign. CENTAF intelligence, however, had only begun to collect material on Iraqi targets in the spring of 1990. It would also appear that in the period before the war, CENTCOM headquarters did not give the needs of the Special Planning Group the highest priority in the requests it forwarded to the national intelligence agencies such as the Central Intelligence Agency (CIA) and the Defense Intelligence Agency (DIA).

When CENTAF intelligence could not, or would not in the minds of the Special Planning Group, supply needed intelligence, the planners turned to other sources. To do so they had merely to pick up their secure telephones (Secure Telephone Unit Mark III [STU–III]) and call U.S. military contacts anywhere in the world. The STU–III's ease of use, hardly more difficult than that of an ordinary phone, and its wide distribution troughout the DoD had an almost subversive influence on war planning and execution. It gave junior officers and subordinate units the ability to circumvent the chain of command and inconvenient, slow, or misunderstood practices at the cost of creating confusion. All command elements no longer precisely understood what information other elements had used as the basis for their actions. In earlier wars, the existence of a single command message center, which encrypted and decrypted all traffic and catered to authorized users only, severely limited the access of junior staff members to overseas communication. Because Special Planning Group members came from all parts of the service, many alternatives to CENTAF intelligence were available. One was Checkmate, which was only too eager to ensure that its air power philosophy had a thorough trial. For instance, the Special Planning Group first learned of the intracacies and vulnerabilities of the Iraqi integrated air defense system from an Air Force intelligence report passed to it by Checkmate, not from CENTAF intelligence.

The problems of timeliness and interpretation of intelligence, especially BDAs,* became far more acute during the conflict and created a mutual aura of distrust and bitterness between CENTAF intelligence and its customers. Before

---

*BDA assays the effects of all air attacks (bombs, rockets, and strafing) on a target. The prime source of BDA is photographic intelligence, usually from various means of collection such as satellite imagery, reconnaissance aircraft, and strike aircraft gun cameras or video tape recorders. All aircraft capable of designating for laser guided bombs carried video tape recorders as part of their weapons delivery system. As one would expect, poor weather over a target can seriously degrade or delay BDA.

the conflict and even more directly during the war, air planners and officers in the air combat wings bypassed CENTAF intelligence by developing an unofficial back-channel network among themselves for the rapid daily exchange of video tape recordings of air strikes and bomb damage. Moreover, the air planners established direct (if informal) links with the DIA and Checkmate in Washington, D.C. Checkmate arranged informal representation in DIA BDA cells, with the CIA, and with other intelligence agencies in Washington. It cast a wide net for information, even bringing in numerous outside consultants, including the former ambassador to Iraq April Glaspie, in an attempt to learn all it could about Iraq and its potential vulnerabilities. This broad range of reliable information and Checkmate's expertise in offensive planning allowed it to serve as an intelligence fusion center for the Black Hole.[58] The shortage of timely, accurate BDAs that informed the air planners on what to retarget threatened to cripple the strategic bombardment campaign and force planning of it to a hit-or-miss basis. This difficulty revealed a damning, two-way lack of prewar communication between intelligence and operations, though in fairness the lack of hard and current data (forcing the hurried rush to discover it once the crisis broke in August 1990) was also a natural outgrowth of the shift in American intelligence assets away from the Persian Gulf in the summer of 1988 after the close of the Iran-Iraq War. Fortunately, the system functioned well enough not to impede either the strategic or tactical air campaigns.

The Persian Gulf War

On January 16, 1991, the U.S. Air Force had stationed in Saudi Arabia approximately 25 percent (700) of its combat aircraft, including 90 percent of the service's precision bomb droppers: 46 F–15Es (more than half equuipped with low-altitude navigation and targeting infrared for night [LANTIRN]* targeting pods), 64 F–111Fs with PAVE TACK pods,** and 36 (soon to be 42) F–117As. These aircraft formed the heart and soul of the strategic air campaign against Iraq. The Air Force supplied another 450 support aircraft and the Navy and Marine Corps furnished 724, of which approximately 400 were on carriers in or on their way to the theater. Arab allies contributed 490 combat aircraft while NATO Coalition members chipped in 146.[59] The bulk of these 2,096 combat aircraft had experienced pilots, sophisticated weapons, and other

---

*LANTIRN was a two-pod electronics system that increased aircraft capability in night flying. The navigation pod, carried by the F–15Es and some F–16s, allowed for precise nighttime location and navigation. The targeting pod, just coming into service in mid-1990, gave a self-designating precision-guided munitions–capability to its carrying aircraft. There were too few targeting pods for both the F–15E and F–16, so all available pods were reserved for the more capable F–15Es.

**PAVE TACK was a belly-mounted laser and infrared sensor system that allowed F–111Fs to designate targets for their own laser-guided bombs.

technological advantages that outclassed those of the Iraqi Air Force.

Iraq had purchased both an air force and an air defense force designed to overawe and defend it from its neighbors. Its air force performed reasonably well in the Iran-Iraq War, but ultracautious tactics and a reluctance to press home attacks demonstrated an aversion to incurring casualties. Iraq apparently regarded air power as a resource better preserved for future intimidation than as an asset to be expended in combat for immediate gain. The Iraqi Air Force in 1991, sixth largest in the world, had 24 main operating bases; 30 dispersal bases; 750 to 800 fixed-wing combat-capable aircraft; and 200 support aircraft.[60] Combat planes included 265 ground attack aircraft (some could deliver chemical weapons), 405 fighter interceptors, and 130 combat-capable trainer aircraft.[61]

Aging, short-range, Soviet-designed MiG–21s made up 204 of Iraq's 405 interceptors. Lacking modern avionics and carrying antiquated Soviet air-to-air missiles (the Vietnam-era, heat-seeking AA–2 Atoll), they had little utility beyond point defense. More capable Soviet-built MiG–23s and MiG–25s (the latter having an early generation look-down shoot-down weapon system) accounted for another 99 aircraft. Finally, the Iraqis had 35 MiG–29s that had a modern look-down shoot-down capability and 65 French-built Mirage F–1Es, an aircraft roughly comparable with the U.S. F–4 Phantom, the mainstay of the U.S. Navy and U.S. Air Force forces deployed in Vietnam in the 1960s and 1970s. But all these aircraft were less capable export versions of the supplying nation's frontline counterparts.

The Iraqis further sabotaged themselves by insisting that pilot transition training to the MiG–29s be done in Iraq rather than in the Soviet Union, and the MiG–29 pilots had not yet mastered their machines when the conflict erupted. The F–1E pilots, to whom their French instructors gave mixed reviews but of whom Israeli Air Force analysts cautioned the Coalition forces not to dismiss lightly, were the best in the Iraqi Air Force. Iraqi doctrine emphasized point, not area, defense and the use of fighters rather than AAA and SAMs. Finally, some suppliers compromised Iraq's inventory by turning over their equipment specifications to Coalition intelligence organizations. Altogether, the Iraqi Air Force possessed inferior aircraft and pilots, all in inferior numbers, with weapons systems that were an open book to their opponents.

The Iraqi air defense system suffered similar defects. Most of its French and Soviet radars were no longer top-of-the-line, and equipment performance (and countermeasures) were known to the Coalition. On the plus side, the five-month sitzkrieg had permitted Iraqi technicians to reorient the air defense system from east–west to north–south. Iraq possessed approximately 970 AAA sites with 7,000 AAA pieces (4,000 of them 23-mm or less), most without radar direction, and 16,000 SAMs (not counting shoulder-held and other missiles in the hands of the Iraqi ground forces), many concentrated in the Baghdad area.[62] Altogether, Baghdad was twice as heavily defended as the most heavily defended Eastern European target at the height of the Cold War, and seven times as heavily defended as Hanoi at the height of LINEBACKER II. From the outset of the war until its end, the SAM threat, particularly from hand-held and other missiles with infrared-seeking warheads, and the barrage threat from AAA denied low altitudes to attackers. The air defense system's design optimized AAA to defend altitudes up to 10,000 feet, with radar-guided missiles employed for higher altitudes. The Iraqis relied on aging, though still dangerous, export versions of Soviet SAMs (e.g., SA–2s and SA–3s), more modern Soviet missiles (e.g., SA–6s and SA–8s), and the low-altitude French Roland for strategic air defense. The initial Coalition attacks on the air defense system were designed in part to suppress the radar-directed SAMs in order to minimize their impact on medium and higher altitude operations.

Defects in the air defense command and control system added to Iraqi military shortcomings. The Iraqi-designed Kari air defense system,* built after the Israeli air raid on Iraq's nuclear reactor in 1981, exhibited a flaw typical of many items built for its dictator: overcentralization. Kari divided the country into four defense sectors, with Kuwait added as a fifth. Each Sector Operations Center (SOC) reported to the national Air Defense Operations Center (ADOC) in Baghdad, and each had two to five subsidiary Intercept Operations Centers (IOCs) reporting directly to it. Each IOC, in turn, had several early warning radar posts. Information flowed from the IOCs to the SOCs and thence to Baghdad, but information did not flow laterally to other sector headquarters. The Iraqis designed the system to repel its worst-case threats, its regional enemies — Iran, Israel, or Saudi Arabia — and the system was well-adapted for this use. It simply could not, however, handle the sheer numbers of Coalition aircraft. The SOCs or IOCs made the decision as to which weapons system would engage intruders — SAMs or fighter interceptors. American intelligence judged Iraqi pilots to be inordinately dependent on ground control interception information received from Kari and probably incapable of managing an aerial intercept on their own (though in fact some did so on the opening night of the war after Kari shut down). SAM and AAA batteries used information from Kari to avoid turning on their own fire control radars until the last possible moment, which helped them negate or diminish allied countermeasures. In the prewar period the Coalition detected a daily average of 1,300 to 1,700 individual SAM/AAA and early warning radar emitters.[63]

The trench fighting of the 1980–1988 Iran-Iraq War left Iraqi military leaders with a preference for fixed defenses, not just in the field but for shielding strategic points as well. Much like the French Army before World War II, the Iraqi military suffered from a Maginot Line complex that led it to rely on concrete and steel rather than mobility. This made good sense in the face of regional threats; indeed, many in the U.S. defense establishment in the

*The French, who built the system, named it "Kari," Iraq spelled backward in French.

fall of 1990 argued that Iraqi forces would have to be dug out of Kuwait, foxhole by foxhole. Proportionately speaking, Iraq's enormous number of hardened concrete facilities gave it the most highly protected target base in the world. The array of hardened facilities could conceivably withstand even an enemy employing tactical nuclear weapons. Nevertheless, the U.S. Air Force's monopoly on hard target–penetrating precision-guided munitions made the Coalition's attack feasible, for facilities hardened to withstand blast overpressures nonetheless proved vulnerable to the scalpel-like lancing of laser-guided bombs. Almost all of the Iraqis' air defense command and control facilities rested within massive concrete structures whose distinctive configurations easily identified them to Coalition pilots. The Iraqi Air Force hid its aircraft in 594 hardened air shelters and had numerous additional hardened personnel shelters for their crews.[64]

Scant hours before the outbreak of hostilities, General Schwarzkopf issued the initial operations order (CENTCOM OPORD 91–001) for the upcoming campaign, restating the Coalition's key military objectives as follows:

1. Attack Iraqi political-military and leadership and C^2 [command and control].
2. Gain and maintain air supremacy.
3. Sever Iraqi supply lines.
4. Destroy known nuclear, biological, and chemical production, storage, and delivery capabilities.
5. Destroy Republican Guard forces in the KTO.
6. Liberate Kuwait City with Arab forces.[65]

The strategic air campaign against Iraq would play a direct role in all but the last of Schwarzkopf's objectives.[66]

Coalition air forces pursued five specific objectives related to those of the theater commander:

1. Gain and maintain air supremacy to permit unhindered air and ground operations.
2. Isolate and incapacitate the Iraqi regime.
3. Destroy Iraq's known nuclear biological and chemical warfare capability.
4. Eliminate Iraq's offensive military capability by destroying key military production, infrastructure, and power capabilities.
5. Render the Iraqi Army and its mechanized equipment in Kuwait ineffective, causing its collapse.[67]

The strategic air campaign against Iraq, which had principal responsibility for each of the first four specific air objectives, applied the twelve strategic target sets developed by the Warden group and the CENTAF Special Planning Group within the broader context of President Bush's announced political objectives, the theater commander's overall military objectives, and the specified air objectives.

The Initial Attacks

On the evening of January 16/17 at air bases throughout the Arabian Peninsula, American air and ground crews approached the night's mission with special care and anticipation. Unless Coalition leaders had a last-minute change of mind, the largest multination combat air operation since the invasion of Normandy would begin in a few short hours. This time the crew chiefs, the armaments specialists, and the electronics technicians were especially painstaking in their work. This time each pilot would be flying into battle, not into practice.* The aviators absolutely depended on the guns and missiles, chaff and flare ejectors, radar and communication systems, and all of the other complicated pieces of their military aircraft to function perfectly. In wing meetings, pilots and planning officers fleshed out the final details of the ATO, plotted ingress and egress routes one more time, made attack approaches clear to all, and pointed out details contained in the latest intelligence. This time, too, the pilots paid special attention to their survival kits, sidearms, water bottles, and last letters home. Each shared, in some fashion, an anticipation of possible death and the rush of adrenalin and sweaty palms that accompanies it. Despite their high level of professional training and well-honed skills, most of the U.S. Air Force pilots in the Persian Gulf were about to embark on their first combat mission.

At 0230L (L = Baghdad time) January 17, 1991, thirty minutes before H-Hour (H–30),** at a time of the night when Iraqi radar operators had reached a physiological and psychological low ebb, ten F–117As dropped off the booms of their tankers near the Saudi-Iraqi border, went "stealthy," and headed for key air defense, command, and communications targets within Iraq, including Baghdad. At the same time, Task Force NORMANDY, consisting of nine Army AH–64 Apache gunships led by three Air Force MH–53J PAVE LOW Pathfinders, hedgehopped across the Iraqi-Saudi border and attacked two Iraqi early warning sites at 0239L (H–21) 400 miles west of Kuwait City with rockets and cannon fire. At least one of the two stations got off a warning. AAA fire immediately filled the air over Baghdad, even though no aircraft as yet were overhead. At 0251L (H–9) two F–117A Nighthawks dropped two GBU–27s on the Nukhayb sector air defense center. The Black Hole had designed these two

---

*Throughout the conflict, the U.S. Air Force provided its units in-theater with excellent combat logistics and support. A reliable local airlift composed of Air Force C–141 and C–130 transports carried vital spare parts from logistics facilities in the United States and distributed them to the field, permitting ground crews to turn around F–15Es in twenty-two minutes.

**Local Baghdad time, not Zulu (Greenwich Mean Time), is used here, as it was by the strategic planners. H-Hour was originally set for 0300 Baghdad time, but in the many modifications of the plan the pre–H-Hour attacks were added. It seemed easier to leave them pre–H-Hour than to rearrange the times for all the other units, at least one of which could be counted on not to get it right at the worst possible time. Washington, D.C., on Eastern Standard Time (EST), was seven hours behind Baghdad time.

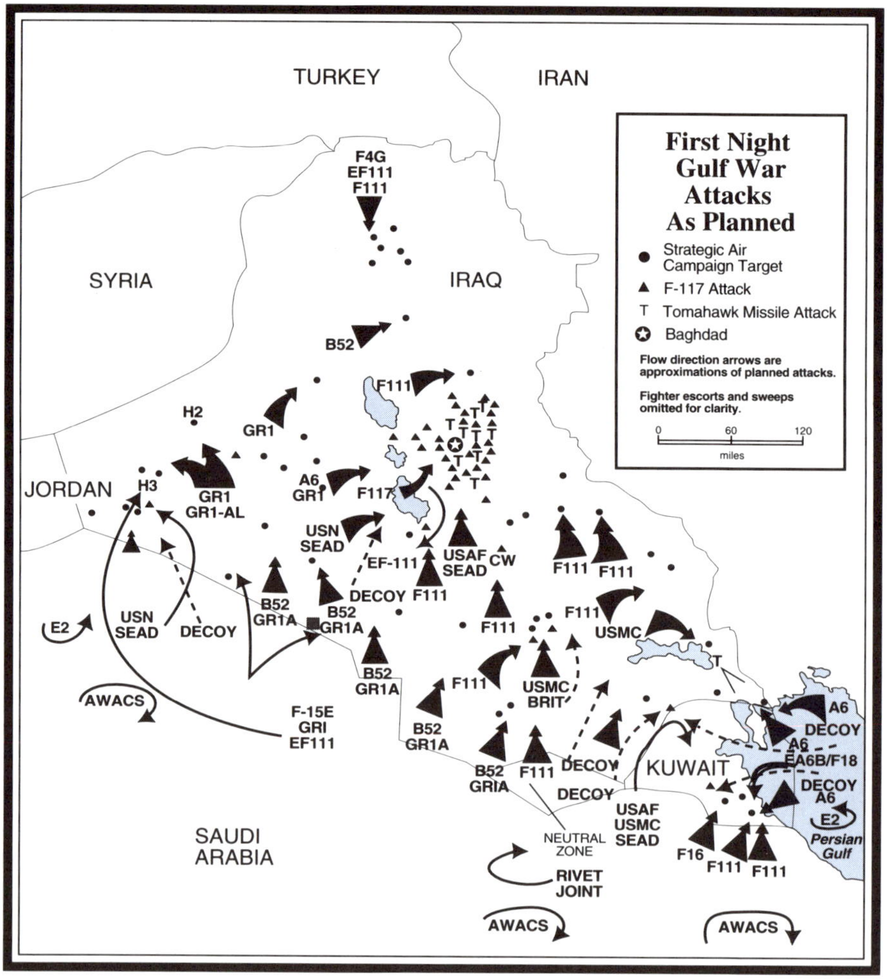

attacks to punch an opening through Iraqi air defenses in order to conceal the subsequent entrance into Iraq of three EF–111A electronic warfare aircraft, meant to assist the F–117s nearing Baghdad,* and a force of nineteen F–15E fighter-bombers headed for fixed SRBM or Scud sites in western Iraq. The execution of the CENTAF Offensive Plan — Phase I, the strategic air campaign

*This electronic assistance for the attacking F–117As was unique in the campaign. The F–117 wing, over the objections of the Iraqi Target Cell, insisted on EF–111A support for this one attack, which was to be the first and only attack against the all-up, undegraded defenses of Baghdad. At no other time in the conflict did the F–117s request or receive dedicated electronic warfare support for a mission. On occasion, of course, the F–117s gained some benefit from electronic warfare aircraft supporting other force packages operating in their proximity.

against Iraq, had begun.

For the Air Force, the war had commenced hours before in the United States with the takeoff of seven B–52Gs bound for Iraq. The big bombers left the ground from Barksdale AFB, Louisiana, at 0735 (EST) on January 16 to begin a 14,000 mile round-trip flight, carrying air-launched cruise missiles (ALCMs) on their first combat test.[68] Meanwhile, the first attack wave of F–117A and F–15E fighter-bombers based in Saudi Arabia, and of Tomahawk land-attack missiles launched from U.S. Navy ships in the Persian Gulf and the Red Sea,* had the task of disrupting the Kari command and control network, hitting permanent Scud launchers directed at Israel, damaging landline communications, and halting the flow of electricity into Baghdad. Southwest of Baghdad, the three EF–111As came through the gap blown in Iraqi radar coverage and headed for their stations south of Baghdad. The F–15Es, flying at very low level, also came through the gap and continued west to strike fixed Scud launchers within range of Israel. At 0305L (H+5) they delivered their weapons, Mk–20 Rockeyes (a munitions canister containing 247 bomblets), on the fixed Scud launchers at low level to insure accuracy. The air planners assumed that when the first bomb dropped on Iraq, Hussein would order Scud launches toward Israel. They scheduled this mission to thwart that action.

Then, at 0300L (H-Hour), two F–117As dropped the first bombs on Baghdad when they attacked communications targets including the so-called AT&T building, which American intelligence fingered as the nerve center for 60 percent of the Iraqi military landline communications capacity. Its loss would disrupt Baghdad's air defenses, which relied heavily on landlines. One of the aircraft also struck North Taji Military Facility No. 1, reputedly a shelter for high-level members of the Iraqi leadership, but the bomb failed to penetrate the bunker. At 0305L (H+5) five more F–117As bombed the sleeping quarters of the Abu Guryahb Presidential Palace (possible quarters of Hussein and the wartime home of the Iraqi General Staff), restruck the AT&T building, and hit the Baghdad SOC at Al Taji, just north of Baghdad.[69]

On the heels of the F–117As, between 0306L and 0311L (H+6 to H+11), Navy Tomahawk land-attack missiles, which had been fired ninety minutes earlier, hit national Baath Party Headquarters and the Presidential Office Complex, known as the Republican Palace. The planners hoped that visible damage to Baath Party Headquarters would demonstrate to the citizens of Baghdad that the Coalition meant to attack the regime, not the people. Damage to the headquarters might also wreck key files and equipment in one of Hussein's organs of internal security. The Republican Palace was not only a psychological target, but was the administrative center of Saddam Hussein's personal govern

*The Tomahawk land-attack missile, a U.S. Navy ship-launched cruise missile containing stealth design elements, featured inertial digital precision guidance and could carry either a single warhead or one capable of dispensing multiple submunitions. It did not mount a penetrating or hard target–killing warhead.

While Pathfinders struck to knock out early warning sites, tactical air-launched decoys (TALDs) (*left*) and drones were used to deceive Iraqi radar during the initial strikes on Baghdad. Through the gap blown in radar coverage came the EF–111s (*below*) and F–15s (*bottom*).

Two F–117A aircraft dropped the first bombs on Baghdad when they attacked communications targets at 3 o'clock in the morning of January 17, 1991.

ment. It included several large office buildings and facilities for the Republican Guard. Additional Tomahawk land-attack missiles, at a cost of $1.1 million apiece,[70] landed on each of the six electrical power plants supplying Baghdad, carrying warheads that reputedly dispensed spools of carbon fibers that festooned the lines and transformer yards at the facilities, causing massive short circuits and automatic shutdowns.[71]

The electrical attacks proved extremely effective. By 0310L (H+10) CNN television reported Baghdad had completely lost commercial power. Few, if any, electrons flowed through Baghdad's power grids for the remainder of the six-week conflict. The loss of electricity also shut down the Iraqi capital's water treatment plants which eventually led to public health problems from raw sewage being dumped in the Tigris River, one of the ramifications of the interconnection of electric power with modern life. Because Kari used commercial electricity as its primary power source, loss of electrical power disrupted this system and reduced its efficiency by forcing defenders to resort to backup generators. The fluctuating output from emergency generators, as air planners knew, would play hob with sensitive electronic equipment and computers. Loss of electricity further hampered daily governmental functions and literally put Iraq's leaders in the dark. In the following week, Tomahawk land-attack missiles and Coalition aircraft reduced every major city in Iraq to the same unhappy situation. At Al Taji, the Iraqis had built a mammoth, centralized logistical plant with maintenance, overhaul, construction, research, and assembly facilities for every item in their military inventory. At 0330L (H+30) twenty

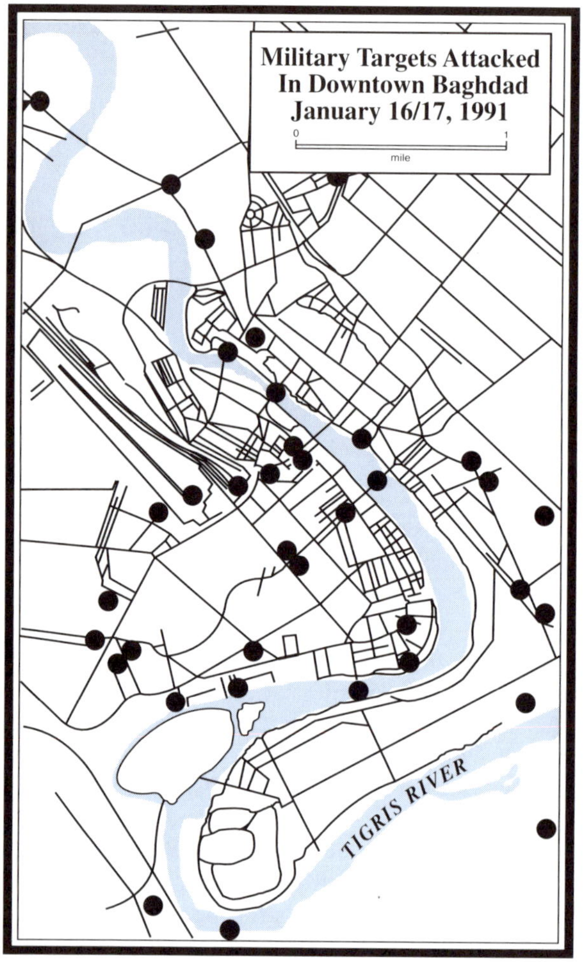

Compare with the map on page 231 showing the Allied bombing of Dresden during World War II.

Tomahawk land-attack missiles crashed into the Scud assembly plant there. Iraq could no longer assemble Scuds. Coalition aircraft would attack Taji repeatedly throughout the hostilities, while striking other key production and support targets.

After the F–117s and cruise missiles came conventional aircraft. From 0355L to 0420L (H+55 to H+80) large numbers of U.S. Air Force, U.S. Navy, U.S. Marine Corps, Royal Saudi Air Force, and RAF aircraft smashed Iraqi air defenses and airfields from H–3, an airfield located in western Iraq, to Ahmed Al Jaber, an airfield in occupied Kuwait. Two packages of aircraft, one a U.S. Navy package from the Red Sea carriers and the other a U.S. Air Force package from the south, pointed directly at Baghdad.[72] The air planners hoped that these

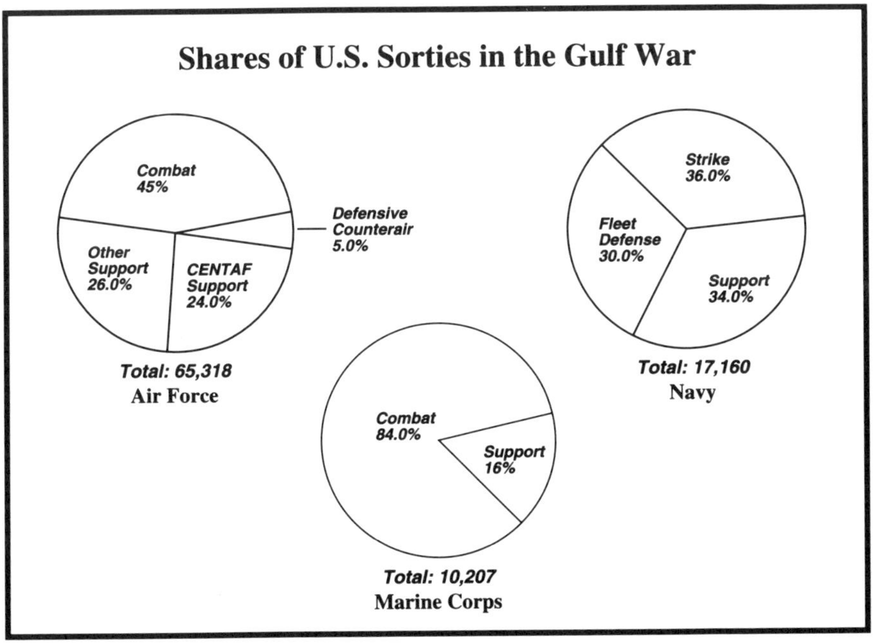

"gorilla" packages would seem particularly threatening to the Iraqi defenders and force them to hurl their air defense resources at them. Air Force ground-launched BQM–34 and Navy tactical air-launched pilotless decoys, which mimicked the radar return of conventional aircraft, further alarmed Iraqi radar operators, many already confused by the absence of central control from Kari.[73] Finally, radar-jamming aircraft accompaning the Coalition planes radiated blanketing electronic emissions that drove Iraqi radar operators to go to full power in an attempt to break through the interference. Then the two incoming Coalition flights revealed their true nature and pounced, for they were a shrewd and devastating ruse.[74]

Instead of bomb-carrying fighter-bombers, they were radar-killing electronic warriors carrying AGM–88 high-speed antiradiation missiles (HARMs) designed to home in on SAM and AAA radar. U.S. Air Force F–4G Wild Weasels alone expended dozens of HARMs in twenty minutes, while U.S. Navy and U.S. Marine Corps F/A–18s fired one hundred for the night. HARMs filled the air over the Baghdad area, the site of more than half of Iraq's SAM and AAA batteries. The stunned Iraqis did not turn off their radars, even when the HARMs fireballed in their midst; as one U.S. Air Force flight leader averred, "the emitters came on and stayed on for the entire flight of the missiles."[75] This deadly surprise not only destroyed many Iraqi radars, it also sensitized radar operators to their own vulnerability. For the rest of the conflict, the Iraqis showed an extreme reluctance to use radar and often chose to launch their SAMs with optical or even no guidance. The initial HARM attack and the F–117 bombings of the Kari system left Iraq's integrated air defense system shattered, opening the country so completely that, within days, Coalition air-to-air tankers regularly operated in Iraqi airspace.

Other nonstealthy aircraft, meanwhile, pummeled Iraqi airfields. At 0400L, B–52s from Diego Garcia, flying at less than 400 feet and behind the cover of the ongoing HARM mission, made shallow penetrations a few miles deep to strike Iraqi forward airfields and runways that might service aircraft waiting to strike at high-value Coalition aircraft, such as AWACS aircraft, or deliver chemical weapons on Coalition troops. The B–52s carried CBU–89s and a special 1,000-lb bomb and delayed fuze combination borrowed from the RAF. Not to be outdone, RAF Tornadoes dispensed JP–233 runway denial munitions that cratered runways at Al Taqqadum, Iraq's largest air base, while Saudi Tornadoes attacked Shaibah in Iraq, a few miles north of Kuwait.

This sudden, overwhelming aerial onslaught against Iraq vividly illustrated the changed roles of U.S. Air Force tactical and strategic aircraft. The large, eight-engined B–52G, once the queen of the strategic nuclear bomber fleet, made shallow, low-level conventional assaults against weakly defended *tactical* targets, while waves of friendly aircraft that had preceded them attracted atten-

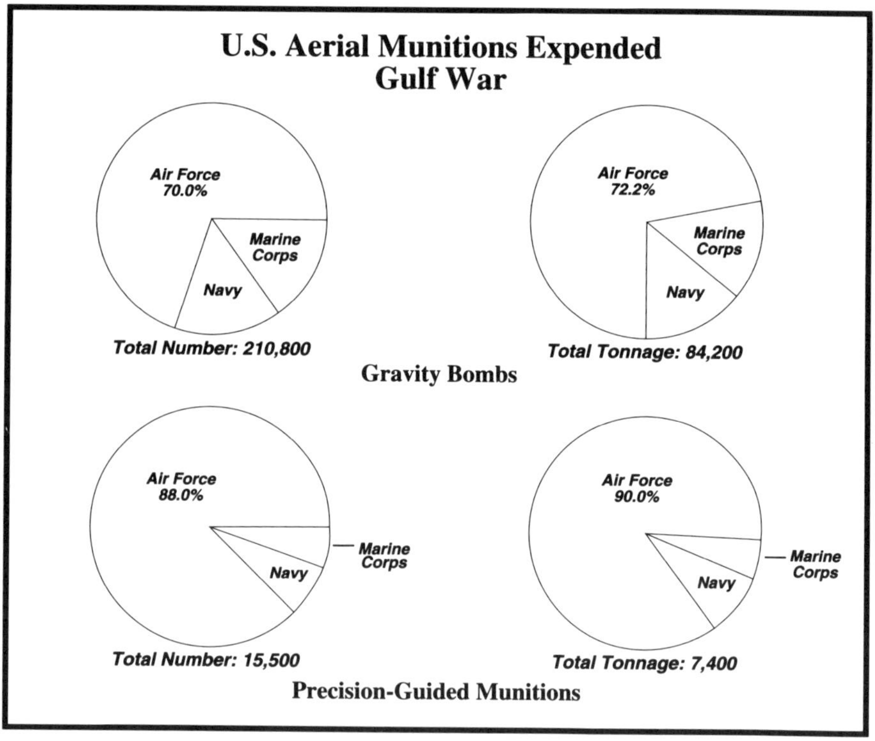

tion to themselves. Light, single-place, twin-engined fighter-bombers flew into the teeth of enemy defenses to strike at *strategic* targets in the Iraqi capital. Late twentieth century technology, both offensive and defensive, had dramatically changed the definition of what constituted a strategic bomber.*

F–117A Nighthawks returned to Baghdad at 0420L, January 17, 1991, where they blasted the Kari IOC at Al Taqqadum airfield, the Baghdad area Kari SOC at Taji, the North Taji Military Related Facility No. 2, and the new Iraqi Air Force headquarters complex. Weather prevented a strike on Baghdad's primary TV transmitter, which the air planners hoped to take off the air, thereby separating Hussein's propaganda machine from any audience still retaining electrical power.

Seven F–117As attacked bunkers holding possible biological warfare material in central Iraq in the last strike of the night. This material required distinctively configured refrigerated bunkers for stability and safety. Because of the potentially horrific consequences of breaching these storage units, such as possibly sending a cloud of airborne anthrax or other toxins over a highly populated area or contaminating the Tigris or Euphrates Rivers, air planners had debated not only the method of attack, but whether or not to hit them at all. Not until late December 1990 did Cheney and Powell acquiesce to Schwarzkopf's recommendation to strike the biological warfare bunkers.[76] The Americans decided to attack the bunkers at an hour before dawn, the day's low ebb for wind and a time that provided for subsequent maximum exposure to sunlight.** First, F–117As would make precision strikes to crack open the bunkers. Then F–111Fs carrying CBU–87s and –89s would make additional passes, to ignite fires and mine the area to prevent salvage.[77] The mission, at 0500L on the morning of January 17, demonstrated once again that Clausewitz's friction in war applies to any military operation, just as does Murphy's Law. The F–117As, followed by sixteen F–111Fs, attacked thirteen biological warfare bunkers. But late-night fog, a common January occurrence near the Euphrates River, obscured the targets and prevented many pilots from dropping bombs or identifying their targets. Of all the bombs carried into Iraq against this target on that night, almost 80 percent could not be dropped because of poor weather.[78]

In sum, the first night's Coalition air attack severed Baghdad from the national power grid, disrupted and heavily damaged key elements of the national air defense network, cut a significant percentage of the state's landline commu-

---

*The distinction between tactical and strategic airframes had begun to blur some twenty-five years earlier in Southeast Asia. B–52s flew many missions against communist troop concentrations in South Vietnam (ARC LIGHT sorties), while U.S. Air Force F–105s (a fighter-bomber designed to deliver nuclear weapons) conducted missions with iron bombs against strategic targets in North Vietnam. The B–52s' strong identification with nuclear strategic war made their later use over North Vietnam during LINEBACKER II politically significant, with a marked effect on world public opinion.

**General Horner and General Glosson had received expert advice that anthrax would decompose rapidly when exposed to sunlight.

nications system, suppressed some Iraqi airfields, and struck the Scud assembly and launching complexes. After the initial punches of one night of aerial action, Iraq's military was well on the way to an eventual knockout. A few hours later, at 0830L on January 17, 1991, the first B–52G air-launched cruise missiles fired in combat exploded in Iraq. The missiles struck not only targets in northern Iraq, beyond the range of Coalition strike aircraft based in Saudi Arabia, but also other targets near Baghdad.[79] These B–52s from Barksdale AFB, Louisianna, completed the longest combat mission ever flown (exceeding thirty-five hours) and demonstrated that the U.S. Air Force could respond forcefully and within hours to any crisis anywhere.

Intense preplanned aerial attacks against strategic targets continued for the next two days. Still later in the morning of January 17, 1991, U.S. aircraft struck the oil refinery at Samawah, halfway between Baghdad and Basra. This inaugurated a series of attacks against refineries and oil storage centers with the goal of eventually limiting fuel to both the civilian and military sectors of the Iraqi economy. These attacks damaged the portion of the industry supplying fuel directly to the Iraqi economy — refineries and oil storage facilities — while sparing oil extraction and export facilities that would permit Iraq to maintain itself and pay reparations after the war. Turning off the Iraqi oil

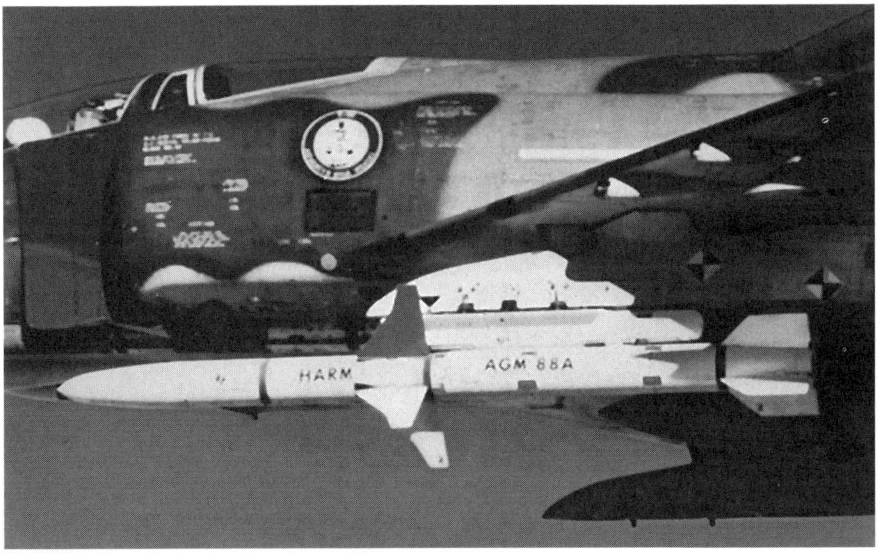

A HARM missile can be seen attached to the wing pylon of an F–4G Wild Weasel aircraft (*above*). HARM missiles homed in on Iraqi radars to virtually shatter these defenses, such as the Fan Song radar (the NATO designation for the Soviet-designed SAM radar tracking and guidance system) (*opposite page*).

supply would in the long run not only severely curtail the military's mobility, it would also deny civilians fuel for heating and cooking, halt internal transport and distribution, and heighten popular dissatisfaction with the regime.

With Phase I (the strategic air offensive against Iraq) of General Schwarzkopf's overall war plan well under way at daylight on the morning of January 17, 1991, Coalition air power simultaneously executed Phase II (suppression of Iraqi air defenses in Kuwait) and Phase III (preparation of the battlefield) of the Commander-in-Chief's plan. Henceforth the three phases ran concurrently, with Schwarzkopf directing shifts of emphasis or effort among them. During the day, A–10As destroyed radar stations in southern Iraq and Kuwait in pinpoint attacks, while B–52G bombers began an around-the-clock attack on Republican Guard targets. Horner promised Schwarzkopf that the big bombers would hit the Republican Guard every three hours for the remainder of the conflict. The B–52Gs, dropping iron bombs from medium and higher altitudes, proved better employed against large diffuse targets, such as dug-in military units, logistics dumps, and spread-out industrial complexes, than against individual ADOCs and assembly plants.*

---

*A SAC briefing, circa 1991, claimed the following circular error probables for SAC bombers: less than 300 feet for offensive avionics systems (radar bombsights and supporting systems) and less

Smoke pours from a burning oil refinery hit by Coalition bombs during DESERT STORM.

B–52s were employed almost exclusively against ground support targets in Kuwait.* Sixty-eight B–52s launched 1,175 strikes against the Iraqi ground

than 100 feet using GPS. These statistics do not state if the figures apply to the B–52 or the much more modern B–1, or to both. Nor do they indicate at what altitude the bombs were dropped. The accuracy of high-altitude drops, like most of those in the Gulf War, is inherently less than it is with low-altitude delivery. Even with GPS, B–52-dropped iron bombs were an order of magnitude less accurate than were precision-guided munitions. Briefing Slide, "Bomber Capabilities: Precision and Mass," in Briefing, "Strategic Air Command," ca 1991, found in SAC Historian's Office, Historians Working Files for CY 1991 Annual History. Bombing computer software or personnel training problems (such as those caused in switching from low- to high-altitude attacks) might further affect bombing accuracy and introduce additional errors in bomb delivery. A shortage of BDA data might compound any delivery problems by allowing missions to repeat mistakes until planners were told what corrections to make.

*SAC figures indicate that B–52s employed in the Gulf War flew 85 percent of their sorties against Iraqi ground units or ground support targets and only 15 percent against "strategic targets." The Republican Guard alone absorbed 37 percent of the B–52 sorties. DoD, *Conduct of the Gulf War*, pp 675–676, presents a fact-filled, but distorted, picture of the B–52s' performance in the war:

> B–52s flew 954 air interdiction sorties against strategic targets (industrial facilities, C^3 facilities, nuclear/chemical/biological facilities, and short-range ballistic missiles), interdiction targets including Republican Guard units as well as fixed installations such as petroleum, oil, and lubricant storage facilities, and railroads. . . .

After the conflict, Iraqi troops in debriefings revealed that they feared the B–52 more than any other Coalition aircraft, mostly from the psychological impact, a result of the concussion of large amounts of detonating ordnance rather than from the actual damage inflicted.

order of battle, compared with 42 F–117s launching only 49 strikes against the same target system.[80] In fact, CENTCOM ground commanders from Schwarzkopf on down so valued the B–52 strikes on Iraqi ground forces that they routinely opposed any suggestion from the air planners of employing the B–52s against targets outside Kuwait.[81] After the conflict, Iraqi troops in debriefings revealed that they feared the B–52 more than any other Coalition aircraft, albeit mostly from psychological impact and concussion of ordnance falling on or near adjacent units than from actual damage inflicted.[82]

Even when the big bombers went deep into Iraq they never employed more than a dozen aircraft with approximately 400 tons of bombs. Still, the B–52s compiled impressive performance figures while flying more than 1,600 sorties. Constituting only 3 percent of the Coalition's total combat aircraft, they delivered 30 percent of the total tonnage of air munitions (72,000 bombs or

---

> B–52Gs flew 527 BAI [Battlefield Air Interdiction] sorties striking armor, mechanized, and infantry units with a variety of general purpose and cluster bomb munitions.

The inclusion of Republican Guard "armor, mechanized, and infantry units," not to mention Iraqi ground forces logistics dumps (fixed installations), in the "strategic" category stands the traditional definition of strategic targets on its head. If these targets are placed, where they should be, in the BAI or ground support categories, one will quickly arrive at the 15 percent to 85 percent ratio of strategic to tactical targets previously noted. It would appear that the authors of *Conduct of the Gulf War* were loath to admit to the B–52s' overwhelming *tactical* role.

cluster bomb units totaling 27,000 tons of munitions), most from high altitude, using radar ground mapping for target acquisition.[83] Because of their long flights from the United States, from Diego Garcia, and from Spain and England, they consumed 25 percent of the theater's air-to-air tanker fuel offloadings and tied up 40 percent of the U.S. Air Force's most modern tankers. Nonetheless, only 4,000 tons of the bombs (fewer than 1 in 6) dropped by B–52s during the conflict were released over targets identified by the Black Hole as "strategic." B–52s going into Iraq also required the same large packages of supporting aircraft that they did when they earlier had participated in LINEBACKER II against the Hanoi-Haiphong area of North Vietnam. In contrast, the F–117A Nighthawks flew 1,296 sorties (2 percent of the total attack sorties) yet struck 40 percent of the strategic targets. Only F–117s hit targets in downtown Baghdad;[84] they delivered more than 3,000 tons of precision-guided munitions and claimed to have placed 80 percent of the bombs released* on target while exposing no support aircraft to enemy fire.[85]

Since the middle of the Great War, in 1916, aircraft designed to bomb strategic targets (high-value objectives usually located deep within the enemy's heartland) had to carry to their destination the weight of an adequate bombload, the weight of fuel, and the weight of defensive systems, and then they had to return to base. The only aircraft possessing the requisite payload, radius, and survivability was the heavy multiengined bomber. For the next seventy-five years, the heavy bomber and strategic bombing were synonymous. All this changed in the first two months of 1991. The U.S. Air Force carried strategic bombing aircraft on its books in the Gulf War (primarily the F–117A, the F–111F, and the F–15E) as fighter-bombers — an aircraft of fighter size** with

---

*Not all F–117 sorties released their bombs. Likewise, not all F–117s dropped their bombs on their primary targets. Orders encouraged pilots to seek secondary targets when they could not strike their primaries. As a result, the bombs actually impacting on a target may not necessarily accurately reflect the number of precision munitions intended for that target by mission planners. In addition, to ensure the necessary damage to a target, planners would often assign multiple precision weapons to the same aiming point, either because of its extreme protection (i.e., its hardening) or because of the absolute necessity of knocking it out. Two or more aircraft assigned to a single target lessened the likelihood of failure due to enemy action or to Murphy's Law.

**The F–117A, F–111F, and the F–15E have takeoff weights (aircraft, gas, bombs, missiles, and crew) of 52,500, 100,000, and 73,000 pounds, respectively. The B–29 (the largest World War II operational bomber) and the B–52G had takeoff weights of 140,000 and 488,000 pounds, respectively. For all of 1944, B–17s and B–24s of the U.S. Eighth Air Force in Europe averaged only 13.1 percent of their bombs within 500 feet of the aiming point, when bombing under "conditions of good visibility." Rpt, HQ 8th AF, Ops Anlys Sec, "Report on Bombing Accuracy, Eighth Air Force," Apr 20, 1945. During the Gulf War precision weapons achieved circular error probables of far greater accuracy at several times a rate of 13.1 percent. The significance of this quantum leap in accuracy and its effect on planning and operations cannot be overstressed. Critics of the use of precision munitions in the Gulf War, especially the authors of a Government Accounting Office report, have exploded the accuracy claims of U.S. Air Force contractors and the initial exaggerated U.S. Air Force claims, but they fail to understand this fundamental change in accuracy and its implications. (Rpt, "Operation DESERT STORM: Evaluation of the Air War," GAO/PEMD–96–10, Jul 1996) Even if the claim of "one bomb, one target" or an 80 percent hit rate are halved, a 40 percent hit rate to within a few dozen feet of the aiming point is a revolutionary advance when compared to

An Air Force KC–135 tanker and a U.S. Navy A–6 refuel a group of Navy A–7s during DESERT STORM.

the principal mission of bombing enemy targets rather than dog-fighting. These fighter-bombers employed new technologies, such as air-to-air refueling that extended their range and reduced the weight of the fuel load required, to meet classical strategic specifications.* Precision-guided munitions multiplied the effectiveness of an aircraft's bombload by increasing the amount of high-explosives delivered exactly on target — an aircraft with a small bombload of precise weapons could equal the effectiveness of a large aircraft with a large bombload of dumb, or conventional iron, bombs by eliminating the need to carpet a target area with bombs to ensure at least one or two hits. Indeed, as Gulf War strike records showed, *a single* F–117 with *two* laser-guided bombs could in some situations achieve the same degree of target destruction that in World War II had required *108* B–17s dropping *648* bombs. These changes also enabled the F–111F and the F–15E to hit all but the most heavily defended strategic targets. Cloaked by stealth, the radical F–117A could strike *any* target. The twin-engined strike aircraft thus replaced the traditional heavy bomber as the "strategic bomber" of choice.**

---

all previous experience in conventional bombing accuracy.

*Within a few days of the conflict's opening, unarmed U.S. Air Force tanker aircraft began to cross into Iraq to fuel aircraft for deep missions or support aircraft in trouble. By the end of the conflict, tanker refueling operations over Iraq had become routine.

**In the unlikely event of a full-scale, nuclear exchange with Russia, China, or the Ukraine, and whatever the role of intercontinental ballistic missiles, U.S. nuclear-capable bombers (B–52s and

At random intervals throughout the first and second days, more Tomahawk land-attack missiles roiled the pot and strained Iraqi nerves, hitting power plants, oil pumping stations, and the Ministry of Defense in Baghdad. The second night, January 17/18, 1991, F–117As hit more ADOCs and targets in the Baghdad area, including the two modified Il–76 Iraqi AWACS aircraft, Iraqi Air Force headquarters, Taji Command Bunker No. 2, the Ministry of Defense Computer Center, Iraqi Intelligence Service Headquarters, and a VIP bunker — the possible location of Saddam Hussein and other leaders. The strike on Iraqi Intelligence Service Headquarters was intended to destroy key files and computer equipment and discourage its personnel. Air planners hoped that the loss of files and morale would hamper the ability of this key organ of the regime's internal control apparatus to perform its duties. The attack showed the citizens of Baghdad that Hussein could not protect his government's most sinister and powerful organizations from Coalition attack, presumably further weakening the hold of the regime on its populace. Also on this second night, two B–52Gs conducted a low-level raid on Al Sahra Undergraduate Pilot Training airfield north of Baghdad. The base also housed the Iraqi Air Force Academy and was within sight and sound of the city of Tikrit, Saddam's hometown and base of the clan that supplied the bulk of his political support.[86] The planners hoped this raid would bring the war home to Saddam Husein's closest supporters.

That same evening Marine air and Air Force F–15Es using conventional bombs commenced attacks on highways and bridges in the Basra region. Iraq's second largest city, Basra, served as a major supply and reinforcement terminus for the Iraqi Army in Kuwait, and the bridge-busting campaign sought to isolate Iraqi forces in Kuwait from their logistics bases in Iraq. More importantly, the bridge-busting was to ensure that Iraqi ground forces, especially the Republican Guard, could not retreat from the theater. Generals Schwarzkopf and Powell wanted the Republican Guard fixed in place so that subsequent air and ground operations might destroy it. Soon Coalition aircraft went after rail bridges as well. But without precision-guided munitions, these and later bridge attacks failed, forcing diversion of U.S. and RAF aircraft equipped with precision-guided munitions to this task. By February 6 the Air Staff informed Secretary Cheney that Coalition bombing had destroyed twenty-two of twenty-four critical highway bridges, a feat unparalleled in air power annals and one made possible entirely by precision-guided munitions.[87] As in LINEBACKER I, the destruction of so many key bridges in such a short time applied a chokehold to the logistics system supplying Iraqi troops in Kuwait. Coalition aircraft revisit-

---

B–1s) might bomb targets deep within those countries and return. Such nuclear missions likely would suffer heavy casualties and could not be sustained for a lengthy period. But the hydrogen bomb is both the ultimate area-bombing attack and ultimate smart weapon. Even if it misses by a few hundred feet, one device is almost guaranteed to deliver more than enough destructive energy on the aiming point (and on everything else in the surrounding area).

ed the bridges whenever necessary, and by the end of the conflict they had dropped forty-one of fifty-four rail and highway bridges between Baghdad and Basra, as well as thirty-two pontoon bridges the Iraqis built as replacements.[88]

At the same time, on the night of January 17/18, 1991, Joint Task Force PROVEN FORCE, consisting of approximately one hundred U.S. Air Force aircraft, none precision-guided munition–capable,* began air operations from the Republic of Turkey. These aircraft remained under the operational control of EUCOM, but Schwarzkopf maintained tactical control. Striking from the northwest, they compounded the Iraqis' air defense problems and placed all of that nation in reach of Coalition air power.** Daylight attacks on January 18, 1991, continued the assault on the Iraqi air defense system (after forty-eight hours of aerial attacks, the number of active Iraqi SAM and early warning radars detected operating by Coalition forces declined by almost 70 percent, from 700 to 200 per day[89]). Other attacks continued against Iraqi air base runways and facilities, blasted chemical and biological warfare bunkers, and further damaged Iraqi communications systems. Tomahawk land-attack missiles rained down on the Ministry of Defense, power plants, refineries, and the Abu Guryahb Presidential Grounds. But bad weather now intervened. On the night of January 18/19, weather completely disrupted an F–117A attack on Baghdad and the nearby nuclear facilities at Tuwaitha. A second F–117A mission to Baghdad also dropped no bombs because of unsuitable weather.

Altogether, in the first two days, Coalition air attackers struck 169 of 298 potential strategic targets, rendered air defenses ineffective (allowing CENTAF to order all strike aircraft to operate with relative impunity at medium and high altitudes***), drove the Iraqi Air Force from the skies (Iraqi night sorties ceased

---

*These were F–15Cs, F–16s, F–4Gs, and F–111Es, plus electronic and tanker support aircraft. At the end of the conflict on February 25, four F–4Es of the 3d TFW (three were fully mission-capable) joined PROVEN FORCE. These precision-capable F–4Es from the Philippines flew only two unsuccessful sorties. Trip Rpt, Maj Gen G. B. Harrison, JTF PROVEN FORCE, AF Center for Studies and Analysis, Mar 7, 1991, in GWAPS files. Initially, to reduce coordination problems and the chances of fratricide (engagement of friendly aircraft by other friendly aircraft), JTF PROVEN FORCE confined its operations to north of the 35th parallel, while planes based in Saudi Arabia operated south of that line. CENTAF air planners supplied JTF PROVEN FORCE's targets but did not include them in the CENTAF daily ATO. JTF PROVEN FORCE operated somewhat autonomously, but it remained handicapped by its lack of precision-guided munition–capable aircraft. Lack of precision aircraft either forced precision-guided munition aircraft from Saudi Arabia to undertake difficult missions to northern Iraq or else that some important targets were not attacked. (The Turkish Air Force did not participate in operations against Iraq.)

**The Turkish government, in part because of confusion within its military General Staff, did not grant permission for U.S. Air Force air strikes against Iraq from Turkish territory until after the start of Coalition operations from Saudi Arabia.

***B–52s, for example, began to conduct strikes from 31,000 feet and above. F–15Es, F–16s, and F–111Fs all switched to medium and high altitudes. The change in attack altitudes saved allied lives by placing attacking aircraft beyond the effective range of low-level Iraqi AAA, but it further increased the effect of poor weather in that the increase in strike altitudes meant higher minimum weather conditions. Clouds cause mission cancellation more often at 10,000 feet and above than they do at 2,500 feet. Higher altitude also adversely affected bombing accuracy by requiring a higher bomb release altitude. The longer the distance a bomb must travel, the greater the magnification of on the night of January 18/19, 1991), established air superiority, shut down the Iraqi electrical grid, and began to dry up the internal oil supply of one of the world's largest petroleum producers. Coalition aircraft erased fifty years of Iraqi development in as many hours.[90] Because of weaknesses in prewar intelligence, however, significant Iraqi biological warfare, chemical warfare, and SRBM capabilities remained. Furthermore, Iraqi troops in Kuwait had not yet suffered significant damage. Their time would come.

The Strategic Air Campaign Concluded

In the first two days the Coalition devoted approximately 2,400 combat sorties to strategic targets in Iraq; from this peak the total dropped daily. By day five, January 21, 1991, Coalition aircraft flew 550 strategic sorties. By January 25, 1991, the number of tactical combat missions directed against Iraqi forces in Kuwait and in direct support in southern Iraq equaled the total for strategic sorties. By January 29, 1991, the thirteenth day of the war, the number of sorties directed at strategic targets within Iraq had declined to 250 per day, of which Coalition air forces expended approximately 75 on Scud-related targets, while the number of tactical sorties increased to some 800 per day and continued to rise. This strategic daily total did not include approximately one hundred sorties flown from Turkey. In February 1991 sorties directed toward Kuwait averaged more than 1,200 per day.[91] For the last two-thirds of the Gulf War, the Coalition conducted the strategic air campaign at a minimum level.

Not unexpectedly, Iraq's Scud ballistic missiles forcefully intruded into Coalition military and political calculations. Beginning on the night of January 17/18, 1991, a volley of seven SRBMs exploded in the nearby state of Israel.[92] Attacking this neutral power with the cynical purpose of provoking a knee-jerk military retaliation, Hussein hoped to exploit the decades-long history of mutual antipathy between the Jewish nation and its Arab neighbors. He hoped to use an Israeli counterstrike as a wedge to split some or all of the Coalition's Arab member states from the rest of the alliance, thereby disrupting or even ending the Coalition's attacks. Whatever the validity of that calculation, Coalition leaders could take no chances, especially since their intelligence services credited Iraq with chemical warheads for its Scuds. For the first time since the founding of Israel, American combat units (U.S. Army Patriot air defense missile batteries) deployed on Israeli soil, subsequently firing Patriot missiles at incoming Scuds. The stationing of U.S. armed forces within the state of Israel and their use against Arab forces set a precedent that few in the Middle

---

any error. Lack of high-altitude training for the crews and delays in reprogramming onboard aircraft bomb-aiming software to compensate for the new height variables also contributed to bombing inaccuracies. In the first ten days of the war, foul weather forced CENTAF to cancel 15 percent of all attack sorties. DoD, *Conduct of the Gulf War*, p 169.

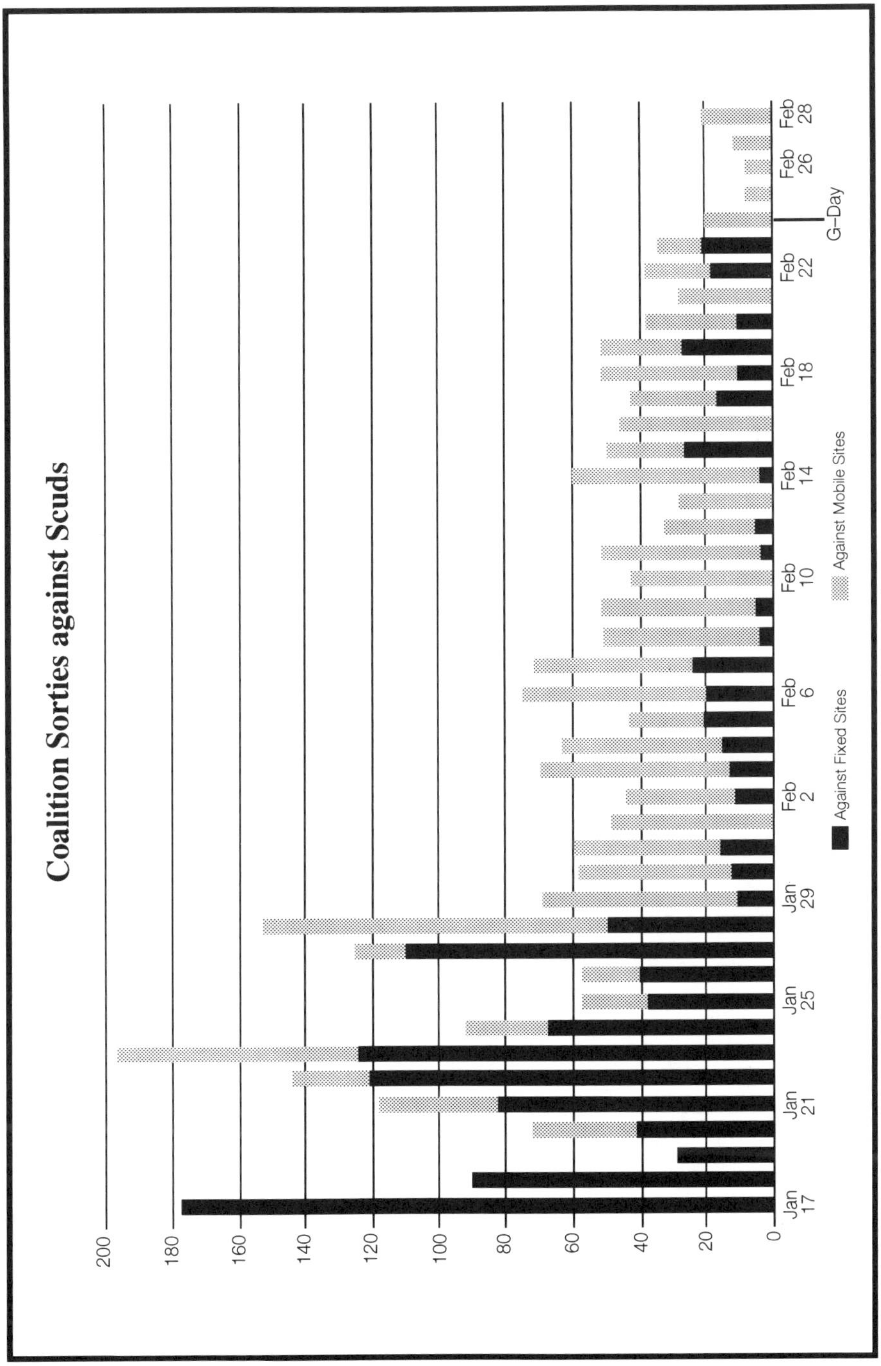
Coalition Sorties against Scuds
0
20
40
60
80
100
120
140
160
180
200
Jan 17
Jan 21
Jan 25
Jan 29
Feb 2
Feb 6
Feb 10
Feb 14
Feb 18
Feb 22
G–Day
Feb 26
Feb 28
Against Fixed Sites
Against Mobile Sites

East failed to notice: when called upon, the United States would physically defend the Jewish state. Three days after the first Scud attack on January 21, 1991, senior U.S. officials, such as Deputy Secretary of State Lawrence S. Eagleburger and the CENTAF second-in-command, Maj. Gen. Thomas R. Olsen, hurried to Tel Aviv, where their impressions confirmed intelligence information that the Israelis were preparing a military response. The obvious Israeli military preparations, of course, increased apprehensions among U.S. leaders that Israel might counterattack, possibly fracturing the Coalition. This pressure on Washington which was immediately passed to CENTCOM translated into a doubling of insistence by senior American commanders in Riyadh that the U.S. Air Force suppress the Scud firings. The demand for this aerial action and for a change in priorities to emphasize Scud targeting, General Horner subsequently stated, generated more intense pressure on him than that occasioned by any other issue.[93]

The political dimension lent urgency to the anti-Scud air campaign and caused the Coalition (much to the frustration of Black Hole planners) to divert air resources from other strategic targets to attack Scud manufacturing facilities and go on search and destroy missions against mobile Scud launchers.[94] Rarely has one nation worked so hard to prevent another from entering a war as its ally. By January 25, 1991, the daily rate of Iraqi missile launches declined from a peak of ten (the average was four), to just one. Scuds launched toward Saudi Arabia did not create the same political problems, but the threat of chemical warheads, which never materialized, occasioned much anxiety and time-consuming effort in the donning of protective clothing and the scurrying to slit

On the night of January 17/18, seven Scud ballistic missiles exploded in nearby Israel. One explosive ordnance disposal team recovers the remains of a Scud (opposite page, *below*). To counter the Iraqi Scud effort, LANTIRN-equipped F–16s and F–15Es were diverted from the strategic air campaign to a Scud hunt (*above*).

trenches as Scud warnings arrived. Altogether, with its increased priority, CENTAF devoted 15 percent of its strategic air effort against Scud manufacturing, assembly, storage, and launchers, with a high of 200 sorties on January 21, 1991. Scud launches further forced CENTAF to hold a squadron of F–15Es and LANTIRN-equipped F–16s on anti-Scud alert at all times. This counter-Scud effort represented a significant diversion of force from a strategic air campaign already greatly attenuated by the increasing concentration on tactical operations against Iraqi ground forces and related targets in the KTO as preparation for the ground offensive commenced. The Iraqi ballistic missiles forced a diversion of aerial resources, but Coalition leaders thwarted the main Iraqi goal: Israel stayed out of the war and did not exercise its unquestioned right of retaliation. The Coalition remained intact, and international contempt for the Hussein regime grew.

On day five, January 22, 1991, CENTAF discontinued airfield suppression

The Coalition had destroyed or severely damaged 375 of Iraqis 594 hardened air shelters and majority of Iraq's hardened maintenance hangers (double-sized, hardened air shelters). An aerial view of the shelters is shown *above* as is the extent of destruction obtained (*next page*).

attacks and switched to using penetrating GBU–10 and GBU–24 precision-guided munitions on the hardened aircraft shelters, now concealing the bulk of the Iraqi Air Force. This effort was intended to prevent an Iraqi Air Tet or, more appropriately, an Operation BODENPLATTE–like response (as when, in a last-ditch effort near the end of World War II, the *Luftwaffe* lashed out at Allied airfields in northern Europe on January 1, 1945). Swing-wing F–111Fs, which could carry four laser-guided bombs to the F117As' two, served as the principal airfield-busters. F–117s continued to fly against Baghdad and other more heavily defended targets where their stealth features were critical for success. For three days the Iraqi Air Force stayed underground instead of aboveground — its entombed airplanes, one after another, dissolving into fireballs. It mounted not a single fighter sortie on January 25, 1991, as if it were standing down to assess what was happening. By the end of the conflict, the Coalition and had destroyed or severely damaged 375 of the Iraqis' 594 hardened air shelters.[95] Shelter strikes also destroyed most of Iraq's hardened maintenance hangars

(double-sized, hardened air shelters). Their loss cost Iraq vital spare parts, specialized ground equipment, and unique shop equipment that would greatly delay reconstitution of a fully operative air force.

For the Iraqi Air Force and Saddam Hussein, if confronting Coalition air forces aloft meant certain destruction, so did staying in protective holes. Only the option of escape remained. As early as January 21, 1991, the Iraqis had sent twenty-five large aircraft, including fourteen looted from Kuwait, to Iran, ostensibly Iraq's mortal enemy. Its other neutral neighbor, Jordan, apparently refused to accept the aircraft. In any case, routing the aircraft east to Iran sent them away from Coalition air operations, whereas directing them west to Jordan would have put them squarely in the path of American F–15Cs protecting ongoing Coalition air combat operations. On January 26, 1991, first-line Iraqi combat planes began to flee to Iran. By January 28, 1991, nearly eighty aircraft rested on Iranian airfields; more than forty additional aircraft, avoiding barrier air patrols hastily established by the Coalition, arrived by February 10, 1991.[96] The Iranians were the big winners in this affair. If Iraq somehow won the war or gained a political stalemate, Iran gained its favor (for whatever that was worth); if Iraq lost the conflict, Iran gained its enemy's air force. Subsequently, the hard currency–starved Soviet Union and its successor states were easily persuaded to sign maintenance and training contracts for Iran's new Soviet-made inventory (the French proved less helpful concerning the F–1s). Included in the 148 interned Iraqi aircraft were the following squadron-sized blocks: 24 Mirage F–1 interceptors, 24 Sukhoi Su–24 Fencer strike aircraft, and 40 Sukhoi Su–22 Fitter–H fighter-bombers.[97] As late as September 1997 the Iraqis were pleading in vain to the United Nations for help in regaining these planes.[98]

At the end of eleven days, on January 27, 1991, CENTAF declared it held *air supremacy*, as distinguished from the *air superiority* it had seized in the first moment of the conflict. The announcement confirmed the ongoing shift of the three-quarters or more of the Coalition air effort to the tactical bombing of Iraqi ground force targets in Kuwait. With a rump effort remaining of approximately 175 daily strategic sorties, CENTAF strategic planners attempted to keep the pressure on the Iraqi regime. Planners could direct about a hundred precision-capable U.S. Air Force aircraft plus some Navy A–6s* and a few RAF Buccaneers equipped with PAVE SPIKE target designators used to buddy lase for RAF Tornadoes until the Tornadoes received their own British-made, thermal-

*As of February 1, 1991, the U.S. Navy had ninety-six A–6Es onboard carriers in the Arabian Gulf and Red Sea, while the U.S. Marine Corps had twenty A–6s at Shaik Isa. During the conflict A–6s flew 2,617 strikes of all types, including 307 precision strikes. The Navy expended 149 precision-guided munition strikes against the Iraqi ground order of battle (tank plinking) and thirty-nine against naval targets, leaving 119 strikes (an average of three a day) against all other targets. GWAPS vol 5, *Statistics*, tables 187, "Strikes by AIF Categories," 193, "PGM Strikes by AIF Categories," pp 436, 532. Throughout the war, naval aviation forces labored under shortages of precision weapons, insufficient ship-based tankers and electronic warfare aircraft, and the inability to provide more than sequential (as opposed to simultaneous) carrier strike packages.

imaging laser designation pods. As the strategic air campaign progressed, support of the ground effort and Scud diversions siphoned off more and more of these aircraft: a squadron of F–15Es, twenty-four aircraft, was assigned to Scud targets and Scud alert; a second squadron of F–15Es was directed to perform tank plinking.* By February 5, 1991, two-thirds of the F–111Fs were committed against targets in Kuwait. On January 25, 1991, to recoup some of this loss of strategic sorties, CENTAF planners required the F–117A wing to begin flying three missions each night. Six additional F–117A Nighthawks arrived from the United States to complement this effort.[98]

During the second week of the strategic air campaign CENTAF began to bomb, not just for *effect* (which initial planning had emphasized), but for *destruction* (to follow up the initial disruption). Squadron-sized or larger missions of precision-guided munition aircraft pummeled the Tuwiatha Nuclear Center, the Latifaya solid-propellant plant, chemical warfare bunkers, and hardened air shelters. Single or paired aircraft repeatedly attacked TV and radio transmitters throughout Iraq, while other single aircraft used their sharpshooting skills to wreck telephone exchanges in the lesser cities and also to deafen the Iraqi National Security Agency by toppling its monitoring sites. The last scheduled attacks on electrical generating facilities, directed against nine smaller facilities, occurred during this period.

In this February 6–14, 1991, time, Coalition air forces struck 24 Iraqi airfields; numerous chemical warfare bunkers; 11 chemical research, production, and development targets; 16 oil targets (including 10 storage facilities); and 53 separate communications targets. After F–117As smashed SAM sites, B–52Gs based in Diego Garcia** began to hit the Taji logistics center, especially the armored fighting vehicle and missile repair facilities. Coalition aircraft struck this large Iraqi military industrial support complex repeatedly.[99] Starting on the night of February 11/12, 1991, F–117As began intensively reattacking strategic Iraqi leadership targets in and around Baghdad. The first wave hit the Iraqi Intelligence Service Headquarters, the Ministry of Information (an organ of internal control), and the large complex that formed Baath Party Headquarters. The second wave struck the Ministry of Defense, the Abu Guryahb Presidential C^3 Bunker, and the Ministry of Information. The first wave over Baghdad on the night of February 12/13, 1991, unloaded its weapons on Iraqi Air Force Headquarters, the Ministry of Defense, the Taji Governmental Command Bunker, and the Baghdad Conference Center, a prestige project

---

*Tank plinking, a term coined by U.S. Air Force pilots, described the act of destroying individual Iraqi tanks, usually located under cover in revetments or bunkers. Using the thermal signatures of the tanks to locate them and subsequently illuminating them with a laser, the pilots loosed 500-lb GBU–12 laser-guided bombs to destroy them. Armor adherents took great umbridge at the term tank plinking, which, of course, ensured its widespread use among aviators!

**During this period, most B–52Gs based in Diego Garcia continued to strike Iraqi ground targets in Kuwait. B–52Gs based in the Gulf region were directed exclusively at these targets.

F–117s destroyed this transport aircraft and small jet, suspected to be Saddam's escape from Baghdad.

built to host the conference of the world's nonaligned nations. The second wave bombed the international TV and press buildings (employed for propaganda broadcasts), the Baghdad Director of Military Intelligence Headquarters, Baath Party Headquarters, and Iraqi Intelligence Service Headquarters. Wave three sustained the pressure, assailing two Baghdad bridges spanning the Tigris River, Baghdad Baath Party Headquarters, Iraqi Intelligence Service Headquarters, the Baghdad Director of General Security Headquarters, the Baghdad Director of Military Intelligence Headquarters, the Baghdad Presidential Residence and Bunker, Camp Taji Presidential Retreat (one of Hussein's favorite residences), and the Al Firdos District Bunker. During these two nights, 90 percent of the bombs from five attacking waves of aircraft struck their primary targets.[100]

The attack on the Al Firdos District Bunker would prove to be the most controversial raid of the war. Two F–117As dropped one bomb each on the bunker: one clipped the outside of the facility; the other penetrated the bunker and exploded inside, killing as many as 300 Iraqi civilians who had taken shelter on its upper floor.* Earlier in the month, air planners had received

*The Al Firdos District Bunker was enclosed in high fences topped with barbed wire and clearly

information that the Al Firdos District Bunker had been "activated and its communications capabilities were being used by senior Iraqi military officials."[101] That made the bunker, previously off-limits as a probable civilian bomb shelter,* a legitimate target of war. After some deliberation, air planners added it to the list of important leadership targets.[102] If the bomb had hit the bunker's fuel tanks as intended, no one would have survived.[103] Water from ruptured internal water tanks, however, flooded the basement, where intelligence activities occurred. The real impact was political, for coverage of the incident by the international press, aided by a spokeswoman from the Ministry of Information, ignited a firestorm of criticism of the Coalition bombing campaign throughout the world. U.S. leaders feared that another instance like this one might undermine both international and U.S. domestic support for the war at a time when Coalition forces clearly had the overwhelming advantage. As a consequence, Schwarzkopf, on instructions from General Powell in Washington, informed CENTAF that it could no longer strike any targets in Baghdad, including the Tigris River bridges, without his explicit permission.[104] This order amounted to a mini–bombing halt on the Iraqi leadership and its internal instruments of control; it abruptly canceled the air planners' highest priority strikes. From February 13, 1991, to the night of February 22/23, 1991, Baghdad was off-limits. Coalition raids on NBC, Scud, and military support targets at the city's periphery, however, continued to shake the city's inhabitants.

In retrospect the attack on the Al Firdos District Bunker may have had a historical impact far greater than merely limiting strikes against Baghdad. Much like the Doolittle Raid on Toyko in April 1942, which helped put the Midway Campaign in train, or the initial small RAF raid on Berlin in 1940, which led Hitler to switch targeting priorities from industry and the RAF to London and other British cities, the Al Firdos mission may have sparked a dramatic and disproportionate change in the enemy's assessment of the wartime situation. Although all available U.S. Air Force evidence demonstrates that the U.S. strike on the high-ranking civilian families in the bunker was inadvertent, Saddam apparently drew the opposite conclusion. At the same time that Saddam Hussein felt the stick of Al Firdos, a visiting Soviet official offered a carrot — Moscow's good offices in brokering a peace. From mid-February 1991 onward the diplomatic tempo of the conflict changed radically as the Iraqi

---

marked as off-limits to the general public. Apparently the civilian occupants of the facility were family members belonging to either high-ranking Iraqi government officials or the intelligence personnel working in the lower level of the bunker.

*Baghdad contained twenty-five bunkers similar to the Al Firdos District Bunker. Before the war, the purpose of these structures, given the Iraqi proclivity of placing many important functions in hardened facilities, confused the Special Planning Group because these buildings appeared to have military capabilities beyond those of an ordinary bomb shelter. For instance, the Iraqis had painted them in camouflage colors. Air planners, therefore, carried them on their prewar target lists. Shortly before hostilities, air planners learned that intelligence could not confirm any function for these structures other than as civilian bomb shelters. Therefore, the air planners, in accordance with U.S. policy, placed them off-limits to bombing.

leader finally sought to extract himself from his predicament while the Coalition kept its terms as stiff as possible and Washington increased its pressure on Schwarzkopf to launch his ground offensive. All this, perhaps, spurred by one precision weapon.

From February 15 to 23, 1991, as the furor over Al Firdos simmered, the strategic air campaign pressed on, striking at targets throughout Iraq including 17 airfields (to keep the Iraqi Air Force suppressed), 17 nuclear and chemical targets, 31 military support facilities, 14 C^3 facilities, 13 highway bridges, and mobile Scud launching sites.[105] On the night of February 22/23, 1991, F–117As returned to Baghdad, striking leadership targets, the Special Operations Headquarters, and some intelligence headquarters, to aid in the imminent ground operation. The next night, when Coalition ground forces began an almost unopposed advance into Kuwait and southern Iraq, simultaneous strategic air missions struck (in one of the worst periods of weather during the war) Iraqi airfields housing ground attack and chemical-capable aircraft near the front and any remaining highway bridges. With targets in Baghdad approved, F–117As attacked the Baghdad bomb assembly plant, chemical warfare bunkers, and the Iskandariyah ammunition plant south of the capital. Seven bombs hit Baghdad Special Security Headquarters and three hit Iraqi Baghdad Regional Intelligence Service Headquarters.[106] The next night F–117As released their laser-guided bombs on the guard facilities of Abu Guryahb Presidential Complex, the Baghdad Special Security Services, and military support targets.

On the night of February 25/26, 1991, for the only time during the campaign, weather completely halted F–117A operations. In the final two nights of the conflict, the strategic campaign continued to strike leadership targets and important industrial facilities. To increase the pressure on Baghdad, strategic planners sent the F–111Es southward from Turkey both nights to hit the Taji complex. On the night of February 26/27, 1991, the 37th TFW launched 63 F–117A Nighthawk sorties with 43 weapons intended for Baath Party Headquarters in Baghdad, 15 for two of Hussein's residences in Baghdad, and 6 for his home in Tikrit. With the Iraqi Army in Kuwait in total disarray, the strategic air planners hoped that this strike, and others like it aimed at the symbols of Hussein's power, would supply the final push to what they suspected must be an already tottering regime. Other F–117As targeted important industrial facilities. Weather prevented the release of all but 12 weapons; only 2 fell on leadership targets. On the last night of the conflict, a wave of 21 F–117As headed for Baghdad and at least 18 hit Baath Party Headquarters. A second wave of 10 aircraft attacked the Alteena Nuclear Center, a site associated with the Al Musayyib missile research and development (R&D) and production complex. The Iraqis had designated the center as the construction and assembly facility for their atomic bomb project.[107] Fourteen of 18 bombs hit their primary target. General Schwarzkopf canceled the night's third wave and suspended further air operations.

One of CENTAF's last strategic missions employed a new laser-guided weapon, the GBU–28, which was literally hot off the assembly line. Just before the cease-fire, two F–111Fs of the 48th TFW each carried a single GBU–28 to Command Leadership Bunker No. 2 at Al Taji. The specially developed bombs, machined from the barrels of surplus Army 8-inch (203-mm) howitzers and tested and deployed in a mere seventeen days, weighed 4,700 pounds each.[108] They arrived from the United States, still hot from the molten high-explosive mixture poured into them just before departure, and went straight from transport aircraft to strike aircraft, remaining warm to the touch as the F–111s taxied out. The Iraqi command bunker had successfully withstood repeated attacks with BLU–109 penetrators from the first night of the conflict onward; however on this night, one GBU–28 missed, but the second scored a direct hit, with debris and smoke spewing from the bunker's entrances, a sure sign of penetration and destruction. The strategic air campaign thus ended with the Nighthawks and Aardvarks demonstrating to the Baath regime that it no longer had anywhere to hide.

Analysis

At the end of any strategic bombing campaign, one confronts three fundamental questions. Did it expend its efforts on targets vital to the enemy's conduct of the war? Did it select targets vulnerable to friendly air action? And did it contribute decisively to the overall success of air, ground, and sea operations and the achievement of national political objectives? For air power in the Gulf War, the quick answer to all of these questions is "yes." It is profitable, however, to examine each aspect in greater detail to learn what succeeded and what failed.

The Core Strategic Target Sets*

In the Gulf War, the core strategic target sets — those target sets most important to maintaining current and a future military capability — consisted of Iraq's

1. National leadership,
2. Military and civil C^3,
3. Electric power generation,
4. Oil refineries, distribution, and storage,
5. NBC weapons research, development, and production,

---

*The term *core* comes from GWAPS, vol 2, *Effects*, p 269. The chief authors of that volume, Barry D. Watts and Thomas A. Keaney, use it to define what they consider "the eight 'strategic' target categories that were perceived by those who planned and executed the DESERT STORM air campaign as constituting the core of Iraq's current and future military power." I deleted one of their eight categories, Rail and Highway Bridges, from the core and placed it into what I designate the "mixed" target category. Since much of Iraq's political status in the region rested on its powerful military capabilities, an attack on those capabilities can also be considered an attack on its political position as well.

6. Military support (R&D, production, and storage of conventional armaments), and
7. Scud ballistic missiles.

To incapacitate the Hussein regime* and slow the capability of the Iraqi leadership to react and transmit decisions, the strategic air campaign targeted Iraqi leadership and C^3 targets. From January 16 to February 28, 1991, the number of leadership targets grew from 33 to 44 while the command and control targets grew from 56 to 146.[109] The later target set grew as the air planner sought to shut down alternate means of civil and military communications.

In the past half-century the speed and flow of information exchanged between a modern military and its leadership has greatly expanded. Restricting a foe's information flow delays his reaction time and causes him to fall further and further behind an attacker's actions until he is knocked out. The inclusion of these target sets in a strategic air attack stemmed directly from the ideas of the Warden group. The promise of total incapacitation of the regime and the severing of all its communications with the forces in Kuwait depended upon what proved to be inadequate intelligence. Thus, some important communication means remained unknown until a few days before the war, some were too closely associated with civilian facilities and so declared off-limits, some were more damage-resistant than had been realized, some were difficult targets even for precision-guided munitions, and some remained hidden throughout the entire course of the war.

Other than images of damaged facilities, such as the well-known strike footage of the Iraqi Air Force Headquarters Building and news footage of the destruction of the minaret-shaped microwave tower of the Al Karakh telephone exchange building, *no solid data is available to connect bombing of leadership and of command and control with specific consequences*. Nevertheless, the bombing of primary facilities did force the Iraqi leadership to resort to far less secure means of communication,[110] a circumstance welcomed by Coalition forces, especially during the 100-hour ground war.[111] Prisoner accounts were replete with reports that Iraqi units were dependent for information on messages delivered by bicycle or motorcycle. The bombing of security and intelligence ministries may have disrupted their operations, perhaps causing a decline in their effectiveness, a loss of files not already evacuated, and lessened control

---

*Warden expected that bombing of the Iraqi leadership would result in Saddam Hussein's elimination or overthrow. Most others, including some of Warden's own planners, thought not. But they did not deny that Saddam Hussein's overthrow would be a welcome and positive result of bombing leadership targets, even if the intended result was not a given. In September 1990, the Black Hole proposed to "Decapitate the Saddam Regime," to bring about a change in regime. Briefing, Gen Glosson to CJSC, Slide 17, Sep 13, 1990. When Glosson briefed President Bush on October 11, 1990, he stated that the strategic air campaign would destroy the Iraqi leadership's command and communications and disrupt its ability to communicate with the Iraqi people. Glosson also received the impression the President did not want to go on the record as targeting Saddam Hussein. GWAPS vol 2, *Effects*, p 281.

over the populace. The widespread and unprecedented criticism of the regime by Iraqi citizens immediately after the war may offer some evidence to that effect. The Shiite and Kurdish rebellions against the Iraqi central government that erupted after the war underscored the regime's obvious loss of control.* The February 12–14, 1991, visit of Soviet Special Envoy Yevgeny Primikov to Saddam Hussein in Baghdad highlighted Hussein's communications difficulties. Primikov brought Soviet satellite imagery that showed the surprised Iraqi President the full extent of the damage Coalition air strikes had inflicted.[112] He also met with the Iraqi leader in a private residence rather than a government facility. From that point onward,** using Soviet good offices, Hussein began actively to seek a way to extricate himself from his predicament.

The C^3 bombing apparently had little effect on Saddam Hussein's control of his Scuds. A DoD intelligence memorandum of March 1991 stated,

> The Coalition's inability to permanently degrade SRBM command and control is also significant, despite determined efforts to incapacitate Iraqi military and civilian national networks. Even in the last days of the war, Baghdad retained a sufficient capability to initiate firings from new launch areas and to retarget SRBMs from urban to mititary and high-value targets, such as the Dimona nuclear reactor.[112]

Precision-guided munitions proved ideal weapons for attacking government buildings and communications centers. Even so, as the war progressed, planners realized they had selected a more elusive and redundant target system than they had suspected. Fiber-optic networks and computerized switching systems were difficult targets to eliminate completely, and some ran along the Baghdad-Tigris bridges. Stealth bombers cut the spans on these two bridges, but CNN news reports of the bombing, which provoked a critical international reaction, coupled with the Al Firdos District Bunker aftermath led Washington*** to place the remaining bridges off-limits in mid-February.[113] The leadership and commu-

---

*The timing and noncoordination of these rebellions, after the Coalition-Iraqi cease-fire of March 3, 1991, allowed Saddam Hussein to crush them separately. Apparently the leaders of the uprisings delayed action in hopes that the Coalition would destroy the Hussein regime for them. If the rebels had taken action during hostilities and appealed for Coalition aid at that point, they might have had a greater chance for success. They would have had a good claim to Coalition aid, and Coalition forces would have continued to tie up the bulk of the Iraqi armed forces. A scenario of this sort would have made the bombing of leadership and command and control a clear success. Instead, the Coalition appears to have left itself open to the same charge as was leveled against western nations in the aftermath of the Hungarian revolution of 1956 — egging on the rebels, with no intention of providing aid.

**The bombing of the Al Firdos District Bunker, which occurred during Primyakov's visit, may also have affected Saddam Hussein's calculations.

***The origin of the stand-downs on Baghdad targeting is a problem that requires further investigation. General Schwarzkopf placed the Baghdad bridges off-limits in early February, and after the Al Firdos District Bunker incident on February 13, General Horner had to obtain General Schwarzkopf's approval for any targets in downtown Baghdad. General Schwarzkopf, in turn, discussed the targets with General Powell in Washington before granting his permission to attack. General Powell seems to have imposed this restriction by his own authority, but there is some evidence that the target "hold" may have come from the White House. Secretary Cheney apparently went to the White House to get the bombing halt removed on February 21, 1991. GWAPS vol 2, *Operations*, pp 249–251.

nications targets placed off-limits in Baghdad were the only target systems subjected to detailed approval and review by higher authorities — Generals Schwarzkopf and Powell, among others. That review prevented full execution of the attack on these targets. INSTANT THUNDER planners, as part of their integrated attack on the regime, also hoped to use another weapon to separate Saddam Hussein from the populace — psychological warfare operations, which they thought might be effectively employed against such a highly centralized state. Bureaucratic competition among American intelligence agencies unable to agree on methods and tactics, however, and the Arabian peninsula Coalition host nations' concerns about confining the campaign's potentailly destabilizing influence solely to Iraq combined to stymie the implementation of such operations and deny a possibly devastating follow-up punch to the attacks on Iraqi leadership.

TABLE 7–2
Strategic Targets: Level of Effort

Strategic Target Sets	Number of Strategic Combat Sorties Flown	Percentage of Strategic Effort	
		Including Republican Guard	Excluding Republican Guard
Electrical Power	215	1	2
Naval	247	2	2
Leadership	429	2	3
Air Defense	436	2	3
Oil	518	3	4
C^3	601	3	5
Railroads & Bridges	712	4	6
NBC	902	5	7
Military Support	2,756	15	22
Scud	2,767	15	22
Airfields	3,047	17	24
Republican Guard	5,646	31	—
Total Strategic Combat Sorties:			
Including Republican Guard	18,276		
Excluding Republican Guard	12,630		
Percentage of Total Coalition Strategic Effort		36	24

Source: DoD, *Conduct of the Gulf War*, table, "Strategic Targets Level of Effort," p 159.

The Hussein regime did not fall, and it retained, at least, a minimal ability to communicate with its forces in Kuwait and its organs of control in other areas. The disappointment of air planners stemmed from overly ambitious goals and circumstances beyond their control. Still, the bombing of the leadership and communications network served a vital purpose. It caused the regime untold inconveniences, possibly slowed effective communication, thus interferring with Hussein's responses to the Coalition onslaught; forced an expenditure of valuable and not easily replaced spares; consumed specialized repair efforts; and damaged government buildings and key communications facilities.

The bombing of C^3 and leadership targets consumed approximately a thousand sorties — 8 percent of the entire strategic effort and only 2 percent of overall Coalition air effort. The result of that bombing may have been minimal, as some critics have claimed, or extensive, as some air power advocates have hoped. Until, or if ever, hard evidence of its effects is obtained from unbiased Iraqi sources, the outside analyst can form no certain opinion as to its actual effectiveness. However, given the potentially great benefits of such bombing, an investment of a small fraction of the air effort toward it seems reasonable and justifiable both in the Gulf War and for future strategic efforts. Conversely, given the fact that no hard data is available concerning acutal results, building an air campaign around or devoting a substantial portion of an air campaing against these target systems in the future would be too great a leap of faith into an uncertain area and would rest on the unsubstantiated claim and the wishes of only one faction of air power activists. The eventual place of C^3 and leadership bombing and their respective priority in the scheme of air campaign operations is, at best, yet to be determined. More experience, testing, and hard information is required.

Chairman of JCS Colin Powell communicates with the Pentagon from Saudi Arabia; Secretary of Defense Richard Cheney addresses the media from the war zone.

Iraq's twenty-five major electrical plants constituted a compact, highly leveraged target system. Loss of their services would force the Iraqi military to use backup generators, vastly complicating its operations. Coalition air attacks by manned strike aircraft and Tomahawk land-attack missiles shut down the southern and central Iraqi power grids within hours, reaffirming the adage that selecting a weapon appropriate to a target's vulnerabilities produces outstanding results. In January 1991, Iraq was a potential major oil producer, controlling 10 percent of the world's oil production capacity and 20 percent of the world's known oil reserves. Three large refineries produced 90 percent of Iraq's refined petroleum products. Tomahawk land-attack missiles hit the distillation towers of two of those refineries in the first two days.[114] Aircraft equipped for both precision-guided munitions and non–precision-guided munitions also conducted extensive raids on Iraqi oil facilities, and 500 sorties delivered 1,200 tons of bombs on 28 different refineries. In ten days, Coalition air attacks shattered Iraq's oil refinery capacity. Ironically, the war's short duration and the Iraqi forces' static posture worked against the effectiveness of oil strikes. The Iraqi military had considerable refined petroleum stocks spread throughout its military logistics system and supply dumps, thanks to prewar preparations. Given these supplies, the oil campaign had little measurable impact on the outcome of the war as fought. Iraqi air and ground operations would appear to have never been affected by any shortages of fuel. The planners, of course, could not foresee the conflict's speedy conclusion, and military prudence dictated an attack on this system that, in the long run, controlled Iraqi mobility.

As an added effect of hitting these two related targets — power and fuel — strategic air planners hoped that lack of heat, hot water, cooking fuel, private automobile fuel, and labor-saving electric appliances would further alienate the populace from their leadership and help to bring about its change. This, added to the normal stress of everyday living, certainly angered and frustrated Iraqi civilians and fostered some feeling of malaise. One might speculate that when added to traditional grievances, it may have helped to spark or perhaps facilitate the Shiite and Kurdish mutinies. But the air planners underestimated the stake of Saddam Hussein's Sunni followers in the status quo as well as the strength of the Iraqi internal security apparatus and its ability to bank or deflect from the regime the fires of popular dissatisfaction.

This underestimation of the hold of a police state on its thralls was not unique to U.S. Air Force planners. Outside observers also underestimated the hold of both the Nazi and Soviet states on their populace before World War II. One might also suggest that the fall of the Soviet Union in the late 1980s resulted more from the recognition of economic failure among the regime's ruling elite than from disapproval of the general population. As to the strength of an internal security apparatus, the continued survival of the KGB and its successors provides an object lesson in their durability and strength.

Planners sought to limit the damage incurred by the power and fuel systems to facilitate their eventual and speedy repair.* To prevent electricity from flowing into the Iraqi national power grid, Coalition air power did not have to bomb the electrical generating plants into rubble. Rather, their operations needed merely to be stopped for a few weeks or months, during the time of the actual fighting. To encourage eventual oil exports, the bombing of oil fields or leveling of refineries was avoided, in contrast to what had been done at Ploesti or Balikpapan in World War II. It was for this reason that planners naturally took umbrage in the immediate postwar period when the Iraqi government and some international visitors exaggerated the public health and economic consequences of the damage inflicted on the electrical power and oil systems. In fact, despite a continuing international embargo and noncooperation from Saddam Hussein's regime, Iraq recovered much of its electrical generating capacity by mid-1992, and by October 1992 Iraq was again actually *exporting* finished petroleum products.[115] Because these targets were located beyond city areas, the bombing of them produced little collateral damage — an almost perfect example of the Warden group's theories of bombing for *effect*, not for *destruction*.

The Coalition attempted to destroy Hussein's NBC weapons research, development, and production because these capabilities were key to his efforts to expand Iraqi control and influence in the Persian Gulf region. They formed part of the original INSTANT THUNDER plan, which promised to destroy, i.e., obliterate, these target sets. That plan, however, was based on limited intelligence and intended for execution in late August 1990. Although more intelligence did become available, the Iraqis gained five months to take countermeasures, such as moving or more thoroughly hiding their programs. Thus, by January 16, 1991, the earlier promise of destruction** of these targets had become more a planning goal than a realistic objective. Iraq's NBC capabilities proved far more diverse than air planners, dependent on flawed intelligence, had imagined. After the 1981 Israeli air raid on their reactor complex at Tuwaitha, just south of Baghdad, the Iraqis dispersed and duplicated all the important segments of their nuclear program. Before the Gulf War began, they removed all fissionable material, equipment, and documentation from these facilities and dispersed or buried it.[116] Thus, U.S. intelligence failed to locate and identify the bulk of Iraq's nuclear effort.

The strategic air campaign planners on January 16, 1991, carried only two nuclear targets: Tuwaitha and the Al Qaim uranium mine near the Syrian

---

*Air planners assumed Saddam Hussein would fall and import restrictions on the new regime would be few if any. By making oil and electricity unavailable during the conflict but speedily available to the new regime, planners hoped to a strengthen Hussein's successors in the eyes of the Iraqi people.

**The author discussed the use of the word *destroy* in relation to Iraqi NBC capabilities with members of Checkmate and the Black Hole. Their intention was to eliminate totally all NBC capability, not just damage it.

border. The list grew to eight (five of which were destroyed and two damaged) by war's end. Facilities known to the planners were struck hard, but the Air Force could not hit, nor be expected to hit, what it did not know about. The Coalition air effort and the Iraqi's own dispersion efforts interrupted their atomic bomb project to an unknown extent, and certainly delayed it for many months.* After the conflict UN inspection teams dismantled many of Iraq's weapons of mass destruction and related programs. By October 1991, the teams had uncovered twenty-one nuclear weapons–related facilities.[117]

Attacks on Iraq's chemical and biological production and research fared better than those on Iraq's nuclear effort. The Coalition destroyed or heavily damaged 75 percent of Iraq's known chemical warfare research and production and almost all of its known biological capability.[118] Nonetheless, in addition to facilities it had concealed or those that had otherwise escaped Coalition intelligence, Iraq had a fair-sized pharmaceutical industry that, on relatively short notice, could convert to chemical warfare production and produce sufficient material to threaten its neighbors. As for chemical and biological warfare weapons, UN inspectors located more than 150,000 chemical artillery shells, but they found no evidence of biological weapons.[119] The strategic bombing campaign also delayed production and reconstitution of this target system for an indeterminate time, certainly for many months if not for years. As with the nuclear campaign, intelligence limitations ensured that the planners' pledge to destroy them was a promise beyond their capability to deliver.

Altogether, NBC targets accounted for only 7 percent of the strategic air campaign's total sorties. Perhaps not surprisingly, little evidence exists to show that strategic air planners were pressured by higher authorities to increase their efforts against any of these systems. After all, the top command levels worked with the same intelligence limitations that the air planners did. They, too, did not realize until after the war's end that Iraq's NBC complexes were far more sophisticated and extensive than previously supposed.

The Coalition's inability to strike such weapons from Hussein's hand raises a problem of disturbing magnitude for international political and military leaders alike. Ownership of such weapons is an all-or-nothing proposition. If a potential foe possesses just a few of them or if he has the ability to produce them quickly, his political/diplomatic/military position is almost as good as if he had them by the dozens or hundreds. But one can never be certain that the enemy has none unless the enemy's land is thoroughly searched (thus obtaining highly accurate data). In fact, the Iraqi experience demonstrated that even the most sensitive components of such weapons can be relocated at will. The lesson is as ominous now as at any time since 1945 — any nation wishing to expend

*Evidence indicates that Iraq had launched a crash effort to produce one A-bomb in August 1990. The outbreak of hostilities apparently halted this program.

the resources can acquire and maintain nuclear and other terror weapons. Without a basic change in the philosophy of a nation's ruling class, such as occurred in West Germany and Japan after World War II, physical destruction of existing facilities is not the entire answer. Military power alone cannot remove the desire for these weapons from a leader's psyche, nor can it remove the knowledge of them from the heads of physicists, biologists, and chemists.

Today, more than ever, intelligence is key to the successful conclusion of any military campaign. In the Gulf War it was adequate, indeed better than in most previous wars, but it was certainly far from perfect. For example, the U.S. decision to devote minimal national intelligence priorities to Iraq and its NBC weapons programs before August 1990 produced crucial failures in identifying the extent and diversity of these programs. This shortfall rippled through the system and caused a breakdown among intelligence, targeting, and campaign planning functions.

The strategic air campaign planners struck at Iraq's massive military storage and production network in order to reduce Hussein's ability to field and sustain his armed forces. Twenty-two percent, or 2,756, of all sorties in the strategic air campaign (and probably an even greater proportion of the strategic air campaigns' total bomb tonnage) went into this target system. This essentially equaled the number of sorties expended on Scuds (2,767). Coalition aircraft destroyed or heavily damaged many physical plants and vulnerable equipment too bulky to move. A DoD assessment affirmed that "at least 30 percent of Iraq's conventional weapons production capability, which made small arms, artillery, small- and large-caliber ammunition, electronic and optical systems, and repaired armored vehicles, was damaged or destroyed."[120] The damage to specific key maintenance and manufacturing bottlenecks as well as to specialty metallurgy and aircraft engine repair facilities meant that the effect on overall Iraqi arms production was probably even greater than the direct loss of 30 percent of production capability. Nonetheless, given its size, diffusion, lack of high priority, and the short duration of the conflict, the bulk of this target system survived, though not in any condition to resume prewar production rates.

The Scud diversion likely saved this large, easily located target system from even more damage, as did the concentration of air effort on Iraqi ground forces in the KTO. A greater use of nonprecision B–52s and F–16s, and fewer precision sorties for tank plinking, would have resulted in far more damage to military stores and production facilities, with the possible disadvantage of also producing greater collateral damage. What must not be forgotten, of course, is that the strategic air campaign inflicted expensive and inconvenient damage to this target system, damage unrepairable without assistance from the outside world. Air power had the capacity to virtually destroy Iraq's military production system, but only with a commitment of resources greater than those made available to the strategic air campaign. However, the unanticipated requirements for the "Great Scud Hunt" and the decision to provide massive, pre–ground warfare, frontline strikes in support of General Schwarzkopf's three corps commanders prevented execution of that option.

The strategic air campaign planners also directed the attack on Scud missile support facilities and their communications, testing centers, and launchers to protect Coalition forces and procure stability in the Persian Gulf region. They targeted the Scuds directly to prevent Hussein from using them to attack Israel and thus provoke Israeli retaliation.* The anti-Scud effort eventually consumed 22 percent (2,767) of the strategic air campaign's total sorties, the same amount as was expended by the Coalition on all the rest of Iraq's conventional military manufacturing. Coalition bombing heavily damaged almost all known production and research facilities, but subsequent UN inspection revealed that the Iraqis had removed most production equipment, components, and documents before the start of the air campaign. The DoD Final Report ruefully stated, "actual damage to Scud production and storage facilities is less than previously thought."[121]

The attack on Scud mobile launchers failed to destroy them in significant numbers. The Iraqis never employed the exposed fixed launchers, though bombs expended on them prevented their future use. By the end of August 1990, the Iraqis had dispersed their mobile launchers to areas within range of Israel and Saudi Arabia, where they continued to operate until the end of the conflict. Mobile launchers proved a difficult target for strike aircraft, for within ten minutes a mobile launcher could move more than five miles from its firing site, and this without traveling on roads. To mislead Coalition aircraft, the Iraqis employed numerous high- and low-quality decoys, some indistinguishable from an actual launcher at more than twenty-five yards. Launcher crews practiced extreme electronic emissions control and nighttime light discipline, and they had streamlined Soviet launch procedures, shaving launch time from hours to minutes. The crews received launching instructions via landlines and couriers, which made them impossible for Coalition electronic intelligence to intercept.

In late 1990, U.S. Air Force exploitation flights against a borrowed launcher and crew showed that U.S. strike aircraft had difficulty in visually or electronically finding launchers in daytime and that these sensory devices proved even less effective at night.[122] The Iraqi mobile Scuds benefitted from a mismatch of available air weapons. The missiles could not be located readily with existing air technology, although the Coalition made use of special operations forces such as the British Special Air Service and the American Delta Force to physically locate and fix the mobile missiles with laser

---

*The Iraqis possessed an inventory of several hundred missiles, several fixed launchers (simple rails on a concrete pad), and probably no more than forty mobile launchers. The Iraqi SCUD had a range of approximately 437 miles, a 500-lb warhead, and a circular error probable of 2,000 meters. For a detailed discussion, see GWAPS vol 2, *Effects*.

designators, thus illuminating them as targets for laser-guided weapons. As of early February, after Coalition air forces had already expended more than half of their total effort against Scuds, there were no confirmed kills of Scud mobile launchers, though postwar Special Operations Force accounts suggest otherwise. Nevertheless, whatever successes air may have had against Scuds served to illustrate the general relative immunity of the mobile Scud ballistic missile to air attack at the time of the Gulf War.[123] After the war, UN inspectors oversaw the destruction of the nineteen mobile launchers and several decoys, which the Iraqis admitted they still possessed.[124]

The Coalition's anti-Scud air effort cannot be considered a dead loss to the overall war effort. The high level of resources expended against the system probably convinced the Israelis that their own efforts could add little to the outcome. Israeli counterattacks might have given Hussein his one chance to end the war on favorable terms. The comparative success of the Scuds, however, presented a troubling new problem for air power. Scuds were relatively cheap to purchase or produce, easy to hide, and accurate enough to inflict great, if indiscriminate, damage with NBC warheads. The Scud confronts modern air forces with many of the problems of guerrilla warfare, with all its complications. Militarily, the anti-Scud effort was a successful strategic diversion imposed by the Iraqis on the Coalition — strategic sorties expended on those missiles might have damaged targets of more lasting significance. Nevertheless, in the political and diplomatic sense, the Iraqis failed. Scuds neither drove Saudi Arabia from the conflict nor dragged Israel into it, and the negative publicity this device generated further alienated the Hussein regime from the world community.

The core target sets and the methods and means of attacking them constituted the centerpiece of the strategic air campaign against Iraq in the Gulf War. Sixty-five percent of the strategic air campaign, a total of 8,188 sorties, flew against them. These were targets of vital importance to the Iraqi war effort, and their vulnerability to air attack varied. The attack on the Iraqi leadership and C^3 produced mixed results. It failed to depose Hussein, tainting the minds of some against the rest of what was, undeniably, an extraordinary air campaign. It also greatly complicated the Iraqis' ability to prosecute their war effort and constituted an important and encouraging lesson for future strategic air operations.

In sum, the attacks on electrical and oil targets (especially the electrical ones) were highly successful and produced immediate and damaging results. They validated the Warden group's methodology of searching for centers of gravity and bombing for effect. Attacks on NBC objectives were effective against known targets. They inflicted significant delays on Iraqi weapons programs, but they did not destroy them because intelligence shortfalls denied air power the opportunity to strike these target systems in their entirety.

Also, the air strikes on the Iraqi military support structure caused serious delays in production and a consequent delay in full reconstitution of the Iraqi armed forces, though not entirely destroying Iraqi capabilities. Delay in and of itself benefitted Iraq's neighbors who, following the war, launched aggressive acquisition and training programs to prepare their defenses against the threat of a renascent Saddam Hussein or his successor. Overall, however, the results of the support-structure strikes strongly suggest that if the goal is solely to prevent current and immediate future production instead of bombing the many diffuse factories that produce arms directly for the military, it is far more economical to disrupt the power sources and transportation nets feeding them.

Finally, the Scud target system illustrated the intimate link between military and political power. When considered alone, the anti-Scud effort was a disappointment that raises disturbing problems for future air power campaigns. When one examines the anti-Scud effort in the Clausewitzian sense, i.e., viewing warfare as extension of the state's politics, then the anti-Scud effort justified the military resources invested in it, for it kept Israel neutral. It demonstrated yet again that, at the strategic level, nearly every target system includes both political and military considerations. In this instance, the political objective was paramount, and the assault on the target system, therefore, was ironically successful.

Self-Defense Targets

The achievement of air superiority and the protection of friendly bases and forces are vital for a successful strategic air campaign. To guarantee those two prerequisites, Coalition air forces attacked the following three strategic target sets dealing with Iraqi defenses capable of harming Coalition air and naval forces or of protecting the core strategic targets:

1. The Iraqi Integrated Air Defense System (IADS),
2. The Iraqi Air Force, and
3. The Iraqi Navy, with its associated port facilities and antishipping missiles.

The Coalition directed 29 percent of its strategic air effort to the suppression or destruction of the aforementioned target sets vital to both the Iraqi and the Coalition war efforts. All three proved extremely vulnerable to Coalition air action.

CENTAF air planners directed the first air actions against Iraq's strategic IADS. Coalition aircraft with antiradiation missiles intimidated Iraqi SAM and AAA radar operators, who hesitated to operate their equipment lest their signal lock-on attract a destructive, beam-riding response. Because of the anti-IADS strikes, within minutes of the start of hostilities on the night of January 16/17, 1991, Coalition aircraft could operate with impunity at high and medium altitudes.[125] Coalition aircraft losses came primarily during the last weeks of the conflict as low-level attacks on Iraqi ground forces paved the way for Coalition ground operations and close air supported the liberation of Kuwait that began on February 24, 1991. (Iraqi ground units, especially the Republican Guard, had ample AAA and numerous shoulder-held SAMs to establish a dangerous, low-altitude environment.) In brief, the destruction and defeat of the Iraqi integrated air defense network gave the Coalition freedom of the air. This was a success for the strategic air campaign and set the tone and tempo for the rest of the war.

Iraqi airfields and the aircraft they housed absorbed 24 percent (3,047) of the strategic air campaign's total sorties. Initial Coalition air attacks concentrated on runway denial before they turned to hardened air and maintenance shelters. These attacks destroyed or forced out of the country nearly half of the Iraqi Air Force and damaged its essential support facilities and equipment. The remaining Iraqi Air Force suffered from spare parts, maintenance, and training problems until it could regain access to the international arms market. Until then, it was limited in its ability to suppress internal revolts and inadequate to threaten its neighbors. Given the Coalition air forces' lopsided advantages in quality and quantity of men and materiel, the destruction of the Iraqi Air Force was but a question of time and blood, and how much of each the Coalition wished to sacrifice. In fact, the Coalition lost only one F/A–18 and pilot in air-to-air combat (to a look-down shoot-down MiG–25), and only handful to ground fire or SAMs in missions against airfields. It achieved air superiority the moment the first F–15C crossed into Iraqi airspace; it achieved air supremacy soon after. The U.S. Air Force had trained for counterair operations, and its use of precision-guided munitions denied the Iraqis a secure sanctuary. Coalition air achieved a more complete success even sooner than anticipated. Its achievement ranks with the opening Israeli Air Force effort in the Six-Day War of 1967 as the most rapid and overwhelming seizure of air supremacy over a hostile nation with a significant air force in military aviation history.

The U.S. Navy, relying heavily on its air arm, overwhelmed the Iraqi Navy. The only target environment offering less concealment for enemy forces from air power than the desert is the water's surface. Naval aircraft and helicopters destroyed 11 of 13 antiship missile boats, destroyed or damaged 143 of 165 Iraqi combatant vessels, and eliminated 3 of 7 shore-based, antiship missile sites, as well as heavily damaging Iraq's two naval facilities.[126] This, plus the absence of any air threat, allowed the Coalition to bring its American aircraft carriers closer to Iraq, easing the air refueling workload that hampered naval aviation operations.* The elimination of the Iraqi Navy also increased the effectiveness of a major Coalition deception plan — the threat of a U.S. Marine Corps amphibious invasion near Kuwait City. This target system absorbed 2 percent (247 sorties) of the total strategic effort.

Because it posed the simplest strategic problem, the attack on air defenses was by far the most successful portion of the strategic air campaign against

---

*U.S. Navy aircraft burned a less volatile aviation fuel than did Air Force aircraft and could be refueled only with the probe-and-drogue method instead of the flying boom, all of which demanded specialized aerial tanker assistance. Configuration of U.S. Air Force tankers for Navy use effectively denied those tankers to the Air Force.

Iraq. American intelligence could easily locate the massive facilities (hardened air operations centers, airfields, and naval ports) supporting these defenses, while the Iraqis could hardly dismantle, disperse, and bury them. SAMs and AAA without central control rapidly lost effectiveness, as did aircraft and ships separated from their permanent bases.

"Mixed" Target Sets

The last two target systems of the strategic air campaign — rail and highway bridges and the Republican Guard — shared political and strictly military considerations belonging to both the tactical air campaign conducted by the Coalition in the KTO and the strategic air campaign against Iraq. The bombing of bridges for the purpose of interdicting Iraqi lines of communication* into Kuwait, thereby depriving Iraqi forces there the means to fight, was strictly a military objective. Destroying bridges to prevent the escape from Kuwait of the Iraqi ground forces, particularly the Republican Guard, however, involved military *and* political considerations. The strategic air campaign expended 6 percent of its effort against bridges,** a vital target system (but less vital than had been anticipated by the air planners) to the Iraqi war effort.

As evident from historical experience, bridges are highly resistant to the effects of dumb bombs but are quite vulnerable to precision-guided munitions. The strategic air campaign destroyed or heavily damaged three-fourths of the major bridges between Baghdad and Basra, including all nine major railroad bridges.[127] This single-track railroad carried the bulk of Iraqi Army and Republican Guard heavy equipment when these units moved for other than tactical or battlefield operations. The complete loss of Iraqi rail capacity in and *out* of the KTO inhibited the speedy retreat of Iraqi armored and mechanized units, and it constricted Iraqi lines of supply to Az Zubayr on the Iraqi-Kuwaiti border, which served as the Iraqi supply head for troops in Kuwait. Loss of the bridges reduced truck traffic to Az Zubayr and southward on the four-lane superhighway as well as rail traffic on the temporary line that linked Az Zubayr to Kuwait City and points west and south. Supplies for the occupation forces in Kuwait dropped to a small fraction of their prewar level, and indeed, proved insufficient for offensive *or* defensive operations. At war's end, the amount of

---

*Lines of communications are all the routes—land, water and air—that connect an operating military force with a base of operations and along which supplies and military forces move. JCS Pub 1.

**I count all bridge sorties, even the 55 percent conducted by F–16s, F/A–18s, and other non–precision bombing aircraft, as belonging to the strategic air campaign. The aircraft serving as the bombing platform, whether a B–17 during World War II or an F–15E today, does not define the target as strategic or tactical, for aircraft can do (and in the Gulf War, *did*) both. Depending on the situation, as in the mixed Gulf War target sets, a given set can have aspects of both. In the "mixed" case, I categorize bridges under strategic bombing because the physical targeting and planning was handled by the CENTAF strategic air planners, many of the targets lay outside the KTO, and the political objective of destroying the Republican Guard may have been of more practical significance than merely constricting Iraqi supply lines.

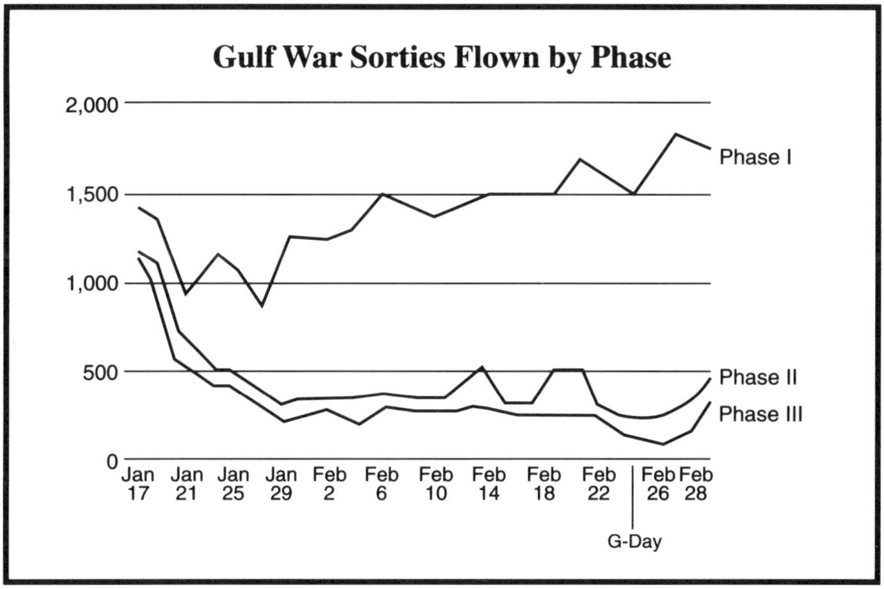

supplies getting through was insufficient to meet the average Iraqi's long-term survival needs.[128]

Bridge bombing also produced traffic jams vulnerable to Coalition air action and increased wear and tear on the Iraqi motor transport fleet, which was forced to travel greater distances over alternative routes. In response, the Iraqis built numerous pontoon bridges (of much lesser capacity than the permanent bridges they replaced) and devised other work-around solutions, such as earthen causeways. In the precision weapons era, however, Coalition air easily countered these historically useful defensive measures by destroying many of them and by mounting frequent "river reconnaissance" patrols to disrupt other Iraqi efforts. Thus, as in LINEBACKER I, the precision-guided munitions proved the most effective weapon against bridges, constituting 45 percent of the Coalition bomb tonnage expended on bridges and inflicting most of the damage incurred. In this instance, the strategic air campaign achieved an ideal match between weapons capability and its target system.

Anticipating the interdiction campaign, Iraq's forces in Kuwait attempted to offset the effect of Coalition air power's severe constriction of their lines of communication by stockpiling large amounts of supply and material in the months between their invasion of Kuwait in August 1990 and mid-January 1991. In fact, the vast physical extent of the numerous, revetted, and highly dispersed Iraqi supply dumps so discouraged Coalition tactical air planners that they "never attempted a coherent campaign to interdict the flow of supplies into the theater."[129] Iraqi forces in Kuwait consumed only a fraction of the stores available before the ground offensive commenced on February 24, 1991. Yet the overwhelming air presence over the front denied mobility to Iraq's forces and prevented their using the stores they had so carefully hoarded. Thus, some frontline forces were virtually on the brink of starvation even though prewar stockpiles were easily within reach. How was this so? The Coalition tactical air campaign in Kuwait succeeded in greatly complicating Iraqi supply distribution from their dumps in Kuwait to the unit logistics centers and thence to the troops in the field. This they achieved by attacking supply convoys and even individual trucks. Although the Iraqis had adequate supplies in Kuwait, they could not get them to the front where some units had few or no deliveries of rations and water after the air offensive began. This interference with Iraqi supply distribution resulted more from the pattern of Coalition tactical air attack than from a consistent assault on the supply system. Coalition aircraft operating in their assigned kill boxes routinely struck at any moving target (supply, fuel, and water trucks are more vulnerable to enemy fire than armored fighting vehicles) or at any target in the open (the Iraqis did not routinely park their trucks in revetments). Reports from Iraqi prisoners of war clearly reveal that tactical air power caused serious logistical difficulties for the Iraqi frontline forces.[130] If the ground war had lasted for an extended period of intense combat, the lack of resupply capacity would have fatally handicapped Iraqi the defenders.

In addition to the the destruction of bridges, a vital target system in the Iraqi war effort, oil tanker loading docks were also struck, as the severed pier with the associated fuel lines shows below.

General Schwarzkopf's Operations Order of January 17, 1991, reflecting his instructions from Secretary Cheney, identified the Republican Guard as an "Iraqi center of gravity,"[131] i.e., a target essential to the Iraqi conduct of the war and survival of the Hussein regime. The DoD Final Report claimed 31 percent (5,646) of the total *strategic air campaign* sorties flew against these units. Although this number of strategic sorties expended on the Republican Guard is official, it is most misleading in two ways. First, the inclusion of all Coalition air sorties flown against the Republican Guard in the summary count for strategic target sets (it is the largest number of sorties flown against any of the twelve strategic target sets) overstates by one-third the overall size and level of the strategic effort compared with both the Coalition air effort in Kuwait and the overall Coalition air effort.* Second, at the war's end, the percentage of Republican Guard units and their heavy equipment that avoided destruction in Kuwait was the highest of any portion of the Iraqi Army.** Applying all sorties flown against the Republican Guard against the strategic air campaign appears to place the blame, if any, for a failure to destroy the Republican Guard on the strategic air campaign alone, to its detriment. It also capriciously absolves tactical air power, Coalition ground forces, and American military and political decision-makers of their responsibility for the failure to destroy Saddam Hussein's political lifeguards.

The fact that much of the Republican Guard survived the war intact has prompted an increasingly controversial and heated debate that has already generated revisionist interpretations.[132] The survival of the Republican Guard echoed the successful German and Italian evacuation from Sicily in August 1943 (which took place in the face of overwhelming Allied land, sea, and air superiority) and, to some degree, the escape of German forces after the near-disaster of the Falaise Gap in 1944. Echoing Falaise, the survival of the Republican Guard rested as much on a complicated series of decisions taken by Coalition ground commanders and political leaders as it did on Iraqi initiatives.*** Hussein committed eight divisions of the Republican Guard to

*The GWAPS, a study paid for by the U.S. Air Force and much less widely circulated than the unclassified DoD Title V Final Report, defines the term *strike* as the delivery of a weapon against a specific target. Using the CENTAF strategic air planners' Master Target List target categories, study originators record only 33 of 9,731 strategic strikes against Republican Guard targets (the least number for any target category). Also noted is that the strategic campaign flew only 9,731 of the 41,309 Coalition strikes, 24 percent of the Coalition air effort. GWAPS vol 5, *A Statistical Compendium*, table 195, "Strikes by Master Target List Categories," p 535.

**Some of these surviving Republican Guard units that escaped through Basra — as well as other Republican Guard formations that were never committed to the KTO and were held in Baghdad during the conflict (to provide security for the Hussein regime) — proved instrumental in suppressing the postwar Shiite and Kurdish insurrections.

***This relatively unsatisfactory outcome did not result from any weakness of air power; it was a product of a decision insisted on by the U.S. Army Corps Commanders and concurred in by General Schwarzkopf that gave higher air priority to the frontline Iraqi divisions rather than the Republican Guard. It derived from a failure to clearly define campaign objectives. For DoD's breakdown of campaign results by target system, see DoD, *Conduct of the Gulf War*, pp 149–159.

Kuwait.* The three heavy divisions — the Tawakalna (We Trust in God) Mechanized Division and the Hammurabi and the Madinah Armored Divisions — held second-echelon or strategic reserve positions on the Kuwaiti-Iraqi border. Traditionally, after an operation (such as the invasion of Kuwait) the Iraqis withdrew the Republican Guard from the front to rest, rehabilitate, retrain for future operations, and sustain their elite edge by avoiding the daily depletion of human and material resources associated with frontline duty. Their internal security functions made it essential for Hussein's regime to preserve their fighting ability at all times. The southern and westernmost, as well as the most powerful, of these three divisions was the Tawakalna, covering the Wadi Al-Batin where the Iraqis anticipated a major Coalition ground attack. Its placement also gave it the potential to swing westward to confront a Coalition flanking attack (the actual Coalition maneuver), which made it the most militarily threatening of the Republican Guard divisions. Consequently, it received the heaviest aerial bombardment of any Republican Guard division, and it suffered the heaviest losses.[133] The Madinah Division had dug in and dispersed at the point where the Kuwaiti-Iraqi border began to curve to the south, while the Hammurabi Division stood on the border midway between the Persian Gulf and the Madinah Division.[134]

When Coalition ground and air commanders referred to the Republican Guard, they invariably meant these three units. The three kill boxes containing the heavy Republican Guard Divisions and parts or all of eight of twelve of the regular Iraqi Army heavy divisions ranked as the top three boxes in numbers of Coalition air strikes, and they absorbed almost one-third of the total fixed-wing strike sorties allotted to the KTO.[135] Prior to the start of the ground war, the Republican Guard heavy divisions had lost, by the most conservative estimates, at least 24 percent of their armored fighting vehicles to air attack.[136] By the end of the war, the Iraqi Army deployed in the KTO lost approximately 76 percent of its tanks, 55 percent of its armored personnel carriers, and 90 percent of its artillery, while the Republican Guard divisions suffered only a 50 percent loss in the same categories.[137]

Several factors account for the lighter loss among the Republican Guard heavy divisions. First was the nature of the desert itself. Many analysts have remarked that the open spaces of the desert served air power well by making concealment more difficult and by revealing enemy movement. On the other hand, the desert sand gave the defenders a somewhat compensatory advantage, for sand absorbed and muffled high-explosive effects and concussion of bombs and shells. Unlike more compact soils, which help to spread blast and

---

At the end of the conflict, the U.S. Army refused to allow Coalition aircraft to attack the fleeing Guards, but the VIIth Corps proved unable to halt their retreat into Iraq.

*Two to four independent Republican Guard brigades and some lesser units remained deep in Iraq, where they performed internal security duties, e.g., serving as the only troops allowed to garrison Baghdad.

fragments over wide areas, sand limits a shell's impact.* This factor reduced secondary bomb damage.

The Republican Guard's geographic position in the Iraqi theater, fifty or more miles removed from the front line, enabled it to exit the theater or avoid combat without having to abandon its vehicles to Coalition ground forces.[138] The heavy divisions' geographic position also put them closer to the Iraqi military's excellent combat engineering corps, which apparently constructed more permanent and bombproof revetments for the Republican Guard heavy divisions' equipment than for that of the other Iraqi formations closer to the front. Moreover, their distance from the front line meant that the heavy divisions did not have to deploy themselves into relatively tight tactical defensive positions meant to repel immediate Coalition ground attacks. Instead, they could disperse their formations over a much wider physical area, taking advantage of the blast protection of sand, which further attenuated the effect of tactical bombing done in the bulk with nonprecision, dumb ordnance. This extra protection made them a more difficult target to attack. Their distance from the front also increased the logistical effort needed to mount an air package against them, while their heavier air defenses made it more costly for Coalition aircraft to approach them.** The Coalition's policy of avoiding casualties also contributed the Republican Guard's survival. When CENTAF lost two A–10As to ground air defenses in Republican Guard areas on February 15, 1991, it limited that aircraft to shallower penetrations at higher altitudes. At the same time, at the insistence of Coalition ground force corps commanders, General Schwarzkopf concentrated tactical air efforts upon the Iraqi frontline divisions, which lessened the effort applied against the Republican Guard.

During the Coalition ground assault, the U.S. Army VII Corps caught up with the Tawakalna Division, which apparently served as a rear guard, inflicting severe casualties on it. The VII Corps also encountered elements of the Madinah Division (which suffered about a 50 percent loss of equipment from all Coalition ground and air action) as it retreated into an assembly of Iraqi forces near Basra, an area known as the Basra pocket. The Hammurabi Division (which sustained a 25 percent loss of equipment to all Coalition action) also retreated into the Basra pocket. Unlike the desert, the Basra pocket contained a large city, its suburbs, and numerous farming villages. Imagery shows that the

---

*Conversely, correctly timed airbursts and proximity-fuzed shells or bombs can be very effective in the desert because their blast is still spread over a wide area, and the defender finds it more difficult to prepare shelters with adequate overhead covering in the absence of proper, easily available local building materials.

**Regular Iraqi Army division air defense elements had a few Soviet SA–2s or –3s (maximum altitude, 25,000 meters), a few, more modern SA–9s (effective maximum altitude, only 4,500 meters), and some shoulder-launched SAMs. Republican Guard division air defense elements possessed more plentiful numbers of the more modern Soviet SA–6s (maximum effective altitude, 10,000–15,000 meters) and SA–13s (maximum effective altitude, 5,500 meters), as well as more modern SA–16 shoulder-launched missiles. All Iraqi heavy divisions had plentiful AAA.

Republican Guard and other Iraqi ground forces were well aware of the Coalition policy of limiting collateral damage and took advantage of this fact to huddle as near as they could to civilian structures within the pocket. This situation, plus poor weather that lowered bombing accuracy, frustrated air operations. The close proximity of Coalition ground forces heightened the chance of losses to friendly fire and required careful identification of ground targets. Finally, General Schwarzkopf placed the Iraqi territory near the border of the Islamic Republic of Iran off-limits in order to avoid airspace incursions and unnecessary international incidents.[139] Nonetheless, by the last day of the war, Coalition air power had damaged or destroyed all of the bridges leading

from the pocket.[140] On February 28, 1991, both the Hammurabi and Madinah Divisions had reached Iraqi-controlled territory in the Basra pocket.

The tale of the Republican Guard infantry* and special forces divisions is quickly told. They occupied positions at least ten to fifteen miles behind the Republican Guard heavy divisions in an arc stretching from An Nasiriyah to the Persian Gulf, which made them, of all the Iraqi ground forces in the KTO,

*U.S. intelligence sources differed as to the exact number of Republican Guard infantry divisions in the theater and variously placed that number at two, three, or four divisions. GWAPS uses the higher figure. I see no reason to disagree. GWAPS places the infantry divisions, from west to east within the KTO, in this order: Nebuchadnezzar, Baghdad, Adnan, and Al-Faw. The Special Forces Division appears to have stationed one brigade at each end of this line.

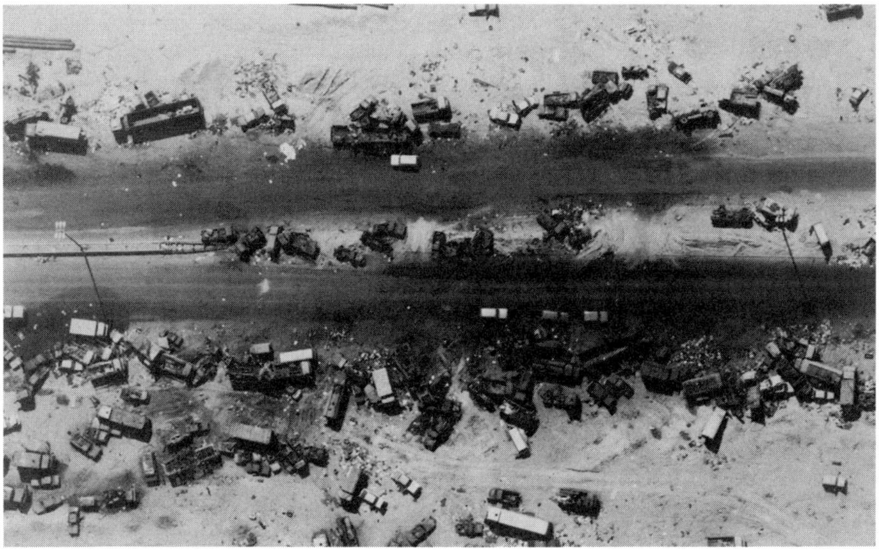

Bridge-busting produced traffic jams that inhibited hasty withdrawal from Kuwait City and made Iraqi vehicles vulnerable to Coalition bombing. The ensuing carnage amounted to what the media misleadingly termed the "Highway of Death," though actual Iraqi casualties were mercifully small.

closest to escape routes (the Euphrates River and Basra bridges) and the furthest removed from Coalition ground and air forces.[141] As some have pointed out, they not only occupied reserve positions, but positions that blocked the retreat of individual Iraqi deserters and all other Iraqi ground forces. These units (infantry forces far from the potential ground battle areas) offered little military threat to Coalition ground operations. If they moved forward, they would have to depart their prepared defensive positions and oppose heavy Coalition units on the march, a prospect which offered little hope for long-term survival. As infantry divisions far from the ground battle zones, they also received far less tactical air bombardment than Republican Guard and regular army heavy divisions. These units made no attempt to engage Coalition ground forces. By the end of the first twenty-four hours of the ground war, three had left the theater.[142] The remaining two divisions either left the theater soon afterward or retreated to the relative safety of the Basra pocket. Although not unscathed by Coalition air operations, the casualties of all five divisions is unknown.[143] It seems probable that they suffered light casualties in the course of the war. These units, assisted by troops left in Iraq and other relatively intact units escaping from the theater, would prove decisive in suppressing the later Kurdish and Shiite uprisings.*

The theater commander directed air and ground forces against the Republican Guard, identifying it as a center of gravity in a fashion that overemphasized its military threat to Coalition ground operations, and did not fully consider its political function of maintaining the Hussein regime. He concentrated the ground and air forces on the three Republican Guard heavy divisions, making little effort against the Republican Guard infantry, which was militarily insignificant and a difficult target for both air and ground operations. Given that the commander's direction of effort accented the *tactical* rather than the *strategic* implications of the target system and that CENTAF tactical air planners handled the physical targeting and planning for all strikes against that target system,** the Republican Guard ground forces in practice and actuality were not, and could not be, a strategic target system. Survival of the Republican Guard should not be used to discredit the concept of strategic bombing as a method of waging war. Indeed, the most remarkable aspect of the Gulf War was that no armored battle transpired equivalent to El Alamein, Kursk, Mortain,

---

*The lightly armed, haphazardly organized and trained, and internationally isolated rebels would stand little chance of sustained resistance against regular Iraqi Army units, *let alone* the Republican Guard. The Iraqis also had large stockpiles of second-line armored and other equipment, notably around Tikrit, that they could and did use to reequip their forces. Thus, they could readily field repatriated prisoners or units that escaped with their personnel but abandoned their equipment.

**In fact, the internal records of the CENTAF Strategic Planning Cell — such as a complete listing of planned strategic missions for the entire war, the "Missions, Day-by-Day Log" — do not list a single mission against the Republican Guard in Kuwait, only for the Guard headquarters in Baghdad, as part of the campaign against leadership and C^3 targets. The CENTAF strategic planners obviously did not consider the Guard to be a strategic target.

or those of the 1973 Arab-Israeli War.

In spite of the well-deserved praise for the U.S. Air Force's superlative performance in the Gulf War, that conflict did reveal two potentially grave organizational flaws: the difficulty of melding the U.S. Air Force planning, operations, and intelligence functions into a smoothly functioning team; and the lack of an organization at the operational level charged with strategic targeting and planning. Looking back on the war General Horner stated,

> One final area that requires significant attention and change is intelligence. In peacetime, we train our intelligence personnel to hedge, to be mediocre rather than wrong. In wartime, it all starts and ends with intelligence. Intel defines what you need to do and how you will do it. Each of the various communities that make up our military staffs — A-2 Intelligence, A-3 Operations, A-4 Logistics and A-6 Communications — is important. But each develops self-serving organizational walls and formal processes of interaction designed to protect its own prerogatives. This is inappropriate in war, where success depends on detailed interaction among all of them. Where air power is concerned, staffs should be reorganized so that the artificial walls of the past are eliminated and cooperation is enhanced. Perhaps the way to do this is to create a functional staff along the lines of "strategy" needed to plan the campaign and "execution" needed to carry out the ATO once it is created [146]

As of 1997 the service had not addressed the problems observed by General Horner. After the war the Black Hole, always an ad hoc group, disappeared as its members returned to regular duties. Within two years of the end of the war, an air staff reorganization eliminated the Deputy Directorate of Warfighting Concepts. As of 1997 within the U.S. Air Force's numbered air forces, which will form the core of the U.S. Air Force components of any Joint Command, strategic thinking has been instituted a haphazard fashion. What should be done is yet to be decided, but the longer the problems continue, the less likely their solution and the more likely their reoccurrence in a future conflict.

Especially when compared with earlier strategic bombing, the Gulf War strategic air campaign had a final and significant achievement: it avoided inflicting large numbers of civilian (and military) casualties. Although an initial (but later disavowed) DIA estimate reported 100,000 dead and 300,000 wounded in Kuwait — a sum larger than all Iraqi forces in Kuwait — by 1993 the figure of Iraqi military casualties had shrunk to 700–2,000 dead and 3,000–7,000 wounded. John G. Heidenrich determined the latter figures, basing his analysis on the absence of mass Iraqi graves in Kuwait, lack of mammoth field hospitals containing Iraqi wounded, and the small number of wounded Iraqi prisoners. He argued convincingly that few wounded means even fewer dead. He placed the number of Iraqi civilian dead at less than 1,000.[144] Both figures are a testimonial to the extent to which Coalition air strikes concentrated on equipment rather than on men and to the care that the Coalition exercised in avoiding collateral damage.

Conclusion

The planners who assembled the strategic air campaign plan against Iraq could take satisfaction in the results. The swift, devastating attacks against the Iraqi air defense network, the Iraqi air force, and communications, electrical power, and transportation targets set the stage for the rapid and overwhelming destruction of a nation that had, on the eve of the war, possessed a very large and reasonably proficient military force and a record of inflicting punishing losses on an attacker. In any case, Iraq had no hope of defeating the Coalition; after the first night's aerial blitz, its ability to defend itself began a precipitous decline. The war reaffirmed the most important lesson of air power — without air superiority, a nation loses the ability to exercise its national prerogatives. The success of the air campaign against emplaced and fielded Iraqi forces and the Iraqi Navy likewise demonstrated that in the current era when one side attains and retains air superiority, surface forces of the enemy are held hostage by friendly aerial forces. The war also confirmed the powerful synergy of high technology, particularly the important interrelationships between tankers and airlift and between airlift, tankers, and strike aircraft and precision munitions. In particular, the highly controversial stealth fighter (the F–117A) proved its overwhelming value, ending a debate on the relative merits of stealth that raged throughout the 1980s. The war emphasized that high-technology systems with precision munitions in sufficient numbers can offset more numerous, less-sophisticated ones. It highlighted advances in SAMs and AAA that have essentially made the low-altitude environment off-limits to conventional strike aircraft, and it forced a resurgence of interest in greater distance, stand-off attacks employing autonomous or near-autonomous precision munitions. Future advances of such antiaircraft weapons that would enable them to operate at higher altitudes and longer ranges may force less reliance on expensive, slow-to-produce manned aircraft and more on mass-produced unmanned aerial vehicles. In the air campaign, the greatest disappointment was the Great Scud Hunt. This is an area that will require greater attention in the future as Scud-like ballistic rockets and cruise missile systems proliferate.

If the strategic air campaign against Iraq did not achieve all of the lofty goals of the CHECKMATE and CENTAF Special Planning Group — particularly, the collapse of the Hussein regime — its inability to do so echoed Robert Burns' aphorism that "a man's reach should exceed his grasp, or what's a Heaven for?" It certainly fulfilled the theater commander's expectations of what air power should do, and it played a crucial role in fulfilling President George Bush's political objectives. (After the war, speaking at the Air Force Academy, Bush declared that "Gulf Lesson One is the value of air power.") Those who fault the strategic air campaign in the Gulf War because it failed to achieve 100 percent success argue for perfection, which no work of mankind has yet attained. What can be said is this: The *strategic* air campaign against Iraq was *a* decisive factor in Iraq's defeat. More important, when joined to the *tactical* air effort against Iraqi forces — the element that consumed nearly 75 percent of the total air effort — strategic and tactical air power combined constituted *the* decisive factor in the Coalition's quick and almost bloodless victory in the Persian Gulf War.

NOTES

Although the author consulted classified sources in the preparation of this chapter, all informaiton contained herein has been declassified.

1. Briefing Slide, "Instant Thunder," from briefing, Col John A. Warden, U.S. Air Force, Chief of the Deputy Directorate of Warfighting Concepts, Air Staff, to Gen Schwarzkopf, MacDill AFB, Fla., Aug 17, 1990.

2. Maj A. J. C. Lavalle, ed., *The Tale of Two Bridges*, U.S. Air Force Southeast Asia Monograph Series (Washington, D.C.: GPO, 1976), vol 1, monogr 1, pp 84–86.

3. Richard P Hallion, *Storm over Iraq: Air Power and the Gulf War* (Washington, D.C.: Smithsonian Institution Press, 1992), app E, "The 'Smart' Bomb," pp 303–307. This is a useful short history of GBU use and development. See also Ray Braybrook, "On Target!: A Review of Precision Air Attacks in the Gulf War," *Air International*, vol 41, no 4 (Oct 1991), pp 177–182; David R. Mets, *Quest for a Surgical Strike: The United States Air Force and Laser Guided Bombs* (Eglin AFB, Fla.: Armaments Division History Office, 1987).

4. For the F–117 figure, see slide 741622, Air Force Stealth Technology Review, Jun 10–14, 1991, F–117 folder, DESERT STORM files, Deputy Directorate of Warfighting Concepts. As of mid-1990 the F–15E Strike Eagles, 4th TFW, Seymour Johnson AFB, N.C., had not yet received their low-altitude navagation and targeting infrared for night (LANTIRN) targeting pods and were not precision-guided munition–capable. By the beginning of 1991 some of the F–15Es did have targeting pods.

5. Capt Jeffery A. Paulk, "Unguided BLU–109/B," *U.S. Air Force Fighter Weapons Review*, spring 1990, p 32.

6. *Ibid.*, p 33.

7. For a day-by-day description of the campaign, see Brig Gen James R. McCarthy and Lt Col George B. Allison, *LINEBACKER II: A View from the Rock*, U.S. Air Force Southeast Asia Monograph Series (Maxwell AFB, Ala.: Airpower Research Institute, Air War College, 1979), vol 6, monogr 8.

8. Earl H. Tilford, Jr., *Set Up: What the Air Force Did in Vietnam and Why* (Maxwell AFB, Ala.: Air University Press, 1991), p 263. For probably the best historical overview, see Karl J. Eschmann, *LINEBACKER: The Untold Story of the Air Raids over North Vietnam* (New York: Ballantine Books, 1989).

9. U.S. Air Force Statistical Digest, Fiscal Year 1973 (28th ed.), Jul 31, 1974, p 59.

10. Richard H. Shultz, Jr., and Robert L. Pfaltzgraff, Jr., eds., *The Future of Air Power in the Aftermath of the Gulf War* (Maxwell AFB, Ala.: Air University Press, Jul 1992), p 220. In September 1990 the Strategic Air Command had approximately 300 bombers, a little more than 200 of which were B–52Gs and Hs.

11. McCarthy and Allison, *LINEBACKER II*, p 101.

12. For a thumbnail sketch of the Stealth Fighter's development, see Hallion, *Storm over Iraq*, pp 293–294.

13. DoD, *Conduct of the Persian Gulf Conflict: An Interim Report to Congress* (Washington, D.C.: DoD, Jul 1991), p 6–2.

14. Briefing Slide, "F–117 Program," Air Force Stealth Technology Review, Jun 10–14, 1991.

15. Marcelle Size Knaack, *Post–World War II Fighters, 1945–1973*, U.S. Air Force Reference Series (Washington, D.C.: Office of Air Force History, 1986), p 259 n 113.

16. Public lecture, Richard P. Hallion, "The Stealth Revolution," Smithsonian Institution, Washing-

ton, D.C., Apr 4, 1992.

17. Thomas A. Keaney and Eliot A. Cohen, *Gulf War Air Power Survey: Summary Report* (hereafter Keaney and Cohen, *GWAPS Summary Report*) (Washington, D.C.: GPO, 1993), p 194.

18. For a detailed survey of satellite contributions, see the Gulf War Air Power Survey (GWAPS) vol 4, *Space Operations*. See also Gen Donald J. Kutyna, U.S. Air Force, "The Military Space Program and DESERT STORM," *The Space Times*, Sep–Oct 1991, pp 3–7.

19. For descriptions of AirLand Battle Doctrine and the U.S. Air Force's participation in it, see John L. Romjue, *From Active Defense to Air Land Battle: The Development of Army Doctrine, 1973–1982* (Fort Monroe, Va.: TRADOC Historical Office, Jun 1984), pp 51–86, and Richard G. Davis, *The 31 Initiatives: A Study in Air Force–Army Cooperation* (Washington, D.C.: Office of Air Force History, 1987), pp 24–33.

20. Intvw, Diane T. Putney, Richard G. Davis, and Perry Jamieson, Center for Air Force History, with Col John A. Warden III, Special Advisor to the Vice President for Policy Studies and National Security Affairs, Office of the Air Force Historian, Bolling AFB, D.C., Feb 6, 1992.

21. John A. Warden III, *The Air Campaign Plan: Planning for Air Combat* (Washington, D.C.: National Defense University Press, 1988).

22. Intvw tape, Richard G. Davis, Center for Air Force History, with Col David R. Tretler, Deputy Air Force Historian, Bolling AFB, D.C., Jan 14, 1992. Colonel Tretler served in the Deputy Directorate of Warfighting Concepts in 1988 and 1989.

23. *Ibid.*

24. Tape recording, Col John A. Warden III, Commandant, Air Command and Staff School, "Review and Comments on Draft Manuscript," ACSC Commandant's Office, Maxwell AFB, Ala., Mar 2, 1993.

25. Speech, Col John A. Warden, "Strategic Warfare: The Enemy as a System," Jan 3, 1993. The text of this speech contains Colonel Warden's expansion of the five rings and his analysis of the Persian Gulf War in terms of the rings. He presents another mature version of the five rings in Shultz and Pfaltzgraff eds., *Future of Air Power*, pp 64–68.

26. Anthony H. Cordesman and Abraham R. Wagner, *The Lessons of Modern War*, vol 2, *The Iran-Iraq War* (Boulder, Colo.: Westview Press, 1990), table 1.2, World Oil Reserves, pp 6–7.

27. Don Oberdorfer, "Missed Signals in the Middle East," *Washington Post Sunday Magazine*, Mar 17, 1991, p 20.

28. Rick Atkinson, *Crusade: The Untold Story of the Persian Gulf War* (Boston, Mass.: Houghton, Mifflin & Co., 1993), pp 50–51.

29. William Scott Malone, "Did the U.S. Teach Iraq to Hide Its Terror Arms?" *The Washington Post*, Nov 3, 1991, p C1.

30. George Lardner, Jr., "Gonzalez's Iraq Expose: Hill Chairman Details U.S. Prewar Courtship," *The Washington Post*, Mar 22, 1992, p A1. Rep Henry B. Gonzalas (D–Tex), Chairman of the House Banking Committee, conducted a series of hearings on prewar U.S.-Iraq relations. His entre was the scandal involving the Atlanta Branch of the Italian government–owned *Banco de Nazionale del Lavoro*, which apparently served as a front for the Iraqi government. During the course of his hearings in the spring of 1992, Chairman Gonzalez read dozens of classified documents concerning prewar relations into the *Congressional Record*. They serve as useful (if out of context) sources.

31. Malone, "Did the U.S. Teach Iraq?" p C4. Mr. Gates made this statement while testifying before the Senate Intelligence Committee during his confirmation hearings to become Director of the CIA.

32. Gen H. Norman Schwarzkopf, *The Autobiography: It Doesn't Take A Hero* (New York, New York: Bantam Books, 1992), p 299.

33. DoD, *Conduct of the Persian Gulf War: Final Report to Congress* (Washington, D.C.: DoD, Apr 1992) (hereafter DoD, *Conduct of the Gulf War*), p 65.

34. *Ibid.*, p 19.

35. *Ibid.*, p 20.

36. *Ibid.*, p 65.

37. *Ibid.*, p 91.

38. Telephone conversation, Diane T. Putney, Center for Air Force History, with Gen H. Norman Schwarzkopf, USA (Ret), May 5, 1992.

39. Atkinson, *Crusade*, p 60.

40. DoD, *Conduct of the Gulf War*, p 94.

41. Briefing Slide, "Strategic Air Campaign INSTANT THUNDER," from briefing, Col John A.

Warden III to the Center for Air Force History, Feb 6, 1992.

42. DoD, *Conduct of the Gulf War*, pp 92–93.

43. Briefing Slide, 14: "Expected Results," Warden briefing to CAFH, Feb 6, 1992.

44. For a fuller discussion of the target sets, see DoD, *Conduct of the Gulf War*, pp 95–98.

45. *Ibid.*, pp 93–94.

46. Atkinson, *Crusade*, pp 61–62.

47. Operations Order, COMUSCENTAF, "Offensive Campaign — Phase I," Sep 2, 1990, box 12, papers of Lt Col David A. Deptula, U.S. Air Force History Support Office.

48. DoD, *Conduct of the Gulf War*, p 66.

49. *Ibid.*, p 101.

50. Briefing Slides, HO31931BC: "Deployment of Selected U.S. Air Force Assets" and HO31931BD: "Coalition Air Strength on Eve of War," from briefing, Richard Hallion, Langley AFB, Va., Sep 28, 1993.

51. Benjamin F. Schemmer, "Six Navy Carriers Launch Only 17% of Attack Missions in DESERT STORM," *Armed Forces Journal International*, Jan 1992, chart, p 13.

52. DoD, *Conduct of the Gulf War*, p 102.

53. *Ibid.*, p 94.

54. *Ibid.*, p 102.

55. Atkinson, *Crusade*, p 63.

56. *Ibid.*

57. Briefing Slide, 10: "THE AIR CAMPAIGN: Planning and Execution," Lt Col David A. Deptula, U.S. Air Force, Secretary of the Air Force Working Group, 1st revision, n.d. [*ca.* Nov 1, 1991], in Deptula Papers, Gulf War Air Power Survey Collection, U.S. Air Force Historical Research Agency, Maxwell AFB, Ala.

58. Memo, Lt Comdr D. J. Muir, subj: The Targeting Process in Operation DESERT STORM, Mar 4, 1991. Lt. Comdr. Muir served as one of the Navy planners assigned to the CENTAF planning shop.

59. Capt Steven B. Michael, comp., *The Persian Gulf War: An Air Staff Chronology of DESERT SHIELD–DESERT STORM* (Washington, D.C.: Center for Air Force History, 1992), pp 219, 228.

60. DoD, *Conduct of the Gulf War*, p 154.

61. Briefing Slide, HO31931BF: "Iraqi Fighter-Interceptor Threat," Hallion briefing, Langley AFB, Va.

62. DoD, *Conduct of the Gulf War*, chart, SAM/AAA Threat, January 1991, p 177.

63. *Ibid.*, chart, Iraqi SAM/EW Radar Activity, p 150.

64. *Ibid.*, p 154.

65. *Ibid.*, p 74, citing Schwarzkopf's Operations Order 91–001, Jan 17, 1991. Also see GWAPS vol 2, *Effects and Effectiveness*, table 23, DESERT STORM Objectives and Constraints, p 277.

66. DoD, *Conduct of the Persian Gulf War*, chart, Theater Campaign Plan and Military Objectives, p 75.

67. *Ibid.*, p 75.

68. Msg, SecDef, Washington DC, to AIG 8798 *et al.*, subj: DoD Press Conference of Jan 16, 92, 170831Z Jan 92.

69. Printout, F–117A Mission Results Days 1–34, p 1.

70. GWAPS vol 5, *A Statistical Compendium and Chronology*, table 204, DESERT SHIELD/STORM: Total U.S. Air Force, USN, USMC Weapons Cost and Utilization (FY 90/91$), p 582. In all, the U.S. Navy fired 298 TLAMs during the conflict.

71. David A. Fulgham, "Secret Carbon-Fiber Warheads Blinded Iraqi Air Defenses," *Aviation Week & Space Technology*, Apr 27, 1992, pp 18–20. For first-hand accounts of the supposed effects of this warhead, see the Harvard Study Team report, "Public Health in Iraq After the Gulf War," May 1991, pp 22–24. This group made a close inspection of Iraqi power plants and interviewed plant employees.

72. DoD, *Conduct of the Gulf War*, map, The First Wave (Planned) 0239 (L)–0525 (L), p 117.

73. *Ibid.*, p 120.

74. Paper, Richard G. Davis, "Dismantling Iraq: The Coalition's Strategic Air Campaign," Nov 1993, p 11. This paper provided documentation and additional data to accompany the author's DESERT STORM Road Show presentation.

75. For discussion of this disarming attack on Iraqi physical air defenses, see James P. Coyne, *Airpower in the Gulf War* (Arlington, Va.: The Air Force Association, 1992), p 71, and Keaney and Cohen, *GWAPS Summary Report*, p 12.

76. Atkinson, *Crusade*, pp 88–89.
77. DoD, *Conduct of the Gulf War*, pp 154–155; Atkinson, *Crusade*, pp 88–89.
78. Printout, F–117A Mission Results Days 1–34, p 1.
79. DoD, *Conduct of the Gulf War*, p 119.
80. GWAPS vol 5, *Statistical Compendium*, table 187, Strikes by AIF [Automated Intelligence Installation File] Categories, p 436. GWAPS states that it considers the AIF the "single authoritative source for U.S. targeting efforts." It defines strike as "the delivery of a weapon or weapons against a specific target, with no subjective assessment as to the degree of success of that strike." For example, if an F–117, which normally carries two precision-guided munitions, drops both bombs on one target, that would count as one "strike" and one attack "sortie." However, if that same aircraft had dropped its bombs on two different targets, that would count as two "strikes" but still only one "sortie." For the Gulf War, 95 percent of the sorties (almost all non-precision guided munition weapons) equaled strikes. *Ibid.*, p 421.
81. GWAPS vol 2, *Effects and Effectiveness*, p 99.
82. GWAPS vol 2, *Operations*, pp 335–337.
83. DoD, *Conduct of the Gulf War*, p 676.
84. *Ibid.*, p 703.
85. Working paper, "F–117 Summary Data," n.d., based on GWAPS Missions Database, Aug 26, 1993.
86. Briefing Slide, "1700–1735{Z}," from briefing, Comdr Donald McSwain, USN, to Richard G. Davis, USAF Historian, Pentagon, Apr 12, 1992.
87. Briefing Slide, "Strategic Resupply, Baghdad to Southern Iraq," briefing, Col John A. Warden III to Secretary Richard Cheney, Checkmate War Room, Pentagon, Feb 6, 1991.
88. Briefing Slide, HO31931BY: "What the War Accomplished," Hallion briefing at Langley AFB, Va.
89. DoD, *Conduct of the Gulf War*, chart, Iraqi SAM/EW Radar Activity, p 150. By January 22, the average fell to one hundred emitters per day and stayed at that level for the remainder of the war. The hundred emitters represented mostly early warning radars; few SAM/AAA gunners risked turning their search equipment on.
90. Atkinson says that by February 11th the war had reduced Baghdad to "nineteenth-century deprivations." *Crusade*, p 282.
91. DoD, *Conduct of the Gulf War*, chart, Air Campaign — Sorties by Phase, p 101; Coyne, *Airpower in the Gulf War*, table, Sorties against Scuds, p 57.
92. DoD, *Conduct of the Gulf War*, chart, Dedicated Scud Sorties/Scuds Launched, p 165.
93. Intvw, Richard Davis, Perry Jamieson, and Barry Barlow with Gen Charles A. Horner, Comdr 9th AF, Shaw AFB, S.C., Mar 4, 1992.
94. Intvw, Richard Davis, Perry Jamieson, and Diane Putney with Lt Col David A. Deptula, Secretary of the Air Force Working Group, Pentagon, Washington, D.C., Jan 8, 1992.
95. DoD, *Conduct of the Gulf War*, p 154.
96. *Ibid.*, p 128.
97. Anthony H. Cordesman, *After the Storm: The Changing Military Balance in the Middle East* (Boulder, Colo.: Westview Press, 1993), p 537.
98. Transcript, National Public Radio, *Morning Show*, Sep 24, 1997.
99. Contingency Hist, 37th TFW(P), Jan 20–26, 1991, p 23.
100. Targets Attacked Day by Day by Aircraft, Iraqi Cell, CENTAF HQ, Jan – Mar 1991.
101. Contingency Hist, 37th TFW(P), Feb 7–14, 1991.
102. DoD, *Conduct of the Gulf War*, p 141. See also Atkinson, *Crusade*, pp 275–277. Atkinson claims the United States had definite electronic intercepts emanating from an Iraqi gestapo unit within the bunker and claims confirmation by a very reliable human intelligence source.
103. Intvws, Richard Davis, Perry Jamieson, and Diane Putney with Maj Gen Buster C. Glosson, CAFH Office, Bolling AFB, Dec 11, 1991, and with Lt Col David Deptula, Pentagon, Washington, D.C., Jan 8, 1992.
104. Intvw, Donna Clark with Maj Michael B. Hoyes and Capt John B. Wilcox, Historian's Office, USAF Fighter Weapons Center, Nellis AFB, Nev., Mar 29, 1991, p 40.
105. Notes, Lt Gen Buster C. Glosson, "CINCCENT Daily Priorities," Feb 12, 1991. Both General Glosson and Colonel Deptula insist that the original date is in error and that the new orders came down after the Al Firdos District Bunker incident.
106. Targets Attacked by Day by Aircraft, Iraqi Cell, CENTAF HQ, Feb 1991.

107. Contingency Hist, 37th TFW(P), Feb 24, 1991.
108. Gary Milhollin, "The Iraqi Bomb," *The New Yorker*, Feb 1, 1993, pp 47–56.
109. DoD, *Conduct of the Gulf War*, p 148.
110. GWAPS vol 2, *Effects*, fig 30, Total Strikes against L and CCC during DESERT STORM, p 284.
111. DoD, *Conduct of the Gulf War*, p 150.
112. *Ibid.*, p 152.
113. Norman Cigar, "Iraq's Strategic Mindset and the Gulf War," *The Journal of Strategic Studies*, Mar 1992, p 25.
114. Memo, DIA, "Iraqi Short-Range Ballistic Missiles in the Persian Gulf War: Lessons and Prospects," Mar 1991, cited in William M. Arkin, "Baghdad: The Urban Sanctuary in DESERT STORM?" *Airpower Journal*, vol 11, no. 1 (spring 1997), p 16.
115. GWAPS vol 2, *Effects*, p 291.
116. *Ibid.*, p 303.
117. *Ibid.*, pp 312, 315.
118. GWAPS vol 2, *Operations and Effects and Effectiveness*, pp 318–320.
119. *Ibid.*, p 333.
120. DoD, *Conduct of the Gulf War*, p 155.
121. GWAPS vol 2, *Effects*, p 330 n 146. Documentation turned over to the UN by Iraq in August 1995 affirms that Iraq had biological weapons ready to deploy at the start of hostilities.
122. DoD, *The Conduct of the Gulf War*, p 159.
123. *Ibid.*, p 156.
124. GWAPS vol 2, *Effects*, pp 340–341.
125. DoD, *Conduct of the Gulf War*, p 168.
126. GWAPS vol 2, *Effects*, pp 339, 341.
127. DoD, *Conduct of the Gulf War*, p 154.
128. *Ibid.*, p 157.
129. *Ibid.*, p 159. GWAPS offers a more pessimistic view, noting that by Feb 28, 1991, bombing had rendered all the railroad bridges unusable but only a little less than one-third of the highway bridges unusable. GWAPS vol 2, *Effects*, p 181. The difference in the two figures stems from differing definitions of bridges. DoD, like CENTAF and Checkmate, counted only the fifty-four "major" bridges, while GWAPS counted all bridges.
130. GWAPS vol 2, *Effects*, p 189.
131. GWAPS vol 2, *Operations*, p 332.
132. *Ibid.*, p 334.
133. DoD, *Conduct of the Gulf War*, p 158.
134. For an example of the controversy beginning to surround this subject, see William S. Lind, Letter to the Editor, *U.S. Naval Institute Proceedings*, vol 119, no 12 (Dec 1993), pp 23–24, and J. G. Burton "Pushing Them Out the Back Door," *U.S. Naval Institute Proceedings*, vol 119, no 6 (Jun 1993), pp 37–42.
135. The two strike boxes (AE6 and AF6) that contained the Tawakalna Division received almost one-quarter of all Coalition fixed-wing strike sorties directed at kill boxes in the KTO. GWAPS vol 5, *Statistical Compendium*, table 191, Kill Box Strikes, p 487.
136. 24th Infantry Division (Mech), *24th Mechanized Infantry Division Combat Team, Operation DESERT STORM, Attack Plan OPLAN 91–3* (Fort Stewart, Ga.: 24th Infantry Division [Mech], Apr 1991), Offensive Operations, Jan 17, 1991, app 1, Intelligence Estimate, tab C, Enemy Situation, to annex B, Intelligence, pp B-1–C-3 and C-4.
137. DESERT STORM Iraqi vehicle loss summary from Maj Richard King, AF/XOXW–G, Jun 27, 1991. See also GWAPS vol 5, *Statistical Compendium*, tables 189, KTO Ground Order of Battle (as Feb 21, 91), and 191, Kill Box Strikes, pp 482–483, 487. Table 191 notes that the three boxes (AF7, AG7, and AE6) received 6,517 strikes of the 21,973 sent against kill boxes.
138. Keaney and Cohen, *GWAPS Summary Report*, p 106.
139. *Ibid.*
140. GWAPS vol 2, *Effects*, table 15, Equipment Destroyed or Abandoned in Republican Guard Heavy Division Areas, p 217. The division lost approximately 93 percent of its tanks, 76 percent of its armored personnel carriers, and 50 percent of its artillery.
141. GWAPS vol 2, *Operations*, p 323.
142. GWAPS vol 2, *Effects*, pp 258–259.
143. *Ibid.*, map 8, Cumulative Air Strikes by Kill Box Jan 17–Feb 28, 1991 (Total Strikes, 21,391),

p 224.

144. *Ibid.*, p 253.

145. *Ibid.*

146. General Charles A. Horner, "Air Power 'Growing Beyond DESERT STORM,'" *Aviation Week & Space Technology*, vol 146, no 16 (Apr 16, 1997), p 74.

147. John G. Heidenrich, "The Gulf War: How Many Died?" *Foreign Policy*, no 90 (spring 1993), pp 108–125. The figures cited indicate Heidenrich's best guess at Iraqi military casualties. He does speculate (p 125) that Iraqi battle deaths might go as high as 6,000. His article, because of the low estimate, has generated some controversy. See subsequent issues of *Foreign Policy* for a lively dispute. None of Heidenrich's critics has yet overcome the absence of Iraqi bodies and scarcity of Iraqi wounded. On the so-called Highway of Death between Kuwait and Basra, for example, observers found as many as 300 Iraqi dead and no fewer than 1,400 destroyed vehicles. The scene was clearly a vehicular graveyard, not a human one.

Bibliographic Essay

Primary and Official Sources

This chapter was submitted for publication in May of 1994. This essay reflects the fact the author finished the bulk of his research at that time. Two earlier versions of this chapter have been published by the USAF History and Museums Program: the monograph *Strategic Air Power in DESERT STORM*, and the commemorative pamphlet in the 50th Anniversary of the USAF series, *Decisive Force: Strategic Bombing in the Gulf War*. These earlier versions illustrate the difficulty of writing contemporaneous history before all the details have emerged. This chapter corrects errors of interpretation and fact made by the author and eliminates other interpolations as of 1997.

The Persian Gulf War, a brief, single-theater conflict involving a half million American servicemen and servicewomen, produced an enormous amount of documentation. U.S. forces generated paper at a rate of approximately 42,000 pages per day, compared with about 17,000 pages per day for all U.S. forces in all theaters in World War II. Clearly, the historian has a plethora of potential source material. Because of the current nature of much of the tactics, weapons, intelligence methods, and relations with friendly powers, however, the bulk of the material prepared during the war is still security classified and is likely to remain so for a decade or two. What is more, the Gulf War historian is in a difficult position, much like World War II historians in the postwar era, who "didn't know what they didn't know." The "secret history of the secret history" remains to be written.

The U.S. Air Force Historical Research Agency (AFHRA) at Maxwell AFB, Alabama, is the repository of most of the historically significant U.S. Air Force documentation on the war, most of which is security classified. In addition to unit and command annual histories for the period, AFHRA's holdings include the records of the Black Hole/Special Campaign Planning Group and several hundred Hollinger boxes containing records collected by the U.S. Air Force Historical Program in-theater during the conflict. Thanks to the efforts of hard-working U.S. Air Force enlisted historians at bases throughout the Persian Gulf and at CENTAF headquarters, in Riyadh, the U.S. Air Force History Program collected and preserved thousands of key documents. Future historians will find U.S. Air Force activities for DESERT SHIELD/STORM by far the best documented among the military services. Many of AFHRA's Gulf War holdings are now on microfilm.

The primary official source in the open literature cited in this chapter is DoD, *Conduct of the Persian Gulf War* (Washington, D.C.: DoD, unclassified version, Apr 1992) written for the U.S. Congress in Response to Title V of the Persian Gulf Conflict Supplemental Authorization and Personnel Benefits Act of 1991 (Public Law 102–25). Several versions of this work exist. The interim (unclassified) version of July 1991 is of interest because it has not yet been homogenized to fit the joint perspective and, therefore, contains sharper disagreements between the services. Two final editions of the work, one security classified and one unclassified, bear the publication date of April 1992. Approximately 30 percent of this chapter's notes refer to *Conduct of the War*, primarily because this work, although somewhat uncritical of service shortcomings, is an excellent, fact-filled narrative/compendium covering the war. The author also employed *Conduct of the War* because, on numerous occasions, it provided an approved "unclassified" source for information the author already possessed in security classified form.

The Gulf War Air Power Survey (GWAPS), paid for and published under the aegis of the Secretary of the Air Force, Donald B. Rice, is an independent study of the conduct and results of the Gulf War, and the U.S. Air Force does not consider it an official work. The multivolume GWAPS study is itself a useful source of information amassed from a variety of sources. The U.S. Government Printing office has published an unclassified single-volume summary of the GWAPS findings written by Eliot A. Cohen and Thomas A. Keaney, patterned on the summary reports of the U.S. Strategic Bombing Survey of World War II. (Keaney and Cohen, *Gulf War Air Power Survey Summary Report* [Washington, D.C.: GPO, 1993].) All GWAPS-collected and -created documentation has been retired to the AFHRA at Maxwell for preservation and microfilming. Its documentary collection includes several dozen feet of significant and unique material generated and gathered by Checkmate from August 1990 through May 1991. For a brief summary of GWAPS' views by the coauthor of the GWAPS volume on effects and effectiveness, see Thomas Keaney (Col USAF), "Surveying Gulf War Airpower," *Joint Forces Quarterly*, no. 2 (autumn 1993), pp 25–36.

Three Air Force colonels, Edward C. Mann III, Suzanne B. Gheri, and Richard T. Reynolds at the Air University, Maxwell AFB, provided a valuable addition to Gulf War historiography by interviewing several senior officers for their unclassified work on air planning in the Gulf War. Richard G. Davis' unclassified work, *Roots of Conflict: A Military Perspective on the Middle East and Persian Gulf Crisis* (Washington, D.C.: Center for Air Force History, 1993) is an introduction for students beginning their studies of the region.

Published Sources

American wars invariably produce numerous works of varied viewpoint and quality. Two works by reporters associated or once associated with the *Washington Post* are well written, well researched, and well informed: Rick Atkinson, *Crusade: The Untold Story of the Persian Gulf War* (Boston, Mass.: Houghton-Mifflin, 1993) and Bob Woodward, *The Commanders* (New York: Simon & Schuster, 1991). Norman Schwarzkopf's memoirs, *General H. Norman Schwarzkopf the Autobiography: It Doesn't Take a Hero* (New York: Bantam Books, 1992) became the first of a major U.S. commander in the war to appear in print. From the air point of view it reveals his concentration on the land campaign and, unfortunately, shows a propensity to mix up dates and events. On the other hand he acknowledges some of air power's achievements and credits Colonel Warden with the ideas behind the strategic campaign.

Several histories of the conflict have appeared since its end. Norman Friedman's *Desert Victory: The War for Kuwait* (Annapolis, Md.: Naval Institute Press, 1991) was obviously rushed into print shortly after the war and, as a result, must be used with caution. Maj. James Blackwell (USA, Ret) *Thunder in the Desert: the Strategy and Tactics of the Persian Gulf War* (New York: Bantam Books, 1991) presents a relatively balanced picture of the war itself. Richard P. Hallion's *Storm over Iraq: Air Power and the Gulf War* (Washington, D.C.: Smithsonian Institution Press, 1992) offers a background in air power history, a pro–U.S. Air Force overview of the air war, and appendices on various weapons systems. James P. Coyne's *Airpower in the Gulf* (Washington, D.C.: Air Force Association, 1992) contains useful interviews and other colorful items. Lawrence Freedman and Efraim

Karsh, *The Gulf Conflict, 1990–1991: Diplomacy and War in the New World Order* (Princeton, N.J.: Princeton University Press, 1993) offers excellent insights into the diplomacy of the Coalition and the conflict. Perhaps the best overall history of the war, which appeared too late to incorporate into this chapter, is Michael R. Gordon and Bernard E. Trainor, *The Generals' War: The Inside Story of the Conflict in the Gulf* (Boston: Little, Brown, 1995). For additional regional background see Trevor Mostyn, *Major Political Events in Iran, Iraq and the Arabian Peninsula 1945–1990* (New York: Facts on File, 1991). Two other useful background works are Samir al-Khalil, *The Republic of Fear: The Inside Story of Hussein's Iraq* (New York: Pantheon Books, 1989), which demonstrates the level of repression that Saddam Hussein's organs of internal control have achieved over the Iraqi populace, and Stephen C. Pelletiere and Douglas V. Johnson II's, *Lessons Learned: The Iran-Iraq War* (Carlisle, Pa.: Strategic Studies Institute, U.S. Army War College, 1991).

The war spawned numerous newspaper and journal articles, essays, and so on. As his newspaper source, the author relied on *The Washington Post* as well as on the big three weekly news magazines: *Time*, *Newsweek*, and *U.S. News and World Report*. The number of Iraqi civilian and military dead remains an open question. The most authoritative and, perhaps, the most well reasoned article on Iraqi casualties is John G. Heidenrich, "The Gulf War: How Many Died?" *Foreign Policy*, no. 90 (spring 1993), pp 108–125. *Air Power History*, in its fall 1991 and spring 1992 issues carried a number of articles by Air Force Major Command History Offices on the roles of their commands in the war. For a controversial article on the escape of the Republican Guard, see James G. Burton, "Pushing Them Out the Back Door," *U.S. Naval Institute Proceedings*, vol 119, no. 6 (Jun 1993), pp 37–42. For one author's opinion of the ineffectiveness of the strategic air campaign against the city of Baghdad, see William M. Arkin, "Baghdad: The Urban Sanctuary in DESERT STORM?" *Airpower Journal*, vol 11, no. 1 (spring 1997), pp 4–20.

Strategic Bombardment: A Retrospective

Lee Kennett

In the spring of 1899, an extraordinary international conference convened at The Hague. The delegates set about to make significant changes in the laws of war and peace, and they undertook, with less success, to limit armaments. The Hague conference of 1899 also had the distinction of being the first international body to discuss aerial armaments. No airplane had yet flown and the dirigible still seemed of doubtful utility, but the delegates sensed that a viable aerial weapon would soon appear. Spokesmen for the land and seas services, who dominated the proceedings, felt that the weapons already in their arsenals were more than sufficient for the warfare of the day. More than one delegate expressed the opinion that attacks from the air smacked of "perfidy," and considerable sentiment existed for imposing a permanent ban on such acts, which would thus destroy air power in the egg.

An American delegate, Captain William Crozier of the U.S. Army, argued against the prohibition. He held out hope that the new weapon might enhance the "efficacy" of war, reducing its horrors by hastening the decision. He suggested that the air weapon could do this by directing destructive force at a "critical point" inaccessible to conventional weaponry.[1] This proposition, elaborated and refined, would become the *raison d'être* of strategic aerial bombardment. Captain Crozier won his argument. Within two decades strategic bombing moved from hypothesis to reality; by midcentury the bomber had become the supreme strategic weapon.* The chapters in this anthology treat the

*The heavy bomber with atomic weapons and air refueling that gave it intercontinental range would undergird President Dwight D. Eisenhower's New Look American defense policy in the 1950s.

most significant stages in the evolution of this form of warfare, and the author of each has written a conclusion incorporating what is most important to grasp and retain. Yet it might be well in these final pages to convey something of the sum and burden of our experience with this remarkable weapon, following some of the threads that have run persistently through its history, and may well reappear in any future development.

The most visible of these threads, and probably the most essential, was the long evolution of the bomber itself; everything else rests ultimately on the capabilities of a machine. In its basic function — the carrying of heavy burdens over long distances — there is a clear line of improvement, with progress accelerating during wartime. In 1914 the aircraft of the Royal Flying Corps could take aloft several pounds of bombs; by 1918 Royal Air Force aircraft could take off with several thousand pounds of bombs and carry them all the way to Germany. But the bombs themselves were not correspondingly improved, for high explosives evolved but little. As a consequence, an increase in destructive force was possible only by using larger bombers and more of them, until the flight of the *Enola Gay*. There were similar disparities in other lines of progression: the ability of the bomber to find its way to its objective in any kind of weather scarcely improved from 1919 to 1939, and then it was spectacularly enhanced by a variety of electronic means. On the other hand, the Second World War did not bring a spectacular improvement in the bombardier's ability to direct a bomb from high altitude to a target of limited size; that would wait for the passing of what Will Jacobs calls "the age of dumb iron bombs."

Whatever its limitations, the heavy bomber of the 1930s and 1940s was probably the ultimate challenge to aircraft designers and builders, a challenge that only a handful of technologically sophisticated countries could meet, and here Alvin Coox reminds us that Japan in the 1930s was not of that number. Even where expertise and technological bases were present, success in heavy bomber development was not automatic; if the Americans had the good fortune to bring out the B–17 well before the war, the Germans encountered insuperable difficulties with their He 177. Even if a country had the technology to build a heavy bomber, it might not have the treasure to field a fleet of them. Both the Italians and the Soviets developed satisfactory heavy bomber designs for use in the Second World War, but the Piaggio P–108 was built in only a handful of copies, and Tupolev TB–7 production did not reach one hundred planes. Sir Arthur Harris at one point hoped to double the size of Bomber Command to 4,000 planes, but Britain's resources would not permit it. Strategic bombing, then, was a rich country's game; not surprisingly the Americans built and operated their bomber fleets with the least concern over resources. In the fall of 1944, when the *Luftwaffe* in Europe found it difficult to obtain aviation gasoline even for the most essential functions, the American XXI Bomber Command in the Orient might burn eleven gallons of aviation gasoline just to airlift one gallon from India to its staging base in Chengtu, China.

Well into the 1930s there seemed to be a solution of sorts in what was called the multipurpose airplane, ideally a twin-engine craft that could perform a limited strategic bombing role and fulfill other functions as well. The Polish government, for example, attempted to achieve this compromise with their small fleet of Łódź bombers. But the strategic bomber was evolving into a highly specialized weapon. As it became more perfectly adapted to its principal calling, the greater difficulty it had in performing other tasks — this the Allies discovered when they called on their strategic air fleets for help with Operations CROSSBOW and OVERLORD.

With the end of World War II, the rules in strategic bombing changed radically. Now the heavy bomber was paired with the nuclear bomb, a combination that remained an American monopoly in the immediate postwar era. The Soviets, the British, and the French eventually succeeded in marrying the bomber with nuclear weapons, but by 1960, when the French conducted their first atomic test, another revolution was in the making, one that would eventually spell the end of strategic bombing as it had been known. Both the Americans and the Soviets were developing an extensive panoply of intercontinental ballistic missiles, while the French and British had to content themselves with much more modest efforts.

Technological and material factors counted for much in the decision to develop a strategic bombing force, but some element of choice or inclination also came into play. Richard Overy is on solid ground when he suggests that this decision depended upon the weight that the army had in determining interarms doctrine. This weight would be greatest in a continental country with potential enemies nearby; there the army would have special status, fortified by history, as the nation's sword and buckler (here France and Germany spring to mind). In such states, the army could insist that all the nation's effort be concentrated on the land battle. Its argument was irrefutable: when the infantry loses, everybody loses.

In each country the interplay of forces was different, the solution a distinctly national one. Sometimes it was the navy rather than the army that blocked the independence or the strategic mission of the air force — as was the case with Japan in 1936. Sometimes the political leadership threw its weight decisively to one viewpoint or the other; in countries with representative governments, the wishes and fears of the public carried great weight.

Among airmen themselves there was an understandable attraction to any role that embraced offensive action independent of the land battle; beyond that they were attracted to the idea of an independent air force. Such an arrangement accorded well with the distinctiveness, indeed the uniqueness, they saw in themselves and in their function. As General Spaatz put it: "I guess we considered ourselves a different breed of cat right in the beginning. We flew through the air and other people walked on the ground. It was as simple as that."[2] Although the same aspirations could probably be found in every air force, these had to achieve a certain "critical mass" before they came out in the open, or perhaps the partisans of air power simply needed someone to speak for them.

This brings us to that series of vigorous and outspoken air leaders — from Douhet to LeMay — whose professional lives were bound up with strategic aerial warfare. David MacIsaac characterizes airmen as "congenitally aggressive optimists," and if this is true, then the champions of the strategic bomber were a particularly hardy strain. They could galvanize others who thought as they did, airmen or not, and lead them in what was very often an uphill fight. The intensity of conviction of some of these men is striking. Sir Arthur Harris fought his superiors to the point of insubordination; Billy Mitchell went beyond that point. Douhet was relentless and unyielding in his convictions about the strategic air offensive; if anything he grew more categorical as he elaborated his beliefs. Perhaps this is what made such men so redoubtable: it was not that they were advocates, or even enthusiasts, they were zealots.

They took their cause beyond purely military circles and perpetuated their views at the highest policy levels. Douhet and Mitchell, who could both write vigorous prose, preached their cause in the media. In a country where the views of the man in the street strongly influenced policy and strategy, such as in the United States, air power won its case. In 1935, when Billy Mitchell's campaign was moving to its climax, the Gallup organization polled the American people on defense policy. The results showed that half of the public was in favor of increasing the Army, and the same proportion favored enlarging the Navy; but of every four Americans polled, three said they would like to see an expansion of the "air force."[3]

If Richard Overy's chapter explains the role of a small number of exceptional men in bringing the great air fleets into existence, then the three chapters dedicated to the operational history of the Second World War show how men of the same mettle made the strategic air campaigns work. None of the Allied strategic bombing efforts evolved quite as intended; there were bureaucratic or military checks, unanticipated technical complications, and sometimes defeats. In response there would be abrupt changes in tactics and targets, but the bombers continued to fly. There were also changes, sometimes brutally abrupt ones, in command. A leader not obtaining results would be replaced by one who could, quite often by a "hard charger," a man who was driven and who knew how to drive others.

The air leadership, British and American, that was collectively responsible for the achievements of the strategic bombing fleets employed against the Axis powers in World War II must also bear its share of the responsibility for the failures and shortcomings of strategic bombing. If there is a fundamental criticism to be made of the strategic bombing campaigns, it lies in what Will Jacobs calls the "disparity between strategic intentions and the means to carry them out"; scarcely a chapter in this anthology does not express that complaint in some form or other. This may well be the chief charge that both land and sea services have leveled at airmen over the years: Aviators envisaged and promised great things, but their aircraft continually let them down.*

Strategic bombardment, particularly high-altitude precision bombing, does seem to have been rather consistently oversold by its enthusiasts. Moreover, most advocates were men of action rather than of reflection, confident in the future of the new weapon and secure in the clarity and consistency of their doctrine. In the case of the British, the emergence of the RAF in 1918 was the beginning of a long struggle with the two older services, in which at times even the continued existence of an independent air force seemed in doubt. The airmen saw the strategic bombing offensive as a guarantee of survival and growth, provided it was accepted as a viable weapon by the government and the public. The RAF's vast and ambitious history, *The War in the Air*,[4] helped sell strategic bombing by giving it prominent coverage. In addition, the Air Ministry had made a survey of the effectiveness of the World War I bombing campaign against Germany. That report offered convincing testimony of the bomber's power, but it minimized the problems encountered. Most recently a charge has been made that the report bore little relationship to the facts and figures from which it was supposedly drawn; yet this report was the foundation for British bombing policy between the wars.[5] If this allegation is true, then what started as deception ended in self-deception.

With the Americans, the formulation of strategic bombing doctrine encountered other problems. The U.S. Air Service did not have the fund of British experience; in the Great War, the Americans dropped only 138 tons of bombs. With war's end and the United States's retreat into isolation, Americans had little contact with the foreign air services or air intelligence from abroad. A few typed translations of Douhet's *Command of the Air* circulated, but no clear proof exists that it had any impression on American bombing doctrine in the postwar decade (Douhet's name was mentioned in official publication for the first time in an Air Corps tactical school text in 1929). The evidence indicates, then, that American bombing doctrine was formulated largely *in vacuo*. With no fund of wartime experience and with little or no knowledge of developments elsewhere, its formulators tended to believe in the invulnerability of unescorted heavy bombers and to conceive strategic bombardment in a "best-case" scenario.

Even had British and American air leaders tried to explore more aggressively the prospects and limitations of strategic bombing fleets, they would have run into a wall of public opposition. In the mid-1930s, notably, offensive weapons in general and the bomber in particular had a bad image. The RAF

*During the Second World War, the guidance technology available and unanticipated winds at high altitudes, not to mention foul weather, precluded the "precision" bombing of strategic targets to which American airmen, especially, subscribed.

took the habit in communications of referring to its bombing forces as "Wessex area," while the Army Air Corps represented the B–17 as most useful in defending America's far-flung sea frontiers. In such a climate an extensive effort to probe the offensive capacities of the strategic bomber was manifestly impossible.

By the late 1930s, when conflicts broke out in China, Ethiopia, and Spain in which air power played a hand, bombing doctrine had already jelled; whatever intelligence did reach British and American airmen, they tended to fit it into their doctrinal preconceptions. Once war came, some of those preconceptions died hard deaths. The Americans had to discover for themselves what the British had already learned two bombing campaigns earlier about the hazards of deep, unescorted aerial incursions over enemy territory conducted in the daytime. It is sometimes said that the American military accepts lessons from its enemies more quickly than it does from its allies; if this is so, it does not apply only to our airmen.

Before leaving this matter of the "disparity" between strategic intentions and the means to execute them, we should note that some air forces *incapable* of carrying out their strategic goals have, nonetheless, won victories of sorts. In the mid-1930s Hitler used propaganda and posturing to inflate his *Luftwaffe* far beyond its actual size and power (at one point he counted all Lufthansa transports as bombers). The visions of vast cities in flames that this phantom force engendered counted for more than a little in the *Reichschancellor's* foreign policy achievements; in 1938, the shadow of the bomber notably hung over the French and British delegations at Munich. A decade later, American policymakers used the menace of atomic bomb–carrying B–29s to keep the Soviets in awe, even though for the most part this force was a "hollow threat" as one historian has described it, chiefly because of a dearth of nuclear weapons and of bombers modified to carry them. In both cases the threats succeeded because they seemed credible to the other side. When the threat was not credible, as when American authorities dispatched B–17s to the Philippines on the eve of Pearl Harbor, the subsequent loss of aircraft was scarcely noticed in the general catastrophe.

At the operational level, the Anglo-American bombing fleets of World War II faced two basic problems: one was penetrating to the objective and returning with an acceptable loss ratio; the other was delivering sufficient destructive force on a target bombed from high altitude. For the first, a solution of sorts lay in night operations, but a better one was to beat German fighter forces to the ground. The second problem was overcome chiefly by enlarging the target, for example, from a single factory to an entire industrial area, or by lowering the bomber's altitude of operation. Both Allied air forces began the war with the intention of delivering precise, telling strokes on carefully selected objectives whose destruction would most hamper the enemy's ability to wage war, what are today called high-value targets. When they found their weapon incapable of the desired knifelike precision, they were forced to use it as a hammer.

After the initial wartime bombing plan was abandoned, the strategic effort often conveyed the impression of groping and indecision. British and American air leaders seemed to direct their bombing fleets much as an impatient fireman might direct a stream of water first at one point in a conflagration, then at another. Chokepoint and bottleneck targets never lost their attractiveness, especially to the Americans. They would emphasize one, then shift to another. The RAF was increasingly drawn to "area attacks," or "city busting," with its double dividend in physical destruction and lowered morale. And for a time, both air forces were mobilized for interdiction in support of the Normandy invasion.

What emerges very clearly in the chapters of this book is the difference among air leaders over the strategy that put large numbers of enemy civilians at hazard. Sir Arthur Harris remained the only outspoken champion of the policy, an advocacy in which he was for a time seconded by Winston Churchill. It was often described in euphemistic terms, a certain sign that people were uncomfortable with it. Lord Cherwell, Churchill's scientific adviser, spoke of "dehousing" as if one could destroy a habitation without doing harm to those who lived within. Stephen McFarland and Wesley Newton relate the concern of Ira Eaker, that history might convict American airmen of choosing as a target "the man in the street."

The intellectual origins of "morale" or "terror" bombing are difficult to establish. Even in the 1930s, when it was a major preoccupation, remarkably little of a "clinical" nature was written on it. Of all the target options for the strategic bomber, this one seems to be the least carefully studied. And on the basis of experience in the Second World War and subsequent conflicts, this type of bombing seems to have been an indifferent success at best. Such is the impression one gains for example, from that vast postmortem found in the U.S. Strategic Bombing Survey. In addition to the moral questions such warfare raised and will continue to raise, it seems to have a low "efficacy." Civilians — living in London, Hamburg, Osaka, or Chungking — proved surprisingly tough animals.

At the other end of the scale, in the precise, systematic attack on critical strategic targets or "chokepoints," what Sir Arthur Harris disparaged as "panacea targets," the Bombing Survey revealed that the fundamental principle was sound; virtually any of the "plans" — oil, transport, electric power, or any other — could have brought Germany and Japan to their knees with *accurate* and *persistent* bombing. These findings might have carried more weight in postwar thinking and planning had the strategic air offensive not become linked, as General Hap Arnold feared it would, to the atomic bomb, a weapon of awesome but indiscriminate destruction.

The massed B–29 raids against Japan on August 14, 1945, may well have marked the final appearance of the "classical" strategic bombing fleets in that vast, apocalyptic scenario imagined by Douhet and brought to reality by British and American airmen. Though the great bomber airplanes themselves remained in service of the RAF Bomber Command and the U.S. Air Force Strategic Air Command, and for a time reigned supreme as the chosen instrument of intercontinental atomic warfare (the B–52 may go down as the final and supreme expression of the genus), they could find only occasional and limited employment in post-1945 conventional conflicts; this is the message of Steven Rearden's and Thomas Hone's chapters. One cannot read these last chapters without wondering whether they constitute a sort of postlude, writing finis to a form of aerial warfare whose day has come and gone.

In the 1950s when SAC measured tactical warning in hours, Soviet air defenses remained thin, and intercontinental ballistic missiles were still experimental, nearly everyone regarded strategic bombers as indispensable. Indeed, the term "strategic bomber" defined a nuclear-armed heavy bomber. In the 1960s, after advances in electronics and rocket technology altered conditions markedly, informed opinion no longer judged bombers in this role with the same reverence. Though Air Force regents could protest, President John Kennedy and his Secretary of Defense Robert McNamara judged the less costly ballistic missile better suited to the nation's strategic needs, rating its time-to-target and likelihood of arrival as vastly superior to the manned bomber's. "What is the role of a [strategic] bomber," McNamara asked in February 1964, "after you place 1,000 to 2,000 missiles on the Soviet Union? What do you have left to mop up? This is *the* question. If it is not a mop-up operation, what is the role of the bomber?"[6]

The technology of air refueling that gave strategic bombers intercontinental range in the 1950s also conferred the same range to air refuelable fighter bombers.* Furthermore, technology enabling attack aircraft of all kinds to identify quickly and strike accurately tactical *or* strategic targets at any time in virtually any weather improved rapidly throughout the 1970s and 1980s. In the "contingency" or "low intensity" conflicts that did occur, fighter-bombers attacked strategic targets such as communication centers, power stations, petroleum refineries, or transportation nodes far more frequently than heavy bombers did. As Richard Davis makes plain, the recent war in the Persian Gulf offered a striking reversal in aircraft roles: Air Force fighter-bombers directed most of the laser-guided ordnance against strategic targets in Iraq; B–52s, unequipped to make such strikes, dropped iron bombs on tactical targets such as Iraqi troop concentrations in the desert.**

The lessons of the Gulf War were not lost on the senior leadership of the

*For the solitary fighter pilot unable to leave his seat, however, fatigue, or sheer physical exhaustion, determines the length of any mission.

**B–1s, meanwhile, remained on nuclear alert in the continental United States. A handful of modified B–52s, it must be said, did fire air-launched cruise missiles in DESERT STORM. Altogether, the B–52 bombers flew only 3 percent of all U.S. combat sorties in that conflict; however, they accounted for 30 percent of the total U.S. bomb tonnage. Stealth F–117 fighter bombers, on the other hand, comprised 2 percent of the Gulf War air contingent but flew against 30 percent of all strategic targets.

Air Force; nor did that leadership overlook the air power implications inherent in the disintegration of the Soviet monolith and the end of the Cold War. By early 1991, it was clear that if heavy bombers were to have a significant role in future conflicts, they could no longer remain simply a vehicle to deliver nuclear weapons. The institutional changes that followed were profound. On June 1, 1992, SAC passed into history, its intercontinental air and missile operations divided between two new commands. Bomber operations were combined with those of the former Tactical Air Command into a single force, Air Combat Command, which had as its first priority fitting the heavy bomber into conventional warfare scenarios.

If the currents of change have been strong in the recent history of strategic bombing, some have also been constant. The list of strategic air objectives in the Gulf War was strongly reminiscent: transport, communications, electrical power, fuel, and lubricants. A half-century ago British and American airmen were battering away at such targets with little evidence of success after weeks and months of effort. In Iraq, such objectives appear to have been "taken out" within a few days, with almost surgical precision, with one or two aircraft doing the work of hundreds of B–17s and Lancasters. Videotapes of the bombing in DESERT STORM detailed what appears to be a quantum leap in the ability of aircraft to hit very small targets, targets approaching in size the proverbial "pickle barrel."

A half-century has passed since the incineration of Hiroshima and Nagasaki. Although it certainly would be premature to speak of the current epoch as "postnuclear," the menace of this form of city killing no longer looms so heavily as once it did. During this period, land- and sea-based ballistic missiles largely replaced the bomber in the strategic nuclear role, while fighter-bombers increasingly displaced it in the precision bombardment role. Configured for the contemporary come-as-you-are contingency war, however, the strategic bomber may yet know something of a renaissance and achieve some of the hopes held for it when it was still only an idea. It possesses features and capabilities that promise much: an internal bomb bay offers heavier, more diverse weapon loads; a multiperson crew permits extended, long-range missions; stealth technology masks its presence to the enemy; and by means of air-launched cruise missiles and laser-guided bombs, in recent years it has acquired the ability to direct incredible destructive power with extreme accuracy. We still have a great deal to learn from the Gulf War, but it has offered us a glimpse of a strategic air weaponry of extraordinary "efficacy," to use Captain Crozier's nineteenth-century term. Perhaps a century later, in whatever guise, this type of aerial weapon is capable of that higher form of warfare he envisaged.

NOTES

1. Russell J. Parkinson, "Aeronautics at The Hague Conference of 1899," *Airpower Historian*, vol 7, no. 2 (Apr 1960), p 109; see also James Brown Scott, ed., *The Hague Conventions of 1899 and 1907* (London: Oxford University Press, 1915).

2. James P. Tate, "The Army and Its Air Corps: A Study of the Evolution of Army Policy toward Aviation, 1919–1941" (Ph.D. diss. Indiana University, 1976), p 1.

3. "Military Appropriations," poll published Dec 29, 1935, reprinted in George H. Gallup, *The Gallup Poll: Public Opinion 1935–1971*, 3 vols (New York: 1972), vol 1, p 6. Indicative of the effort of American airmen to make their case, and doubtless influencing the popular mind, Arnold and Eaker also coauthored three volumes during this period: *The Flying Game* (New York: Funk & Wagnalls, 1936); *Winged Warfare* (New York: Harper & Brothers, 1941); and *Army Flyer* (New York: Harper & Brothers, 1942).

4. W. Raleigh and H. A. Jones, *The War in the Air*, 6 vols, (London: Oxford University Press, 1922).

5. "Results of Air Raids on Germany Carried out by British Aircraft," A.P. Jan 25, 1990, PRO Air 1/2104/207/36; George Kent Williams, "Statistics and Strategic Bombardment: Operations and Records of the British Long-range Bombing Force during World War I and Their Implications for the Development of the Postwar Royal Air Force, 1917–1923" (Ph.D. diss., History, Merton College, Oxford, 1987).

6. U.S. Congress, Senate, Subcommittee of the Committee on Appropriations and the Committee on Armed Services, *Department of Defense Appropriations: Hearings on H.R. 10939*, 88th Cong, 2d sess, 1965, pt 1, p 57. Emphasis added.

Contributors

ALVIN D. COOX is a professor of history and director of the Japan Studies Institute at San Diego State University. He earned his Ph.D. from Harvard University and served as historian with the Japanese Research Division of the U.S. Army in Japan, as an analyst with the U.S. Air Force, and as managing editor of Orient/West Press in Tokyo. His published works include four books, among them *Nomonham: Japan Against Russia, 1939* (Stanford University Press, 1986). Coox resided in Japan for more than fifteen years since the end of World War II, most recently as a Fulbright Senior Research Fellow in 1988–1989, when he reexamined Japan's decision to attack the United States in 1941. In recognition of his sustained contributions to U.S.-Japanese relations, in 1992 the Japanese Emperor awarded Coox *The Order of the Rising Sun, Gold Rays with Neck Ribbon.*

RICHARD G. DAVIS is a senior historian at the Air Force History Support Office in Washington, D.C. He earned a B.A. from the University of Virginia, an M.A. from the University of North Carolina at Chapel Hill, and a Ph.D. from the George Washington University. Before joining the Air Force Historical Program in 1980, he was an archivist at the National Archives and Records Administration. He is the author of *Carl A. Spaatz and the Air War in Europe* (Washington, D.C.: Center for Air Force History, 1993) and the Air Staff Historical Studies, *The 31 Initiatives: A Study in Air Force–Army Cooperation* (Washington, D.C.: Office of Air Force History, 1987) and *Tempering the Blade: General Carl Spaatz and American Tactical Air Power in North Africa, November 8, 1942–May 14, 1943* (Washington, D.C.: Office of Air Force History, 1989). Dr. Davis is currently researching aspects of U.S. air strategy in the Persian Gulf War of 1991.

R. CARGILL HALL, chief historian of National Reconnaissance Office, arranged and edited this volume while he managed the contract histories program for the Air Force History Support Office. Previously, he served as chief of the research division and deputy director of the U.S. Air Force Historical Research Agency and as a historian at Headquarters Military Airlift Command and Headquarters Strategic Air Command. He also served the National Aeronautics and Space Administration history program as historian at the Jet Propulsion Laboratory, California Institute of Technology. Hall is the author of *Lunar Impact: A History of Project Ranger* (Washington, D.C.: National Aeronautics and Space Administration, 1977), editor and contributor to *Lightning Over Bougainville* (Washington, D.C.: Smithsonian Institution Press, 1991). He has also contributed chapters to and published numerous articles on space law and the history of aeronautics and astronautics in diverse books and journals, and serves on the board of advisers for the Smithsonian Institution Press's history of aviation book series.

THOMAS C. HONE is a member of the staff of the Commander, Naval Air Systems Command. He holds a Ph.D. in political science from the University of Wisconsin at Madison and served as a task force leader on the Gulf War Air Power Survey sponsored by the Secretary of the Air Force. He contributed to *Case Studies in the Achievement of Air Superiority* (Washington, D.C.: Center for Air Force History, 1994), and is the author of *Power and Change: The Administrative History of the Office of the Chief of Naval Operations, 1946–1986* (Washington, D.C.: Naval Historical Center). His research articles have been published in the *Naval War College Review*, *Armed Forces and Society*, *Defense Analysis*, the *Journal of Strategic Studies*, and other professional publications.

WILLIAM A. JACOBS is Professor and Chair of Political Science at the University of Alaska, Anchorage, where he has taught history and political science since 1973. He is a graduate of Wisconsin State University, Eau Claire, and holds M.A. and Ph.D. degrees from the University of Oregon. Originally trained as a social and economic historian, he later turned to the history of warfare and to the field of strategic studies. Jacobs is the author of articles on close air support doctrine and organization in the U.S. Army Air Forces and the Royal Air Force, the development of strategic bombing in American national strategy in the Second World War, and the organizational politics of Anglo-American air force high commands. He is currently working on an analysis of American security policy in Northeast Asia after the Cold War.

LEE KENNETT, who received his doctoral degree at the University of Virginia, spent most of his teaching career at the University of Georgia, retiring from that institution as a Research Professor of History in 1993. In the course of his teaching career, he held three Fulbright professorships to France, taught at the *Ecole Pratique des Hautes Etudes* (Section IV) in 1978–1979, and was Lindbergh Professor at the National Air and Space Museum in 1989–1990. Among Kennett's more recent books ar *A History of Strategic Bombing* (New York: Charles Scribner's Sons, 1982); *For the Duration: America Goes to War, Pearl Harbor — 1942* (New York: Charles Scribner's Sons, 1985); *G.I.: The American Soldier in World War II* (Charles Scribner's Sons, 1987); *The First Air War, 1914–1918* (New York: The Free Press, 1991); and *Marching through Georgia: The Story of Soldiers and Civilians in Sherman's Campaign* (New York: Harper Collins, 1995).

STEPHEN L. McFARLAND is an associate professor of history at Auburn University and a visiting professor of military history at Air University. He holds a Ph.D. from the University of Texas, and his specialties are the history of flight, air power, and the modern Middle East. He is the coauthor (with Wesley Phillips Newton) of *To Command the Sky: The Battle for Air Superiority over Germany, 1942–1944* (Smithsonian Institution Press, 1991). McFarland has also authored articles on air power, chemical and biological warfare, and Iranian history for *Diplomatic History*, the *International Journal of Middle East Studies*, *Defense Analysis*, *Literature East and West*, and the *Journal of Strategic Studies*. He recently completed a manuscript, *America's Pursuit of Precision Bombing, 1910–1945*. He is currently working on a history of high-octane aviation gasolines and their effect on flight improvement between the two world wars.

DAVID MacISAAC, Lt. Col. (Ret., USAF), holds degrees in history from Trinity College, Yale University, and Duke University. He taught military history and strategy at the U.S. Air Force Academy as well as at the Naval War College and the Air War College. From August 1978 through September 1979 he was a Fellow at the Woodrow Wilson International Center for Scholars at the Smithsonian Institution. On retirement from active service, he served as Professor of Military History at Air University. MacIsaac is the author of *Strategic Bombing in World War II: The Story of the U.S. Strategic Bombing Survey* (New York & London: Garland Publishing Co., 1976), and editor of Garland Publishing's ten-volume series, *The U.S. Strategic Bombing Survey: Selected Reports* (1976). He has contributed chapters to Peter Paret, ed., *Makers of Modern Strategy from Machiavelli to the Nuclear Age* (Princeton, N.J.: Princeton University Press, 1986); R. A. Mason, ed., *War in the Third*

Dimension (Washington and London: Brassey's Defence Publishers, 1986); Sam C. Sarkesian and John Allen Williams, *The U.S. Army in a New Security Era* (Boulder, Colo.: Lynne Rienner Publishers, 1990); and L. Robineau, ed., *Precurseurs et Prophetes de L'Aviation Militaire* (Paris, 1992).

WESLEY PHILLIPS NEWTON is Professor Emeritus of History, Auburn University. He earned his Ph.D. from the University of Alabama. His books include *To Command the Sky: The Battle for Air Superiority Over Germany, 1942–1944* (Washington, D.C.: Smithsonian Institute Press, 1991), with Stephen L. McFarland; *The Perilous Sky: U.S. Aviation Diplomacy and Latin America, 1919–1931* (University of Miami Press, 1978); *Delta: The History of an Airline* (University of Georgia Press, 1979), with W. David Lewis; *Wings of Gold: The [World War II] . . . Correspondence of Aviation Cadet/Ensign Robert R. Rea* (University of Alabama Press, 1987), with Robert R. Rea. He authored and co-authored various articles, including ones on Pan American Airways and Delta (with W. David Lewis) in *The Airline Industry* volume of *The Encyclopedia of American Business History and Biography* (Columbia, S.C.: Facts on File, 1992), edited by William M. Leary. He has completed a history of Montgomery, Alabama, in World War II and is working on a history of Maxwell AFB with Jerome Ennels.

RICHARD J. OVERY is Professor of Modern History at King's College, London, where he has taught since 1980 after teaching at Queens' College, Cambridge. He has written extensively on modern German and airpower history. His books include *The Air War 1939–1945* (London: Europa, 1980), *Goering: The 'Iron Man'* (London & Boston: Routledge & Kegan Paul, 1984); (with Andrew Wheatcroft) *The Road to War* (New York, Random House, 1989); *War and Economy in the Third Reich* (New York: Oxford University Press, 1994); and most recently, *Why the Allies Won* (New York: Norton, 1996). He gave the Harmon Memorial Lecture at the Air Force Academy in 1989 and was winner of the T. S. Ashton Prize in 1983 and the Cass Prize in 1987 for his work on the Nazi economy. He is currently writing the Oxford history of World War II and a history of the Nazi Economy.

STEVEN L. REARDEN is an independent consultant in Washington, D.C., who specializes in contemporary foreign affairs and national security issues. He completed a book-length study of Congress and the Strategic Defense Initiative and is working on a study of the role of the secretary of defense in foreign affairs. Rearden earned his Ph.D. in history from Harvard University and has been an adviser to the Departments of State and Defense and to the U.S.

General Accounting Office and has held appointments as Guest Scholar at the Nitze School of Advanced International Studies of the Johns Hopkins University and as Research Fellow at the John F. Kennedy School of Government, Harvard University. His published works include *The Evolution of American Strategic Doctrine: Paul H. Nitze and the Soviet Challenge* (Boulder, Colo.: Westview Press, 1984), *History of the Office of the Secretary of Defense: The Formative Years, 1947–1950* (Washington, D.C.: Office of the Secretary of Defense History Office, 1984), and *The Origins of U.S. Nuclear Strategy, 1945–1953* (New York, St. Martin's Press, 1993), which he co-authored with Samuel R. Williamson, Jr.

Index

ISBN 0-16-049781-7